T0280619

Dynamic Energy Budget Theory for Metabolic Organisation

Dynamic Energy Budget (DEB) theory is a formal theory for the uptake and use of substrates (food, nutrients and light) by organisms and their use for maintenance, growth, maturation and propagation. It applies to all organisms (micro-organisms, animals and plants). The primary focus is at individual level, from a life cycle perspective, with many implications for sub- and supra-individual levels. The theory is based on sound chemical and physical principles, and is axiomatic in set-up to facilitate testing against data. It includes effects of temperature and chemical compounds; ageing is discussed as an effect of reactive oxygen species, with tight links to energetics. The theory also includes rules for the covariation of parameter values, better known as body size scaling and quantitative structure–activity relationships. Many well-known empirical models turn out to be special cases of DEB theory and provide empirical support. Many additional applications are illustrated using a wide variety of data and species. After 30 years of research on DEB theory, this third edition presents a fresh update. Since the second edition in 2000, some 140 papers have appeared in journals with a strong focus on DEB theory.

A lot of supporting material for this book meanwhile has been developed and is freely available, such as the software package DEBtool; one of its toolboxes contains code that generates the figures of this book: setting of data, model specification, parameter estimation, and plotting. By replacing data of your own, you have a convenient tool for applying the theory.

This edition includes a new chapter on evolutionary aspects; discusses methods to quantify entropy for living individuals; isotope dynamics; a mechanism behind reserve dynamics and toxicity of complex mixtures of compounds. An updated ageing module is now also applied to demand systems; there are new methods for parameter estimation; adaptation of substrate uptake; the use of otoliths for reconstruction of food level trajectories; the differentiated growth of body parts (such as tumours and organs) linked to their function, and many more topics are new to this edition.

BAS KOOIJMAN has been Professor of Applied Theoretical Biology at the Vrije Universiteit in Amsterdam since 1985 and is head of the department.

Third Edition

Dynamic Energy Budget Theory for Metabolic Organisation

S. A. L. M. Kooijman, Vrije Universiteit, Amsterdam

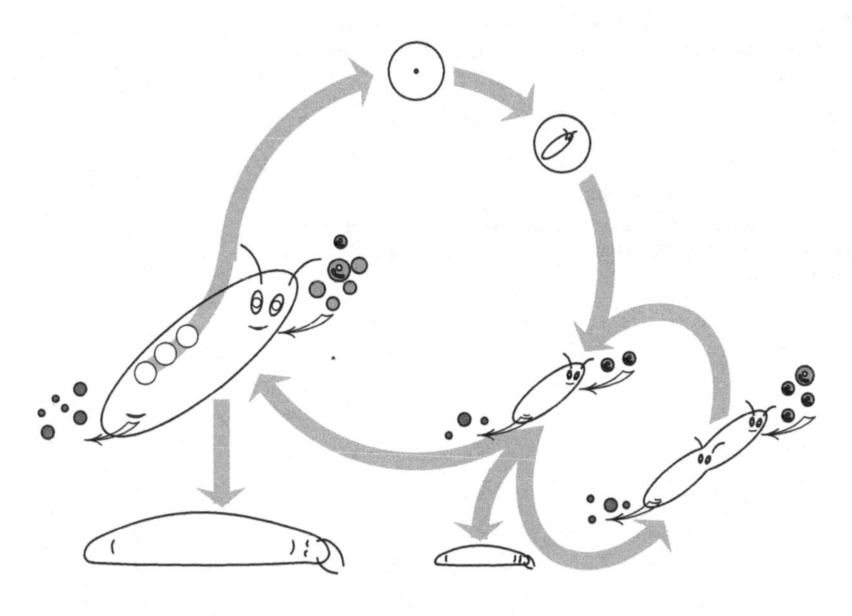

CAMBRIDGE
UNIVERSITY PRESS

CAMBRIDGE
UNIVERSITY PRESS

University Printing House, Cambridge CB2 8BS, United Kingdom

One Liberty Plaza, 20th Floor, New York, NY 10006, USA

477 Williamstown Road, Port Melbourne, VIC 3207, Australia

314-321, 3rd Floor, Plot 3, Splendor Forum, Jasola District Centre, New Delhi - 110025, India

79 Anson Road, #06-04/06, Singapore 079906

Cambridge University Press is part of the University of Cambridge.

It furthers the University's mission by disseminating knowledge in the pursuit of education, learning and research at the highest international levels of excellence.

www.cambridge.org
Information on this title: www.cambridge.org/9780521131919

First published 1993 as *Dynamic Energy Budgets in Biological Systems*
Second edition published 2000 as *Dynamic Energy and Mass Budgets in Biological Systems*
Third edition published 2010

A catalogue record for this publication is available from the British Library

ISBN 978-0-521-13191-9 Paperback

Summary of contents

Contents

Preface

What to expect in this book?

This book is about a formal consistent and coherent theory for the processes of substrate uptake and use by organisms, which I called the Dynamic Energy Budget (DEB) theory. Over the 30 years of research on this theory, it became well established; some 140 papers on DEB theory have appeared since the second edition in 2000. The application of the theory by the international research group AQUAdeb, http://www.ifremer.fr/aquadeb/, and of this book in the DEB tele-courses, http://www.bio.vu.nl/thb/deb/course/, urged for a new edition. This book gives a fresh update of the present state of the theory. In view of its accelerating development, this update will probably not be the last one. To accommodate all new material, I had to cut out most methodological parts of the previous edition, which is a pity because opponents of DEB theory typically seem to differ in opinion on 'details', but actually differ in opinion on the role of models in research and related methodological issues. I wrote a document on methods in theoretical biology, which also summarises the mathematics that is used in this book, see http://www.bio.vu.nl/thb/deb/.

Many empirical models, ranging from Lavoisier's model for indirect calorimetry, Kleiber's model for the respiration as function of body weight, von Bertalanffy's model for animal growth and Droop's model for nutrient-limited algal growth, turned out to be special cases of DEB models that follow from the theory. This means that DEB theory is the best tested quantitative theory in biology.

Support of this book

Although I have tried hard to avoid errors, experience tells me that they are unavoidable. A list of detected errors can be found at the DEB information page http://www.bio.vu.nl/thb/deb/, and I offer my apologies for any inconvenience. The errata, and all support material mentioned below, will frequently be updated.

I have tried to emphasise the concepts in this book, and to reduce on technicalities a section-wise summary of concepts gives an overview, skipping tests against realism. Mathematical derivations of results are important, however, especially for people who want to contribute to the further development of the theory. These derivations are collected in the comments on DEB theory, which can also be found at the DEB information page. These comments also give further background information and summarise the developments of the theory and its applications since this book appeared.

Software package DEBtool can be downloaded freely from the electronic DEB laboratory, http://www.bio.vu.nl/thb/deb/deblab/; the manual is included (run file *index.html* in subdirectory *manual* in a browser, such as *firefox*). DEBtool is written in Octave and in Matlab. The purpose of this package is a mix of demonstrations of what the theory can do, routines that can be used to fit DEB models to data, to calculate quantities such as the initial amount of reserve in eggs, to reconstruct food-density trajectories from observations on an individual, to run numerical simulations for plant growth, etc. Toolbox *fig* collects the files that have been used to create the figures of this book. Here you can see how the data are set, how the model is specified and how it is fitted to the data. The files give the figures in colours, and also present standard deviations of parameters that are estimated. These standard deviations are not presented in this book. If you want to apply the theory to your own data, and your application resembles one of the figures, an efficient way to do this is to go the file that produces the figure, replace the data with your own data and rerun the file.

This book is used in the international biannual DEB tele-course. Its set-up can be found on http://www.bio.vu.nl/thb/deb/course/; starting in 2009, the course will be linked to an international symposium on DEB theory, organised by previous participants of the tele-courses. The DEB information page also gives access to other material that is used and produced in this course. This includes collections quizzes, exercises and solutions, Powerpoint sheets, questions and answers, essays written by participants and, typically, later used in publications.

The book mentions many names of taxa; I have collected recent ideas on the evolutionary relationships between living organisms. This document can be found on the DEB information page.

The DEB information page also presents a number of papers that introduce DEB theory, ongoing activities, job opportunities, etc. Examples of application of DEB theory can be found in the special issues of the *Journal of Sea Research*, **56** (2006), issue 2 and **62** (2009), issue 1/2.

Set-up of this book

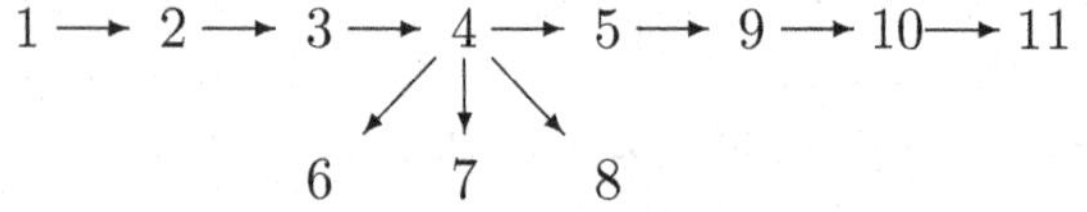

$1 \rightarrow 2 \rightarrow 3 \rightarrow 4 \rightarrow 5 \rightarrow 9 \rightarrow 10 \rightarrow 11$; $4 \rightarrow 6$, $4 \rightarrow 7$, $4 \rightarrow 8$

The logical structure of the chapters is indicated in the diagram (left). *A first quick glance through the section on notation and symbols, page* {494}, *saves time and annoyance.*
A glossary at {487} explains technical terms.

Chapter 1 gives introductory concepts, namely the notion of the individual, its life-stages, the various varieties of homeostasis and the effects of temperature on metabolic rates. The choice of topics is based on their relevance for the standard DEB model.

Chapter 2 specifies the standard DEB model, which represents the simplest non-degenerated model in DEB theory, so a canonical form, which uses one type of substrate (food), one reserve, one structure for an isomorph, i.e. an organism that does not change in shape during growth. It neglects all sorts of complications for educational purposes, to illustrate the basic DEB concepts in action; a summary section presents the list of assumptions from which the standard DEB model follows.

Chapter 3 discusses the relationships between energy, compounds and biomass, and presents basic concepts on metabolism. It presents the actions of synthesising units (SUs), i.e. a generalised form of enzymes that basically follow enzyme kinetics, but with an important modification: SU kinetics is based on fluxes, not on concentrations. The material in this chapter prepares for the next one.

Chapter 4 describes univariate DEB models, which have one type of substrate (food), one reserve and one structure, and starts with a discussion on changes in food density and extends the standard DEB model of Chapter 2 by accounting for changing shapes. The various chemical compounds, isotopes and energies are followed, product formation is specified and respiration is discussed in some detail. The quantification of entropy of living biomass is discussed. The parameter estimation section of Chapter 2 is now extended to include mass, energy and entropy parameters. The final section shows the use of observations on individuals to reconstruct how the food availability and temperature changed in time. This is useful e.g. to study size-dependent food selection.

Chapter 5 extends the theory to include several substrates, reserves and structural masses to increase the metabolic versatility that is found in organisms that acquire nutrients and light independently, and have to negotiate the problem of simultaneous limitation caused by stoichiometric coupling. The various ways in which substrate can take part in metabolic transformations are discussed. The processes of photosynthesis and calcification are discussed; the implications for plant development are evaluated.

Chapter 6 starts with a discussion of ageing that is caused by the effects of reactive oxygen species, which are formed as side-products of respiration, and its links with energetics. This chapter considers the uptake and effects of non-essential compounds, such as toxicants. The significance of effects of toxicants for energetics is that DEB parameter values are changed and the response to these changes reveals the metabolic organisation of individuals.

Chapter 7 extends DEB theory to include more detail for the various applications, especially if the shorter timescales need to be included to link to developments in molecular biology. The purpose of this chapter is to show how DEB theory fits into a wider context of biological research. Some parameter values turn out to change sometimes during the development of an organism and are discussed; it prepares the topic of the next chapter.

Chapter 8 analyses the intra- and inter-specific variation of parameter values among individuals. It compares the energetics of different species by studying the implications of DEB theory for the covariation of parameter values among species. The chapter shows how, for a wide variety of biological variables, body size scaling relationships can be derived from first principles rather than established empirically. This approach to body size scaling relationships is fundamentally different from that of existing studies.

Chapter 9 considers interactions between individuals and develops population consequences. The population, after its introduction as a collection of individuals, is considered as a new entity in terms of systems analysis, with its own relationships between input, output and state. These new relationships are expressed in terms of those for individuals. The coupling between mass and energy fluxes at the population level is studied and the behaviour of food chains and of canonical communities is discussed briefly.

Chapter 10 presents scenarios for the evolution of metabolic organisation and the gradual coupling and uncoupling of the dynamics of partners in symbiotic interactions; it also aims to make DEB theory biologically explicit. Apart from showing how DEB theory fits into an evolutionary context, this chapter demonstrates a key issue: two species that follow DEB rules can merge such that the merged new species again follows DEB rules. The process that life became increasingly dependent on life is discussed and illustrated with examples.

Chapter 11 places the approach taken by the DEB theory in existing eco-energetic research, and highlights some differences in concepts. A collection of well-known empirical models is presented that turn out to be special cases of DEB theory and their empirical support also supports DEB theory.

Acknowledgements

Many people have contributed to the development of the theory and to this book in different ways. I would specifically like to thank Bob Kooi and Tjalling Jager for the continuing support for this work at the Theoretical Biology Department in Amsterdam. Years of intensive discussions with Tânia Sousa, Tiago Domingo and Jean-Christophe Poggiale sharpened my mind to the extent of directing theoretical developments. The 30 Ph.D. students who have graduated on DEB-related topics also contributed a lot; it has been a real pleasure to work with them. I also want to thank people of the AQUAdeb group, and especially Marianne Alunno-Bruscia, for their work and enthusiasm, and Ifremer for supporting this group. Anna Hodson was of great help in the production of this book and Yues Descatoire created the banner on the cover.

Present behind all aspects of the 30 years of work on the theory is the critical interest of Truus Meijer, whose loving patience is unprecedented. The significance of her contribution is beyond words.

1

Basic concepts

The purpose of this chapter is to introduce some general concepts to prepare for the development of the simplest version of DEB theory, which is discussed in the next chapter. I start with the explanation why the organisation level of the individual plays a key role in DEB theory, followed by homeostasis concepts. Mechanisms for homeostasis and evolutionary aspects are discussed later. Then we need to introduce the notion of life stages and effects of temperature in some detail.

1.1 Individuals as dynamic systems

1.1.1 The basic level of metabolic organisation

From a systems analysis point of view, individuals are special for metabolic organisation because at this organisational level it is relatively easy to make energy and mass balance. This is important, because the conservation law for energy and mass is one of the few hard laws available in biology. At the sub- and supra-individual levels it is much more difficult to measure and model mass and energy flows.

Life started as an individual in evolutionary history, see {386}, and individuals are the units of selection and the survival machines of life; differences between individuals are, in combination with selection, key to evolution. DEB theory captures these differences by parameter values, which can differ between individuals, see Chapter 8.

Individuals are also special because behaviour is key to food intake and food selection (food fuels metabolism) and to mate selection; reproduction controls survival across generations in many species.

The analysis of metabolic organisation should, therefore, start at the level of the individual. Many species are unicellular, which links subcellular organisation directly to the individual level.

1.1.2 Vague boundaries: the cell–population continuum

The emphasis on individuals should not mask that the boundaries between cells, individuals, colonies, societies and populations are not always sharp. Fungal mycelia can cover up to 15 hectares as in the basidiomycete *Armillaria bulbosa*, but they can also fragment easily. Cellular slime moulds (dictyostelids) have a single-celled free-living amoeboid stage, as well as a multicellular one; the cell boundaries dissolve in the multicellular stage of acellular slime moulds (Eumycetozoa), which can now creep as a multi-nucleated plasmodium over the soil surface.

The mycetozoans are not the only amoebas with multi-nuclear stages; *Mastigamoeba* (a pelobiont) is another example [88]. Many other taxa also evolved multi-nucleated cells, plasmodia or stages, e.g. ciliates, xenophyophores, actinophryids, *Biomyxa*, loukozoans, diplomonads, Gymnosphaerida, haplosporids, Microsporidia, nephridiophagids, Nucleariidae, plasmodiophorids, Pseudospora, Xanthophyta (e.g. *Vaucheria*), most classes of Chlorophyta (Chlorophyceae, Ulvophyceae, Charophyceae (in mature cells) and all Cladophoryceae, Bryopsidophyceae and Dasycladophyceae) [516, 871]. Many higher fungi have hyphen where cells are fused in a multi-nucleated plasma, and nuclei of several rhodophytes can crawl from one cell into another. The Paramyxea have cells inside cells. The Myxozoa have multicellular spores, but a single-cellular adult stage. Some bacteria have multicellular tendencies [1044].

Certain plants, such as grasses and sedges, can form runners that give off many sprouts and cover substantial surface areas; sometimes, these runners remain functional in transporting and storing resources as tubers, whereas in other cases they soon disintegrate. A similar situation can be found in, for example, corals and bryozoans, where the tiny polyps can exchange resources through stolons.

Behavioural differentiation between individuals, such as between those in syphonophorans, invites to consider the whole colony an integrated individual, whereas the differentiation in colonial insects and mammals is still so loose that it is recognised as a group of coordinated individuals. Schools of fish {348}, bacterial colonies and forests {134} can behave as a super-individual,

These examples illustrate the vague boundaries of multicellularity, and even those of individuality. A sharpening of definitions or concepts may reduce the number of transition cases to some extent, but this cannot hide the fact that we are dealing here with a continuum of metabolic integration in the twilight zone between individuals and populations. This illustrates that organisms, and especially eukaryotes, need each other metabolically.

1.1.3 Why reserves apart from structure?

DEB theory partitions biomass into one or more reserves and one or more structures. Reserves complicate the dynamics of the individual and the application of the model considerably, so it makes sense to think about its necessity and become motivated to deal with this more complex dynamics.

We need reserve for the following reasons:

- to include metabolic memory. A variable substrate (food) supply does not combine easily with constant maintenance needs. Organisms use reserve(s) to smooth out fluctuations. The metabolic behaviour of an individual does not depend on the actual food availability, but of that of the (recent) past. Individuals react slowly to changes in their feeding conditions. This cannot be described realistically with the digestive system as a buffer, because its relaxation time is too short. Spectacular examples of prolonged action without food intake are the European, North American and New Zealand eels, *Anguilla*, which stop feeding at a certain moment. Their alimentary canal even degenerates, before the 3000-km-long journey to their breeding grounds where they spawn. The male emperor penguin *Aptenodytes forsteri* breeds its egg in Antarctic midwinter for 2 months and feeds the newly hatched chick with milky secretions from the stomach without access to food. The male loses some 40% of its body weight before assistance from the female arrives.

- to smooth out fluctuations in resource availability to make sure that no essential type of resource is temporarily absent {394}; growth can only proceed if all essential resources are available in certain relative amounts. This argument concerns a different form of memory that is used by multiple reserve systems. Single reserve systems evolved from multiple reserve systems. This will be discussed in Chapter 10. Non-limiting reserves can dam up, which causes strong changes in the composition of biomass, see {197}.

- the chemical composition of the individual depends on the growth rate. This can only be captured if biomass has more than one component.

- fluxes (e.g. dioxygen, carbon dioxide, nitrogen waste, heat) are linear sums of three basic energy fluxes: assimilation, dissipation and growth (as we will see). The method of indirect calorimetry is based on this fact. Without reserve, using a single structure only, two rather than three basic energy fluxes would suffice, while experimental evidence shows that this is not true.

- to explain observed patterns in respiration and in body size scaling relationships. Eggs decrease in mass during development, but increase in respiration, while juveniles increase in mass as well as in respiration. This cannot be understood without reserve. A freshly laid egg consists (almost) fully of reserve and hardly respires; a simple and direct empirical support for the DEB assumption that structure requires maintenance, but reserve does not. We will see that reserve plays a key role in body size scaling relationships, and to understand, for instance, why respiration increases approximately with weight to the power 3/4 among species.

- to understand how the cell decides on the use of a particular (organic) substrate, as building block or as source of energy. This problem will be discussed in the section on organelle–cytoplasm interactions at {282}.

The term reserve does *not* mean 'set apart for later use'; reserve 'molecules' can have active metabolic functions while 'waiting' for being used. Ribosomal RNA, for instance, turns out to belong to the reserve, see {143}; it is used for peptide elongation. The primary difference between reserve and structure is in their dynamics:

all chemical compounds in the reserve have the same turnover time, in the structure they can be different. Reserves are used to fuel all metabolic needs of the individual.

Most metabolic behaviour of animals, i.e. organisms that live off other organisms, can be understood using a single reserve, but autotrophs, which obtain nutrients independently from the environment, require the delineation of more reserves, as will be discussed in Chapter 5.

1.1.4 Metabolic switching is linked to maturation

Metabolic switches occur, for instance, at the start of development of an individual, the moment at which age is initiated in DEB theory. Another switch occurs when assimilation is initiated, a moment called birth, or when allocation to maturation is redirected to reproduction, a moment called puberty, or when cell division occurs or at which DNA duplication is initiated. The age at which such switches occur differs widely among individuals of the same species, depending on the food uptake in the past. The size at which the switches occur differs already much less, but still shows some scatter.

DEB theory links the occurrence of such metabolic switches to the level of maturity, i.e. the set of regulation systems that control metabolic performance. Although allocation to reproduction does not occur as long as maturity is still increasing, this does not imply that maturity directly relates to preparation of the reproductive machinery only. I see maturity as a much more general investment to prepare the body for the adult state, which involves, among other things, extensive gene regulation switching and cell and tissue differentiation. Its formal status is information, not energy, mass or entropy. The building up of maturity costs energy, and maturity is quantified as the cumulated energy or amount of reserve that is invested in maturity. After being used to build up maturity, this energy becomes lost. Maturation can be conceived as metabolic learning and can be compared with reading a book or a newspaper; this costs considerable energy but forgetting the information does not give an extra release of heat or an extra carbon dioxide emission.

In multicellular organisms birth typically precedes puberty, which naturally leads to three life-stages: embryo, juvenile and adult.

Embryo

The first stage is the embryonic one, which is defined as a state early in the development of the individual, when no food is ingested. The embryo relies on stored energy supplies. Freshly laid eggs consist, almost entirely, of stored energy, and for all practical purposes the initial structural volume of the embryo can realistically be assumed to be negligibly small. At this stage it hardly respires, i.e. it uses no dioxygen and does not produce carbon dioxide. (The shells of bird eggs initially produce a little carbon dioxide [125, 469].) In many species, this is a resting stage. This especially holds for plants, where seeds are equivalent to eggs; seeds can be dormant for many years and the number of dormant seeds greatly exceeds the number of non-dormant individuals [462]. Many seeds (particularly berries) require to be treated by the digestive juices of a particular animal species for germination; others need fire, see {428}. Although the seed or egg exchanges gas and water with the environment, it is otherwise a rather closed system.

Foetal development represents a variation on embryo development, where the mother provides the embryo with reserve material, such as in the Placentalia, some species of velvet worm *Peripatus* and the Devonian placoderm *Materpiscus* [717]. Complicated intermediates between reproduction by eggs and fetuses exist in fish [956, 1275, 1276], reptiles and amphibians [110, 899, 1112]. The evolutionary transition from egg to foetal development occurred several times independently. From the viewpoint of energetics, fetuses are embryos because they do not take food. The digestive system is not functional and the embryo does not have a direct impact on food supplies in an ecological sense. The crucial difference from an energetics point of view is the supply of energy to the embryo. In lecithotrophic species, nutrients are provided by the yolk of the ovum, whereas in matrotrophic species nutrients are provided by the mother as the foetus grows, not just in vitellogenesis. The fact that eggs are kept in the body (viviparity) or deposited in the environment (oviparity) is of no importance from an energetic perspective. (The difference is important in a wider evolutionary setting, of course.) As in eggs, a number of species of mammal have a developmental delay just after fertilisation, called diapause [1052].

Juvenile

The second stage in life history is the juvenile one, in which food is taken but resources are not yet allocated to the reproductive process. In some species, the developing juvenile takes a sequence of types of food or sizes of food particles. Most herbivores, for instance, initially require protein-rich diets that provide nitrogen for growth, see {185}. Some species, such as *Oikopleura*, seem to skip the juvenile stage. It does not feed as a larva, a condition known as lecithotrophy, and it starts allocating energy to reproduction at the moment it starts feeding. A larva is a morphologically defined stage, rather than an energy defined one. If the larva feeds, it is treated as a juvenile; if not, it is considered to be an embryo. So, the tadpole of the gastric-brooding frog *Rheobatrachus*, which develops into a frog within the stomach of the parent, should for energy purposes be classified as an embryo, because it does not feed. The switch from feeding to non-feeding as a larva seems to be made easily, from an evolutionary perspective. Sea urchins have developed a complex pattern of species that do or do not feed as a larva, even within the same genus, which comes with dramatic differences in larval morphology [1277, 1278, 1279]. Sperm of the sea urchin *Heliocidaris tuberculata*, which has feeding larvae, can fertilise eggs of *H. erythrogramma*, which has non-feeding larvae; the zygote develops into feeding hybrid larvae that resemble starfish larvae, similar to that of the distant ancestor of sea urchins and starfishes, some 450 Ma ago [928].

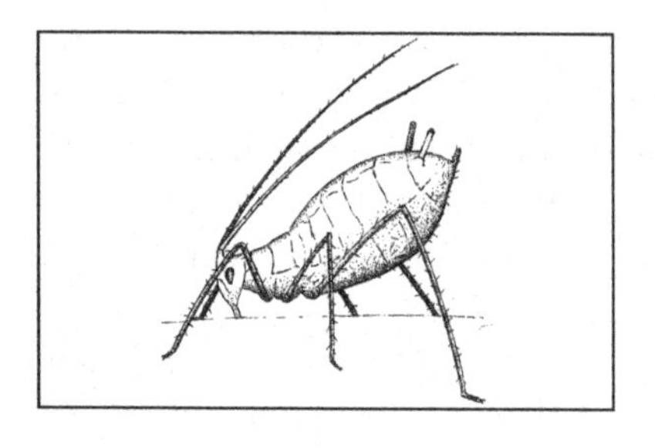

Parthenogenetic aphids have a spectacular mode of reproduction: embryos producing new embryos [596], see {358}. Since aphids are ovoviviparous, females carry daughters and granddaughters at the same time. From a formal point of view, the juvenile period is negative; the embryonic stage overlaps with the adult one. Aphids illustrate that the metabolic events of switching on feeding and reproduction matter, rather than the stages.

The word 'mammal' refers to the fact that the young usually receive milk from the mother during the first stage after birth, called the baby stage. Pigeons, flamingos and penguins also do this. The length of the baby stage varies considerably. If adequate food is available, the guinea-pig *Cavia* can do without milk [1052]. At weaning the young experience a dramatic change in diet, and after weaning the growth rate frequently drops substantially. Few biochemical transformations are required from milk to building blocks for new tissue. The baby, therefore, represents a transition stage between embryo and juvenile. The baby stage relates to the diet in the first instance, see {185}, and not directly to a stage in energetic development, such as embryo and juvenile. This can best be illustrated by the stoat *Mustela erminea*. Although blind for some 35–45 days, the female offspring reaches sexual maturity when only 42–56 days of age, before they are weaned. Copulation occurs whilst they are still in the nest [597, 1052].

Asexually propagating unicellular organisms take food from their environment, though they do not reproduce in a way comparable to the production of eggs or young by most multicellular organisms. For this reason, I treat them as juveniles in this energy-based classification of stages. Although I realise that this does not fit into standard biological nomenclature, it is a logical consequence of the present delineations. I do not know of better terms to indicate energy-defined stages, which highlights the lack of literature dealing with the individual-based energetics of both micro- and multicellular organisms. This book shows that both groups share enough features to try to place them in a single theoretical framework. Some multicellular organisms, such as some annelids, triclads and sea cucumbers (e.g. *Holothuria parvula* [323]), also propagate by division. Some of them sport sexual reproduction as well, causing the distinction between both groups to become less sharp and the present approach perhaps more amenable.

The eukaryotic cell cycle is usually partitioned into the interphase and mitotic phases; the latter is here taken to be infinitesimally short. The interphase is further decomposed into the first gap-phase, the synthesis phase (of DNA) and the second gap-phase. Most cell components are made continuously through the interphase, so that this distinction is less relevant for energetics. The second gap-phase is usually negligibly short in prokaryotes. Since the synthesis phase is initiated upon exceeding a certain cell size, size at division depends on growth conditions and affects the population growth rate. These phenomena are discussed in some detail on {279}.

In many species, the switch from the juvenile to the adult stage is hardly noticeable, but in the paradoxical frog, for instance, the switch comes with a dramatic change in morphology and a substantial reduction in size from 20 to 2 cm; the energy parameters differ between the stages. Holometabolic insects are unique in having a pupal stage between the juvenile and adult ones. It closely resembles the embryonic stage from an energetics point of view, see {284}. Pupae do not take food, and start synthesising (adult) tissue from tiny imaginal disks. A comparable situation occurs in echinoderms, bryozoans, sipunculans and echiurans, where the adult stage develops from a few undifferentiated cells of the morphologically totally different larva. In some cases, the larval tissues are resorbed, and so converted to storage materials; in other cases the new stage develops independently. When *Luidia sarsi* steps off its bipinnaria larva as a tiny starfish, the relatively large larva may continue to swim actively for another 3 months, [1136] in [1257]. Some jelly fishes (Scyphomedusae) alternate between an asexual stage, i.e. small sessile polyps, and a sexual stage, i.e. large free-swimming medusae. Many parasitic trematodes push this alternation of generations to the extreme. Mosses, ferns and relatives alternate between a gametophyte and a sporophyte stage; the former is almost completely suppressed in flowering plants. From an energetics perspective, the sequence embryo, juvenile is followed by a new sequence, embryo, juvenile, adult, with different values for energy parameters for the two sequences. The coupling between parameter values is discussed on {295}.

Adult

The third stage is the adult one, in which energy is allocated to the reproduction process. The switch from the juvenile to the adult stage, puberty, is here taken to be infinitesimally short. The actual length differs from species to species and behavioural changes are also involved. The energy flow to reproduction is continuous and usually quite slow, while reproduction itself is almost instantaneous. This can be modelled by the introduction of a buffer, which is emptied or partly emptied upon reproduction. The energy flow in females is usually larger than that in males, and differs considerably from species to species.

Some Florideophyceae (red algae) and Ascomyceta (fungi) have three sexes; most animals and plants have two, male and female, but even within a set of related taxa, an amazing variety of implementations can occur. Some species of mollusc and annelid, and most plants, are hermaphroditic, being male and female at the same time; some species of fish and shrimp are male during one part of their life and female during another part; plants such as the bog myrtle *Myrica gale* can change sex yearly; some have very similar sexes while other species show substantial differences between males and females; see Figure 1.1. The male can be bigger than the female, as in many mammals, especially sea elephants, or the reverse can occur, as in spiders and birds of prey. Males of some fish, rotifers and some echiurans are very tiny, compared to the female, and parasitise in or on the female or do not feed at all. The latter group combines the embryo stage with the adult one, not unlike aphids. Differences in ultimate size reflect differences in values for energy parameters, see {299}. Parameter values, however, are tied to each other, because it is not possible to grow rapidly without eating a lot (in the long run). Differences in energy budgets between sexes are here treated in the same way as differences between species.

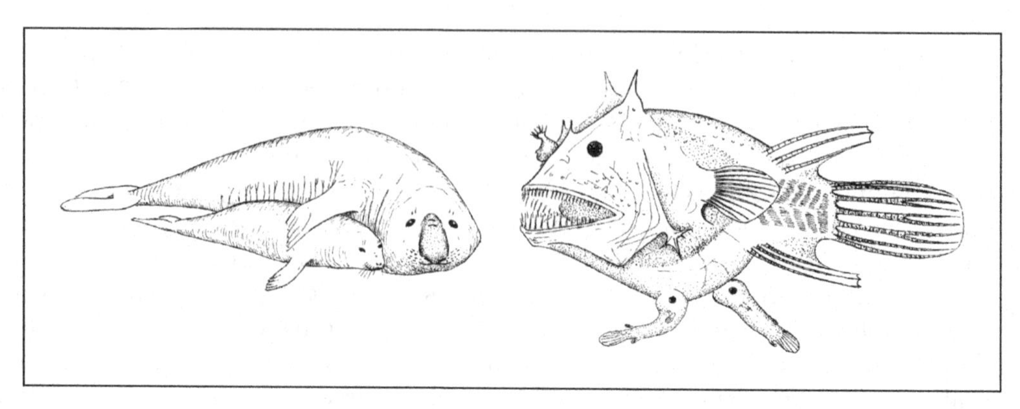

Figure 1.1 Sexual dimorphy can be extreme. The male of the southern sea elephant *Mirounga leonina* is 10 times as heavy as the female, while the parasitic males of the angler fish *Haplophryne mollis* are just pustules on the female's belly.

Reproduction, in terms of the production of offspring, does not always have a simple relationship with gamete production. All oocytes are already present at birth for future ovulations in birds and mammals, where they are arrested at prophase I of meiosis [798] (which occurs at the transition from the second gap-phase to the mitotic phase). In some species of tapeworm, wasp and at least 18 species of mammal (e.g. armadillo) there is a mode, called polyembryony, in which a sexually produced embryo splits into several genetically identical offspring. The opposite also occurs in several species of mammal (e.g. pronghorn, elephant shrews, bats, viscacha), where the mother reduces a considerable number of ova to usually two, early in the development, but also later on, by killing embryos [108]. Cannibalism among juveniles inside the mother has been described for *Salamandra*, some sharks and the sea star *Patiriella* {185}. Parent coots, *Fulica*, are known to drown some hatchlings of large litters, possibly to increase the likelihood of the healthy survival of the remaining ones.

In some species, e.g. humans, a senile stage exists, where reproduction diminishes or even ceases. This relates to the process of ageing, see {214}. An argument is presented for why this stage cannot be considered as a natural next stage within the context of DEB theory.

The summary of the nomenclature used here reads:

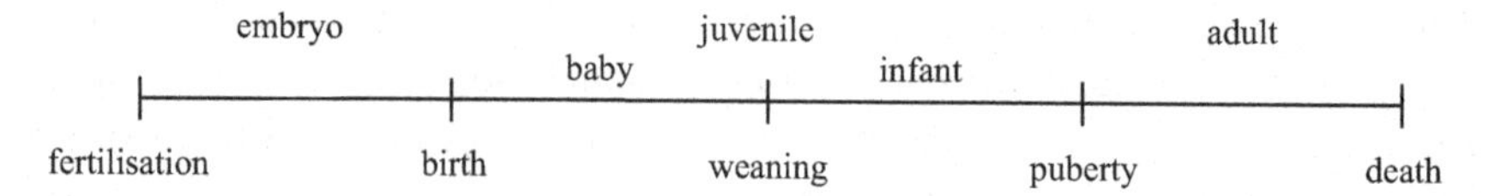

1.2 Homeostasis is key to life

Homeostasis is the ability to run metabolism independent of the (fluctuating) environment. All living systems do this to some extent and to capture this extent DEB theory makes use of several homeostasis concepts, which are discussed in this section.

The compounds that cells use to drive metabolism require enzymes for their chemical transformation. Compounds that react spontaneously are excluded or stored such that this cannot occur. In this way cells achieve full control over all transformations, because they synthesise enzymes, consisting of protein, themselves. No reaction runs without the assistance of enzymes. The properties of enzymes depend on their micro-environment. So homeostasis is essential for full control. Changes in the environment in terms of resource availability, both spatial and temporal, require the formation of reserve pools to ensure a continuous supply of essential compounds for metabolism. This implies a deviation from homeostasis for cells (or individuals) as a whole. The cell's solution to this problem is to make use of polymers that are not soluble. In this way these reserves do not change the osmotic value, and neither do they affect the capacity of monomers to do chemical work (see {79}). In many cases cells encapsulate the polymers in membranes, to reduce interference even further, at the same time increasing access, as many cellular activities are membrane bound.

1.2.1 Strong homeostasis: stoichiometric constraints

The chemical composition in small volumes, such as in bacterial cells and eukaryotic cell compartments is intrinsically stochastic, see {430}, and therefore fluctuates. So homeostasis is never perfect. DEB theory assumes that the chemical composition of reserve(s) and structure(s) are constant, an assumption called the strong homeostasis assumption. The basic idea is to delineate enough reserve(s) and structure(s) to approximate this situation, but for animals a single reserve and a single structure already captures most variation in the chemical composition of biomass, mainly because the variation in the chemical composition of their food is limited. The amounts of reserve(s) and structure(s) can vary, but not their chemical composition. The mixture of chemical compounds that make up these pools can, therefore, be considered as a single generalised compound.

To produce a compound of constant chemical composition, substrates for this production are required in particular relative amounts, which gives what is called stoichiometric constraints on production. A lot of ecological literature focuses on the availability of chemical elements [1110], but the production is from chemical compounds, however, not from chemical elements directly and the problem is that compounds can be transformed into other compounds, which complicates matters considerably. Primary production on Earth is mainly limited by nitrogen, for instance, while 70% of the atmospheric gases consist of dinitrogen; only a few organisms can use this nitrogen, however. The problem of specifying the constraints on production is one of the main tasks of DEB theory, which is a rather complex one because some compounds can partly replace others.

Reserve materials can be distinguished from materials of the structural mass by a change in relative abundance if resource levels change. This defining property breaks down in case of extreme starvation, when structural materials are degraded as well when reserves are exhausted. An example of this is the breakdown of muscle tissue during extreme starvation. Even if food intake is resumed, the structural component of muscle tissue does not recover in mammals such as ourselves.

Since the amount of reserves can change relative to the amount of structural material, the chemical composition of the whole body can change. That is, it can change in a particular way. This is a consequence of choosing energy as a state variable rather than the complete catalogue of all compounds.

1.2.2 Weak homeostasis: restrictions on dynamics

DEB theory also uses another homeostasis assumption: weak homeostasis. Its definition is that *if food density does not change* reserve density, i.e. the ratio between the amounts of reserve and structure, becomes constant *even when growth continues*; reserve and structure grow in harmony and biomass no longer changes in composition. This makes sense only if reserve and structure obey strong homeostasis, so weak homeostasis implies strong homeostasis, but is more restrictive. The fact that growth still can continue is essential for the weak homeostasis concept. Weak homeostasis applies to the whole body, not to its compartments, but under steady-state conditions only. Strong homeostasis has nothing to do with reserve dynamics, but weak homeostasis turns out to fully determine reserve dynamics, see {38}.

1.2.3 Structural homeostasis: isomorphy

Structural homeostasis is about shapes, not about chemical composition. For an understanding of energetics, only two aspects of size and shape are relevant, as is explained later: surface areas for acquisition processes and volumes for maintenance processes. The shape defines how these measures relate to each other. If an individual does not change in shape during growth, it is called an isomorph. Isomorphism is an important property that applies to the majority of species on Earth by approximation. The shape can be any shape and the comparison is only between the shapes that a single individual takes during its development. If organisms have a permanent exoskeleton, however, there are stringent constraints on their shape [637].

Two bodies of a different size are isomorphic if it is possible to transform one body into the other by a simple geometric scaling in three-dimensional space: scaling involves only multiplication, translation and rotation. This implies, as Archimedes already knew, that if two bodies have the same shape and if a particular length takes value L_1 and L_2 in the different bodies, the ratio of their surface areas is $(L_1/L_2)^2$ and that of their volumes $(L_1/L_2)^3$, irrespective of their actual shape. It is, therefore, possible to make assertions about the surface area and the volume of the body relative to some standard, on the basis of lengths only. One only needs to measure the surface area or volume if absolute values are required. This property is used extensively in this book.

Structural homeostasis is an assumption of the standard DEB model, but not of DEB models generally. Notice that length itself does not play a role in DEB theory and everywhere where it occurs while isomorphy is assumed, length actually stands for the ratio of volume and surface area. Section 4.2 at {124} considers changes in shape and its consequences for energetics.

Shape coefficients convert physical to volumetric lengths

Each length measure L_w needs a definition of how the length is taken. If we would relate quantities about the performance of an individual to its length, the parameter values in the description can differ substantially between two species, not because they would differ in performance, but because they differ in shape. To eliminate this effect, I typically work with volumetric length, being the cubic root of the volume, $L = V^{1/3}$; it is independent of the shape.

The shape coefficient $\delta_{\mathcal{M}}$ is defined as volume$^{1/3}$ length^{-1}, so the physical volume is given by $V_w = (\delta_{\mathcal{M}} L_w)^3$. The practical purpose of shape coefficients is to convert shape-specific length measures to volumetric lengths: $L = \delta_{\mathcal{M}} L_w$. It is specific for the particular way the length measure has been chosen. Thus the inclusion or exclusion of a tail in the length of an organism results in different shape coefficients. A simple way to obtain an approximate value for the shape coefficient belonging to length measure L_w is via the wet weight W_w, i.e. the weight of a living organism without adhering water, and the specific density d_{Vw} to convert weight into volume: $\delta_{\mathcal{M}} = (\frac{W_w}{d_{Vw}})^{1/3} L_w^{-1}$; the specific density d_{Vw} is typically close to 1 g cm^{-3}. So $W_w = d_{Vw} V_w$.

The following considerations may help in getting acquainted with the shape coefficient. For a sphere of diameter L_w and volume $L_w^3 \pi/6$, the shape coefficient is 0.806 with respect to the diameter. For a cube with edge L_w, the shape coefficient takes the value 1, with respect to this edge. The shape coefficient for a cylinder with length L_w and diameter L_ϕ is $(\frac{\pi}{4})^{1/3} (L_w/L_\phi)^{-2/3}$ with respect to the length.

The shapes of organisms can be compared in a crude way on the basis of shape coefficients. Figure 1.2 shows the distributions of shape coefficients among European birds and Neotropical mammals; they fit the normal distribution closely. Summarising statistics are given in Table 1.1, which includes European mammals as well. Some interesting conclusions can be drawn from the comparison of shape coefficients. They have an amazingly small coefficient of variation (cv), especially in birds (including sphere-like wrens and stick-like flamingos), which probably relates to constraints for

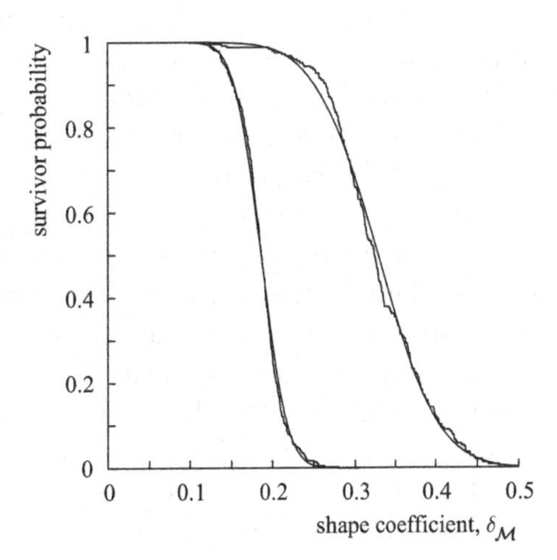

Figure 1.2 The sample survivor function (see Glossary) of shape coefficients for European birds (left) and Neotropical mammals (right). The lengths include the tail for the birds, but not for mammals. Data are from Bergmann and Helb [86] and Emmons and Feer [321]. The fitted survivor functions are those of the normal distribution.

Table 1.1 The means and coefficients of variation of shape coefficients of European birds and mammals and Neotropical mammals.

			tail included		tail excluded	
taxon	source	number	mean	cv	mean	cv
European birds	[167, 321]	418	0.186	0.14		
European mammals	[149]	128	0.233	0.27	0.335	0.28
Neotropical mammals	[86]	246	0.211	0.41	0.328	0.18

flight. Mammals have somewhat larger shape coefficients than birds. They tend to be more spherical, which possibly relates to differences in mechanics. The larger coefficient of variation indicates that the constraints are perhaps less stringent than for birds. The spherical shape is more efficient for energetics because cooling is proportional to surface area and a sphere has the smallest surface area/volume ratio, namely $6/L_\phi$. When the tail is included in the length, European mammals have somewhat larger shape coefficients than Neotropical mammals, but the difference does not arise when the tail is excluded. Neotropical mammals tend to have longer tails, which is probably because most of them are tree-dwellers. The temperature difference between Europe and the Neotropics does not result in mammals in Europe being more spherical to reduce cooling.

Shape at the subcellular level: membrane-cytosol interactions

Surface-area–volume relationships play an essential role in the communication between the extensive variable body size and intensive variables such as concentrations of compounds and rates of reaction between compounds. It is not difficult to imagine the physiological significance of isomorphism. Process-regulating substances in the body tend to have a short lifetime to cope with changes, so such substances have to be produced continuously. If some organ secretes at a rate proportional to its volume (i.e. number of cells), isomorphism will result in a constant concentration of the substance in the body. The way the substance exercises its influence does not have to change with changing body volume in order to obtain the same effect in isomorphs. Organisms and cells monitor their size, but the way they do this is considered to be an open problem [1213, p 123]; the argument in Figure 1.3 shows that organisms and cells do not need to accumulate compounds with increasing size to monitor their size.

Most enzymes can be conceived of as fluffy structures, with performance depending on the shape of the molecule's outer surface and the electrical charge distribution over it. If bound to a membrane, the outer shape of the enzyme changes into the shape required for the catalysis of the reaction specific to the enzyme. Membranes thus play a central role in cellular physiology [395, 465, 1239]. The change in surface area/volume ratios has important kinetic implications at all scales, including the micro-scale .

Many pathways require a series of transformations and so involve a number of enzymes. The binding sites of these enzymes on the membrane are close to each other,

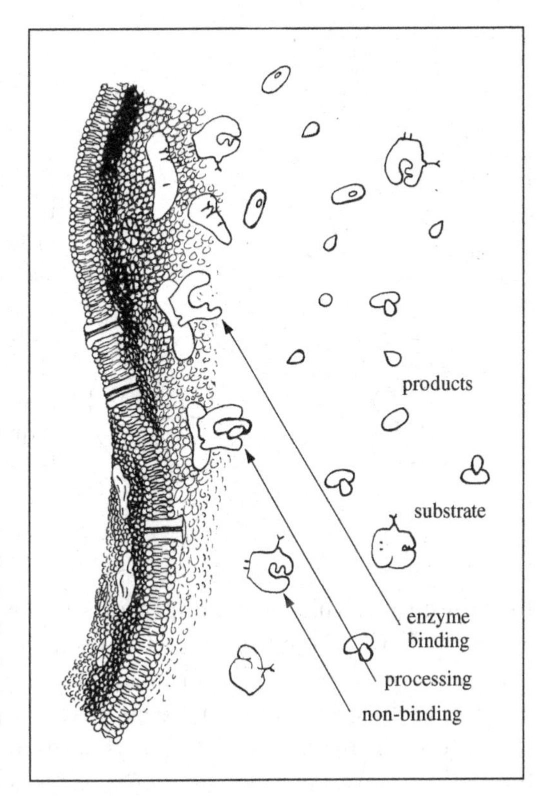

Figure 1.3 Each cell in the body 'knows' its volume by the ratio between its volume and the surface area of its membranes. This is because most enzymes only function if bound to a membrane, with their substrates and products in the cell volume as illustrated; the production of enzymes is a relatively slow process, while regulation is fast.

so that the product of one reaction does not disperse in the cytosol before being processed further. The product is just handed over to the neighbouring enzyme in a process called piping. Interplay between surface areas and volumes is basic to life, not only at the level of the individual, but also at the molecular level.

Reserve–structure interface at subcellular level

Figure 1.4 illustrates how isomorphy works out at the subcellular level for the distribution of reserve material in a matrix of structural material. Since monomers are osmotically active, their concentration in cells is typically very low, and reserve and structure are mainly present in form of polymers (carbohydrates, proteins, RNA) and lipids. The surface area of their interface is proportional to $EV^{-1/3} = E/L$, where E stands for the energy in reserve and V for the volume of structure of the cell and L for (volumetric) length. This implies that the number of reserve vesicles reduces, if the structural cell volume grows, but not the amount of reserves. This property is used in the mechanism for reserve dynamics, {41}.

1.2.4 Thermal homeostasis: ecto-, homeo- and endothermy

Temperature affects metabolic rates, see {17}, so control over metabolic rates requires control over body temperature. Heat is a side product of all uses of energy, see {160}.

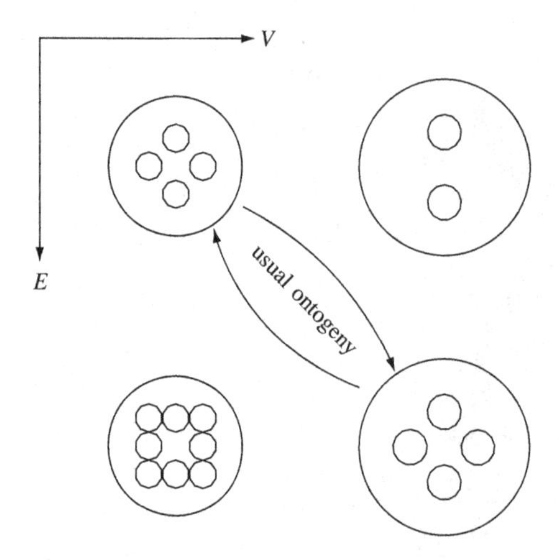

Figure 1.4 The structural cell volume V is growing to the right by a factor two, i.e. the cell diameter is growing by a factor $2^{1/3}$. The reserve E is growing to the bottom by a factor two, i.e. the number of blobs of E in V is growing by a factor two. V and E typically grow simultaneously; if E/V remains constant, the blobs of E in V grow as if we look through a magnification glass of increasing strength. The consequence is that the surface area of the interface between E and V is proportional to $EV^{-1/3}$.

In ectotherms, this heat simply dissipates without increasing the body temperature above that of the environment to any noticeable amount as long as the temperature is sufficiently low, especially in the aquatic environment. If the environmental temperature is high, as in incubated bird eggs just prior to hatching, metabolic rates are high as well, see {17}, releasing a lot more energy in the form of heat, which increases the body temperature even further, see {173}. This is called positive feedback in cybernetics. The rate of heat dissipation obviously depends on the degree of insulation and is directly related to surface area.

A small number of species, known as endotherms, use energy to maintain their body temperature at a predetermined high level, 27 °C in sloths (*Bradypus*), 34 °C in monotremes, 37 °C in most mammals, 39 °C in non-passerine birds, 41 °C in passerine birds. Mammals and birds change from ectothermy to endothermy during the first few days of their juvenile stage. Some species temporarily return to the ectothermic state or partly so at night (hummingbirds, insectivores) or during hibernation (poorwills [705], rodents, bats) or dry seasons (tenrecs, see {122}). Not all parts of the body are kept at the target temperature, especially not the extremities. The naked mole rat *Heterocephalus glaber* (see Figure 1.5) has a body temperature that is almost equal to that of the environment [719] and actually behaves as an ectotherm. Huddling in the nest plays an important role in the thermoregulation of this colonial species [1263]. The body temperature of Grant's golden mole *Eremitalpa granti* normally matches that of the sand in which it lives, but it is able to maintain the diurnal cycle of its body temperature if the temperature of the sand is kept constant [720].

Many ectotherms can approach the state of homeothermy under favourable conditions by moving from shady to sunny places, and back, in an appropriate way. In an extensive study of 82 species of desert lizards from three continents, Pianka [893] found that body temperature T_b relates to ambient air temperature T_e as

$$T_b = 311.8 + (1 - \beta)(T_e - 311.8) \tag{1.1}$$

Figure 1.5 The naked mole rat *Heterocephalus glaber* (30 gram) is one of the few mammals that are essentially ectothermic. They live underground in colonies of some 60 individuals. The single breeding female suppresses reproductive development of all 'frequent working' females and of most 'infrequent working' females, a social system that reminds us of termites [721].

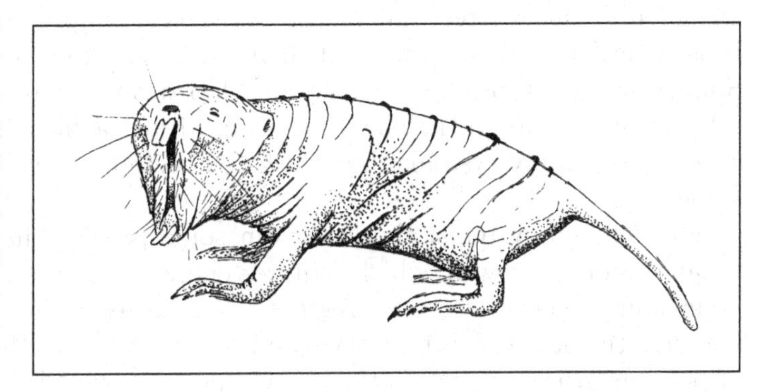

where β stands for the species-specific thermoregulatory capacity, spanning the full range from perfect regulation, $\beta = 1$, for active diurnal heliothermic species, to no regulation, $\beta = 0$, for nocturnal thigmothermic species. The target temperature of 311.8 K or 38.8 °C varies somewhat between the different sub-groups and is remarkably close to that of mammals. Many species of plants enhance the interception of radiation by turning their flowers to the moving sun. The parabolic shape of flowers helps to focus radiation on the developing ovum. Sunflowers, *Helianthus annuus*, follow the sun with their leaves and developing inflorescence, but when the flowers open they are oriented towards the east [682]. This probably relates to thermoregulation.

Several species can raise their temperature over 10 °C above that of the environment (bumble-bees and moths [487], tuna fish, mackerel shark, leatherback turtles *Dermochelys*). Some species of *Arum*, which live in dark forests, heat their flowers metabolically. These examples do make it clear that energy investment in heating is species-specific and that the regulation of body temperature is a different problem.

The 'advantages' of homeothermy are that enzymes can be used that have a narrow tolerance range for temperatures and that activity can be maintained at a high level independent of environmental temperature. At low temperatures ectotherms are easy prey for endotherms. Development and reproduction are enhanced, which opens niches in areas with short growing seasons that are closed to ectotherms. The costs depend on the environmental temperature, insulation and body size. If temperature is high and/or insulation is excellent and/or body size is large, there may be hardly any additional costs of heating; the range of temperatures to which this applies is called the thermo-neutral zone.

1.2.5 Acquisition homeostasis: supply and demand

Individuals can be ranked on the spectrum from supply to demand systems. For supply systems, the lead is in the feeding process, which offers an energy input to the individual. The available energy flows to different destinations, more or less as water flows through a river delta. A sea anemone is an example of a 'supply' type of animal.

It is extremely flexible in terms of growth and shrinkage, which depend on feeding conditions. It can survive a broad range of food densities. Japanese bonsai cultures cannot illustrate better that plants are typical supply systems as well. Supply organisms typically move less and find their food via a kind of (activated) diffusion process. Supply systems typically have less developed sensors and are metabolically more flexible and vary more in the chemical composition of their bodies. By far the majority of species are supply systems, but the few demand systems get relatively more research attention.

For demand systems, the lead is in some process that uses energy, such as maintenance and/or growth, which requires energy intake of matching size. Food-searching behaviour is then subjected to regulation processes in the sense that an animal eats what it needs; the nervous system plays an important role in this regulation [812]. The range of possible growth curves is thus much more restricted. Demand systems evolved from supply systems, see {422}, and froze existing metabolic rules, lost metabolic flexibility (to deal with extreme starvation conditions), but increased in behavioural flexibility. All demand systems are animals, i.e. organisms that feed on other organisms; they are often mobile and move to where the food is. Hence they less frequently encounter extreme starvation conditions; they typically cannot shrink during starvation, but die. The increased behavioural flexibility gives them the possibility to specialise on one type of food species and translates in a small value for the half saturation coefficient for demand systems. They also have a relatively large difference between the peak and the standard metabolic rate, and have typically closed circulation systems (efficient transport under extreme metabolic performance), some developed endothermy (birds and mammals) and many have highly developed sensors. Closed circulation systems developed in vertebrates, echinoderms, cephalopods and annelids. Altogether only a small fraction of the animal species have a closed circulatory system and most of these species still have a position near the supply end of the spectrum. Demand systems typically sport accelerated ageing, where a high survival probability during the juvenile period is combined with a relatively fast ageing during the adult stage, see {219}.

Even in the 'supply' case, growth is typically regulated carefully by hormonal control systems (see Figure 2.14). This is because growth should not proceed faster than the rate at which the energy and elementary compounds necessary to build the new structures can be mobilised, nor should growth proceed slower, else metabolites accumulate locally which gives several problems. Models that describe growth as a result of hormonal regulation should deal with the problem of what determines the hormone levels. This requires studying organisation at the individual level. The conceptual role of hormones is linked to the similarity of growth patterns despite the diversity of regulating systems, see Figure 2.11. In the DEB theory, messengers such as hormones are part of the physiological machinery that an organism uses to regulate its growth. Their functional aspects can only be understood by looking at other variables and compounds.

The development of demand systems from supply systems can be seen as a step up in the degree of homeostasis.

1.3 Temperature affects metabolic rates

1.3.1 Arrhenius temperature

All metabolic rates depend on the body temperature. For a species-specific range of temperatures, the description proposed by S. Arrhenius in 1889, see, e.g. [410], usually fits well

$$\dot{k}(T) = \dot{k}_1 \exp\left(\frac{T_A}{T_1} - \frac{T_A}{T}\right) \tag{1.2}$$

with T the absolute temperature (in Kelvin), T_1 a chosen reference temperature, the parameter T_A the Arrhenius temperature, $\dot{k}$ a (metabolic) reaction rate and $\dot{k}_1$ its value at temperature T_1. So, when $\ln \dot{k}$ is plotted against T^{-1}, a straight line results with slope $-T_A$, as Figure 1.6 illustrates.

Arrhenius based this formulation on the van't Hoff equation for the temperature coefficient of the equilibrium constant and amounts to $\dot{k}(T) = \dot{k}_\infty \exp(\frac{-E_a}{RT})$, where $\dot{k}_\infty$ is known as the frequency factor, R is the gas constant $8.31441\,\mathrm{J\,K^{-1}\,mol^{-1}}$, and E_a is called the activation energy. Justification rests on the collision frequency which obeys the law of mass action, i.e. it is proportional to the product of the concentrations of the reactants. The Boltzmann factor $\exp(\frac{-E_a}{RT})$ stands for the fraction of molecules that manage to obtain the critical energy E_a to react.

In chemistry, the activation energy is known to differ widely between different reactions. Processes such as the incorporation of [^{14}C]leucine into protein by membrane-bound rat-liver ribosomes have an activation energy of 180 $\mathrm{kJ\,mol^{-1}}$ in the range 8–20 °C and 67 $\mathrm{kJ\,mol^{-1}}$ in the range 22–37 °C. The difference is due to a phase transition of the membrane lipids, [1167] after [15]. Many biochemical reactions seem to have an activation energy in this range [1097]. This supports the idea that the value of activation energy is a constraint for functional enzymes in cells.

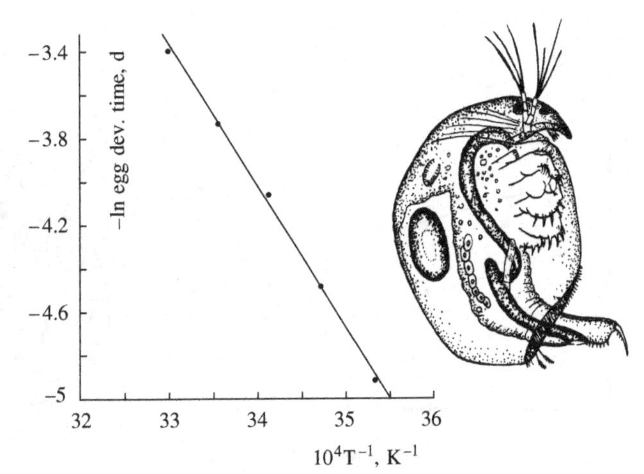

Figure 1.6 The Arrhenius plot for the development rate for eggs of the waterflea *Chydorus sphaericus*, i.e. the inverse time between egg laying and hatching. Data from Meyers [790].

Glasstone *et al.* [410] studied the thermodynamic basis of the Arrhenius relationship in more detail. They came to the conclusion that this relationship is approximate for bimolecular reactions in the gas phase. Their absolute rate theory for chemical reactions proposes a more accurate description where the reaction rate is proportional to the absolute temperature times the Boltzmann factor. This description, however, is still an approximation [410, 510].

The step from a single reaction between two types of particles in the gas phase to metabolic rates where many compounds are involved and gas kinetics do not apply is, of course, enormous. Due to the somewhat nebulous application of thermodynamics to describe how metabolic rates depend on temperature, I prefer to work with the Arrhenius temperature, rather than the activation energy. I even refrain from the improvement offered by Glasstone's theory, because the small correction does not balance the increase in complexity of the interpretation of the parameters for biological applications. The Arrhenius relationship seems to describe the effect of temperature on metabolic rates with acceptable accuracy in the range of relevant temperatures.

1.3.2 Coupling of rates in single reserve systems

Figure 1.7 shows that the Arrhenius temperatures for different rates in a single species are practically the same. If each reaction depended in a different way on temperature, cells or individuals would have a hard time coordinating the different processes at fluctuating temperatures. Metabolism is about the conversion of chemical compounds by organisms, for which they use a particular biochemical machinery that operates with a particular efficiency in ways that do not depend on temperature. Obviously, animals cannot respire more without eating more. As a first approximation it is realistic to assume that all metabolic rates in a single individual are affected by temperature in the same way, so that a change in temperature amounts to a simple transformation of time.

Table 1.2 gives Arrhenius temperatures, T_A, for several species. It ranges somewhere between 6 and 12.5 kK. Many experiments do not allow for an adaptation

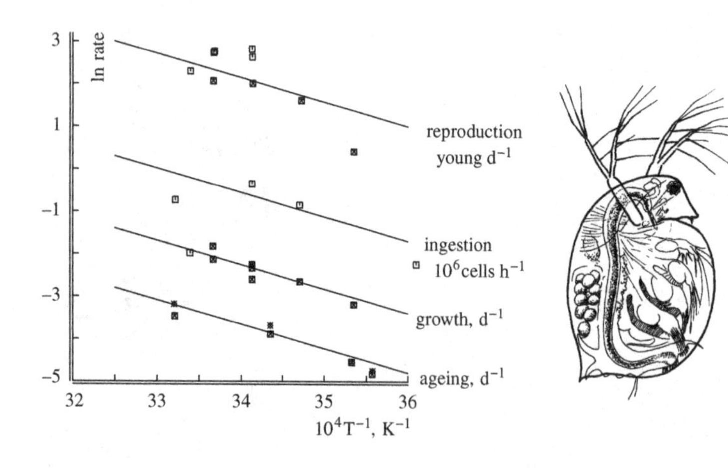

Figure 1.7 The Arrhenius plot for reproduction, ingestion, von Bertalanffy growth and Weibull ageing of *Daphnia magna*; from [658]. The Arrhenius temperature is 6400 K. ◇ males, □ females. Food: the algae *Scenedesmus subspicatus* (open symbols) or *Chlorella pyrenoidosa* (filled symbols). The ingestion and reproduction rates refer to 4-mm individuals.

Table 1.2 Arrhenius temperatures as calculated from literature data on the growth of ectothermic organisms. The values for the mouse cells are obtained from Pirt [901]. The other values were obtained using linear regressions.

species	range (°C)	T_A (K)	type of data	source
Escherichia coli	23–37	6 590	population growth	[802]
Escherichia coli	26–37	5 031	population growth	[545]
Escherichia coli	12–26	14 388	population growth	[545]
psychrophilic pseudomonad	12–30	6 339	population growth	[545]
psychrophilic pseudomonad	2–12	11 973	population growth	[545]
Klebsiella aerogenes	20–40	7 159	population growth	[1166]
Aspergillus nidulans	20–37	7 043	population growth	[1168]
nine species of algae	13.5–39	6 842	population growth	[418]
mouse tissue cells	31–38	13 834	population growth	[1220]
Nais variabilis	14–29	9 380	population growth	[569]
Pleurobrachia pileus	5–20	10 000	Bertalanffy growth	[433]
Mya arenaria	7–15	13 000	Bertalanffy growth	[30]
Daphnia magna	10–26.5	6 400	Bertalanffy growth	[636]
Ceriodaphnia reticulata	20–26.5	6 400	Bertalanffy growth	[636]
Calliopius laeviusculus	6.5–15	11 400	Bertalanffy growth	[241]
Perna canaliculus	7–17	5 530	lin. growth	[506]
Mytilus edulis	6.5–18	8 460	lin. growth larvae	[1095]
Cardium edule and *C. glaucum*	10–30	8 400	lin. growth larvae	[598]
Scophthalmus maximum	8–15	15 000	lin. growth larvae	[560]
25 species of fish	6–29	11 190	embryonic period	[769]
Brachionus calyciflorus	15–25	7 800	embryonic period	[448]
Chydorus sphaericus	10–30	6 600	embryonic period	[790]
Canthocampus staphylinus	3–12	10 000	embryonic period	[1012]
Moraria mrazeki	7–16.2	13 000	embryonic period	[1012]

period, which affects the resulting value. The problem is that alloenzymes are produced with somewhat different temperature–activity relationships when temperature changes. This takes time, depending on species and body size. Without an adaptation period, the performance of enzymes adapted to one temperature is measured at another temperature, which affects the apparent Arrhenius temperature.

1.3.3 States can depend on temperature via rates

The standard DEB model implies that ultimate size results from a ratio of two rates, see {52}, so it should not depend on the temperature, as all rates are affected in the same way. Table 1.3 confirms this for two species of daphnids cultured under well-standardised conditions and abundant food [636]. It is also consistent with the observation by Beverton, see appendix to [202], that the walleye *Stizostedion vitreum* matures at 2 years at the southern end of its range in Texas and at 7 or 8 years in

Table 1.3 The von Bertalanffy growth rate for the waterfleas *Ceriodaphnia reticulata* and *Daphnia magna*, reared at different temperatures in the laboratory both having abundant food. The length at birth is 0.3 and 0.8 mm respectively

	Ceriodaphnia reticulata				*Daphnia magna*			
temp °C	growth rate a^{-1}	sd a^{-1}	ultimate length mm	sd mm	growth rate a^{-1}	sd a^{-1}	ultimate length mm	s.d mm
10					15.3	1.4	4.16	0.16
15	20.4	4.0	1.14	0.11	25.9	1.3	4.27	0.06
20	49.3	3.3	1.04	0.09	38.7	2.2	4.44	0.09
24	57.3	2.6	1.06	0.01	44.5	1.8	4.51	0.06
26.5	74.1	4.4	0.95	0.02	53.3	2.2	4.29	0.06

northern Canada, while the size at maturation of this fish is the same throughout its range.

Although ultimate sizes are not rates, they are frequently found to decrease with increasing temperature. The reason may well be that the feeding rate increases with temperature, so, at higher temperatures, food supplies are likely to become limited, which reduces ultimate size. I discuss this phenomenon in more detail in relation to the Bergmann rule, {297}. For a study of the effects of temperature on size, it is essential to test for the equality of food density. This requires special precautions.

Apart from effects on rate parameters, temperature can affect egg size [327] and sex. High temperatures produce males in lizards and crocodiles, and females in turtles [255, 1049], within a range of a few degrees.

We will see that the fractionation of isotopes depends on temperature, {155}, not because the fractionation mechanism itself does, {97}, but the rates that generate fractionation depend on temperature.

1.3.4 Patterns in Arrhenius temperatures

The catalysing rate of enzymes in metabolic transformations can be adapted by the individual to the current temperature by changing the tertiary configuration. This takes time, up to days to weeks depending on the detailed nature of the adaptation. This time-dependence is frequently a reason for conflicting results on the effects of temperature on rates as reported in the literature.

Species living in habitats that typically sport large (and rapid) temperature fluctuations (e.g. juvenile and adult bivalves that live in the intertidal zones of sea coasts) have to use enzymes that function well in a broad temperature range, with the result that they have a relatively low Arrhenius temperature (around 6 kK). Species that live in habitats with a rather constant temperature (e.g. the larvae of the intertidal bivalves that live in the pelagic, or the deep ocean) typically have a high Arrhenius temperature (around 12 kK). The Arrhenius temperature can thus change with the stage in some species.

1.3.5 van't Hoff coefficient

A common way to correct for temperature differences in physiology is on the basis of Q_{10} values, known as van't Hoff coefficients. The Q_{10} is the factor that should be applied to rates for every 10 °C increase in temperature: $\dot{k}(T) = \dot{k}(T_1)Q_{10}^{(T-T_1)/10}$. The relationship with the Arrhenius temperature is thus $Q_{10} = \exp(\frac{10T_A}{T(T+10)})$. Because the range of relevant temperatures is only from about 0 to 40 °C, the two ways to correct for temperature differences are indistinguishable for practical purposes.

1.3.6 Temperature tolerance range

At low temperatures, the actual rate of interest is usually lower than expected on the basis of (1.2). If the organism survives, it usually remains in a kind of resting phase, until the temperature rises again. For many seawater species, this lower boundary is between 0 and 10 °C, but for terrestrial species it can be much higher; caterpillars of the large-blue butterfly *Maculinea rebeli*, for instance, cease to grow below 14 °C [318]. The lower boundary of the temperature tolerance range frequently sets boundaries for geographical distribution. Tropical reef-building corals only occur in waters where the temperature never drops below 18 °C; cold deep-water reefs have different species. Plants can experience chilling injury if the temperature drops below a species-specific threshold.

At temperatures that are too high, the organism usually dies. At 27 °C, *Daphnia magna* grows very fast, but at 29 °C it dies almost instantaneously. The tolerance range is sharply defined at the upper boundary. A few degrees rise of the seawater temperature, due to the intense 1998 El Niño event, caused death and the subsequent bleaching of vast areas of coral reef. It will take them decades to recover. Nisbet [838] gives upper temperature limits for 46 species of Protozoa, ranging from 33 to 58 °C. Thermophilic bacteria and organisms living in deep ocean thermal vents thrive at temperatures of 100 °C or more. The width of the tolerance range depends on the species; many endotherms have an extremely small one around a body temperature of 38 °C.

The existence of a tolerance range for temperatures is of major evolutionary importance; many extinctions are thought to be related to changes in temperature. This is the conclusion of an extensive study by Prothero, Berggren and others [924] on the change in fauna during the middle–late Eocene (40–41 Ma ago). This can most easily be understood if the ambient temperature makes excursions outside the tolerance range of a species. If a leading species in a food chain is the primary victim, many species that depend on it will follow. The wide variety of indirect effects of changes in temperature complicates a detailed analysis of climate-related changes in faunas. Grant and Porter [426] discuss in more detail the geographical limitations for lizards set by temperature, if feeding during daytime is only possible when the temperature is in the tolerance range, which leads to constraints on ectotherm energy budgets.

1.3.7 Outside the temperature tolerance range

Sharpe *et al.* [1029, 1045] proposed a quantitative formulation for the reduction of rates at low and high temperatures, based on the idea that the rate is controlled by an enzyme that has an inactive configuration at low and high temperatures. The reaction to these two inactive configurations is taken to be reversible with rates depending on temperature in the same way as the reaction that is catalysed by the enzyme, however the Arrhenius temperatures might differ. This means that the reaction rate has to be multiplied by the enzyme fraction that is in its active state, which is assumed to be at its equilibrium value. This fraction is

$$s(T)/s(T_1), \quad \text{with } s(T) = \left(1 + \exp\left(\frac{T_{AL}}{T} - \frac{T_{AL}}{T_L}\right) + \exp\left(\frac{T_{AH}}{T_H} - \frac{T_{AH}}{T}\right)\right)^{-1} \qquad (1.3)$$

where T_L and T_H relate to the lower and upper boundaries of the tolerance range and T_{AL} and T_{AH} are the Arrhenius temperatures for the rate of decrease at both boundaries. All are taken to be positive and all have dimension temperature. We usually find $T_{AH} \gg T_{AL} \gg T_A$. Figure 1.8 illustrates the quantitative effect of applying the correction factor.

The effects of chemical compounds on individuals can be captured using three ranges of internal concentrations: 'too little', 'enough' and 'too much', {235}. This approach has a nice similarity with the effects of temperature using the temperature tolerance range.

1.3.8 Uncoupling of rates in multiple reserve systems

The interception of photons by chlorophyll is less effected by temperature than dioxygen or carbon dioxide binding by Rubisco, which implies an enhanced electron leak at low

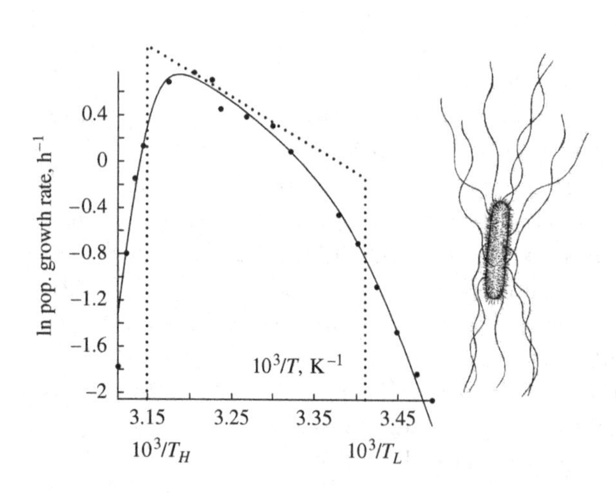

Figure 1.8 The Arrhenius plot for the population growth rate of *Escherichia coli* B/r on rich complex medium. Data from Herendeen *et al.* [496]. The Arrhenius temperatures for the growth rate, and for both deactivation rates are T_A = 4 370 K, T_{AL} = 20 110 K, and T_{AH} = 69 490 K. The dotted line shows the Arrhenius relationship with the same value for T_A and the population growth rate (1.94 h^{-1} at T_1 = 310 K), but without accounting for the deactivation, between the upper and lower boundaries of the tolerance range, T_L = 293 K and T_H = 318 K.

temperatures. Photosynthesis is known to depend on temperature at high light levels, but hardly so at low light levels [111, 755]. A build-up of carbohydrate reserve in a multiple reserve system can occur at low temperature, so a shift in the composition of biomass, which affects its nutritional value for consumers of this biomass. In the case of a single generalised reserve, this flexibility is absent, and the other rates (growth, reproduction, etc.) must follow the temperature dependence of the assimilation process to avoid changes in conversion efficiencies.

The solubility of dioxygen in water decreases less with temperature than that of carbon dioxide, which means that the compensation point, see {191}, i.e. the ratio of the carbon dioxide to the dioxygen partial pressures for which photorespiration balances photosynthesis, increases with temperature [680]. This leads to an optimum relationship of photosynthesis with temperature, but the location of the optimum is highly adaptable, and can change during the season in a single individual.

1.4 Summary

This chapter dealt with some basic concepts that are required to set up the DEB theory systematically, without too many asides.

The individual is introduced as the basic level of organisation with typically three stages (embryo, juvenile and adult) for multicellulars and one (juvenile) for unicellulars. The delineation is motivated of structure that requires maintenance, of reserve that quantifies metabolic memory and of maturity that controls metabolic switching.

The concept of homeostasis is discussed, which is subtle because homeostasis is not perfect and takes several qualities: strong, weak, structural, thermal and acquisition. Structural homeostasis is discussed in some detail because it controls surface area to volume relationships. This is important because uptake is coupled to surface area, and maintenance to volume. I argued that changes in surface area to volume relationships inform molecules about the size of the structure.

The effects of temperature on metabolic rates are quantified and I argued why the different rates in a single reserve systems depend on temperature in the same way, while multiple reserves allow for more degrees of freedom.

2

Standard DEB model in time, length and energy

This chapter discusses the simplest non-degenerated DEB model that is implied by DEB theory, the standard or canonical DEB model, to show the concepts of the previous chapter in action. The next chapter introduces more concepts on chemical transformations to deal with more complex situations. The standard DEB model assumes isomorphy and has a single reserve and a single structure, which is appropriate for many aspects of the metabolic performance of animals; other organisms typically require more reserves, and some (e.g. plants) also more structures. So in this chapter, we keep an animal in mind as a reference, which helps to simplify the phrasing. In this chapter substrate (food) has a constant composition that matches the needs of the individual. Food density in the environment might vary in the standard DEB model, but the discussion of what happens during prolonged starvation is delayed to Chapter 4. The discussion of mass aspects is also delayed and we here only use time t, length L and energy E. Length L is the volumetric length, $L = V^{1/3}$, where V is the structural volume. We use energy only conceptually, and typically in scaled form, and also delay the discussion of its quantification. The discussion of energy aspects does not imply that the individual should be energy-limited.

The logic of the energy flows will be discussed in this chapter and we start with a brief overview in this introductory section. We here keep the amount of detail to a minimum, neglecting all fast process, and focusing on the slow ones that matter on a life-cycle basis. Since reserve dynamics is slow relative to gut content dynamics, see {263}, the latter is not part of the standard DEB model. The dynamics of blood composition is linked to the dynamics of gut contents, so we neglect the blood compartment as well. Most aspects of behaviour are even faster than the dynamics of gut contents, so behaviour is here treated in very much reduced form.

The relationships between the different processes are schematically summarised in Figure 2.1. Food is ingested by a post-embryonic animal, transformed into faeces and egested; defecation is a special case of product formation. The feeding rate depends on food availability and the amount of structure. Energy, in the form of metabolites, is derived from food and added to the reserve. The reserve is mobilised at a rate that

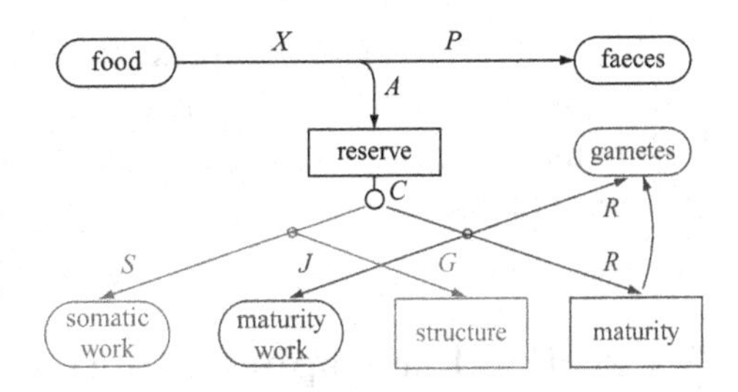

Figure 2.1 Energy fluxes in the standard DEB model. The rounded boxes indicate sources or sinks. The symbols stand for: X food intake; P defecation; A assimilation; C mobilisation; S somatic maintenance; J maturity maintenance; G growth; R reproduction. See Figure 10.4 for an evolutionary setting of this diagram.

depends on the amount of reserve and the amount of structure, and used for metabolic purposes. A fraction κ of the mobilised reserve flux is used for somatic maintenance plus growth, the rest for maturity maintenance plus maturation (in juveniles) or reproduction (in adults). Somatic maintenance has priority over growth, so growth ceases if all energy available for maintenance plus growth is used for somatic maintenance. Likewise maturity maintenance has priority over maturation or reproduction. Reserve that is allocated to reproduction is first collected in a buffer; the reproduction buffer has species-specific handling rules for transformation into eggs that typically leave the body upon formation. Eggs consist initially almost exclusively of reserve, the amount of structure and the level of maturity being negligibly small. The reserve density at birth (hatching) equals that of the mother at egg formation, a maternal effect.

Each of these processes will be quantified in the following sections on the basis of a set of simple assumptions that are collected in Table 2.4.

2.1 Feeding

Feeding is part of the behavioural repertoire and, therefore, notoriously erratic compared with other processes involved in energetics. The three main factors that determine feeding rates are food availability, body size and temperature.

2.1.1 Food availability is per volume or surface area of environment

For some species it is sensible to express food availability per surface area of environment, for others food per volume makes more sense, and intermediates also exist. The body size of the organism and spatial heterogeneity of the environment hold the keys to the classification. Food availability for krill, which feed on algae, is best expressed in terms of biomass or biovolume per volume of water, because this links up with processes that determine filtering rates. The spatial scale at which algal densities differ is large with respect to the body size of the krill. Baleen whales, which feed on krill, are intermediate between surface and volume feeders because some dive below the top layer, where most algae and krill are located, and sweep the entire column to the surface; so it does not matter where the krill is in the column. Cows and lions are typically surface feeders and food availability is most appropriately expressed in terms of biomass per surface area.

To avoid notational complexity, we here express food density X relative to the value that results in half of the maximum food uptake rate, K, and treat $x = X/K$ as a dimensionless quantity. K is known as the half-saturation coefficient or Michaelis–Menten constant.

2.1.2 Food transport is across surface area of individual

Organisms use many methods to obtain their meal; some sit and wait for the food to pass by, others search actively. Figure 2.2 illustrates a small sample of methods, roughly classified with respect to active movements by prey and predator. The food items can be very small with respect to the body size of the individual and rather evenly distributed over the environment, or the food can occur in a few big chunks. This section briefly mentions some feeding strategies and explains why feeding rates tend to be proportional to the surface area when a small individual is compared to a large one of the same species. (Comparisons between species are made in Chapter 8, {295}.) The examples illustrate a simple physical principle: mass transport from one environment to another, namely to the organism, must be across a surface, so the ingestion rate in watts is

$$\dot{p}_X = \{\dot{p}_X\}L^2 \tag{2.1}$$

where L is the volumetric length of the individual, and the specific ingestion rate $\{\dot{p}_X\}$ is some function of food density. Notice that not all of the surface of an isomorph needs to be involved in food acquisition, it might use a fixed fraction of it, such as gut surface area.

Marine polychaetes, sea anemones, sea lilies and other species that feed on blind prey are rather apathetic. Sea lilies simply orient their arms perpendicular to an existing current (if mild) at an exposed edge of a reef and take small zooplankters by grasping them one by one with many tiny feet. The arms form a rather closed fan in mild currents, so the active area is proportional to the surface area of the animal. Sea gooseberries stick plankters to the side branches of their two tentacles using cells that are among the most complex in the animal kingdom. Since the length of the side branches as well as the tentacles are proportional to the length of the animal, the encounter probability is proportional to a surface area.

Filter feeders, such as daphnids, copepods and larvaceans, generate water currents of a strength that is proportional to their surface area [158], because the flapping frequency of their limbs or tails is about the same for small and large individuals [911], and the current is proportional to the surface area of these extremities. (Allometric regressions of currents give a proportionality with length to the power 1.74 [147], or 1.77 [312] in daphnids. In view of the scatter, they agree well with a proportionality with squared length.) The ingestion rate is proportional to the current, so to squared length. Allometric regressions of ingestion rates resulted in a proportionality with length to the power 2.2 [771], 1 [912], 2.4–3 [262] and 2.4 [863] in daphnids. This wide range of values illustrates the limited degree of replicatability of these types of measurements. This is partly due to the inherent variability of the feeding process, and partly to the technical complications of measurement. Feeding rate depends on food density, as is discussed on {33}, while most measurement methods make use of changes in food densities so that the feeding rate changes during measurement. Figure 2.10 illustrates results obtained with an advanced technique that circumvents this problem [330].

The details of the filtering process differ from group to group. Larvaceans are filterers in the strict sense; they remove the big particles first with a coarse filter and collect the small ones with a fine mesh, see Figure 2.19. The collected particles are transported to the mouth in a mucous stream generated by a special organ, the endostyle. Copepods take their minute food particles out of the water, one by one with grasping movements [1182]. Daphnids exploit centrifugal force and collect them in a groove; Figure 2.10 shows that the resulting feeding rate is proportional to squared length. Ciliates, bryozoans, brachiopods, bivalves and ascidians generate currents, not by flapping extremities but by beating cilia on part of their surface area. The ciliated part is a fixed portion of the total surface area [366], and this again results in a filtering rate proportional to squared length; see Figure 2.3.

Some surface feeding animals, such as crab spiders, trapdoor spiders, praying mantis, scorpion fish and frogs, lie in ambush; their prey will be snatched upon arrival within reach, i.e. within a distance that is proportional to the length of a leg, jaw or tongue. The catching probability is proportional to the surface area of the predatory isomorphs. When aiming at a prey with rather keen eyesight, they must hide or apply camouflage.

Many animals search actively for their meal, be it plant or animal, dead or alive. The standard cruising rate of surface feeders tends to be proportional to their length, because the energy investment in movement as part of the maintenance costs tends to be proportional to volume, while the energy costs of transport are proportional to surface area; see {32}. Proportionality of cruising rate to length also occurs if limb movement frequency is more or less constant [922]. The width of the path searched for food by cows or snails is proportional to length if head movements perpendicular to the walking direction scale isomorphically. So feeding rate is again proportional to surface area, which is illustrated in Figure 2.4 for the pond snail.

The duration of a dive for the sperm whale *Physeter macrocephalus*, which primarily feeds on squid, is proportional to its length, as is well known to whalers [1221]. This can be understood, since the respiration rate of this endotherm is approximately proportional to surface area, as I argue on {147}, and the amount of reserve dioxygen is proportional to volume on the basis of a homeostasis argument. It is not really obvious how this translates into the feeding rate, if at all; large individuals tend to feed on large prey, which occur less frequently than small prey. Moreover, time investment in hunting can depend on size as well. If the daily swimming distance during hunting were independent of size, the searched water volume would be about proportional to surface area for a volume feeder such as the sperm whale. If the total volume of squid per volume of water is about constant, this would imply that feeding rate is about proportional to surface area.

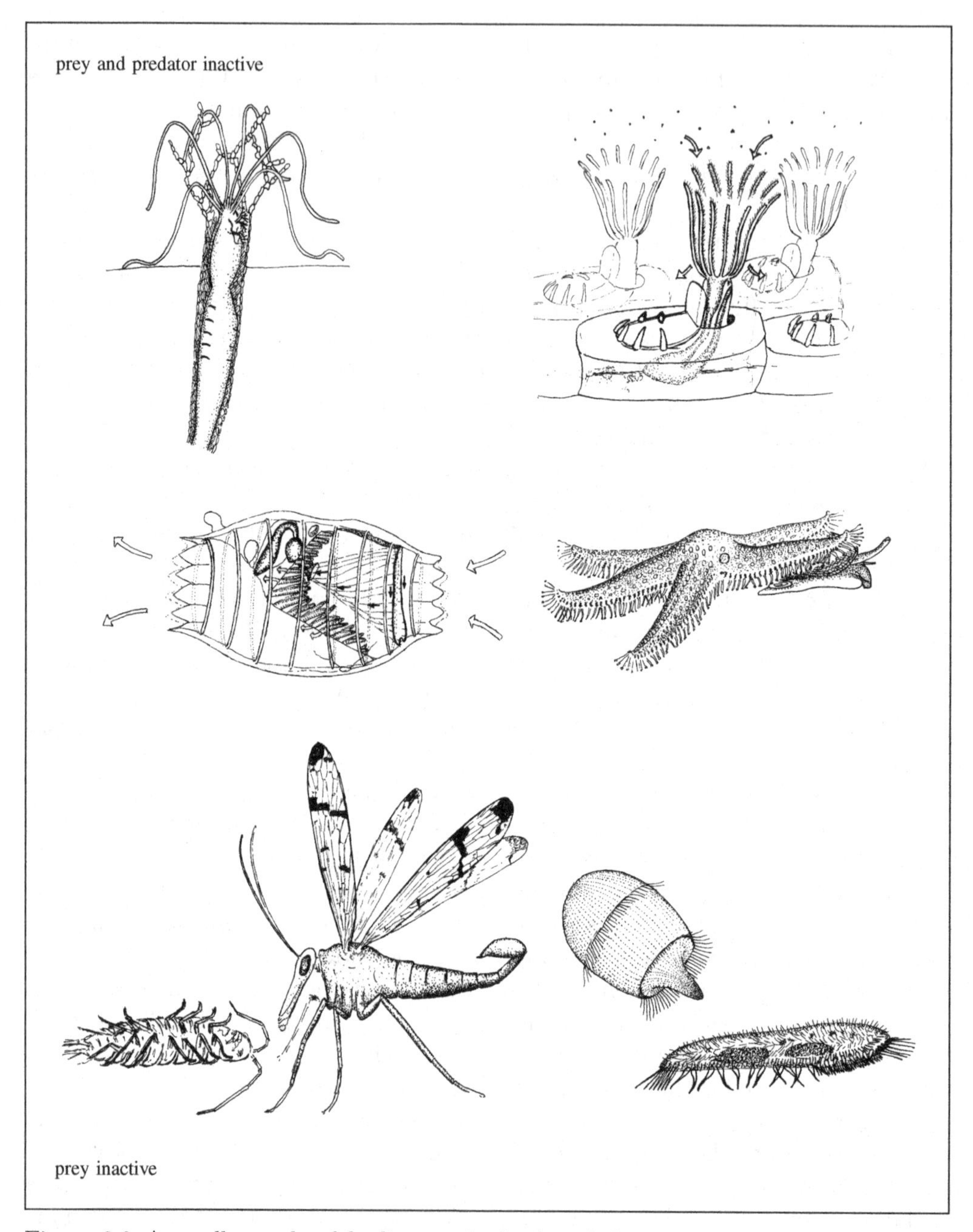

Figure 2.2 A small sample of feeding methods classified with respect to the moving activities of prey and predator.

The amount of food parent birds feed per nestling relates to the requirements of the nestling, which is proportional to surface area; Figure 2.5 illustrates this for chickadees. This is only possible if the nestlings can make their needs clear to the parents, by crying louder: demand systems in the strict sense of the word.

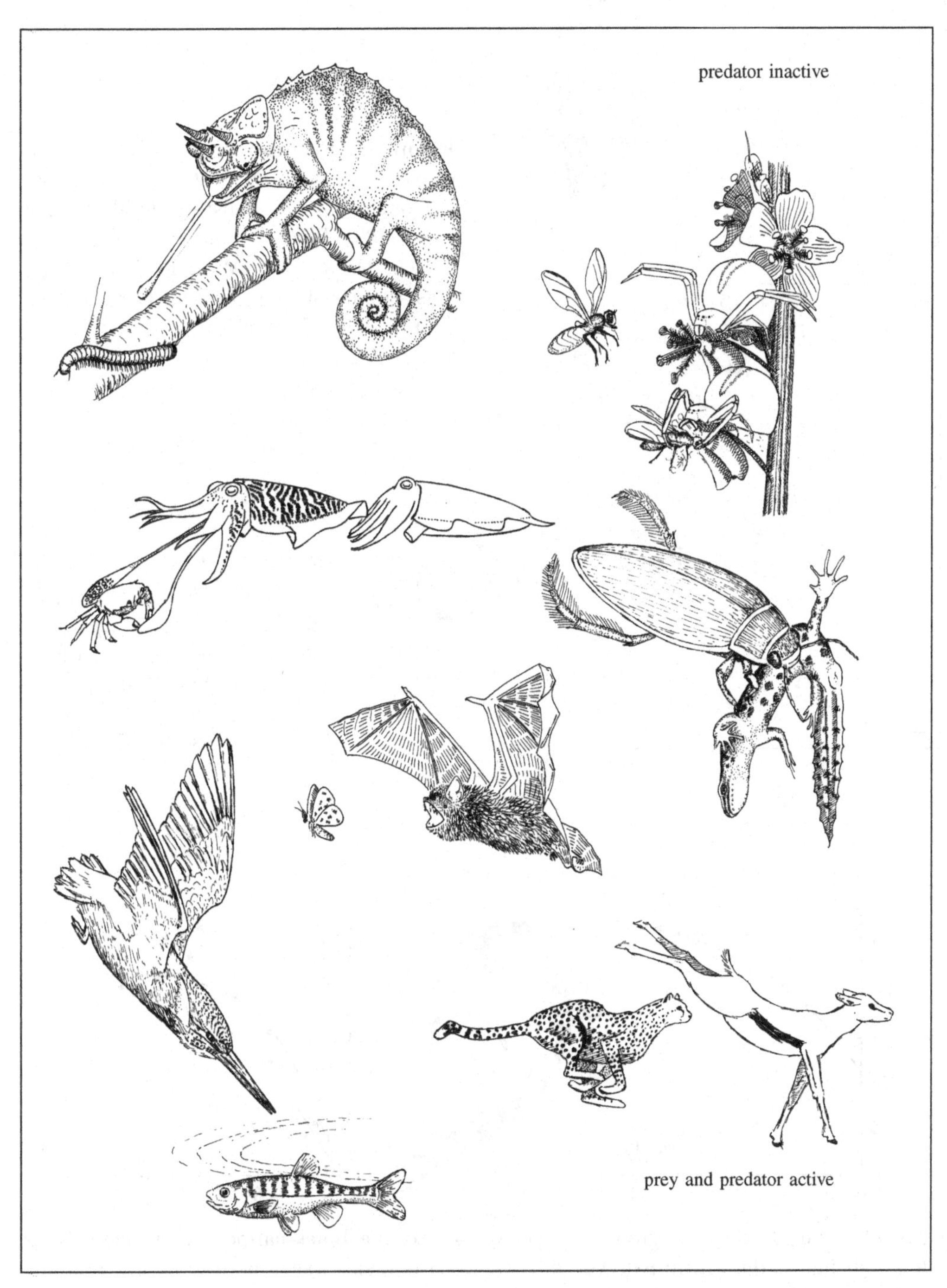

Figure 2.2 (cont.)

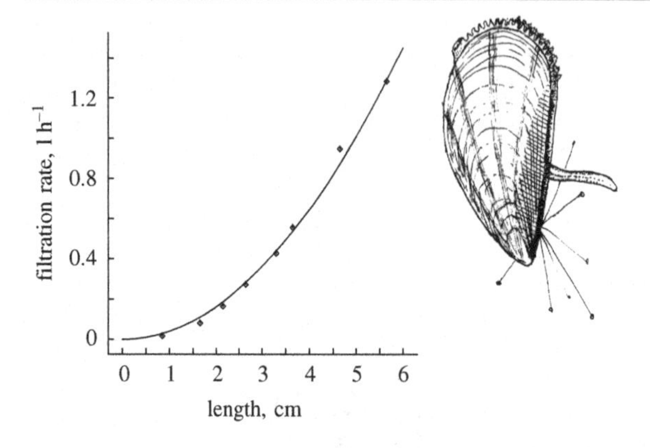

Figure 2.3 Filtration rate as a function of shell length, L, of the blue mussel *Mytilus edulis* at constant food density (40×10^6 cells l^{-1} *Dunaliella marina*) at 12 °C. Data from Winter [1260]. The least-squares-fitted curve is $\{\dot{F}\}L^2$, with $\{\dot{F}\} = 0.041\,\mathrm{l\,h^{-1}\,cm^{-2}}$.

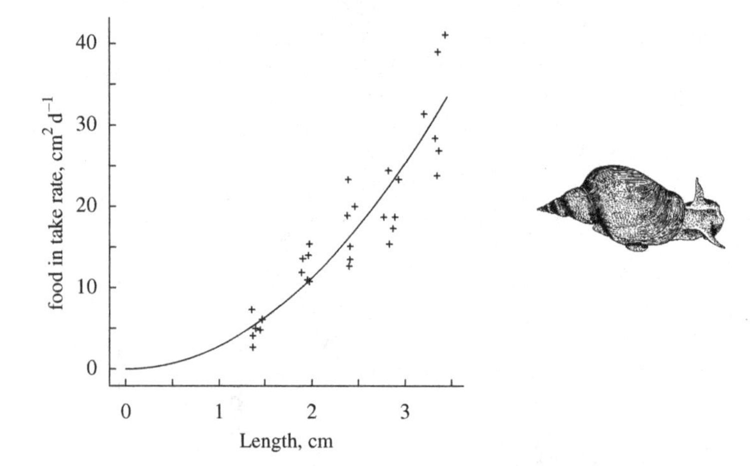

Figure 2.4 Lettuce intake as a function of shell length, L, in the pond snail *Lymnaea stagnalis* at 20 °C [1294]. The weighted least-squares-fitted curve is $\{\dot{J}_{XA}\}L^2$, with $\{\dot{J}_{XA}\} = 2.81\,\mathrm{cm^2\,d^{-1}\,cm^{-2}}$.

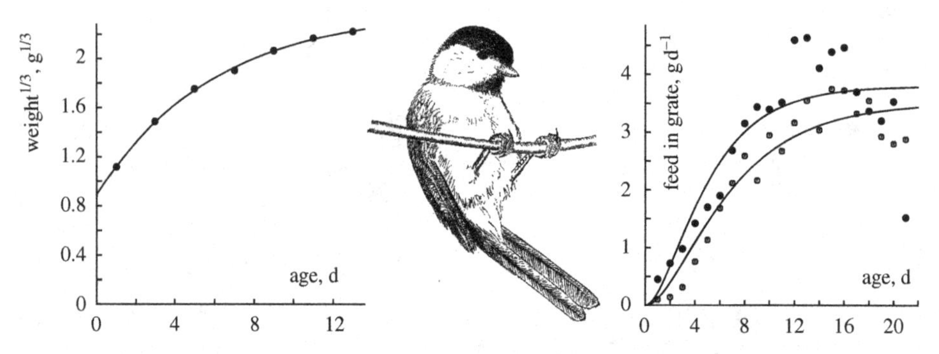

Figure 2.5 The von Bertalanffy growth curve applies to the black-capped chickadee, *Parus atricapillus* (left figure, data from Kluyver [607, 1082]. Brood size was a modest 5.) The amount of food fed per male (•) or female (○) nestling in the closely related mountain chickadee, *P. gambeli*, is proportional to weight$^{2/3}$ (right figure), as might be expected for individuals that grow in a von Bertalanffy way. Data from Grundel [439, 1082]. The last five data points were not included in the fit; the parents stop feeding, and the young still have to learn to gather food while rapidly losing weight.

Catching devices, such as spider or pteropod webs and larvacean filter houses [19], have effective surface areas that are proportional to the surface area of the owner.

Bacteria, floating freely in water, are transported even by the smallest current, which implies that the current relative to the cell wall is effectively nil. Thus bacteria must obtain substrates through diffusion, {266}, or attach to hard surfaces (films) or each other (flocs, {136}) to profit from convection, which can be a much faster process. Some species develop more flagellae at low substrate densities, which probably reduces diffusion limitation (L. Dijkhuizen, pers. comm.). Uptake rate is directly proportional to surface area, if the carriers that bind substrate and transport it into the cell have a constant frequency per unit surface area of the cell membrane [8, 177]. *Arthrobacter* changes from a rod shape into a small coccus at low substrate densities to improve its surface area/volume ratio. Caulobacters do the same by enhancing the development of stalks under those conditions [904].

Some fungi, slime moulds and bacteria glide over or through the substrate, releasing enzymes and collecting elementary compounds via diffusion. Upon arrival at the cell surface, the compounds are taken up actively. The bakers' yeast *Saccharomyces cerevisiae* typically lives as a free-floating, budding unicellular, but under nitrogen starvation it can switch to a filamentous multicellular phase, which can penetrate solids [519]. Many protozoans engulf particles (a process known as phagocytosis) with their outer membrane (again a surface), encapsulate them into a feeding vacuole and digest them via fusion with bodies that contain enzymes (lysosomes). Such organisms are usually also able to take up dissolved organic material, which is much easier to quantify. In giant cells, such as the Antarctic foraminiferan *Notodendrodes*, the uptake rate can be measured directly and is found to be proportional to surface area [257]. Ciliates use a specialised part of their surface for feeding, which is called the cytostome; isomorphic growth here makes feeding rate proportional to surface area again.

All these different feeding processes relate to surface areas in comparisons between different body sizes within a species at a constant low food density. At high food densities, the encounter probabilities are no longer rate-limiting, this becomes the domain of digestion and other food processing activities involving other surface areas, for example the mouth opening and the gut wall. The gradual switch in the leading processes becomes apparent in the functional response, i.e. the ingestion rate as function of food density, {33}.

2.1.3 Feeding costs are paid from food directly

As feeding methods are rather species-specific, costs of feeding will also be species-specific if they contribute substantially to the energy budget. I argue here that costs of feeding and movements that are part of the routine repertoire are usually insignificant with respect to the total energy budget. For this reason this subsection does not do justice to the voluminous amount of work that has been done on the energetics of movements [878], a field that is of considerable interest in other contexts. Alexander [14] gives a most readable and entertaining introduction to the subject of energetics and biomechanics of animal movement. Differences in respiration between active and non-active individuals give a measure for the energy costs of activity, but metabolic activity might be linked to behavioural activity more generally, which complicates the interpretation. The resting metabolic rate is a measure that excludes active movement. The standard or basal metabolic rate includes a low level of movement only. The field metabolic rate is the daily energy expenditure for free-ranging individuals. Karasov [576] found that the field metabolic rate is about twice the standard metabolic rate for several species of mammal, and that the costs of locomotion range at 2–15% of the field metabolic rate. Mammals are among the more active species. The respiration rate associated with filtering in animals such as larvaceans and ascidians was found to be less than 2% of the total dioxygen consumption [356]. Energy investment in feeding is generally small, which does not encourage the introduction of many parameters to describe this investment. Feeding costs can be accommodated in two ways within the DEB theory without the introduction of new parameters, and this subsection aims to explore to what extent this accommodation is realistic.

The first way to accommodate feeding costs is when they are proportional to feeding rate. They then show up as a reduction of the energy gain per unit of food. One can, however, argue that feeding costs per unit of food should increase with decreasing food density, because of the increased effort of extracting it from the environment. This type of cost can only be accommodated without complicating the model structure if these costs cancel against increased digestion efficiency, caused by the increased gut residence time, see {273}.

The second way to accommodate feeding costs without complicating the model structure applies if the feeding costs are independent of the feeding rate and proportional to body volume. They then show up as part of the maintenance costs, see {44}. This argument can be used to understand how feeding rates for some species tend to be proportional to surface area if transportation costs are also proportional to surface area, so that the cruising rate is proportional to length, {27}. In this case feeding costs can be combined with costs of other types of movement that are part of the routine repertoire. A fixed (but generally small) fraction of the maintenance costs then relates to movement.

Schmidt-Nielsen [1022] calculated 0.65 $\text{ml}\,O_2\,\text{cm}^{-2}\,\text{km}^{-1}$ to be the surface-area-specific transportation costs for swimming salmon, on the basis of Brett's work [148]. (He found that transportation costs are proportional to weight to the power 0.746, but respiration was not linear with speed. No check was made for anaerobic metabolism of the salmon. Schmidt-Nielsen obtained, for a variety of fish, a power of 0.7, but 0.67 also fits well.) Fedak and Seeherman [339] found that the surface-area-specific transportation cost for walking birds, mammals and lizards is about 5.39 $\text{ml}\,O_2\,\text{cm}^{-2}\,\text{km}^{-1} \simeq 118\ \text{J}\,\text{cm}^{-2}\,\text{km}^{-1}$. (They actually report that the transportation costs are proportional to weight to the power 0.72 as the best fitting allometric relationship, but the scatter is such that 0.67 fits as well.) This is consistent with data from Taylor *et al.* [1137] and implies that the costs of swimming are some 12% of the costs of running. Their data also indicate that the costs of flying are between those of swimming and running and amount to some 1.87 $\text{ml}\ O_2\,\text{cm}^{-2}\,\text{km}^{-1}$.

The energy costs of swimming are frequently taken to be proportional to squared speed on sound mechanical grounds [672], which questions the usefulness of the above-mentioned costs and comparisons because the costs of transportation become dependent on speed. If the inter-species relationship that speeds scale with the square root of volumetric length, see {310}, also applies to intra-species comparisons, the transportation costs are proportional to volume if the travelling time is independent of size.

The energy required for walking and running is found to be proportional to velocity for a wide diversity of terrestrial animals including mammals, birds, lizards, amphibians, crustaceans and insects [388]. This means that the energy costs of walking or running a certain distance are independent of speed and just proportional to distance. If the costs of covering a certain distance are dependent on speed, and temperature affects speeds, these costs would work out in a really complex way at the population and community levels.

The conclusion is that, for the purposes of studying how energy budgets change during the life span, transportation costs either show up as a reduction of energy gain from food, or as a fixed fraction of the somatic maintenance costs when these costs are proportional to structural mass.

2.1.4 Functional response converts food availability to ingestion rate

The feeding or ingestion rate, $\dot{p}_X$, of an organism as a function of scaled food density, $x = X/K$, expressed in watts, is described well by the hyperbolic functional response

$$\dot{p}_X = f\dot{p}_{Xm} = f\{\dot{p}_{Xm}\}L^2 \quad \text{with} \quad f = \frac{x}{1+x} \tag{2.2}$$

where $\dot{p}_{Xm}$ is the maximum ingestion rate, $\{\dot{p}_{Xm}\}$ the specific maximum ingestion rate and L the volumetric length. The Michaelis–Menten function f is referred to as the scaled functional response. Holling [523] called it type II functional response, and it is illustrated in Figure 2.6. It applies to the uptake of organic particles by

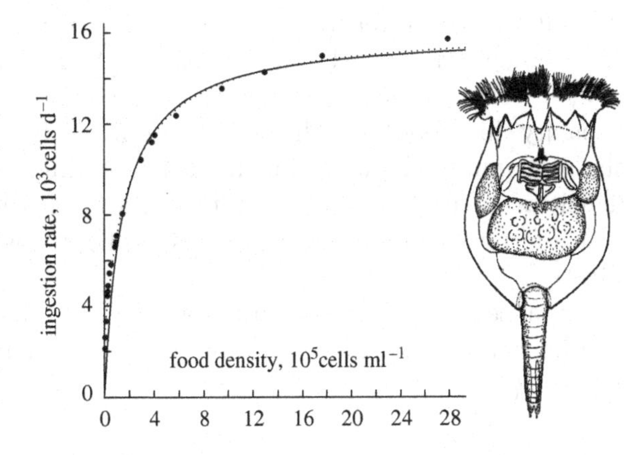

Figure 2.6 The ingestion rate, $\dot{p}_X$, of an individual (female) rotifer *Brachionus rubens*, feeding on the green alga *Chlorella* at 20 °C, as a function of food density, X. Data from Pilarska [898]. The curve is the hyperbola (2.2), with maximum feeding rate $15.97\,10^3$ cells d^{-1} and saturation coefficient $K = 1.47\,10^5$ cells ml^{-1}. The stippled curve allows for an additive error in the measurement of the algal density of $0.35\,10^5$ cells ml^{-1}.

ciliates (phagocytosis), the filtering of algae by daphnids, the catching of flies by mantis, the uptake of substrates by bacteria, the uptake of nutrients by algae and plants, and the transformation of substrates by enzymes. Although these processes differ considerably in detail, some common abstract principle gives rise to the hyperbolic functional response: the busy period, which is characteristic of the Synthesising Unit, see {104}.

All behaviour is classified into just two categories: food acquisition and food processing, which not only includes food handling, but also digestion and other metabolic steps that keep the individual away from food acquisition. These two behavioural components compete for time allocation by the individual.

Let $\dot{F}$ denote the filtering rate (in volume per time), a rate that is taken to depend on mean particle density only, and not on particle density at a particular moment. The arrival rate of food particles (in number per time), present in density N (in number per volume), equals $\dot{h} = \dot{F}N$. Notice that we have to multiply N with the mass per particle M_X to arrive at the mass of food per volume of environment $X = NM_X$, and that we have to multiply $\dot{h}$ with M_X to arrive at the ingestion rate $\dot{J}_{XA} = M_X\dot{h}$ in terms of C-moles per time. The mean time between the end of a handling period and the next arrival (the binding period) is $t_b = \dot{h}^{-1}$. The mean handling period is $t_p = \dot{h}_m^{-1}$, which is the maximum value of $\dot{h}$, where no time is lost in waiting for particles. The time required to find and eat one particle is thus given by $t_c = t_b + t_p$ and the mean ingestion rate is $\dot{h} = t_c^{-1} = (\dot{h}_m^{-1} + \dot{h}^{-1})^{-1} = \dot{h}_m N(\dot{h}_m/\dot{F} + N)^{-1}$, which is hyperbolic in the density X. The (half) saturation coefficient is inversely proportional to the product of the handling time and the filtering (or searching) rate, i.e. $K = M_X(t_p\dot{F})^{-1} = \dot{J}_{XAm}/\dot{F}$, where $\dot{J}_{XAm}$ is the food uptake rate in moles per time.

Filter feeders, such as rotifers, daphnids and mussels, reduce filtering rate with increasing food density [365, 911, 972, 973], rather than maintain a constant rate, which would imply the rejection of some food particles. They reduce the rate by such an amount that no rejection occurs because of the handling (processing) of particles. If all incoming water is swept clear, the filtering rate is found from $\dot{F}(X) = \dot{h}/N$, which reaches a maximum if no food is around (temporarily), so that $\dot{F}_m = \{\dot{h}_m\}L^2/N$, and $\dot{F}$ approaches zero for high food densities. An alternative interpretation of the saturation coefficient in this case would be $K = \dot{J}_{XAm}/\dot{F}_m = \{\dot{J}_{XAm}\}/\{\dot{F}_m\}$, which is independent of the size of the animal, as long as only intra-specific comparisons are made. It combines the maximum capacity for food searching behaviour, only relevant at low food densities, with the maximum capacity for food processing, which is only relevant at high food densities. This mechanistic interpretion of the saturation coefficient is a special case of the dynamics of Synthesising Units, which will be discussed in Chapter 3, {100}.

Because the specific searching rate $\{\dot{F}_m\}$ is closer to the underlying feeding process, it will be treated as primary parameter, rather than the descriptive saturation coefficient $K = \{\dot{J}_{XAm}\}/\{\dot{F}_m\}$.

A most interesting property of the hyperbolic functional response is that it is the only one with a finite number of parameters that maps onto itself. For instance, an exponential function of an exponential function is not again an exponential function. A polynomial (of degree higher than one) of a polynomial is also a polynomial, but it is of an increasingly higher degree if the mapping is repeated over and over again. The ratio of two linear functions of a ratio of two linear functions as in (2.2) is, however, again such a ratio; the linear function is a special case of this. In a metabolic pathway each product serves as a substrate for the next step. Neither the cell nor the modeller needs to know the exact number of intermediate steps to relate the production rate to the original substrate density, if and only if the functional responses of the subsequent intermediate steps are of the hyperbolic type. If, during evolution, an extra step is inserted in a metabolic pathway the performance of the whole chain does not change in functional form. This is a crucial point because each pathway has to be integrated with other pathways to ensure the proper functioning of the individual as a whole. If an insert in a metabolic pathway simultaneously required a qualitative change in regulation at a higher level, the probability of its occurrence during the evolutionary process would be remote. This suggests that complex regulation systems in metabolic pathways fix and optimise the kinetics that originate from the simpler kinetics on which Synthesising Units are based.

A most useful property of the hyperbolic functional response is that it has only two parameters that serve as simple scaling factors on the food density and ingestion rate axis. So if food density is expressed in terms of the saturation coefficient, and ingestion rate in terms of maximum ingestion rate, the functional response no longer has dimensions or parameters.

When starved animals are fed, they often ingest at a higher rate for a short time [1222], but this is here neglected. Starved daphnids, for instance, are able to fill their guts within 7.5 minutes [393].

2.1.5 Generalisations: differences in size of food particles

The derivation of the functional response can be generalised in different ways without changing the model. Each arriving particle can have an attribute that stands for its probability of becoming caught. The ith particle has some fixed probability ρ_i of being caught upon encountering an animal if the animal is not busy handling particles, and a probability of 0 if it is. The mass of each particle does not need to be the same. The flux $\dot{J}_X$ should be interpreted as the mean mass flux (in C-moles per time), where the mass of each particle represents a random trial from some frequency distribution.

It is not essential for the handling time to be the same for all particles; handling time can be conceived of as a second attribute attached to each particle, but it must be independent of food density. The amount of time required for food processing is taken to be proportional to the amount of mass of the food item to ensure that the maximum uptake capacity is not exceeded.

The condition of zero catching probability when the animal is busy can be relaxed. Metz and van Batenburg [783, 784] and Heijmans [482] tied catching probability to satiation (thought to be related to gut content in the mantis). An essential condition for hyperbolic functional responses is that catching probability equals zero if satiation (gut content) is maximal.

When offered different food items, individuals can select for size. Shelbourne [1046] reports that the mean length of *Oikopleura* eaten by plaice larvae increases with the size of the larvae. Copepods appear to select the larger algal cells [1125]. Daphnids do not collect very small particles, $<0.9\,\mu m$ cross-section [421], or large ones, >27 and $>71\,\mu m$; the latter values were measured for daphnids of length 1 and 3 mm respectively [175]. Kersting and Holterman [593] found no size selectivity between 15 and $105\,\mu m^3$ (and probably $165\,\mu m^3$) for daphnids. Size selection is rarely found in daphnids [970], or in mussels [365, 1250], but selection of food type does occur frequently [132].

Deviations from the hyperbolic functional response can be expected if the mass per particle is large, while the intensity of the arrival process is small. They can also result from, e.g. more behavioural traits, see {263}, social interactions, see {264}, transport processes of resources, see {266}.

2.2 Assimilation

The term 'assimilated' energy here denotes the free energy fixed into reserves; it equals the intake minus free energy in faeces and in all losses in relation to digestion. Unlike in typical static budgets, see {433}, the energy in urine is included in assimilation energy, because urine does not directly derive from food and is excreted by the organism, see {86} and {151}. (Faeces are not excreted, because they have never been inside the organism.)

The assimilation efficiency of food is here taken to be independent of the feeding rate. This makes the assimilation rate proportional to the ingestion rate, which seems to be realistic, see Figure 7.17. I later discuss the consistency of this simple assumption with more detailed models for enzymatic digestion, {273}. The conversion efficiency of food into assimilated energy is denoted by κ_X, so that $\{\dot{p}_{Am}\} = \kappa_X\{\dot{p}_{Xm}\}$, where $\{\dot{p}_{Am}\}$ stands for the maximum surface-area-specific assimilation rate. Both κ_X and $\{\dot{p}_{Am}\}$ are diet-specific parameters. The assimilated energy that comes in at food density X is now given by

$$\dot{p}_A = \kappa_X \dot{p}_X = \{\dot{p}_{Am}\} f L^2 \quad \text{where } f = \frac{X}{K+X} \tag{2.3}$$

and L is the volumetric length. It does not involve the parameter $\{\dot{p}_{Xm}\}$ in the notation, which turns out to be useful in the discussion of processes of energy allocation in the next few sections.

2.3 Reserve dynamics

The change of the reserve energy E in time t can be written as $\frac{d}{dt}E = \dot{p}_A - \dot{p}_C$, where the assumption is that the mobilisation rate of reserve, $\dot{p}_C$, is some function of the amount of reserve energy E and of structural volume V only. This function is fully determined by the assumptions of strong and weak homeostasis, see {10}, but its derivation is not the easiest part of this book. Since mobilised reserve fuels metabolism, reserve dynamics is discussed in some detail. For the sake of parameter estimation from data that has no energy in its units, I will use the scaled reserve $U_E = E/\{\dot{p}_{Am}\}$, which has the un-intuitive dimension time × length^2, and treat it as 'something that is proportional to reserve energy'.

The dynamics for the reserve density has to be set up first, in general form. It can be written as the difference between the volume-specific assimilation rate, $[\dot{p}_A] = \dot{p}_A/V = f\{\dot{p}_{Am}\}/L$, and some function of the state variables: the reserve density $[E] = E/V$ and the structural volume V. The freedom of choice for this function is greatly restricted by the requirement that $[E]$ at steady state does not depend on size, while $[\dot{p}_A] \propto L^{-1}$. It implies that the dynamics can be written as

$$\frac{d}{dt}[E] = [\dot{p}_A] - L^{-1}\dot{H}([E]|\boldsymbol{\theta}) + ([E]^* - [E])\dot{G}([E], L) \tag{2.4}$$

where $\dot{H}([E]|\boldsymbol{\theta})$ is some function of $[E]$ and a set of parameters $\boldsymbol{\theta}$, that does not depend on L, and $\dot{G}([E], L)$ some function of $[E]$ and L. The value $[E]^*$ represents the steady-state reserve density, which can be found from $\frac{d}{dt}[E] = 0$. Since $[E]^*$ depends on food density via the assimilation power $[\dot{p}_A]$, the requirement that the rate of use of reserves should not depend on food density implies that $\dot{G}([E], L) = 0$ and (2.4) reduces to

$$\frac{d}{dt}[E] = [\dot{p}_A] - \dot{H}([E]|\boldsymbol{\theta})/L \tag{2.5}$$

The mass balance for the reserve density can be written as

$$\frac{d}{dt}[E] = [\dot{p}_A] - [\dot{p}_C] - [E]\dot{r} \tag{2.6}$$

where $\dot{r} = \frac{d}{dt}\ln V$ stands for the specific growth rate; the third term stands for the dilution by growth, which directly follows from the chain rule for differentiation of E/V. Because maintenance (work) and growth are among the destinations of mobilised reserve, the volume-specific mobilisation rate $[\dot{p}_C] = \dot{p}_C/V$ relates to these fluxes as $\kappa([E], V)[\dot{p}_C] = [\dot{p}_S] + [\dot{p}_G]$, or $\dot{r} = [\dot{p}_G]/[E_G] = (\kappa([E], V)[\dot{p}_C] - [\dot{p}_S])/[E_G]$, where the specific somatic maintenance cost $[\dot{p}_S]$ is some function of V and the volume-specific cost of structure $[E_G]$ is constant, in keeping with the homeostasis assumption for structural mass. The allocation fraction $\kappa([E], V)$ is, at this stage in the reasoning, some function of the state variables $[E]$ and V. Substitution of the expression for growth into (2.6) results in

$$\frac{d}{dt}[E] = [\dot{p}_A] - [\dot{p}_C](1 + \kappa[E]/[E_G]) + [E][\dot{p}_S](V)/[E_G] \tag{2.7}$$

Substitution of (2.5) leads to the volume-specific mobilisation rate

$$[\dot{p}_C] = \frac{\dot{H}([E]|\boldsymbol{\theta}^\circ)/L + [E][\dot{p}_S](V)/[E_G]}{1 + \kappa([E], V|\boldsymbol{\theta}^\circ)[E]/[E_G]} \tag{2.8}$$

where $\boldsymbol{\theta}^\circ$ is a subset of $\boldsymbol{\theta} = (\boldsymbol{\theta}^\circ, [\dot{p}_S], [E_G])$. Note that the functions $\dot{H}$ and κ cannot depend on $[\dot{p}_S]$ or $[E_G]$ because allocation occurs after mobilisation.

2.3.1 Partitionability follows from weak homeostasis

The next step in the derivation of reserve dynamics follows from the partitionability of reserve kinetics, meaning that the partitioning of reserve should not affect its dynamics, i.e. the sum of the dynamics of the partitioned reserves should be identical to that of the lumped one in terms of growth, maintenance, development and reproduction. Partitionability is implied by weak (and strong) homeostasis, as shown in [1087]. This originates from the fact that the reserves are generalised compounds, i.e. mixtures of various kinds of proteins, lipids, etc. Each of these compounds follows its own kinetics, which are functions of the amounts of that compound and of structural mass, but their relative amounts do not change; all reserve compounds have identical kinetics. The strong homeostasis assumption ensures that the amount of any particular compound of the reserve is a fixed fraction, say κ_A, of the total amount of reserve. This compound must account for a fraction κ_A of the maintenance costs and growth investment, see Figure 2.7.

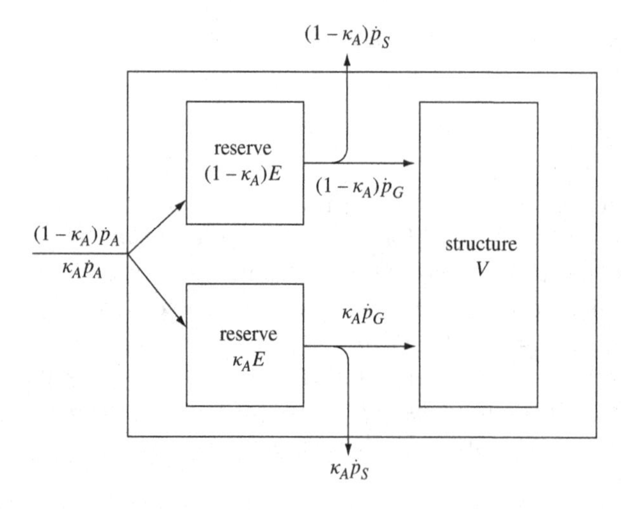

Figure 2.7 When the reserve is partitioned in two parts, the somatic maintenance costs and the costs for structure also need to be partitioned to arrive at a situation where the partitioning remains without effects for the individual.

In quantitative terms, partitionability means

$$\kappa_A[\dot{p}_C]([E], V|[\dot{p}_S], [E_G], \boldsymbol{\theta}^\circ) = [\dot{p}_C](\kappa_A[E], V|\kappa_A[\dot{p}_S], \kappa_A[E_G], \boldsymbol{\theta}^\circ) \tag{2.9}$$

for an arbitrary factor κ_A in the interval $(0, 1)$. This factor not only applies to $[E]$, but also to the specific maintenance $[\dot{p}_S]$ and structure costs $[E_G]$, because the different fractions of the reserve contribute to these costs. The factor does not apply to V and the parameters $\boldsymbol{\theta}^\circ$. We can check in (2.8) that $[\dot{p}_C]$ is partitionable if

- the function $\dot{H}$ is a first-degree homogeneous function, which means that $\kappa_A\dot{H}([E]|\boldsymbol{\theta}^\circ) = \dot{H}(\kappa_A[E]|\boldsymbol{\theta}^\circ)$. It follows that this function can be written as $\dot{H}([E]) = \dot{v}[E]$, for some constant $\dot{v}$, which will be called the *energy conductance* (dimension length per time). The inverse, $\dot{v}^{-1}$, has the interpretation of a resistance. Conductances are often used in applied physics. Therefore, it is remarkable that the biological use of conductance measures seems to be restricted to plant physiology [562, 842] and neurobiology [674].
- the function κ is a zero-th degree homogeneous function in E, which means that $\kappa(\kappa_A[E], V) = \kappa([E], V)$. In other words: κ may depend on V, but not on $[E]$. Later, I argue that $\kappa(V)$ is a rather rudimentary function of V, namely a constant, see {42}.

Substitution of the function $\dot{H}$ into (2.5) gives

$$\frac{d}{dt}[E] = [\dot{p}_A] - [E]\dot{v}/L = (\{\dot{p}_{Am}\}f - [E]\dot{v})/L \tag{2.10}$$

The reserve density at steady state is $[E]^* = L[\dot{p}_A]/\dot{v} = f\{\dot{p}_{Am}\}/\dot{v}$. The maximum reserve density at steady state occurs at $f = 1$, which gives the relationship $[E_m] = \{\dot{p}_{Am}\}/\dot{v}$. So the reserve capacity $[E_m]$ represents the ratio of the assimilation and mobilisation fluxes. The scaled reserve density $e = [E]/[E_m]$ is a dimensionless quantity and has the simple dynamics

$$\frac{d}{dt}e = (f - e)\dot{v}/L \tag{2.11}$$

Notice that, if f is constant, e converges to f.

The reserve dynamics (2.10) results in the specific mobilisation and growth rates

$$[\dot{p}_C] = [\dot{p}_A] - \frac{d}{dt}[E] - [E]\dot{r} = [E](\dot{v}/L - \dot{r}) = [E]\frac{[E_G]\dot{v}/L + [\dot{p}_S]}{\kappa[E] + [E_G]} \tag{2.12}$$

$$\dot{r} = \frac{[E]\dot{v}/L - [\dot{p}_S]/\kappa}{[E] + [E_G]/\kappa} \tag{2.13}$$

After further specification of κ and $[\dot{p}_S]$, the mobilisation and specific growth rates are fully specified.

The mean time compounds stay in the reserve amounts to $t_E = E/\dot{p}_C$, which increases with length. For metabolically active compounds that lose their activity spontaneously, this phenomenon might contribute to the ageing process, to be discussed at {214}. This storage residence time must be large with respect to that of the stomach and the gut to justify neglecting the smoothing effect of the digestive tract.

If the reserve capacity, $[E_m]$, is extremely small, the dynamics of the reserves degenerates to $[E] = f[E_m]$, while both $[E]$ and $[E_m]$ tend to 0. The mobilisation rate then becomes $\dot{p}_C = \{\dot{p}_{Am}\}fV^{2/3}$. This case has been studied by Metz and Diekmann [786], but some consistency problems arise in variable environments where maintenance costs cannot always be paid.

2.3.2 Mergeability is almost equivalent to partitionability

Mergeability means that reserves can be added without effects on the reserve (density) dynamics if assimilation of the resources to synthesise the reserves is coupled and the total intake is constant. This is also implied by weak homeostasis; partitionable dynamics is also mergeable. This property is essential to understand the gradual reduction of the number of reserves during evolution, see {399}. The mergeability is also essential to understand symbiogenesis in a DEB theory context, see {406}: Given that species 1 and 2 each follow DEB rules, and species 1 evolves into an endosymbiont of species 1, the new symbiosis again should follow the DEB rules [648] (otherwise the theory becomes species-specific), see {406}. Evolution might have found several mechanisms to obtain mergeability of reserves, but the fact that they are mergeable is essential for evolution, see {395}.

A quantitative definition of mergeability is as follows. Given $\frac{d}{dt}[E_i] = [\dot{p}_{A_i}] - \dot{F}([E_i], V)$ for $i = 1, 2, \ldots$ and $[\dot{p}_{A_i}] = \kappa_{A_i}[\dot{p}_A]$ with $\sum_i \kappa_{A_i} = 1$, two reserves E_1 and E_2 are mergeable if $\frac{d}{dt}\sum_i [E_i] = [\dot{p}_A] - \dot{F}(\sum_i [E_i], V)$. The mergeability condition summarises to $\sum_i \dot{F}([E_i], V) = \dot{F}(\sum_i [E_i], V)$.

Weak homeostasis implies that $\dot{F}([E], V) = V^{-1/3}\dot{H}([E])$, see (2.5), so together with the mergeability requirement this translates into the requirement that $\sum_i \dot{H}([E_i]) = \dot{H}(\sum_i [E_i])$ or $\kappa_A \dot{H}([E]) = \dot{H}(\kappa_A [E])$ for an arbitrary positive value of κ_A. In other words: H must be first-degree homogeneous in $[E]$. From this follows directly $\dot{H}([E]) = \dot{v}[E]$.

Since from partitionability it also follows that κ is a zero-th degree homogeneous function in E, while this does not follow from mergeability, the latter requirement is less restrictive. In other words, partitionability imposes constraints on the fate of mobilised reserve, mergeability does not. More specifically, partitionability involves maintenance explicitly, mergeability does not.

To demonstrate the difference I now translate the mergeability constraint on $\dot{F}$ to a constraint on the mobilisation flux $\dot{p}_C$. These two fluxes relate to each other as $\dot{F} = [\dot{p}_C] + [E]\dot{r}$, where the specific growth rate $\dot{r} = [\dot{p}_G]/[E_G] = (\kappa[\dot{p}_C] - [\dot{p}_S])/[E_G]$. So

$$\dot{F} = [\dot{p}_C] + (\kappa[\dot{p}_C] - [\dot{p}_S])[E]/[E_G] = \left(1 + \frac{\kappa[E]}{[E_G]}\right)[\dot{p}_C] - \frac{[E]}{[E_G]}[\dot{p}_S] \quad (2.14)$$

The mergeability constraint $\kappa_A \dot{F}([E], V) = \dot{F}(\kappa_A[E], V)$ can be written as

$$\kappa_A[\dot{p}_C]([E], V) = [\dot{p}_C](\kappa_A[E], V)\frac{[E_G] + \kappa_A\kappa[E]}{[E_G] + \kappa[E]} \quad (2.15)$$

which can now be compared with the partitionability definition (2.9).

2.3.3 Mechanism for mobilisation and weak homeostasis

Although Lemesle and Mailleret [696] correctly observed that the use of food and reserve are both Michaelis–Menten functions of their densities, and the mechanism behind the use of food is known, the use of reserve must have a different mechanism because the parameters have a specific interpretation and reserve and structure are not homogeneously mixed. Also first-order kinetics, which is very popular in chemistry, cannot apply because it is not partitionable for isomorphs. Finding a mechanism for reserve mobilisation has been a challenge; a elegant and realistic mechanism for the reserve dynamics rests on the structural homeostasis concept, see {10}. The arguments are as follows.

Since reserve primarily consists of polymers (RNA, proteins, carbohydrates) and lipids, an interface exists between reserve and structure and the gross mobilisation rate of reserve is now taken to be proportional to the surface area of the reserve–structure interface, so $E\dot{v}/L$, see Figure 1.4, and allocated to the mobilisation Synthesising Units (SUs). This flux is partitioned for $\dot{r} \geq 0$ into a net mobilisation flux $\dot{p}_C = E\dot{v}/L - E\dot{r}$ that is accepted by the mobilisation SUs and used for metabolism, and a flux $E\dot{r}$ is rejected and fed back to the reserve. This particular partitioning follows from the weak homeostasis argument, where the ratio of formation of reserve, $\dot{r}E$, and structure, $\dot{r}V$, equals the existing ratio of both amounts, E/V, and the rejected flux is formally considered as a 'synthesis' of reserve. From (2.6) now directly follows (2.10).

A beautiful property of this mechanism is that the correct partitioning of the gross mobilisation flux automatically follows from SUs kinetics, see {101}, if the specific number of mobilisation SUs (in C-moles) equals $\frac{y_{EV}\dot{v}}{L\dot{k}}$, where y_{EV} is the yield of reserve on structure (in C-moles), and $\dot{k}$ is the (constant) dissociation rate of the SUs [662]. An increased deviation from this value results in an increased deviation from weak homeostasis, and the selection of the proper value possibly links directly to the evolution of weak homeostatis. If membranes wrap reserve vesicles, the SU density in these membranes would be constant. Strong homeostasis can only apply strictly if mobilisation SUs switch to the active state if bound to these membranes.

This mechanism has a problem for embryos (eggs, seeds), because L is initially very small, so the gross mobilisation rate as well as the rejected flux are very large. This is unrealistic, because the absence of respiration in the early embryo excludes substantial metabolic activity. This problem hardly exists for organisms that propagate by division, so it became a problem when embryos evolved during evolution. Also organisms with a large body size suffer from the problem, because they have a relatively large amount of reserve, see {300}. This problem can be solved by a self-inhibition mechanism for monomerisation; polymers as such do not take part in metabolism as substrates. The SUs take their substrate from a small pool of reserve monomers, and the pool size of the reserve monomers is proportional to that of the reserve polymers; this is implied by the strong homeostasis assumption.

The self-inhibition mechanism might be by rapid interconversion of the first-order type, but this comes with metabolic costs. A more likely possibility is that monomerisation is product inhibited and ceases if the monomers per polymer reach a threshold. The monomerisation cost is then covered by maintenance and growth. For an individual with an amount of structure M_V and reserve M_E, the kinetics of the amount of monomers M_F could be

$$\frac{d}{dt}M_E = -M_E(\dot{k}_{EF} - \frac{m_F}{m_E}\dot{k}_{FE}); \quad \frac{d}{dt}M_F = y_{FE}M_E(\dot{k}_{EF} - \frac{m_F}{m_E}\dot{k}_{FE}) \tag{2.16}$$

with $m_E = M_E/M_V$ and $m_F = M_F/M_V$. This kinetics makes that in steady state $\frac{m_F^*}{m_E^*} = \frac{\dot{k}_{EF}}{\dot{k}_{FE}}$. The monomerisation occurs at the E–V interface, which has a surface area proportional to E/L in isomorphs. This means that k_{EF} and k_{EF} are proportional to L^{-1} as well.

2.4 The κ-rule for allocation to soma

The motivation for a κ-rule originates from the maturation concept, see {47}, which implies four destinies of mobilised reserve: growth plus somatic maintenance, summarised by the term *soma*, and maturation (or reproduction) plus maturity maintenance. The partitionability requirement states that the fraction κ of mobilised reserve that is allocated to the soma cannot depend on the amount of reserve (or on the reserve density) but still can depend on the amount of structure. The simplest assumption is that κ is constant and does also not depend on the amount of structure. This assumption is used by the standard DEB model.

The empirical evidence for a constant value of κ is that then the von Bertalanffy growth curve results at constant food density, see {52}, which typically fits data from many species very well. Even stronger support is provided by the resulting body size scaling relationships, which are discussed in Chapter 8. Moreover Huxley's allometric model for relative growth of body parts closely links up with multivariate extensions of this κ-rule, see {202}. Strong support for the κ-rule also comes from situations where the value for κ is changed to a new fixed value. Such a simple change affects reproduction as well as growth and so food intake in a very special way. Parasites such as the trematode *Schistosoma* in snails harvest all energy to reproduction and increase κ to maximise the energy flow they can consume; see [564] for a detailed physiological discussion.

Parasite-induced gigantism, coupled to a reduction of the reproductive output, is also known from trematode-infested chaetognaths (*Sagitta*) [821], for instance. The daily light cycle also affects the value for κ in snails, and the allocation behaviour during prolonged starvation; see {118}. The effect of some toxic compounds can be understood as an effect on κ, as is discussed on {240}.

A possible mechanism for a constant κ is as follows. At separated sites along the path the blood follows, somatic cells and ovary cells pick up energy. The only information the cells have is the energy content of the blood and body size, see {12}. They do not have information about each other's activities in a direct way. This also holds for the mechanism by which energy is added to or taken from energy reserves. The organism only has information on the energy density of the blood, and on size, but not on which cells have removed energy from the blood. This is why the parameter κ does not show up in the dynamics of reserve density. The activity of all carriers that remove energy from the body fluid and transport it across the cell membrane depends, in the same way, on the energy density of the fluid. Somatic cells and ovary cells both may use the same carriers, but the concentration in their membranes may differ so that $1 - \kappa$ may differ from the ratio of ovary to body weight. This concentration of active carriers is controlled, by hormones for example, and depends on age, size and environment. Once in a somatic cell, energy is first used for maintenance, the rest is used for growth. This makes maintenance and growth compete directly, while development and reproduction compete with growth plus maintenance at a higher level. The κ-rule makes growth and development parallel processes that interfere only indirectly, as is discussed by Bernardo [89], for instance.

If conditions are poor, the system can block allocation to reproduction, while somatic maintenance and growth continue to compete in the same way, see {118}.

The κ-rule solves quite a few problems from which other allocation rules suffer. Although it is generally true that reproduction is maximal when growth ceases, a simple allocation shift from growth to reproduction leaves similarity of growth between different sexes unexplained, since the reproductive effort of males is usually much less than that of females. The κ-rule implies that size control is the same for males and females and for organisms such as yeasts and ciliates, which do not spend energy on reproduction but do grow in a way that is comparable to species that reproduce; see Figure 2.14. It is important to realise that although the fraction of mobilised reserve that is allocated to the soma remains constant, the absolute size of the flow tends to increase during growth at constant food densities.

The value of κ can be extracted from combined observations on growth and reproduction. The observed value of 0.8 for *D. magna* in Figure 2.10 is well above the value that would maximise the reproduction rate, given the other parameter values. This questions the validity of maximisation of fitness without much attention for mechanisms, a line of thinking that has become popular among evolutionary biologists, e.g. [1101, 982]. The reason why the measured value of κ is so high is an open problem, but might be linked with the length at birth and at puberty, which increases sharply with κ in the standard DEB model.

2.5 Dissipation excludes overheads of assimilation and growth

Dissipation is defined as the use of reserve that is not coupled to net production, where reproduction is seen as an excretion process, rather than a production process in the strict sense of the word. Assimilation and growth have overheads that also appear in the environment, and are excluded from dissipation. So dissipation is not all that dissipates, but less than that. The reason to group these fluxes together under one label, dissipation, becomes clear in the discussion on the organisation of mass fluxes.

Dissipation represents metabolic work. Reserve is metabolised and the metabolites are generally excreted into the environment, frequently in mineralised form (mainly water, carbon dioxide and ammonia), see {165}. Dissipation has four components: somatic and maturity maintenance, maturation and reproduction overhead.

2.5.1 Somatic maintenance is linked to volume and surface area

Maintenance costs can generally be decomposed in contributions that are proportional to structural body volume, and to surface area, which gives the quantification

$$[\dot{p}_S] = [\dot{p}_M] + \{\dot{p}_T\}/L \tag{2.17}$$

where the costs that comprise the volume-linked maintenance costs $[\dot{p}_M]$ and the surface-area-linked maintenance costs $\{\dot{p}_T\}$ are discussed below.

Volume-linked maintenance costs

Maintenance processes include the maintenance of concentration gradients across membranes, the turnover of structural body proteins, a certain (mean) level of muscle tension and movement, and the (continuous) production of hairs, feathers, scales, leaves (of trees), see Figure 2.8.

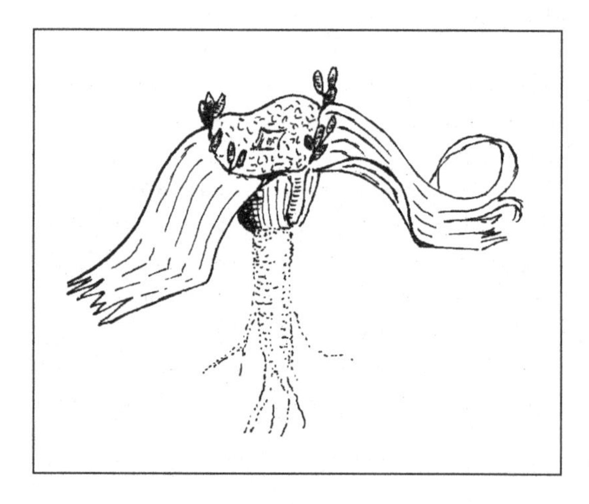

Figure 2.8 The leaves of most plant species grow during a relatively short time period, and are shed yearly, after the plant has recovered useful compounds. The leaves of some species, however, such as *Welwitschia mirabilis*, grow and weather continuously. The life span of this remarkable gymnosperm can exceed 2000 years. Leaves have a very limited functional life span, but plants have found different ways to deal with that problem.

The idea that maintenance costs are proportional to biovolume is simple and rests on strong homeostasis: a metazoan of twice the volume of a conspecific has twice as many cells, which each use a fixed amount of energy for maintenance. A unicellular of twice the original volume has twice as many proteins to turn over. Protein turnover seems to be low in prokaryotes [613]. Another major contribution to maintenance costs relates to the maintenance of concentration gradients across membranes. Eukaryotic cells are filled with membranes, and this ties the energy costs for concentration gradients to volume. (The argument for membrane-bound food uptake works out differently in isomorphs, because feeding involves only the outer membrane directly.) Working with mammals, Porter and Brand [913] argued that proton leak in mitochondria represents 25% of the basal respiration in isolated hepatocytes and may contribute significantly to the standard metabolic rate of the whole animal.

The energy costs of movement are also taken to be proportional to volume if averaged over a sufficiently long period. Costs of muscle tension in isomorphs are likely to be proportional to volume, because they involve a certain energy investment per unit volume of muscle. In the section on feeding, I discuss briefly the energy involved in movement, {32}, which has a standard level that includes feeding. This can safely be assumed to be a small fraction of the total maintenance costs. Sustained powered movement such as in migration requires special treatment. Such activities involve temporarily enhanced metabolism and feeding. The occasional burst of powered movement hardly contributes to the general level of maintenance energy requirements. Sustained voluntary powered movement seems to be restricted to humans and even this seems of little help in getting rid of weight!

There are many examples of species-specific maintenance costs. Daphnids produce moults every other day at 20 °C. The synthesis of new moults occurs in the intermoult period and is a continuous and slow process. The moults tend to be thicker in the larger sizes. The exact costs are difficult to pin down, because some of the weight refers to inorganic compounds, which might be free of energy cost. Larvaceans produce new feeding houses every 2 hours at 23 °C [340], and this contributes substantially to organic matter fluxes in oceans [17, 18, 248]. These costs are taken to be proportional to volume. The inclusion of costs of moults and houses in maintenance costs is motivated by the observation that these rates do not depend on feeding rate [340, 634], but only on temperature.

It will be convenient to introduce the maintenance rate coefficient $\dot{k}_M = [\dot{p}_M]/[E_G]$ as compound parameter. It stands for the ratio of the costs of somatic maintenance and of structure and has dimension time^{-1}. It was introduced by Marr *et al.* [746] for the first time and publicised by Pirt [900].

Surface-area-linked maintenance costs

Some specialised maintenance costs relate to surface areas of individuals.

Aquatic insects are chemically fairly well isolated from the environment. Euryhaline fishes, however, have to invest energy in osmoregulation when in waters that are not iso-osmotic. The cichlid *Oreochromis niloticus* is iso-osmotic at 11.6‰ and 29% of the respiration rate at 30‰ can be linked to osmoregulation [1273]. Similar results have been obtained for the brook trout *Salvelinus fontinalis* [377].

Endotherms (birds, mammals) use reserve to heat their body such that a particular part of the body (the head and the heart region in humans) has a constant temperature during the post-embryonic stages; embryos don't do this. Heat loss is not only proportional to surface area but, according to I. Newton, also to the temperature difference between body and environment. This is incorporated in the concept of thermal conductance $\{\dot{p}_T\}/(T_e - T_b)$, where T_e and T_b denote the temperature of the environment and the body. It is about 5.43 $\mathrm{J\,cm^{-2}\,h^{-1}\,{}^\circ C^{-1}}$ in birds and 7.4–9.86 $\mathrm{J\,cm^{-2}\,h^{-1}\,{}^\circ C^{-1}}$ in mammals, as calculated from [497]. The unit $\mathrm{cm^{-2}}$ refers to volumetric squared length, not to real surface area which involves shape. The values represent crude means in still air. The thermal conductance is roughly proportional to the square root of wind speed.

This is a simplified presentation. Birds and mammals moult at least twice a year, to replace their hair and feathers which suffer from wear, and change the thick winter coat for the thin summer one. Cat owners can easily observe that when their pet is sitting in the warm sun, it will pull its hair into tufts, especially behind the ears, to facilitate heat loss. Many species have control over blood flow through extremities to regulate temperature. People living in temperate regions are familiar with the change in the shape of birds in winter to almost perfect spheres. This increases insulation and generates heat from the associated tension of the feather muscles. These phenomena point to the variability of thermal conductance.

There are also other sources of heat exchange, through ingoing and outgoing radiation and cooling through evaporation. Radiation can be modulated by changes in colour, which chameleons and tree frogs apply to regulate body temperature [719]. Evaporation obviously depends on humidity and temperature. For animals that do not sweat, evaporation is tied to respiration and occurs via the lungs. Most non-sweaters pant when hot and lose heat by enhanced evaporation from the mouth cavity. A detailed discussion of heat balances would involve a considerable number of coefficients [803, 1092], and would obscure the main line of reasoning. I discuss heating in connection with the water and energy balances on {160}. It is important to realize that all these processes are proportional to surface area, and so affect the heating rate $\{\dot{p}_T\}$ and in particular its relationship with the temperature difference between body and environment.

It turns out to be convenient to introduce the heating length $L_T \equiv \{\dot{p}_T\}/[\dot{p}_M]$ or the heating volume $V_T \equiv L_T^3$ as compound parameters of interest. Heating volume stands for the reduction in volume endotherms experience due to the energy costs of heating. It can be treated as a simple parameter as long as the environmental temperature remains constant. Sometimes, it will prove to be convenient to work with the scaled heating length $l_T \equiv \{\dot{p}_T\}/\kappa\{\dot{p}_{Am}\}$ as a compound parameter. If the temperature changes slowly relative to the growth rate, the heating volume is a function of time. If environmental temperature changes rapidly, body temperature can be taken to be constant again while the effect contributes to the stochastic nature of the growth process, see {114}.

2.5.2 Maturation for embryos and juveniles

The κ-rule means that growth and development are parallel processes, which links up beautifully with the concepts of acceleration and retardation of developmental phenomena such as sexual maturity [423]. These concepts are used to describe relative rates of development in species that are similar in other respects.

The ideas on maturation and maturation maintenance in the DEB context rest on four observations:

- Contrary to age, the volume at the first appearance of eggs hardly depends on food density, typically; see Figure 2.9. The same holds for the volume at birth.
- Some species, such as daphnids, continue to grow after the onset of reproduction. *Daphnia magna* starts to reproduce at a length of 2.5 mm, while its ultimate size is 5 mm, if well fed. This means an increase of well over a factor eight in volume during the reproductive period. Other species, however, such as birds, only reproduce well after the growth period. The giant petrel wanders 7 years over Antarctic waters before it starts to breed for the first time. This means that stage transitions cannot be linked to size.
- The total cumulative energy investment in development at any given size of the individual depends on food density. Indeed, if feeding conditions are so poor that the ultimate volume is less than the threshold for allocation to reproduction, the cumulated energy investment in development becomes infinitely large if survival allows.
- If food density is constant, the reproduction rate at ultimate size is a continuous function of the food density; it is zero for low food densities, and increases from zero for increasing food densities. So it does not make a big jump if the various food densities differ sufficiently little.

The combination of the four graphs in Figure 2.10 illustrates a basic problem for the rules of allocating energy quantitatively. The problem becomes visible as

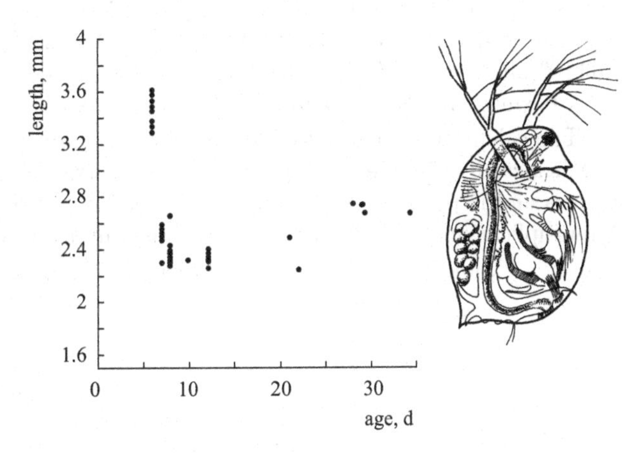

Figure 2.9 The carapace length of the daphnid *Daphnia magna* at 20 °C for five different food levels at the moment of egg deposition in the brood pouch. Data from Baltus [57]. The data points for short juvenile periods correspond with high food density and growth rate. They are difficult to interpret because length increase is only possible at moulting in daphnids.

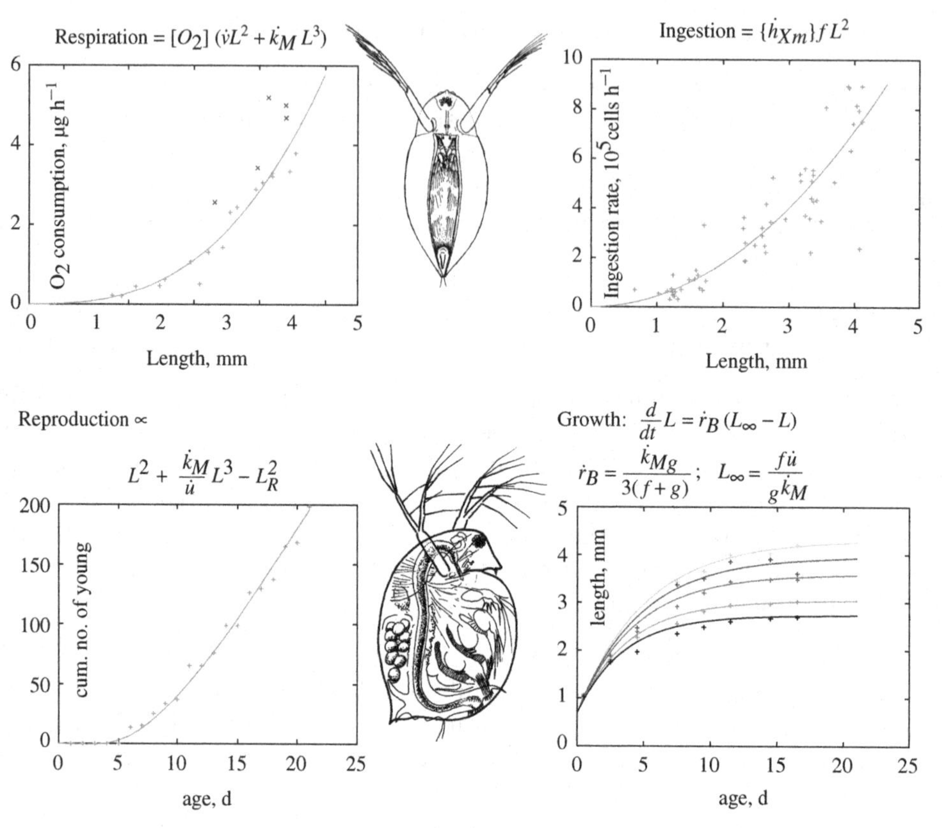

Figure 2.10 Respiration (upper left), and ingestion (upper right) as a function of body length, and reproduction (lower left) at high food level and body length (lower right) as function of age at different food levels in the waterflea *Daphnia magna* at 20 °C. Original data and from [330, 652]; the DEB model specifies the curves. The reproduction curve shows that *D. magna* starts to reproduce at the age of 7 d, i.e. when its length exceeds 2.5 mm. However, respiration and ingestion do not increase steeply at this size, nor does growth decrease. Where did the substantial reproductive energy come from? The κ-rule gives the explanation. The open symbols in the graph for respiration relate to individuals with eggs in their brood pouch. Parameter values: $\delta_{\mathcal{M}} = 0.54$ (fixed), $\kappa = 0.799$, $\kappa_R = 0.95$ (fixed), $g = 0.15$, $\dot{k}_J = 3.57\,\mathrm{d}^{-1}$, $\dot{k}_M = 4.06\,\mathrm{d}^{-1}$, $\dot{v} = 1.62\,\mathrm{mm\,d}^{-1}$, $U_H^b = 0.001\,\mathrm{mm}^2\mathrm{d}$, $U_H^p = 0.049\,\mathrm{mm}^2\mathrm{d}$, $[O_2] = 2.033\,\mu\mathrm{g\,mm}^{-3}$, $\{\dot{h}_{Xm}\} = 1.53\,10^5\mathrm{cells\,h}^{-1}\mathrm{mm}^{-2}$, $f = 0.88, 0.81, 0.73, 0.63, 0.56$ in lower-right graph. The observation that $L_b = 0.8$ mm and $L_p = 2.5$ mm at $f = 1$ and 0.5 has been used to stabilise the estimate for $\dot{k}_J$. The calculated lengths are $L_b = 0.686, 0.685$ mm and $L_p = 2.46, 2.45$ mm, respectively. All length measures in the parameters are volumetric. The shape coefficient $\delta_{\mathcal{M}}$ affects physical lengths and is probably too large because of the water pockets inside the daphnid's carapace. A multiplication of $\delta_{\mathcal{M}}$ by a factor z means a multiplication of $\dot{v}$ by z, of U_H^b and U_H^p by z^2, of $[O_2]$ by z^{-3}, and of $\{\dot{h}_{Xm}\}$ by z^{-2}. This can be seen from the units of the parameters.

soon as one realises that a considerable amount of energy is invested in reproductive output. The volume of young produced exceeds one-quarter of that of the mother each day. The problem is that growth is not retarded in animals crossing the 2.5 mm barrier; they do not feed much more and simply follow the surface area rule with a fixed proportionality constant at constant food densities; they do not change sharply in respiration, so it seems unlikely that they digest their food much more efficiently. So where does the energy allocated to reproduction come from?

These observations fit naturally if stage transitions are linked to maturity and a maturity maintenance flux exists that is proportional to the level of maturity. The recognition of the problem and its solution is the cornerstone of DEB theory.

A solution to this problem can be found in maturation. Juvenile animals have to mature and become more complex. They have to develop new organs and install regulation systems. The increase in size (somatic growth) of the adult does not include an increase in complexity. The energy spent on development in juveniles is spent on reproduction in adults. This switch does not affect growth and suggests the κ-rule: a fixed proportion κ of energy mobilised from the reserve is spent on somatic maintenance plus growth, the remaining portion $1 - \kappa$ on maturity maintenance plus maturation plus reproduction. The partitionability of reserve kinetics has led to the conclusion that κ cannot depend on the reserve density, see {39}. The argument that allocation is an intensive process, not an extensive one, suggests that κ is independent of V as well.

If the maturity maintenance ratio equals the somatic one, the maturity density, so the ratio of the maturity and the amount of structure, remains constant and stage transitions then also occur at fixed amounts of structure, see {54}. Some species do show variations in the size at first reproduction, however, see [89]. Little is known about the molecular machinery that is involved in the transition from the juvenile to the adult stage. Recent evidence points to a trigger role of the hormone leptin in mice, which is excreted by the adipose tissue [203]. This finding supports the direct link between the transition and energetics.

The increase in the level of maturity, quantified as cumulative investment of reserve into maturity equals

$$\frac{d}{dt}E_H = \dot{p}_R(E_H < E_H^p) \quad \text{with} \quad \dot{p}_R = (1-\kappa)\dot{p}_C - \dot{p}_J \qquad (2.18)$$

where E_H^p is the maturity threshold at puberty. Because of the arbitrariness of a unit for information, I refrain from an explicit conversion to information, but the mass as well as the energy in this investment is typically excreted into the enviroment in the form of metabolites and heat.

The literature distinguishes determinate growers, which cease growth during the adult stage, and indeterminate growers, which continue growth. In DEB theory, this is just a matter of the value of E_H^p relative to other DEB parameters; even ecdysozoans that reproduce only in their final moult typically follow von Bertalanffy growth curves, see {52}. The only real determinate growers in the DEB context are the holometabolic insects, which insert a pupal stage between the juvenile and adult stages, see {284}, and don't grow as adults; their growth as larvae to pupation is not asymptotic.

2.5.3 Maturity maintenance: defence systems

The maturity maintenance is assumed to be proportional to the maturity level

$$\dot{p}_J = \dot{k}_J E_H \tag{2.19}$$

where $\dot{k}_J$ is the maturity maintenance rate coefficient. It can be compared with the somatic maintenance rate coefficient $\dot{k}_M$, and we will see, {54}, that if the maintenance ratio $k = \dot{k}_J/\dot{k}_M = 1$, stage transitions also occur at fixed structural volumes. Notice that $\dot{k}_M$ expresses the maintenance costs relative to the cost for structure; likewise $\dot{k}_J$ expresses the maintenance costs relative to the cost of a unit of maturity, but since we quantify maturity as the cumulative reserve investment into maturity this unit equals 1 by definition; we don't make the conversion to maturity as information explicitly.

An observation that strongly supports the existence of maturity maintenance concerns pond snails, where the day/night cycle affects the fraction of utilised energy spent on maintenance plus growth [1294] such that κ at equal day/equal night, κ_{md}, is larger than that at long day/short night, κ_{ld}. Apart from the apparent effects on growth and reproduction rates, volume at the transition to adulthood is also affected. If the cumulated energy investment in the increase of maturity does not depend on the value for κ and if the maturity maintenance costs are $\dot{p}_J = \frac{1-\kappa}{\kappa}\dot{p}_M$, the expected effect is $\frac{V_{p,ld}}{V_{p,md}} = \frac{\kappa_{ld}(1-\kappa_{md})}{\kappa_{md}(1-\kappa_{ld})}$, which is consistent with the observations on the coupling of growth and reproduction investments to size at puberty [1294].

Maturity maintenance can be thought to relate to the maintenance of regulating mechanisms and concentration gradients, such as those found in *Hydra*, that maintain head/foot differentiation [406].

During extreme forms of starvation, many organisms shrink {121}. They can only recover enough energy from the degradation of structural mass to pay the somatic maintenance costs if they can reduce the maturity maintenance costs under those conditions. Thomas and Ikeda [1150] concluded from studies on laboratory populations of *Euphausia superba* that female krill can regress from the adult to the juvenile state during starvation.

Organisms also become more vulnerable to diseases during starvation. This suggests that defence systems, see {397}, such as the immune system of vertebrates, are fuelled from maturity maintenance, and that maturity maintenance is more facultative than somatic maintenance. All species have several defence systems, also to protect themsevels against effects of toxicants. This also explains why maturity maintenance can be substantial.

2.5.4 Reproduction overhead

Since embryos initially consist almost exclusively of reserve, the allocation to reproduction consists of reserve, and the strong homeostasis assumption means that reserve cannot change in composition, little metabolic work is involved in reproduction. Yet some work is involved in the conversion of (part of) the buffer of reserve that is allocated to reproduction into eggs. A fraction κ_R of the reproduction flux, called the reproduction efficiency, is assumed to be fixed in embryo reserve, and the rest, a fraction $1-\kappa_R$, is used as reproduction overhead.

2.6 Growth: increase of structure

Now that the allocation fraction κ and the specific maintenance costs $[\dot{p}_S]$ are specified, the specific mobilisation (2.12) and growth rates (2.13) can be written as

$$[\dot{p}_C] = [E_m](\dot{v}/L + \dot{k}_M(1 + L_T/L))\frac{eg}{e+g} \tag{2.20}$$

$$\frac{d}{dt}V = \dot{r}V \quad \text{with} \quad \dot{r} = \dot{v}\frac{e/L - (1 + L_T/L)/L_m}{e+g} \tag{2.21}$$

where the energy investment ratio $g = [E_G]/\kappa[E_m]$ stands for the costs of new biovolume relative to the maximum potentially available energy for growth plus maintenance. It is dimensionless.

The maximum length $L_m = \kappa\{\dot{p}_{Am}\}/[\dot{p}_M] = \dot{v}/\dot{k}_M g$ represents the ratio of the part of the assimilation flux that is allocated to the soma and the somatic maintenance flux. Notice the conceptual similarity with the reserve capacity $[E_m] = \{\dot{p}_{Am}\}/\dot{v}$, which is also a ratio of in- and out-going fluxes. The maximum volume $V_m = L_m^3$ is also a compound parameter of interest.

On Gough Island, the house mouse *Mus musculus* changed diet and turned to prey on the chicks of the Tristan albatross *Diomedea dabbenena* and the Atlantic petrel *Pterodroma incerta*, despite the fact that these birds are 250 times heavier. From an energy point of view, this had the remarkable effect that the weight of the adult mice is 40 g, rather than the typical 15 g. The reason is probably that the conversion efficiency from birds to mice is higher than their typical conversion efficiency. This supports the idea that ultimate (structural) weight represents the ratio of assimilation and maintenance. From a nature conservation point of view the problem is that 99% of the world population of these two bird species live on this island; the birds are now threatened with extinction.

For some applications, it will be convenient to work in scaled length $l = L/L_m$ and with $\frac{d}{dt}l = l\dot{r}/3$ and scaled reserve density e. From (2.11) and (2.21) we have

$$\frac{d}{dt}l = \dot{k}_M\frac{g}{3}\frac{e - l - l_T}{e+g} \quad \text{and} \quad \frac{d}{dt}e = (f-e)g\dot{k}_M/l \tag{2.22}$$

Animals that have non-permanent exoskeletons, the Ecdysozoa, have to moult to grow. The rapid increase in size during the brief period between two moults relates to the uptake of water or air, not to synthesis of new structural biomass, which is a slow process occurring during the intermoult period. This minor deviation from the DEB model relates more to size measures than to model structure.

2.6.1 Von Bertalanffy growth at constant food

If food density X and, therefore, the scaled functional response f are constant, and if the initial energy density equals $[E] = f[E_m]$, energy density will not change and $e = f$. Volumetric length as a function of time since birth can then be solved from $\frac{d}{dt}L = L\dot{r}/3$ and results in

$$L(t) = L_\infty - (L_\infty - L_b)\exp(-t\dot{r}_B) \quad \text{or} \quad t(L) = \frac{1}{\dot{r}_B}\ln\frac{L_\infty - L_b}{L_\infty - L} \tag{2.23}$$

$$\dot{r}_B = \frac{1}{3/\dot{k}_M + 3fL_m/\dot{v}} = \frac{\dot{k}_M/3}{1+f/g} \tag{2.24}$$

$$L_\infty = fL_m - L_T \tag{2.25}$$

where length at birth $L_b \equiv L(0)$ is a quantity that will be discussed at {63}. See Figure 2.14 for a graphical interpretation of the $L(t)$ curve. I will follow tradition and call this curve the von Bertalanffy growth curve despite its earlier origin and von Bertalanffy's contribution of introducing allometry, which I reject. Equations (2.24) and (2.25) give a physiological interpretation of the von Bertalanffy growth rate $\dot{r}_B$ and the ultimate length L_∞.

The von Bertalanffy growth curve results for post-natal isomorphs at constant food density and temperature and has been fitted successfully to the data of some 270 species from many different phyla, which have very different hormonal systems to control growth; see Figure 2.11 and Table 8.3. The gain in insight since A. Pütter's original formulation in 1920 [926] is in the interpretation of the parameters in terms of underlying processes. It appears that heating cost does not affect the von Bertalanffy growth rate $\dot{r}_B$. Food density affects both the von Bertalanffy growth rate and the ultimate length. The inverse of the von Bertalanffy growth rate is a linear function of the ultimate volumetric length; see Figure 2.10. This is in line with Pütter's original formulation, which took this rate to be inversely proportional to ultimate length, as has been proposed again by Gallucci and Quinn [385]. DEB theory shows, however, that the intercept cannot be zero.

The requirement that food density is constant for a von Bertalanffy curve can be relaxed if food is abundant, because of the hyperbolic functional response. As long as food density is higher than four times the saturation coefficient, food intake is higher than 80% of the maximum possible food intake, which makes it hardly distinguishable from maximum food intake. Since most birds and mammals have a number of behavioural traits aimed at guaranteed adequate food availability, they appear to have a fixed volume–age relationship. This explains the popularity of age-based models

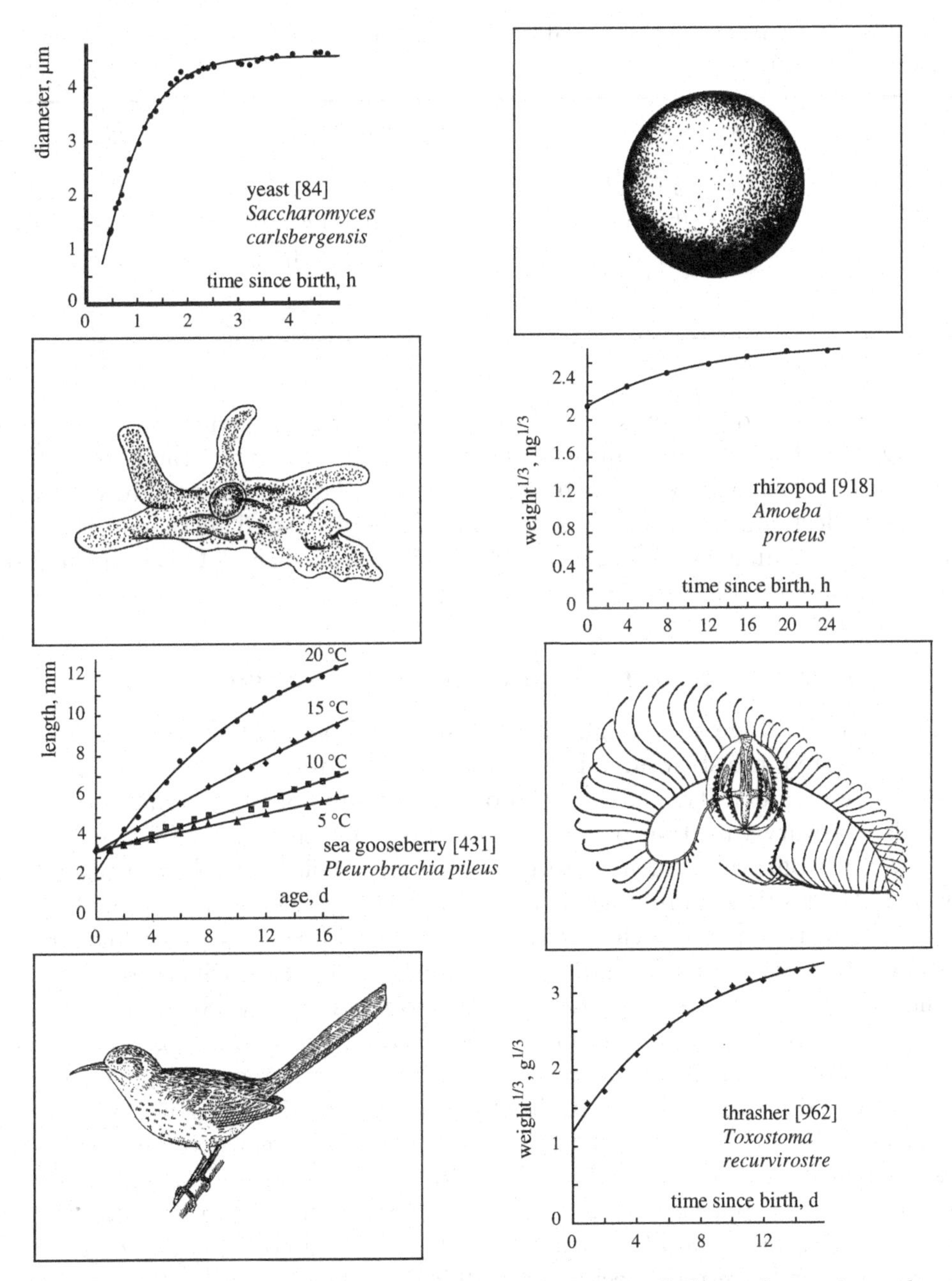

Figure 2.11 These von Bertalanffy growth curves fit very well data of organisms that differ considerably in their growth-regulating systems. This suggests that hormones are used to match local supply and demand of metabolites, but growth is controlled at the level of the individual, including hormonal activity.

Table 2.1 The dimensionless scaled variables and parameters that are used to find the initial amount of scaled reserve.

$\tau = a\dot{k}_M$	$\tau_b = a_b\dot{k}_M$	$l = L/L_m$	$l_b = L_b/L_m$
$u_E = \frac{E}{g[E_m]L_m^3}$	$u_E^0 = \frac{E_0}{g[E_m]L_m^3}$	$u_H = \frac{E_H}{g[E_m]L_m^3}$	$u_H^b = \frac{E_H^b}{g[E_m]L_m^3}$
$e = gu_E/l^3$	$e_b = gu_E^b/l_b^3$	$e_H = gu_H/l^3$	$e_H^b = gu_H^b/l_b^3$
$x = \frac{g}{e+g}$	$x_b = \frac{g}{e_b+g}$	$\alpha = 3gx^{1/3}/l$	$\alpha_b = 3gx_b^{1/3}/l_b$
$y = \frac{xe_H}{1-\kappa}$	$y_b = \frac{x_b e_H^b}{1-\kappa} = gx_b v_H^b l_b^{-3}$	$v_H^b = \frac{u_H^b}{1-\kappa}$	$k = \dot{k}_J/\dot{k}_M$

for growth in 'demand' systems. Later, on {173}, I discuss deviations from the von Bertalanffy growth curve that can be understood in the context of the present theory.

In contrast, at low food densities, fluctuations in food density soon induce deviations from the von Bertalanffy curve. This phenomenon is discussed further in the section on genetics and parameter variation, {296}. Growth ceases, i.e. $\frac{d}{dt}V = 0$, if the reserve density equals a threshold value, $[E] = [\dot{p}_S]L/\kappa\dot{v}$.

2.6.2 States at birth and initial amount of reserve

During the embryo stage, dry weight decreases, but the amount of structure increases: reserve is converted into structure. The initial amount of reserve is not a free parameter because the reserve density, i.e. the ratio of the amounts of reserve and structure, at birth tends to covary with that of the mother at egg production; well-fed mothers give birth to well-fed offspring. Such maternal effects are typical and have been found in e.g. birds [824], reptiles, amphibians [715], fishes [478], insects [770, 994, 993], crustaceans [411], rotifers [1287], echinoderms and bivalves [98]. Maternal effects explain, for instance, why the batch fecundity of the anchovy *Engraulis* increases during the spawning season in response to a decrease in food availability [877]. However, some species seem to produce large eggs under poor feeding conditions, e.g. some poeciliid fishes [955], daphnids [413] and *Sancassania* mites [82]. Moreover, egg size can vary within a clutch [314, 823, 1218], according to geographical distribution [1067], with age [770] and race. Nonetheless, the pattern that the reserve density at birth $[E_b]$ equals the reserve density $[E]$ of the mother at egg formation generally holds; this not only removes a parameter, but also has the nice implication that the von Bertalanffy growth curve applies from birth on, at constant food density. Embryos don't feed, $f = 0$, and even embryos of endothermic species don't allocate to heating, $L_T = 0$.

The state variables (E_H, E, L) evolve from $(0, E_0, 0)$ at age $a = 0$ to $(E_H^b, [E_b]L_b^3, L_b)$ at age $a = a_b$, the age at birth. To find E_0, a_b, and L_b, given E_H^b and $[E_b]$, is a bit of a challenge, which has been overcome only recently [645].

For this purpose, it is most convenient to remove parameters by scaling to dimensionless quantities: (τ, u_E, l, u_H), see Table 2.1. The reformulated problem is now: find τ_b, l_b, u_E^0 given u_H^b, k, g, κ and $u_E^b = e_b l_b^3/g$, where e_b is the scaled reserve density at birth.

For the variable (τ, u_E, l, u_H) evolving from the value $(0, u_E^0, 0, 0)$ to the value $(\tau_b, u_E^b, l_b, u_H^b)$, the scaled model amounts to

$$\frac{d}{d\tau}u_E = -u_E l^2 \frac{g+l}{u_E + l^3} \tag{2.26}$$

$$\frac{d}{d\tau}l = \frac{1}{3}\frac{g u_E - l^4}{u_E + l^3} \tag{2.27}$$

$$\frac{d}{d\tau}u_H = (1-\kappa)u_E l^2 \frac{g+l}{u_E + l^3} - k u_H \tag{2.28}$$

or alternatively for variable (τ, e, l, e_H) evolving from the value $(0, \infty, 0, e_H^0)$ to the value $(\tau_b, e_b, l_b, e_H^b)$

$$\frac{d}{d\tau}e = -g\frac{e}{l} \tag{2.29}$$

$$\frac{d}{d\tau}l = \frac{g}{3}\frac{e-l}{e+g} \tag{2.30}$$

$$\frac{d}{d\tau}e_H = (1-\kappa)\frac{ge}{l}\frac{l+g}{e+g} - e_H\left(k + \frac{g}{l}\frac{e-l}{e+g}\right) \tag{2.31}$$

where the initial scaled maturity density $e_H^0 = (1-\kappa)g$ is such that $\frac{d}{d\tau}e_H(0) = 0$, otherwise $\frac{d}{d\tau}e_H(0) = \pm\infty$.

If $k = 1$, so $\dot{k}_J = \dot{k}_M$, we have $e_H(\tau) = e_H^0$ for all τ and $u_H(\tau) = (1-\kappa)l_b^3$. In other words: the maturity density remains constant, so maturity exceeds threshold values when structure exceeds threshold values. We then have the relationship for the structural volume at birth

$$V_b = L_b^3 = \frac{E_H^b/[E_m]}{(1-\kappa)g} \tag{2.32}$$

For $k > 1$, e_H is decreasing in (scaled) age, and for $k < 1$ increasing.

Figure 2.12 shows that data on embryo weight, yolk and respiration are in close agreement with model expectations. As is discussed later, {147}, respiration is taken to be proportional to the mobilisation rate. The two or three curves per species have been fitted simultaneously by Zonneveld [1296], and the total number of parameters is five excluding, or seven including, respiration. This is fewer than three parameters per curve and thus approaches a straight line for simplicity when measured this way. I have not found comparable data for plant seeds, but I expect a very similar pattern of development.

The examples are representative of the data collected in Table 2.2, which gives parameter estimates of some 40 species of snails, fish, amphibians, reptiles and birds. The model tends to underestimate embryo weight and respiration rate in the early phases of development. This is partly because of deviations in isomorphism, the contributions of extra-embryonic membranes (both in weight and in the mobilisation of energy reserve) and the loss of water content during development. The parameter estimates

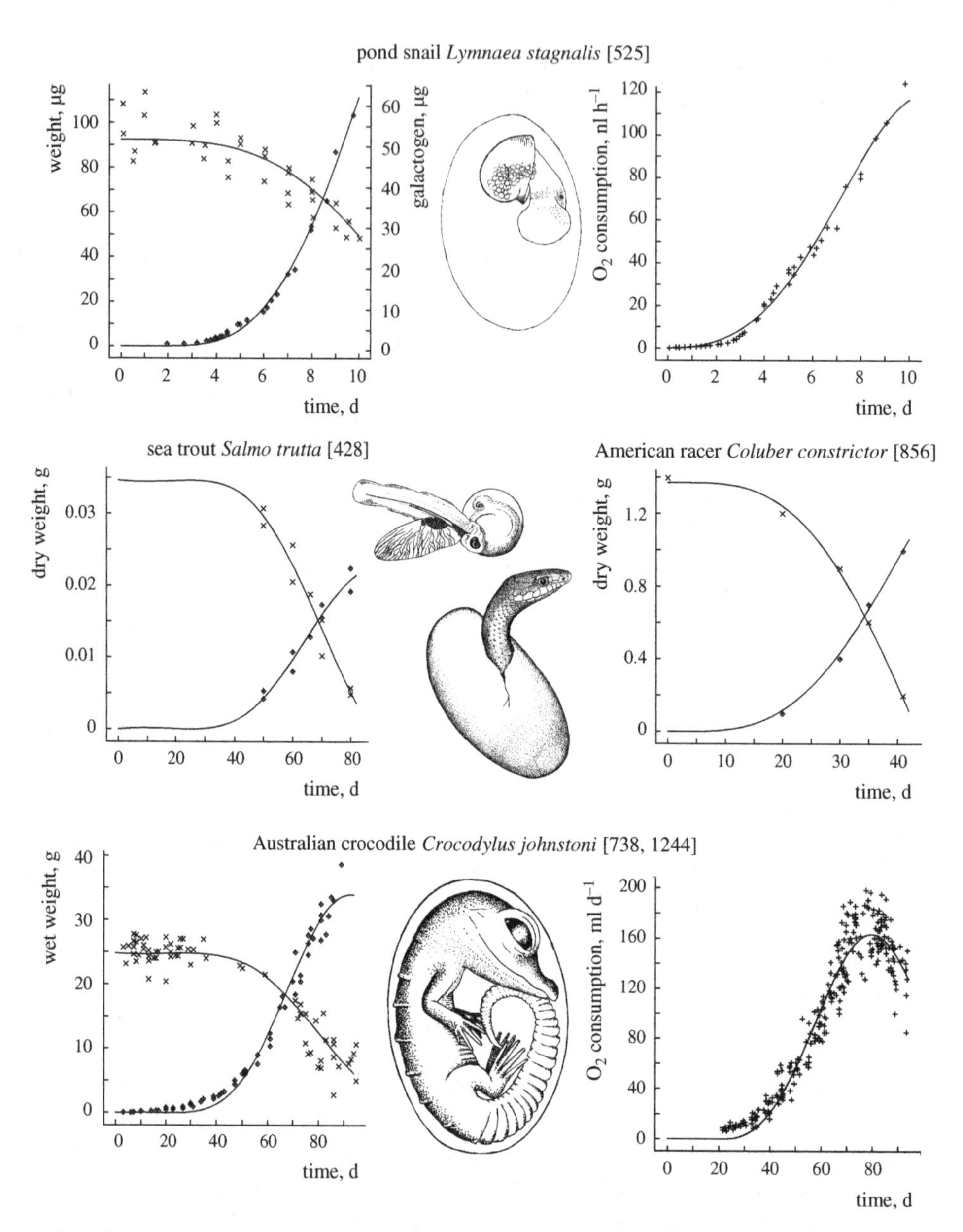

Figure 2.12 Yolk-free embryo weight (◇), yolk weight (×) and respiration rate (+) during embryo development, and fits on the basis of the DEB model. Data sources are indicated.

for altricial birds such as the parrot *Agapornis* should be treated with some reservation, because neglected acceleration caused by the temperature increase during development substantially affects the estimates, as discussed on {173}.

The values for the energy conductance $\dot{v}$, as given in Table 2.2, are in accordance with the average value for post-embryonic development, as given on {312}, which indicates

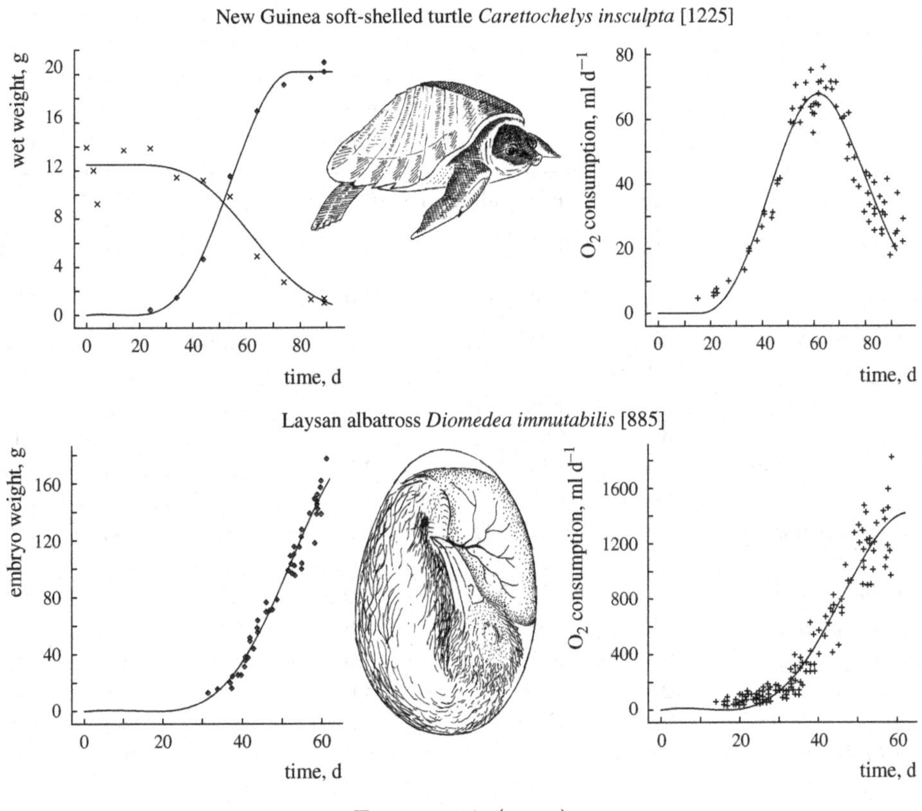

Figure 2.12 (cont.)

that no major changes in energy parameters occur at birth. The maintenance rate constant $\dot{k}_M$ for reptiles and birds is about 0.08 d^{-1} at 30 °C, implying that the energy required to maintain tissue for 12 days at 30 °C is about equal to the energy necessary to synthesise the tissue from the reserve. The maintenance rate constant for freshwater species seems to be much higher, ranging from 0.3 to 2.3 d^{-1}. Data from Smith [1081] on the rainbow trout *Oncorhynchus mykiss* result in 1.8 d^{-1} and Figure 2.10 gives 4.06 d^{-1} at 20 °C, which corresponds to some 8.5 d^{-1} at 30 °C (if it would survive that) for the waterflea *Daphnia magna*. The costs of osmosis might contribute to these high maintenance costs, as has been suggested on {45}, but for Ecdysozoa (to which *Daphnia* belongs), moulting might be costsly. The high value for *Oikopleura*, see Figure 2.19, probably relates to house production. Although information on parameter values is still sparse, it indicates that no (drastic) changes in these values occur at the transition from the embryonic to the juvenile state.

The general pattern of embryo development in eggs is characterised by unrestricted fast development during the first part of the incubation period (once the process has started) due to unlimited energy supply, at a rate that would be impossible to reach if the animal had to refill reserves by feeding. This period is followed by a retardation of

Table 2.2 Survey of re-analysed egg data, and parameter values standardised to a temperature of 30 °C, taken from [1296].

species	temp. °C	type of data	$\dot{v}_{30}$ mm d^{-1}	$\dot{k}_{M\,30}$ d^{-1}	E_b/E_0	reference
Lymnaea stagnalis	23	ED, galac, O	0.80	2.3	0.55	[527]
Salmo trutta	10	ED, YD	3.0	0.31	0.37	[430]
Rana pipiens	20	EW, O	2.5		0.87	[40]
Crocodylus johnstoni	30	EW, YW	1.9	0.060	0.31	[740]
	29, 31	O				[1245]
Crocodylus porosus	30	EW, YW	2.7	0.024	0.19	[1227]
	30	O				*1*
Alligator mississippiensis	30	EW, YW	2.7		0.34	[256]
	30	O				[1153]
Chelydra serpentina	29	ED, YD	1.9		0.35	[859]
	29	O				[399]
Carettochelys insculpta	30	EW, YW, O	1.9	0.040	0.08	[1226]
Emydura macquarii	30	EW, O	1.6	0.14	0.35	[1153]
Caretta caretta	28–30	EW, O	3.0		0.65	[4, 3]
Chelonia mydas	28–30	EW, O	3.0		0.57	[4, 3]
Amphibolurus barbatus	29	ED, YD	0.92	0.061	0.47	[860]
Coluber constrictor	29	ED, YD	1.4		0.69	[861]
Sphenodon punctatus	20	HW, O	0.85	0.062	0.25	*2*
Gallus domesticus	39	EW, O, C	3.2	0.039	0.34	[987]
	38	EW, C	3.4		0.52	[125]
Leipoa ocellata	34	EE, YE, O	1.7	0.031	0.55	[1200]
Pelicanus occidentalis	36.5	EW, O	3.2	0.10	0.77	[61]
Anous stolidus	35	EW, O	2.0	0.11	0.59	[887]
Anous tenuirostris	35	EW, O	1.8	0.20	0.59	[887]
Diomedea immutabilis	35	EW, O	2.5	0.069	0.57	[886]
Diomedea nigripes	35	EW, O	2.5	0.049	0.58	[886]
Puffinus pacificus	38	EW, O	0.92	0.084	0.61	[5]
Pterodroma hypoleuca	34	EW, O	1.9		0.20	[886]
Larus argentatus	38	EW, C	2.7	0.15	0.56	[293]
Gygis alba	35	EW, O	1.4		0.53	[885]
Anas platyrhynchos	37.5	EW	2.5	0.10	0.67	[921]
	37.5	O				[578]
Anser anser	37.5	EW	4.1	0.039	0.23	[986]
	37.5	O				[1198]
Coturnix coturnix	37.5	EW, O	1.7		0.49	[1198]
Agapornis personata	36	EW, O	0.8		0.79	[170]
Agapornis roseicollis	36	EW, O	0.84		0.81	[170]

Table 2.2 (cont.)

species	temp. °C	type of data	$\dot{v}_{30}$ mm d^{-1}	$\dot{k}_{M\,30}$ d^{-1}	E_b/E_0	reference
Troglodytes aëdon	38	EW, O	1.4		0.82	[590]
Columba livia	38	EW	2.7		0.80	[582]
	37.5	O				[1198]

EW: Embryo Wet weight; YW: Yolk Wet weight; ED: Embryo Dry weight
EE: Embryo Energy content; YE: Yolk Energy content; YD: Yolk Dry weight
O: Dioxygen consumption rate; C: Carbon dioxide prod. rate; HW: Hatchling Wet weight
'galac': galactogen content
1 P. J. Whitehead, pers. comm., 1989 ; *2* M. B. Thompson, pers. comm., 1989.

development due to the increasing depletion of energy reserves. Apart from the reserves of the juvenile, the model works out very similar to that of Beer and Anderson [74] for salmonid embryos.

In view of the goodness of fit of the model in species that do not possess shells (see the turtle data), retardation is unlikely to be due to limitation of gas diffusion across the shell, as has been frequently suggested for birds [932]. The altricial and precocial modes of development have been classified as being basically different; the precocials show a plateau in respiration rates towards the end of the incubation period, whereas the altricials do not. Figure 2.13 shows that this difference can be traced back to the simple fact that altricial birds hatch relatively early. A frequently used argument for diffusion limitation is the strong negative correlation between diffusion rates across the egg shell and diffusion resistance, when different egg sizes are compared, ranging from hummingbirds to ostriches; the product of diffusion rate and resistance does not vary a lot. This correlation probably results from a minimisation of water loss by eggs. Data from fossil plants from Greenland show a dramatic drop in stomata frequency at the end of the Triassic (208 Ma ago), which has been linked to a steep rise of the atmospheric carbon dioxide concentration to three times the present values, possibly as a result of the activity of volcanoes that mark the break-up of Pangaea [75, p 98]. The plants no longer needed many stomata, and reduced the number to reduce the water loss by evaporation.

To find τ_b, l_b, u_E^0, I first observe that from Table 2.1 and ODEs (2.26–2.27), we have

$$\frac{d}{d\tau}x = gx\frac{1-x}{l}; \quad \frac{d}{d\tau}l = \frac{g - xg - lx}{3}; \quad \frac{d}{d\tau}\,\alpha = \frac{x^{1/3}}{1-x}\frac{d}{d\tau}x \tag{2.33}$$

so

$$\alpha = 3g(u_E^0)^{-1/3} + B_x\left(\frac{4}{3}, 0\right) \tag{2.34}$$

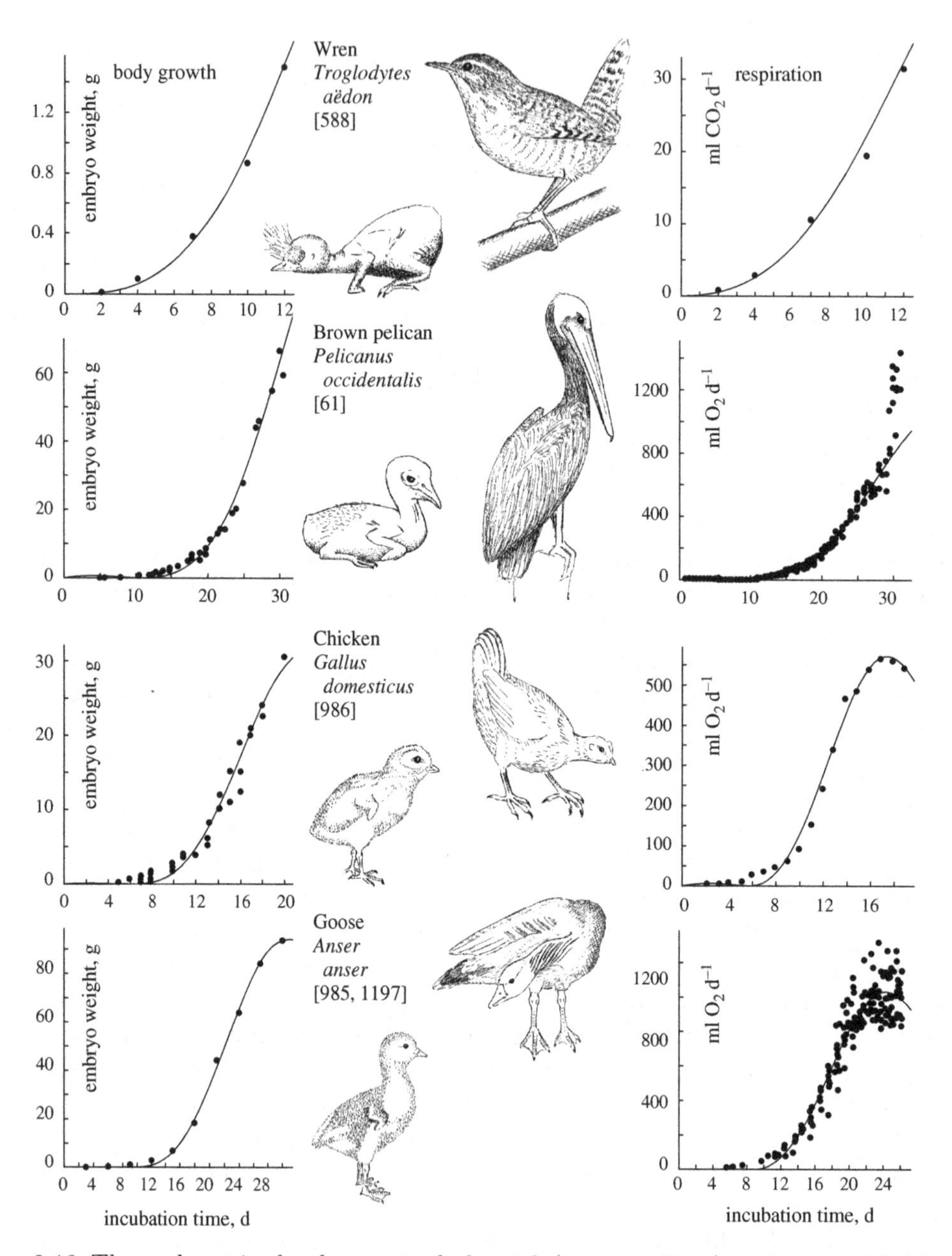

Figure 2.13 The embryonic development of altricial (wren, pelican) and precocial (chicken, goose) birds. Data from the sources indicated; fits are on the basis of the DEB model (parameters in Table 2.2). The underestimation of the initial development possibly relates to embryonic membranes. Pelican's high respiration rate just prior to hatching is attributed to internal pipping, which is not modelled. The drawings show hatchlings and adults.

for the incomplete beta function

$$B_x(\frac{4}{3}, 0) \equiv \int_0^x y^{1/3}(1-y)^{-1}\, dy \tag{2.35}$$

$$= \sqrt{3}\left(\arctan\frac{1+2x^{1/3}}{\sqrt{3}} - \arctan\frac{1}{\sqrt{3}}\right) + \frac{1}{2}\log(1+x^{1/3}+x^{2/3}) - \log(1-x^{1/3}) - 3x^{1/3}$$

Consequently we have

$$\alpha_b - \alpha = B_{x_b}(\frac{4}{3}, 0) - B_x(\frac{4}{3}, 0) \tag{2.36}$$

$$\frac{1}{l} = \frac{1}{l_b}\left(\frac{x_b}{x}\right)^{1/3} - \frac{B_{x_b}(\frac{4}{3}, 0) - B_x(\frac{4}{3}, 0)}{3gx^{1/3}} \tag{2.37}$$

We need this expression for $l(x)$ later in the derivation of l_b.

Scaled age at birth τ_b

The scaled age at birth τ_b follows from (2.33) and (2.37) by separation of variables and integration

$$\tau_b = 3\int_0^{x_b} \frac{dx}{(1-x)x^{2/3}(\alpha_b - B_{x_b}(\frac{4}{3}, 0) + B_x(\frac{4}{3}, 0))} \tag{2.38}$$

Notice that τ_b requires l_b in α_b, which is given below.

The parameter values in Figure 2.10 for *D. magna* imply an incubation time of 0.76 d at $f = 1$ to 0.80 d at $f = 0.5$. The eggs are deposited in the brood pouch just after moulting and develop there till birth just before the next moult, some 1.5 to 2 d later at 20 °C. This suggests a food-level-dependent diapause of around 0.5 d in *D. magna*.

The age at birth simplifies for small g and large $\dot{k}_M$, while $\dot{r}_B = \frac{\dot{k}_M g}{3(e_b+g)}$ remains fixed [661]:

$$a_b = \frac{3}{\dot{k}_M}\int_0^{x_b} \frac{dx}{(1-x)x^{2/3}(3gx_b^{1/3}l_b^{-1} - B_{x_b}(\frac{4}{3}, 0) + B_x(\frac{4}{3}, 0))} \tag{2.39}$$

$$\overset{g,\dot{k}_M^{-1}\text{ small}}{\simeq} \frac{1}{3e_b\dot{r}_B}\int_0^{x_b} \frac{dx}{(1-x)x^{2/3}x_b^{1/3}(l_b^{-1} - (x_b^{4/3} - x^{4/3})/(4e_b))}$$

$$\overset{g,\dot{k}_M^{-1}\text{ very small}}{\simeq} \frac{l_b}{3e_b\dot{r}_B}\int_0^{x_b} \frac{dx}{(1-x)x^{2/3}x_b^{1/3}} \overset{g,\dot{k}_M^{-1}\to 0}{=} \frac{l_b}{e_b\dot{r}_B} \overset{e_b=f}{=} \frac{3l_b}{\dot{k}_M g}\left(1+\frac{g}{f}\right) \tag{2.40}$$

where $x_b \equiv \frac{g}{e_b+g}$. The significance of this result is in the fact that for fixed $\dot{r}_B$, L_b and L_∞, $g \to 0$ while $\dot{k}_M \to \infty$ if a_b is running from 0 to this upper boundary. See Figure 2.14 for a graphical interpretation.

The chameleon *Furcifer labordi* is reported to live 8–9 months as egg and only 4–5 months as juvenile plus adult [577]. The males grow from some 3 cm to 10.2 cm (snout-to-vent length) with a von Bertalanffy growth rate of 0.035 d^{-1}, and the females from

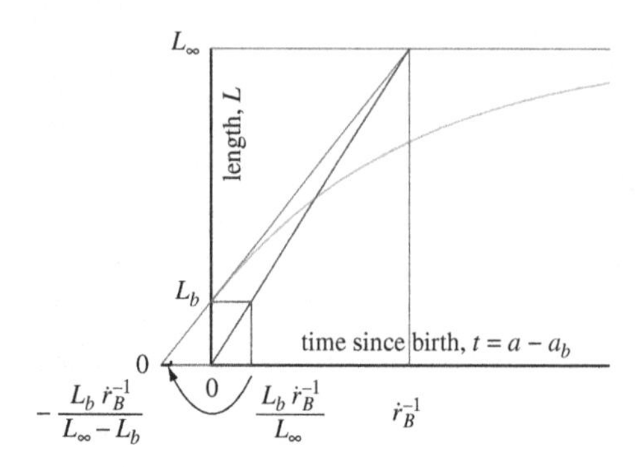

Figure 2.14 The von Bertalanffy growth curve $\frac{d}{dt}L = \dot{r}_B(L_\infty - L)$, with the graphical interpretation of the von Bertalanffy growth rate $\dot{r}_B$, and the maximum possible age at birth $\frac{L_b \dot{r}_B^{-1}}{L_\infty}$ in the context of DEB theory. The tangent line at $t = 0$ intersects the asymptote at time $\dot{r}_B^{-1}$; the line from the origin to this intersection point hits level L_b at time $\frac{L_b}{L_\infty \dot{r}_B}$, which is the maximum possible incubation time.

2.6 cm to 7.8 cm with a von Bertalanffy growth rate of $0.04\,\mathrm{d}^{-1}$. The maximum age at birth that is consistent with DEB theory is thus some 8.3 d, rather than the observed 8–9 months. The eggs are buried in the soil, where it might be some 10 °C cooler than during post-embryonic growth. Even after correction for this difference, this suggests that this species is on metabolic hold for most of the time.

The foetal special case, where $a_b = 3L_b/\dot{v} = \frac{3l_b}{\dot{k}_M g}$, represents a lower boundary for the age at birth (of eggs). So the possible range for a_b is

$$\frac{3l_b}{\dot{k}_M g} < a_b < \frac{3l_b}{\dot{k}_M g}(1 + g/e_b) \quad \text{or} \quad 1 < a_b \frac{\dot{k}_M g}{3l_b} < 1 + g/e_b \tag{2.41}$$

Scaled initial amount of reserve u_E^0

The scaled initial amount of reserve u_E^0 directly follows from (2.34) for $x = x_b$ and $\alpha = \alpha_b$

$$u_E^0 = \left(\frac{3g}{\alpha_b - B_{x_b}\left(\frac{4}{3}, 0\right)} \right)^3 \quad \text{so} \quad E_0 = u_E^0 g [E_m] L_m^3 \tag{2.42}$$

which again requires l_b in α_b.

The parameter values in Figure 2.10 for *D. magna* imply a scaled initial amount of reserve of $U_E^0 = 0.04\,\mathrm{mm}^2\mathrm{d}$ at $f = 1$ and $0.0246\,\mathrm{mm}^2\mathrm{d}$ at $f = 0.5$ at 20 °C. The scaled amount at birth equals $U_E^b = f[E_m]L_b^3 = 0.0313$ and $0.0156\,\mathrm{mm}^2\mathrm{d}$, respectively. So a fraction of 0.79 and 0.63 for $f = 1$ and 0.5, respectively, of the initial reserve is still left at birth. Figure 2.10 has, however, no data on embryo development or reserve, which demonstrates the strength of DEB theory.

Data on embryo development, Table 2.2, also show that about half of the reserves are used during embryonic development. The deviating values for altricial birds are artifacts, caused by the above-mentioned acceleration of development by increasing temperatures. Congdon *et al.* [217] observed that the turtles *Chrysemus picta* and *Emydoidea blandingi* have 0.38 of the initial reserves at birth. Respiration measurements on seabirds by Pettit *et al.* [888] indicate values that are somewhat above the ones reported in the table. The extremely small value for the soft-shelled turtle, see also

Figure 2.12, relates to the fact that these turtles wait for the right conditions to hatch, after which they have to run the gauntlet as a cohort at night from the beach to the water, where a variety of predators wait for them.

Scaled length at birth l_b

For the variable (τ, e_H) evolving from the value $(0, e_H^0)$ to the value (τ_b, e_H^b) we have

$$\frac{d}{d\tau}e_H = (1-\kappa)g(1-x)\left(\frac{g}{l(x)}+1\right) - e_H\left(k - x + g\frac{1-x}{l(x)}\right) \quad (2.43)$$

Now consider the variable (x, e_H) evolving from the value $(0, e_H^0)$ to the value (x_b, e_H^b) or the variable (x, y) evolving from the value $(0, 0)$ to the value (x_b, y_b):

$$\frac{d}{dx}e_H = \frac{e_H^0}{x}\left(\frac{l(x)}{g}+1\right) - \frac{e_H}{x}\left(\frac{k-x}{1-x}\frac{l(x)}{g}+1\right) \quad \text{for} \quad e_H^0 = e_H(0) = (1-\kappa)g$$

$$\frac{d}{dx}y = r(x) - ys(x) \quad \text{for} \quad r(x) = g + l(x); \quad s(x) = \frac{k-x}{1-x}\frac{l(x)}{gx} \quad (2.44)$$

where $l(x)$ is given in (2.37). The ODE for y can be solved to

$$y(x) = v(x)\int_0^x \frac{r(x_1)}{v(x_1)}dx_1 \quad \text{with} \quad v(x) = \exp\left(-\int_0^x s(x_1)\,dx_1\right) \quad (2.45)$$

The quantity l_b must be solved from $y_b = y(x_b) = gx_b v_H^b l_b^{-3}$, see Table 2.1. So we need to find the root of t as function of l_b with

$$t(l_b) = \frac{x_b g v_H^b}{v(x_b)l_b^3} - \int_0^{x_b}\frac{r(x)}{v(x)}dx = 0 \quad (2.46)$$

From this equation it becomes clear that the parameters κ and u_H^b affect l_b only via $v_H^b = \frac{u_H^b}{1-\kappa}$; a conclusion that is more difficult to obtain using the ODE for the scaled maturity density e_H rather than that for abstract variable y. Notice that the solution of l_b (and that of u_E^0 and τ_b) for the boundary value problem for the ODE for (u_E, l, e_H) as given in (2.26–2.28) depends on the four parameters g, k, v_H^b and e_b only. The solution for l_b must be substituted into (2.42) to obtain u_E^0 and in (2.38) to obtain τ_b; the scaled reserve at birth is $u_E^b = e_b l_b^3/g$.

The parameter values in Figure 2.10 for *D. magna* imply a birth length of 0.686 and 0.685 mm at $f = 1$ and 0.5, respectively. These values hardly differ much because $\dot{k}_J$ is close to $\dot{k}_M$.

Special case $e \to \infty$: foetal development

Foetal development differs from that in eggs in that energy reserves are supplied continuously via the placenta. The feeding and digestion processes are not involved. Otherwise, foetal development is taken to be identical to egg development, with initial reserves that can be taken to be infinitely large, for practical purposes. At birth, the neonate receives an amount of reserves from the mother, such that the reserve density of the neonate equals that of the mother. So the approximation $[E] \to \infty$ or $e \to \infty$ for the foetus can be made for the whole gestation period, because the foetus lives on the reserves of the mother. In other words: unlike eggs, the development of fetuses is not restricted by energy reserves. Initially the egg and foetus develop in the same way, but the foetus keeps developing at a rate not restricted by the amount of reserves till the end of the gestation time, while the development of the egg becomes retarded, due to depletion of the reserves.

The special case $e \to \infty$ means that $\frac{d}{d\tau}l = g/3$, or $l(\tau) = g\tau/3$. The foetal structural volume thus behaves as

$$\frac{d}{dt}V = \dot{v}V^{2/3} \quad \text{so} \quad V(t) = (\dot{v}t/3)^3 \tag{2.47}$$

This growth curve was proposed by Huggett and Widdas [536] in 1951. Payne and Wheeler [874] explained it by assuming that the growth rate is determined by the rate at which nutrients are supplied to the foetus across a surface that remains in proportion to the total surface area of the foetus itself. This is consistent with the DEB model, which gives the energy interpretation of the single parameter. The graph of foetal weight against age resembles an exponential growth curve, but in fact it is less steep; the model has the property that subsequent weight doubling times increase by a factor $2^{1/3} = 1.26$, while there is no increase in the case of exponential growth.

The fit is again excellent; see Figure 2.15. It is representative for the data collected in Table 2.3 taken from [1296]. A time lag for the start of foetal growth has to be incorporated, and this diapause may be related to the development of the placenta, which possibly depends on body volume as well. The long diapause for the grey seal *Halichoerus* probably relates to timing with the seasons to ensure adequate food supply for the developing juvenile. Variations in weight at birth are primarily due to variations in gestation period, not in foetal growth rate. For comparative purposes, energy conductance $\dot{v}$ is converted to 30 °C, on the assumption that the Arrhenius

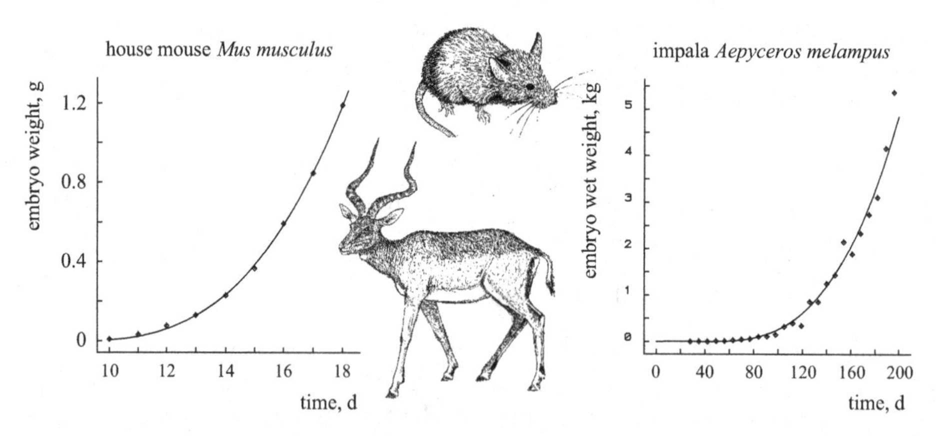

Figure 2.15 Foetal weight development in mammals, see Figure 6.2. Parameters are given in Table 2.3.

Table 2.3 The estimated diapause t_0, age at birth a_b, scaled energy conductance $\dot{v}^* = \dot{v}(1+\omega_w)^{1/3}$, birth weight W_b for mammals; specific density $d_V = 1\,\text{g}\,\text{cm}^{-3}$, $e = 1$.

species	t_0 d	a_b d	$\dot{v}^*$ cm d^{-1}	W_b g	reference
Homo sapiens	26.8	259	0.180	3750	[1230]
Oryctolagus cuniculus	10.7	19.2	0.560	46.2	[695]
Lepus americanus	13.1	21.1	0.573	66	[127]
Cavia porcellus	15.7	48.8	0.269	84	[291]
	19.5	48.5	0.239	101	[541]
Cricetus auratus	9.29	5.8	0.570	1.43	[925]
Mus musculus	8.2	11.8	0.278	1.2	[728]
Rattus norvegicus					
Wistar	11.4	7.7	0.487	2	[346]
	11.8	9.3	0.525	4.3	[536]
albino	12.4	9.5	0.542	5	[355]
Clethrionomys glareolus	8.29	10.4	0.374	2.2	[224]
Aepyceros melampus	39.4	166	0.316	5370	[331]
Odocoileus virginianus	34.9	155	0.296	3577	[977]
Dama dama	9.94	155	0.345	56400	[35]
Cervus canadensis	24.9	217	0.336	14500	[811]
Lama pacus	7.47	90.7	0.120	47.7	[347]
Ovis aries					
Welsh	43.9	105.8	0.482	4913	[536]
	33.3	111	0.433	4120	[213]
Karakul	31.0	116	0.426	4445	[306]
	27.5	121.8	0.403	4384	[568]
Capra hircus	24.3	126.3	0.339	2910	[317]
	31.3	117.3	0.365	2910	[59]
Bos taurus	59.5	204	0.475	33750	[1261]
Equus caballus	37.0	348	0.370	79200	[789]
Sus scrofa	4.73	129	0.266	1500	[1219]
Yorkshire	5.49	121	0.283	1500	[1178]
large white	23.6	90	0.383	1500	[907]
Essex	14.1	107	0.321	1500	[907]
Felix catus	18.8	39.8	0.371	119.5	[223]
Pipistrellus pipistrellus					
1978	9.95	14.2	0.237	1.4	[927]
1979	13.7	18.5	0.181	1.4	[927]
Halichoerus grypus	145	165	0.375	8732	[501]

temperature, T_A, is 10 200 K and the body temperature is 37 °C for all mammals in the table. This is a rather crude conversion because the cat, for instance, has a body temperature of 38.6 °C. Weights were converted to volumes using a specific density of $[W_w] = 1\ \text{g}\,\text{cm}^{-3}$.

One might expect that precocial development is rapid, resulting in advanced development at birth and, therefore, comes with a high value for the energy conductance. The values collected in Table 2.3, however, do not seem to have an obvious relationship with altricial–precocial rankings. The precocial guinea-pig and alpaca as well as the altricial human have relatively low values for the energy conductance. The altricial–precocial ranking seems to relate only to the relative volume at birth V_b/V_m.

We further have

$$\frac{d}{d\tau}u_H = (1-\kappa)l^2(g+l) - ku_H \tag{2.48}$$

$$u_H(\tau) = \frac{g^3(1-\kappa)}{3^3k^4}\left(k^2\tau^2(3k+k\tau-3) + 6(k-1)(1-\tau-\exp(-k\tau))\right) \tag{2.49}$$

The equation $u_H(\tau_b) = u_H^b$ has to be solved numerically for τ_b, but for $k=1$ we have $u_H^b = (1-\kappa)3^{-3}g^3\tau_b^3 = (1-\kappa)l_b^3$. The solution of this equation is stable and fast; the resulting scaled length at birth $l_b = g\tau_b/3$ can be used to start the Newton Raphson procedure. This start is preferable if k is substantially different from 1. From $l_b < 1$, so $\tau_b < 3/g$, we can derive the constraint

$$\frac{k^2u_H^b}{1-\kappa} < k + g(k-1) + g^3\frac{k-1}{k^2}\frac{1-3/g-\exp(-3k/g)}{9/2} \tag{2.50}$$

For $u_E^b = u_E(\tau_b)$, the cost for a foetus amounts to

$$u_E^0 = u_E^b + \kappa l_b^3 + u_H^b + \int_0^{\tau_b}(\kappa l^3(\tau) + ku_H(\tau))\,d\tau = u_E^b + l_b^3 + \frac{3}{4}\frac{l_b^4}{g}; \quad E_0 = u_E^0 g[E_m]L_m^3 \tag{2.51}$$

where the five terms correspond with the costs of reserve, structure, maturity, somatic and maturity maintenance, respectively. The second equality follows from the structure of DEB theory: the investment in maturity plus maturity maintenance equals $\frac{1-\kappa}{\kappa}$ times the investment in structure plus somatic maintenance and $l(\tau) = g\tau/3$. The cost of a foetus is somewhat smaller than that of an egg (2.42).

Foetal development obviously affects the energetics of the mother. This is discussed at {282}.

2.6.3 States at puberty

Puberty occurs as soon as $E_H = E_H^p$, or $U_H = U_H^p = E_H^p/\{\dot{p}_{Am}\} = U_H^p$ or $u_H = u_H^p = \frac{E_H^p}{g[E_m]L_m^3}$. If $\dot{k}_J = \dot{k}_M$, and so maturity density is constant, and $L_T = 0$, we have

$$V_p = \frac{E_H^p/[E_m]}{(1-\kappa)g} \tag{2.52}$$

Otherwise the structural volume at puberty $V_p = L_p^3$ can vary with food density history and even more than V_b can. It must be found numerically from integration of $\frac{d}{dt}V$ to $E_H(t) = E_H^p$.

Generally little can be said about age and length at puberty, but if food density X, and so scaled functional response f, remains constant, (2.23), (2.22) and (2.43) show that age and scaled length at puberty amount to

$$a_p = a_b + \frac{1}{\dot{r}_B} \ln \frac{L_\infty - L_b}{L_\infty - L_p} \tag{2.53}$$

$$l_p = l(v_H^p) \quad \text{with} \quad \frac{d}{dv_H} l = \frac{(f - l - l_T)g/3}{fl^2(g+l) - kv_H(g+f)} \quad \text{and} \quad l(v_H^b) = l_b \tag{2.54}$$

where $k = \dot{k}_J/\dot{k}_M$ as before. This differential equation has to be integrated numerically and results in a satiating function $l_p(f)$ if $k < 1$. We must have that $f > l_p + l_T$ to allow for adolescence.

The parameter values in Figure 2.10 for *D. magna* imply a length at puberty of 2.46 and 2.45 mm at $f = 1$ and 0.5, respectively. These values hardly differ because $\dot{k}_J$ is close to $\dot{k}_M$.

2.6.4 Reduction of the initial amount of reserve

The embryonic period is elongated if the initial amount of reserve is reduced, while the cumulative energy investment to complete the embryonic stage is the same. The mechanism is that reducing the amount of reserve reduces the mobilisation of reserve, so it takes longer to reach a certain maturity level. Large eggs, so large initial energy supplies, result in short incubation times if eggs of one species are compared. Crested penguins, *Eudyptes*, are known for egg dimorphism [1218]; see Figure 2.16. They first lay a small egg and, some days later, a 1.5 times bigger one. As predicted by the DEB

Figure 2.16 Egg dimorphism occurs as standard in crested penguins (genus *Eudyptes*). The small egg is laid first, but it hatches later than the big one, which is 1.5 times as heavy. The DEB theory explains why the large egg requires a shorter incubation period. The illustration shows the Snares crested penguin *E. atratus*.

model, the bigger one hatches first, if fertile, in which case the parents cease incubating the smaller egg, because they are only able to raise one chick. They continue to incubate the small egg only if the big one fails to hatch. This is probably an adaptation to the high frequency of unfertilised eggs or other causes of loss of eggs (aggression [1218]), which occur in this species.

Incubation periods only decrease for increasing egg size if all other parameters are constant. The incubation period is found to increase with egg size in some beetle species, lizards and marine invertebrates [210, 327, 1067]. In these cases, however, the structural biomass at hatching also increases with egg size. This is again consistent with the DEB theory, although the theory does not explain the variation in egg sizes.

Hart [466] studied the effect of separation of the embryonic cells of the sea urchin *Strongylocentrotus droebachiensis* in the two-cell stage on the energetics of larval development. Both the size and the feeding capacity of the resulting larvae were reduced by about one-half, but the time to metamorphosis is about the same (7 d at 8–13 °C). The maximum clearance rate of dwarf and normal larvae was found to be the same function of the ciliated band length. Larvae fed a smaller ration had longer larval periods, but food ration hardly affected size at metamorphosis. Egg size affected juvenile test diameter only slightly.

Armadillos typically separate cells in the four-cell stage of the embryo, giving birth to four identical offspring. Humans rarely do this successfully, then giving birth to four babies of about 1 kg each, rather than the typical 3 kg. In terms of an effect on length this reduction amounts to a factor $(1/3)^{1/3} = 0.69$. The human growth curve fits the von Bertalanffy curve very well, with a von Bertalanffy growth rate of $\dot{r}_B = \frac{\dot{k}_M g}{3(e+g)} = 0.123$ a^{-1}, see {289}. We can safely assume that the scaled reserve density was close to its maximum $e = 1$ for the post-embryonic stages. Moreover the age at birth is $a_b = \frac{3l_b}{g\dot{k}_M} = 0.75$ a for humans. If we take a typical maximum adult weight of 70 kg, then the scaled length at birth equals $l_b = (3/70)^{1/3} = 0.35$. So the energy investment ratio equals $g = \frac{l_b}{a_b \dot{r}_B} - 1 = 2.79$, the somatic maintenance rate coefficient $\dot{k}_M = \frac{3l_b}{g a_b} = 0.5$ a^{-1} and the scaled age at birth $\tau_b = a_b \dot{k}_M = 0.375$. With these values for g, e_b and l_b, the scaled cost amounts to $u_E^0 = 0.062$ from (2.51). In the case of four babies with a length reduced by a factor 0.69, the scaled cost per baby equals $u_E^0 = 0.02$, so summed over the four babies this is 1.3 times the amount of a single baby; not a surprising result, in view of the 4 kg of babies relative to the 3 kg for a single baby.

If the separation of the cells affects the required cumulative investment in development, however, other predictions result. It is then quite well possible that incubation is hardly affected, while size at birth is. Standard DEB theory correctly predicts that a reduction of the feeding level elongates the larval period and hardly affects size at puberty for a particular relationship between the somatic and maturity maintenance costs.

As discussed at {300}, the reserve density capacity $[E_m] = \{\dot{p}_{Am}\}/\dot{v}$ scales with maximum structural length of a species. So species with a larger ultimate body size tend to have a relatively larger reserve capacity. It turned out that for the combination of parameter values as found for *D. magna* we have to apply a zoom factor of at least $z = 1.87$ to arrive at a minimum maximum body size for which cell separation might be successful.

For $k > 1$, the structural volume at birth increases after halving, and decreases for $k < 1$. Since reserve contributes to weight, the weight at birth is close to half of the original weight at birth, irrespective of the value of k. The age of the two-cell stage is probably smaller than $\tau_b/3$, but the results are very similar.

The removal of an amount of reserve at the start of the development, as is frequently done [351, 555, 556, 809, 1067], elongates the incubation time (as observed in the gypsy moth [994]), and reduces the reserve at hatching. This experiment simulates the natural situation, where the nutritional status of the mother affects the initial amount of reserve. The pattern is rather similar to that of the separation of cells at an early stage, because reductions of structure and maturity at an early stage have little effect. The initial amount of reserve is a U-shaped function of the reserve at birth. The right branch is explained by the larger amount of reserve at birth, the left branch by the larger age at birth, which comes with larger cumulative somatic maintenance requirements.

The size of neonates of trout and salmon was found to increase with initial egg size [314, 478], suggesting that $k < 1$ for salmonids. This also applies to the emu *Dromaius novaehollandiae* [305], and probably represents a general pattern.

2.7 Reproduction: excretion of wrapped reserve

Organisms can achieve an increase in numbers in many ways. Sea anemones can split off foot tissue that can grow into a new individual. This is not unlike the strategy of budding yeasts. Colonial species usually have several ways of propagating. Fungi have intricate sexual reproduction patterns involving more than two sexes. Under harsh conditions some animals can switch from parthenogenetic to sexual reproduction, others develop spores or other resting phases. It would not be difficult to fill a book with descriptions of all the possibilities. The standard DEB model assumes propagation via eggs; dividing organisms don't have an embryo or adult stage, and divide when the maturity exceeds a threshold. This resets maturity and reduces the amount of structure and reserve.

Energy allocation to reproduction equals

$$\dot{p}_R = (1 - \kappa)\dot{p}_C - \dot{k}_J E_H^p \tag{2.55}$$

cf. (2.18), where E_H is now replaced by the constant E_H^p, because the flux to maturity is redirected to reproduction at puberty, which means that maturity does not change in adults. The cost of an egg E_0 or a foetus is given in (2.42) or (2.51), so the mean reproduction rate $\dot{R}$ in terms of number of eggs per time equals

$$\dot{R} = \frac{\kappa_R \dot{p}_R}{E_0} = \frac{\kappa_R \dot{k}_M}{v_E^0}\left(\frac{el^2}{e+g}(g + l_T + l) - kv_H^p\right) \tag{2.56}$$

for $v_E^0 = \frac{u_E^0}{1-\kappa}$, see Table 2.1.

At constant food density, where $e = f$, the reproduction rate is, according to (2.56), proportional to

$$\dot{R} \propto L^2 + \frac{\dot{k}_M}{\dot{v}} L^3 - L_R^2 \tag{2.57}$$

where L_R is just a constant, which depends on the nutritional status. Comparison of reproduction rates for different body sizes thus involves three compound parameters, i.e. the proportionality constant, $\dot{k}_M/\dot{v}$ and L_R, if all individuals experience the same food density for a long enough time. Figure 2.17 shows that this relationship is realistic, but that the notorious scatter for reproduction data is so large that access to the parameter $\dot{k}_M/\dot{v}$ is poor. The fits are based on guestimates for the maintenance rate coefficient, $\dot{k}_M = 0.011$ d^{-1}, and the energy conductance, $\dot{v} = 0.433$ mm d^{-1} at 20 °C. The main reason for the substantial scatter in reproduction data is that they are usually collected from the field, where food densities are not constant, and where spatial heterogeneities, social interactions, etc., are common.

The reproduction rate of spirorbid polychaetes has been found to be roughly proportional to body weight [498]. On the assumption by Strathmann and Strathmann [1124] that reproduction rate is proportional to ovary size and that ovary size is proportional to body size (an argument that rests on isomorphy), the reproduction rate is also expected to be proportional to body weight. They observed that reproduction

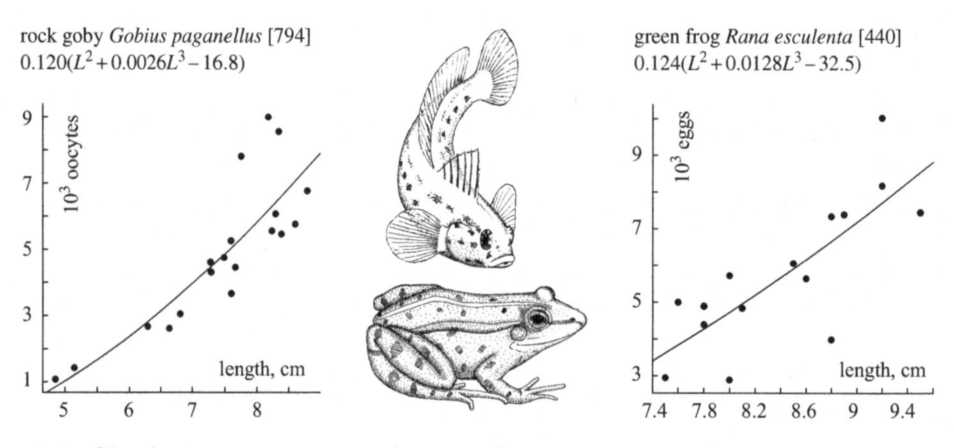

Figure 2.17 Clutch size, as a measure for reproduction rate, as a function of body length L for two randomly selected species. The data sources and DEB-based curves are indicated. The parameter that is multiplied by L^3 in both fits has been guestimated on the basis of common values for the maintenance rate coefficient and the energy conductance, with a shape coefficient of $\delta_{\mathcal{M}} = 0.1$ for the goby and of $\delta_{\mathcal{M}} = 0.5$ for the frog. Both other parameter values represent least-squares estimates.

rate tends to scale with body weight to the power somewhat less than 1 for several other marine invertebrate species, and used their observation to identify a constraint on body size for brooding inside the body cavity. The DEB theory gives no direct support for this constraint; an allometric regression of reproduction rate against body weight would result in a scaling parameter between 2/3 and 1, probably close to 1, depending on parameter values.

The maximum (mean) reproduction rate for an individual of maximum volume, i.e. $l = 1 - l_T$, amounts to

$$\dot{R}_m = \kappa_R \dot{k}_M \frac{(1 - l_T)^2 - k v_H^p}{v_E^0} \tag{2.58}$$

with $v_E^0 = \frac{u_E^0}{1-\kappa}$, where u_E^0 is given in (2.42) for eggs and in (2.51) for fetuses.

Under conditions of prolonged starvation, organisms can deviate from the standard reproduction allocation, as is discussed on {118}.

2.7.1 Cumulative reproduction

Oikopleura sports a heroic way of reproduction which leads to instant death. During its week-long life at 20 °C and abundant food, it accumulates energy for reproduction which is deposited at the posterior end of the trunk; see Figure 2.18. This allows an easy test of the allocation rule against experimental data. Except for this accumulation of material for reproduction, the animal remains isomorphic. The total length of the trunk, L_t, including the gonads, can be partitioned into the true trunk length, L, and the length of the gonads, L_R. Since the reproduction material is deposited on a surface area of the trunk, the length of the gonads is about proportional to the accumulated investment of energy in reproduction divided by the squared true trunk length. Fenaux and Gorsky [341] measured both the true and the total trunk length under

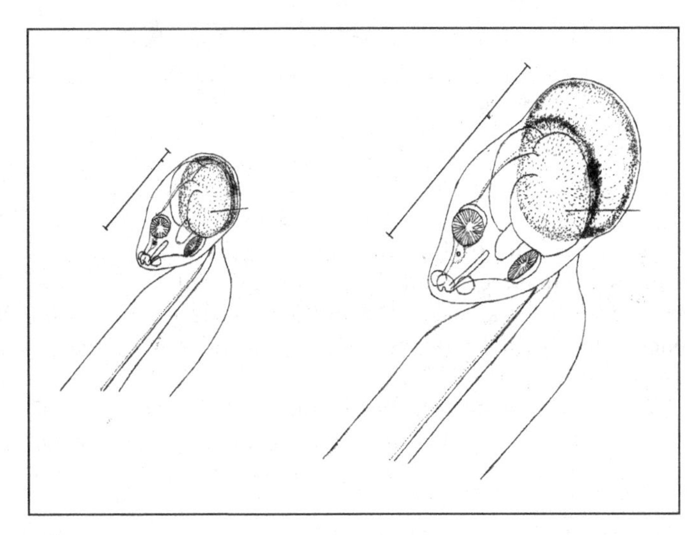

Figure 2.18 The larvacean *Oikopleura* grows isomorphically; during its short life it accumulates reproductive material at the posterior end of the trunk. The energy interpretation of data on total trunk lengths should take account of this. Larvaceans of the genus *Oikopleura* are an important component of the zooplankton of all seas and oceans and have an impact as algal grazers comparable with that of copepods.

laboratory conditions. This allows us to test the consequences of the DEB theory for reproduction.

Let $e_R(t_1, t_2)$ denote the cumulative investment of energy in reproduction between t_1 and t_2, as a fraction of the maximum energy reserves $[E_m]V_m$. From Table 2.5 (see below) we know that this investment amounts for adults to

$$e_R(t_1, t_2) = \kappa_R g \dot{k}_M \int_{t_1}^{t_2} \left((1-\kappa) \frac{g + l(t)}{g + e(t)} e(t) l^2(t) - H k u_H^p \right) dt \tag{2.59}$$

Oikopleura has a non-feeding larval stage and starts investing in reproduction as soon as it starts feeding, so $E_H^b = E_H^p$. From an energetic point of view, it thus lacks a juvenile stage, and the larva should be classified as an embryo. The total trunk length then amounts to $L_t(t) = L(t) + V_R e_R(0,t)/L^2(t)$. The volume V_R is a constant that converts the scaled cumulative reproductive energy per squared trunk length into the contribution to the total length. With abundant food, the true trunk length follows the von Bertalanffy growth curve (2.23) and $e(t) = 1$. If the data set $\{t_i, L(t_i), L_t(t_i)\}_{i=1}^n$ is available, the five parameters L_b, L_m, $\dot{k}_M$, g and $V_R\kappa_R$ can be estimated in principle. Dry weight relates to trunk length and reproductive energy as $W_d(t) = [W_{Ld}]L^3(t) + W_{Rd} e_R(0,t)$, where the two coefficients give the contribution of cubed trunk length and cumulative scaled reproductive energy to dry weight. If dry weight data are available as well, there are seven parameters to be estimated from three curves.

Figure 2.19 gives an example. The data appear to contain too little information to determine both $\dot{k}_M$ and g, so either $\dot{k}_M$ or g has to be fixed. The more

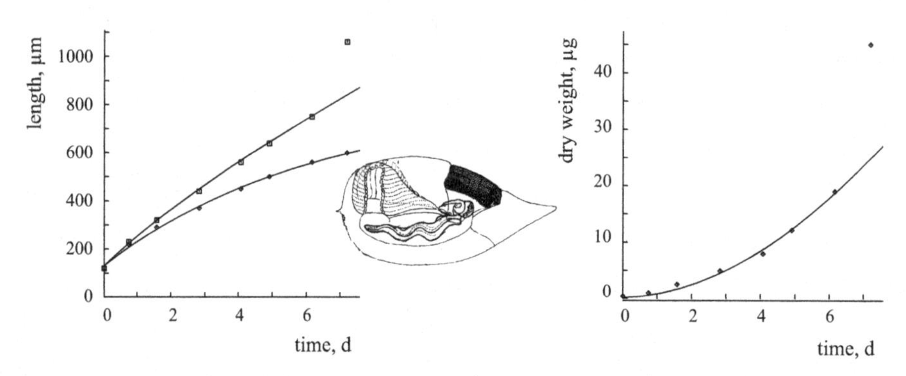

Figure 2.19 The total trunk length, L_t (□ and upper curve, left), the true trunk length, L (◇ and lower curve, left) and the dry weight (right) for *Oikopleura longicauda* at 20 °C. Data from Fenaux and Gorsky [341]. The DEB-based curves account for the contribution of the accumulated energy allocated to reproduction, to total trunk length and to dry weight. The parameter estimates are $L_m = 822\,\mu$m, $l_b = l_p = 0.157$, $\dot{k}_M = 1.64\,\text{d}^{-1}$, $g = 0.4$, $V_R\kappa_R = 0.0379\,\text{mm}^3$. Given these parameters, the weight data give $W_{Ld} = 0.0543\,\text{g}\,\text{cm}^{-3}$, $W_{Rd} = 15.2\,\mu$g. The last data point is excluded in both data sets because here structure is rapidly converted to gametes, and maintenance has probably ceased. This 'detail' is not implemented in the standard DEB model.

or less arbitrary choice $g = 0.4$ is made here. The estimates are tied by the relationship that $\frac{\dot{k}_M g}{1+g}$ is almost constant. The high value for the maintenance rate coefficient $\dot{k}_M$ probably relates to the investment of energy in the frequent synthesis of new filtering houses. The cumulative reproduction fits the data for *D. magna* also very well (see Figure 2.10), where $\dot{k}_M$ is even higher, probably related to moulting.

2.7.2 Buffer handling rules

Individuals are discrete units, which implies the existence of a buffer, where the energy allocated to reproduction is accumulated and converted to eggs at the moment of reproduction. The translation of reproduction rate into number of eggs in Figure 2.17 assumes that this accumulation is over a period of 1 year. The energy content of the buffer is denoted by E_R.

The strategies for handling this buffer are species-specific. Some species (e.g. some rotifers) reproduce when enough energy for a single egg has been accumulated, others wait longer and produce a large clutch. If the reproduction buffer is used completely, the size of the clutch equals the ratio of the buffer content to the energy costs of one young, $\kappa_R E_R / E_0$, where E_0 is given in (2.42). This resets the buffer. So after reproduction $E_R = 0$ and further accumulation continues from there. That is to say, the bit of energy that was not sufficient to build the last egg can become lost or still remain in the buffer; fractional eggs do not exist. In the chapter on population dynamics, {358,361}, I show that this uninteresting detail substantially affects dynamics at low population growth rates, which occur most frequently in nature. If food is abundant, the population will evolve rapidly to a situation in which food per individual is sparse and reproduction low if harvesting processes do not prevent this.

Reproduction is coupled to the moulting cycle in daphnids; neonates in the brood pouch are released, the old moult is shed, the new eggs are deposited in the brood pouch and the new carapace hardens out and the cycle repeats some 1.5 to 2 days later at 20 °C. The moulting cycle is linked to somatic maintenance and, therefore, is independent of nutritional status, while the incubation time is dependent. This means that there must be a variable diapause.

Many species use environmental triggers for spawning at particular times t_R. Many molluscs spawn if temperature exceeds a threshold value, while $[E_R]$ is larger than a threshold value [918]. Other species use food availability as trigger; *Pecten maximus* spawns just after an algal bloom. The spectacular synchronisation of reproduction in corals [1051], the pelagic palolo worms *Eunice viridis*, South East Asian dipterocarps and bamboo forests probably reduce losses, because potential predators have little to eat between the events. Most species are able to synchronise the moment of reproduction with seasonal cycles such that food availability just matches the demand of the offspring. Clutch size in birds typically relates to food supply during a 2-month period prior to egg laying and tends to decrease if breeding is postponed in the season

[779]. The laying date is determined by a rapid increase in food supply. Since feeding conditions tend to improve during the season, internal factors must contribute to the regulation of clutch size. These conclusions result from an extensive study of the energetics of the kestrel *Falco tinnunculus* by Serge Daan and co-workers [269, 754, 778]. I see reproductive behaviour like this for species that cease growth at an early moment in their life span as variations on the general pattern that the DEB theory is aiming to grasp. Aspects of reproduction energetics for species that cease growth are worked out on {284}.

Multiple spawning

Batch preparation in adult anchovy is initiated when the temperature in spring exceeds a threshold (14 °C in anchovy [876, 877]); juveniles that mature after this time point have to wait for batch preparation till the next spring. The batch size expressed to total amount of reserves is proportional to structural volume; dividing by the cost per eggs gives the number of eggs per spawning, but this involves the reserve density of the spawner. This means that if the scaled functional response decreases during the spawning season, the number of eggs increases (if length remains constant).

The rate of batch preparation equals the maximum allocation to reproduction (i.e. as if $e = 1$) and batch preparation ceases for that spawning season if the reproduction buffer is emptied. The rate still depends on length of the individual and is motivated by the avoidance of an unbounded accumulation of the reproduction buffer with abundant food (during the whole year). The spawning season, however, lasts less than a year, so the rate of batch preparation is divided by the fraction of the year that has good spawning conditions, which is about 7/12. Only in the last batch of the spawning season will the batch size be smaller than the target size. If food were abundant, this rule for spawning implies that spawning, once initiated, continues till death.

2.7.3 Post-reproductive period

Many animal species have a post-reproductive period. In the context of DEB theory, this is (or can be) implemented as an aspect of ageing, see {214}, which is taken to be an effect of reactive oxygen species (ROS). Effects of compounds are, in general, implemented as effects on particular DEB parameters, like effects of temperature. This implementation requires more state variables than the standard DEB model has.

2.8 Parameter estimation I: numbers, lengths and time

Parameters with energy or moles in their dimension require measurements of energy and moles, respectively. The compound parameter g is dimensionless, and $\dot{k}_M$ has only time in its dimension, so they don't need such measurements to be estimated. The DEB parameters that can be extracted from observations on length and reproduction over time at several food densities are: κ, g, $\dot{k}_J$, $\dot{k}_M$, $\dot{v}$, $U_H^b = E_H^b/\{\dot{p}_{Am}\}$ and $U_H^p = E_H^p/\{\dot{p}_{Am}\}$, see [661] and Figure 2.10. These parameters also determine the initial scaled reserve $U_E^0 = E_0/\{\dot{p}_{Am}\}$ as a function of scaled functional response f. A data set using a single measured time-varying food density would be enough, in principle, but the estimation process is more complex, compared to a set (>1) of constant food densities. The food densities only have to be measured in the latter case if $\{\dot{F}_m\}$ needs to be obtained. The parameter κ_R must be obtained from mass balances; a default value of $\kappa_R = 0.95$ would probably be appropriate in most cases.

Common practice is that these observations are not always available, and the question is: what can be done with fewer data? If observations on just a single food density are available, we don't know how size at birth and/or puberty depends on food availability, and we are forced to assume that $\dot{k}_J = \dot{k}_M$.

The (compound) DEB parameters that only have time and length in their dimension can be obtained from growth and reproduction data with functions in DEBtool. These parameters don't depend on food level, while the growth and reproduction data do. To emphasise this, the quantities that depend on food level are printed bold below:

Growth at a single food level: debtool/animal/get_pars_g

$$(L_b, \boldsymbol{L_\infty}, \boldsymbol{a_b}, \boldsymbol{\dot{r}_B} \text{ at } \boldsymbol{f_1}) \longrightarrow (g, \dot{k}_M = \dot{k}_J, \dot{v}; \boldsymbol{U_E^0}, \boldsymbol{U_E^b} \text{ at } \boldsymbol{f_1})$$

Growth at several food levels: debtool/animal/get_pars_h

$$\begin{pmatrix} \boldsymbol{L_b}, \boldsymbol{L_\infty}, \boldsymbol{\dot{r}_B} \text{ at } \boldsymbol{f_1} \\ \boldsymbol{L_b}, \boldsymbol{L_\infty}, \boldsymbol{\dot{r}_B} \text{ at } \boldsymbol{f_2} \end{pmatrix} \longrightarrow \begin{pmatrix} g, \dot{k}_M, \dot{k}_J, \dot{v}, & \begin{matrix} \boldsymbol{U_E^0}, \boldsymbol{U_E^b} \text{ at } \boldsymbol{f_1} \\ \boldsymbol{U_E^0}, \boldsymbol{U_E^b} \text{ at } \boldsymbol{f_2} \end{matrix} \end{pmatrix}$$

Growth at several food levels: debtool/animal/get_pars_i

$$\begin{pmatrix} L_b, & \begin{matrix} \boldsymbol{L_\infty}, \boldsymbol{\dot{r}_B} \text{ at } \boldsymbol{f_1} \\ \boldsymbol{L_\infty}, \boldsymbol{\dot{r}_B} \text{ at } \boldsymbol{f_2} \end{matrix} \end{pmatrix} \longrightarrow \begin{pmatrix} g, \dot{k}_M = \dot{k}_J, \dot{v}, & \begin{matrix} \boldsymbol{U_E^0}, \boldsymbol{U_E^b} \text{ at } \boldsymbol{f_1} \\ \boldsymbol{U_E^0}, \boldsymbol{U_E^b} \text{ at } \boldsymbol{f_2} \end{matrix} \end{pmatrix}$$

Growth and reproduction at a single food level: debtool/animal/get_pars_r

$$(L_b, L_p, \boldsymbol{L_\infty}, \boldsymbol{a_b}, \boldsymbol{\dot{r}_B}, \boldsymbol{\dot{R}_\infty} \text{ at } \boldsymbol{f_1}) \overset{\text{given } \kappa_R}{\longrightarrow} (\kappa, g, \dot{k}_J = \dot{k}_M, \dot{v}, U_H^b, U_H^p; \boldsymbol{U_E^0}, \boldsymbol{U_E^b}, \boldsymbol{U_E^p} \text{ at } \boldsymbol{f_1})$$

Growth and reproduction at several food levels: debtool/animal/get_pars_s

$$\begin{pmatrix} \boldsymbol{L}_b, \boldsymbol{L}_p, \boldsymbol{L}_\infty, \dot{\boldsymbol{r}}_B, \dot{\boldsymbol{R}}_\infty \text{ at } \boldsymbol{f}_1 \\ \boldsymbol{L}_b, \boldsymbol{L}_p, \boldsymbol{L}_\infty, \dot{\boldsymbol{r}}_B, \dot{\boldsymbol{R}}_\infty \text{ at } \boldsymbol{f}_2 \end{pmatrix} \overset{\text{given } \kappa_R}{\longrightarrow} \begin{pmatrix} \kappa, g, \dot{k}_J, \dot{k}_M, \dot{v}, U_H^b, U_H^p, & \boldsymbol{U}_E^0, \boldsymbol{U}_E^b, \boldsymbol{U}_E^p \text{ at } \boldsymbol{f}_1 \\ & \boldsymbol{U}_E^0, \boldsymbol{U}_E^b, \boldsymbol{U}_E^p \text{ at } \boldsymbol{f}_2 \end{pmatrix}$$

Growth and reproduction at several food levels: debtool/animal/get_pars_t

$$\begin{pmatrix} L_b, L_p, & \boldsymbol{L}_\infty, \dot{\boldsymbol{r}}_B, \dot{\boldsymbol{R}}_\infty \text{ at } \boldsymbol{f}_1 \\ & \boldsymbol{L}_\infty, \dot{\boldsymbol{r}}_B, \dot{\boldsymbol{R}}_\infty \text{ at } \boldsymbol{f}_2 \end{pmatrix} \overset{\text{given } \kappa_R}{\longrightarrow} \begin{pmatrix} \kappa, g, \dot{k}_J = \dot{k}_M, \dot{v}, U_H^b, U_H^p, & \boldsymbol{U}_E^0, \boldsymbol{U}_E^b, \boldsymbol{U}_E^p \text{ at } \boldsymbol{f}_1 \\ & \boldsymbol{U}_E^0, \boldsymbol{U}_E^b, \boldsymbol{U}_E^p \text{ at } \boldsymbol{f}_2 \end{pmatrix}$$

The unscaled reserve E and maturity E_H require energies, and thus knowledge of $\{\dot{p}_{Am}\}$. This can be obtained from observations on the feeding process and the mass at zero and birth. These extra observations give access to more parameters, and are discussed at {168}. DEBtool also has functions iget_pars that do the inverse mapping from (compound) DEB parameters to easy-to-measure quantities. This can be used for checking the mapping and testing against empirical data.

2.9 Summary of the standard DEB model

The standard DEB model applies to an isomorph that feeds on a single type of food and has a single reserve and a single structure. In this chapter we have used time, length and energy only; this is not always most convenient. The model has three state variables: structural volume V, reserve energy E and maturity, expressed in terms of cumulative energy investment, E_H. Maturity has no mass or energy, however, and represents information. The input variable is food density $X(t)$, and temperature $T(t)$ modifies all rates (which can be recognised by the dots); they typically vary in time in harmony. The model has 12 individual-specific parameters: specific searching rate $\{\dot{F}_m\}$, assimilation efficiency κ_X, maximum specific assimilation rate $\{\dot{p}_{Am}\}$, energy conductance $\dot{v}$, allocation fraction to soma κ, reproduction efficiency κ_R, volume-specific somatic maintenance cost $[\dot{p}_M]$, surface-area-specific somatic maintenance cost $\{\dot{p}_T\}$, maturity maintenance rate coefficient $\dot{k}_J$, specific cost for structure $[E_G]$, maturity at birth E_H^b and maturity at puberty E_H^p. Typical parameter values are discussed in Chapter 8.

The 10 assumptions that fully specify the standard DEB model are listed in Table 2.4. If the somatic and maturity rate coefficients are equal, birth and puberty occur at fixed amounts of structure, so that there is no longer a need for maturity as an explicit state variable; otherwise the scaled length at birth l_b and l_p are not constant. Other DEB models are modified versions of the standard DEB model to include dividing organisms, changing of shapes, multiple types of food, reserve and structure, adaptation, etc. Effects of compounds, such as ageing, need more state variables.

Table 2.4 The assumptions that specify the standard DEB model quantitatively.

1. The amounts of reserve, structure and maturity are the state variables of the individual; reserve and structure have a constant composition (strong homeostasis) and maturity represents information.
2. Substrate (food) uptake is initiated (birth) and allocation to maturity is redirected to reproduction (puberty) if maturity reaches certain threshold values.
3. Food is converted into reserve and reserve is mobilised at a rate that depends on the state variables only to fuel all other metabolic processes.
4. The embryonic stage has initially a negligibly small amount of structure and maturity (but a substantial amount of reserve). The reserve density at birth equals that of the mother at egg formation (maternal effect). Fetuses develop in the same way as embryos in eggs, but at a rate unrestricted by reserve availability.
5. The feeding rate is proportional to the surface area of the individual and the food-handling time is independent of food density.
6. The reserve density *at constant food density* does not depend on the amount of structure (weak homeostasis).
7. Somatic maintenance is proportional to structural volume, but some components (osmosis in aquatic organisms, heating in endotherms) are proportional to structural surface area.
8. Maturity maintenance is proportional to the level of maturity.
9. A fixed fraction of mobilised reserves is allocated to somatic maintenance plus growth, the rest to maturity maintenance plus maturation or reproduction (the κ-rule).
10. The individual does not change in shape during growth (isomorphism). This assumption applies to the standard DEB model only.

Table 2.5 gives the resulting energy fluxes, called powers, as functions of the scaled energy density e and scaled length l, where the scaled length at birth l_b and puberty l_p depend on the food history if $\dot{k}_J \neq \dot{k}_M$. The relationships between compound and primary parameters are summarised in Table 3.3. Notice that all powers are cubic polynomials in the (scaled) length, while the weight coefficients depend on (scaled) reserve density.

The mobilisation power equals the sum of the non-assimilative powers, and κ times the mobilisation power equals the sum of somatic maintenance and growth:

$$\dot{p}_C = \dot{p}_S + \dot{p}_G + \dot{p}_J + \dot{p}_R \quad \text{with} \quad \kappa\dot{p}_C = \dot{p}_S + \dot{p}_G \tag{2.60}$$

A three-stage individual invests either in maturation, or in reproduction. This is why these powers have the same index: the stage determines the destination.

The dissipating power excludes assimilation and somatic growth overheads by definition and amounts to

$$\dot{p}_D = \dot{p}_S + \dot{p}_J + (1 - \kappa_R)\dot{p}_R \tag{2.61}$$

Table 2.5 The scaled powers $\dot{p}_*/\{\dot{p}_{Am}\}L_m^2$ as specified by the standard DEB model for an isomorph of scaled length $l = L/L_m$, scaled reserve density $e = [E]/[E_m]$ and scaled maturity density $u_H = \frac{E_H}{g[E_m]V_m}$ at scaled functional response $f \equiv \frac{X}{K+X}$, where X denotes the food density and K the saturation constant. The powers $\dot{p}_X$ and $\dot{p}_P$ for ingestion and defecation occur in the environment, not in the individual. Assimilation is switched on at scaled length l_b, and allocation to maturation is redirected to reproduction at $l = l_p$. Compound parameters: allocation fraction κ, investment ratio g, somatic maintenance rate coefficient $\dot{k}_M$, scaled heating length l_T. Implied dynamics for $e > l > l_b$: $\frac{d}{dt}e = \dot{k}_M g \frac{f-e}{l}$ and $\frac{d}{dt}l = \frac{\dot{k}_M}{3}\frac{e-l-l_T}{1+e/g}$ and $\frac{d}{dt}u_H = \dot{k}_M(u_H < u_H^p)((1-\kappa)el^2\frac{g+l+l_T}{g+e} - ku_H))$.

$\frac{\text{power}}{\{\dot{p}_{Am}\}L_m^2}$	embryo $0 < l \leq l_b$	juvenile $l_b < l \leq l_p$	adult $l_p < l < 1$
assimilation, $\dot{p}_A$	0	fl^2	fl^2
mobilisation, $\dot{p}_C$	$el^2\frac{g+l}{g+e}$	$el^2\frac{g+l+l_T}{g+e}$	$el^2\frac{g+l+l_T}{g+e}$
somatic maintenance, $\dot{p}_S$	κl^3	$\kappa l^2(l + l_T)$	$\kappa l^2(l + l_T)$
maturity maintenance, $\dot{p}_J$	ku_H	ku_H	ku_H^p
growth, $\dot{p}_G$	$\kappa l^2\frac{e-l}{1+e/g}$	$\kappa l^2\frac{e-l-l_T}{1+e/g}$	$\kappa l^2\frac{e-l-l_T}{1+e/g}$
maturation, $\dot{p}_R$	$(1-\kappa)el^2\frac{g+l}{g+e} - ku_H$	$(1-\kappa)el^2\frac{g+l+l_T}{g+e} - ku_H$	0
reproduction, $\dot{p}_R$	0	0	$(1-\kappa)el^2\frac{g+l+l_T}{g+e} - ku_H^p$

where $\kappa_R = 0$ for the embryo and juvenile stages. Reproduction power $\dot{p}_R$ has a special status because reserve of the adult female are converted into reserve of the embryo(s) which have the same composition; $(1-\kappa_R)\dot{p}_R$ is dissipating and $\kappa_R\dot{p}_R$ returns to the reserve, but now of the embryo.

3

Energy, compounds and metabolism

Metabolism is about the transformation of compounds by organisms; some aspects of this process can only be understood by considering the abundance of the various compounds, and dealing with the links with energy and entropy. This chapter discusses the basic concepts for these links and considers a framework for the quantification of metabolic rates. Like the first chapter, this is a concepts-chapter to prepare for the further development of DEB theory.

3.1 Energy and entropy

Energy fluxes through living systems are difficult to measure and even more difficult to interpret. Let me briefly mention some of the problems.

Although it is possible to measure the enthalpy of food through complete combustion, we need the free energy to quantify the amount of (metabolic) work that can be done with it. Food has a dual role in providing the capacity to do work as well as elementary compounds for anabolism. Another problem is that of digestive efficiency. The difference between the energy contents of food and faeces is just an upper boundary for the uptake by the animal, because there are energy losses in the digestion process. Part of this difference is never used by the organism, but by the gut flora instead. Another part is lost through enhanced respiration coupled to digestion, especially of proteins, called the *heat increment of feeding*, which is discussed on {151}.

Growth involves energy investment, which is partially preserved in the new biomass. In addition to the energy content of the newly formed biomass, energy is invested to give it its structure. Part of this energy is lost during growth and can be measured as dissipating heat. This heat can be thought of as an overhead of the growth process. The energy that is fixed in the new biomass is present partly as energy-bearing compounds.

Cells are highly structured objects and the information contained in their structure is not measured by bomb calorimetry.

The thermodynamics of irreversible or non-equilibrium processes offers a framework for pinpointing the problem; see for instance [446, 669]. While bomb calorimetry measures the change in enthalpy, Gibbs free energy is the more useful concept for quantifying the energy performance of individuals. Enthalpy and Gibbs free energy are coupled by the concept of entropy: the enthalpy of a system equals its Gibbs free energy plus the entropy times the absolute temperature. This basic relationship was formulated by J. W. Gibbs in 1878. The entropy depends on the (local) chemical environment, including the spatial structure and the transformations that are going on. It is, therefore, hard to quantify directly.

The concept of strong homeostasis offers a solution to the problem of defining and measuring free energies and entropies. This solution is based on the assumption that the free energy per C-mole of structural biomass and of reserves is constant, i.e. it does not depend on the (absolute) amounts. Most chemists probably find this assumption offensive, since free energies depend on the concentration of a compound in spatially homogeneous systems. The reason for the dependence is that the molecules interfere, which affects their ability to do work in the thermodynamic sense. Yet, I think that the assumption is more than just a conceptual trick to solve problems; it is the way living cells solve the problem of a compound's capacity to do (chemical) work depending on the concentration. If this capacity changes substantially as a function of the changing cell composition, the cell would have an immensely complex problem to solve when regulating its metabolic processes. It is not just a coincidence that cells use large amounts of polymers (i.e. proteins, carbohydrates and lipids) to store bulk compounds, and small amounts of monomers to run their metabolism. Cells keep the concentration of monomers low and relatively constant, and prevent any interference that makes the monomers' capacity to do work depend on their abundance. They also solve their osmotic problems this way. Their osmotic pressure equals that of seawater, which is frequently seen as a relic of the evolutionary process: life started in the sea.

I assume that the Gibbs relationship still applies in the complex setting of living organisms. If the free energy per C-mole does not change, then neither will the entropy per C-mole, because the enthalpy per C-mole is constant. The Gibbs relationship can be used to obtain the entropy and the free energy of complex organic compounds, such as food, faeces, structural biomass and reserves, as is worked out on {162}. The mean specific Gibbs free energy (i.e. chemical potential) of biomass is -67 kJ C-mol^{-1} (pH $= 7$, 10^5 Pa at 25 °C, thermodynamic reference) or $+474.6$ kJ C-mol^{-1} (pH $= 7$, combustion reference) [485]. Since biomass composition is not constant, such crude statistics are of limited value and a more subtle approach is necessary to quantify dissipating heat. We will quantify entropy of reserve(s) and structure(s) via the entropy balance: the dynamics of what goes in and out from a living individual. Since the entropy balance rests on the energy balance, and the energy balance rests on the mass balance, a detailed discussion is delayed to Chapter 4, see {162}.

3.2 Body mass and composition

3.2.1 Mass quantified as gram

Common practice is to take wet weight proportional to physical volume, $W_w = d_{Vw}V_w$. This mapping in fact assumes that the compositions of structural mass and reserves are identical. Much literature is based on this relationship or on the similar one for dry weights: $W_d = d_{Vd}V_w$.

The contribution made by reserves, relative to that made by structure, to size measures depends on their nature. For example, energy allocated to reproduction, but temporarily stored in a buffer, will contribute to dry weight, but much less to wet weight [394]. While wet weight is usually easier to measure and can be obtained in a non-destructive way, dry weight has a closer link to chemical composition and mass balance implementations. I show on {142} how to separate structural body mass from reserves and determine the relative abundances of the main elements for both categories on the basis of dry weight.

The relationships between physical volume V_w, wet weight W_w and dry weight W_d with *structural* body volume V, non-allocated energy reserves E, and energy reserves allocated to reproduction E_R are

$$V_w = V + (E + E_R)\frac{w_E}{d_E\overline{\mu}_E} \overset{E_R=0}{=} V(1 + \omega_V e) \quad \text{for} \quad \omega_V = \frac{[E_m]}{d_E}\frac{w_E}{\overline{\mu}_E} \tag{3.1}$$

$$W_w = d_V V + (E + E_R)\frac{w_E}{\overline{\mu}_E} \overset{E_R=0}{=} d_V V(1 + \omega_w e) \quad \text{for} \quad \omega_w = \frac{[E_m]}{d_V}\frac{w_E}{\overline{\mu}_E} \tag{3.2}$$

$$W_d = d_{Vd} V + (E + E_R)\frac{w_{Ed}}{\overline{\mu}_E} \overset{E_R=0}{=} d_{Vd} V(1 + \omega_d e) \quad \text{for} \quad \omega_d = \frac{[E_m]}{d_{Vd}}\frac{w_{Ed}}{\overline{\mu}_E} \tag{3.3}$$

where d_* are densities, which convert volumes to weights, $\overline{\mu}_E$ the chemical potential of reserve (energy per C-mole), and w_* are molecular weights (weight per C-mole, see {83}). The parameters ω_* weigh the contribution of reserve to weight.

The contribution of reserves to weight has long been recognised, and is used to indicate the nutritional condition of fish and birds [894]. A series of coefficients has been proposed, e.g. (weight in g)×(length in cm)$^{-1}$, known as the condition factor, Hile's formula or the ponderal index [11, 382, 507, 538].

Although the relationship between weight and reserves plus structural volume is more accurate than a mere proportionality, it is by no means 'exact' and depends on species-specific details. The gut contents of earthworms, shell of molluscs or exoskeleton of crustaceans do not require maintenance and for this reason they should be excluded from biovolume and weight for energetic purposes. The contribution of inorganic salts to the dry weight of small marine invertebrates is frequently substantial. Because weights combine structural and reserve mass, they should not be used to set up a theory of substrate uptake and use, and their role is restricted to link model predictions to data. The problem can be illustrated by the observation that the weight-specific maintenance

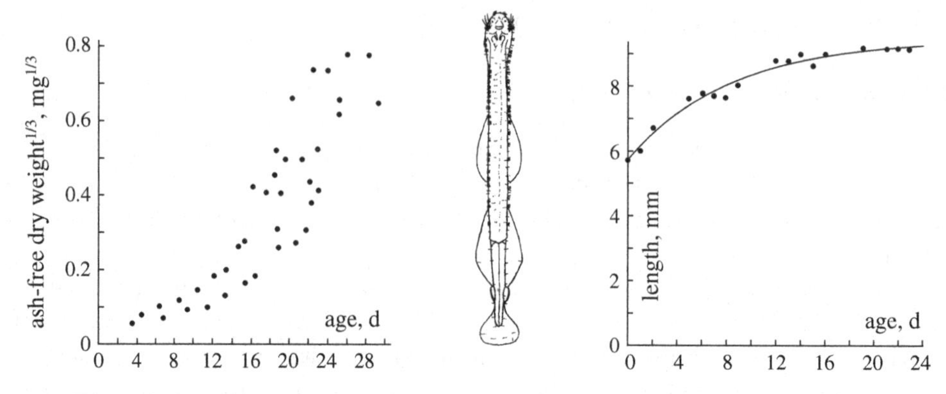

Figure 3.1 The ash-free dry weight and the length of the chaetognath *Sagitta hispida*. Data from Reeve [948, 949]. The curve through the lengths is $L(t) = L_\infty - (L_\infty - L_0)\exp(-\dot{r}_B t)$.

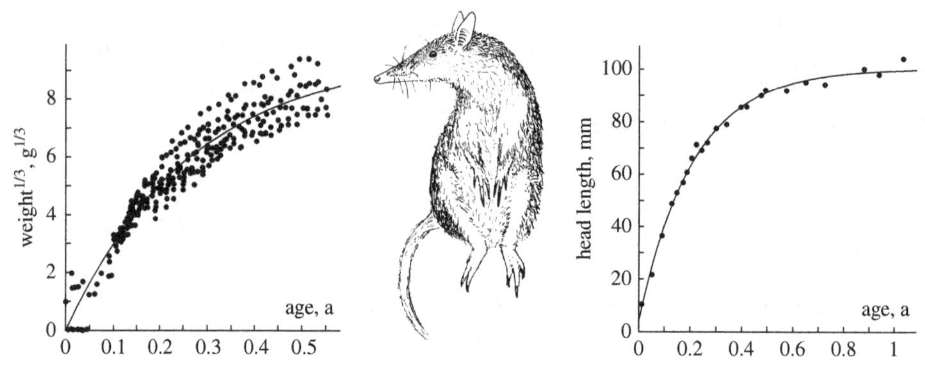

Figure 3.2 The weight to the power 1/3 and the head length of the long-nosed bandicoot *Perameles nasuta*. Data from Lyne [723]. The curves are again $L(t) = L_\infty - (L_\infty - L_0)\exp(-\dot{r}_B t)$.

costs of fungi and trees are extremely low. This does not point, however, to exceptional metabolic qualities, but to the fact that their weights include products (cell wall material, wood), that do not require maintenance. The production rates are quantified by the DEB theory, {165}, which allows weights to be decomposed into the contributions from structure, reserves and products.

Figure 3.1 illustrates an interpretation problem in the measurement of the ash-free dry weight of chaetognaths. Length measurements follow the expected growth pattern closely when food is abundant, while the description of weight requires an *ad hoc* reasoning, possibly involving gut contents. Although quickly said, this is an important argument in the use of measurements within a theoretical context: if an explanation that is not species-specific competes with one that is, the first explanation should be preferred if the arguments are otherwise equally convincing. Since energy reserves contribute to weight and are sensitive to feeding conditions, weights are usually much more scattered, in comparison to length measurements. This is illustrated in Figure 3.2.

The determination of the size of an embryo is complicated by the extensive system of membranes that the embryo develops in order to mobilise stored energy and materials and the decrease in water content during development [1246]. In some species, the embryo can be separated from 'external' yolk. As long as external yolk is abundant, the energy reserves of the embryo without that yolk, if present at all, will, on the basis of DEB theory, turn out to be a fixed fraction of wet and dry weight, so that the embryo volume is proportional to weight. Uncertainty about the proportionality factor will hamper the comparison of parameter values between the embryonic stage and the post-embryonic one.

Weights play no role in the DEB theory itself, but they are important for relating theoretical predictions to measurements.

3.2.2 Mass quantified as C-mole

Microbiologists frequently express the relative abundances n_{*W} of the elements hydrogen, oxygen and nitrogen in dry biomass relative to that of carbon, and conceive the combined compound so expressed as a kind of abstract 'molecule' that can be counted and written as $CH_{n_{HW}}O_{n_{OW}}N_{n_{NW}}$. So one mole of glucose, $C_6H_{12}O_6$, equals 6 C-moles of glucose. As is standard in the microbiological literature, the concept of the C-mole is extended to (simple) substrates, the difference from an ordinary mole being that it always has at most 1 C-atom.

While mass quantified as gram does not consider the chemical composition, mass quantified as C-mole rapidly become less valuable if the chemical composition varies.

I denote structural mass in terms of C-moles by M_V, reserve mass by M_E, and the ratio of reserve to structural mass by $m_E = M_E/M_V$. Table 3.3 on {93} gives useful conversions between volumes, masses and energies.

3.2.3 Composition of biomass

The aqueous fraction of an organism is important in relation to the kinetics of toxicants. Water is treated just like any other compound in the decomposition of biomass. The aqueous weight is the difference between wet weight and dry weight, so $W_H = W_w - W_d$. It can be written as $W_H = [W_H]V$, for

$$[W_H] = d_V - d_{Vd} + (w_E - w_{Ed})(E + E_R)/\overline{\mu}_E = d_V - d_{Vd} + (w_E - w_{Ed})(e + e_R)[M_{Em}] \quad (3.4)$$

where $[M_{Em}] = [E_m]/\overline{\mu}_E$ is the maximum molar reserve density of juveniles and adults. The volume occupied by water is $V_H = W_H/d_H \simeq (d_V - d_{Vd})V/d_H$, where d_H stands for the specific density of water, which is close to 1 g cm^{-3}. The aqueous fraction of body volume V_H/V_w typically takes values between 0.7 and 0.9.

For each C-atom in dry biomass, there are typically $n_{HW} \simeq 1.8$ H-atoms, $n_{OW} \simeq 0.5$ O-atoms and $n_{NW} \simeq 0.2$ N-atoms for a randomly chosen micro-organism [981]. This gives a mean degree of reduction of 4.2 and a 'molecular weight' of $w_W = 24.6$ g mol^{-1}. The latter can be used to convert dry weights into 'C-moles'. The relative abundances of elements in biomass-derived sediments largely remain unaltered on a geological timescale, apart from the excretion of water. The Redfield ratio C : N : P = 105 : 15 : 1 is popular [947] in geology and oceanography, or for silica-bearing organisms such as diatoms, radiolarians, silico-flagellates and (some) sponges C : Si : N : P = 105 : 40 : 15 : 1. This literature usually excludes hydrogen and oxygen, because their abundances in biomass-derived sediments change considerably during geological time. Other bulk elements in organisms are S, Cl, Na, Mg, K and Ca, while some 14 other trace elements play an essential role, as reviewed by Fraústo da Silva and Williams [369]. The ash that remains when dry biomass is burnt away is rich in these elements. Ash weight typically amounts to some 5% of dry weight only, and the elements C, H, O and N comprise more than 95% of the total dry weight. I focus on these four elements only, which happen to be the four lightest of the periodic table that can make covalently bounded compounds [199]. The inclusion of more elements is straightforward; as stated before, some taxa require special attention on this point.

Reserves and structural mass are thought of as generalised compounds: rich mixtures of compounds that do not change in chemical composition. The concept rests fully on the strong homeostasis assumption. If a 'molecule' of structural biomass is denoted by $CH_{n_{HV}}O_{n_{OV}}N_{n_{NV}}$ and a 'molecule' of energy reserves by $CH_{n_{HE}}O_{n_{OE}}N_{n_{NE}}$, then their relative abundances in biomass consisting of structural mass M_V, reserves M_E and reserves allocated to reproduction M_{E_R} are given by

$$n_{*W} = \frac{n_{*V}M_V + n_{*E}(M_E + M_{E_R})}{M_V + M_E + M_{E_R}} = \frac{n_{*V} + n_{*E}(m_E + m_{E_R})}{1 + m_E + m_{E_R}} \tag{3.5}$$

where $*$ stands for H, O or N and $m_E = M_E/M_V$ and $m_{E_R} = M_{E_R}/M_V$ are molar reserve densities.

The molar weights of structural biovolume and energy reserves are given by

$$\begin{aligned} w_V &\simeq 12 + n_{HV} + 16n_{OV} + 14n_{NV} \quad \text{gram mol}^{-1} \\ w_E &\simeq 12 + n_{HE} + 16n_{OE} + 14n_{NE} \quad \text{gram mol}^{-1} \end{aligned}$$

since the contribution of the other elements to weight is negligibly small. The problem of uncovering the relative abundances n_{*V} and n_{*E} from measurements of n_{*W} is discussed on {142}.

Similarly we have $M_E = \overline{\mu}_E^{-1}E$ and $M_V = [M_V]V$, where $[M_V]$ denotes the conversion coefficient from structural volume to C-mole. Using a specific density of wet mass of $d_V = 1$ g cm^{-3}, a wet weight/dry weight ratio of 10 and a molecular weight of $w_V = 24.6$ g mol^{-1} for structure, a typical value for $[M_V]$ would be 4.1 mmol cm^{-3}.

We need the link between generalised and 'pure' chemical compounds in applications of isotopes, for instance. Like generalised compounds, we quantify organic compounds in terms of C-moles. Suppose that compound i is present in M_{V_i} (C-)moles in structure, for example. So structure has mass $M_V = \sum_i M_{V_i} n_{CV_i}$, where n_{CV_i} is either 0 (inorganic compounds) or 1 (organic compounds). The chemical indices of the generalised compound relate to that of the chemical compounds as

$$n_{*V} = \frac{\sum_i M_{V_i} n_{*V_i}}{\sum_i M_{V_i} n_{CV_i}} = \sum_i w_i n_{*V_i} \quad \text{with} \quad w_i = \frac{M_{V_i}}{M_V} \quad \text{and} \quad i \in \{C, H, O, N\} \qquad (3.6)$$

This decomposition can also be done for other generalised compounds, such as reserve and food. Notice that any chemical compound can potentially partake in all generalised compounds and all chemical elements can be included.

The delineation of more than one type of reserve (or structural mass) comes with additional contributions to mass and weight. For n reserves, a single structural mass, and no reserves allocated to reproduction, (total) biomass can be decomposed into the masses (in C-moles) $\{M_{E_1}, M_{E_2}, \ldots, M_{E_n}, M_V\}$ which, in combination with maturity, define the state of the organism. The strong homeostasis assumption states that these masses do not change in chemical composition and, therefore, they can be treated as generalised compounds.

The wet weight of the n-reserves organism amounts to $W_w = w_V M_V + \sum_i w_{E_i} M_{E_i}$, where the ws stand for the C-molar weights. The dry weight of this organism can be expressed similarly as $W_d = w_{Vd} M_V + \sum_i w_{E_i d} M_{E_i}$, where the w_{*d}s represent the C-molar weights, after removal of water. This can be done in this way because the assumption that neither reserves nor structural biomass can change in composition means that their water fractions are constant.

The weight of any particular chemical compound Y in the n-reserves organism can be expressed as $W_Y = w_Y(n_{YV} M_V + \sum_i n_{YE_i} M_{E_i})$, where w_Y is the molar weight of the compound Y and the ns denote the molar amounts of the compound per C-mole of reserve or structural biomass. This again is a consequence of the strong homeostasis assumption. The ns are zero if the compound does not happen to occur in that biomass component. The density of the compound in biomass can be expressed as W_Y/W_d on the basis of weights, or as $W_Y(w_Y M_V + w_Y \sum_i M_{E_i})^{-1}$ on the basis of moles per mole of carbon.

The chemical composition of biomass becomes increasingly flexible with the number of delineated reserves, and depends on the nutritional conditions of the environment. In terms of relative frequencies of chemical elements, all restrictions in the composition of (total) biomass disappear if the number of reserves exceeds the number of chemical elements minus one.

3.3 Classes of compounds in organisms

Chemical elements obey conservation laws, not compounds, and we need this to quantify fluxes of compounds. Two sets of chemical compounds partake in three (sets of) transformations:

compounds →	minerals, $\mathcal{M}$				org. comp. $\mathcal{O}$			
↓ transformations	carbon dioxide	water	dioxygen	nitrogenous waste	food	structure	reserves	faeces
	C	H	O	N	X	V	E	P
assimilation A	+	+	−	+	−		+	+
growth G	+	+	−	+		+	−	
dissipation D	+	+	−	+			−	

The organic compounds V and E constitute the individual, the other organic compounds and the minerals define the chemical environment of the individual. The signs indicate appearance (+) or disappearance (−); a blank indicates that the compound does not partake in that transformation.

The following subsections briefly discuss some features of the different compounds to supplement earlier introductions.

3.3.1 Mineral compounds

Dioxygen and carbon dioxide

Most organisms use dioxygen as an electron acceptor in the respiration chain, the final stage in the oxidation of pyruvate, see Figure 3.11. If an electron acceptor is not available, a substantial amount of energy cannot be extracted from pyruvate, and metabolic products have to be excreted. Photosynthetic bacteria, algae and plants not only use dioxygen, but also produce dioxygen, see {189, 427}. This production exceeds the consumption if light intensity is high enough.

Most organisms consume and excrete carbon dioxide, see {425}; consumption exceeds excretion in photosynthetic organisms in the light, and in bacteria that use methane as substrate.

Water

Water is formed metabolically from other compounds. This rate of water production is studied first; the direct exchange of water with the environment via drinking and evaporation, and its use for transport, are discussed on {153}.

Nitrogenous waste

From an energy perspective, the cheapest form of nitrogenous waste is ammonia. Since ammonia is rather toxic at high concentrations, terrestrial animals usually make use of more expensive, less toxic nitrogenous wastes. Terrestrial isopods are an

Table 3.1 Various nitrogenous wastes that animals use [1262].

nitrogenous waste	formula	solubility (mM)	insects	crustaceans	fish	birds	mammals
ammonia	NH_3	52.4		o	o		
amm. bicarbonate	NH_4HCO_3	1.5		o			
urea	$CO(NH_2)_2$	39.8					o
allantoin	$C_4H_6O_3N_4$	0.015					o
allantoic acid	$C_4H_8O_4N_4$	slight					o
uric acid	$C_5H_4O_3N_4$	0.0015				o	
sodium urate	$C_5H_2O_3N_4Na_2$	0.016	o			o	
potassium urate	$C_5H_2O_3N_4K_2$	slight	o			o	
guanine	$C_4H_5ON_5$	0.0013	o			o	
xanthine	$C_5H_4O_2N_4$	0.068	o			o	
hypoxanthine	$C_5H_4ON_4$	0.021	o			o	
arginine	$C_6H_{14}O_2N_4$	3.4	o			o	

exception; Dutchmen call them ‘pissebed’, a name referring to the smell of ammonia that microbes produce from urine in a bed. Terrestrial eggs have to accumulate the nitrogenous waste during development; they usually make use of even more expensive, less soluble nitrogenous wastes that crystallise outside the body, within the egg shell. Table 3.1 lists the different chemical forms of nitrogenous waste. The nitrogenous waste (urine) includes its water in its chemical ‘composition’, for simplicity's sake.

Nitrogenous waste mainly originates from protein turnover, which is part of somatic maintenance. A second origin of nitrogenous waste can be assimilation, when metazoans feed on protein-rich food, and nitrogen is excreted in the transformation of food to reserves. The (energy/carbon) substrate for micro-organisms can be poor in nitrogen, such that nitrogen must be taken up from the environment, rather than excreted. Though the term nitrogenous waste no longer applies, this does not matter for the analysis; the sign of the flux defines uptake or excretion. Bacteria that live on glucose as an energy source will have negative nitrogenous waste.

3.3.2 Organic compounds

I here briefly introduce the chemical aspects of food, products and storage materials. Observed changes in the elemental composition of the body mass, as a function of growth rate, or starvation time, can be used to obtain the elemental composition of reserves and structure, as is discussed on {142}.

Food

Food for micro-organisms is usually called 'substrate', which can be very simple chemical compounds, such as glucose. Most animals feed on other organisms, i.e. complex substrates. For simplicity's sake, I assume that the composition of food is constant, but this is not essential; the composition of faeces is taken to be constant as a consequence. This condition will be relaxed on {185}.

Products

Faeces are the remains of food after it has passed through the gut. Animals add several products to these remains, such as bile and enzymes that are excreted in the gut, and excreted micro-flora formed in the gut. Mammals in particular also add substantial quantities of methane, which is produced by the bacterial gut flora, see {426}; the Amazonian hoatzins (*Opisthocomus*) smell like cows, because these birds have a similar gut flora and digestion. I include these products in faeces, since these excretions are tightly coupled to the feeding process.

The 'faeces' of micro-organisms are usually called 'metabolic products'. Sometimes, substrate molecules are taken up entirely, and completely metabolised to carbon dioxide and water; in this case no faeces are produced. In other cases products are formed that generally do not originate from substrate directly, but indirectly with a more complex link to the metabolic machinery of the organism. The role of such products is then similar to that of nitrogenous wastes in animals. I cope with these situations by including such products in the overheads of the three basic energy fluxes, the assimilation flux, the dissipating flux and the somatic growth flux. The number of different products can be extended in a straightforward manner, see {165}.

Two chemically related organic products changed the world: chitin $(C_{16}H_{26}O_{10}N_2)_n$ because of its role in the organic carbon pump, see {425}, and cellulose $(C_{12}H_{20}O_{10})_n$ because of its accumulation in soils, see {424}, and its transformation to coal. Both products serve mechanical support functions (carapace and wood respectively) and are difficult to degrade (no animal can degrade cellulose without the help of bacteria and/or fungi), so they accumulate in the environment (and change e.g. the water retention properties of soils). Opisthokonts (= fungi + animals) produce chitin and most bikonts, which comprise all eukaryotes except amoebas and opisthokonts, and some amoebas (*Dictyostelium*) produce cellulose. Eukaryotic cellulose production originates from Cyanobacteria [843]. Two opisthokont taxa produce cellulose, however, rather than chitin: urochordates [756] and *Aspergillus fumigastus* [843]. They got this ability from α-proteobacteria via lateral gene transfer. Deuterostomes don't produce chitin but calcium carbonate and tetrapods produce keratin for mechanical support. Other well-known animals products are mucus, hair, scales, otoliths, see {159, 180}.

If dioxygen is poorly available, a variety of products are formed and released in the environment:

product	chemical formula	rel. freq.	μ
ethanol	CH_3CH_3O	$CH_3O_{0.5}$	657
lactate	$CH_3CH_2OCHO_2$	CH_2O	442
succinate	$CHO_2CH_2CH_2CHO_2$	$CH_{1.5}O$	376
propionate	$CH_3CH_2CHO_2$	$CH_2O_{0.66}$	493
acetate	CH_3CHO_2	CH_2O	442

where the last column gives the Gibbs energy of formation in kJ per C-mole at pH = 7 in the combustion frame of reference [484]. The kind of product depends on the species and the environmental conditions. The quantitative aspects are discussed on {166}.

Storage materials

Storage material can be classified into several categories; see Table 3.2. These categories do not point to separate dynamics. Carbohydrates can be transformed into fats, for instance, see Figure 3.11. Most compounds have a dual function as a reserve pool for both energy and elementary compounds for anabolic processes. For example, protein stores supply energy, amino acids and nitrogen. Ribosomal RNA (rRNA) catalyses protein synthesis. In rapidly growing cells such as those of bacteria in rich media, rRNA makes up to 80% of the dry weight, while the relative abundance in slowly growing cells is much less. For this reason, it should be included in the storage material. I show how this point of view leads to realistic descriptions of peptide elongation rates and growth-rate-related changes in the relative abundance of rRNA, {143}. There is no requirement for storage compounds to be inert.

Waxes can be transformed into fats (triglycerides) and play a role in buoyancy, e.g. of zooplankton in the sea [81]. By increasing their fat/wax ratio, zooplankters can ascend to the surface layers, which offer different food types (phytoplankton), temperatures and currents. Since surface layers frequently flow in different directions than deeper ones, they can travel the Earth by just changing their fat/wax ratio and stepping from one current to another. Wax ester biosynthesis may provide a mechanism for rapidly elaborating lipid stores from amino acid precursors [1011].

Unsaturated lipids, which have one or more double bonds in the hydrocarbon chain, are particularly abundant in cold water species, compared with saturated lipids. This possibly represents a homeoviscous adaptation [1048].

The amount of storage materials depends on the feeding conditions in the (recent) past, see {37}. Storage density, i.e. the amount of storage material per unit volume of structural biomass, tends to be proportional to the volumetric length for different species, if conditions of food (substrate) abundance are compared, as explained on {300} and tested empirically on {312}. This means that the maximum storage density of bacteria is small. However, under conditions of nitrogen limitation, for instance, bacteria can become loaded with energy storage materials such as polyphosphate or polyhydroxybutyrate, depending on the species, see {197}. This property is used in biological plastic production and phosphate removal from sewage water. Intracellular lipids can accumulate up to some 70% of the cell dry weight in oleaginous yeasts, such as *Apiotrichum* [939, 1283]. This property is used in the industrial production of lipids.

Table 3.2 Some frequently used storage materials in heterotrophs.

phosphates	
pyrophosphate	bacteria
polyphosphate	bacteria (*Azotobacter, Acinetobacter*)
polysaccharides	
β-1,3-glucans	
leucosin	Chrysomonadida, Prymnesiida
chrysolaminarin	Chrysomonadida
paramylon	Euglenida
α-1,4-glucans	
starch	Cryptophyceae, Dinozoa, Volvocida, plants
glycogen	cyanobacteria, protozoa, yeasts, molluscs
amylopectin	Eucoccidiida, Trichotomatida, Entodiniomorphida
trehalose	fungi, yeasts
lipoids	
poly-β-hydroxybutyrate	bacteria
triglyceride	oleaginous yeasts, most heterotrophs
wax	marine animals
proteins	most heterotrophs
ovalbumin	egg-white protein
casein	milk protein (mammals)
ferritin	iron storage in spleen (mammals)
cyanophycine	cyanobacteria
phycocyanin	cyanobacteria
ribosomal RNA	all organisms

The excess storage is due to simultaneous nutrient limitation that is associated with what is called 'luxurious' uptake.

Storage deposits

Lipids, in vertebrates, are stored in cell lysosomes in specialised adipose tissue, which occurs in rather well-defined surface areas of the body. The cells themselves are part of the structural mass, but the contents of the vacuole are part of the reserves, see {405}. In molluscs specialised glycogen storage cells are found in the mantle [490]. The areas for storage deposits are usually found scattered over the body and therefore appear to be an integral part of the structural body mass, unless superabundant; see Figure 3.3. The occurrence of massive deposits is usually in preparation for a poor feeding season. The rodent *Glis glis* is called the 'edible dormouse', because of its excessive lipid deposits just prior to dormancy {122}. Stewed in honey and wine, dormice were a gourmet meal for the ancient Romans. Tasmania's yellow wattlebird *Anthochaera paradoxa* accumulates lipid deposits during the rich season to the extent that it has problems with flight; it then becomes exceedingly wary for a good reason [431].

Figure 3.3 Some storage deposits are really eye-catching.

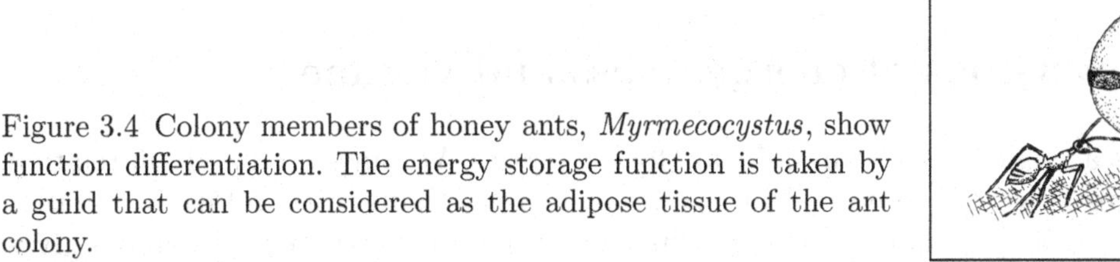

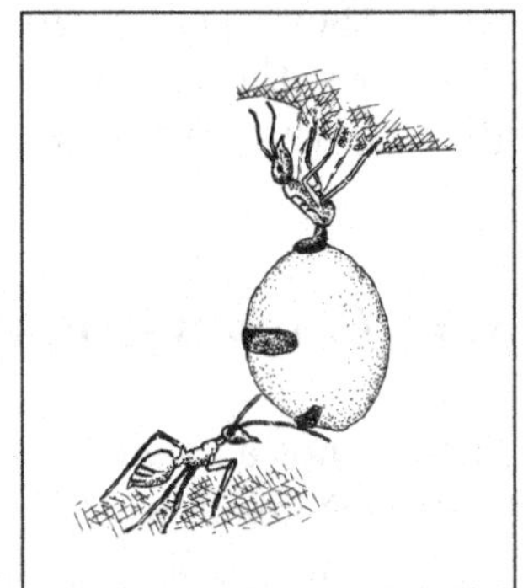

Figure 3.4 Colony members of honey ants, *Myrmecocystus*, show function differentiation. The energy storage function is taken by a guild that can be considered as the adipose tissue of the ant colony.

In most invertebrate groups, storage deposits do not occur in specialised tissues, but only in the cells themselves in a quantity that relates to requirements. So reproductive organs tend to be rich in storage products. The mesoglea of sea anemones, for instance, has mobile cells that are rich in glycogen and lipid, called glycocytes, which migrate to sites of demand during gametogenesis and directly transfer the stored materials to developing oocytes [1048]. Glycogen that is stored for a long time typically occurs in rosettes, and for short time in particles [512, 1048]. A guild of honey ants specialises in the storage function for the colony, not unlike adipose tissue in vertebrates, see Figure 3.4.

The recently discovered anaerobic sulphur bacterium *Thiomargarita namibiensis* [1032] accumulates nitrate up to 0.8 M in a vacuole of up to 750 µm in diameter; it can survive over 2 years without nitrate or sulphur at 5 °C. The bacterium *Acinetobacter calcoaceticus* accumulates polyphosphates to spectacular levels under carbon-limiting aerobic conditions, and releases phosphates under energy-limiting anaerobic conditions, which is used technically in sewage water treatment, see {199}.

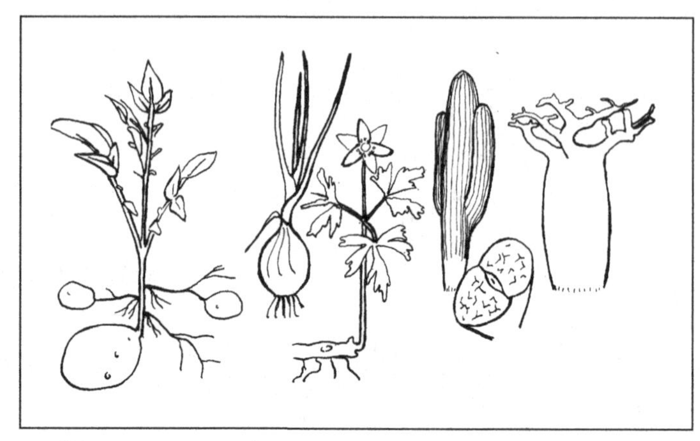

Figure 3.5 Plants can store large amounts of carbohydrates and/or water.

Since autotrophs acquire energy and the various nutrients independently from each other, they usually store possibly limiting substrates independently in specialised organelles: the vacuoles [694]. Carbohydrate (starch) and water storage are most bulky in plants that live in seasonal environments, see Figure 3.5.

3.4 Conversions of energy, mass and volume

There is not a single most useful notation for energetics. Volumes are handy in relation to surface areas, which are needed for the process of food/substrate uptake in the DEB model, while moles are handy for mass fluxes and mass conservation. Table 3.3 gives conversions between volume-based, mole-based and energy-based quantities, some of which are introduced in the next chapter. The specific fluxes j_* relate to the total fluxes $\dot{J}_*$ by $j_* = \dot{J}_*/M_V$, but contrary to the total fluxes, they are always taken to be non-negative. Because of strong homeostasis $[M_V] = M_V/V$ is a constant.

The three state variables for maturity, reserve and structure of the standard DEB model can be written as maturity energy, reserve energy density and structural volume $(E_H, [E], V)$, with dynamics given in (2.18), (2.10) and (2.21), or in dimensionless scaled variables (e_H, e, l) with dynamics given in (2.31), (2.11) and (2.22), or in masses (M_H, m_E, M_V) using Table 3.3:

$$\frac{d}{dt}M_H = (M_H < M_H^p)\Big((1-\kappa)m_E M_V(\frac{\dot{v}}{L} - \dot{r}) - \dot{k}_J M_H\Big) \quad \text{with} \quad L = \left(\frac{M_V}{[M_V]}\right)^{1/3} \tag{3.7}$$

$$\frac{d}{dt}m_E = j_{EAm}(f - m_E/m_{Em}) \quad \text{with} \quad j_{EAm} = \{\dot{J}_{EAm}\}M_V^{-1/3}[M_V]^{-2/3} \tag{3.8}$$

$$\frac{d}{dt}M_V = M_V\dot{r} \quad \text{with} \quad \dot{r} = \frac{j_{EAm}(m_E/m_{Em} - l_T) - j_{EM}/\kappa}{m_E + y_{EV}/\kappa} \tag{3.9}$$

Table 3.3 Conversions and compound parameters.

relationship	unit	description
$K = \frac{\{\dot{J}_{EAm}\}}{y_{EX}\{\dot{F}_m\}}$	$\text{mol}\,\text{m}^{-3}$	half-saturation constant
$\{\dot{J}_{XAm}\} = -\{\dot{J}_{EAm}\}/y_{EX}$	$\text{mol}\,\text{m}^{-2}\text{d}^{-1}$	maximum specific ingestion rate
$M_E = V[M_E] = \frac{E}{\overline{\mu}_E} = M_V \frac{e y_{EV}}{\kappa g}$	mol	mass of reserve
$M_V = V[M_V]$	mol	structural mass
$M_{Vm} = V_m[M_V]$	mol	maximum structural mass
$[M_V] = d_V/w_V$	$\text{mol}\,\text{m}^{-3}$	specific structural mass
$[M_{Em}] = \frac{\{\dot{J}_{EAm}\}}{\dot{v}} = \frac{y_{EV}[M_V]}{g\kappa}$	$\text{mol}\,\text{m}^{-3}$	maximum reserve density
$m_{Em} = \frac{[M_{Em}]}{[M_V]} = \frac{y_{EV}}{g\kappa}$	$\text{mol}\,\text{mol}^{-1}$	maximum reserve density
$m_E = \frac{M_E}{M_V} = \frac{e}{\overline{\mu}_E}\frac{[E_m]}{[M_V]}$	$\text{mol}\,\text{mol}^{-1}$	reserve density
$m_H = M_H/M_V$	$\text{mol}\,\text{mol}^{-1}$	maturity density
$y_{PE} = \overline{\mu}_E/\mu_{AP} = j_{PA}/j_{EA}$	$\text{mol}\,\text{mol}^{-1}$	coupler of faeces and reserve prod.
$y_{PX} = \mu_{AX}/\mu_{AP} = j_{PA}/j_{XA}$	$\text{mol}\,\text{mol}^{-1}$	coupler of faeces prod. and food cons.
$y_{VE} = \frac{\mu_E}{\mu_{GV}} = \frac{j_{VG}}{j_{EG}} = \frac{[M_V]\overline{\mu}_E}{[E_G]}$	$\text{mol}\,\text{mol}^{-1}$	coupler of struct. prod. and res. invest.
$y_{EX} = \frac{\overline{\mu}_E}{\mu_{AX}} = -\frac{y_{EV}[M_V]\dot{v}}{g\kappa\{\dot{J}_{XAm}\}} = -\frac{[M_{Em}]\dot{v}}{\{\dot{J}_{XAm}\}}$	$\text{mol}\,\text{mol}^{-1}$	coupler of food cons. and reserve prod.
$[E_G] = \mu_{GV}[M_V]$	$\text{J}\,\text{m}^{-3}$	specific costs for structure
$[E_m] = \{\dot{p}_{Am}\}/\dot{v} = \overline{\mu}_E[M_{Em}]$	$\text{J}\,\text{m}^{-3}$	maximum reserve density
$L = V^{1/3} = (M_V/[M_V])^{1/3}$	m	structural (volumetric) length
$L_m = \kappa\frac{\{\dot{J}_{EAm}\}}{[\dot{J}_{EM}]} = \kappa\frac{\{\dot{p}_{Am}\}}{[\dot{p}_M]} = \frac{\dot{v}}{\dot{k}_M g}$	m	maximum structural length
$L_T = \{\dot{p}_T\}/[\dot{p}_M]$	m	heating length
$V = L^3 = \frac{M_V}{[M_V]} = \frac{E_V}{\overline{\mu}_V[M_V]}$	m^3	structural volume
$V_m = L_m^3 = M_{Vm}/[M_V]$	m^3	maximum structural volume
$U_E = M_E/\{\dot{J}_{EAm}\} = E/\{\dot{p}_{Am}\}$	$\text{d}\,\text{m}^2$	scaled reserve
$U_H = M_H/\{\dot{J}_{EAm}\} = E_H/\{\dot{p}_{Am}\}$	$\text{d}\,\text{m}^2$	scaled maturity
$\{\dot{p}_{Am}\} = \overline{\mu}_E\{\dot{J}_{EAm}\} = -\mu_{AX}\{\dot{J}_{XAm}\}$	$\text{J}\,\text{d}^{-1}\,\text{m}^{-2}$	maximum specific assim. flux
$\{\dot{p}_T\} = -\{\dot{J}_{ET}\}\overline{\mu}_E$	$\text{J}\,\text{d}^{-1}\,\text{m}^{-3}$	surface-area-specific maint. flux
$[\dot{p}_M] = -[\dot{J}_{EM}]\overline{\mu}_E = \dot{k}_M\mu_{GV}[M_V]$	$\text{J}\,\text{d}^{-1}\,\text{m}^{-3}$	specific somatic maint. flux
$[\dot{p}_J] = -[\dot{J}_{EJ}]\overline{\mu}_E = \dot{k}_J[E_H]$	$\text{J}\,\text{d}^{-1}\,\text{m}^{-3}$	specific maturity maint. flux
$\dot{k}_M = [\dot{p}_M]/[E_G] = j_{EV}\,y_{VE}$	d^{-1}	somatic maintenance rate coefficient
$\dot{J}_{EJ} = -\dot{k}_J M_H$	$\text{mol}\,\text{d}^{-1}$	maturity maintenance flux
$j_{XA} = j_{EA}/y_{EX}$	$\text{mol}\,\text{mol}\,\text{d}^{-1}$	specific food uptake flux
$j_{EV} = -[\dot{J}_{EV}]/[M_V] = \dot{k}_M\,y_{EV}$	$\text{mol}\,\text{mol}\,\text{d}^{-1}$	specific somatic maintenance flux
$E_H = \overline{\mu}_E M_H$	J	maturity
$E = \overline{\mu}_E M_E$	J	reserve energy
$[E_G] = \overline{\mu}_E[M_V]/y_{VE}$	$\text{J}\,\text{m}^{-3}$	energy costs per structural volume
$\overline{\mu}_E = \frac{\{\dot{p}_{Am}\}}{\{\dot{J}_{EAm}\}} = \frac{[E_m]}{[M_{Em}]}$	$\text{J}\,\text{mol}^{-1}$	chemical potential of reserve
$\mu_{AX} = \{\dot{p}_{Am}\}/\{\dot{J}_{XAm}\} = \overline{\mu}_E/y_{XE}$	$\text{J}\,\text{mol}^{-1}$	energy–mass coupler for assimilation
$\mu_{AP} = \{\dot{p}_{Am}\}/\{\dot{J}_{PAm}\} = \overline{\mu}_E/y_{PE}$	$\text{J}\,\text{mol}^{-1}$	energy–mass coupler for defecation
$\mu_{GV} = [E_G]/[M_V] = \overline{\mu}_E/y_{VE}$	$\text{J}\,\text{mol}^{-1}$	energy–mass coupler for growth
$g = \frac{[E_G]}{\kappa[E_m]} = \frac{\dot{v}[M_V]}{\kappa\{\dot{J}_{EAm}\}y_{VE}}$	–	energy investment ratio
$f = X/(K + X)$	–	scaled functional response
$l = (M_V/M_{Vm})^{1/3} = L/L_m$	–	scaled length
$e = \frac{m_E}{m_{Em}} = \frac{[M_E]\dot{v}}{\{\dot{J}_{EAm}\}} = \frac{[E]}{[E_m]}$	–	scaled reserve density

where y_{EV} denotes the moles of reserves required to synthesise a mole of structural mass. Equation (3.8) shows that the parameter m_{Em} can be interpreted as the maximum value of the molar reserve density m_E. The scaled heating length l_T is zero for most ectotherms.

The 12 primary parameters expressed in a mass–length–time frame are: specific searching rate $\{\dot{F}_m\}$, yield of reserve on food y_{EX}, maximum specific assimilation rate $\{\dot{J}_{EAm}\}$, energy conductance $\dot{v}$, allocation fraction to soma κ, reproduction efficiency κ_R, volume-specific somatic maintenance cost $[\dot{J}_{EM}]$, surface-area-specific somatic maintenance cost $\{\dot{J}_{ET}\}$, maturity maintenance rate coefficient $\dot{k}_J$, yield of structure on reserve y_{VE}, maturity at birth M_H^b, maturity at puberty M_H^p.

3.5 Macrochemical reaction equations

Suppose that substrates $\mathcal{S} = (S_1, S_2, \ldots)$ and products $\mathcal{P} = (P_1, P_2, \ldots)$ partake in a transformation k with fluxes $\dot{\boldsymbol{J}}_{\mathcal{S}k} = (\dot{J}_{S_1k}, \dot{J}_{S_2k}, \ldots)^T$ and $\dot{\boldsymbol{J}}_{\mathcal{P}k} = (\dot{J}_{P_1k}, \dot{J}_{P_2k}, \ldots)^T$ collected in column vectors. The rate of the transformation is fully specified by the vector $\dot{\boldsymbol{J}} = (\dot{\boldsymbol{J}}_{\mathcal{S}k}; \dot{\boldsymbol{J}}_{\mathcal{P}k})$. If $\boldsymbol{n}_{\mathcal{S}}$ denotes the matrix of chemical indices for the substrates, with typical element n_{is} for substrate $s \in \mathcal{S}$, and $\boldsymbol{n}_{\mathcal{P}}$ that for the products, the constraint

$$\boldsymbol{0} = \boldsymbol{n}_{\mathcal{S}}\dot{\boldsymbol{J}}_{\mathcal{S}k} + \boldsymbol{n}_{\mathcal{P}}\dot{\boldsymbol{J}}_{\mathcal{P}k} \tag{3.10}$$

applies to the vector of rates to ensure conservation of chemical elements. Substrates disappear, so $\dot{\boldsymbol{J}}_{\mathcal{S}k} < \boldsymbol{0}$, and products appear, so $\dot{\boldsymbol{J}}_{\mathcal{P}k} > \boldsymbol{0}$. For all transformations simultaneously we can write $\boldsymbol{0} = \boldsymbol{n}\dot{\boldsymbol{J}}$, where $\boldsymbol{n} = (\boldsymbol{n}_{\mathcal{S}}, \boldsymbol{n}_{\mathcal{P}})$. The number of constraints equals the number of chemical elements that are followed; the constraints can be used to specify some of the fluxes. If appropriate, these constraints could be extended with e.g. constraints on energy, electrical charge and isotopes.

To separate information on rates from that on stoichiometric coupling, it is frequently useful to divide the fluxes by one of the fluxes, say of reference compound $j \in (\mathcal{S}, \mathcal{P})$, and introduce the yield coefficients $\boldsymbol{Y}_{\mathcal{S}j}^k = \dot{\boldsymbol{J}}_{\mathcal{S}k}/\dot{J}_{jk}$ and $\boldsymbol{Y}_{\mathcal{P}j}^k = \dot{\boldsymbol{J}}_{\mathcal{P}k}/\dot{J}_{jk}$.

If one or more of the compounds stand for some generalised compound, rather than pure compounds, we speak of a macrochemical reaction equation, which can typically be split up into two or more microchemical reaction equations. A macrochemical reaction equation is defined for $Y_{S_*j} < 0$ and $j \in \mathcal{P}$ as

$$-Y_{S_1j}^k S_1 - Y_{S_2j}^k S_1 - \cdots \rightarrow Y_{P_1j}^k P_1 + Y_{P_2j}^k P_2 + \cdots \quad \text{or} \quad 0 = \sum_{i \in \mathcal{S},\mathcal{P}} Y_{ij}^k i \tag{3.11}$$

Notice that the macrochemical reaction equation is not a mathematical equation; i in this equation stands for a label (i.e. a type), not for a concentration or other quantity.

Conservation of chemical elements translates to constraints on yield coefficient for reference compound j as

$$\boldsymbol{0} = \boldsymbol{n}_{\mathcal{S}}\boldsymbol{Y}_{\mathcal{S}j}^k + \boldsymbol{n}_{\mathcal{P}}\boldsymbol{Y}_{\mathcal{P}j}^k \tag{3.12}$$

Suppose that there are not one but several chemical transformations simultaneously. Let $\boldsymbol{M}$ be the column matrix of the masses of all compounds. Then $\frac{d}{dt}\boldsymbol{M} = \dot{\boldsymbol{J}}\boldsymbol{1}$, where the summation is over all transformations.

3.6 Isotopes dynamics: reshuffling and fractionation

Isotope dynamics can be followed in the context of DEB theory, due to the fact that DEB theory specifies *all* mass fluxes. We here derive the dynamics, excluding physiological effects of isotopes. Applications of isotope dynamics could include history reconstructions and monitoring particular fluxes.

We neglect the decay of isotopes, so if this decay can't be neglected (e.g. for some unstable isotopes), the present treatment should be adjusted. Transformations convert substrates into products; the isotope ratios of the substrates are assumed to be known. The isotope ratios of the products are assumed to be known at time zero only, and the task is to specify the trajectory of the ratio given a specification of the transformation rate as function of time. We first discuss the process of reshuffling of atoms in transformations, which leads to a redistribution of isotopes, then we study fractionation.

We take the fluxes of substrate in a transformation to be negative by definition and that of products positive. Since the roles of substrates and products are asymmetrical with respect to isotope transduction, the next section assumes that all substrates and products are specified, even if they happen to be chemically identical. Some transformations might use e.g. water both as substrate and as product; water should then appear twice in the equation for the transformation.

The literature on isotope distributions, see e.g. [231], uses the isotope ratio R, which stands for the ratio of the frequencies of one isotope of a certain element (typically the rare type) and that of another (typically the most common type). Sometimes R is the ratio of masses, rather than frequencies. This ratio relates to the relative isotope frequency of type 0 of element i in compound j, γ_{ij}^0, as $R = \frac{\gamma_{ij}^0}{1-\gamma_{ij}^0}$. Data typically refer to isotope frequencies relative to a standard 'ref' and are denoted by

$$\delta_i = 1000\frac{R_i - R_{\text{ref}}}{R_{\text{ref}}} = 1000\left(\frac{\gamma_{ij}^0}{1-\gamma_{ij}^0}\frac{1-\gamma_{ij}^{0\text{ref}}}{\gamma_{ij}^{0\text{ref}}} - 1\right) \tag{3.13}$$

This notation does not make explicit the compound(s) in which the element i occurs. If the compound occurs in phases A and B, two other frequently used definitions are

$$\Delta_{A-B} = \delta_A - \delta_B; \quad \alpha_{A-B} = \frac{1000 + \delta_A}{1000 + \delta_B} = \frac{\gamma_{ij}^{0A}}{1-\gamma_{ij}^{0A}}\frac{1-\gamma_{ij}^{0B}}{\gamma_{ij}^{0B}} \tag{3.14}$$

We use this notation only in auxiliary theory (to link predictions to measurements) because in the core theory we need more notational detail in compounds and transformations and a closer link to the underlying processes. The notation in the literature is rather natural for selection of isotopes from pools, but not for that from fluxes, as we will see.

3.6.1 Reshuffling

Let $\mathcal{S}$ be the set of substrates and $\mathcal{P}$ be the set of products. The dimensionless reshuffling parameter α_{ps}^{ik}, with $0 \leq \alpha_{ps}^{ik} \leq 1$, specifies what fraction of the atoms of chemical element i in substrate s ends up in product p in transformation k. Given the relative frequency of isotope 0 of element i in substrate $s \in \mathcal{S}$ in transformation k, n_{is}^{0k}, the coefficients n_{ip}^{0k} are given for $p \in \mathcal{P}$ by

$$0 = n_{ip}^{0k} \dot{J}_{pk} + \sum_{s \in \mathcal{S}} \alpha_{ps}^{ik} n_{is}^{0k} \dot{J}_{sk} \quad \text{or} \quad n_{ip}^{0k} = -\sum_{s \in \mathcal{S}} \alpha_{ps}^{ik} n_{is}^{0k} / Y_{ps}^{k} \tag{3.15}$$

with $1 = \sum_{p \in \mathcal{P}} \alpha_{ps}^{ik}$. If n_s substrates and n_p products exist, the number of reshuffling parameters α is $(n_p - 1)n_s$.

In matrix notation we can write

$$\mathbf{0} = \dot{\boldsymbol{J}}_{\mathcal{P}k}^{0i} + \boldsymbol{\alpha}^{ik} \dot{\boldsymbol{J}}_{\mathcal{S}k}^{0i} \quad \text{with} \quad \mathbf{1}^T \boldsymbol{\alpha}^{ik} = \mathbf{1}^T \tag{3.16}$$

where (column) vector $\dot{\boldsymbol{J}}_{\mathcal{S}k}^{0i}$ has elements $n_{is}^{0k} \dot{J}_{sk}$ and vector $\dot{\boldsymbol{J}}_{\mathcal{P}k}^{0i}$ has elements $n_{ip}^{0k} \dot{J}_{pk}$ and matrix $\boldsymbol{\alpha}^{ik}$ has elements α_{ps}^{ik} with $p \in \mathcal{P}$ and $s \in \mathcal{S}$. Notice that the use of the reshuffling parameters is via the product with the chemical indices, $\alpha_{ps}^{ik} n_{ns}^{0k}$, so the requirement that the sum of the rows of each column of $\boldsymbol{\alpha}^{ik}$ equals 1 is only essential for elements that actually occur in that substrate. If element i does not occur in substrate s, the entries of $\boldsymbol{\alpha}^{ik}$ in column s don't matter. From the conservation of elements and isotopes, we must have

$$0 = \mathbf{1}^T \dot{\boldsymbol{J}}_{\mathcal{P}k}^{i} + \mathbf{1}^T \dot{\boldsymbol{J}}_{\mathcal{S}k}^{i} \quad \text{and} \quad 0 = \mathbf{1}^T \dot{\boldsymbol{J}}_{\mathcal{P}k}^{0i} + \mathbf{1}^T \dot{\boldsymbol{J}}_{\mathcal{S}k}^{0i} \tag{3.17}$$

where vector $\dot{\boldsymbol{J}}_{\mathcal{S}k}^{i}$ and $\dot{\boldsymbol{J}}_{\mathcal{P}k}^{i}$ have elements $n_{is} \dot{J}_{sk}$ and $n_{ip} \dot{J}_{pk}$, respectively.

To illustrate the application of the reshuffling matrix, consider the oxygenic photosynthesis L:

$$\mathrm{CO_2} + 2\,\mathrm{H_2O} + \text{light} \rightarrow \mathrm{CH_2O} + \mathrm{H_2O} + \mathrm{O_2} \quad \text{or} \quad C + 2\mathrm{H} \rightarrow \mathrm{X} + H' + \mathrm{O}$$

where the oxygen atoms of dioxygen are known (from biochemistry) to come from water, not from carbon dioxide; this is why water is both substrate and product in this transformation. Water as product is labelled H' because its isotope composition can deviate for water as substrate H. For this transformation L and isotopes ^{13}C, ^{2}H and ^{18}O we have

$$\begin{array}{llllllll}
\alpha_{XC}^{CL} = 1 & & \alpha_{XH}^{HL} = \frac{1}{2} & & \alpha_{XC}^{OL} = \frac{1}{2} & \alpha_{XH}^{OL} = 0 & & & & n_{OX}^{18\,L} = \frac{1}{2} n_{OC}^{18\,L} \\
\alpha_{H'C}^{CL} = 0 & \vdots & \alpha_{H'H}^{HL} = \frac{1}{2} & \vdots & \alpha_{H'C}^{OL} = \frac{1}{2} & \alpha_{H'H}^{OL} = 0 & \vdots \; n_{CX}^{13\,L} = n_{CC}^{13\,L} & \vdots & \begin{array}{l} n_{HX}^{2\,L} = n_{HH}^{2\,L} \\ n_{HH'}^{2\,L} = n_{HH}^{2\,L} \end{array} & \vdots \; n_{OH'}^{18\,L} = \frac{1}{2} n_{OC}^{18\,L} \\
\alpha_{OC}^{CL} = 0 & & \alpha_{OH}^{HL} = 0 & & \alpha_{OC}^{OL} = 0 & \alpha_{OH}^{OL} = 1 & & & & n_{OO}^{18\,L} = 2 n_{OH}^{18\,L}
\end{array} \tag{3.18}$$

$\alpha_{XC}^{OL} = \frac{1}{2}$ tells that half of the oxygen of carbon dioxide ends up in carbohydrate; $\alpha_{OH}^{OL} = 1$ tells that all of the oxygen of water ends up in dioxygen. So the oxygen-isotope distribution in carbon dioxide has no relevance for that in dioxygen (in this transformation).

Suppose we have the absurd reaction mechanism that all substrate atoms of element i are allocated to product molecules after complete randomisation. The isotope ratios of that element are equal in all products, so for product $p \in \mathcal{P}$ we have

$$\frac{n_{ip}^{0k}}{n_{ip}} = \frac{\mathbf{1}^T \dot{\boldsymbol{J}}_{\mathcal{S}k}^{0i}}{\mathbf{1}^T \dot{\boldsymbol{J}}_{\mathcal{S}k}^{i}} \quad \text{or} \quad \dot{\boldsymbol{J}}_{\mathcal{P}k}^{0i} = \dot{\boldsymbol{J}}_{\mathcal{P}k}^{i} \frac{\mathbf{1}^T \dot{\boldsymbol{J}}_{\mathcal{S}k}^{0i}}{\mathbf{1}^T \dot{\boldsymbol{J}}_{\mathcal{S}k}^{i}} \quad \text{and} \quad \alpha_{ps}^{ik} = \frac{\dot{J}_{pk}^{i}}{\mathbf{1}^T \dot{\boldsymbol{J}}_{\mathcal{P}k}^{i}} \quad \text{or} \quad \boldsymbol{\alpha}^{ik} = \frac{\dot{\boldsymbol{J}}_{\mathcal{P}k}^{i} \mathbf{1}^T}{\mathbf{1}^T \dot{\boldsymbol{J}}_{\mathcal{P}k}^{i}} \tag{3.19}$$

Division of the numerator and denumerator by one of the fluxes, typically a flux of substrate, converts fluxes to yield coefficients which are not time-dependent. Although the mechanism is unrealistic, this choice of reshuffle coefficients can serve as baseline to reduce the number of parameters in specific applications where no information about the mechanism is available. If the transformation is really complex, like in living systems, complete reshuffling might be not too far from reality.

Addition of transformations

Macrochemical reaction equations are typically additions of several (or even many) equations. Transfer of isotopes comes with an asymmetry of the roles of substrates and products, which means that a particular compound in a macrochemical reaction equation can play both roles, even if no net synthesis or decay of that compound occurs.

Before adding transformations k and l to a new transformation m, we extend the set of substrates $\mathcal{S}$ and products $\mathcal{P}$, such that these sets include *all* substrates and compounds, and allow that some of the fluxes are zero, and some compounds occur in both sets. Let $\{\dot{\boldsymbol{J}}_{\mathcal{S}m}, \dot{\boldsymbol{J}}_{\mathcal{P}m}\} = \{\dot{\boldsymbol{J}}_{\mathcal{S}k} + \dot{\boldsymbol{J}}_{\mathcal{S}l}, \dot{\boldsymbol{J}}_{\mathcal{P}k} + \dot{\boldsymbol{J}}_{\mathcal{P}l}\}$ be the sets of fluxes of the total transformation. To define transformation m properly, we must have $n_{is}^{0k} = n_{is}^{0l} = n_{is}^{0m}$ for $s \in \mathcal{S}$, so $\dot{\boldsymbol{J}}_{\mathcal{S}m}^{0i} = \dot{\boldsymbol{J}}_{\mathcal{S}k}^{0i} + \dot{\boldsymbol{J}}_{\mathcal{S}l}^{0i}$. Although generally we will have $n_{ip}^{0k} \neq n_{ip}^{0l}$, we still have $\dot{\boldsymbol{J}}_{\mathcal{P}m}^{0i} = \dot{\boldsymbol{J}}_{\mathcal{P}k}^{0i} + \dot{\boldsymbol{J}}_{\mathcal{P}l}^{0i}$. Further $\alpha_{ps}^{im} \dot{J}_{sm}^{i} = \alpha_{ps}^{ik} \dot{J}_{sk}^{i} + \alpha_{ps}^{il} \dot{J}_{sl}^{i}$. We then have

$$\mathbf{0} = \dot{\boldsymbol{J}}_{\mathcal{P}m}^{0i} + \boldsymbol{\alpha}^{im} \dot{\boldsymbol{J}}_{\mathcal{S}m}^{0i} \quad \text{with} \quad \mathbf{1}^T \boldsymbol{\alpha}^{im} = \mathbf{1}^T \tag{3.20}$$

and

$$\boldsymbol{\alpha}^{im} \text{diag}(\dot{\boldsymbol{J}}_{\mathcal{S}m}) = \boldsymbol{\alpha}^{ik} \text{diag}(\dot{\boldsymbol{J}}_{\mathcal{S}k}) + \boldsymbol{\alpha}^{il} \text{diag}(\dot{\boldsymbol{J}}_{\mathcal{S}l}) \tag{3.21}$$

Notice that the reshuffle parameters become time-dependent if the ratio of the rates of transformation k and l changes in time. In practice we will only add fully coupled transformations.

3.6.2 Fractionation

Selection of substrate molecules on the basis of their possession of isotopes of particular elements (at particular positions) can occur from (large) pools and from fluxes, which corresponds to selection from *small* pools by integrating the flux over a time increment. Simple chemical transformations don't allow for fractionation in the transformation (so a flux), because the reaction mechanism determines which atoms of the substrates become each of the atoms of the products. This situation is, however, more complex in macroscopic reactions that involve a network of transformations and alternative routes exist for the intermediary metabolites. These intermediary metabolites serve the role of substrates.

Fractionation from pools

Preamble: Suppose we have m_0 white balls with weight β_0 each and $m_1 = m - m_0$ black balls with weight β_1. The number of white balls in a sample of size n follows a binomial distribution if m_0 and m_1 are large relative to n and selection is random, but proportional to the weight of the balls. For odds ratio $\beta = \beta_0/\beta_1$, the expected number of white balls in the sample is

$$n_0 = \frac{n m_0 \beta}{m_0 \beta + m_1} \qquad (3.22)$$

Notice that this number only depends on $\frac{m_0}{m_1} = \frac{\gamma_{ij}^0}{1-\gamma_{ij}^0}$, and not on m_0 and m_1 separately. Applied to molecules with isotopes, the physical mechanism of differential selection relates to the differential mean velocity of molecules with isotopes. All molecules have the same kinetic energy $mc^2/2$; if the mass m of an isotope is larger, its velocity c is smaller.

Fractionation can occur in the selective uptake of dioxygen and carbon dioxide (mostly by phototrophs) and in the selective release of carbon dioxide, nitrogenous waste and water. The latter might be of some importance for terrestrial organisms, where this release is associated with a phase transition for liquid (= organism) to gas. The isotope frequency in assimilation, dissipation and/or growth has expectation

$$n_{ij}^{0k} = \frac{n_{ij}\beta_{ij}^{0k}}{\beta_{ij}^{0k} - 1 + 1/\gamma_{ij}^0}, \quad \text{so} \quad n_{OO}^{0k} = \frac{2\beta_{OO}^{0k}}{\beta_{OO}^{0k} - 1 + 1/\gamma_{OO}^0} \qquad (3.23)$$

For odds ratio $\beta_{ij}^{0k} = 1$, this gives $n_{ij}^{0k} = n_{ij}\gamma_{ij}^0$. In the case of dioxygen, there is little reason to expect that this relationship depends on the transformation k.

Suppose that the odds ratio equals the ratio of molecular velocities and that ^{18}O and ^{16}O combine randomly in dioxygen. So a fraction $(1-\gamma_{OO}^{18})^2$ of the dioxygen molecules has velocity $\dot{v}_{32}$, a fraction $(\gamma_{OO}^{18})^2$ has velocity $\dot{v}_{36}$, and a fraction $2\gamma_{OO}^{18}(1-\gamma_{OO}^{18})$ has velocity $\dot{v}_{34}$ at some given temperature. All dioxygen molecules have the same kinetic energy so $32\dot{v}_{32}^2 = 34\dot{v}_{34}^2 = 36\dot{v}_{36}^2$. So

$$\beta_{OO}^{18k} = \frac{(1-\gamma_{OO}^{18})\dot{v}_{34} + \gamma_{OO}^{18}\dot{v}_{36}}{(1-\gamma_{OO}^{18})\dot{v}_{32} + \gamma_{OO}^{18}\dot{v}_{34}} = \frac{1-\gamma_{OO}^{18} + \gamma_{OO}^{18}\sqrt{34/36}}{(1-\gamma_{OO}^{18})\sqrt{34/32} + \gamma_{OO}^{18}} \simeq \sqrt{\frac{32}{34}} = 0.97 \qquad (3.24)$$

The latter approximation applies for small γ_{OO}^{18}. However, it is very doubtful that this simple reasoning applies; the link between molecular and macroscopic phenomena is typically less direct.

The observations for ^{13}C in the oxidative photosynthesis L are: $^{13}\delta C = -8$ for CO_2 in the atmosphere, and -28 from carbohydrate in C_3-plants [376, p44]. The

$R_{\rm ref} = 0.01118$ for carbon in the PDB standard [376, p 34]. So $R = 0.011091$ for $^{13}CO_2$ and 0.010750 for $^{13}CH_2O$. This gives $\gamma_{CC}^{13} = R/(1+R) = 0.010969$ for $^{13}CO_2$ and $\gamma_{CX}^{13} = 0.010750$ for $^{13}CH_2O$. The odds ratio for $^{13}CO_2$ is $\beta_{CC}^{13\,L} = \frac{1/\gamma_{CC}^{13}-1}{1/\gamma_{CX}^{13}-1} = 0.97982$; a small deviation from 1 gives a strong fractionation.

Selection from food, reserve and structure as pools is less likely. Food is processed as whole items; at the interface of reserve and structure mobilisation SUs are at work locally with no 'knowledge' of the neighbouring reserve molecules. Selection is more likely in mobilised fluxes that have several fates; isotopes can affect binding strength in a molecule and so the energy required to transform the compound; compounds with light isotopes are more easily degraded, so more likely to be used for catabolic, rather than anabolic purposes.

Fractionation from fluxes

Preamble: Suppose we have m_0 white balls with weight β_0 each and $m_1 = m - m_0$ black balls with weight β_1. The number of white balls in a sample of size n follows Fisher's non-central hypergeometric distribution if selection is random, non-interactive, and proportional to weight. For odds ratio $\beta = \beta_0/\beta_1$ and $n \in (0, m)$, the expected number of white balls in the sample is

$$n_0 = P_1/P_0 \quad \text{with} \quad P_k = \sum_{y=\max(0,n-m_1)}^{\min(n,m_0)} \binom{m_0}{y}\binom{m_1}{n-y} \beta^y y^k \quad \text{or}$$

$$n_0 \approx \frac{-2c}{b - \sqrt{b^2 - 4ac}} = \frac{r m_0 \beta}{r\beta + 1} \quad \text{with} \quad r > 0 \quad \text{such that} \quad n = \frac{r m_0 \beta}{r\beta + 1} + \frac{r m_1}{r + 1} \qquad (3.25)$$

where $a = \beta - 1$, $b = n - m_1 - (m_0 + n)\beta$, $c = m_0 n \beta$. Multivariate extensions are known. For isotope applications we focus on fractions m_0/m in the total flux and n_0/n in the sub-flux (the anabolic flux). For large m_0 and m_1 relative to n, this non-central hypergeometric distribution converges to the binomial distribution. Wallenius' non-central hypergeometric distribution should not be used here, because interactive selection is excluded within a time increment. The physical mechanism is in the differential strength of chemical bounds; the larger the mass, the stronger the bound. This means that light molecules have a preference for the catabolic route.

Suppose that a molecule of a compound has more than one possible fate in a transformation. Selection occurs if the probability on the fate of a molecule depends on the presence of one or more isotopes. We here assume that each molecule in a well-mixed pool has the same probability to be selected to partake in a transformation, independent of its isotope composition; selection only interferes with the fate of the mobilised molecule.

Suppose that the fluxes of substrates $\dot{\boldsymbol{J}}_{Sk}$ are partitioned into two fluxes (e.g. a catabolic and an anabolic one) as $\dot{\boldsymbol{J}}_{Sk_a} = \kappa_k \dot{\boldsymbol{J}}_{Sk}$ and $\dot{\boldsymbol{J}}_{Sk_c} = (1 - \kappa_k)\dot{\boldsymbol{J}}_{Sk}$. The partitioning is, however, selective for the isotope of element i in compound j.

We must have $n_{is}^{0k} \dot{J}_{sk} = n_{is}^{0k_a} \dot{J}_{sk_a} + n_{is}^{0k_c} \dot{J}_{sk_c}$ or $n_{is}^{0k} = n_{is}^{0k_a} \kappa_k + n_{is}^{0k_c}(1 - \kappa_k)$. Again we write $n_{is}^{0k} = \gamma_{is}^{0} n_{is}$ and introduce an odds ratio $\beta_{is}^{0k_a}$ for an isotope of type 0 of element

i in compound s in transformation k_a. The number of isotopes in the anabolic flux times a time increment follows Fisher's non-central hypergeometric distribution with approximate mean for $B = \frac{n_{is}^{0k}}{n_{is}} - (1 - \kappa_k) - (\frac{n_{is}^{0k}}{n_{is}} + \kappa_k)\beta_{is}^{0k_a}$

$$n_{is}^{0k_a} \simeq \frac{2n_{is}^{0k}\beta_{is}^{0k_a}}{\sqrt{B^2 + 4(1 - \beta_{is}^{0k_a})\beta_{is}^{0k_a}\frac{n_{is}^{0k}}{n_{is}}\kappa_k} - B}; \quad n_{is}^{0k_c} = \frac{n_{is}^{0k} - n_{is}^{0k_a}\kappa_k}{1 - \kappa_k} \tag{3.26}$$

If $\beta_{is}^{0k_a} = 1$, we have $n_{is}^{0k_a} = n_{is}^{0k}$ and the process is unselective. We must have

$$n_{is}^{0k} \geq n_{is}^{0k_a}\kappa_k \quad \text{and} \quad B^2 + 4(1 - \beta_{is}^{0k_a})\beta_{is}^{0k_a}n_{is}^{0k}\kappa_k \geq 0 \tag{3.27}$$

Notice that only molecules can be selected on the basis of having a particular isotope of a particular element; the selection is not on elements independently. Once the selective element i is determined for a compound s, $\beta_{hs}^{0k_a} = 1$ for all $h \neq s$. The selection of a single isotope of a particular atom in a particular compound is the simplest possibility; many more complex forms of selection can exist.

Suppose that substrate S is subjected to selection with respect to element I and that $\boldsymbol{\alpha}^{ik_a}$ and $\boldsymbol{\alpha}^{ik_c}$ are the reshuffling parameters of the anabolic and the catabolic sub-fluxes. So the fraction κ_k applies to flux $\dot{J}_S$. Let $n_{IS}^{0k} = n_{IS}\gamma_{IS}^0$. In adding these two fluxes, we should take into account that the anabolic flux experiences a different isotope frequency for element I than the catabolic flux: $\dot{J}_{Sk}^{0I} = \dot{J}_{Sk_a}^{0I} + \dot{J}_{Sk_c}^{0I} = (n_{IS}^{0k_a}\kappa_k + n_{IS}^{0k_c}(1 - \kappa_k))\dot{J}_{Sk}$. Let $\dot{\boldsymbol{J}}_{Sk}^{0I*}$ be $\dot{\boldsymbol{J}}_{Sk}^{0I}$, but with element S replaced by this modified flux $\dot{J}_{Sk}^{0I}$. The reshuffle parameters $\boldsymbol{\alpha}^{Ik}$ are not affected by selection. The coefficients n_{Ip}^{0I} in $\dot{\boldsymbol{J}}_{\mathcal{P}k}^{0I}$ are now given by

$$\mathbf{0} = \dot{\boldsymbol{J}}_{\mathcal{P}k}^{0I} + \boldsymbol{\alpha}^{Ik}\dot{\boldsymbol{J}}_{\mathcal{S}k}^{0I*} \quad \text{with} \quad \boldsymbol{\alpha}^{Ik}\text{diag}(\dot{\boldsymbol{J}}_{\mathcal{S}k}) = \boldsymbol{\alpha}^{Ik_a}\text{diag}(\dot{\boldsymbol{J}}_{\mathcal{S}k_a}) + \boldsymbol{\alpha}^{Ik_c}\text{diag}(\dot{\boldsymbol{J}}_{\mathcal{S}k_c}) \tag{3.28}$$

3.7 Enzyme-mediated transformations based on fluxes

Enzymes are compounds that catalyse a transformation without being transformed themselves. They are typically proteins, frequently only working in combination with an RNA co-factor at the binding site of the enzyme. The protein just enhances the binding of the substrates and brings them into an orientation such that the transformation occurs. The transformation rates are constrained by the time budget of the enzyme. The enzyme can be in a state where it can bind substrate(s) or where it is processing substrates. In the simplest formulations these traits exclude each other, so they compete for the time of the enzyme, but in more advanced formulations they do that only partly.

Classic enzyme kinetics specifies fluxes of product in terms of substrate concentrations. This catenates two different processes, the arrival process of substrate molecules to the binding site(s) of the enzyme molecules and the transformation of bounded substrate into product, which can better be dealt with separately. In homogeneous environments, arrival rates of substrate molecules to the enzyme molecules are proportional to the concentration, on the basis of diffusive transport. The rejected substrate molecules return to the environment, which makes it difficult, if not impossible, to determine their existence. When growth is modelled as a function of mobilised reserve fluxes (see {194}), the situation is different, because this process represents arrival and replaces diffusive transport. Transformations are hard to link to concentrations in those situations.

The concept concentration implies spatially homogeneous mixing at the molecular level; it hardly applies to the living cell, see {430}. Another argument for avoiding the use of concentrations as much as possible is that concentrations should be thought of as states of the system. The inclusion of concentrations of intermediary metabolites in a metabolic pathway increases the number of state variables of the system. A reduction of this number, to simplify the model, is only possible when the amounts are small enough. This problem is avoided by using fluxes, where intermediaries do not accumulate.

Thinking in terms of fluxes, rather than concentrations, allows us to treat light in a similar way to compounds, with stoichiometric coupling coefficients in photochemical reactions, see {189}. This idea may be less wild than might first appear; cells extract a fixed amount of energy from the photons that are able to excite the pigment system, the remaining energy dissipates as heat. The light flux can be quantified in einstein (or mole) per second, i.e. in $6.023\,10^{23}$ photons per second [449].

Synthesising Units (SUs) [640] are generalised enzymes that follow the rules of classic enzyme kinetics with two modifications: transformation is based on fluxes, rather than on concentrations of substrates, and the backward fluxes are assumed to be negligibly small in the transformation. The arrival flux can be taken to be proportional to the density in spatially homogeneous environments. In spatially structured situations, SUs can interfere and handshaking protocols can be formulated to understand the relationships between organelles and the cytosol, see {255}.

The specification of behaviour of SUs has strong parallels with that of individuals from an abstract point of view, see {263, 264}. If we identify the SU with an individual, and the product with reserve, the transformation rate is directly given by the functional response (2.2). We have already encountered the SU fed by a flux in the mechanism behind the reserve dynamics {41}.

3.7.1 From substrate to product

The simplest transformation of substrate S to product P by enzyme $\mathcal{E}$ can be written as

$$S + \mathcal{E} \rightleftharpoons S\mathcal{E} \rightleftharpoons P\mathcal{E} \rightleftharpoons P + \mathcal{E}$$

The backward fluxes (controlled by the rates $\dot{b}_P$, $\dot{k}_{PS}$ and $\dot{k}_S$) might be small, not because of enzyme performance as such, but because of the spatial organisation of the supply of substrate and the removal of product by transporters. The transformation can then be captured in a simple diagram, see Figure 3.6. The differences from classic enzyme kinetics do not affect the simple one-substrate one-product conversion in spatially homogeneous environments, but do affect more complex transformations.

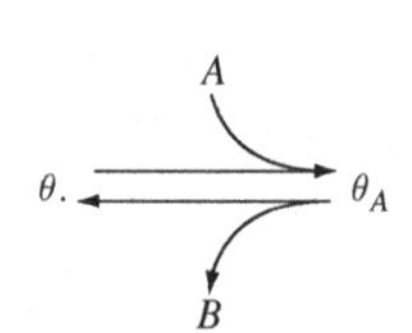

Figure 3.6 Uptake of a single substrate is well quantified on the basis of a fixed handling time of substrate (prey) by the uptake machinery. The time need not be constant, but it must be independent of substrate density [783, 784]. The handling time not only includes mechanical handling but also metabolic processing. This is why eating prey by predators and transformation rate by enzymes depend in a similar way on substrate (food) density.

3.7.2 Rejection vs. Synthesising Units

Let us consider a very simple chemical transformation, where an enzyme requires one copy of each of two substrates, present in concentrations X_A and X_B, to produce a product, present in concentration X_C.

Classic enzyme kinetics states that substrate–enzyme association follows the law of mass action, so the rate is proportional to the product of the concentrations, and dissociation is a first-order process, so the rate is proportional to the concentration of the complex. Given the dissociation rate parameters $\dot{k}_A$, $\dot{k}_B$ and $\dot{k}_C$, and the association parameters $\dot{b}_A$ and $\dot{b}_B$, the change in the fractions of enzyme in the various binding states is given by

$$1 = \theta_{..} + \theta_{A\cdot} + \theta_{\cdot B} + \theta_{AB} \tag{3.29}$$

$$\frac{d}{dt}\theta_{..} = \dot{k}_C\theta_{AB} + \dot{k}_A\theta_{A\cdot} + \dot{k}_B\theta_{\cdot B} - (\dot{b}_A X_A + \dot{b}_B X_B)\theta_{..} \tag{3.30}$$

$$\frac{d}{dt}\theta_{A\cdot} = \dot{k}_B\theta_{AB} + \dot{b}_A X_A\theta_{..} - (\dot{k}_A + \dot{b}_B X_B)\theta_{A\cdot} \tag{3.31}$$

$$\frac{d}{dt}\theta_{\cdot B} = \dot{k}_A\theta_{AB} + \dot{b}_B X_B\theta_{..} - (\dot{k}_B + \dot{b}_A X_A)\theta_{\cdot B} \tag{3.32}$$

where $\theta_{..}$ is the fraction of free enzymes. Steady state is reached when the substrate–enzyme complexes do not change in concentration, so $\frac{d}{dt}\theta_{**} = 0$. The relative abundance of enzyme–substrate complexes is now given by

$$\begin{pmatrix} \theta_{..} \\ \theta_{A\cdot} \\ \theta_{\cdot B} \\ \theta_{AB} \end{pmatrix} = \begin{pmatrix} 1 & 1 & 1 & 1 \\ x_A k_A & -x_B - k_A & 0 & 1 \\ x_B & 0 & -1 - x_A k_A & k_A \\ 0 & x_B & x_A k_A & -1 - k_A - k_C \end{pmatrix}^{-1} \begin{pmatrix} 1 \\ 0 \\ 0 \\ 0 \end{pmatrix} \tag{3.33}$$

with $x_A = X_A\dot{b}_A/\dot{k}_A$, $x_B = X_B\dot{b}_B/\dot{k}_B$, $k_A = \dot{k}_A/\dot{k}_B$, $k_C = \dot{k}_C/\dot{k}_B$. The appearance rate of product is for $\dot{J}_{Cm} = \dot{k}_C X_+$ given by

$$\frac{d}{dt}X_C = \dot{J}_C = \dot{J}_{Cm}\theta_{AB} \tag{3.34}$$

Two limiting cases are of special interest: the Synthesising Unit (SU), where the substrate–enzyme dissociation rates are small, and the Rejection Unit (RU), where these rates are high, but the association rates are high as well. (Another way to obtain the same RU is when the product–enzyme dissociation rate $\dot{k}_C$ is small, and the total amount of enzyme X_+ is high, but this hardly applies to organisms.) These limiting cases give the following results:

SU: $\dot{k}_A, \dot{k}_B \to 0$	RU: $\dot{k}_A, \dot{k}_B, \dot{b}_A, \dot{b}_B \to \infty$ and $\frac{\dot{k}_A}{\dot{b}_A}$, $\frac{\dot{k}_B}{\dot{b}_B}$ constant
for $x_A = X_A \frac{\dot{b}_A}{\dot{k}_C}$ and $x_B = X_B \frac{\dot{b}_B}{\dot{k}_C}$	for $x_A = X_A \frac{\dot{b}_A}{\dot{k}_A}$ and $x_B = X_B \frac{\dot{b}_B}{\dot{k}_B}$
$\dot{j}_C = \frac{\dot{j}_{Cm}}{1+x_A^{-1}+x_B^{-1}-(x_A+x_B)^{-1}}$	$\dot{j}_C = \frac{\dot{j}_{Cm}}{(1+x_A^{-1})(1+x_B^{-1})}$
$\begin{pmatrix} \theta_{..} \\ \theta_{A.} \\ \theta_{.B} \\ \theta_{AB} \end{pmatrix} = \begin{pmatrix} (x_A+x_B)^{-1} \\ x_A x_B^{-1}(x_A+x_B)^{-1} \\ x_B x_A^{-1}(x_A+x_B)^{-1} \\ 1 \end{pmatrix} \frac{\dot{j}_C}{\dot{j}_{Cm}}$	$\begin{pmatrix} \theta_{..} \\ \theta_{A.} \\ \theta_{.B} \\ \theta_{AB} \end{pmatrix} = \begin{pmatrix} (x_A x_B)^{-1} \\ x_B^{-1} \\ x_A^{-1} \\ 1 \end{pmatrix} \frac{\dot{j}_C}{\dot{j}_{Cm}}$

Despite its popularity [66, 336, 337, 846], the RU has a number of problems that make it less attractive than the SU. The first, but perhaps not the most important, problem is a mild form of inconsistency at the molecular level. The law of mass action is used for association between substrate and enzyme. It requires completely homogeneous mixing, which is hard to combine with infinitely large dissociation and association rates; as soon as a substrate molecule is rejected by an enzyme molecule, it becomes attracted again if the mixing rate is not infinitely large, which is obviously not realistic. Moreover, it is hard to see in terms of molecular geometry and electrical charge distributions how a high association rate can combine with a high dissociation rate. The SU is much more natural in this respect, because the binding sites on the enzyme molecule mirror-match the substrates in shape and electrical charge, which makes it likely that the substrate–enzyme dissociation rate is small compared to the product–enzyme dissociation rate, because of the shape and charge changes during the substrates–product transition. Product molecules do not mirror-match the substrate-binding sites in shape and electrical charge, and products, not substrates, are rejected by the enzyme.

For very large concentrations x_B, both SU and RU simplify to what is known as Michaelis–Menten kinetics (MM-kinetics): $\dot{j}_C = \dot{j}_{Cm}(1 + x_A^{-1})^{-1}$, but the convergence for SU is much faster than for RU. In fact, the RU converges really slowly to MM-kinetics, which means that substrate concentrations must exceed the saturation constant by at least an order of magnitude to become (almost) non-limiting. A substrate is defined to be non-limiting if a change in substrate concentration does not affect the production rate. Given the fact that models for uptake and use of nutrients are likely to include only a small subset of the required nutrients and compounds, the implication that compounds that are not included must be really abundant is not acceptable.

Last, but not least, the multiplicative model for nutrient uptake, as implied by the RU, is found to be inconsistent with empirical data [295]. MM-kinetics, and its various generalisations, plays a central role in models for enzyme kinetics and substrate (food, nutrient) acquisition by organisms; it was first described by Henri in 1902 [494].

Given identical production rates if only one substrate is limiting (this is when the other substrate is abundant), the production rate of the RU is always smaller than that of the SU, $\dot{J}_{C,\,\mathrm{SU}} > \dot{J}_{C,\,\mathrm{RU}}$, while their ratio tends to infinity for small substrate concentrations (x_A, $x_B \to 0$).

3.7.3 Four basic classes of transformations

When two substrates are complementary (or supplementary if you wish), i.e. they are both required in fixed stoichiometric proportions, the absence of one substrate prevents the uptake of the other; think, for instance, of ammonia and carbon dioxide as substrates and amino acids as reserves. Empirical evidence frequently indicates that the uptake of the most abundant substrate (relative to the needs) is set by the least abundant substrate: the popular minimum rule of von Liebig [703]. The rule originally related biomass yields to nutrient levels, but was later applied to uptake processes [41]. However, this application becomes complex if reserves are included; the environment may not contain the substrate, but growth is not restricted because of the presence of reserves. If the role of limiting and non-limiting substrate does not switch at the same time for all individuals in the population in a variable environment, it is almost impossible to evaluate population behaviour on the basis of individual behaviour. Moreover, sharp switches are not realistic at the molecular level, because of the intrinsic stochasticity of the substrate arrival process.

Let us characterise the states of the SUs in bounded fractions with vector $\boldsymbol{\theta}$, while $\mathbf{1}^T\boldsymbol{\theta} = 1$ and $0 \leq \theta_i < 1$ for all states i. The change in bounded fractions of SUs can be written as $\frac{d}{dt}\boldsymbol{\theta} = \dot{\boldsymbol{k}}\boldsymbol{\theta}$, for a matrix of rates $\dot{\boldsymbol{k}}$ with diagonal elements $\dot{k}_{ii} = -\sum_{j \neq i} \dot{k}_{ij}$, while $\dot{k}_{ij} \geq 0$, so $\mathbf{1}^T\dot{\boldsymbol{k}} = \mathbf{0}$. Using a timescale separation argument, a flux of metabolite X can be written as $\dot{J}_X = \dot{\boldsymbol{J}}^T\boldsymbol{\theta}^*$, with weight coefficients $\dot{\boldsymbol{J}}$ and fractions $\boldsymbol{\theta}^*$ such that $\mathbf{0} = \dot{\boldsymbol{k}}\boldsymbol{\theta}^*$. Substrates can be classified as substitutable or complementary and binding schemes as sequential or parallel. These four classes comprise the standard SU kinetics (see Figure 3.7), and, in retrospect, have direct links with the waiting time theory derived by O'Neill *et al.* [852].

Mixtures of the four classes of standard kinetics have the property that $\dot{\boldsymbol{k}} = \sum_i \dot{\boldsymbol{k}}_i$, where $\dot{\boldsymbol{k}}$ is the matrix of rates of the mixture, and $\dot{\boldsymbol{k}}_i$ that of a standard type.

Cows eat lots of grass because of the low protein content of grass. The microorganisms in their stomachs transform cellulose into products such as acetate, propionate, butyrate and valerate [1017], which they cannot use as substrate in this anaerobic environment. The cow absorbs these products only partly, so energy supply is *ad libitum.* Feeding the cow some extra protein allows reduction of the required grass intake by a factor of 10. Grass and protein are to some extent substitutable for a cow; the conversion from grass and protein to cow must be described by a mixture between

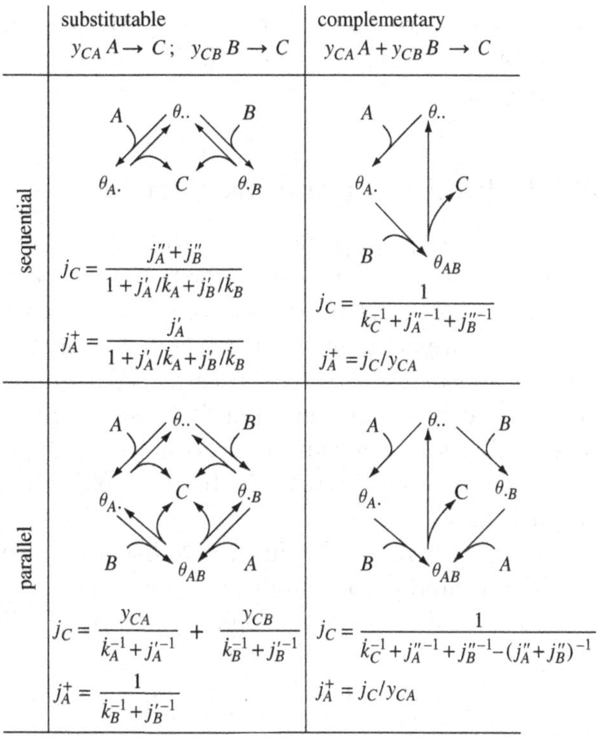

	substitutable $y_{CA} A \rightarrow C$; $y_{CB} B \rightarrow C$	complementary $y_{CA} A + y_{CB} B \rightarrow C$
sequential	$j_C = \frac{j''_A + j''_B}{1 + j'_A/\dot{k}_A + j'_B/\dot{k}_B}$ $j^+_A = \frac{j'_A}{1 + j'_A/\dot{k}_A + j'_B/\dot{k}_B}$	$j_C = \frac{1}{\dot{k}_C^{-1} + j''^{-1}_A + j''^{-1}_B}$ $j^+_A = j_C/y_{CA}$
parallel	$j_C = \frac{y_{CA}}{\dot{k}_A^{-1} + j'^{-1}_A} + \frac{y_{CB}}{\dot{k}_B^{-1} + j'^{-1}_B}$ $j^+_A = \frac{1}{\dot{k}_B^{-1} + j'^{-1}_B}$	$j_C = \frac{1}{\dot{k}_C^{-1} + j''^{-1}_A + j''^{-1}_B - (j''_A + j''_B)^{-1}}$ $j^+_A = j_C/y_{CA}$

Figure 3.7 The interaction of substrates A and B in transformations into product C can be understood on the basis of a classification of substrates into substitutable and complementary, and of binding into sequential or parallel. The symbol $\theta_{*_1 *_2}$ represents an SU that is bound to the substrates $*_1$ and $*_2$, the dot representing no substrate, so $\theta_{..}$ represents an unbounded SU. The symbol $y_{*_1 *_2}$ denotes a stoichiometric coupling coefficient. j^+_A is the accepted flux of A; $j'_* = \rho_* j_*$, where $\rho_*$is the binding probability; $j''_* = y_{C*} j'_*$; $\dot{k}_*$ is the dissociation rate. Modified from Kooijman [642].

substitutable and complementary compounds to ensure that the yield of the combination of grass and protein is much higher, especially if proteins are extracted from the cow in the form of milk. It is a consequence of work with generalised compounds; some chemical compounds serve as energy substrate, others as nutrient (building blocks). Grass and protein weigh these functions differently. Another example of the use of mixtures between substitutable and complementary compounds is in the gradual transition between these basic types which occurs in evolution of symbiosis that is based on syntrophy, see {409}.

Number of SUs

When a flux of substrate arrives at a set of N SUs, it depends on the local spatial organisation how this translates into the arrival rate of substrate for each SU. Classic enzyme kinetics works with the law of mass action, which takes the meeting frequency proportional to the product of the concentrations of substrate and enzyme, using a diffusion argument; the arrival rate is proportional to the concentration of substrate. The enzyme molecules don't interact and the arrival rate per SU is the substrate flux divided by N. SUs dynamics, however, allows for interaction.

Suppose that the SUs are localised on a membrane and can bind substrate molecules that are within a threshold distance from the binding site. If the distance between the SUs is smaller than two times this threshold distance, they start to interact, and the accepted substrate flux becomes a satiating function of the number of SUs. Such a situation occurs, for instance, in the case of carriers for the uptake of substrate in the outer membrane.

3.7.4 Inhibition and preference

Preference frequently occurs in the uptake of substrates, sometimes for nutritional requirements, or when a predator becomes specialised on particular prey species, or to minimise risks of injury (selection of old or weak prey individuals). Inhibition occurs in gene expression, where carriers for a particular substrate are not synthesised as long as another substrate is present. We here deal with interacting substitutable substrates that are bound in a parallel fashion. Standard inhibition makes part of the SUs unavailable for catalysing transformations (Figure 3.8). Stronger forms of interaction can occur if one substrate is able to replace another that is already bound to an SU (Figure 3.8).

Let j_{S_1} and j_{S_2} be the fluxes of substrate S_1 and S_2 that arrive at an SU, and ρ_{S_1} and ρ_{S_2} be the binding probabilities. The binding kinetics, i.e. the changes in the bounded fractions of SUs, for scaled fluxes $j'_{S_1} = \rho_{S_1} j_{S_1}$, $j'_{S_2} = \rho_{S_2} j_{S_2}$ and $1 = \theta_{\cdot} + \theta_{S_1} + \theta_{S_2}$ are

$$\frac{d}{dt}\theta_{S_2} = j'_{S_2}\theta_{\cdot} - (j'_{S_1} + \dot{k}_{S_2})\theta_{S_2}; \quad \frac{d}{dt}\theta_{S_1} = j'_{S_1}(\theta_{\cdot} + \theta_{S_2}) - \dot{k}_{S_1}\theta_{S_1} \tag{3.35}$$

where $\dot{k}_{S_1}$ and $\dot{k}_{S_2}$ are the dissociation constants of the SU–substrate complexes.

Supply kinetics

For the binding fraction at steady state, the production flux of P equals $j_P = y_{PS_1} j^+_{S_1} + y_{PS_2} j^+_{S_2}$, while the fluxes of S_1 and S_2 that are used are

$$j^+_{S_1} = \dot{k}_{S_1}\theta^*_{S_1} = \frac{\dot{k}_{S_1} j'_{S_1}}{\dot{k}_{S_1} + j'_{S_1}}; \quad j^+_{S_2} = \dot{k}_{S_2}\theta^*_{S_2} = \frac{\dot{k}_{S_1}\dot{k}_{S_2} j'_{S_2}}{\dot{k}_{S_2} + j'_{S_1} + j'_{S_2}} \tag{3.36}$$

Although their derivation has been set up slightly differently, this formulation is used in [138] to model substrate preference and diauxic growth in micro-organisms, see {291}.

Figure 3.8 Left: Interaction between the conversions $S_1 \to P$ and $S_2 \to P$, with preference for the first transformation. θ_* indicates the fraction of Synthesising Units that are bound to substrates. Right: The standard inhibition scheme, where S_2 inhibits the transformation $S_1 \to P$.

Demand kinetics

If the flux of P is given (and constant), we require that

$$j_P = y_{PS_1}\dot{k}_{S_1}\theta_{S_1} + y_{PS_2}\dot{k}_{S_2}\theta_{S_2} \tag{3.37}$$

is constant at value $\dot{k}_P$, say, by allowing $\dot{k}_{S_1}$ and $\dot{k}_{S_2}$ to depend on θ_*. The following rates fulfil the constraint:

$$\dot{k}_{S_1} = \dot{k}_P/\theta \text{ and } \dot{k}_{S_2} = \rho\dot{k}_P/\theta \quad \text{with } \theta = y_{PS_1}\theta_{S_1} + y_{PS_2}w\theta_{S_2} \tag{3.38}$$

where the preference parameter $\rho_{S_2} = \dot{k}_{S_2}/\dot{k}_{S_1}$ has the interpretation of the ratio of dissociation rates. For the fractions in steady state, the fluxes of S_1 and S_2 that are used to produce P are

$$j^+_{S_1} = (\dot{k}_P - y_{PS_2}j^+_{S_2})y_{S_1P} \quad \text{and} \quad j^+_{S_2} = \rho\dot{k}_P\frac{\theta^*_{S_2}}{\theta^*} = \frac{2a\dot{k}_P/y_{PS_2}}{2A + y_{PS_1}(\sqrt{B^2 - 4AC} - B)} \tag{3.39}$$

with $A = \rho_{S_2}j'_{S_2}\dot{k}_P y_{PS_2}$, $B = y_{PS_1}C + ((1-\rho_{S_2})j'_{S_1} + j'_{S_2})\dot{k}_P$, $C = -j'_{S_1}(j'_{S_1} + j'_{S_2})$.

Tolla [1164] proposed this model to quantify the preference to pay maintenance (flux j_P) from reserve (flux j_{S_1}) rather than from structure (flux j_{S_2}).

Another variation on the demand version of (partly) substitutable compounds was studied by Kuijper [675], where carbohydrate reserve is preferred above protein reserve for paying the energy maintenance in zooplankton, but protein reserve is required to pay the building-block maintenance. This increase in metabolic flexibility has the consequence that a nutrient–light–phytoplankton–zooplankton system evolves to a situation in which it becomes both energy and nutrient limited, rather than a single limitation only.

3.7.5 Co-metabolism

The biodegradation of organic pollutants in soils can sometimes be enhanced by adding readily degradable substrates. This is a special case of a more general phenomenon that the processing of one substrate affects that of another.

Suppose that substrates A and B are substitutable and are bounded in parallel and that the binding probability of each substrate depends on binding with the other substrate as described and applied by Brandt [140]. We study the process $1\,A \rightarrow y_{CA}\,C$ and $1\,B \rightarrow y_{CB}\,C$. So we have three binding probabilities of each substrate; for substrate A we have the binding probabilities 0 if A is already bounded; ρ_A if A and B are not bounded; ρ_{AB} if B is bounded, but A is not.

No interaction occurs if $\rho_A = \rho_{AB}$; full co-metabolism occurs if $\rho_A = 0$, see Figure 3.9. Sequential processing occurs if $\rho_{AB} = \rho_{BA} = 0$. The dissociation rates $\dot{k}_A$ and $\dot{k}_B$ of product C, and the stoichiometric coefficients y_{AC} and y_{BC}, might differ

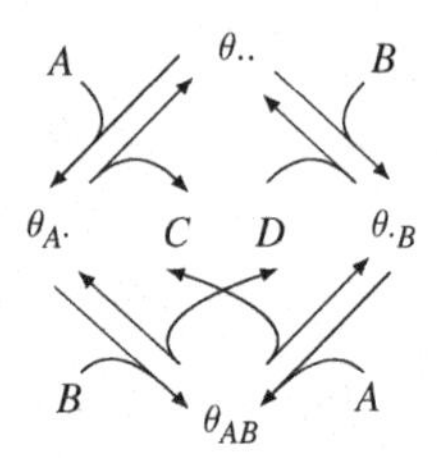

Figure 3.9 The scheme for general co-metabolism of the transformations $A \to C$ with $B \to D$.

for both substrates. The binding period is measured as the period between arrival of substrate and dissociation of product, so it includes the production period.

For $j'_A = j_A\rho_A$, $j''_A = j_A\rho_{AB}$, $j'_B = j_B\rho_B$, $j''_B = j_B\rho_{BA}$, the fractions of bounded SUs follow the dynamics

$$1 = \theta_{..} + \theta_{A\cdot} + \theta_{\cdot B} + \theta_{AB}; \quad \frac{d}{dt}\theta_{..} = -(j'_A + j'_B)\theta_{..} + \dot{k}_A\theta_{A\cdot} + \dot{k}_B\theta_{\cdot B} \tag{3.40}$$

$$\frac{d}{dt}\theta_{A\cdot} = j'_A\theta_{..} - (\dot{k}_A + j''_B)\theta_{A\cdot} + \dot{k}_B\theta_{AB}; \quad \frac{d}{dt}\theta_{\cdot B} = j'_B\theta_{..} - (\dot{k}_B + j''_A)\theta_{\cdot B} + \dot{k}_A\theta_{AB}$$

Assuming pseudo steady state (i.e. $\frac{d}{dt}\theta_{**} = 0$ for $\theta_{**} = \theta^*_{**}$), the production flux amounts to

$$j_C = j_{C,A} + j_{C,B} = y_{CA}\dot{k}_A(\theta^*_{A\cdot} + \theta^*_{AB}) + y_{CB}\dot{k}_B(\theta^*_{\cdot B} + \theta^*_{AB}) \tag{3.41}$$

$$= \frac{y_{CA}\dot{k}_A\left(j'_A\dot{k}_B + j''_A\frac{j'_B(\dot{k}_A+\dot{k}_B)+j''_B(j'_A+j'_B)}{j''_A+j''_B+\dot{k}_A+\dot{k}_B}\right) + y_{CB}\dot{k}_B\left(j'_B\dot{k}_A + j''_B\frac{j'_A(\dot{k}_A+\dot{k}_B)+j''_A(j'_A+j'_B)}{j''_A+j''_B+\dot{k}_A+\dot{k}_B}\right)}{j'_A\dot{k}_B + j'_B\dot{k}_A + \dot{k}_A\dot{k}_B + \frac{j''_Aj'_B\dot{k}_B+j''_Bj'_A\dot{k}_A+j''_Aj''_B(j'_A+j'_B)}{j''_A+j''_B+\dot{k}_A+\dot{k}_B}} \tag{3.42}$$

If B represents a xenobiotic substrate, and A a natural one, the case $\rho_A = \rho_{AB}$ and $\rho_B = 0$ is of special interest. The use of A is not affected by B, but B can only be processed if A is present. The expression for the product flux simplifies for $j'_A = j''_A$ and $j'_B = 0$ to

$$j_C = \frac{y_{CA}\dot{k}_A}{1 + \dot{k}_A j'^{-1}_A} + \frac{y_{CB}\dot{k}_B}{1 + \dot{k}_A j'^{-1}_A}\frac{j''_B(j'_A + \dot{k}_A + \dot{k}_B)}{j''_B(j'_A + \dot{k}_B) + \dot{k}_B(j'_A + \dot{k}_A + \dot{k}_B)} \tag{3.43}$$

The accepted flux of substrate B, so the specific biodegradation rate of B, is $j^+_B = y_{BC}j_{C,B}$ with $y_{BC} = y^{-1}_{CB}$, and $j_{C,B}$ is given by the second term in the expression for j_C.

3.8 Metabolism

3.8.1 Trophic modes: auto-, hetero- and mixotrophy

Trophic strategies are labelled with respect to the energy and the carbon source, as indicated in Table 3.4. Animals typically feed on other organisms, which makes them organochemotrophs, and so heterotrophs. If these organisms are only animals, we call them carnivores, if they are only veridiplants (glaucophytes, rhodophytes or chlorophytes, including plants), we call them herbivores, and in all other cases we call them omnivores. The implication is that daphnids, which also feed on heterokonts, ciliates and dinoflagellates, should be classified as omnivores, although many authors call them herbivores. Many trophic classifications are very imprecise and sensitive to the context.

Animals' food has a complex composition (mixtures of polysaccharides, lipids and proteins), from which animals extract energy, electrons, as well as all necessary building blocks: carbon, nitrogen, vitamins, etc. As is the case in many other organisms, some of the amino acids, purines and pyrimidines in food are taken up and used as building blocks directly, while other amino acids are synthesised *de novo* if not available in food. They thus obtain energy from oxidation–reduction reactions, and carbon from organic compounds. This classifies them as chemo-organotrophs (chemo- is opposite to photo-; organo- is opposite to litho-; the latter dichotomy is synonymous with hetero- versus auto-). They frequently use dioxygen as an electron acceptor. As a consequence, they excrete carbon dioxide and nitrogen waste, such as ammonia or urea, see Figure 3.11.

Most plants, in contrast, use light energy, and take carbon dioxide as a carbon source. This classifies them as photolithotrophs. Energy that comes from light is usually stored in polysaccharides and/or lipids, which also serve as carbon reserves. Plants use water, rather than organic compounds, as an electron donor, and, with carbon dioxide as the electron acceptor, dioxygen is produced in the light. Most plants can synthesise all the compounds they need from very simple minerals (nitrate, phosphate, etc.), but some plants also use complex organic compounds (for instance the parasitic plants that lack chlorophyll, or the hemi-parasites that still have chlorophyll). Quite a few species of plants in unrelated families are purely heterotrophic. So plants combine chemo-organotrophic with photolithotrophic properties, which classifies them as mixotrophs.

Bacteria, as a group, use a wide range of metabolic modes, some resembling those of animals or plants. The purple non-sulphur bacteria Rhodospirallacea use light as their energy source, but different kinds of organic compounds as the electron donor and acceptor. This classifies them as photo-organotrophs. Most photo-assimilable organic

Table 3.4 The classification of trophic modes among organisms.

trophy	hetero-	auto-
energy source	chemo	photo
carbon source	organo	litho

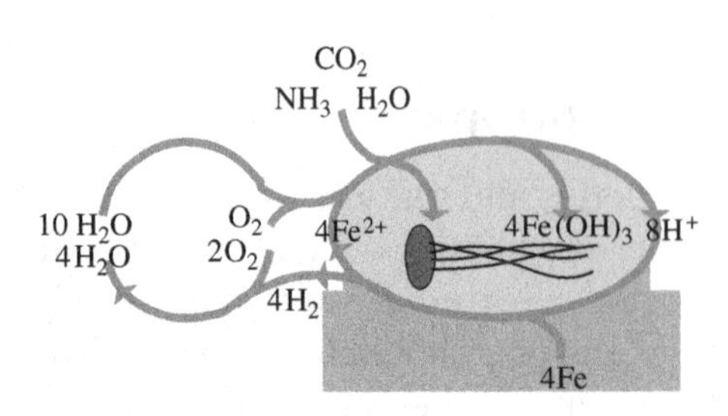

Figure 3.10 The next time you look at your car or bike, you will remember that corrosion is an example of chemolithotrophy. Most corrosion is microbe-mediated and the main culprit is the iron bacterium *Gallionella*; it uses 220 g of iron-II and produces 430 g of rust to make just 1 g of cells from carbon dioxide, water and ammonia [737]. It excretes long strands of rust at one side of the cell.

compounds can also be respired, but benzoate, for instance, can be used in the light, but cannot be respired [1099]. Sulphur bacteria use light as an energy source, carbon dioxide as a carbon source, and H_2S, elemental sulphur or H_2 as an electron donor. Like plants, they classify as photolithotrophs. Most bacteria are chemotrophs, however, which use oxidation–reduction or fermentation reactions to fuel energy-demanding reactions. Figure 3.10 gives an example of chemolithotrophy.

Chemolithotrophy is mostly confined to the bacteria. However, several eukaryotes can respire nitrate non-symbiotically. The ciliate *Loxodes* (Karyorelicta) reduces nitrate to nitrite; the fungi *Fusarium oxysporum* and *Cylindrocarpon tonkinense* reduce nitrate to nitrous oxide; the foraminifera *Globobulimina* and *Nonionella* live in anoxic marine sediments and are able to denitrify nitrate completely to N_2 [975].

Individuals of many phototrophic prokaryotes and protoctists can also activate the chemotrophic mode, depending on the environmental conditions, which somewhat degrades the usefulness of the classification. They are, therefore, mixotrophs. Figure 3.11 illustrates that organisms can differ in their assimilation strategies, but otherwise have substantial similarities in the organisation of metabolism.

Some organisms, like ourselves, rapidly die when dioxygen is not available. Intertidal animals (crustaceans, molluscs), animals in sediments, parasitic animals, yeasts and goldfish can survive its absence for some time, by switching from respiration to fermentation (see {166}), see [168] for a review. (At some stage, all need some dioxygen to synthesise steroids or collagen [343]). Some bacteria do not need dioxygen, but can survive in its presence, but others rapidly die when exposed to dioxygen. This is because dioxygen is rather reactive and can form free radicals in the cell, which are extremely reactive. Organisms can only survive in the presence of dioxygen (aerobic conditions) if they 'catch' these free radicals efficiently with specialised enzymes, called superoxide dismutases (some prokaryotes use high concentrations of Mn^{2+} or other means), to convert the radicals to the highly toxic hydrogen peroxide, and subsequently back to dioxygen, using the enzyme catalase. The handling of dioxygen remains rather tricky, however, and is at the basis of the process of ageing, see {214}.

From a dynamic point of view, it is important to realise that the availability of the various nutrients and light can fluctuate wildly, while autotrophs must couple them to synthesise structural biomass with a constant chemical composition. This requires the installation of reserves, one for each nutrient (mineral) that has to be taken up, with rules for the use of these reserves and their replenishment. This is less necessary for chemo-organotrophs such as animals; an imbalance between the composition of food and their needs to synthesise structural biomass can be modelled realistically, as a first approximation, by a conversion of food into reserves that is not very efficient. The match is perfect for animals that feed on closely related species, and explains why they evolved in many taxa: mammal-eating mammals, starfish-eating starfish, comb-jelly-eating comb jellies, etc.

Animals can buffer varying availabilities of food with a single reserve, because all required nutrients covary, while plants also need auxiliary reserves, because mineral nutrients and light vary independently. Since growth of structural biomass can change, the machinery to synthesise biomass would face very busy and very quiet periods if it were a fixed part of the structural biomass. (The part must be fixed on the basis of the homeostasis assumption.) If the synthesis machinery is part of the reserves, however, the fluctuations in activity would be much less, and the amount of required machinery could be 'chosen' much more economically, see {143}. This is because growth tends to increase with the reserves, as we will see. Auxiliary reserves for a particular nutrient, in contrast, can increase considerably if growth is limited by other nutrients or energy, see {197}. This is how large (auxiliary) reserves can accompany low growth rates. The homeostasis assumption also applies to each auxiliary reserve. Homeostasis for the organism as a whole decreases with an increasing number of reserves, and the composition of the body increases in flexibility.

3.8.2 Central metabolism

Central metabolism is the core of metabolism that deals with energy extraction from glucose, and the formation of building blocks for other main compounds, such as lipids, animo acids and RNA. Its evolution is discussed at {388}.

The central metabolic pathway of many prokaryotes and almost all eukaryotes (Figure 3.11) consists of four main modules [657].

The pentose phosphate (PP) cycle comprises a series of extra-mitochondrial transformations by which glucose-6-phosphate is oxidised with the formation of carbon dioxide, reduced NADP and ribulose 5-phosphate. Some of this latter compound is subsequently transformed to sugar phosphates with 3 to 7 or 8 carbon atoms, whereby glucose-6-phosphate is regenerated. Some ribulose 5-phosphate is also used in the synthesis of nucleotides and amino acids. Higher plants can also use the same enzymes in reverse, thus running the reductive pentose phosphate cycle. The PP cycle is primarily used to inter-convert sugars as a source of precursor metabolites and to produce reductive power. Theoretical combinatorial optimisation analysis has indicated that the number of steps in the PP cycle is evolutionarily minimised [781, 780], which maximises the flux capacity [488, 1206].

The glycolytic pathway converts glucose-6-phosphate (aerobically) to pyruvate or (anaerobically) to lactate, ethanol or glycerol, with the formation of 2 ATP and 2 NADH. The transformations occur extra-mitochondrially in the free cytoplasm. However, in kinetoplastids they are localised in an organelle, the glycosome, which is probably homologous to the peroxisome of other organisms [51, 194]. The flux through this pathway is controlled by phospho fructokinase and by hormones.

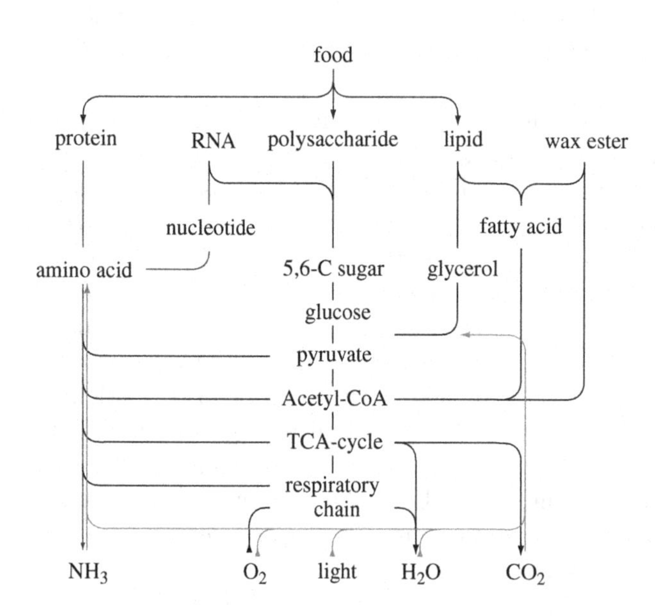

Figure 3.11 A simplified map of metabolism. The second line gives the main polymers that are used as reserves, below that are the monomers that play an active role in metabolism. The bottom line gives the main end products and external sources. Dioxygen is used as an electron acceptor by the respiratory chain, but sometimes other electron acceptors are used. Most pathways are reversible, although different sets of enzymes are usually involved. Most animals can synthesise lipids from polysaccharides, but not vice versa. Heterotrophs use food to supply the reserve polymers, autotrophs use light and minerals to synthesise sugar and animo acids (in grey), mixotrophs do both. TCA, tricarboxylic acid.

Heinrich and Schuster [488] studied some design aspects of the glycolytic pathway. Most pyruvate is converted to acetyl and bound to coenzyme A.

The tricarboxylic acid (TCA) cycle, also known as the citric acid or the Krebs cycle, oxidises (without the use of dioxygen) the acetyl group of acetyl coenzyme A to two carbon dioxide molecules, under the reduction of four molecules of NAD(P) to NAD(P)H. In eukaryotes that contain them, these transformations occur within their mitochondria. Some plants and micro-organisms have a variant of the TCA cycle, the glyoxylate cycle, which converts pyruvate to glyoxylate and to malate (hence a carbohydrate) with another pyruvate. Since pyruvate can also be obtained from fatty acids, this route is used for converting fatty acids originating from lipids into carbohydrates. Some plants possess the enzymes of the glyoxylate cycle in specialised organelles, the glyoxysomes.

The respiratory chain oxidises the reduced coenzyme NAD(P)H, and succinate with dioxygen, which leads to ATP formation through oxidative phosphorylation. Similarly to the TCA cycle it occurs inside mitochondria. Amitochondriate eukaryotes process pyruvate through pyruvate-ferredoxin oxidoreductase, rather than through the pyruvate dehydrogenase complex. If the species can live anaerobically, the respiratory chain can use fumarate, nitrate or nitrite as electron acceptors in the absence of dioxygen [1162].

The glycolysis, TCA cycle and respiratory chain in series convert

$$\text{anaerobic:glucose} + 2\,\text{ADP} + 2\,\text{P} \rightarrow 2\,\text{ethanol} + 2\,\text{CO}_2 + 2\,\text{ATP} + 2\,\text{H}_2\text{O}$$
$$\text{aerobic:glucose} + 6\,\text{O}_2 + 30\,\text{ADP} + 30\,\text{P} \rightarrow 6\,\text{CO}_2 + 30\,\text{ATP} + 36\,\text{H}_2\text{O}$$

Many intermediary metabolites escape further conversion, however. In combination with nutrients (phosphates, sulphates, ammonia, iron oxides, etc.), the first three pathways of the central metabolic pathway provide almost all the essential cellular building blocks, including proteins, lipids and RNA. The universality of this central metabolic pathway is partly superficial or, if you like, the result of convergent evolution because the enzymes running it can differ substantially. This diversity in enzymes partly results from the modular make-up of the enzymes themselves. Some variation occurs in the intermediary metabolites as well.

Obviously, glucose plays a pivotal role in the central metabolism. However, its accumulation as a monomer for providing a metabolism with a permanent source of substrate would give all sorts of problems, such as osmotic ones. This also applies to metabolic products. To solve these problems, cells typically store the supplies in polymeric form (polyglucose (i.e. glycogen), starch, polyhydroxyalkanoate, polyphosphate, sulphur, proteins, RNA), which are osmotically neutral. Their storage involves so-called inclusion bodies, the inherent solid/liquid interface of which controlls their utilisation dynamics (see reserve dynamics at {41}).

3.9 Summary

The quantification of the amount and composition of biomass is discussed, including the energy and entropy aspects. Classes of organic and inorganic compounds are briefly introduced for later use and conversions between volumes, masses and energies are discussed.

Macrochemical reaction equations are presented and applied in the dynamics of isotopes. This dynamics has mixing and fractionation aspects. The dynamics of fractionation from pools and from fluxes differ. Selection from pools is on the basis of differential velocities of molecules; anabolic versus catabolic fluxes select on the basis of differential strength of bounds.

DEB theory makes frequent use of Synthesising Units (SUs), generalised enzymes that follow the rules of classic enzyme kinetics with two modifications: their dynamics is specified in terms of substrate fluxes, rather than substrate concentrations, and the backward fluxes are taken to be negligibly small as a result of the spatial organisation of transport processes. Processing can be classified into sequential and parallel and compounds into complementary and substitutable. Mixtures of these basic types are possible, and variations are discussed to deal with e.g. co-metabolism and inhibition. Demand processes can be modelled with SUs by letting dissociation rates depend on the relative frequency of SUs in the various binding states. As a result of working with fluxes, SUs dynamics can deal with spatial structure, with flexibility for how the rates depend on the number of copies. SUs can interact on the basis of handshaking protocols.

The trophic modes auto-, hetero- and mixotrophy are described and the four modules of central metabolism are summarised.

4

Univariate DEB models

This chapter discusses the fluxes of compounds in univariate DEB models (one type of substrate, one reserve and one structure). Univariate DEB models follow directly from the assumptions of Table 2.4 for the standard DEB model, but the assumption of isomorphy is no longer used.

Figure 4.16 shows the example of *Klebsiella* which lives on glycerol. It must have many, rather than a single reserve. Multiple reserve systems, which are discussed in the next chapter, can behave as single reserve systems in the context of DEB theory, if growth is limited by a single nutrient and all rejected reserve fluxes are excreted.

I start with a more detailed discussion of phenomena at varying food densities, followed by effects of changes in shape during growth. The rest of the chapter discusses mass and energy aspects that are implied by the assumptions of Table 2.4 and show, for instance, why the fluxes of essential compounds, as well as the dissipating heat, are weighted sums of the three basic powers assimilation, dissipation and growth. Therefore, dissipating heat can also be written as a weighted sum of three mineral fluxes: carbon dioxide, dioxygen and nitrogenous waste. This relationship is the basis of the method of indirect calorimetry. After half a century of wide application, this empirical method has finally been underpinned theoretically. Simple extensions of univariate DEB models can deal with drinking by terrestrial organisms.

4.1 Changing feeding conditions

Food density, as experienced by an individual, is never really constant and feeding frequently takes the form of meals. The next subsections analyse phenomena of changing food availability at an increasing timescale. Adaptations to seasonal variations in food availability are further discussed at {296}.

4.1.1 Scatter structure of weight data

For simplicity's sake, the processes of feeding and growth have been modelled deterministically, so far. This is not very realistic, as (feeding) behaviour especially is notoriously erratic. This subsection discusses growth if feeding follows a special type of random process, known as an alternating Poisson process or a random telegraph process. Because of the resulting complexity, I rely here on computer simulation studies.

Suppose that feeding occurs in meals that last an exponentially distributed time interval $\underline{t}_1$ with parameter $\dot{\lambda}_1$, so $\Pr\{\underline{t}_1 > t\} = \exp(-t\dot{\lambda}_1)$. The mean length of a meal is then $\dot{\lambda}_1^{-1}$. The time intervals of fasting between the meals are also exponentially distributed, but with parameter $\dot{\lambda}_0$. Food intake during a meal is copious, so the scaled functional response switches back and forth between $f = 1$ and $f = 0$. The mean value for f is $\mathcal{E}\underline{f} = \dot{\lambda}_0(\dot{\lambda}_0 + \dot{\lambda}_1)^{-1}$. This on/off process is usually smoothed out by the digestive system, but let us here assume that this is of minor importance. In the change of scaled length and scaled reserve density in (2.22), time can be scaled out as well, using $\tau = t\dot{k}_M$ and a single parameter, g, is involved in this growth process, while two others, λ_0 and λ_1, occur in the description of the on/off process of f. (Note that the λs do not have dots, because scaled time is dimensionless.) The process is initiated with $l(0) = l_b$ and $e(0)$ equals the scaled energy density of a randomly chosen adult.

Figure 4.1 shows the results of a computer simulation study, where scaled weight relates to scaled length and scaled energy density, according to (3.2) as

$$W_w(d_V V_m)^{-1} = (1 + e w_E[M_{Em}]/d_V) l^3 \tag{4.1}$$

The resemblance of the scatter structure with experimental data is striking, see for instance Figure 3.2. This does not imply, however, that the feeding process is the

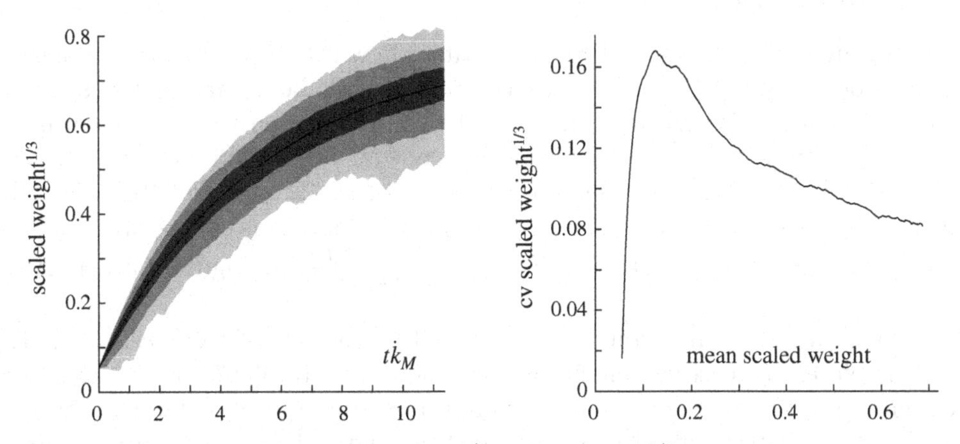

Figure 4.1 Computer-simulated scaled weight$^{1/3}$, $(W_w/d_V V_m)^{1/3}$, is plotted against scaled time in the left figure, if feeding follows an alternating Poisson process. The shade areas give frequency intervals of 99%, 90% and 50%, the drawn curve gives the mean and the dotted one gives the deterministic growth curve, if feeding is constant at the same mean level. The coefficient of variation is given in the right figure. The parameters are $\lambda_0 = 11.666$, $\lambda_1 = 5$, $g = 1$, $l_b = 0.05$ and $w_E[M_{Em}]/d_V = 0.5$. The small difference between the mean and deterministic curves relates to the step size of the numerical integration (Mrs F. D. L. Kelpin, pers. comm.).

only source of scatter. Differences of parameter values between individuals are usually important as well. The results do suggest a mechanism behind the generally observed phenomenon that scatter in weights increases with the mean.

4.1.2 Step up/down in food availability

The difference between age-based and size-based models becomes apparent in situations of changing food densities. As long as food density remains constant, size-based models can always be converted into age-based ones, which makes it impossible to tell the difference.

Figure 4.2 shows the result of an experiment with *Daphnia magna* at 20 °C, exposed to constant high food densities with a single instantaneous switch to a lower food density at 1, 2 or 3 weeks. The reverse experiment with a single switch from low to high food densities has also been done, together with continuous exposure to both food densities. Figure 2.10 has already shown that the maintenance rate coefficient $\dot{k}_M$ and energy conductance $\dot{v}$ can be obtained by comparing growth at different constant food densities. These compound parameters, together with ultimate and maximum lengths and the common length at birth, have been obtained from the present experiment without a switch. These five parameters completely determine growth with a switch, both up and down, leaving no free parameters to fit in this situation. The excellent fit strongly supports the DEB theory.

4.1.3 Mild starvation

If a growing individual is starved for some time, it will (like the embryo) continue to grow (at a decreasing rate) till it hits the non-growth boundary of the state space ($e = l$). Equation (2.37) describes the e, l-path. Depending on the amount of reserves, the change in volume will be small for animals not far from maximum size. Strömgren and Cary [1126] found that mussels in the range of 12–22 mm grew 0.75 mm. If the change in size is neglected, the scaled reserve density changes as $e(\tau) = e(0)\exp(-g\tau/l)$ and the growth of scaled length is $\frac{d}{d\tau}l = \frac{g}{3}\frac{\exp(-g\tau/l) - l/e(0)}{\exp(-g\tau/l) + g/e(0)}$. Figure 4.3 confirms this prediction.

Respiration during starvation is proportional to the use of reserves; see {147}. It should, therefore, decrease exponentially in time at a rate of $\dot{v}/L$ if size changes can be neglected; see (2.10). Figure 4.4 confirms this prediction for a daphnid. If a shape coefficient of $\delta_{\mathcal{M}} = 0.6$ is used to transform the length of *D. pulex* into a volumetric one, the energy conductance becomes $\dot{v} = 0.6 \times 1.62 \times 0.23 = 0.22\,\mathrm{mm\,d^{-1}}$. This value seems to be somewhat small in comparison with that for *D. magna*, cf. Figure 2.10, and the mean energy conductance of many species, see {312}. The next section suggests an explanation in terms of changes in allocation rules to reproduction during starvation.

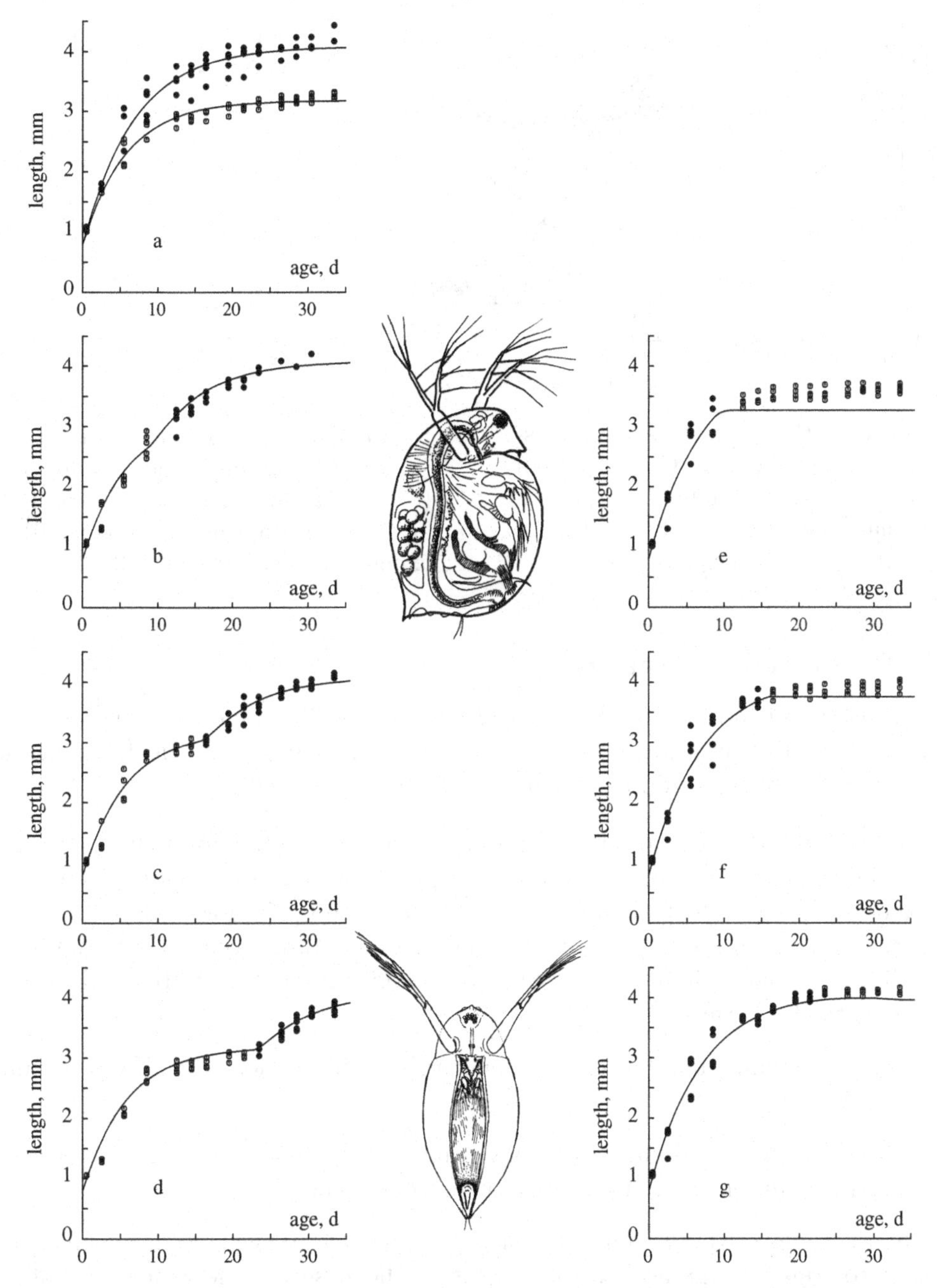

Figure 4.2 Length-at-age for the waterflea *Daphnia magna* at 20 °C feeding at a high (•) and a low (∘) constant density of the green alga *Chlorella pyrenoidosa* (a), and with a single interchange of these two densities at 1 (b,e), 2 (c,f) or 3 (d,g) weeks. The curves b to g describe the slow adaptation to the new feeding regime. They are completely based on the five parameters obtained from (a), so no additional parameters were estimated. From [634].

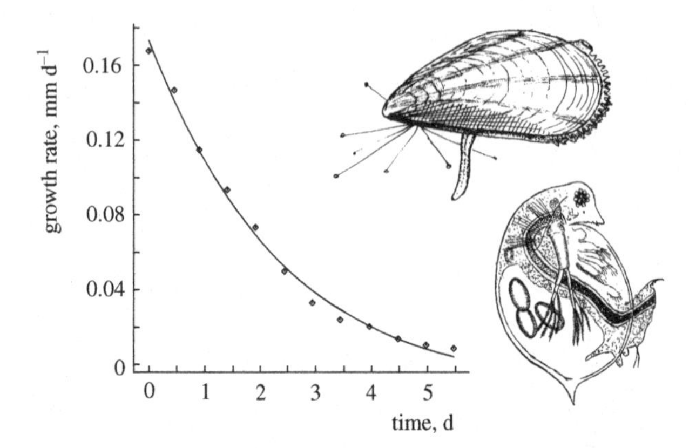

Figure 4.3 Growth rate in the starved mussel *Mytilus edulis* at 21.8 °C. Data from Strömgren and Cary [1126]. The parameter estimates are $\frac{g}{e(0)} = 12.59$, $\dot{k}_M = 2.36\,10^{-3}\,\mathrm{d}^{-1}$ and $\dot{v} = 0.252\,\mathrm{cm\,d}^{-1}$.

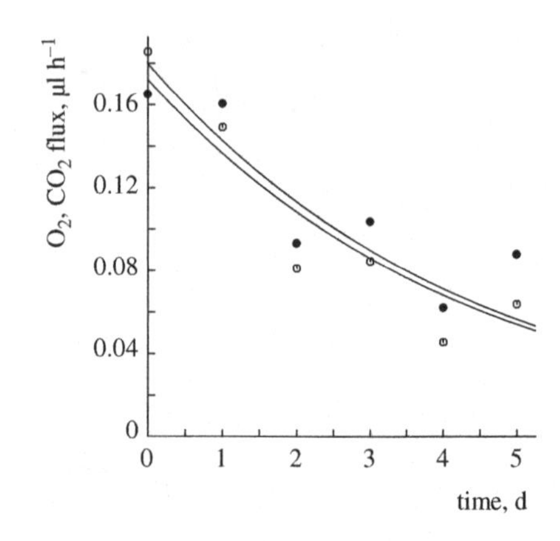

Figure 4.4 The dioxygen consumption rate (•) and the carbon dioxide production rate (◦) in starved *Daphnia pulex* of 1.62 mm at 20 °C. Data from Richman [961]. The exponential decay rate is $0.23\,\mathrm{d}^{-1}$.

4.1.4 Prolonged starvation

As long as growth is non-negative, i.e. $e \leq l + l_T$, see (2.22), standard dynamics applies. If (full or partial) starvation continues, the response can be at the following levels depending on the species and environmental factors:

1 Continue the standard reserve dynamics till death follows; don't change the κ-rule for allocation; use the buffer for reproduction (few data are available to tell us how exactly, but see [309] for studies on polychaetes and [876] for studies on anchovy); if necessary shrink (i.e. pay somatic maintenance from structure). Variant 1: partially reduce somatic maintenance costs. Variant 2: migrate to better locations or switch to the dormant state.

2 Like the previous rule, but change the κ-rule for allocation when the reproduction buffer is empty.

3 Change the reserve dynamics to pay somatic maintenance only; no allocation to maturity maintenance, maturation or reproduction.

4 Change the reserve dynamics by converting reserve to eggs (seeds); convert structure to eggs (as far as possible). This is the case of emergency reproduction, typically followed by death. It is a popular strategy among plants.

Sometimes systems start to respond at level 1, but then continue to levels 2, 3 and 4.

Pond snails seem to continue energy allocation to reproduction during prolonged starvation under a light : dark 16 : 8 cycle (summer conditions, denoted by LD, long day), but they cease reproduction under a 12 : 12 cycle (spring/autumn conditions, denoted by MD, mid-day) [124, 1294]. This makes sense because under summer conditions, an individual can expect high primary production, so, if it has consumed a plant, it will probably find another one in the direct neighbourhood. Under spring/autumn conditions, however, it can expect a long starvation period. By ceasing allocation to reproduction, it can increase its survival period by a factor two; see Figure 4.6. Another aspect is that offspring have a remote survival probability if there is no food around. They are more vulnerable than the parent, as follows from energy reserve dynamics. These dynamics can be followed on the basis of the assumption that LD snails do not change the rule for utilisation of energy from the reserves, and neither MD nor LD snails cut somatic maintenance.

This example shows that the diurnal cycle also affects the allocation under non-starvation conditions in the pond snail *Lymnaea stagnalis*. This is obvious from the ultimate length. Snails kept under a 12 : 12 cycle (MD conditions) have a larger ultimate length than under a 16 : 8 cycle (LD conditions) [1294]. MD snails also have a smaller von Bertalanffy growth rate and a smaller volume at puberty, but MD and LD snails are found to have the same energy conductance of $\dot{v} \simeq 1.55$ mm d^{-1} at 20 °C. This is a strong indication that the photoperiod only affects the partition coefficient κ.

If starvation is complete and volume does not change, i.e. $f = 0$ and l is constant, the energy reserves will be $e(t) = e(0)\exp(-g\dot{k}_M t/l)$; see (2.11). Dry weight is a weighted sum of volume and energy reserves, so according to (3.3) for LD snails we must have

$$W_d(l,t) = V_m l^3 (d_{Vd} + w_{Ed}[M_{Em}]e(0)\exp(-g\dot{k}_M t/l)) \tag{4.2}$$

if the buffer of energy allocated to reproduction is emptied frequently enough (E_R small). For MD snails, where $e(t) = e(0) - ([\dot{p}_M]/[E_m])t$, dry weight becomes

$$W_d(l,t) = V_m l^3 (d_{Vd} + w_{Ed}[M_{Em}](e(0) - t[\dot{p}_M]/[E_m])) \tag{4.3}$$

So the dry weight of LD snails decreases exponentially and that of MD snails linearly. Figure 4.5 confirms this. It also supports the length dependence of the exponent.

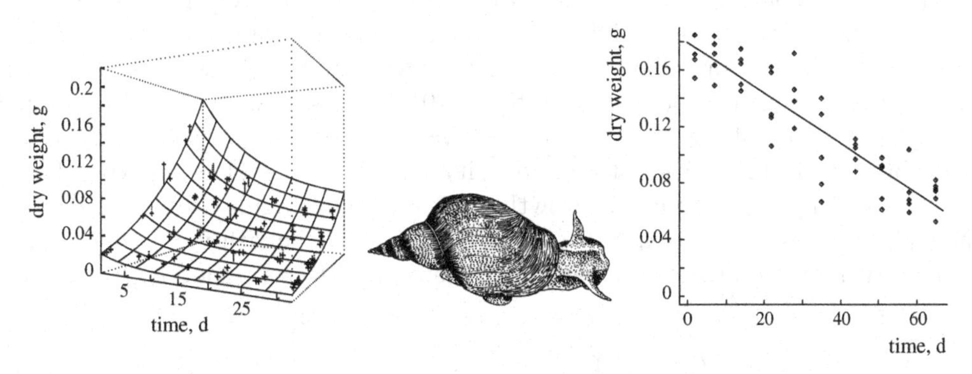

Figure 4.5 Dry weight during starvation of long-day (LD, left) and mid-day (MD, right) pond snails *Lymnaea stagnalis* at 20 °C. The left figure gives dry weights (z-axis) as a function of starvation time (x-axis) and length (y-axis: 1.6–3.3 cm). In the right figure, the length of the MD pond snails was 3 cm. From Zonneveld and Kooijman [1294]. The surface and curve are fitted DEB-based expectations.

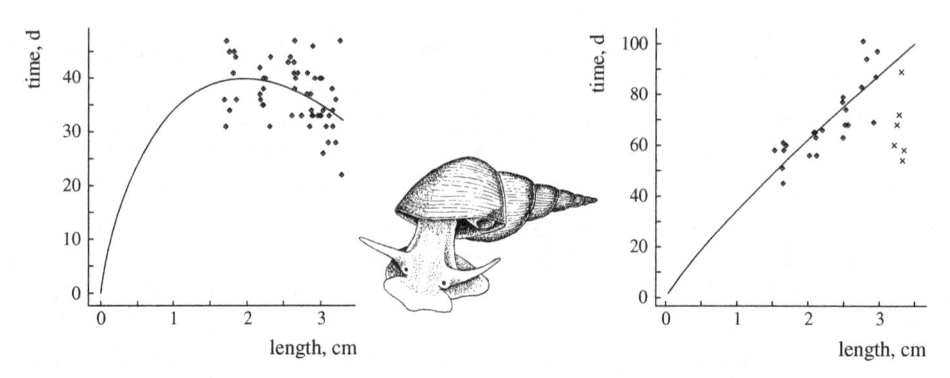

Figure 4.6 Survival time during starvation of LD (left) and MD (right) pond snails as a function of length. From Zonneveld and Kooijman [1294]. The data points $\times$ in the right figure are not included in the DEB-based fit. These large individuals had deformations of the shell.

If we exclude the possibility of prolonging life through decomposition of structural body mass, and if death strikes when the utilisation rate drops below the maintenance level, the time till death by starvation can be evaluated.

In animals such as LD snails, which do not change storage dynamics, the utilisation rate, $-\frac{d}{dt}[E]$, equals the maintenance rate, $[\dot{p}_M]$, for $[E]/[E_m] = V^{1/3}[\dot{p}_M]/\{\dot{p}_{Am}\}$ or $e = \kappa l$. Since $e(t) = e(0)\exp(-\dot{k}_M tg/l)$, death strikes at $t_\dagger = \frac{l}{\dot{k}_M g} \ln \frac{e(0)}{\kappa l}$. This only holds if the length increase is negligibly small.

In animals such as MD snails, which change storage dynamics to $\frac{d}{dt}e = -[\dot{p}_M]/[E_m]$ or $e(t) = e(0) - t[\dot{p}_M]/[E_m]$, death strikes when $e = 0$, that is at $t_\dagger = e(0)[E_m]/[\dot{p}_M] = \frac{e(0)}{\kappa \dot{k}_M g}$. This only holds as long as there is no growth, so $e(0) < l$. In practice, this is a more stringent condition than the previous one. The first part of the starvation period usually includes a period where growth continues, because $e > l$. This complicates the analysis of starvation data, as illustrated in the following example. In a starvation experiment with MD snails, individuals were taken from a standardised culture and initially fed *ad libitum* for 4 days prior to complete starvation. If we assume that food density in the culture has been constant, so $e(0) = f_c$, say, with f_c being about 0.7, and $f = 1$ during the 4 days prior to the starvation experiment, the change in length is negligibly small. The initial storage density is $e(0) = 1 - (1 - f_c)\exp(-4\dot{k}_M g/l)$, according to (2.10). The time till growth ceases is found again from (2.10) and the boundary condition $l = e(0)\exp(-t\dot{k}_M g/l)$. (Although the length increase is negligibly small, energy allocation to growth can be substantial.) After a period $l(\kappa \dot{k}_M g)^{-1}$ death will strike, so

$$t_\dagger = \frac{l}{\dot{k}_M g}\left(\frac{1}{\kappa} + \ln\left\{l^{-1}\left(1 - (1 - f_c)\exp(-4\dot{k}_M g/l)\right)\right\}\right) \tag{4.4}$$

Figure 4.6 confirms model predictions for the way survival time depends on length in LD and MD snails, and shows that MD snails can prolong life by a factor of two by not reproducing during starvation. In contrast to the situation concerning embryonic growth, this confirmation gives little support to the theory, because the shape of the survival time–length curve is very flexible for the LD case, although there are only two free parameters. The upper size class of the MD snails has been left out of the model fit, because the shape of their shell suggested a high age, which probably affected energy dynamics.

When storage levels become too low for maintenance, some species can decompose their structural biomass to some extent. If feeding conditions then become less adverse, recovery may be only partial. The distinction between structural biomass and energy reserves fades at extreme starvation. The priority of storage materials over structural biomass is perhaps even less strict in species that shrink during starvation. Species with (permanent or non-permanent) exoskeletons usually do not shrink in physical dimensions, but the volume-specific energy content nonetheless decreases during starvation.

If an individual ceases reproduction during starvation, any consistent specification of $\frac{d}{dt}e$ must be continuous in f, e and l. One possibility is by first obeying maturity maintenance requirements, then switching on reproduction gradually if food intake increases from a low level.

4.1.5 Shrinking and the turnover of structure

Many species can, to some extent, shrink in structural mass during starvation, as a way to pay their somatic maintenance costs. Even animals with a skeleton, such as shrews of the genus *Sorex*, can exhibit a geographically varying winter size depression, known as the Dehnel phenomenon [396]. Molluscs seem be to able to reduce shell size [290].

The turnover of structure as part of the somatic maintenance process directly relates to the quantification of shrinking during starvation. This turnover implies a mobilisation of structure at a fixed specific rate, say $\dot{J}_{VC} = \dot{j}_{VC} M_V = [\dot{J}_{VC}]V$ and I suppose that this flux is large enough to pay somatic maintenance costs if necessary. Normally this mobilised flux equals the synthesised flux, as part of the turnover process, but not during shrinking. Somatic maintenance is normally paid from reserve at rate $\dot{J}_{ES} = [\dot{J}_{EM}]L^3 + \{\dot{J}_{ET}\}L^2 = (\dot{j}_{EM} + \{\dot{J}_{ET}\}M_V^{-1/3}[M_V]^{-2/3})M_V$, and if it is fully paid from structure the costs are $\dot{J}_{VS} = [\dot{J}_{VM}]L^3 + \{\dot{J}_{VT}\}L^2 = (\dot{j}_{VM} + \{\dot{J}_{VT}\}M_V^{-1/3}[M_V]^{-2/3})M_V$. The surface-linked component is only paid after birth, so a worst-case scenario gives $\dot{j}_{VC} \geq \dot{j}_{VM} + \{\dot{J}_{VT}\}(M_V^b)^{-1/3}[M_V]^{-2/3}$.

When shrinking becomes opportune, the somatic maintenance SUs receive a reserve flux $\dot{J}_{EC}$ and a structure flux $\dot{J}_{VC}$; reserve and structure are substitutable compounds for somatic maintenance, with a strong preference for reserve. The demand version of the preference case for SUs, {107}, specifies the somatic maintenance fluxes for reserve and structure

$$\dot{j}_E^S = \dot{j}_{ES}(1 - \dot{j}_V^S/\dot{j}_{VS}); \quad \dot{j}_V^S = \dot{j}_{VS}\frac{2A}{2A + \sqrt{B^2 - 4AC} - B} \tag{4.5}$$

with $A = \rho_V j_{VC} j_{ES}^2 / j_{VS}$, $B = C + ((1 - \rho_V) j_{EC} + y_{EV} j_{VC}) j_{ES}$, $C = -j_{EC}(j_{EC} + y_{EV} j_{VC})$. So shrinking occurs at rate $\frac{d}{dt} M_V = -j_V^S M_V$. Given appropriate parameter values shrinking hardly occurs if maintenance can be paid from reserve.

Shrinking thus comes with four extra parameters: j_{VC}, $[\dot{J}_{VM}]$, $\{\dot{J}_{VT}\}$ and preference parameter ρ_V. The relationships $[\dot{J}_{VM}]/[\dot{J}_{EM}] = \{\dot{J}_{VT}\}/\{\dot{J}_{ET}\} \geq y_{VE}$ seem reasonable. The latter inequality is based on thermodynamic considerations, which imply losses for each transformation; payment via structure involves an extra transformation, so extra losses. The equality sign can considered to be the thermodynamic edge for shrinking. Close to this edge, two shrinking parameters are left. For absolute priority for reserve as substrate for somatic maintenance, $\rho_V = 0$, we have

$$j_E^S = \min\{j_{ES}, j_{EC}\} \quad \text{and} \quad j_V^S = j_{VS}(1 - j_E^S / j_{ES}) \tag{4.6}$$

so that j_{VC} is no longer relevant for the mass dynamics; it still is for the dynamics of isotopes. I will call this the switch special case for shrinking.

If the turnover rates of compounds in the structure differ between compounds, shrinking gives deviations from homeostasis with this mechanism. Compounds with no turnover can be treated as products and formally excluded from structure to preserve strong homeostasis for structure.

4.1.6 Migration

Seasonal forcing of food availability induces quite a few animal species (vertebrates, butterflies) to migrate. Long-distance migration requires quite a bit of physiological preparation. Many bird species manage to increase the size of their guts temporarily, fatten up, reduce the size of their guts and increase the size of their muscles before they go. All these changes occur in a matter of days [835]. Such hormonally controlled adaptations of organ size can only be captured by multiple structure versions of DEB models, see {202}.

4.1.7 Dormancy

Some species manage to escape adverse feeding conditions (and/or extreme temperature or drought) by switching to a torpor state in which growth and reproduction cease, while maintenance (and heating) costs greatly diminish. The finding that metabolic rate in homeotherms is proportional to body weight during hibernation [585] suggests that maintenance costs are reduced by a fixed proportion.

As heating is costly, a reduction in the body temperature of endotherms saves a lot of energy. Bats and hummingbirds lower their body temperature in a daily cycle. This probably relates to the relatively long life span of bats (for their size) [350]. Although most bird embryos have a narrow temperature tolerance range, swifts survive significant cooling. This relates to the food-gathering behaviour of the parents. Dutch swifts are known to collect mosquitoes above Paris at a distance of 500 km, if necessary. During hibernation, not only is the body temperature lowered, but other maintenance costs are reduced as well.

Hochachka and Guppy [513] found that the African lungfish *Protopterus* and the South American lungfish *Lepidosiren* reduce maintenance costs during torpor in the dry season, by removing ion channels from the membranes. This saves energy expenses for maintaining concentration gradients over membranes, which proves to be a significant part of the routine metabolic costs. This metabolic arrest also halts ageing. The life span of lungfish living permanently submerged, so always active, equals the cumulative submerged periods for lungfish that are regularly subjected to desiccation. This is consistent with the DEB interpretation of ageing, see {214}.

If maintenance cannot be reduced completely in a torpor state, it is essential that some reserves are present, {90}. This partly explains why individuals frequently survive adverse conditions as freshly laid eggs, because the infinitesimally small embryo requires little maintenance; it only has to delay development. The start of the pupal stage in holo-metabolic insects is also very suitable for inserting a diapause in order to survive adverse conditions, {284}.

4.1.8 Emergency reproduction

The determination of sex in some species is coupled to dormancy in a way that can be understood in the context of the DEB model. Daphnids use special winter eggs, packed in an ephippium. The diploid female daphnids usually develop diploid eggs that hatch into new diploid females. If food densities rapidly switch from a high level to a low one and the energy reserves are initially high, the eggs hatch into diploid males, which fertilise females that now produce haploid eggs [1073]. After fertilisation, the 'winter eggs' or resting eggs develop into new diploid females. The energy reserves of a well-fed starving female are just sufficient to produce males, to wait for their maturity and to produce winter eggs.

The trigger for male/winter egg development is not food density itself, but a change of food density. If food density drops gradually, females do not switch to the sexual cycle [633], see Figure 9.13. Sex determination in species such as daphnids is controlled by environmental factors, so that both sexes are genetically identical [172, 480]. Mrs D. van Drongelen and Mrs J. Kaufmann informed me that a randomly assembled cohort of neonates from a batch moved to one room proved to consist almost exclusively of males after some days of growth, while in another cohort from the same batch moved to a different room all individuals developed into females as usual. This implies that sex determination in *Daphnia magna*, and probably in all other daphnids and most rotifers as well, can be affected even after hatching. More observations are needed. Male production does not seem to be a strict prerequisite for winter egg production [604]. Kleiven, Larsson and Hobæk [604] found that crowding and shortening of day length also affect male production in combination with a decrease in food availability at low food densities. The females that hatch from winter eggs grow faster, mature earlier and reproduce at a higher rate than those from subitaneous eggs [32]; the size at maturation and the ultimate body size are also larger for the exephippial generation. The physiological nature of these interesting differences is still unknown.

The switch to sexual reproduction as a reaction to adverse feeding conditions frequently occurs in unrelated species, such as slime moulds, myxobacteria, oligochaetes (*Nais*) and plants. The difference between emergency and suicide reproduction, see {289}, is that the individual can still switch back to standard behaviour if the conditions improve.

4.2 Changing shapes

The structural volume is of interest because of maintenance processes, and surface area for acquisition processes; this gives a focus on the scaling between volumes and surface areas that are involved in uptake. I will argue that not only the shape itself matters, but also the local environment that affects uptake.

The fact that wing development, for instance, is delayed in birds is of little relevance to whole body growth. Some species such as echinoderms, molluscs and some insects change shape over different life-stages. Plants are extreme in this, and environmental factors contribute substantially to changes in shape. Some of these changes do not cause problems because food intake is sometimes restricted to one stage only. If the shape changes considerably during development, and if volume has been chosen as the basis for size comparisons, the processes related to surface area should be corrected for these changes in shape.

Surface areas are only proportional to volume to the power 2/3 for isomorphs. If organisms change in shape during growth, surface areas relate to volume in different ways, which can be captured by the dimensionless shape correction function $\mathcal{M}(V)$, which stands for the actual surface area relative to the isomorphic one for a body with volume V, where a particular shape has been chosen as the reference. So $\mathcal{M}(V) = 1$ for an isomorph. The derivation of this function will be illustrated for what I call V0- and V1-morphs: idealised morphs that change in shape during growth in a particular way. Many organisms approach these idealised changes quite accurately, others can be conceived as static or dynamic mixtures of two or more of these idealised growing morphs, as will be shown.

4.2.1 V0-morphs

The surface area of a V0-morph is, per definition, proportional to volume^0, so it remains constant. Only the surface area matters that is involved in the uptake process. A biofilm on a plane, diatoms and dinoflagellates are examples, see Figure 4.8. The outer dimensions do not increase during the synthesis of cytoplasm. The vacuoles shrink during growth of the cell, and should be excluded from the structural volume that requires maintenance costs. The surface area of a V0-morph is A_d, say. An isomorph has surface area $A_d(V/V_d)^{2/3}$. The value V_d is a reference that is required to compare both types of morphs; at this volume they have the same surface area. The shape correction function for a V0-morph is

$$\mathcal{M}(V) = (V/V_d)^{-2/3} \tag{4.7}$$

In the section on diffusion limitation on {266}, I discuss situations where the outer boundary of the stagnant water mantle around a small organism restricts uptake. If the mantle is thick enough, the uptake will resemble that of a V0-morph, whatever the actual changes in shape of the organism.

The ingestion rate, storage dynamics and growth for V0-morphs can be found from that of isomorphs by multiplying $\{\dot{p}_{Am}\}$ and $\dot{v}$ with the shape correction function $\mathcal{M}(V)$ in (4.7). This results in

$$\dot{J}_{XA} = \{\dot{J}_{XAm}\} V_d^{2/3} f \tag{4.8}$$

$$\frac{d}{dt} e = (f - e)\dot{v} V_d^{2/3}/V \tag{4.9}$$

$$\frac{d}{dt} V = \frac{\dot{v}}{e+g}\left(e V_d^{2/3} - V V_m^{-1/3}\right) \tag{4.10}$$

where V_d is the volume at division and V_m is defined by $V_m^{1/3} = L_m = \frac{\dot{v}}{g \dot{k}_M}$.

Figure 4.7 shows that, in the situation of feeding laboratory male Sprague–Dawley rats, *Rattus norvegicus*, a fixed amount of food each day, irrespective of their size, isomorphs can numerically behave similar to V0-morphs, although reserve and structure change differently; the data points almost hide the curves. Possibly due to social interaction, rats in a group in the laboratory eat all there is [693]; this implies hyperphagia in the *ad libitum* cohort {262}. The details of food supply matter; fish in a tank that daily receive a fixed number of food particles rapidly develop substantial size differences as a result of social interaction [646].

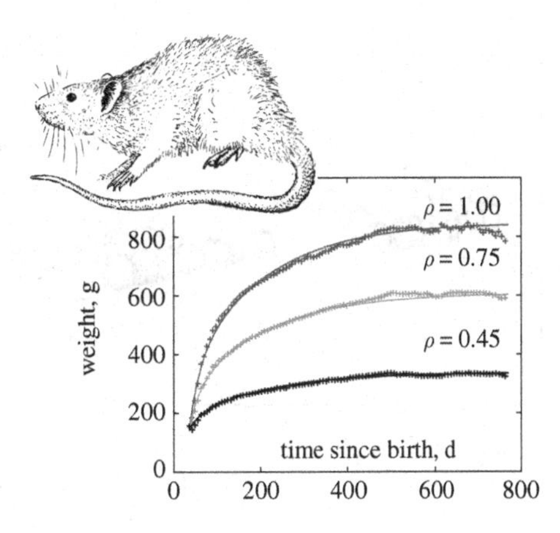

Figure 4.7 Male Sprague–Dawley rats eat at constant rate, irrespective of their size, in a particular laboratory situation. Their specific growth rate is given by (2.21) and the scaled reserve density changes as $\frac{d}{dt}e = \frac{\dot{v}}{L}\left(\rho \frac{(L_m - L_T)^2}{L^2} - e\right)$. Body weight is given by (3.2) with $d_V = 1\,\text{g}\,\text{cm}^{-3}$ and $\omega_w = 0.94$. Data from Hubert *et al.* [534], who used feeding levels $\rho = 1, 0.75$ and 0.45. Parameter values: $e(0) = 1$ (fixed), $V(0) = 73.7\,\text{cm}^3$, $L_T = 0\,\text{cm}$ (fixed), $\dot{k}_M = 0.0059\,\text{d}^{-1}$, $\dot{v} = 0.317\,\text{cm}\,\text{d}^{-1}$, $g = 7.1$.

4.2.2 V1-morphs

The surface area of a V1-morph is, per definition, proportional to volume[1]. It (usually) grows in one dimension only, and it is possible to orient the body such that the direction of growth is along the x-axis, while no growth occurs along the y- and z-axes. The different body sizes can be obtained by multiplying the x-values by some scalar l. An example of a V1-morph is the filamentous hyphae of a fungus with variable length, and thus variable volume V, but a fixed diameter, see Figure 4.8. Its surface area equals $A(V) = A_d V/V_d$, where A_d denotes the surface area at $V = V_d$. The surface area of an isomorph equals $A(V) = A_d(V/V_d)^{2/3}$. So the shape correction function for V1-morphs becomes

$$\mathcal{M}(V) = \frac{A_d V/V_d}{A_d(V/V_d)^{2/3}} = (V/V_d)^{1/3} \tag{4.11}$$

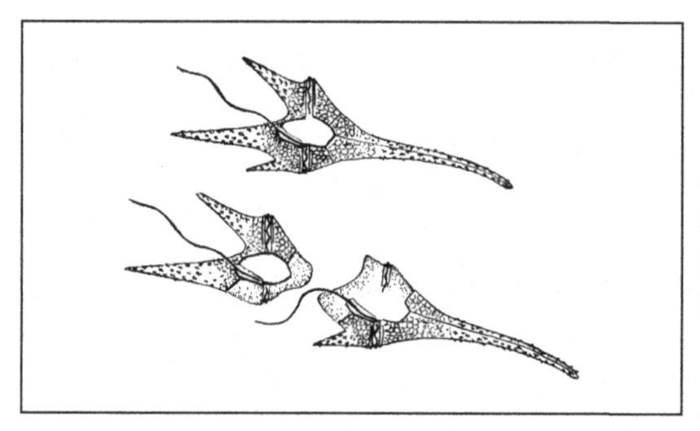

V0-morph. The dinoflagellate *Ceratium* has a rigid cell wall, which does not grow during the cell cycle, nor does the adjacent outer membrane that takes up nutrients. Cytoplasm growth is at the expense of internal vacuoles.

V1-morph. A mycelium of a fungus, such as *Mucor*, can be conceived as a branching filament, with a constant diameter. If the mycelium becomes dense, uptake is usually no longer proportional to the total filament length or number of growing tips of branches.

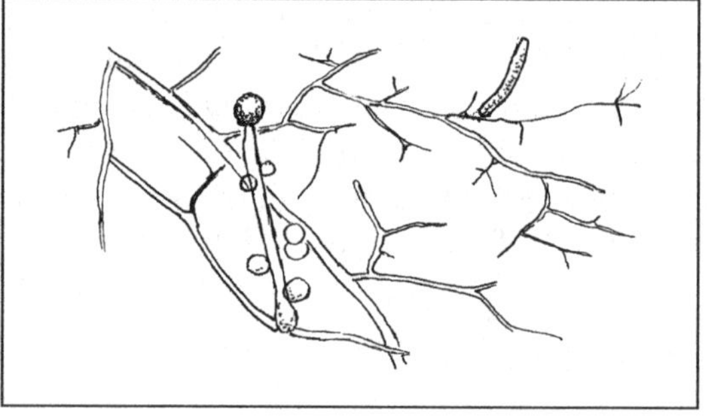

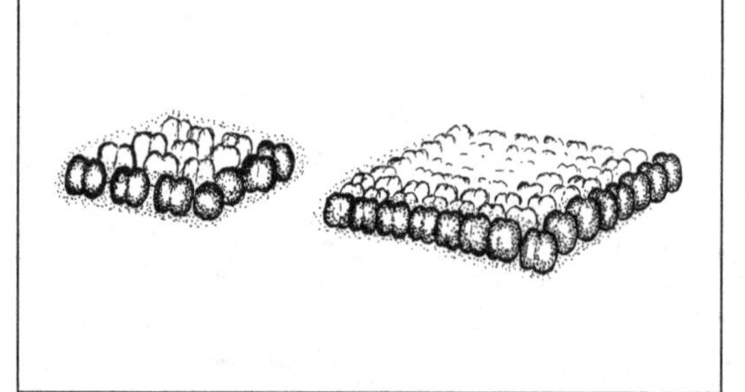

V1-morph. The cyanobacterial colony *Merismopedia* is only one cell layer thick. Although this sheet grows in two dimensions, it is a V1-morph. The arrangement of the cells requires an almost perfect synchronisation of the cell cycles.

Figure 4.8 A sample of organisms that change in shape during growth in very particular ways.

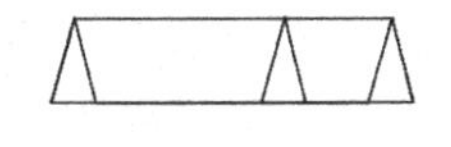

It is not essential that the cross-section through a filament is circular; it can be any shape, as long as it does not change during growth.

A V1-morph can also grow in two dimensions, however, as is illustrated by sheets, i.e. flat bodies with a constant, but small, height. The archaebacterium *Methanoplanus*, and Walsby's bacterium [594, 1216] fit this description. Several colonies, such as the sulphur bacterium *Thiopedia*, the cyanobacterium *Merismopedia* and the green alga *Pediastrum*, also fall into this category; see Figure 4.8. How sheets grow in two dimensions does not matter: they may change wildly in shape during growth. Height must be small to neglect the contribution of the sides to the total surface area. The surface area of the sheet relates to its volume as $A(V) = 2VL_h^{-1}$, where L_h denotes the height of the sheet and the factor 2 accounts for the upper and lower surface areas of the sheet. Division by the isomorphic surface area $A(V_d)(V/V_d)^{2/3}$ gives $\mathcal{M}(V) = (V/V_d)^{1/3}$, as for filaments, i.e. V1-morphs.

V1-morphs play an important role in DEB theory because of their simple dynamics, especially for organisms that divide into two daughter cells. Even if the actual changes in shape differ from V1-morphy, the dynamics can be approximated very well by that of V1-morphs, because of the narrow range in body sizes. From a population perspective the detailed morphology of the growth curve hardly matters, only the time it takes to double the initial volume. All individuals converge to the same reserve density in homogeneous space and the change of the total structure in a population behaves in a similar way to that of any individual. In other words a population of many small individuals behaves identically to one of few big individuals, as long as the parameters are identical. Size at division is irrelevant at the population level, which means that maturity only needs to be evaluated in connection with details of the cell cycle. Moreover, the somatic maintenance costs are proportional to volume (because the surface-area-linked maintenance now scales with volume as well) and can be added to the maturity maintenance cost (because dividers are always in the juvenile stage and the maturity does not stick at some threshold value). This is also the reason why the maturity costs can be combined with that of structure. Thanks to the κ-rule, there is no need to evaluate maturity explicitly. Whether or not unicellulars and particularly prokaryotes invest in cell differentiation during the cell cycle is still open to debate. Dworkin [304] reviewed development in prokaryotes and pointed to the striking similarities between myxobacteria and cellular slime moulds and between *Actinomyceta* and some fungi.

To avoid excessive notation by introducing new symbols for combinations of fluxes and amounts, I take $\kappa = 1$, $\{\dot{p}_T\} = 0$, so $[\dot{p}_S] = [\dot{p}_M]$, and $\dot{k}_J = 0$ for V1-morphs and don't discuss maturity dynamics. The consequence of these choices is that in fact the interpretation of some parameters changes for V1-morphs.

The ingestion rate, storage dynamics and growth for V1-morphs can be found from those of isomorphs by multiplying $\{\dot{p}_{Am}\}$, $\{\dot{F}_m\}$, $\{\dot{p}_T\}$ and $\dot{v}$ with the shape correction function $\mathcal{M}(V)$ in (4.11). The result is for $[\dot{J}_{XAm}] \equiv \{\dot{J}_{XAm}\}/L_d$ and $[\dot{F}_m] \equiv \{\dot{F}_m\}/L_d$

$$\dot{J}_{XA} = [\dot{J}_{XAm}]fV \quad \text{with} \quad f = \frac{X}{K+X} \quad \text{and} \quad K = [\dot{J}_{XAm}]/[\dot{F}_m] \tag{4.12}$$

$$\frac{d}{dt}e = (f-e)\dot{k}_E \quad \text{or} \quad \frac{d}{dt}m_E = \dot{\jmath}_{EA} - \dot{k}_E m_E \tag{4.13}$$

$$\frac{d}{dt}V = \dot{r}V \quad \text{with} \quad \dot{r} = \frac{m_E\dot{k}_E - \dot{\jmath}_E^M - y_{EV}\dot{\jmath}_V^M}{m_E + y_{EV}} \tag{4.14}$$

The reserve turnover rate $\dot{k}_E = \dot{v}/L_d = [\dot{p}_{Am}]/[E_m]$, with $[\dot{p}_{Am}] = \{\dot{p}_{Am}\}V_d^{-1/3}$, has dimension 'per time'. The expressions for $\dot{\jmath}_E^M$ and $\dot{\jmath}_V^M$ are given in (4.5) or (4.6), where the volume-linked somatic maintenance process M for V1-morphs plays the role of total maintenance, S and J for isomorphs. The smallest turnover rate of structure that covers the maintenance costs is $\dot{\jmath}_{VC} = \dot{\jmath}_{VM}$. The specific reserve mobilisation rate is found from (2.12) and (4.13). The switch case amounts to

$$[\dot{p}_C] = [\dot{p}_{Am}](e - \dot{r}/\dot{k}_E) \quad \text{or} \quad \dot{\jmath}_{EC} = \dot{\jmath}_{EAm}(e - \dot{r}/\dot{k}_E) \tag{4.15}$$

$$\dot{\jmath}_E^M = \min\{\dot{\jmath}_{EM}, \dot{\jmath}_{EC}\} \quad \text{and} \quad \dot{\jmath}_V^M = \dot{\jmath}_{VM}(1 - \dot{\jmath}_E^M/\dot{\jmath}_{EM}) \tag{4.16}$$

while the more general preference formulation gives for $\dot{\jmath}_{VC} = \dot{\jmath}_{VM}$

$$\dot{\jmath}_E^M = \dot{\jmath}_{EM}(1 - \dot{\jmath}_V^M/\dot{\jmath}_{VM}); \quad \dot{\jmath}_V^M = \dot{\jmath}_{VM}\frac{2A}{2A + \sqrt{B^2 - 4AC} - B} \tag{4.17}$$

with $A = \rho_V \dot{\jmath}_{EM}^2$, $B = C + ((1-\rho_V)\dot{\jmath}_{EC} + y_{EV}\dot{\jmath}_{VM})\dot{\jmath}_{EM}$, $C = -\dot{\jmath}_{EC}(\dot{\jmath}_{EC} + y_{EV}\dot{\jmath}_{VM})$. As before (4.17) reduces to (4.16) for $\rho_V \to 0$. Substitution into (4.14) gives

$$\dot{r} = \frac{m_E\dot{k}_E - \dot{\jmath}_{EM}}{m_E + y_{EV}} = \dot{k}_E\frac{e - l_d}{e + g} \quad \text{for} \quad m_E \geq \frac{\dot{\jmath}_{EM}}{\dot{k}_E} \quad \text{or} \quad e \geq l_d \tag{4.18}$$

$$\dot{r} = \frac{m_E\dot{k}_E - \dot{\jmath}_{EM}}{m_E + \dot{\jmath}_{EM}/\dot{\jmath}_{VM}} = \dot{k}_E\frac{e - l_d}{e + g_V} \quad \text{for} \quad m_E < \frac{\dot{\jmath}_{EM}}{\dot{k}_E} \quad \text{or} \quad e < l_d \tag{4.19}$$

where $g_V = \frac{l_d\dot{k}_E}{\dot{\jmath}_{VM}}$ and $l_d = (V_d/V_m)^{1/3} = L_d/L_m = \frac{\dot{\jmath}_{EM}}{\dot{k}_E m_{Em}} = g\dot{k}_M/\dot{k}_E$ and V_d is the volume at division, $V_m = L_m^3$ is defined by $L_m = \frac{\dot{v}}{g\dot{k}_M}$. The latter compound parameter lost its interpretation as maximum length.

Table 4.1 summarises the basic powers for V1-morphs. The implication is that the residence time of compounds in the reserve of a V1-morph is independent of the amount of structure. The maximum reserve density is $m_{Em} = \dot{\jmath}_{EAm}/\dot{k}_E$.

If no structure is used to pay (somatic) maintenance costs the specific gross growth rate $\dot{\jmath}_{VG}$ equals the specific net growth rate $\dot{r}$, but if (some) maintenance is paid from structure at rate $\dot{\jmath}_V^M$, then the growth rates relate to each other as $\dot{r} = \dot{\jmath}_{VG} - \dot{\jmath}_V^M$.

Figure 4.9 provides empirical support for the reserve dynamics; notice that the reserve *density* decays exponentially, as expected, not the reserve.

Table 4.1 The powers as specified by the DEB model for a dividing V1-morph of scaled length l and scaled reserve density e at scaled functional response f; cf Table 2.5 for reproducing isomorphs. An individual of structural volume $V \equiv M_V/[M_V]$ takes up substrate at rate $[\dot{J}_{Xm}]fV$. The implied dynamics for e and l: $\frac{d}{dt}e = \frac{f-e}{l_d}\dot{k}_M g$ and $\frac{d}{\times}dtl = l\frac{e/l_d-1}{e/g+1}\frac{\dot{k}_M}{3}$; division occurs when $l = l_d$.

power $\frac{}{\{\dot{p}_{Am}\}L_m^2}$	juvenile
assimilation, $\dot{p}_A$	fl^3/l_d
mobilisation, $\dot{p}_C$	$el^3\frac{1+g/l_d}{g+e}$
somatic maintenance, $\dot{p}_S$	κl^3
maturity maintenance, $\dot{p}_J$	$(1-\kappa)l^3$
somatic growth, $\dot{p}_G$	$\kappa l^3\frac{e/l_d-1}{1+e/g}$
maturity growth, $\dot{p}_R$	$(1-\kappa)l^3\frac{e/l_d-1}{1+e/g}$

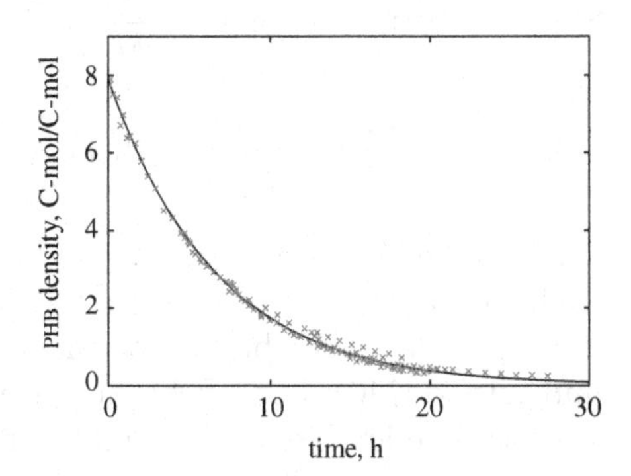

Figure 4.9 The poly-β-hydroxybutyrate (PHB) density (on the basis of C-mol/C-mol) in aerobic activated sludge at 20 °C. The fitted curve is an exponential one with parameter 0.15 h. Data from Beun [100]. She pointed in her thesis to [26] who found that the number of PHB granules per cell is fixed at the earliest stage of polymer accumulation. This supports the structural homeostasis hypothesis.

Exponential growth at constant food density

If substrate density X and, therefore, the scaled functional response f are constant long enough, energy density tends to $[E] = f[E_m]$ and the volume of V1-morphs as a function of time since division becomes for $V(0) = V_d/2$

$$V(t) = \frac{1}{2}V_d \exp(t\dot{r}) \quad \text{or} \quad t(V) = \dot{r}^{-1}\ln\{2V/V_d\} \quad \text{with} \quad \dot{r} \equiv \dot{k}_E\frac{f-l_d}{f+g} \tag{4.20}$$

The time taken to grow from $V_d/2$ to V_d is thus $t(V_d) = \dot{r}^{-1}\ln 2$.

The maximum specific growth rate is

$$\dot{r}_m = \frac{m_{Em}\dot{k}_E - j_{EM}}{m_{Em} + y_{EV}} = \frac{j_{EAm} - j_{EM}}{j_{EAm}/\dot{k}_E + y_{EV}} = \frac{\dot{k}_E - \dot{k}_M g}{1+g} = \dot{k}_E\frac{1-l_d}{1+g} \tag{4.21}$$

so that $g = \frac{\dot{k}_E - \dot{r}_m}{\dot{k}_M + \dot{r}_m}$ and $j_{EAm} = y_{EV}\frac{\dot{r}_m + \dot{k}_M}{1-\dot{r}_m/\dot{k}_E} = y_{EV}\dot{k}_E/g$ for $j_{EM} = y_{EV}\dot{k}_M$.

Exponential growth can be expected if the surface area at which nutrients are taken up is proportional to volume. For V1-morphs, this happens when the total surface area is involved, or a fixed fraction of it. If uptake only takes place at tips, the number of tips should increase with total filament length to ensure exponential growth. This has been found for the fungi *Fusarium* [1169] and *Penicillium* [832, 902], which do not divide; see Figure 4.11. The ascomycetous fungus *Neurospora* does not branch this way [320]; it has a mycelium that grows like a crust, see {134}.

Exponential growth of individuals should not be confused with that of populations. All populations grow exponentially at resource densities that are constant for long enough, whatever the growth pattern of individuals; see {347}. This is simply because the progeny repeats the growth/reproduction behaviour of the parents. Only for V1-morphs is it unnecessary to distinguish between the individual and the population level. This is a characteristic property of exponential growth of individuals and is discussed on {351}.

Yield of biomass on substrate at constant food

The yield of structure on substrate is given by $Y = \frac{\dot{r}}{f\dot{\jmath}_{XAm}}$ on a mole per mole basis. Simple substitution of $f = \frac{\dot{k}_E l_d + \dot{r}g}{\dot{k}_E - \dot{f}}$ from (4.20) gives

$$Y^{-1} = Y_g^{-1}\frac{1+\dot{k}_M/\dot{r}}{1-\dot{r}/\dot{k}_E} \quad \text{with } Y_g = \kappa\frac{\mu_{AX}}{\mu_{GV}} = \frac{\dot{k}_E}{g\dot{\jmath}_{XAm}} = \frac{\kappa}{y_{VX}} \text{ and } y_{VX} = \frac{y_{VE}}{y_{XE}} \tag{4.22}$$

which is known as the 'true' yield in microbiology. This specifies a three-parameter U-shaped relationship between Y^{-1} and $\dot{r}^{-1}$. It typically fits data very well; see Figure 4.10. The right branch of the U-shaped curve relates to maintenance, as is well understood [746]. The left branch relates to reserve in DEB theory, while popular explanations in microbiology speculate on mechanisms for enhanced maintenance at high growth rates [999].

Only y_{EX} is likely to depend on the chemical potential of the substrate, i.e. $y_{VX} \simeq \eta_{VA}\bar{\mu}_X$ with $\eta_{VA} = 0.001$ C-mol kJ^{-1}. Since animals are biotrophs, so their food mainly

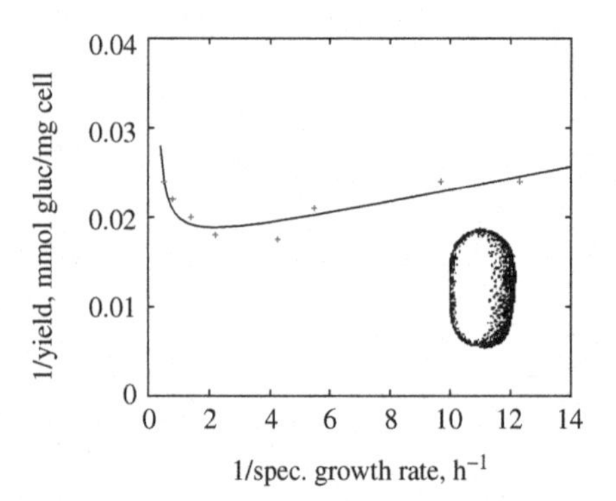

Figure 4.10 When the inverse yield of *Streptococcus bovis* on glucose is plotted against the inverse growth rate, a U-shaped curve results. Data from Russell and Baldwin (1979) as reported in [999] at 37 °C (I assume). The parameters are $Y_g = 62.84$ mg cell/mmol glucose, $\dot{k}_E = 5.93\,\text{h}^{-1}$ and $\dot{k}_M = 0.042\,\text{h}^{-1}$.

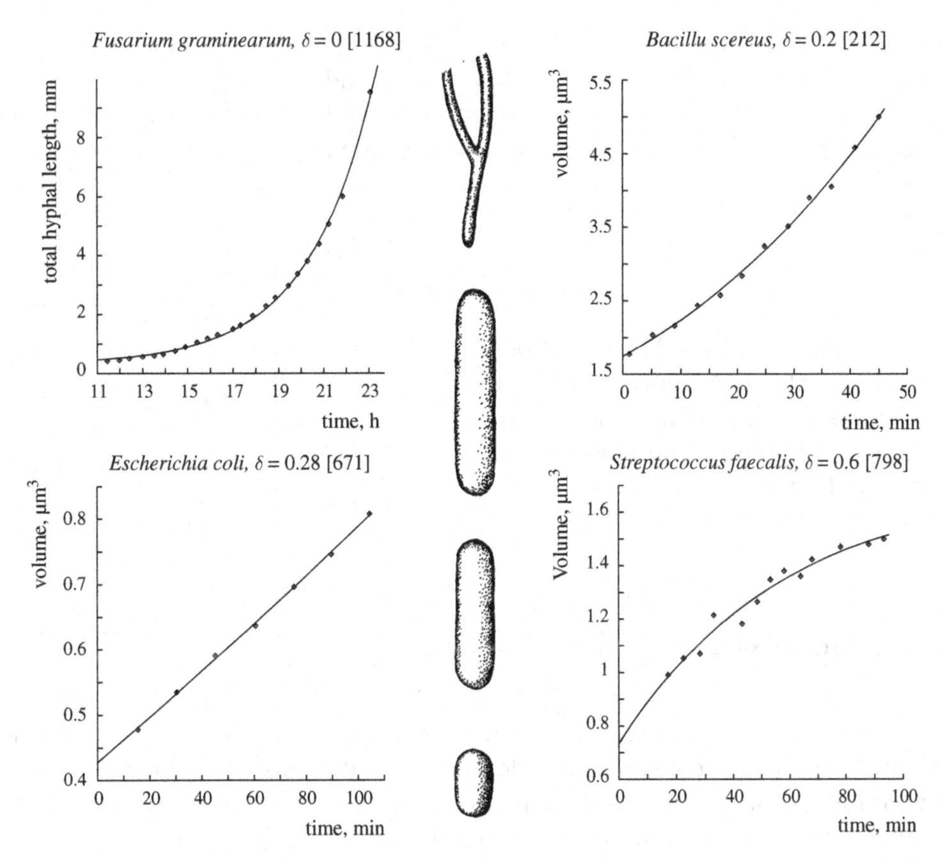

Figure 4.11 DEB-based growth curves for cells of static mixtures between V1- and V0-morphs. The larger the aspect ratio, δ, the more the growth curve turns from the V1-type (exponential) to the V0-type (satiation), reflecting the different surface area/volume relationships and supporting the assumption that uptake is linked to the surface area.

consists of polysaccharides, lipids and proteins, we expect that $y_{VX} = 0.4$ to 0.6 C-mol per C-mol for animals (see the table at {137}).

If no reserve is used to pay (somatic) maintenance, so $j_E^M = 0$ and $j_V^M = j_{VM}$, and the reserve turnover rate is large $\dot{k}_E \to \infty$, the DEB model reduces to the Marr *et al.* [746] and Pirt [900] model. If also $j_{VM} = 0$, the model further reduces to that of Monod [802]. If $j_{EM} = j_{VM} = 0$ but $\dot{k}_E$ is not very large, the model reduces to that of Droop [294, 296]. The Droop model is typically applied for nutrient-limited growth, and the maintenance costs for nutrient reserves might be small.

The Marr–Pirt model does not distinguish between maintenance and death, and is typically applied at the population level. Although the difference between maintenance and death in terms of effects on the population might be small, the difference in products that are formed is substantial; dead biomass is a nutritious substrate for many organisms. Refractory material is formed only during shrinking of bacterium *Alteromonas infernus* [313]. This strongly supports the existence of two maintenance fluxes with different products. Marie Eichinger also found few quantitative differences between the Marr–Pirt and the DEB model under constant environmental conditions, but substantial differences under varying conditions. At small spatial and temporal scales, environmental conditions typically vary substantially.

The yield coefficients of the Monod, Droop, Marr–Pirt and the DEB models relate to each other as at the right. The specific growth rates of these four models are compared in Figure 9.3 and the dynamic behaviour in Figure 9.2.

$[E_m]$ \ $[\dot{p}_M]$	0	>0
0	Monod Y_g	Marr–Pirt $Y_g \frac{f-l_d}{f}$
>0	Droop $Y_g \frac{g}{f+f}$	DEB for V1's $Y_g \frac{g}{f} \frac{f-l_d}{f+g}$

4.2.3 Static mixtures of morphs: rods

Cooper [222] argues that at constant substrate density *Escherichia* grows in length only, while the diameter/length ratio at division remains constant for different substrate densities. This mode of growth and division is typical for most rod-shaped bacteria, and most bacteria are rod-shaped. Shape and volume at division, at a given substrate density, are selected as a reference. The cell then has, say, length L_d, diameter δL_d, surface area A_d and volume V_d. The fraction δ is known as the aspect ratio of a cylinder. The index d will be used to indicate length, surface area and volume at division at a given substrate density. The shape of the rod-shaped bacterium is idealised by a cylinder with hemispheres at both ends and, in contrast to a filament, the caps are now included. Length at division is $L_d = \left(\frac{4V_d}{(1-\delta/3)\delta^2\pi}\right)^{1/3}$, making length $L = \frac{\delta}{3}\left(\frac{4V_d}{(1-\delta/3)\delta^2\pi}\right)^{1/3} + \frac{4V}{\pi\delta^2}\left(\frac{(1-\delta/3)\delta^2\pi}{4V_d}\right)^{2/3}$. Surface area becomes $A = L_d^2\frac{\pi}{3}\delta^2 + \frac{4V}{\delta L_d}$. The surface area of an isomorphically growing rod equals $A_d(V/V_d)^{2/3}$. The shape correction function is the ratio of these surface areas. If volume, rather than length, is used as an argument, the shape correction function becomes

$$\mathcal{M}(V) = \frac{\delta}{3}\left(\frac{V}{V_d}\right)^{-2/3} + \left(1 - \frac{\delta}{3}\right)\left(\frac{V}{V_d}\right)^{1/3} \tag{4.23}$$

When $\delta = 0.6$, the shape just after division is a sphere as in cocci, so this is the upper boundary for the aspect ratio δ. This value is obtained by equating the volume of a cylinder to that of two spheres of the same diameter. When $\delta \to 0$, the shape tends to that of a V1-morph.

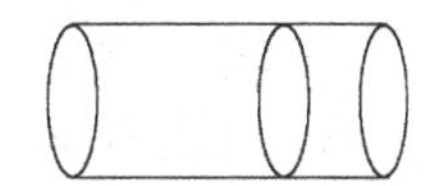

The shape correction function for rods can now be conceived as a weighted sum of those for a V0- and a V1-morph, with a simple geometric interpretation of the weight coefficients. A cylinder that grows in length only, with flat caps and an aspect ratio δ at $V = V_d$, has the shape correction function

$$\mathcal{M}(V) = \frac{\delta}{\delta+2}\left(\frac{V}{V_d}\right)^{-2/3} + \frac{2}{\delta+2}\left(\frac{V}{V_d}\right)^{1/3} \tag{4.24}$$

which is again a weighted sum of correction functions for V0- and V1-morphs. For the aspect ratio $\delta \to \infty$, the shape can become arbitrarily close to that of a V0-morph. The exact geometry of the caps is thus less important for surface area/volume relationships. Rods and cylinders are examples of static mixtures of V0- and V1-morphs, i.e. the weight coefficients do not depend on volume. Crusts are examples of dynamic mixtures of V0- and V1-morphs, and are discussed on {134}.

The growth of rods is on the basis of the shape correction function (4.23)

$$\frac{d}{dt}V = \frac{\delta V_d}{3V_\infty}\frac{\dot{k}_E e}{e+g}(V_\infty - V) \tag{4.25}$$

where $V_\infty \equiv V_d \frac{\delta}{3}(e^{-1}(\frac{V_d}{V_m})^{1/3} - 1 + \frac{\delta}{3})^{-1}$ and, as before, $V_m^{1/3} \equiv \frac{\dot{v}}{g k_M}$. If substrate density X and, therefore, the scaled functional response f are constant long enough, scaled energy density tends to $e = f$ and volume as a function of time since division becomes

$$V(t) = V_\infty - (V_\infty - V_d/2)\exp(-t\dot{r}_r) \tag{4.26}$$

where $\dot{r}_r \equiv \frac{V_d f \dot{k}_E \delta/3}{V_\infty(f+g)}$. The interpretation of V_∞ depends on its value.

- If $V_\infty = \infty$, i.e. if $f(1-\delta/3) = (V_d/V_m)^{1/3}$, the volume of rods grows linearly at rate $\frac{\dot{k}_E f}{f+g}V_d\frac{\delta}{3}$. This is frequently found empirically [49].

- If $0 < V_\infty < \infty$, V_∞ is the ultimate volume if the cell ceases to divide but continues to grow. For these values, $V(t)$ is a satiating function and is of the same type as $V(t)^{1/3}$ for isomorphs, (2.23). Note that volume, and thus cubed length, grows skewly S-shaped for isomorphs. When V_∞ is positive, the cell will only be able to divide when $V_\infty > V_d$, thus when $f > (V_d/V_m)^{1/3}$.

- If $\delta = 0$, $V_\infty = 0$ and the rod behaves as a V1-morph, which grows exponentially.

- For $V_\infty < 0$, $V(t)$ is a concave function, tending to an exponential one. The cell no longer has an ultimate size if it ceases to divide. V_∞ is then no longer interpreted as ultimate size, but this does not invalidate the equations.

The shape of the growth curve, convex, linear or concave, thus depends on substrate density and the aspect ratio. Figure 4.11 illustrates the perfect fit of growth curves (4.26) with only three parameters: volume at 'birth', $V_d/2$, ultimate volume, V_∞, and growth rate, $\dot{r}_r$. The figure beautifully reveals the effect of the aspect ratio; the larger the aspect ratio, the more important the effect of the caps, so a change from V1-morphic behaviour to a V0-morphic behaviour. A sudden irreversible change in morphology from spherical to filamentous cells has been observed in the yeast *Kluyveromyces marxianus* [435], while no other changes could be detected. The associated increase of 30% in the maximum specific growth rate could be related to the observed increase in specific surface area.

The time required to grow from $V_d/2$ to V at constant substrate density is found from (4.26)

$$t(V) = \frac{(f+g)V_\infty}{f\dot{k}_E V_d \delta/3} \ln \frac{V_\infty - V_d/2}{V_\infty - V} \tag{4.27}$$

4.2.4 Dynamic mixtures of morphs

Some organisms change in shape during growth in a complex fashion. Frequently it is still possible, however, to take these changes in shape into account in a rather simple way.

Biofilms on curved surfaces

Figure 4.12 shows that a biofilm on a curved surface can behave somewhere between a V0- and an isomorph.

Crusts

Crusts, i.e. biofilms of limited extent that grow on solid surfaces, are mixtures of V0-morphs in the centre and V1-morphs in the periphery where the new surface is covered. Lichens on rocks or forests behave like crusts and so do bacterial colonies on an agar plate, conceived as super-organisms. A forest or peat of limited extent on a spatially homogeneous plain is a crust. The spatial expansion of geographical distribution areas of species, such as the muskrat in Europe, and of infectious diseases, see

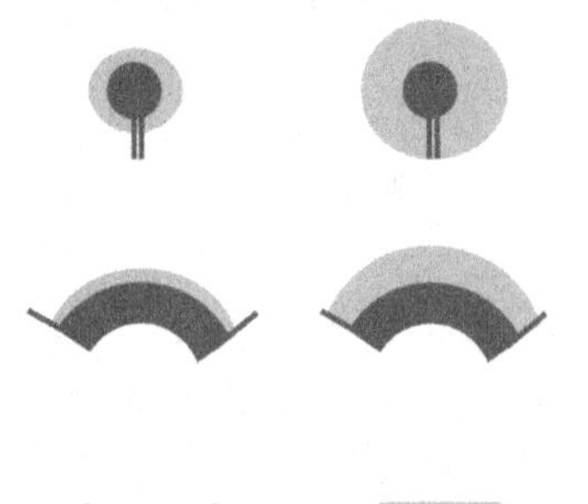

Figure 4.12 Biofilms (light grey) on the surface of a sphere (dark grey) can behave between an isomorph (top; radius of sphere $V_1 = 0$) and a V0-morph (bottom; radius of sphere $V_1 = \infty$). The shape correction function is $\mathcal{M}(V) = \left(\frac{V_d}{V}\frac{V_1+V}{V_1+V_d}\right)^{2/3}$.

[131, 481], closely resembles that of crusts. They all share the property that what happens in the centre has little relevance for the moving border. I will now demonstrate that this has the consequence that, in constant environments, the border moves at a constant rate: biomass in the border area grows exponentially and causes expansion, but that in the inner area settles at a constant density (amount per surface area) and hardly contributes to the expansion. The inner part behaves as a V0-morph and represents an increasing proportion of the biomass; the outer part behaves as a V1-morph, while the width of the annulus is determined by the horizontal transport rate of the limiting resource (in the case of lichens) or individuals (in the case of muskrats).

Let L_ϵ denote the width as well as the thickness of the outer annulus of the circular crust of radius L_r that is growing exponentially in an outward direction. The width and the thickness of the outer annulus remain constant. This biomass thus behaves as a V1-morph; all other biomass in the centre of the crust behaves as a V0-morph. The surface area of the crust is $A_r(t) = \pi L_r^2(t)$, and of the exponentially growing annulus $A_\epsilon(t) = \pi \left(L_r^2(t) - (L_r(t) - L_\epsilon)^2\right) = \pi \left(2L_r(t)L_\epsilon - L_\epsilon^2\right)$. The total surface area is growing at rate $\frac{d}{dt}A_r = \dot{r}A_\epsilon$, so the radius is growing at rate

$$\frac{d}{dt}L_r = \dot{r}L_\epsilon\left(1 - \frac{L_\epsilon}{2L_r}\right) \tag{4.28}$$

from which it follows that the diameter of the crust is growing linearly in time for $L_\epsilon \ll L_r$. Strong empirical support for the linear growth of the diameter of the crustose saxicolous lichen *Caloplaca trachyphylla* is given by [209]. This linear growth in diameter has also been observed experimentally by Fawcett [338], and the linear growth model originates from Emerson [320] in 1950 according to Fredrickson *et al.* [370]. Figure 4.13 shows that this linear growth applies to lichen growth on moraines. Richardson [959] discusses the value of gravestones for the study of lichen growth, because of the reliable dates. Lichen growth rates are characteristic of the species, so the diameter distribution of the circular patches can be translated into arrival times, which can then be linked to environmental factors, for instance.

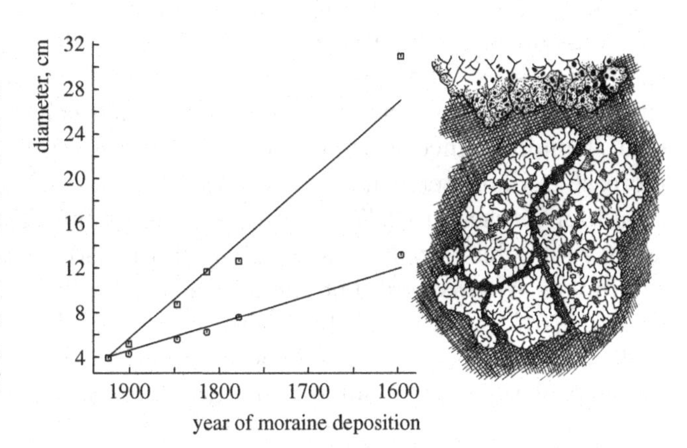

Figure 4.13 The lichens *Aspicilia cinerea* (above) and *Rhizocarpon geographicum* (below) grow almost linearly in a period of more than three centuries on moraine detritus of known age in the European Alps. Data from Richardson [959]. Linear growth is to be expected from the DEB model, when such lichens are conceived as dynamic mixtures of V0- and V1-morphs.

If substrate transport in the vertical direction on the plate is sufficient to cover all maintenance costs, and transport in the horizontal direction is small, the growth rate of the V0-morph on top of an annulus of surface area dA is

$$\frac{d}{dt}V = \frac{f\{\dot{p}_{Am}\}\,dA - [\dot{p}_M]V}{[E_G] + [E_m]f} \tag{4.29}$$

The denominator stands for the volume-specific costs of structural biomass and reserves. Division by the surface area of the annulus gives the change in height L_h of the V0-morph on the top of an annulus of surface area $dA = V_d^{2/3}$; the height is found from (4.10) by substituting $V = L_h V_d^{2/3}$

$$\frac{d}{dt}L_h = \frac{\dot{v}}{e+g}(e - V_m^{-1/3}L_h) \quad \text{or} \quad \frac{d}{dt}l_h = 3\dot{r}_B(f - l_h) \tag{4.30}$$

with the scaled height $l_h \equiv L_h V_m^{-1/3}$. The initial growth rate in scaled height is $3\dot{r}_B(f - l_\epsilon)$. The parameter $l_\epsilon \equiv L_\epsilon V_m^{-1/3}$ can be eliminated, on the assumption that the growth rate in the outward direction equals the initial growth rate in the vertical direction, which gives $l_\epsilon = l_d/2$ for $l_d \ll f$. For $l_d \ll l_r$ with $l_r \equiv L_r V_m^{-1/3}$, the end result amounts to

$$l_h(t, l_r) = f - (f - l_d/2)\exp\left(\frac{l_r}{f - l_d} - 3\dot{r}_B t\right) \tag{4.31}$$

The scaled height of the crust is thus growing asymptotically to f. Different crust shapes can be obtained by accounting for horizontal transport of biomass and diffusion limitation of food transport to the crust.

Flocs and tumours

Growth in the thickness of a biofilm on a plane, which behaves as a V0-morph, is thus similar to that of a spherical biofilm on a small core in suspension, which behaves as an isomorph as long as mass transport in the film is sufficiently large to consider the biomass as homogeneous. Films are growing in a von Bertalanffy way in both situations, if growth via settling of suspended cells on the film is not important. Note that if maintenance is small, so that the asymptotic depth of the film is large, the increase in diameter is linear with time, so that volume increases as time^3, as has been found for fetuses in (2.47) by different reasoning. This mode of growth was called the 'cube root' phase by Emerson [320], who found it applicable to submerged mycelia of the fungus *Neurospora*. The model was originally formulated by Mayneord for tumour growth [761], and frequently applied since then [679, 782, 1103, 1243].

If mass transport in a spherical biofilm on a small core in suspension is not large, the biomass in the centre will become deprived of substrate by the peripheral mass, and die from starvation. Such a film is called a (microbial) floc. A concentration gradient of substrate develops in the living peripheral mass, such that the organisms at the living/dead boundary layer just receive enough substrate to survive, and do not grow. The organisms at the outer edge grow fastest. The thickness of the living layer directly relates to the transport rate of substrate, and so depends on the porosity of the floc. Flocs again behave as dynamic mixtures of V0- and V1-morphs, and, just like crusts, the floc diameter eventually grows linearly in time at constant substrate densities in the environment, if it does not fall apart because of the increasing mechanical instability. This can be seen as follows.

Let L_ϵ denote the thickness of the thin living layer of a spherical floc of radius L_r. The thickness remains constant, while the living mass is growing exponentially at rate $\dot{r}$. The outer layer behaves as a V1-morph, the kernel as a degenerated V0-morph. The total volume of the floc is $V_r(t) = \frac{4}{3}\pi L_r^3(t)$ and of the living layer $V_\epsilon(t) = \frac{4}{3}\pi \left(L_r(t)^3 - (L_r(t) - L_\epsilon)^3\right) = \frac{4}{3}\pi \left(3L_r^2(t)L_\epsilon - 3L_r(t)L_\epsilon^2 + L_\epsilon^3\right)$. The growth of the floc is given by $\frac{d}{dt}V_r = \dot{r}V_\epsilon$, so the radius is growing at rate

$$\frac{d}{dt}L_r = \dot{r}L_\epsilon\left(1 - \frac{L_\epsilon}{L_r} + \frac{L_\epsilon^2}{3L_r^2}\right) \tag{4.32}$$

For $L_r \gg L_\epsilon$, the change in the radius L_r becomes constant, and the floc grows linearly in time. The steady-state population growth rate of flocs can be obtained analytically, given a fixed size at fragmentation into n parts. The dead volume increases with $\frac{d}{dt}V_\dagger(t) = 4\pi(L_r(t) - L_\epsilon)^2\frac{d}{dt}L_r(t)$. Brandt [139] showed that the combination of diffusive transport of substrate into the floc, see {266}, and a hyperbolic functional response leads to a living layer of thickness $\left(\frac{\dot{D}X_k}{2\dot{j}_{Xm}X_1}\right)^{1/2}\int_{x_\dagger}^{x_0}\left(y - x_\dagger + \ln\frac{1+x_\dagger}{1+y}\right)^{-1/2} dy$, where $\dot{D}$ is the diffusion coefficient, the scaled substrate density at the living/ dead boundary is $x_\dagger = \frac{[\dot{p}_M]}{[\dot{p}_{Am}] - [\dot{p}_M]}$ with specific maintenance power $[\dot{p}_M]$ and specific maximum assimilation power $[\dot{p}_{Am}]$, scaled substrate concentration $x_0 = X_0/X_K$ with saturation constant X_K, biomass density in the floc X_1 and maximum specific substrate uptake rate $[\dot{j}_{Xm}]$ [139].

Roots and shoots

The modelling step from algae to plants involves a number of extensions that primarily relate to the fact that plants take up nutrients through roots, while shoots (including leaves) are used for light and carbon dioxide uptake and water transpiration, which affect internal nutrient and metabolite transport. This makes the allocation of resources to root versus shoot growth of special interest, as well as shape changes that affect surface area/volume relationships via the scaling of assimilation and maintenance, respectively, with structural mass. As illustrated in Figure 4.14, most plants naturally develop from a V1-morphic, via an isomorphic, to a V0-morphic growth during their life cycle. Procumbent plants almost skip the isomorphic phase and directly develop from V1- to V0-morphic growth, similar to crusts [105]. Climbing plants seem to stay in the V1-morphic phase.

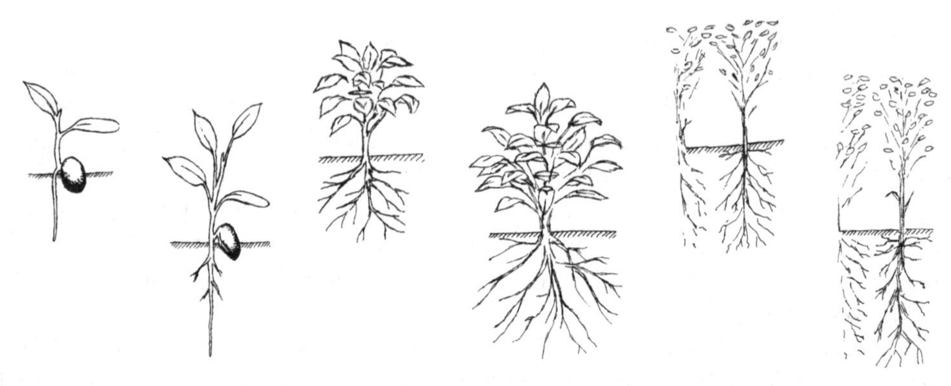

Figure 4.14 Just after germination, plants usually grow as V1-morphs, but when the number of leaves increases, self-shading becomes important, and the plant gradually behaves as an isomorph. If they make contact with other plants, and leaves and roots form a closed layer, they behave as V0-morphs; an increase in mass no longer results in an increase of surface area that is effectively involved in nutrient or light uptake.

These changes in shape can be incorporated using the shape correction function $\mathcal{M}(V)$, which can be chosen differently for roots and shoots. Given the wild diversity of plant shapes and the extreme extent of local adaptations, it is hard to see how a choice can be based on mechanistic arguments. Empirical and convenience arguments can hardly be avoided at this point. A simple choice would be

$$\mathcal{M}(V) = (V/V_d)^{1/3-(V/V_m)^\beta} \tag{4.33}$$

which starts from V1- and ends with V0-morphic growth when it reaches its maximum volume V_m.

4.3 Mass aspects of univariate DEB models

4.3.1 Three basic fluxes

In this subsection I show that all energy and mass fluxes of univariate DEB models are weighted sums of three basic fluxes: assimilation, dissipation and growth. The mineral fluxes follow from the organic fluxes, and the organic fluxes follow from the assumptions of Table 2.4.

The chemical indices of the minerals and the organic compounds are collected in two matrices $\boldsymbol{n}_\mathcal{M}$ and $\boldsymbol{n}_\mathcal{O}$, respectively. A typical element of such a matrix, $n_{*_1 *_2}$, denotes the chemical index of compound $*_2$ with respect to element $*_1$. The chemical indices of the organic compounds for carbon equal 1 by definition. The strong homeostasis assumption amounts to the condition that the chemical indices do not change.

Let $\dot{J}_*$ denote the rate of change of the compound $*$. The conservation of mass amounts to

$$\begin{pmatrix} 0 \\ 0 \\ 0 \\ 0 \end{pmatrix} = \begin{pmatrix} 1 & 0 & 0 & n_{CN} \\ 0 & 2 & 0 & n_{HN} \\ 2 & 1 & 2 & n_{ON} \\ 0 & 0 & 0 & n_{NN} \end{pmatrix} \begin{pmatrix} \dot{J}_C \\ \dot{J}_H \\ \dot{J}_O \\ \dot{J}_N \end{pmatrix} + \begin{pmatrix} n_{CX} & n_{CV} & n_{CE} & n_{CP} \\ n_{HX} & n_{HV} & n_{HE} & n_{HP} \\ n_{OX} & n_{OV} & n_{OE} & n_{OP} \\ n_{NX} & n_{NV} & n_{NE} & n_{NP} \end{pmatrix} \begin{pmatrix} \dot{J}_X \\ \dot{J}_V \\ \dot{J}_E + \dot{J}_{E_R} \\ \dot{J}_P \end{pmatrix} \tag{4.34}$$

This can be summarised in matrix form as $\mathbf{0} = \boldsymbol{n}_{\mathcal{M}} \dot{\boldsymbol{J}}_{\mathcal{M}} + \boldsymbol{n}_{\mathcal{O}} \dot{\boldsymbol{J}}_{\mathcal{O}}$ or $\mathbf{0} = \boldsymbol{n}\dot{\boldsymbol{J}}$, for $\boldsymbol{n} = (\boldsymbol{n}_{\mathcal{M}} \vdots \boldsymbol{n}_{\mathcal{O}})$ and $\dot{\boldsymbol{J}} = (\dot{\boldsymbol{J}}_{\mathcal{M}}; \dot{\boldsymbol{J}}_{\mathcal{O}})$; the only difference from (3.10) is in the grouping of the compounds. Thus the fluxes for the 'mineral' compounds $\dot{\boldsymbol{J}}_{\mathcal{M}}$ can be written as a weighted sum of the fluxes of the organic compounds $\dot{\boldsymbol{J}}_{\mathcal{O}}$

$$\dot{\boldsymbol{J}}_{\mathcal{M}} = -\boldsymbol{n}_{\mathcal{M}}^{-1} \boldsymbol{n}_{\mathcal{O}} \dot{\boldsymbol{J}}_{\mathcal{O}} \tag{4.35}$$

with

$$\boldsymbol{n}_{\mathcal{M}}^{-1} = \begin{pmatrix} 1 & 0 & 0 & -\frac{n_{CN}}{n_{NN}} \\ 0 & 2^{-1} & 0 & -\frac{n_{HN}}{2n_{NN}} \\ -1 & -4^{-1} & 2^{-1} & \frac{n}{4n_{NN}} \\ 0 & 0 & 0 & n_{NN}^{-1} \end{pmatrix}; \quad n \equiv 4n_{CN} + n_{HN} - 2n_{ON} \tag{4.36}$$

I will now explain why the 'organic' fluxes $\dot{\boldsymbol{J}}_{\mathcal{O}}$ relate to the basic powers $\dot{\boldsymbol{p}}$ as

$$\begin{pmatrix} \dot{J}_X \\ \dot{J}_V \\ \dot{J}_E + \dot{J}_{E_R} \\ \dot{J}_P \end{pmatrix} = \begin{pmatrix} -\eta_{XA} & 0 & 0 \\ 0 & 0 & \eta_{VG} \\ \overline{\mu}_E^{-1} & -\overline{\mu}_E^{-1} & -\overline{\mu}_E^{-1} \\ \eta_{PA} & \eta_{PD} & \eta_{PG} \end{pmatrix} \begin{pmatrix} \dot{p}_A \\ \dot{p}_D \\ \dot{p}_G \end{pmatrix}, \quad \text{or} \quad \dot{\boldsymbol{J}}_{\mathcal{O}} = \boldsymbol{\eta}_{\mathcal{O}} \dot{\boldsymbol{p}} \tag{4.37}$$

where $\overline{\mu}_E$ is the chemical potential of the reserve, and $\eta_{*_1 *_2}$ the mass flux of compound $*_1$ per unit of power $*_2$, i.e. the coupling between mass and energy fluxes. The latter coefficients serve as model parameters, and are collected in matrix $\boldsymbol{\eta}$.

The fluxes $\dot{J}_X = -\eta_{XA}\dot{p}_A$ and $\dot{J}_P$ follow from the strong homeostasis assumption and from (2.2). Assimilation energy is quantified by its fixation in reserves, so reserves are formed at a rate $\dot{p}_A/\overline{\mu}_E$, and the yield of food on reserve, $y_{XE} = \overline{\mu}_E/\mu_{AX}$, stands for the C-moles of food ingested per C-mole of reserves formed, where $\mu_{AX} = \eta_{XA}^{-1}$. The rate at which work can be done by ingested food is $\overline{\mu}_X \dot{J}_X$; the flux $\dot{p}_A$ is fixed in reserves, the flux $\dot{p}_A \overline{\mu}_P \eta_{PA}$ is fixed in product, the rest dissipates as heat and mineral fluxes that are associated with this conversion. The coefficient $y_{PX} = \mu_{AX}\eta_{PA}$ stands for the C-mole of product that is derived directly from food per C-mole of food ingested (products can also be formed indirectly from assimilated energy).

If the individual happens to be a metazoan and the product is interpreted as faeces, then $\eta_{PD} = \eta_{PG} = 0$. Faeces production is coupled to food intake only. Alcohol production by yeasts that live on glucose is an example of product formation where $\eta_{PD} \neq \eta_{PG} \neq 0$. At this point there is no need for molecular details about the process of digestion being intra- or extracellular. This knowledge only affects details in the interpretation of the coefficients in $\boldsymbol{\eta}$.

The flux $\dot{J}_V = \dot{p}_G \eta_{VG}$ indicates that $\mu_{GV} = \eta_{VG}^{-1}$ is the invested energy per C-mole of structural biomass, which directly follows from assumption 1 in Table 2.4. Note that $\overline{\mu}_V$ is the energy that is actually fixed in a C-mole of structural biomass, so $\mu_{GV} - \overline{\mu}_V$ dissipates (as heat or via products that are coupled to growth) per C-mole.

The flux of reserves is given by $\dot{J}_E = \overline{\mu}_E^{-1}(\dot{p}_A - \dot{p}_C)$: reserve energy is generated by assimilation and mobilised for further use. The flux of embryonic reserves (i.e. reproduction), $\dot{J}_{E_R} = \overline{\mu}_E^{-1} \kappa_R \dot{p}_R$, appears as a return flux to the reserve because embryonic reserves have the same composition as adult reserves because of the strong homeostasis assumption. Since $\dot{p}_C = \dot{p}_D + \dot{p}_G + \kappa_R \dot{p}_R$, see (2.60) and (2.61), we have $\dot{J}_E + \dot{J}_{E_R} = \overline{\mu}_E^{-1}(\dot{p}_A - \dot{p}_D - \dot{p}_G)$, which is the relationship given in (4.37). So $\dot{J}_E + \dot{J}_{E_R}$ is a weighted sum of three powers, but $\dot{J}_E$ and $\dot{J}_{E_R}$ themselves are not.

Substitution of (4.37) into (4.35) shows that the mass balance equation can be re-formulated as $\mathbf{0} = \boldsymbol{n}_{\mathcal{M}} \boldsymbol{\eta}_{\mathcal{M}} + \boldsymbol{n}_{\mathcal{O}} \boldsymbol{\eta}_{\mathcal{O}}$, which provides the matrix of energy–mineral coupling coefficients $\boldsymbol{\eta}_{\mathcal{M}} = -\boldsymbol{n}_{\mathcal{M}}^{-1} \boldsymbol{n}_{\mathcal{O}} \boldsymbol{\eta}_{\mathcal{O}}$ and the mineral fluxes $\dot{\boldsymbol{J}}_{\mathcal{M}} = \boldsymbol{\eta}_{\mathcal{M}} \dot{\boldsymbol{p}}$.

Table 2.5 shows that all basic powers are cubic polynomials in length, from which it follows that all mass fluxes are also cubic polynomials in length.

The matrix $\boldsymbol{n}_{\mathcal{M}}^{-1}$ of coefficients (4.36) has an odd interpretation in terms of reduction degrees if the nitrogenous waste is ammonia. The third row, i.e. the one that relates to dioxygen, represents the ratio of the reduction degree of the elements C, H, O, N to that of O_2, which is -4. That is to say, N atoms account for -3 of these reduction degrees, whatever their real values in the rich mixture of components that are present. The third row of the matrix $\boldsymbol{n}_{\mathcal{M}}^{-1} \boldsymbol{n}_{\mathcal{O}}$ thus represents the ratio of the reduction degrees of X, V, E and P to that of O. Sandler and Orbey [1010] discuss the concept of generalised degree of reduction.

Figure 4.15 illustrates $\dot{\boldsymbol{J}}_{\mathcal{O}}$ and $\dot{\boldsymbol{J}}_{\mathcal{M}}$ of the DEB model as a function of the structural biomass (i.e. scaled length, see next section), when food is abundant. The embryonic reserve flux is negative, because embryos do not eat. The growth just prior to birth is reduced, because the reserves become depleted. The switch from juvenile to adult, so from development to reproduction, implies a discontinuity in the mineral fluxes, but this discontinuity is negligibly small.

Partitioning of mass fluxes

The mineral and organic fluxes can be decomposed into contributions from assimilation, dissipation power and growth. Let $\dot{J}_* = \dot{J}_{*A} + \dot{J}_{*D} + \dot{J}_{*G}$ for $* \in \{\mathcal{M}, \mathcal{O}\}$, and let us collect these fluxes in two matrices, then

$$\dot{\boldsymbol{J}}_{\mathcal{O}*} = \boldsymbol{\eta}_{\mathcal{O}} \, \mathbf{diag}(\dot{\boldsymbol{p}}) \quad \text{and} \quad \dot{\boldsymbol{J}}_{\mathcal{M}*} = \boldsymbol{\eta}_{\mathcal{M}} \, \mathbf{diag}(\dot{\boldsymbol{p}}) \tag{4.38}$$

where $\mathbf{diag}(\dot{\boldsymbol{p}})$ represents a diagonal matrix with the elements of $\dot{\boldsymbol{p}}$ on the diagonal, so that $\mathbf{diag}(\dot{\boldsymbol{p}})\mathbf{1} = \dot{\boldsymbol{p}}$, and $\dot{\boldsymbol{J}}_{\mathcal{M}*}\mathbf{1} = \dot{\boldsymbol{J}}_{\mathcal{M}}$, $\dot{\boldsymbol{J}}_{\mathcal{O}*}\mathbf{1} = \dot{\boldsymbol{J}}_{\mathcal{O}}$. These results are used in later sections.

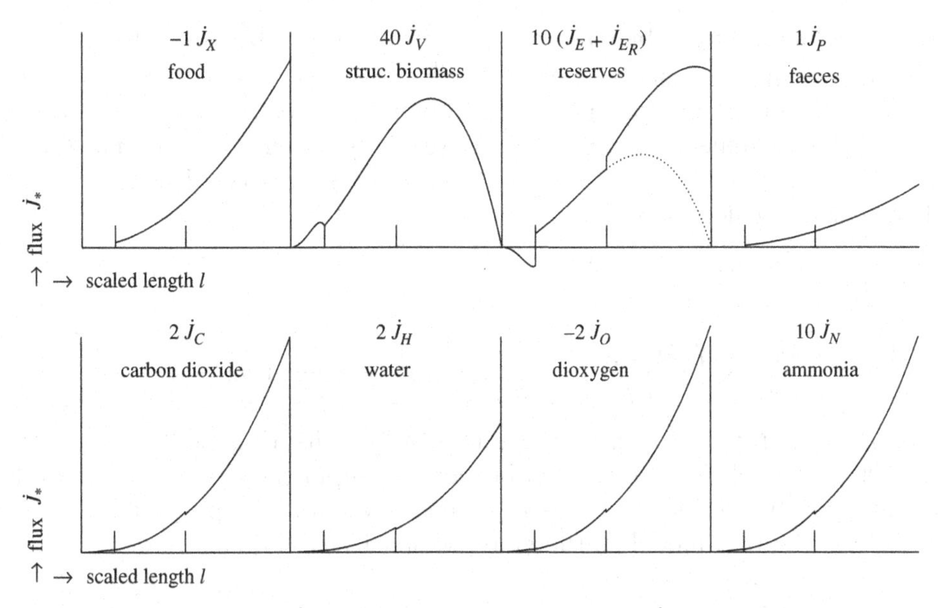

Figure 4.15 The organic fluxes $\dot{\boldsymbol{J}}_{\mathcal{O}}$ (top) and the mineral fluxes $\dot{\boldsymbol{J}}_{\mathcal{M}}$ (bottom) for the DEB model as functions of the scaled length l at abundant food ($e = 1$ for $l > l_b$; $0 < l < 1$). The various fluxes are multiplied by the indicated scaling factors for graphical purposes, while a common scaling factor involves model parameters. The parameters: scaled length at birth $l_b = 0.16$, scaled length at puberty $l_p = 0.5$ (both indicated on the abscissa), scaled heating length $l_T = 0$ (ectotherm), energy investment ratio $g = 1$, partition coefficient $\kappa = 0.8$, reproduction efficiency $\kappa_R = 0.8$. The coefficient matrices are

$$\boldsymbol{\eta}_{\mathcal{O}} = \begin{pmatrix} -1.5 & 0 & 0 \\ 0 & 0 & 0.5 \\ 1 & -1 & -1 \\ 0.5 & 0 & 0 \end{pmatrix}, \; \boldsymbol{n}_{\mathcal{M}} = \begin{pmatrix} 1 & 0 & 0 & 0 \\ 0 & 2 & 0 & 3 \\ 2 & 1 & 2 & 0 \\ 0 & 0 & 0 & 1 \end{pmatrix}, \; \boldsymbol{n}_{\mathcal{O}} = \begin{pmatrix} 1 & 1 & 1 & 1 \\ 1.8 & 1.8 & 1.8 & 1.8 \\ 0.5 & 0.5 & 0.5 & 0.5 \\ 0.2 & 0.2 & 0.2 & 0.2 \end{pmatrix}.$$

The fluxes assimilation, dissipation and growth can be further subdivided into a catabolic and an anabolic aspect, even for the dissipation flux. This is because somatic maintenance includes the turnover of structure, where structure serves the function of both substrate and product. We need this for the specification of fractionation of isotopes.

4.3.2 State versus flux

The mass of reserves and the structural biomass relate to the fluxes as $M_E(a) = M_E^0 + \int_0^a \dot{J}_E(t)\, dt$ and $M_V(a) = \int_0^a \dot{J}_V(t)\, dt$ (the initial structural mass is negligibly small). The mass of reserves of an embryo in C-moles at age 0 is $M_E^0 = E_0/\overline{\mu}_E$, where E_0 is given in (2.42).

The change in structural biomass M_V and reserve mass M_E relate to the powers as $\frac{d}{dt}M_V = \dot{J}_V = \dot{p}_G\eta_{VG}$ and $\frac{d}{dt}M_E = \dot{J}_E = \frac{\dot{p}_A - \dot{p}_C}{\overline{\mu}_E}$. If the model for these powers implies the existence of a maximum for the structural biomass, M_{Vm}, and for the reserve mass, M_{Em}, it can be convenient to replace the state of the individual, M_V and M_E, by the scaled length $l \equiv (M_V/M_{Vm})^{1/3}$ and the scaled energy reserve density $e \equiv \frac{M_E M_{Vm}}{M_V M_{Em}}$. The change of the scaled state then becomes

$$\frac{d}{dt}l = \frac{\dot{p}_G\eta_{VG}}{3M_V^{2/3}M_{Vm}^{1/3}} = \frac{\dot{p}_G}{3l^2\kappa g E_m} \tag{4.39}$$

$$\frac{d}{dt}e = \frac{M_{Vm}}{M_V M_{Em}}\left(\frac{\dot{p}_A - \dot{p}_C}{\overline{\mu}_E} - \frac{M_E}{M_V}\dot{p}_G\eta_{VG}\right) = \frac{1}{E_m l^3}\left(\dot{p}_A - \dot{p}_C - \dot{p}_G\frac{e}{\kappa g}\right) \tag{4.40}$$

The reproduction rate, in terms of the number of offspring per time, is given by $\dot{R} = \dot{J}_{E_R}/M_E^0$. Therefore, the three basic powers, supplemented by the reproductive power, fully specify the individual as a dynamic system. The purpose of the specific assumptions of the DEB model is, therefore, to specify these three powers.

4.3.3 Mass investment in neonates

Several simple expressions can be obtained for changes over the whole incubation period that are useful for practical work. The initial weight (age $a = 0$) and the weight at birth (i.e. hatching, age $a = a_b$), excluding membranes and nitrogenous waste, are

$$\begin{pmatrix} W_w(0) & W_w(a_b) \end{pmatrix} = V_m \begin{pmatrix} w_E & w_V \end{pmatrix} \begin{pmatrix} [M_{Em}]e_0 & [M_{Em}]e_b l_b^3 \\ 0 & [M_V]l_b^3 \end{pmatrix} \tag{4.41}$$

where w_E and w_V denote the molecular weights of reserves and structural biomass. The scaled reserve densities e_0 and e_b are defined as $e_* \equiv E_*([E_m]V_m)^{-1}$, where E_* denotes the initial amount of reserves or the amount at hatching.

The relative weight at hatching is

$$W_w(a_b)/W_w(0) = (e_b + w_V/w_E)l_b^3/e_0. \tag{4.42}$$

The total production of 'minerals' during incubation, $\boldsymbol{M}_{\mathcal{M}}(a_b)$, amounts to

$$\boldsymbol{M}_{\mathcal{M}}(a_b) \equiv \int_0^{a_b} \dot{\boldsymbol{J}}_{\mathcal{M}}(a)\,da = -\boldsymbol{n}_{\mathcal{M}}^{-1}\boldsymbol{n}_{\mathcal{O}}\begin{pmatrix} 0 & -[M_V]V_b & \overline{\mu}_E^{-1}(E_0 - E_b) & 0 \end{pmatrix}^T \tag{4.43}$$

4.3.4 Composition of reserves and structural mass

Figure 4.16 illustrates that the change in composition of biomass for increasing growth rates can be used to obtain the composition of the reserves and of the structural mass. This method can be applied not only to elements but also to any chemical compound that can be measured in organisms. Indirect evidence can be used to obtain the amounts, without separating structure and reserve physically, see Figure 9.7.

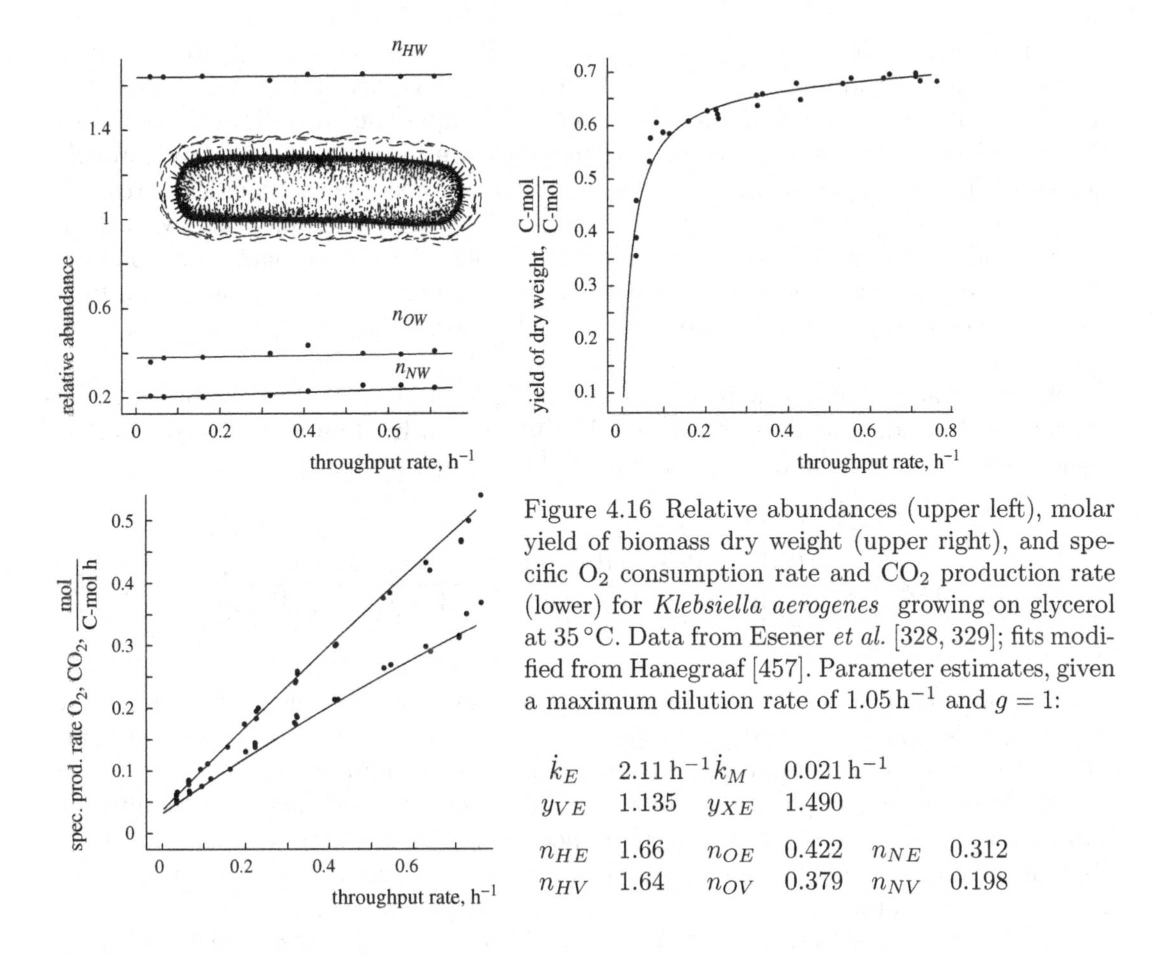

Figure 4.16 Relative abundances (upper left), molar yield of biomass dry weight (upper right), and specific O_2 consumption rate and CO_2 production rate (lower) for *Klebsiella aerogenes* growing on glycerol at 35 °C. Data from Esener *et al.* [328, 329]; fits modified from Hanegraaf [457]. Parameter estimates, given a maximum dilution rate of 1.05 h^{-1} and $g = 1$:

$\dot{k}_E$	2.11 h^{-1}	$\dot{k}_M$	0.021 h^{-1}		
y_{VE}	1.135	y_{XE}	1.490		
n_{HE}	1.66	n_{OE}	0.422	n_{NE}	0.312
n_{HV}	1.64	n_{OV}	0.379	n_{NV}	0.198

The relative contributions of the three basic powers to the mass conversions depend on the substrate density, and therefore on throughput rate of a chemostat, as is illustrated in Figure 4.16 for the conversion process of glycerol into the bacterium *Klebsiella aerogenes* at steady state. The data on the elemental composition, and on the yield of dry weight and the specific dioxygen and carbon dioxide fluxes, lead to the following relationship between mineral fluxes and the three basic powers for $\dot{\boldsymbol{J}}_{\mathcal{M}+} = \dot{\boldsymbol{J}}_{\mathcal{M}*}\mathbf{1} = \dot{\boldsymbol{J}}_{\mathcal{M}}$ and $\dot{\boldsymbol{p}}_+ = \mathbf{diag}(\dot{\boldsymbol{p}})\mathbf{1} = \dot{\boldsymbol{p}}$

$$\dot{\boldsymbol{J}}_{\mathcal{M}+} = \begin{pmatrix} 0.14 & 1.00 & -0.49 \\ 1.15 & 0.36 & -0.42 \\ -0.35 & -0.97 & 0.63 \\ -0.31 & 0.31 & 0.02 \end{pmatrix} \dot{\boldsymbol{p}}_+/\overline{\mu}_E \tag{4.44}$$

rRNA belongs to reserve

RNA, mainly consisting of ribosomal RNA, is an example of a compound known to increase in abundance with the growth rate [609]. This property is used to measure the growth rate of fish, for example [171, 531]. In prokaryotes, which can grow much faster, the increase in rRNA is much stronger. Within the DEB model, we can only account for this relationship when (part of the) RNA is included in the energy reserves. This does not seem unrealistic, because when cells experience a decline in substrate density and thus a decline in energy reserves, they are likely to gain energy through the degradation of ribosomes [249]. It also makes sense, because the kinetics of reserve energy density is first order, which implies that the use of reserves increases with their density. The connection between the abundance of rRNA, i.e. the apparatus for protein synthesis, and energy density is, therefore, a logical one. No assumption of the DEB model implies that the energy reserves should be inert materials that only wait for further use.

RNA as a fraction of dry weight is given in Figure 4.17. If the weight of RNA is a fraction θ_v of the dry weight of structural biomass and a fraction θ_e of the dry weight of the energy reserves, the fraction of dry weight that is RNA equals

$$W_{RNA}/W_d = \frac{\theta_v[W_{Vd}]V + \theta_e[W_{Ed}]fV}{[W_{Vd}]V + [W_{Ed}]fV} = \frac{\theta_v + \theta_e f[W_{Ed}]/[W_{Vd}]}{1 + f[W_{Ed}]/[W_{Vd}]} \qquad (4.45)$$

The data are consistent with the assumption that all rRNA belongs to the reserve in *E. coli* and also most RNA, and that about half the energy reserves consist of RNA. The rate of RNA turnover is completely determined by this assumption.

It also has strong implications for the translation rate and the total number of translations made from a particular RNA molecule. The mean translation rate of a ribosome, known as the peptide elongation rate, is proportional to the ratio of the rate of protein synthesis to the energy reserves, E. The rate of protein synthesis is proportional to the growth rate plus part of the maintenance rate, which is higher the lower the growth rate in bacteria [1119]. The peptide elongation rate is plotted in Figure 4.18 for *E. coli* at 37 °C. If the contribution of maintenance to protein synthesis can be neglected, the elongation rate at constant substrate density is proportional to the ratio of the growth rate $\frac{d}{dt}V$ to the stored energy $[E_m]fV$. As shown by (4.25), the elongation rate in a rod of mean volume should be proportional to $\dot{r}/f$ at population growth rate $\dot{r}$. The relationship allows the estimation of the parameter l_d, which is hard to obtain in any other way.

The lifetime of a compound in the reserve of a rod is exponentially distributed with a mean residence time of $(\dot{k}_E(\frac{\delta}{3}\frac{V_d}{V} + 1 - \frac{\delta}{3}))^{-1}$. The mean residence time thus increases during the cell cycle. At division it is $\dot{k}_E^{-1}$, independent of the (population) growth rate. The total number of transcriptions of a ribosome, in consequence, increases with the population growth rate. Outside the cell, RNA is rather stable. The fact that the RNA fraction of dry weight depends on feeding conditions indicates that an RNA molecule has a restricted life span inside the cell.

Analyses of this type are required to see if the conclusion that rRNA belongs to the reserve also holds more generally for (isomorphic) eukaryotes. If so, rRNA can be used as a proxy for reserve.

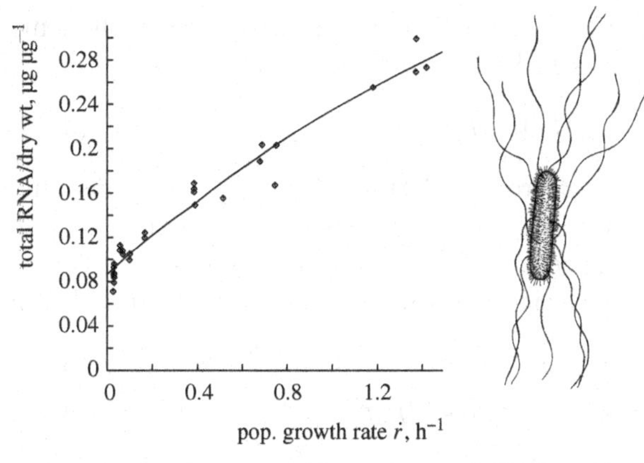

Figure 4.17 The concentration of RNA as a function of the population growth rate in *E. coli*. Data from Koch [609]. The least-squares estimates of the parameters are $\theta_e = 0.44$, $\theta_v = 0.087$ and $[W_{Ed}]/[W_{Vd}] = 20.7$.

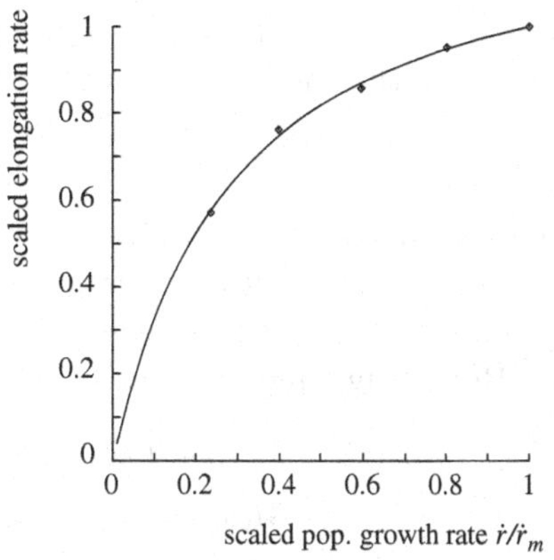

Figure 4.18 Elongation rate in *E. coli* for $\delta = 0.3$, $l_d = 0.24$, $g = 32.4$. Data from Bremer and Dennis [146]. Both elongation rate and population growth rate are expressed as fractions of their maximum value of $\dot{r}_m = 1.73\,\mathrm{h}^{-1}$ with an elongation rate of $21\,\mathrm{aa\,s^{-1}rib^{-1}}$.

DNA belongs to structure

Nuclear DNA belongs to structure, not because of any chemical property or function it has, but because of its dynamics. Each cell has just a single set of DNA molecules, irrespective of its nutritional state. This means that its amount follows that of structure and the amount of DNA per cell weight decreases as a function of the growth rate, because the weight of the cell increases due to the contribution of the reserve. A slightly more complex situation exists in some bacteria that grow fast, such as fast-growing *Escherichia coli*, which can reduce their division interval to some 20 minutes, while it takes an hour to copy their DNA. Their amount of DNA per cell increases with the growth rate, but otherwise the amount of DNA per cell hardly varies, as discussed at {279}. This simple reasoning shows that DNA can be used as a proxy for structure.

If mitochondrial DNA follows the dynamics of mitochondrial activity, it must at least partly belong to the reserve; to my knowledge little is known about its dynamics.

Composition changes during starvation

The linear decrease of compounds during starvation can be used to gain information on the composition of reserve and structure, using the following reasoning.

We first try to understand the decrease of a compound C in an organism during starvation, having measurements of how the amount M_C (in C-mol) changes in time t. At the start of the experiment, the organism has amounts of structure M_V and reserve M_E. Suppose that reserve mobilisation during starvation is just enough to cover the somatic maintenance costs. The amount of structure M_V remains constant, so if we focus on some compound C, e.g. protein, and follow it backward in time, with the time origin at the moment on which the reserve is fully depleted, we have

$$M_C(t^*) = M_{CV} + (M_{CE}/M_E)t^* \dot{J}_{EM} \tag{4.46}$$

where $M_C(t^*)$ is the amount of compound at reverted time t^*, M_{CV} the (constant) amount of compound in structure, M_{CE}/M_E the constant density of the compound in reserve and $\dot{J}_{EM}$ the (constant) rate of use of reserve for somatic maintenance purposes.

Reverting time back into the standard direction, we substitute $t = t_0 - t^*$ and obtain

$$M_C(t) = M_{CV} + (M_{CE}/M_E)(t_0 - t)\dot{J}_{EM} \quad \text{and for } M_{C0} = M_{CV} + \dot{J}_{CM} t_0 \tag{4.47}$$

$$= M_{C0} - t\dot{J}_{CM} \quad \text{with } \dot{J}_{CM} = (M_{CE}/M_E)\dot{J}_{EM} \tag{4.48}$$

This shows that each compound can decrease linearly at its own rate, even under the strong homeostasis assumption, which prescribes that the densities of the compound in reserve M_{CE}/M_E and in structure M_{CV}/M_V remain constant.

It also shows that, if we only know how the compound changes in time, we have access to M_{C0} and $\dot{J}_{CM}$, but not to the more informative M_{CV} and M_{CE} (i.e. information on the composition of structure and reserve).

We do have some relative information on the composition of reserve, if we know the time trajectories of several compounds: $\dot{J}_{C_1M}/\dot{J}_{C_2M} = M_{C_1E}/M_{C_2E}$. If we know when the reserve is depleted (namely at time t_0), we have access to the composition of structure M_{CV}/M_V, since $M_C(t_0) = M_{CV}$, but the individual will probably

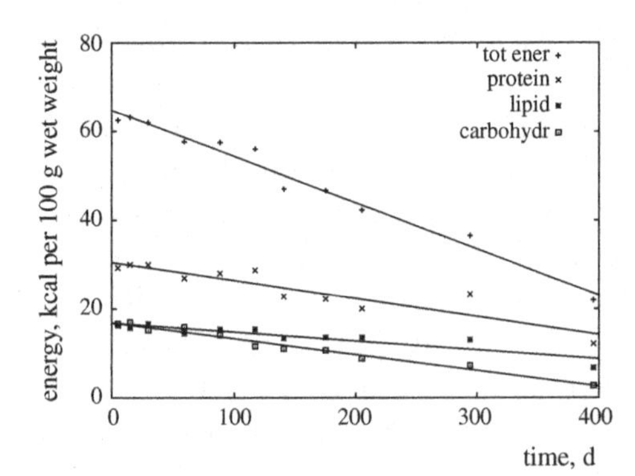

Figure 4.19 The amounts of energy in starving oyster *Crassostrea gigas*. Data from Whyte *et al.* [1248]. The parameter estimates (right), conversions, and their translation into composition information for three choices for the time t_0 at which the reserve is depleted. The caloric values $\overline{\mu}_C$ are from Table 4.2; 1 cal = 4.184 J.

100 g wet weight	t_0, d	total	protein	lipid	carbohydrate
$\overline{\mu}_C M_{C0}$, kcal		64.81	30.54	16.80	16.87
$\overline{\mu}_C \dot{J}_{CM}$, kcal d^{-1}		0.1042	0.0408	0.0200	0.0358
M_{C0}, mol		0.570	0.319	0.114	0.137
$\dot{J}_{CM}$, mmol d^{-1}			0.426	0.136	0.290
M_{CE}/M_E, $\frac{\text{mol}}{\text{mol}}$			0.500	0.159	0.341
M_{CV}/M_V, $\frac{\text{mol}}{\text{mol}}$	200		0.546	0.191	0.263
M_{CV}/M_V, $\frac{\text{mol}}{\text{mol}}$	400		0.537	0.185	0.278
M_{CV}/M_V, $\frac{\text{mol}}{\text{mol}}$	600		0.531	0.181	0.288

start to use structure to pay maintenance costs during prolonged starvation (causing deviations from linear decrease). Moreover it is likely that the reserve buffer that is allocated to reproduction is used under extreme starvation. This makes it difficult to have access to t_0.

Suppose now that we have information for *all* compounds, that is $\sum_i M_{C_iV} = M_V$ and $\sum_i M_{C_iE} = M_E$. Although the actual number of chemical compounds is formidable, they can be grouped into a limited number of chemical categories (e.g. proteins, lipids, etc). We have $\sum_i \dot{J}_{C_iM} = \dot{J}_{EM}$, so $\dot{J}_{C_iM}/\sum_j \dot{J}_{C_jM} = M_{C_iE}/M_E$. We also have $\sum_i M_{C_i0} = M_V + \dot{J}_{EM}t_0$, so $M_V = \sum_i M_{C_i0} - t_0 \sum_i \dot{J}_{C_iM}$, which we know if we have an estimate for t_0. We obviously must have that $t_0 < \sum_i M_{C_i0}/\sum_i \dot{J}_{C_iM}$. The composition of structure is then found from $M_{C_iV}/M_V = (M_{C_i0} - t_0 J_{C_iM})/M_V$.

Figure 4.19 gives an example of application. The composition of reserve and structure turns out to insensitive for the unknown moment of reserve depletion, and reserve of oyster is rich in carbohydrates, compared to structure.

4.4 Respiration

Respiration, i.e. the use of dioxygen or the production of carbon dioxide, is usually taken to represent the total metabolic rate in an organism. The latter is a rather vague concept, however. The conceptual relationship between respiration and use of energy has changed with time. Von Bertalanffy identified it with anabolic processes, while the Scope For Growth concept, {435}, relates it to catabolic processes. The respiration rate can now be defined concisely as the dioxygen flux $\dot{J}_O = \eta_{OA}\dot{p}_A + \eta_{OD}\dot{p}_D + \eta_{OG}\dot{p}_G$, or the carbon dioxide flux $\dot{J}_C = \eta_{CA}\dot{p}_A + \eta_{CD}\dot{p}_D + \eta_{CG}\dot{p}_G$.

If product formation, such as faeces, is only linked to assimilation, the carbon dioxide production rate that is not associated with assimilation, $\dot{J}_C$ for $\dot{p}_A = 0$, follows from (4.35), (4.36) and (4.37)

$$\dot{J}_{CD} + \dot{J}_{CG} = \left(1 - n_{NE}\frac{n_{CN}}{n_{NN}}\right)\overline{\mu}_E^{-1}(\dot{p}_D + \dot{p}_G) - \left(1 - n_{NE}\frac{n_{CN}}{n_{NN}}\right)\eta_{VG}\dot{p}_G \tag{4.49}$$

where the second term represents the carbon from the reserve flux that is allocated to growth and actually fixed into new tissue. The relationship simplifies if the nitrogenous waste contains no carbon ($n_{CN} = 0$). For embryos and juveniles we have $\dot{p}_G + \dot{p}_D = \dot{p}_C$, but adults fix carbon in embryonic reserves. This change at puberty results in a stepwise decrease in carbon dioxide production as illustrated in Figure 4.15. Table 2.5 gives the required powers: for adults we have the growth power $\dot{p}_G = V_m[\dot{p}_M]l^2\frac{e-l-l_T}{1+e/g}$ and the dissipating power

$$\dot{p}_D = V_m[\dot{p}_M]\left(l^3 + \left(\frac{1}{\kappa} - 1\right)l_p^3 + l^2 l_T + (1 - \kappa_R)\left(\frac{1}{\kappa} - 1\right)\left(l^2\frac{e - l + l_T/g}{1 + e/g} + l^3 - l_p^3\right)\right) \tag{4.50}$$

Initially, eggs hardly use dioxygen, but dioxygen consumption rapidly increases during development; see Figure 4.20. In juveniles and adults, dioxygen consumption is

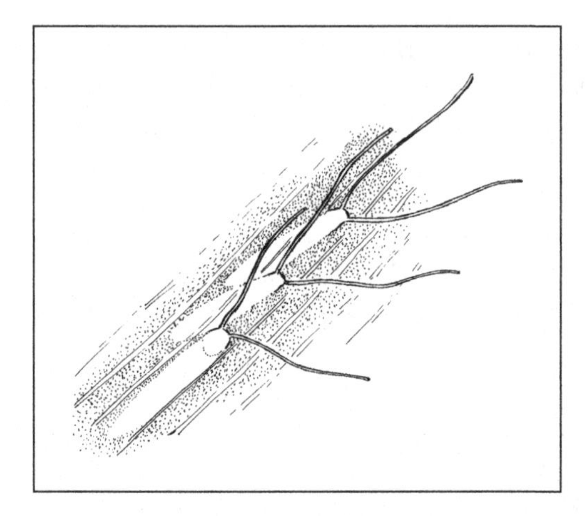

Figure 4.20 The water stick insect *Ranatra linearis* deposits its eggs in floating decaying plant material, where dioxygen availability is usually poor. The eggs are easily spotted by the special respiratory organs that peek out of the plant. Just prior to hatching, eggs typically need a lot of dioxygen, see Figure 2.12.

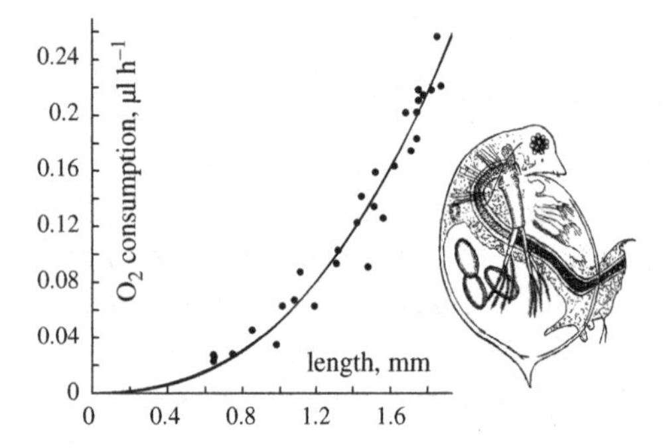

Figure 4.21 The respiration rate of *Daphnia pulex* with few eggs at 20 °C as a function of length. Data from Richman [961]. The DEB-based curve $0.0336L^2 + 0.01845L^3$ as well as the standard allometric curve $0.0516L^{2.437}$ are plotted on top of each other, but they are so similar that this is hardly visible. If you look hard, you will notice that the line width varies a little.

usually measured in individuals that have been starved for some time, to avoid interpretation problems related to digestion. (For micro-organisms this is not possible without a substantial decrease of reserves.) The expression for the dissipating power is consistent with the observation that respiration rate increases with reserve density [602], while reserves themselves do not use dioxygen. Moreover, it explains the reduction of respiration during starvation; see {118}.

The following subsection shows that respiration is a weighted sum of volume and surface area in steady-state conditions for the reserves. This is, for all practical purposes, numerically indistinguishable from the well-known Kleiber's rule, which takes respiration to be proportional to weight to the power 0.75 or length to the power 2.25; see Figure 4.21. There are three major improvements in comparison to Kleiber's rule. This model does not suffer from dimensional problems, it provides an explanation rather than a description and it accommodates species that deviate from Kleiber's rule; endotherms respire in proportion to surface area (approximately), which has given rise to Rubner's surface law.

As already mentioned, this result solves the long-standing problem of why the volume-specific respiration of ectotherms decreases with increasing size when organisms of the same species are compared. This problem has been identified as one of the central problems of biology [1262]. Many theories have been proposed, see e.g. [950] for a discussion, but all use arguments that are too specific to be really satisfactory: heating (but many species are ectothermic), muscle power (but movement costs are relatively unimportant), gravity (but aquatic species escape gravity), branching transport systems (but open circulatory systems are frequent). Peters [884] even argued that we should stop looking for a general explanation. The DEB theory, however, does offer a general explanation: the overhead of growth. A comparison of different species is covered in a later chapter, {295}, where it is shown that inter-species comparisons work out a bit differently.

Since embryos do not assimilate, their respiration has contributions from growth plus maturation and maintenance only. The observation that respiration is proportional to a weighted sum of volume and change in volume goes back to the 1957 Smith study [1081] of salmon eggs. At constant food density, the change in volume is of the von Bertalanffy type, which makes respiration proportional to $3\dot{r}_B(V_\infty^{1/3}V^{2/3} - V) + \dot{k}_M V$. This gives five parameters to be estimated from two data sets on respiration and growth: V_b, V_∞, $\dot{r}_B$, a proportionality constant for respiration and the maintenance rate coefficient, $\dot{k}_M$. This gives 2.5 parameters per data set, which is acceptable if the scatter is not too large.

4.4.1 Respiration Quotient

The Respiration Quotient (RQ) is of practical interest because it yields information on the relative contributions of protein, carbohydrates and lipids. The RQ for a particular compound X with chemical indices $\boldsymbol{n}_X$ can be obtained by decomposing the compound into minerals with chemical indices $\boldsymbol{n}_\mathcal{M}$. The composition of the nitrogenous waste (N), which can also contain C and O, affects the RQ if the compound contains N. The stoichiometric coefficients are $\boldsymbol{y}_{\mathcal{M}X} = \begin{pmatrix} y_{CX} & y_{HX} & y_{OX} & y_{NX} \end{pmatrix}^T = \boldsymbol{n}_\mathcal{M}^{-1}\boldsymbol{n}_X$, and RQ$= y_{CX}/y_{OX}$.

The RQ value can be used to make inferences about the composition of reserves, see Table 4.2. Proteins are by far the most diverse polymers; the composition (and function) of protein differs over the taxa, the RQ varying between 0.8 and 0.9.

The chemical indices of the structural biomass and the reserves relate to that of the three groups of polymers as

$$n_{*_1*_2} = n_{*_1P_s}Y_{P_s,*_2} + n_{*_1L_i}Y_{L_i,*_2} + n_{*_1P_r}Y_{P_r,*_2} \quad *_1 \in \{C,H,O,N\}\ ,\ *_2 \in \{V,E\} \quad (4.51)$$

where $Y_{*_3*_2}$ is the molar yield of $*_3 \in \{P_s, L_i, P_r\}$, on $*_2$, and $1 = Y_{P_s,*_2} + Y_{L_i,*_2} + Y_{P_r,*_2}$. Given the composition of the three polymers, the composition of structural biomass and that of reserves have two degrees of freedom each. The constraint that the RQ is independent of the state of the individual eliminates all degrees of freedom and the value of the RQ can be directly translated into the composition of reserves and structure in terms of the three groups of polymer.

For living organisms, the situation is a bit more complex, because the ratio between the produced carbon dioxide and the consumed dioxygen is not necessarily constant.

Table 4.2 Structural biomass and mainly reserves consist of three groups of polymers. The Respiration Quotient (RQ) value for protein relates to urea as nitrogenous waste. The formula for lipid refers to tripalmitin; octanol ($C_8H_{18}O$, or $CH_{2.25}O_{0.125}$) is frequently used as a chemical model for a typical animal fat, see {327}.

compound	symbol	formula	RQ	kJ/ g	kJ/ C-mol
polysaccharides	*Ps*	CH_2O	1.00	17.2	516
lipids	*Li*	$CH_{1.92}O_{0.12}$	0.67	38.9	616
proteins	*Pr*	$CH_{1.61}O_{0.33}N_{0.28}$	0.84	17.6	401

The standard assumption in animal physiology that the RQ is constant imposes constraints on the composition of reserve relative to structure that I now evaluate, assuming that the gas fluxes that are associated with the assimilation process, and so with feeding, are excluded, as usual, from the measurements of the RQ, by starving the individual prior to the measurement. An explicit expression for the RQ can be obtained from the relationships $\dot{\boldsymbol{J}}_\mathcal{M} = \boldsymbol{\eta}_\mathcal{M}\dot{\boldsymbol{p}}$ and $\boldsymbol{\eta}_\mathcal{M} = -\boldsymbol{n}_\mathcal{M}^{-1}\boldsymbol{n}_\mathcal{O}\boldsymbol{\eta}_\mathcal{O}$. As is usually done, we set the first row of $\boldsymbol{n}_\mathcal{O}$ equal to $\mathbf{1}^T$, set $\eta_{PD} = \eta_{PG} = 0$, and obtain

$$\text{RQ} = -\frac{\dot{J}_{CD} + \dot{J}_{CG}}{\dot{J}_{OD} + \dot{J}_{OG}} = -\frac{(\boldsymbol{n}_\mathcal{M}^{-1})_C\,\boldsymbol{n}_\mathcal{O}\begin{pmatrix} 0 & \dot{p}_G\eta_{VG} & -\dfrac{\dot{p}_D + \dot{p}_G}{\overline{\mu}_E} & 0\end{pmatrix}^T}{(\boldsymbol{n}_\mathcal{M}^{-1})_O\,\boldsymbol{n}_\mathcal{O}\begin{pmatrix} 0 & \dot{p}_G\eta_{VG} & -\dfrac{\dot{p}_D + \dot{p}_G}{\overline{\mu}_E} & 0\end{pmatrix}^T} \tag{4.52}$$

$$= \frac{1 - n_{NV}\dfrac{n_{CN}}{n_{NN}} - \left(1 - n_{NE}\dfrac{n_{CN}}{n_{NN}}\right)\dfrac{\mu_{GV}}{\overline{\mu}_E}\left(1 + \dfrac{\dot{p}_D}{\dot{p}_G}\right)}{1 + \dfrac{n_{HV}}{4} - \dfrac{n_{OV}}{2} - \dfrac{n}{4}\dfrac{n_{NV}}{n_{NN}} - \left(1 + \dfrac{n_{HE}}{4} - \dfrac{n_{OE}}{2} - \dfrac{n}{4}\dfrac{n_{NE}}{n_{NN}}\right)\dfrac{\mu_{GV}}{ol\mu_E}\left(1 + \dfrac{\dot{p}_D}{\dot{p}_G}\right)} \tag{4.53}$$

where $(\boldsymbol{n}_\mathcal{M}^{-1})_*$ denotes the row of $\boldsymbol{n}_\mathcal{M}^{-1}$ that corresponds to compound $*$. The contribution of energetics to the RQ is thus via the ratio of growth to dissipation power. The RQ is in practice usually taken to be a constant for a particular species. Within the DEB model, the RQ is independent of the state of the animal (size l and reserve density e) if the following condition on the composition of E, V and N holds

$$\frac{1 + \dfrac{n_{HE}}{4} - \dfrac{n_{OE}}{2} - \dfrac{n}{4}\dfrac{n_{NE}}{n_{NN}}}{1 + \dfrac{n_{HV}}{4} - \dfrac{n_{OV}}{2} - \dfrac{n}{4}\dfrac{n_{NV}}{n_{NN}}} = \frac{1 - n_{NE}\dfrac{n_{CN}}{n_{NN}}}{1 - n_{NV}\dfrac{n_{CN}}{n_{NN}}} \tag{4.54}$$

in which case

$$\mathrm{RQ} = \frac{1 - n_{NE}\dfrac{n_{CN}}{n_{NN}}}{1 + \dfrac{n_{HE}}{4} - \dfrac{n_{OE}}{2} - \dfrac{n}{4}\dfrac{n_{NE}}{n_{NN}}} = \frac{1 - n_{NV}\dfrac{n_{CN}}{n_{NN}}}{1 + \dfrac{n_{HV}}{4} - \dfrac{n_{OV}}{2} - \dfrac{n}{4}\dfrac{n_{NV}}{n_{NN}}} \tag{4.55}$$

The respiration rate (the dioxygen consumption rate as well as the carbon dioxide production rate) is then proportional to the mobilisation power if the contribution via assimilation is excluded. The literature (which typically treats biomass as being homogeneous) frequently identifies respiration with catabolism; this has been the motivation for the notation $\dot{p}_C$. This link is not correct in the context of DEB theory. Condition (4.54) simplifies considerably if the Urination Quotient (UQ) is constant as well, see (4.61). The elemental composition of the reserves has to be equal to that of the structural biomass, if the Watering Quotient (WQ) is also independent of the state of the animal, see {153}.

4.4.2 Heat increment of feeding

The heat increment of feeding, also known as 'specific dynamic action', and many other terms, is defined (strangely enough) as the *dioxygen* consumption that is associated with the feeding process. Apart from a small part that relates to the processing of proteins, the heat increment of feeding is little understood [1262]. It can be obtained, however, from the conservation law for mass. The dioxygen consumption per C-mole of food is independent of the states of the animal (reserve e and size l) as

$$\frac{\dot{J}_{OA}}{\dot{J}_X} = (\boldsymbol{n}_{\mathcal{M}}^{-1})_{O*}\boldsymbol{n}_{\mathcal{O}} \begin{pmatrix} 1 \\ 0 \\ -\mu_{AX}\overline{\mu}_E^{-1} \\ -\mu_{AX}\eta_{PA} \end{pmatrix} = \begin{pmatrix} -1 \\ -\frac{1}{4} \\ \frac{1}{2} \\ \frac{n}{4n_{NN}} \end{pmatrix}^T \begin{pmatrix} n_{CX} & n_{CE} & n_{CP} \\ n_{HX} & n_{HE} & n_{HP} \\ n_{OX} & n_{OE} & n_{OP} \\ n_{NX} & n_{NE} & n_{NP} \end{pmatrix} \begin{pmatrix} 1 \\ -\frac{\mu_{AX}}{\overline{\mu}_E} \\ -\frac{\mu_{AX}}{\mu_{AP}} \end{pmatrix} \tag{4.56}$$

where $(\boldsymbol{n}_{\mathcal{M}}^{-1})_{O*}$ denotes the row of $\boldsymbol{n}_{\mathcal{M}}^{-1}$ that relates to O, which is the third row. The expression shows how assimilation-associated dioxygen consumption depends on the composition of food, faeces, reserves and nitrogenous waste, and the digestion efficiency through the parameters μ_{AX}, μ_{AP} and $\overline{\mu}_E$.

4.5 Nitrogen balance

Standard 'static' energy budget studies treat energy in urine similar to energy in faeces, by subtracting both from energy contained in food to arrive as metabolisable energy that is available to the animal, see {433}. Since the gut contents still belong to the 'outside world', this is reasonable for energy in faeces, but not for energy in urine. The

DEB model leads to a different point of view, where dissipating power and anabolic power also contribute to the nitrogenous waste. The energy (and nitrogen) in urine originates from all powers, where the contributions to urine appear as overhead costs. Without reserves, the two points of view can be translated into each other, but with reserves the two become essentially different.

If $n_{NE} < \frac{[M_V]}{\overline{\mu}_E^{-1}} \frac{n_{NV}}{[E_G]}$, the flux of nitrogenous waste that relates to anabolic power, $\dot{J}_{NG}$, is negative, meaning that nitrogen is built in rather than wasted in the transformation of reserves to structural biomass. The flux of nitrogenous waste that relates to dissipating power amounts to $\dot{J}_{ND} = \frac{\dot{p}_D}{\overline{\mu}_E} \frac{n_{NE}}{n_{NN}}$, which can be a substantial part of the total flux of nitrogenous waste.

4.5.1 Urination Quotient

Analogous to the Respiration Quotient, we can define the Urination Quotient (UQ) as

$$\mathrm{UQ} = -\frac{\dot{J}_{ND} + \dot{J}_{NG}}{\dot{J}_{OD} + \dot{J}_{OG}} = -\frac{(\boldsymbol{n}_{\mathcal{M}}^{-1})_N \, \boldsymbol{n}_{\mathcal{O}} \begin{pmatrix} 0 & \dot{p}_G \eta_{VG} & -\frac{\dot{p}_D + \dot{p}_G}{\overline{\mu}_E} & 0 \end{pmatrix}^T}{(\boldsymbol{n}_{\mathcal{M}}^{-1})_O \, \boldsymbol{n}_{\mathcal{O}} \begin{pmatrix} 0 & \dot{p}_G \eta_{VG} & -\frac{\dot{p}_D + \dot{p}_G}{\overline{\mu}_E} & 0 \end{pmatrix}^T} \tag{4.57}$$

$$= \frac{\frac{n_{NV}}{n_{NN}} - \frac{n_{NE}}{n_{NN}} \frac{\mu_{GV}}{\overline{\mu}_E} \left(1 + \frac{\dot{p}_D}{\dot{p}_G}\right)}{1 + \frac{n_{HV}}{4} - \frac{n_{OV}}{2} - \frac{n\, n_{NV}}{4\, n_{NN}} - \left(1 + \frac{n_{HE}}{4} - \frac{n_{OE}}{2} - \frac{n\, n_{NE}}{4\, n_{NN}}\right) \frac{\mu_{GV}}{\overline{\mu}_E} \left(1 + \frac{\dot{p}_D}{\dot{p}_G}\right)} \tag{4.58}$$

The UQ is independent of the states of the animal (size l and reserve density e) if the following condition on the composition of E, V and N holds

$$\frac{1 + \frac{n_{HE}}{4} - \frac{n_{OE}}{2} - \frac{n\, n_{NE}}{4\, n_{NN}}}{1 + \frac{n_{HV}}{4} - \frac{n_{OV}}{2} - \frac{n\, n_{NV}}{4\, n_{NN}}} = \frac{n_{NE}}{n_{NV}} \tag{4.59}$$

in which case

$$\mathrm{UQ} = \frac{\frac{n_{NE}}{n_{NN}}}{1 + \frac{n_{HE}}{4} - \frac{n_{OE}}{2} - \frac{n\, n_{NE}}{4\, n_{NN}}} = \frac{\frac{n_{NE}}{n_{NN}}}{1 + \frac{n_{HV}}{4} - \frac{n_{OV}}{2} - \frac{n\, n_{NV}}{4\, n_{NN}}} \tag{4.60}$$

The UQ and the RQ are both constant if

$$n_{NE} = n_{NV} \tag{4.61}$$

$$n_{HE} - 2n_{OE} = n_{HV} - 2n_{OV} \tag{4.62}$$

Analogous to the RQ and UQ, we can define a Watering Quotient WQ $= -\frac{\dot{J}_{HD}+\dot{J}_{HG}}{\dot{J}_{OD}+\dot{J}_{OG}}$: the ratio of the water production to dioxygen consumption that relates to dissipation and growth. (For terrestrial animals, the evaporation of water invokes a drinking behaviour, which is discussed on {153}.) The condition that the RQ, UQ and WQ are all independent of the state of the animal directly translates to the condition that the reserves and the structural biomass have the same elemental composition. The dioxygen consumption, the carbon dioxide production, the nitrogenous waste production and the water production that relate to dissipation and growth are all proportional to the reserve mobilisation rate, comparing individuals of the same species (i.e. the same parameter values), but different states (structural biomass and/or reserves).

If the RQ and the UQ are both constant, the ratio of the carbon dioxide to the nitrogenous waste production equals $\frac{\text{RQ}}{\text{UQ}} = \frac{n_{NN}}{n_{NE}} - n_{CN}$, excluding contributions via assimilation as before. If the WQ is constant as well, the ratio of the water to the nitrogenous waste production equals $\frac{\text{WQ}}{\text{UQ}} = \frac{n_{HE}}{2}\frac{n_{NN}}{n_{NE}} - \frac{n_{HN}}{2}$.

4.5.2 Ammonia excretion

Many algae take up nitrogenous compounds, such as ammonia, from the environment, but even algae also excrete ammonia, associated with maintenance and growth. This follows from the balance equation for nitrogen, given the composition of reserves and structural mass. Ammonia excretion can be quantified for V1-morphs as follows.

Let n_{NE} and n_{NV} denote the chemical indices for nitrogen in reserves and structural mass. The ammonia excretion that is associated with maintenance and growth can then be written as

$$\dot{J}_{N_H,D} + \dot{J}_{N_H,G} = (j_{N_H,D} + j_{N_H,G})M_V = n_{NE}(\dot{p}_D + \dot{p}_G)/\overline{\mu}_E - n_{NV}\dot{p}_G/\mu_{GV} \quad (4.63)$$

$$j_{N_H,D} + j_{N_H,G} = n_{NE}y_{EV}(\dot{k}_E + \dot{r}) - n_{NV}\dot{r} \quad (4.64)$$

with dissipating power $\dot{p}_D = M_V\mu_{GV}\dot{k}_M$ and growth power $\dot{p}_G = M_V\mu_{GV}\dot{r}$ (see Table 4.1); the mass–mass coupler y_{EV} is the ratio of two energy–mass couplers, $y_{EV} = \mu_{GV}/\overline{\mu}_E$, where μ_{GV} is the reserve energy investment per unit increase of structural mass.

The flux of nitrogenous waste that relates to assimilation amounts to $\dot{J}_{N,A} = \eta_{NA}\dot{p}_A$, with $\eta_{NA}n_{NN} = -\eta_{XA}n_{NX} + n_{NE}/\overline{\mu}_E + \eta_{PA}n_{NP}$.

4.6 Water balance

The drinking rate equals the water flux $\dot{J}_{HX} = \dot{J}_H$ for aquatic animals, but terrestrial animals have to deal with evaporation of water. The water balance implies that the sum of the water fluxes by metabolism, evaporation and drinking amounts to zero. Embryos usually do not drink and are 'designed' such that evaporation takes care of water outflux, although small changes in water content have been found. The water content of tissues in birds gradually decreases during growth, which led Ricklefs and Webb [967] and Konarzewski [616] to model juvenile growth on the basis of the water content of the tissue. Here, we idealise the process by assuming strict homeostasis for both the structural biomass and the reserves, while focusing on juveniles and adults. Note that water emission via urine is incorporated in the composition of the nitrogenous waste, which could be large enough to let the water outflux $\dot{J}_H$ be negative and turn it into a water influx.

Evaporation has two main routes, one via water loss linked to respiration, $\dot{J}_{HO}$, and one via transpiration, $\dot{J}_{HH}$. Water loss via respiration is proportional to dioxygen consumption via the amount of inhaled air, so $\dot{J}_{HO} = \dot{J}_O y_{HO}$, while transpiration is proportional to surface area, so $\dot{J}_{HH} = \{\dot{J}_{HH}\} V_m^{2/3} l^2$, where $\{\dot{J}_{HH}\}$ does not depend on the state of the animal. Both loss rates depend on water pressure in the air, temperature, wind speed and behavioural components. The DEB model leads to a drinking rate of

$$\dot{J}_{HX} = \begin{pmatrix} 0 & 1 & y_{HO} & 0 \end{pmatrix} \dot{\boldsymbol{J}}_{\mathcal{M}} + \{\dot{J}_{HH}\} V_m^{2/3} l^2 \tag{4.65}$$

This two-parameter model for the drinking process is, of course, an idealised picture which pushes the concept of homeostasis to the extreme. The water content of urine is actually rather variable, depending on environmental and behavioural factors. However, the model might be helpful as a first approximation to reveal the coupling that must exist between drinking and energetics.

Water plays an essential role in the transport of nutrients from the environment to terrestrial plants, and in the translocation of their metabolites. Its quantitative role can only be understood in a multivariate setting, see next subsection.

4.6.1 Plant–water relationships

Terrestrial plants have intimate relationships with water, and total biomass production is found to vary almost proportionally to the annual precipitation across the globe [837, p 124]. Since plants cannot move, the local availability of water is the main factor determining the distribution of plant species [1272]. Like all organisms they need water for metabolic purposes, as autotrophs they need it as electron donor, but, above all, they need it for transport [935]. From a geophysiological perspective, plants are structures that pump water from the soil into the atmosphere. The evaporation of water from the leaves generates a water flux from the roots to the shoot, which is used for internal transport and for nutrient uptake from the soil. Factors that control evaporation include temperature, relative humidity, wind speed and water supply in the soil [1007, 1071]. Plants can modify evaporation by stomata in the leaves, but this regulation is limited by the need to acquire carbon dioxide. Jones [562], Nobel [842] and Lambers *et al.* [680] give an excellent discussion of quantitative aspects.

Suppose that the arrival rate of nutrients at the receptor in the root, $\dot{J}_r$, is proportional to its concentration in the water, X_n, and the water flux per receptor. The water flux is proportional to the shoot area where transpiration takes place, which controls nutrient transport, and to the availability of water in the soil, X_H. The proportionality factor includes the regulation of stomata opening by the plant, and atmospheric factors (temperature, wind, humidity). The number of receptors is proportional to the surface area of the root. The surface areas of roots and shoot are proportional to $A_r = \mathcal{M}_r(V_r)V_r^{2/3}$ and $A_s = \mathcal{M}_s(V_s)V_s^{2/3}$, respectively. The uptake rate of nutrient is proportional to the number of receptors times $\frac{\dot{k}\rho\dot{J}_r}{\dot{k}+\rho\dot{J}_r}$, where ρ is the binding probability, and $\dot{k}$ the dissociation rate between receptor and bound nutrient. This leads to the uptake rate of nutrient

$$\dot{J}_N = \{\dot{J}_{NAm}\}A_r(1 + X_{KN}/X_N)^{-1}, \quad X_{KN} \propto (X_H A_s/A_r)^{-1} \tag{4.66}$$

The surface area of the shoot appears in the saturation 'constant' X_{KN}, which is no longer constant.

Nutrient uptake is arrested by lack of water transport in this formulation, because the saturation constant becomes very large. This mechanism gives a direct coupling between nutrient uptake and precipitation. In water-rich soils, the control of transport on nutrient uptake might be less, and in subaquatic conditions even absent. This boils down to an additive term X_0, which relates to diffusive transport of the nutrient: $X_{KN} \propto (X_0 + X_H A_s/A_r)^{-1}$.

4.7 Isotope dynamics in the standard DEB model

The isotope dynamics for macrochemical reactions as discussed at {95} can now be applied to the standard DEB model, where dioxygen is a non-limiting substrate. This excludes applications in micro-aerobic environments (e.g. parasites inside hosts), where we have to deal with transitions from aerobic metabolism to fermentation. The focus is on the isotope dynamics of reserve E and structure V with food X as substrate.

4.7.1 Three contributions to isotope fluxes

The three basic fluxes assimilation, dissipation and growth each have a catabolic and an anabolic aspect. In the catabolic aspect energy is generated by oxidation of substrate that is used in the anabolic aspect where the same substrate is used to provide building blocks for products. For simplicity's sake, I now assume that the atoms of the mineral products all originate from the organic substrate or from dioxygen and that the carbon dioxide production in the anabolic aspect is negligibly small. Since it is known that e.g. carbon dioxide is both product and substrate, at least in some transformations, this assumption need not be correct and applications might urge to change this assumption.

Under extreme starvation conditions shrinking might occur; the anatomy of this transformation is basically identical to that of dissipation.

The chemical indices for the minerals $\mathcal{M} = (C, H, O, N)$ and the organic compounds $\mathcal{O} = (X, V, E, P)$ are assumed to be known and $\mathbf{0} = \boldsymbol{n}_\mathcal{M}\mathbf{Y}_{\mathcal{M}s}^k + \boldsymbol{n}_\mathcal{O}\mathbf{Y}_{\mathcal{O}s}^k$ so $\mathbf{Y}_{\mathcal{M}s}^k = -\boldsymbol{n}_\mathcal{M}^{-1}\boldsymbol{n}_\mathcal{O}\mathbf{Y}_{\mathcal{O}s}^k$ for any choice of organic substrate s. Many aquatic organisms use ammonia as nitrogenous waste, so $n_{CN} = 0$, $n_{HN} = 3$, $n_{ON} = 0$, $n_{NN} = 1$. Since $\boldsymbol{n}_{\mathcal{M}}^{-1}$ is well defined, $\boldsymbol{Y}_{\mathcal{M}s}^{k}$ is known, once $\boldsymbol{Y}_{\mathcal{O}s}^{k}$ is given.

Assimilation

Assimilation A is defined as the transformation

$$Y_{XE}^A X + Y_{OE}^A O \to Y_{EE}^A E + Y_{PE}^A P + Y_{HE}^A H + Y_{NE}^A N + Y_{CE}^A C$$

for food X, dioxygen O, reserve E, faeces P, water H, nitrogenous waste N, carbon dioxide C. The organic yield coefficients are

$$\boldsymbol{Y}_{\mathcal{O}E}^A = (\ Y_{XE}^A \quad Y_{VE}^A \quad Y_{EE}^A \quad Y_{PE}^A\ )^T = (\ -\tfrac{1}{y_{EX}} \quad 0 \quad 1 \quad \tfrac{y_{PX}}{y_{EX}}\ )^T \tag{4.67}$$

from which follow the mineral yield coefficients $\boldsymbol{Y}_{\mathcal{M}E}^A = (\ Y_{CE}^A \quad Y_{HE}^A \quad Y_{OE}^A \quad Y_{NE}^A\ )^T = -\boldsymbol{n}_{\mathcal{M}}^{-1}\boldsymbol{n}_{\mathcal{O}}\boldsymbol{Y}_{\mathcal{O}E}^A$; the assimilation flux $\dot{J}_{EA}$ is determined by DEB theory.

The anabolic fraction is $\kappa_A^a = y_{EX} = 1/y_{XE}$, so $\kappa_A^c = 1 - y_{EX}$. If $y_{EX} + y_{PX} = 1$, we have $Y_{CX}^A = 0$.

If selection occurs for isotope 0 of element i of food X in assimilation A, we need to use the apparent coefficient $n_{iX}^{0A_a}$ for reserve, rather than the actual coefficient n_{iX}^{0A}, using $\beta_{iX}^{0A_a}$. Likewise we need to use $n_{iX}^{0A_c}$ for faeces with $\beta_{iX}^{0A_c}$. We have isotope flux $\dot{j}_{XA}^{0i} = (n_{iX}^{0A_a}\kappa_A^a + n_{iX}^{0A_c}\kappa_A^c)\dot{J}_{XA}$.

Dissipation

The catabolic aspect of dissipation just oxidises reserve into minerals. Somatic maintenance, $\dot{J}_{EM}$, which is one of the components of the dissipation flux, is partly used for the turnover of structure, which means that structure is both a substrate and a product. No net synthesis of structure occurs in association with dissipation.

Dissipation D is defined as the transformation

$$Y_{EE}^D E + Y_{VE}^D V + Y_{OE}^D O \to -Y_{VE}^D V + Y_{HE}^D H + Y_{NE}^D N + Y_{CE}^D C$$

for reserve E, structure V, dioxygen O, water H, nitrogenous waste N, carbon dioxide C.

$$\boldsymbol{Y}_{\mathcal{O}E}^D = (\ Y_{XE}^D \quad (Y_{VE}^D - Y_{VE}^D) \quad Y_{EE}^D \quad Y_{PE}^D\ )^T = (\ 0 \quad 0 \quad 1 \quad 0\ )^T \tag{4.68}$$

from which follow the mineral yield coefficients $\boldsymbol{Y}_{\mathcal{M}E}^D = (\ Y_{CE}^D \quad Y_{HE}^D \quad Y_{OE}^D \quad Y_{NE}^D\ )^T = -\boldsymbol{n}_{\mathcal{M}}^{-1}\boldsymbol{n}_{\mathcal{O}}\boldsymbol{Y}_{\mathcal{O}E}^D$; the dissipation flux $\dot{J}_{ED}$ is given by DEB theory.

A fixed fraction of the somatic maintenance flux is used for the turnover of structure; a fraction of this flux is used to generate energy to drive the turnover; the flux $\dot{J}_{EM_a} = \kappa_M \dot{J}_{EM}$ is used as building blocks for turnover, where κ_M is a new parameter. Let $\dot{J}_{VD}$ denote the turnover flux of structure, which enters as well as leaves the pool of structure. Using $\dot{J}_{EM_a}$ as a reference flux, the formation and use of structure in the turnover process read

$$E + y_{VE}^{M_a} V \to (1 + y_{VE}^{M_a}) V + Y_{HE}^{M_a} H + Y_{OE}^{M_a} O + Y_{NE}^{M_a} N \qquad (4.69)$$

$$(1 + y_{VE}^{M_a}) V \to y_{VE}^{M_a} V + Y_{CE}^{M_c} C + Y_{HE}^{M_c} H + Y_{OE}^{M_c} O + Y_{NE}^{M_c} N \qquad (4.70)$$

where $y_{VE}^{M_a} = \dot{J}_{VD}/\dot{J}_{EM_a} - 1$ is a model parameter. The carbon dioxide that is produced in the turnover can be associated with the catabolic reserve flux that generates the energy to drive turnover or with the catabolic structure flux that leaves the structure. This is why it is not implemented in the formation of structure. Fractionation in the turnover process can thus occur in the reserve flux that is used for synthesis, and in the structure flux that partitions in an anabolic and a catabolic aspect. The fraction of the mobilised structure that re-enters the structure pool is $\frac{y_{VE}^{D_a}}{1+y_{VE}^{D_a}}$.

The simplest assumption is that all atoms of the flux $\dot{J}_{EM_a}$ are fixed into new structure. So for selection of isotope 0 of element i in reserve E in the dissipation process D, we might use the apparent coefficient $n_{iE}^{0D_a}$ for structure with $\beta_{iE}^{0D_a}$. We have isotope flux $\dot{J}_{ED}^{0i} = (n_{iE}^{0D_a} \kappa_D^a + n_{iE}^{0D_c} \kappa_D^c) \dot{J}_{ED}$.

During shrinking, the product yield of structure is less than the substrate yield, but otherwise also some synthesis of structure still occurs and the equations remain the same.

Growth

Growth G is defined as the transformation

$$Y_{EV}^G E + Y_{OV}^G O \to Y_{VV}^G V + Y_{HV}^G H + Y_{NV}^G N + Y_{CV}^G C \qquad (4.71)$$

for reserve E, dioxygen O, structure V, water H, nitrogenous waste N, carbon dioxide C.

$$\mathbf{Y}_{\mathcal{O}E}^G = (\ Y_{XE}^G \quad Y_{VE}^G \quad Y_{EE}^G \quad Y_{PE}^G\ )^T = (\ 0 \quad -y_{VE} \quad 1 \quad 0\ )^T \qquad (4.72)$$

from which follow the mineral yield coefficients $\mathbf{Y}_{\mathcal{M}E}^G = (\ Y_{CE}^G \quad Y_{HE}^G \quad Y_{OE}^G \quad Y_{NE}^G\ )^T = -\boldsymbol{n}_{\mathcal{M}}^{-1} \boldsymbol{n}_{\mathcal{O}} \mathbf{Y}_{\mathcal{O}E}^G$; the growth flux $\dot{J}_{EG}$ is determined by DEB theory.

The anabolic fraction is $\kappa_G^a = y_{VE} = 1/y_{EV}$, so $\kappa_G^c = 1 - y_{VE}$. We must have $y_{EV} > 1$. Since all structure originates from reserve in the anabolic route. If selection occurs on reserve with isotope 0 in element i in reserve E in growth G, we need to use the apparent coefficient $n_{iE}^{0G_a}$ for structure, rather than the actual coefficient n_{iE}^{0G}, using $\beta_{iE}^{0G_a}$. We have isotope flux $\dot{J}_{EG}^{0i} = (n_{iE}^{0G_a} \kappa_G^a + n_{iE}^{0G_c} \kappa_G^c) \dot{J}_{EG}$.

4.7.2 Changes in isotope fractions

The coefficients n_{ij}^{0k}, i.e. the isotope frequency in element i of compound j, relative to the carbon frequency in that compound in the various fluxes are quantified in (3.15). Now we focus on the dynamics of the fraction of isotopes in the pools.

Let fraction γ_{ij}^0 denote the amount of isotopes of element i in the pool of compound j as a fraction of the amount of element i in that pool, i.e. $n_{ij}M_j$ and let $n_{ij}^0 M_j$ denote the amount of isotopes of type 0 of element i in the pool of compound j. So

$$\frac{d}{dt}\gamma_{ij}^0 = \frac{d}{dt}\frac{n_{ij}^0 M_j}{n_{ij} M_j} = \frac{\frac{d}{dt} n_{ij}^0 M_j}{n_{ij} M_j} - \gamma_{ij}^0 \frac{\frac{d}{dt} M_j}{M_j} = \frac{\sum_k n_{ij}^{0k} \dot{J}_{jk}}{n_{ij} M_j} - \gamma_{ij}^0 \frac{\sum_k \dot{J}_{jk}}{M_j} \tag{4.73}$$

$$= \sum_{k \in \mathcal{P}} \left(\frac{n_{ij}^{0k}}{n_{ij}} - \gamma_{ij}^0 \right) \frac{\dot{J}_{jk}}{M_j} \tag{4.74}$$

The last equality holds because for the processes with $\dot{J}_{jk} < 0$, so for which compound j serves as substrate rather than as product, we have $n_{is}^{0k} = \gamma_{is}^0 n_{is} = n_{is}^0$.

We now apply this for $j = E, V$ and $k = A, D_a, G$ to the standard DEB model. The changes in isotope fractions γ_{iE}^0 and γ_{iV}^0, given those in the substrates $\gamma_{iX}(t)$ and $\gamma_{OO}(t)$, are

$$\frac{d}{dt}\gamma_{iE}^0 = \left(\frac{n_{iE}^{0A}}{n_{iE}} - \gamma_{iE}^0 \right) \frac{\dot{J}_{EA}}{M_E}; \quad \frac{d}{dt} M_E = \dot{J}_{EA} + \dot{J}_{EC} \quad \text{for} \quad \dot{J}_{EC} < 0 \tag{4.75}$$

$$\frac{d}{dt}\gamma_{iV}^0 = \left(\frac{n_{iV}^{0G}}{n_{iV}} - \gamma_{iV}^0 \right) \frac{\dot{J}_{VG}}{M_V} + \left(\frac{n_{iV}^{0D_a}}{n_{iV}} - \gamma_{iV}^0 \right) \frac{\dot{J}_{VD_a}}{M_V}; \quad \frac{d}{dt} M_V = \dot{J}_{VG} + \dot{J}_{VD_s} \tag{4.76}$$

where $\dot{J}_{VD_a}$ represents the (positive) flux of structure turnover as part of the somatic maintenance process and the (negative) flux $\dot{J}_{VD_s}$ the shrinking, which only occurs during extreme starvation.

4.7.3 Effects of temperature

Temperature affects rates, and selection depends on odds ratios, which are dimensionless. So effects of temperature on fractionation are only indirect, via effects on metabolic rates (assimilation, dissipation, growth). An increase in temperature causes an increase in dissipation, so an increase in the rate at which the isotope fraction in structure increases. Isotope enrichment in the food chain has several components: (1) the isotope fraction of food increases, which cause an increase in the isotope fraction of reserve and structure of the predator, (2) body size typically increases with the trophic level, so the life span and mean age, which means that dissipation-linked enrichment has more time to proceed (independent of food characteristics). So the observation that isotope fractions increase with the trophic level does not imply an enrichment in the assimilation process.

If the trajectory of isotope enrichment is well captured with enrichment in dissipation only (including responses to changes in food availability and temperature), this would give support for the position of maintenance in the metabolic organisation within the context of DEB theory. Notice that the Marr–Pirt model (for prokaryotes) specifies that structure is used for maintenance, rather than reserve, so it would be impossible to obtain enrichment linked to dissipation with this model.

4.7.4 Persistent products and reconstruction

Products that accumulate in solid form (hair, nails, shells, bones, earplugs, otoliths, wood) 'write' a record of the food–temperature history, which can be reconstructed using chemical identifiers, including the isotope signal. DEB theory specifies that these products are formed in association with assimilation, dissipation and/or growth. Collaborative work with Laure Pecquerie on otoliths of anchovy [876] indicates that the contribution of assimilation is negligible, and that those of dissipation and of growth differ in opacity, which produces the typical banded pattern in otoliths. Opacity as a function of length in slices of otoliths can be used to reconstruct (scaled) food intake.

Given that fractionation occurs at the separation of the anabolic and catabolic subfluxes of assimilation, dissipation and growth, the isotope frequency of an element in the products of any of these three fluxes might equal that before separation, that of the catabolic or that of the anabolic flux depending from which flux the product is actually formed.

4.7.5 Doubly labelled water

An ingenious method to measure the carbon dioxide flux indirectly is via the differential loss of isotopes of (injected) doubly labelled water. The method overcomes the problem that direct measurement of the carbon dioxide flux gives an instantaneous value only, and its measurement affects (the behaviour of) the animal. The interest in carbon dioxide fluxes stems from their relationship with energy fluxes, which is discussed on {162}. The method is based on the assumptions that labelled oxygen of water is exchanged (rapidly) with dioxygen of carbon dioxide, and that the loss of deuterium reflects the loss of water. A few additional simplifying assumptions are also useful for obtaining a simple interpretation of the results, such as that labelled and unlabelled body water are completely mixed, and that loss of label other than via water and carbon dioxide loss is negligible [704].

The total water flux equals $\dot{J}_{HL} = \dot{J}_{HX} + y_{HN}\dot{J}_N$, where y_{HN} denotes the moles of water in the nitrogenous waste, per mole of nitrogenous waste. The amount of body water equals $M_H = y_{HV}M_V + y_{HE}M_E$, so that the specific rate at which deuterium is lost equals $\dot{h}_H = \dot{J}_{HL}/M_H$. An estimate for M_H can be obtained by back-extrapolation of the oxygen label density at time zero, given a known amount of injected label. The specific loss rate of deuterium, combined with the total amount of body water, leads to an estimate for total water flux $\dot{J}_{HL}$. The specific loss rate of oxygen label equals $\dot{h}_O = (\dot{J}_{HL} + 2\dot{J}_C)/M_H$, which can be used to obtain $\dot{J}_C$, when $\dot{J}_{HL}$ and M_H are known.

4.8 Enthalpy, entropy and free energy balances

Thornton's rule [1154] relates dissipating heat to dioxygen consumption, by a fixed conversion of 519 ($\pm$13) kJ(mol $O_2)^{-1}$ [49]. This choice is not fully satisfactory, because it lacks a mechanistic underpinning, and because it is obviously not applicable to anaerobic conditions. The correlation between dissipating heat and carbon dioxide production has been found to be reduced by variations in the type of substrate [221]. Heijnen [484] related dissipating heat to C-moles of formed biomass. This choice is problematic because of maintenance. If substrate density is low enough, no new biomass will be produced but heat will still dissipate. Given the chemical coefficients, the proportionality between heat dissipation and dioxygen consumption can be translated into a condition on the specific enthalpies within the context of DEB theory.

The strong homeostasis assumption for structural biomass and reserves implies a direct link between the dissipating heat and the free energies and entropies of structural mass and reserve as worked out in [1088] and provides a theoretical underpinning of the method of indirect calorimetry. We first need to study the energy balance of the system 'individual plus relevant compounds'.

4.8.1 Energy balance: dissipating heat

Work that is involved in changes in volumes is typically negligibly small at the surface of the Earth, but in the deep ocean, this work has profound effects on energetics and biochemistry [401, 1038]. Neglecting this effect, the dissipating heat $\dot{p}_{T+}$ follows from the energy balance equation using (4.35) and (4.37)

$$0 = \dot{p}^\circ_{T+} + \overline{\boldsymbol{h}}^T_{\mathcal{M}} \dot{\boldsymbol{J}}_{\mathcal{M}} + \overline{\boldsymbol{h}}^T_{\mathcal{O}} \dot{\boldsymbol{J}}_{\mathcal{O}} = \dot{p}^\circ_{T+} + (\overline{\boldsymbol{h}}^T_{\mathcal{O}} - \overline{\boldsymbol{h}}^T_{\mathcal{M}} \boldsymbol{n}^{-1}_{\mathcal{M}} \boldsymbol{n}_{\mathcal{O}}) \boldsymbol{\eta}_{\mathcal{O}} \dot{\boldsymbol{p}} \tag{4.77}$$

where $\overline{\boldsymbol{h}}^T_{\mathcal{M}} \equiv (\, \overline{h}_C \;\; \overline{h}_H \;\; \overline{h}_O \;\; \overline{h}_N \,)$ and $\overline{\boldsymbol{h}}^T_{\mathcal{O}} \equiv (\, \overline{h}_X \;\; \overline{h}_V \;\; \overline{h}_E \;\; \overline{h}_P \,)$ are the specific enthalpies of the minerals and the organic compounds, respectively, and $\dot{p}^\circ_{T+}$ is the net heat release by all chemical reactions. If the temperature of the organism is constant, the net heat release $\dot{p}^\circ_{T+}$ is equal to the net heat dissipated by the organism $\dot{p}_{T+}$.

This balance equation can be used to obtain the molar enthalpies of the organic compounds, $\overline{\boldsymbol{h}}^T_{\mathcal{O}}$, given the molar enthalpies for the minerals, $\overline{\boldsymbol{h}}^T_{\mathcal{M}}$ from the literature (see Table 4.3), and the measured dissipating heat. This heat can be negative if heat from the environment is required to keep the temperature of the individual constant. Generally measurements of dissipating heat at four different food levels are required to obtain the four enthalpies for the organic compounds; if the enthalpies of food X and faeces P are known then only measurements of dissipated heat at two different food densities are required. The specific enthalpy of biomass equals $\overline{h}_W = \frac{m_E \overline{h}_E + \overline{h}_V}{m_E + 1}$.

The chemical potentials $\overline{\mu}$ have to be computed simultaneously with the molar entropies $\overline{s}$. The chemical potential and entropies of food $\overline{\mu}_X$ and $\overline{s}_X$, structure $\overline{\mu}_V$

Table 4.3 Formation enthalpies and absolute entropies of CO_2, H_2O and O_2 at 25°C were taken from Dean [252]. The formation enthalpy and absolute entropy for NH_3 at 25°C were taken from Atkins [39].

formula	state	enthalpy kcal mol^{-1}	entropy cal mol^{-1} K^{-1}
CO_2	g	−94.05	51.07
H_2O	l	−68.32	16.71
O_2	g	0	49.00
NH_3	aq.	−19.20	26.63

and $\overline{s}_V$, reserve $\overline{\mu}_E$ and $\overline{s}_E$, and faeces $\overline{\mu}_P$ and $\overline{s}_P$ can be obtained from (4.77) with

$$0 = (\overline{\boldsymbol{\mu}}_{\mathcal{M}} + T\overline{\boldsymbol{s}}_{\mathcal{M}})^T \dot{\boldsymbol{J}}_{\mathcal{M}} + (\overline{\boldsymbol{\mu}}_{\mathcal{O}} + T\overline{\boldsymbol{s}}_{\mathcal{O}})^T \dot{\boldsymbol{J}}_{\mathcal{O}} + \dot{p}^\circ_{T+} \tag{4.78}$$

$$= \left((\overline{\boldsymbol{h}}_{\mathcal{M}} - \overline{\boldsymbol{\mu}}_{\mathcal{M}} - T\overline{\boldsymbol{s}}_{\mathcal{M}}) \, \boldsymbol{n}_{\mathcal{M}}^{-1} \boldsymbol{n}_{\mathcal{O}} - \overline{\boldsymbol{h}}_{\mathcal{O}} + \overline{\boldsymbol{\mu}}_{\mathcal{O}} + T\overline{\boldsymbol{s}}_{\mathcal{O}} \right)^T \dot{\boldsymbol{J}}_{\mathcal{O}}, \tag{4.79}$$

by measuring the dissipated heat $\dot{p}_{T+} \simeq \dot{p}^\circ_{T+}$ and computing the organic and mineral flows at eight different food densities (or four different food densities if molar entropies and chemical potentials of food X and faeces P are known), where $\overline{\boldsymbol{\mu}}_{\mathcal{M}}$ and $\overline{\boldsymbol{s}}_{\mathcal{M}}$ collect the values of the molar chemical potentials and molar entropies for the four minerals, while $\overline{\boldsymbol{\mu}}_{\mathcal{O}}$ and $\overline{\boldsymbol{s}}_{\mathcal{O}}$ do that for the organic compounds, as before. Rather than measuring dissipating heat, the method of indirect calorimetry can be used, see {162}. The specific entropy of biomass equals $\overline{s}_W = \frac{m_E \overline{s}_E + \overline{s}_V}{m_E + 1}$.

Convection and radiation

The dissipating heat contributes to the thermal fluxes to and from the individual. The individual loses heat via convection and radiation at a rate $\dot{p}_{TT} = \{\dot{\pi}_T\}(T_b - T_e)V^{2/3} + \{\dot{\pi}_R\}(T_b^4 - T_e^4)V^{2/3}$. Here T_e denotes the absolute temperature in the environment, including a relatively large sphere that encloses the individual. For radiation considerations, the sphere and individual are assumed to have grey, opaque diffuse surfaces. T_b is the absolute temperature of the body; $V^{2/3}$ is the body surface area; $\{\dot{\pi}_T\}$ is the thermal conductance and $\{\dot{\pi}_R\} = \epsilon\dot{\sigma}$ is the emissivity times the Stefan–Boltzmann constant $\dot{\sigma} = 5.6710^{-8}$ J m^{-2} s^{-1} K^{-4}; see for instance [668]. The body temperature does not change if the heat loss via convection and radiation matches the dissipating heat, $\dot{p}_{T+} = \dot{p}_{TT}$. This relationship can be used to obtain the body temperature or the heating costs, given knowledge about the other components. It specifies, for instance, how a temporary increase in activity reduces heating costs, using complementary physiological information about activity efficiencies [166, 1228, 1252].

Most animals, especially the aquatic ones, have a high thermal conductance, which gives body temperatures only slightly above the environmental ones. Endotherms, however, heat their body to a fixed target value, usually some $T_b = 312$ K, and have a thermal conductance as small as $\{\dot{\pi}_T\} = 5.43$ J cm^{-2} h^{-1} K^{-1} in birds and 7.4–9.86 J cm^{-2} h^{-1} K^{-1} in mammals, as calculated from [497]. Thermal conductance can be modified by environmental and behaviour factors, see e.g. [895, 896].

Heat loss by evaporation and thermo-neutral zone

Most endotherms are terrestrial and lose heat also via evaporation of water at a rate $\dot{p}_{TH}$, say. The relationship $\dot{p}_T > \dot{p}_{TH} + \dot{p}_{TT}$ determines the lower boundary of the thermo-neutral zone: the minimum environmental temperature at which no endothermic heating is required. It also specifies the heating requirement at a given environmental temperature. To see how, we first have to consider the water balance in more detail, to quantify the heat $\dot{p}_{TH}$ that goes into the evaporation of water. The individual loses water via respiration at a rate proportional to the use of dioxygen, i.e. $\dot{J}_{HO} = y_{HO}\dot{J}_O$, see [638, 1187], and via transpiration, i.e. cutaneous losses. The latter route varies between 2% and 84% of the total water loss in birds, despite the lack of sweat glands [250]. Water loss, $\dot{J}_{HH}$, via transpiration is proportional to body surface area, to the difference in vapour pressure of water in the skin and the ambient air and to the square root of the wind speed, and depends on behavioural components. The heat loss by evaporation amounts to $\dot{p}_{TH} = \mu_{TH}(\dot{J}_H + \dot{J}_{HO} + \dot{J}_{HH})$, with $\mu_{TH} = 6$ kJ mol^{-1}. Within the thermo-neutral zone, endotherms control their body temperature by evaporation, through panting or sweating, which affects the water balance via enhanced drinking.

Entropy production

The rate of entropy production by the organism $\dot{\sigma}$ is a measure of the amount of dissipation that is occurring. It can be quantified for each food density if the temperature of the organism and the entropies of the organic compounds are known:

$$0 = \dot{\sigma} + \frac{\dot{p}_{T+}}{T} + \overline{\boldsymbol{s}}_{\mathcal{M}}^T \dot{\boldsymbol{J}}_{\mathcal{M}} + \overline{\boldsymbol{s}}_{\mathcal{O}}^T \dot{\boldsymbol{J}}_{\mathcal{O}} = T\dot{\sigma} + \overline{\boldsymbol{\mu}}_{\mathcal{M}}^T \dot{\boldsymbol{J}}_{\mathcal{M}} + \overline{\boldsymbol{\mu}}_{\mathcal{O}}^T \dot{\boldsymbol{J}}_{\mathcal{O}} \tag{4.80}$$

4.8.2 Indirect calorimetry: aerobic conditions

The relationships between enthalpies, entropies and free energies are simpler for aerobic conditions because for most important reactions in aerobic biological systems $T\,\Delta\overline{s}$ is very small compared to $\Delta\overline{h}$ and therefore the enthalpy of the reaction $\Delta\overline{h}_+$ is approximated using its Gibbs energy $\Delta\overline{\mu}_+$, since at constant temperature we have $\Delta\overline{\mu} = \Delta\overline{h} - T\,\Delta\overline{s} \simeq \Delta\overline{h}$ [386]. Consequently we have

$$-T\dot{\sigma} = \dot{p}_{T+} \quad \text{and} \quad 0 = \dot{p}_{T+}^{\circ} + \overline{\boldsymbol{\mu}}_{\mathcal{M}}^T \dot{\boldsymbol{J}}_{\mathcal{M}} + \overline{\boldsymbol{\mu}}_{\mathcal{O}}^T \dot{\boldsymbol{J}}_{\mathcal{O}} \tag{4.81}$$

see [1088] and (4.78), (4.80).

The method of computing entropy from (4.78) simplifies under aerobic conditions and has been applied to the data and the fitted DEB model reported in Figure 4.16 on *Klebsiella* [1088], which resulted in a molar entropy of reserve of 74.8 J/C-mol K and of structure of 52.0 J/C-mol K. This value gives an entropy for biomass that is almost two times higher than the value obtained using the biochemical method of Battley [65], which does not account for spatial structure or the processes of life.

Indirect calorimetry uses measurements of dioxygen consumption, carbon dioxide and nitrogen production to estimate dissipating heat $\dot{p}_{T+}$:

$$\dot{p}_{T+} = \boldsymbol{\mu}_T^T \dot{\boldsymbol{J}}_{\mathcal{M}} \quad \text{with} \quad \boldsymbol{\mu}_T^T \equiv \left(\mu_{TC} \quad \mu_{TH} \quad \mu_{TO} \quad \mu_{TN} \right) \tag{4.82}$$

Its basis is just empirical when applied to individuals, rather than pure compounds, and has ancient roots, {162}. Examples are: $\mu_{TC} = 60$ kJ mol^{-1}, $\mu_{TH} = 0$, $\mu_{TO} = -350$ kJ mol^{-1} and $\mu_{TN} = -590$ kJ mol^{-1} in aquatic animals [136] that excrete ammonia as nitrogenous waste, or $-86\frac{n_{CN}}{n_{NN}}$ kJ mol^{-1} in birds [116]. For mammals, corrections for methane production have been proposed [161]. The coefficients $\boldsymbol{\mu}_T$ can be obtained by direct calorimetry, using multiple regression. The mass fluxes prove to be a weighted sum of the three basic powers, see {138}. Dissipating heat is again a weighted sum of the three powers and so of (three) mass fluxes, which justifies the method of indirect calorimetry.

Now we can reverse the argument and wonder how measurements of heat dissipation can be used to obtain the chemical potentials of the organic compounds. Substitution of (4.81) and (4.82) into (4.80) results in

$$\overline{\boldsymbol{\mu}}_{\mathcal{O}}^T = (\boldsymbol{\mu}_T^T + \overline{\boldsymbol{\mu}}_{\mathcal{M}}^T)\mathbf{n}_{\mathcal{M}}^{-1}\mathbf{n}_{\mathcal{O}} \tag{4.83}$$

Under anaerobic conditions, the amount of metabolic work substrates can do is typically very much reduced, not because of substantial changes in specific entropy, but because the products are not carbon dioxide and water, but fermentation products, such as ethanol and acetate. These products should be taken into account in (4.81), but otherwise the way to obtain the chemical potentials is similar. In special situations changes in specific entropies cannot be neglected, but then the detailed chemical composition of the environment should be taken into account as well.

4.8.3 Substrate-dependent heat dissipation

When different substrates are compared, the conversion efficiency of substrate to biomass tends to be proportional to the chemical potential $\overline{\mu}_X$, on the basis of C-moles. It seems reasonable to assume that μ_{AX} is proportional to the chemical potential of substrate, and consequently to the yield, cf (4.22). This is confirmed in Figure 4.22.

The dissipating heat $\dot{p}_{T+}$ from a chemostat at steady state, with total structural mass M_{V+} and reserve density e, is found from (4.82), (4.35) and (4.37):

$$0 = \dot{p}_{T+} + (\boldsymbol{\mu}_{\mathcal{O}}^T - \boldsymbol{\mu}_{\mathcal{M}}^T \boldsymbol{n}_{\mathcal{M}}^{-1} \boldsymbol{n}_{\mathcal{O}})\boldsymbol{\eta}_{\mathcal{O}} \dot{\boldsymbol{p}}(e, 1) M_{V+}/M_{Vm} \tag{4.84}$$

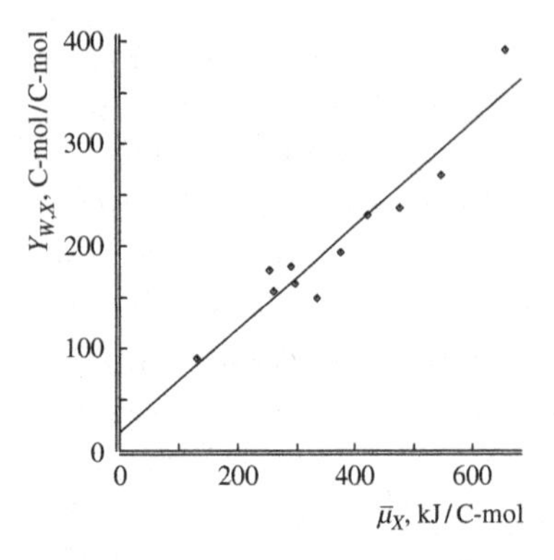

Figure 4.22 The molar yield of biomass corrected for a fixed population growth rate of $\dot{r} = 0.2\,\text{h}^{-1}$ is proportional to the chemical potential of substrate, expressed per C-mole in combustion reference. Data from Rutgers [1005] for *Pseudomonas oxalaticus* (•) and from van Verseveld, Stouthamer and others [777, 1190, 1191, 1192, 1193] for *Paracoccus denitrificans* (◦) under aerobic conditions with NH_4^+ as the nitrogen source, corrected for a temperature of 30 °C. No product, or a negligible amount, is formed during these experiments [1190].

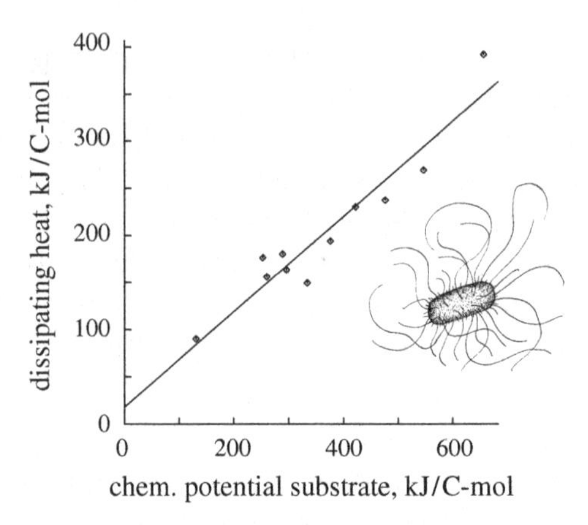

Figure 4.23 The amount of dissipating heat at maximum population growth rate is linear in the free energy per C-mole of substrate on the basis of combustion reference (pH = 7). Data from Rutgers [1005] and Heijnen and van Dijken [484, 485] for *Pseudomonas oxalaticus*, growing aerobically at 30 °C on a variety of substrates.

where $\dot{\boldsymbol{p}}(e,l) = (\ \dot{p}_A \quad \dot{p}_S + \dot{p}_J \quad \dot{p}_G + \dot{p}_R\ )^T$ as further specified in Table 4.1. If the population is growing at maximum rate, we have that $f = e = 1$, and

$$\dot{\boldsymbol{p}}(1,1) = \left(\ \frac{\dot{k}_E}{g} \quad \dot{k}_M \quad \frac{\dot{k}_E - \dot{k}_M g}{1+g}\ \right) \mu_{GV} M_{Vm} \tag{4.85}$$

When different substrates are compared, the dissipating heat tends to increase with the free energy of substrate. This is to be expected, because the maximum volume-specific assimilation rate $[\dot{p}_{Am}]$ and the maximum reserve capacity $[E_m]$ are proportional to the free energy per C-mole of substrate $\overline{\mu}_X$, see {163}, so the reserve turnover rate $\dot{k}_E$ is independent of $\overline{\mu}_X$, $g \propto \overline{\mu}_X^{-1}$, and the dissipating heat at maximum population growth rate is approximately linear in $\overline{\mu}_X$ if the combustion frame of reference is used. This frame of reference is necessary because a high free energy of substrate corresponds with a high degree of reduction, which requires more dioxygen to release the energy. In the combustion reference, this extra use of oxygen does not affect the relationship between free energy of substrate and heat dissipation. This is confirmed by the data of Rutgers [1005]; see Figure 4.23.

The idea that the type of substrate and environmental conditions affect the substrate/energy conversion μ_{AX} (and $[E_m]$) but nothing else is consistent with analyses of data from Pirt [900]. V1-morphs with small reserve capacities $[E_m]$ have $\frac{1}{Y_{WX}} = \frac{\mu_{GV}}{\mu_{AX}}(1 + \frac{\dot{k}_M}{\dot{r}})$, see (4.22). As S. J. Pirt noted, this relationship is linear in $\dot{r}^{-1}$, but the slope depends on the substrate–energy conversion μ_{AX}. Pirt found a wide range of

0.083–0.55 h^{-1} on a weight basis for two species of bacteria (*Aerobacter* species and a lipolytic bacterium) growing on two substrates at 37 °C, aerobically and anaerobically. The ratio of the slope to the intercept equals the maintenance rate coefficient, $\dot{k}_M$, which does not depend on the substrate–energy conversion. Pirt's data fall in the narrow range of $\dot{k}_M = 0.0393 - -0.0418\,h^{-1}$ [634]. These findings support the funnel concept, which states that a wide variety of substrates is decomposed to a limited variety of building blocks, which depend of course on the nature of the substrate and environmental conditions; these products are then built into biomass, which only depends on internal physiological conditions, subject to homeostasis.

4.9 Products

From a dynamic systems point of view, minerals can be considered as products, with contributions from the basic powers, apart from the fact that their fluxes can become negative (e.g. dioxygen for heterotrophs). Faeces are a product as well, where the contributions from dissipating and growth powers are zero, which ties faeces production directly to assimilation. Many micro-organisms produce a variety of products via several routes. If the DEB model still applies in the strict sense, the mere fact that product formation costs energy implies that product formation must be a weighted sum of the basic powers: assimilation, dissipation (maintenance) and growth. The energy drain to product formation can then be considered as an overhead cost in these three processes.

The necessity to tie product formation to all the three energy fluxes in general becomes obvious in a closer analysis of fermentation. If product formation is independent of one or more energy fluxes, mass balance equations dictate that more than one product must be made under anaerobic conditions, and that the relative amounts of these products must depend on the (population) growth rate in a very special way. In the Monod model, which does not include maintenance and reserves (see {351}), assimilation is proportional to growth investment, which leaves just a single energy flux available to couple to product formation. In the Marr–Pirt model, which does not include reserves, assimilation is proportional to maintenance plus growth investment, which leaves two energy fluxes available to couple to product formation. Maintenance and reserves together allow for a three-dimensional base for product formation: $\dot{J}_P = \dot{p}_A \eta_{PA} + \dot{p}_D \eta_{PD} + \dot{p}_G \eta_{PG}$, see (4.37). The quantitative aspects of products only differ from those of 'minerals' in that the weight coefficients for products are free parameters, while those for 'minerals' follow from mass conservation.

Since most unicellulars behave approximately as V1-morphs, assimilation rate and maintenance are both proportional to biomass, with constant proportionality coefficients at steady state. Leudeking and Piret [698] proposed in 1959 that product formation is a weighted sum of biomass and change in biomass (growth). They studied lactic acid fermentation by *Lactobacillus delbrueckii*. The Leudeking–Piret kinetics has proved extremely useful and versatile in fitting product formation data for many different fermentations [49]. It now turns out to be a special case of the DEB theory, where the biomass component links to maintenance.

For practical applications where no energies are measured, it might be useful to convert powers to mass fluxes via the coefficients $\zeta_{*_1*_2} = \eta_{*_1*_2}\overline{\mu}_E m_{Em}$, which leads to the specific production flux for V1-morphs

$$j_P = \zeta_{PM}\dot{k}_M g + \zeta_{PA}\dot{k}_E f + \zeta_{PG}\dot{r}g. \tag{4.86}$$

Milk of female mammals is an example of a product that is coupled to maintenance, which requires a temporal change in parameter values to describe its production in association with giving birth. The same holds for plant secretions (e.g. resin), in response to wounds, for example.

4.9.1 Fermentation

Many organisms can live in anaerobic environments, partly as a relic from their evolutionary history, as life originated in a world without free oxygen. Most parasites [1160, 1161], as well as gut and sediment dwellers [343, 345] do not usually encounter much dioxygen, and aquatic environments can also be low in dioxygen. Some fish [1223, 1224] and molluscs [159] survive periods without dioxygen. Parasitic helminths sport anaerobic metabolism in the core of their bodies, and aerobic metabolism in the peripheral layers, which become relatively less important during growth [1159].

The mass balance equation reveals that such organisms must produce at least one product, with an elemental composition that is independent (in the sense of linear algebra) of the composition of the other 'minerals' (carbon dioxide, water and nitrogenous waste). Usually, several products are formed. Under anoxic conditions, lipids cannot be metabolised, because their degree of reduction is too high, and the respiration chain cannot be used.

Fermentation is an anaerobic process in which organic compounds act as electron donor as well as electron acceptor. Usually several products are made rather than just one. These products can be valuable substrates under aerobic conditions, but under anaerobic conditions mass balances force organisms to leave them untouched. Under anaerobic conditions we have the constraints that

$$\begin{pmatrix} \mu_{AO}^{-1} & \mu_{DO}^{-1} & \mu_{GO}^{-1} \end{pmatrix} = \mathbf{0} \tag{4.87}$$

The practical implementation of these constraints in non-linear regressions is via Lagrange multipliers, which can be found in standard texts on calculus. An interesting consequence of these constraints is that there are no free parameters for product formation if just one product is made. Figure 4.24 shows that the DEB model accurately describes the fermentation process (biomass composition, substrate and product fluxes) with only $17/11 = 1.5$ parameters per curve. The experimental data do not obey the mass balance for carbon and oxygen in detail. Measurements of the volatile ethanol seem to be less reliable. The mass-balance-based model fit of Figure 4.24 suggests that the measured values represent 75% of the real ones when the measurement error is considered as a free parameter. The saturation coefficient X_K was poorly fixed by the data, and the chosen value should be considered as an educated guess.

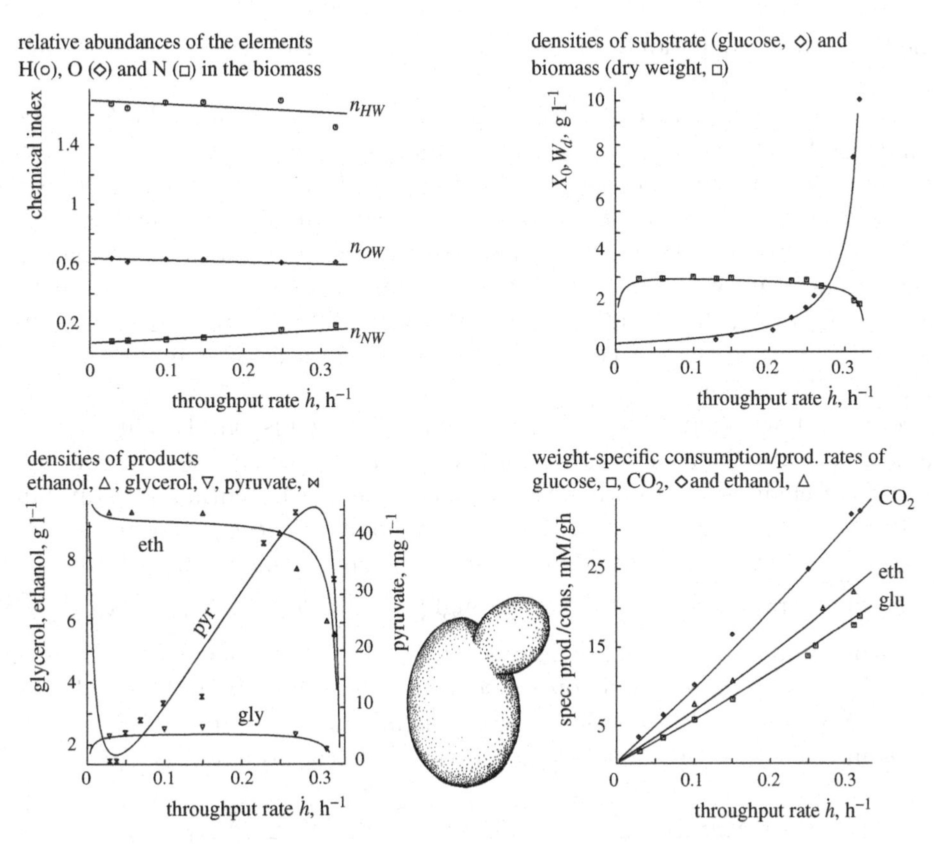

Figure 4.24 All these functions of population growth rate of *Saccharomyces cerevisiae* at 30 °C and a glucose concentration of 30 g l^{-1} in the feed have been fitted simultaneously [457]. The observation that the maximum throughput rate is 0.34 h^{-1} has also been used. Data from Schatzmann [1016]. The curves are based on expectations of the DEB model for V1-morphs, with parameters

$\dot{k}_E$	=	0.461 h^{-1}	g	=	0.385	$\dot{k}_M$	=	0.0030 h^{-1}
y_{VE}	=	1.206	y_{XE}	=	10.28	X_K	=	1.79 g l^{-1}
n_{HV}	=	1.70	n_{OV}	=	0.637	n_{NV}	=	0.071
n_{HE}	=	1.55	n_{OE}	=	0.572	n_{NE}	=	0.205
	ethanol			glycerol			pyruvate	
$\zeta_{P_1,A}$	=	8.047	$\zeta_{P_2,A}$	=	7.398	$\zeta_{P_3,A}$	=	0.0313
$\zeta_{P_1,D}$	=	3.019	$\zeta_{P_2,D}$	=	2.711	$\zeta_{P_3,D}$	=	0.0062
$\zeta_{P_1,G}$	=	0.336	$\zeta_{P_2,G}$	=	0.972	$\zeta_{P_3,G}$	=	−0.0365

Yeasts appear to be relatively rich in proteins when they grow fast, but their maximum growth rate is about half that of *Klebsiella*. Three products are made by the yeast: glycerol ($n_{HP_1} = 8/3$, $n_{OP_1} = 1$), ethanol ($n_{HP_2} = 3$, $n_{OP_2} = 0.5$) and pyruvate ($n_{HP_3} = 4/3$, $n_{OP_3} = 1$). A negative parameter for product formation means that the product is consumed, rather than produced, in the corresponding energy flux. So it is possible that compounds are produced at a rate proportional to one energy flux and consumed at a rate proportional to another energy flux. No theoretical problems occur as long as there is an overall net production.

Note that the maintenance rate coefficient $\dot{k}_M$ for *Klebsiella* at 35 °C is about 10 times that for *Saccharomyces* at 30 °C. The maintenance rate coefficient for fungi is usually found to be much smaller in the literature [92], which Bulthuis [174] explained by the fact that fungi make a lot of protein at high population growth rates, which costs a lot of energy. As the maintenance rate coefficient is the ratio of maintenance to structure costs, its value for fungi is low. Since protein density is coupled to the growth rate, however, the assumption of homeostasis dictates that most protein must be conceived as part of the reserves, so the costs of synthesis of structural biomass are not higher for this reason.

Figure 4.24 shows that biomass density hardly depends on the throughput rate. In practice, this also holds for most other compounds, except for the concentration of substrate. If changes in concentrations affect chemical potentials substantially, the chemical potential for substrate will be the first point to check (although substrate is usually processed intracellularly, rather than in the environment). The extremes of the substrate concentration are found for throughput rate $\dot{h} = 0$, where $X_0 = \frac{X_K g \dot{k}_M}{\dot{k}_E - g\dot{k}_M}$, and for throughput rate $\dot{h} = \dot{h}_m$, where $X_0 = X_r$ if death is negligible. The chemical potential of a compound depends on its concentration X as $\overline{\mu} = \overline{\mu}_{\text{ref}} + RT \ln X/X_{\text{ref}}$, where $R = 8.31441$ $\text{JK}^{-1}\text{mol}^{-1}$ is the gas constant. The maximum relative effect of differences in concentrations of substrate on the chemical potential is

$$\frac{\overline{\mu}_{X0,\text{max}} - \overline{\mu}_{X0,\text{min}}}{\overline{\mu}_{X0,\text{ref}}} = \frac{RT}{\overline{\mu}_{X_0,\text{ref}}} \ln \left\{ \frac{\dot{k}_E - g\dot{k}_M}{g\dot{k}_M} \frac{X_r}{X_K} \right\} \tag{4.88}$$

In the example of Figure 4.24, where the chemical potential of glucose is 2856 kJ mol^{-1} in the combustion frame of reference, the maximum relative effect amounts to 0.00777, which is negligibly small in view of many other uncertainties. However, the effect of changes in concentrations should be tested in each practical application.

4.10 Parameter estimation II: mass, energy and entropy

In the initial stages of estimation of DEB parameters [661], it is useful to avoid the use of moles and energies, which motivates the use of scaled reserve U_E and scaled maturity U_H. The initial scaled reserve U_E^0 can be known from g, $\dot{k}_M$, $\dot{k}_J$, $\dot{v}$, and f, or from L_b, L_∞, a_b, $\dot{r}_B$ and f, using the assumption $\dot{k}_M = \dot{k}_J$.

Suppose that the amount of carbon in a freshly laid egg M_E^0 and in a neonate $M_W^b = M_E^b + M_V^b$ are known, in combination with U_E^0. We first use the information in M_E^0 and obtain $\{\dot{J}_{EAm}\} = M_E^0/U_E^0$, and then, for $\{\dot{J}_{XAm}\} < 0$, $y_{EX} = -\{\dot{J}_{EAm}\}/\{\dot{J}_{XAm}\}$, $M_H^b = \{\dot{J}_{EAm}\}/U_H^b$, $M_H^p = \{\dot{J}_{EAm}\}/U_H^p$, $M_E^b = -U_E^b\{\dot{J}_{XAm}\}$. We then use the information in M_W^b, and obtain $M_V^b = M_W^b - M_E^b$ and $[M_V] = M_V^b L_b^{-3}$ (in actual length, if L_b is in actual length), $y_{VE} = \dot{v}[M_V](\kappa\{\dot{J}_{EAm}\}g)^{-1}$, $[\dot{J}_{EM}] = -\dot{k}_M[M_V]/y_{VE}$.

If the weight of a freshly laid egg W_0 and of a neonate W_b are known, as well as the moles of carbon in a freshly laid egg M_E^0, we can obtain the molecular weights of reserve and structure: $w_E = W_0/M_E^0$, $W_V^b = W_b - w_E M_E^b$ and $w_V = W_V^b/M_V^b$. The shape coefficient is $\delta_{\mathcal{M}} = (M_V^b/[M_V])^{1/3}/L_b = (d_V^{-1}W_V^b)^{1/3}/L_b$.

The parameter y_{VE}, and so $m_{Em} = \kappa g y_{VE}$ can also be obtained from the gonado-somatic index $Q = \frac{M_{E_R}}{M_E+M_V}$ where the reproduction buffer M_{E_R} has accumulated over a period t_1 in an individual that is fully grown at constant scaled functional response f. For an individual of structural volume $V = V_\infty = f^3L_m^3$ (or $L = fL_m$) and reserve mass $M_E = [E]V/\overline{\mu}_E = [E_m]f^4L_m^3/\overline{\mu}_E$, the index relates to κ, g, $\dot{k}_M$, $\dot{k}_J$, U_H^p (see {74}) as

$$Q = \frac{t_1\dot{k}_M g/f^3}{f + \kappa g y_{VE}}((1-\kappa)f^2 - \dot{v}^{-2}\dot{k}_M^2 g^2 \dot{k}_J U_H^p) \tag{4.89}$$

4.10.1 Composition parameters

If the elemental composition of a freshly laid egg (so of reserve) and that of a neonate is known, the chemical index of structure, i.e. the frequency of element $*$ in structure, relative to carbon, is given by

$$n_{*V} = n_{*W}m_W^b - n_{*E}m_E^b \quad \text{for} \quad * = H, O, N, \cdots \tag{4.90}$$

where $m_W^b = M_W^b/M_V^b$.

This is just one of a series of related techniques to unravel the composition of reserve and structure using measurements of biomass. Suppose that we have the elemental frequencies of two individuals of the same length (so the same amount of structure) at two scaled functional responses. We have $M_W = M_V + M_E$, and $M_E = fm_{Em}M_V$, where $m_{Em} = (m_W - 1)/f$ is the maximum reserve density. The structural mass M_V of an individual of total mass M_W equals $M_V = M_W/(1 + m_E)$. Moreover, if an organism has physical length L and structural mass M_V, the shape coefficient is $\delta_{\mathcal{M}} = (M_V/[M_V])^{1/3}/L$.

We also have

$$M_W\, n_{*W} = M_V\, n_{*V} + M_E\, n_{*E} \tag{4.91}$$

so the chemical indices of reserve and structure are

$$n_{*E} = \frac{f_1}{m_{W1}-1}\,\frac{m_{W1}-m_{W2}}{f_1-f_2}; \quad n_{*V} = m_{W1}n_{*W} - f_1\frac{m_{W1}-m_{W2}}{f_1-f_2} \tag{4.92}$$

This technique for computing the concentrations in reserve and structure can also be applied to compounds rather than chemical elements. The contribution of the reproduction buffer in the weight (and composition) of adults should be taken into account, but for juveniles we don't have these complications.

Knowledge about the chemical indices can be used to determine the molecular weights of reserve and structure, so to link masses and weights. A pertinent question is to include or exclude water in mass, volume and weight measurements. If water replaces reserve in starving organisms (likely in aquatic arthropods and other taxa with exoskeletons), strong homeostasis can only apply when we exclude water. In many other cases the inclusion of water is more handy.

4.10.2 Thermodynamic parameters

The estimation of the specific enthalpies, entropies, and chemical potentials is discussed below the balance equations (4.77) and (4.79). These equations make full use of the mass balances for all (generalised) compounds, which makes these thermodynamic parameters difficult to access.

The specific chemical potential $\overline{\mu}$ of a compound converts a flux of this compound (in moles per time) into a flux of Gibbs energy, for instance the assimilation energy flux is $\dot{p}_A = \overline{\mu}_E \dot{J}_{EA}$. The chemical potentials of organic compounds are essential to obtain the energy parameters $\{\dot{p}_{Am}\}$, $[E_G]$, $\{\dot{p}_T\}$, $[\dot{p}_M]$ and $[\dot{p}_J]$, see Table 3.3.

4.11 Trajectory reconstruction

4.11.1 Reconstruction of food intake from growth data

Many data sets on growth in the literature do not provide adequate information about food intake. Sometimes it is really difficult to gain access to this type of information experimentally. The blue mussel *Mytilus edulis* filters what is called particulate organic matter (POM). Apart from the problem of monitoring the POM concentration relevant to a particular individual, its characterisation in terms of nutritional value is problematic. The relative abundances of inert matter, bacteria and algae change continuously [644]. In the search for useful characterisations, it can be helpful to invert the argument: given an observed size and temperature pattern, can the assimilation energy be reconstructed in order to relate it to measurements of POM? The practical gain of such a reconstruction is in the use of correlation measures to determine the nutrition value of bacteria, algae, etc. Since the correlation coefficient is a linear measure, a direct correlation between bacteria numbers and mussel growth, for instance, only has limited value because assimilation and growth are related in a non-linear way.

Kautsky [583] measured mussels from four size classes kept individually in cages (diameter 10 cm) at a depth of 15 m in the Baltic at a salinity of 7 ‰. Suppose that (the mean) food density changes slowly enough to allow an approximation of the energy reserves with $e = f$. The growth equation (2.22) then reduces for a reference temperature T_{ref} to

$$\frac{d}{dt}l = \frac{(f(t)-l)_+}{3(f(t)+g)} g\dot{k}_M (T(t) > T_0) \exp\left(T_A\left(\frac{1}{T_{\text{ref}}} - \frac{1}{T(t)}\right)\right) \qquad (4.93)$$

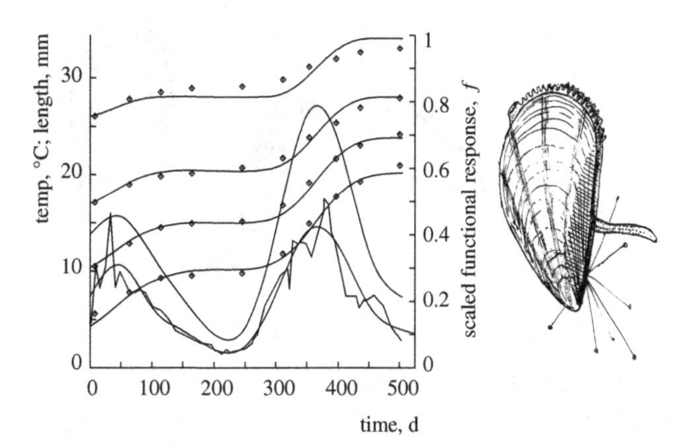

Figure 4.25 The reconstruction of the scaled functional response since 1st August from mean length–time data for four length classes of the mussel *Mytilus edulis* as reported by Kautsky [583] (upper four curves). The reconstruction (the curve in the middle with two peaks) is based on a cubic spline description of the measured temperature (lower curve and capricious line) and the parameter values $L_m = 100$ mm, $g = 0.13$, $\dot{k}_M = 0.03\,\text{d}^{-1}$ at 15 °C and $T_A = 7600$ K.

where T_0 is at the lower end of the tolerance range. The next step is to choose cubic spline functions to describe the observed temperature pattern $T(t)$ and the unobserved scaled functional response $f(t)$. The reconstruction of $f(t)$ from length–time data then amounts to the estimation of the knot values of the spline at chosen time points, given realistic choices for the growth parameters. Figure 4.25 shows that the simultaneous least-squares fit of the numerically integrated growth description (4.93) is acceptable in view of the scatter in the length data (not shown), which increases in time in the upper size class in the original data. The scaled functional response (i.e. the hyperbolically transformed food abundance in terms of its nutritional value) appears to follow the temperature cycle during the year. Such a reconstructed food abundance can be correlated with POM and chlorophyll measurements to evaluate their significance for the mussel.

If food intake changes too fast to approximate the reserve density with its equilibrium value, the reserve density should be reconstructed as well. Figure 4.26 illustrates this for the penguin. The von Bertalanffy growth is shown to apply to the adelie penguin, which indicates that body temperature is constant and food is abundant. The deviation at the end of the growth period probably relates to the refusal of the parents to feed the chicks in order to motivate them to enter the sea. The small-bodied adelie penguin manages to synchronise its breeding cycle with the local peaks in plankton density in such a way that it is able to offer the chicks abundant food. Typically there are two such peaks a year in northern and southern cold and temperate seas. The plankton density drops sharply when the chicks are just ready to migrate to better places. This means that a larger species, such as the king penguin, is not able to offer its chicks this continuous wealth of food, because its chicks require a longer growth period (see Chapter 8 on comparison of species for an explanation, {295}). So they have to face the meagre period between plankton peaks. (Food for king penguins, squid and fish, follows plankton in abundance.) The parents do not synchronise their breeding season with the calendar; they follow a 14–17-month breeding cycle [1064]. The largest living penguin, the emperor penguin, also has to use both plankton peaks for

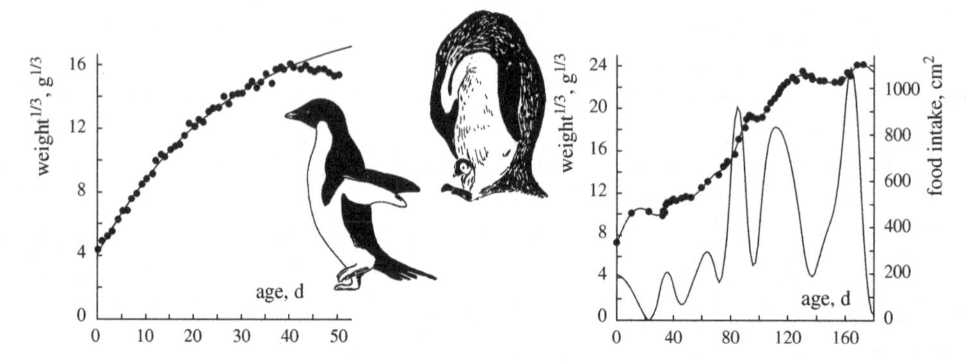

Figure 4.26 Weight ontogeny of the small adelie penguin *Pygoscelis adeliae* (left) and the large emperor penguin *Aptenodytes forsteri* (right). Data from Taylor [1143] and Stonehouse [1115]. The adelie data follow the fitted von Bertalanffy growth curve, which suggests food abundance during the nursery period. The cubic spline through the emperor data is used to reconstruct food intake $fV^{2/3} = \dot{J}_X/\{\dot{J}_{Xm}\}$. $d_V = 0.3\,\mathrm{g\,cm^{-3}}$, $w_E[M_{Em}] = 0.7\,\mathrm{g\,cm^{-3}}$, $g = 0.1$, $\dot{v} = 0.6\,\mathrm{cm\,d^{-1}}$, $l_T = 0.01$, $V_m = 6000\,\mathrm{cm^3}$, $e_0 = 0.6$.

one brood, which implies a structural deviation from a simple von Bertalanffy growth curve.

Given weight–time data, food intake can be reconstructed on the basis of the DEB theory. The relationship between (wet) weights, volumes and energy reserves is given in (3.2). For juveniles, where $E_R = 0$, we have $[W_w] = d_V + w_E[M_{Em}]e$ and specific wet weight is thus not considered to be a constant. Growth according to (2.21) and (2.10) is given by

$$\frac{d}{dt}W_w = [W_w]\frac{d}{dt}V + w_E[M_{Em}]V\frac{d}{dt}e \tag{4.94}$$

$$= \dot{v}V^{2/3}\left(\frac{[W_w]}{e+g}\left(e - l_T - (V/V_m)^{1/3}\right) + [W_{Ew}](f-e)\right) \tag{4.95}$$

Solution of f and substitution of (2.10) gives

$$f = e + \frac{[W_w]^{2/3}}{\dot{v}[W_{Ew}]W_w^{2/3}}\frac{d}{dt}W_w - \frac{[W_w]/[W_{Ew}]}{g+e}\left(e - l_T - \left(\frac{W_w}{V_m[W_w]}\right)^{1/3}\right) \tag{4.96}$$

$$\frac{[M_{Em}]w_E}{[W_w]}\frac{d}{dt}e = \frac{d}{dt}\ln W_w - \frac{\dot{v}}{g+e}\left((e - l_T)\left(\frac{[W_w]}{W_w}\right)^{1/3} - V_m^{-1/3}\right) \tag{4.97}$$

The steps to reconstruct feeding are as follows: first fit a cubic spline through the weight data, which gives $W_w(t)$ and so $\frac{d}{dt}W_w(t)$. Use realistic values for $e(0)$, d_V, $w_E[M_{Em}]$, g, V_m, l_T and $\dot{v}$ and recover $e(t)$ through numerical integration of (4.97) and then $f(t)$ by substitution. Figure 4.26 gives an example. The peaks in the reconstruction will probably be much sharper if the chick's stomach contents are taken into account. This reconstruction can be useful in cases where feeding behaviour that is hard to observe directly is studied and knowledge concerning energetics from captive specimens is available. The significance of this example is to show that the DEB theory hardly poses constraints for growth curves in general. The simple von Bertalanffy growth curve only emerges under the conditions of constant food density and temperature.

4.11.2 Reconstruction of body temperature from growth data

Empirical growth curves of birds frequently deviate from the von Bertalanffy growth curve, even if food is abundant. The body temperature of endotherms can be well above the environmental temperature. If insulation or heat transfer from mother to chick changes in time, deviations from the von Bertalanffy growth curve are to be expected. Altricial birds provide an excellent case to illustrate the problem of the energy interpretation of growth measurements in the case of an unknown body temperature.

Birds become endothermic around hatching; precocial species usually make the transition just before hatching, and altricial ones some days after. The ability to keep the body temperature at some fixed level is far from perfect at the start, so the body temperature depends on that of the environment and the behaviour of the parent(s) during that period. Unless insulation of the nest is perfect, the parents cannot heat the egg to their own body temperature. There will be a few degrees difference, but this is still a high temperature, which means that the metabolic rate of the embryo is high. So it produces an increasing amount of heat as a by-product of its general metabolism before the start of endothermic heating.

The process of pre-endothermic heating can be described by: $\frac{d}{dt}T_b = \alpha_T \dot{p}_{T+} - \dot{k}_{be}(T_b - T_e)$, where T_b is the body temperature of the embryo, T_e the temperature of the environment, α_T the heat generated per unit of utilised energy and $\dot{k}_{be}$ the specific heat flux from the egg to the environment. The latter is here taken to be independent of the body size of the embryo, because the contents of the egg are assumed to be homogeneous with respect to the temperature. (Brunnich's guillemot seems to need a 40 °C temperature difference between one side of the egg and the other to develop [954].)

Figure 4.27 illustrates the development of the lovebird *Agapornis*, with changing body temperature (T_A = 10 kK). The curves hardly differ from those with a constant temperature, but the parameter estimates differ substantially. The magnitude of the predicted temperature rise depends strongly on the parameter values chosen. The information contained in the data of Figure 4.27 did not allow a reliable estimation of all parameters; the predicted temperature difference of 4 °C is arbitrary, but not unrealistic.

It is interesting that the red-headed lovebird *A. pullaria* from Africa and at least 11 other parrot species in South America, Australia and New Guinea breed in termite nests, where they profit from the heat generated by the termites. Breeding golden-shouldered parrots *Psephotus chrysopterygius* in captivity failed frequently, until it became known that one has to heat the nest to 33 °C for some days before hatching and for 2 weeks after.

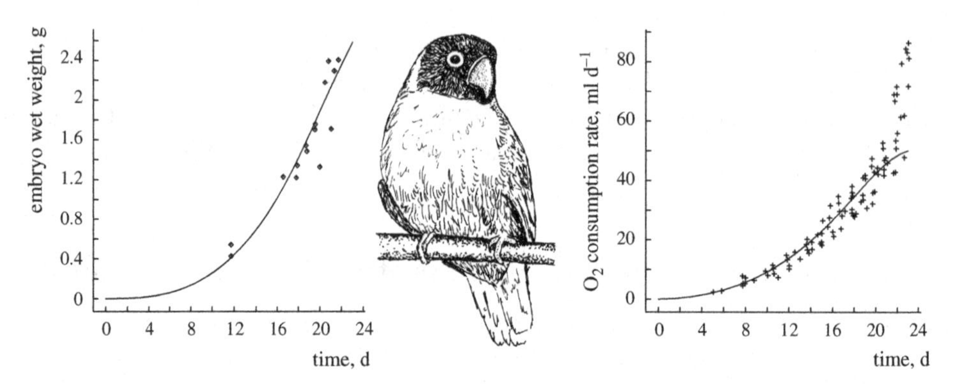

Figure 4.27 Embryo weight and respiration ontogeny in the parrot *Agapornis personata*. Data from Bucher [170]. The curves are DEB model predictions accounting for a temperature increase of 4 °C during development; see text. The temporary respiration increase at day 23 relates to hatching. This detail is not part of the model.

The significance of this exercise is the following: the least-squares-fitted curves remain almost exactly the same, whether or not the body temperature changes, but the parameter estimates for, for example, the energy conductance differ considerably. It follows that these data are not suitable for estimating energy parameters unless the temperature is known as a function of time. This holds specially for altricial birds because they hatch too early to show the reduction in respiration rate that gives valuable information about parameter values. The few studies on bird development that include temperature measurements indicate that the temperature change during incubation is not negligibly small. Drent [293] found an increase from 37.6 to 39 °C in the precocial herring gull *Larus argentatus*.

The main reason why the (empirical) logistic growth curve sometimes fits bird data better than the von Bertalanffy curve at food abundance is probably because body temperature changed. The ontogeny of body temperature can be reconstructed as follows.

At abundant food, (2.22) reduces to $\frac{d}{dt}l = \dot{r}_B(1-l)$, where the von Bertalanffy growth rate $\dot{r}_B = \frac{\dot{k}_M}{3}\frac{g}{1+g}$ is now considered not as a constant but as a function of time, since the temperature and thus the maintenance rate coefficient $\dot{k}_M$ change. Integration gives

$$l(t) = 1 - (1 - l(0)) \exp\left(-\int_0^t \dot{r}_B(t_1)\, dt_1\right) \text{ with} \tag{4.98}$$

$$\dot{r}_B(t) = \dot{r}_{B\infty} \exp(T_A(T_\infty^{-1} - T_b(t)^{-1})) \tag{4.99}$$

where $\dot{r}_{B\infty}$ is the ultimate growth rate when the body temperature is kept constant at some target temperature in the range 39–41 °C, or $T_\infty = 312$ (non-passerines) or 314 K (passerine birds). Body temperature is thus given by

$$T_b(t) = \left(\frac{1}{T_\infty} - \frac{1}{T_A} \ln \frac{\frac{d}{dt} l}{\dot{r}_{B\infty}(1-l)} \right)^{-1} \tag{4.100}$$

Given an observed growth and size pattern, this equation tells us how to reconstruct the temperature. The reconstruction of body temperature, therefore, rests on the assumption of (time-inhomogeneous) von Bertalanffy growth (4.98) and an empirical description of the observed growth pattern. It is a problem, however, that both the growth rate and the length difference with its asymptote 1 vanish, which means that their ratio becomes undetermined if inevitable scatter is present. General purpose functions such as polynomials or splines to describe size-at-age are not suitable in this case.

A useful choice for an empirical description of growth is

$$\frac{d}{dt} l = \frac{\dot{r}_{B\infty}}{\delta_l} (l^{-\delta_l} - 1) l \quad \text{or} \quad l(t) = (1 - (1 - l(0)^{\delta_l}) \exp(-\dot{r}_{B\infty} t))^{1/\delta_l} \tag{4.101}$$

because it covers both von Bertalanffy growth (shape parameter $\delta_l = 1$), and the frequently applied logistic growth ($\delta_l = -3$) and all shapes in between. For the shape parameter $\delta_l = 0$, the well-known Gompertz curve arises: $l(t) = l(0)^{\exp(-\dot{r}_B t)}$. Nelder [830] called this model the generalised logistic equation. It was originally proposed by Richards [957] to describe plant growth. The graph of volume as a function of age is skewly sigmoid, with an inflection point at $V/V_\infty = (1 - \delta_l/3)^{3/\delta_l}$ for $\delta_l \leq 3$. Substitution of (4.101) into (4.100) gives

$$T_b(t) = \left(\frac{1}{T_\infty} - \frac{1}{T_A} \ln \frac{1}{\delta_l} \frac{1 - l^{-\delta_l}}{1 - l^{-1}} \right)^{-1} \tag{4.102}$$

Note that if growth is of the von Bertalanffy type, so $\delta_l = 1$, this reconstruction amounts to $T_b(t) = T_\infty$, which does not come as a surprise. This interpretation of growth data implies that the growth parameters of the logistic, Gompertz and von Bertalanffy growth curves are comparable in their interpretation and refer to the target body temperature. The DEB theory gives the physiological backgrounds. Figure 4.28 gives examples of reconstructions, which indicate that the body temperature at hatching can be some 10 °C below the target and it increases almost as long as growth lasts. The reconstruction method has been tested on several data sets where the body temperature has been measured during growth [1295]. It has been found to be quite accurate given the scatter in the temperature data. Figure 4.28 gives one example. Although the Arrhenius temperature can be estimated from combined weight/temperature data, its value proved to be poorly defined.

4.11.3 Reconstruction from reproduction data

Food intake can also be reconstructed from reproduction data of e.g. *Daphnia hyalina*. Data provided by Stella Berger include body length, egg length, width and number of eggs in the brood pouch in weekly hauls from enclosures. The general idea is to

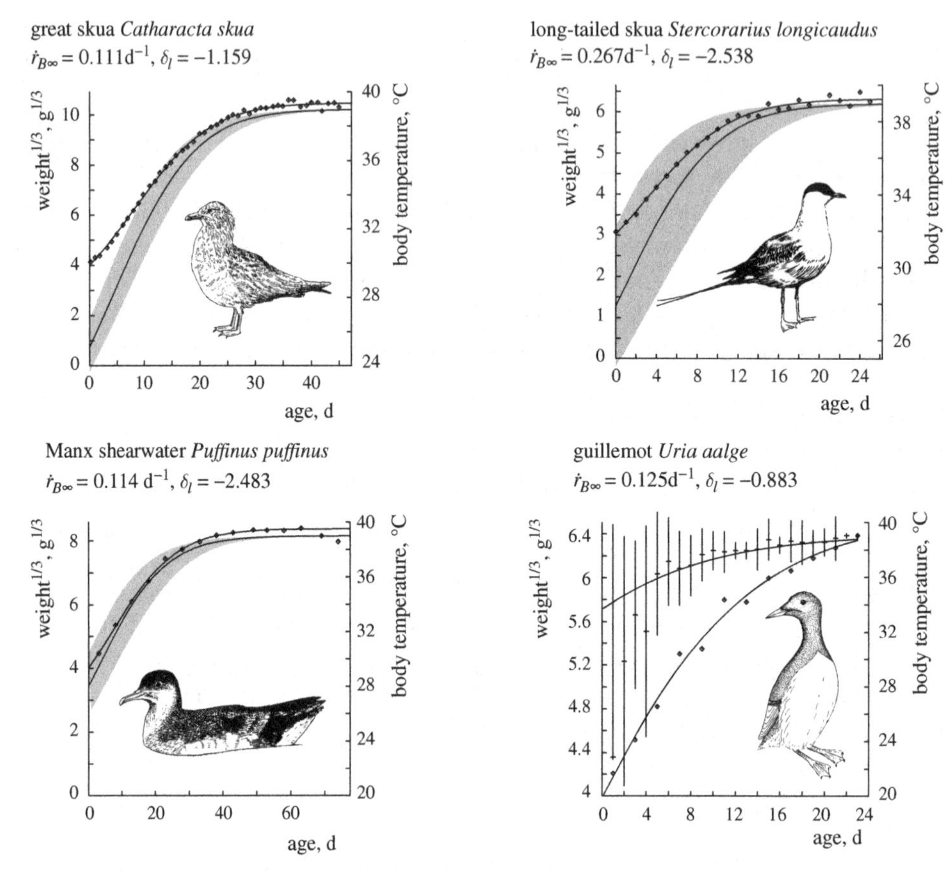

Figure 4.28 The empirical, generalised, logistic growth curves have been fitted to measured data for some birds. The von Bertalanffy growth rate $\dot{r}_{B\infty}$ at the ultimate body temperature and shape parameter δ_l are given. On the basis of these fits the body temperature was reconstructed, on the assumption that $T_\infty = 312\,\text{K}$ and $T_A = 10\,\text{kK}$. The shaded areas around the body temperature curves indicate the 95% confidence interval based on the marginal distribution for k. The reconstruction method is tested on the guillemot data (lower right figure) where measured body temperatures were available. The bars indicate the standard deviation. Both temperature parameters, $T_\infty = 312.3\,\text{K}$ and $T_A = 8.225\,\text{kK}$, have been estimated from the combined weight/temperature data. Data from Furness, de Korte in [384], Thompson in [157] and [735] respectively.

reconstruct food density and then try to link measured quantities in the enclosures, such as chlorophyll concentration, POM and DOC to this reconstructed food density to learn more about the nutritional value of these quantities for daphnids. These links are less than direct (daphnids cannot digest chlorophyll or cellulose) and involve the (unknown) half-saturation constant.

These data also allow the study of maternal effects: is the reserve density at birth indeed equal to the reserve density of the mother at egg formation as the DEB theory assumes? Eggs initially fully consist of reserve. If reserve density at birth is small, initial egg size will be small as well, but less than linear: a low amount of initial reserve leads to low maturation, so long incubation and high cumulated maintenance costs. Hatching (which coincides with start of feeding in *Daphnia*) occurs if maturity exceeds a threshold value. The differences in egg size are small only since only half of the initial reserve is used during the embryo stage in daphnids [661].

I evaluate two different ideas on the main sources of scatter.

Scenario 1: Individuals are idential, local environments are different Each individual experienced a different food history and I use the observed number of eggs N to estimate the scaled functional response f for each individual. To find f, $N = t_R\dot{R}$ with reproduction rate $\dot{R}$ given in (2.56) was solved numerically for each individual, starting from the analytical solution using the scaled reserve U_E^0 for $f = 1$. Using these values of f, the sum of squared deviations between observed egg volumes and expected egg volume $\dot{v}_0 U_E^0$ was minimised to find an estimate for the conversion factor $\dot{v}_0$. In this scenario all scatter is in the local food density of individual daphnids. The environment is supposed to be spatially and temporary heterogeneous. These individuals have identical parameter values. Since eggs grow in volume during incubation (see below), we need to correct the measured egg volumes for growth during development. The 'observed' initial egg volume V_{Oi} of individual i is estimated by $\sum_j w_j V_{Oj} / \sum_j w_j$, where $w_{ij} = \exp(-c_f(f_i - f_j)^2 - c_V(V_{Oj} - V_{Om})^2)$ and V_{Om} is the minimum observed egg volume. So the closer the reconstructed functional response is to the individual at hand and the smaller the egg volume is, the larger is the weight coefficient for the estimated initial egg volume.

Scenario 2: Individuals are different, local environments are identical Individuals in a single haul experienced the same food history and I use the different individuals in one haul to estimate a common scaled functional response. To find f, the sum of squared deviations was minimised between the observed number of eggs N and the expected number of eggs $t_R\dot{R}$ with $\dot{R}$ given in (2.56) for individuals of different lengths, simultaneously with that between observed and expected egg volumes. In this scenario part of the scatter is in the translation of food to eggs, and part in difference of parameter values amount individuals. Since eggs grow in volume during incubation (see below), the smallest egg volume in each haul represents the best estimate for the initial volume for that scaled functional response if individuals do not synchronise moulting cycles.

Since sampling is weekly only, reserve is supposed to be in pseudo-equilibrium and the scaled reserve density e in $\dot{p}_C$ is replaced by the scaled functional response f in reproduction rate is given in (2.56).

The volume of an ellipse of radii a, b, c equals $4\pi abc/3$. Expressed in egg length $L_l = 2a$ and egg width $L_w = 2b = 2c$, egg volume equals $V_e = L_l L_w^2 \pi/6$.

We have no reproduction, $\dot{R} = 0$, if $(1-\kappa)S_C = \dot{k}_J U_H^p$, which happens for

$$f_R^0 = \left((1-\kappa)(L^2 + L^3\dot{k}_M/\dot{v})/(\dot{k}_J U_H^p) - g^{-1}\right)^{-1} \tag{4.103}$$

Using only individuals with eggs, we know that the reconstructed f must be in the interval $(f_R^0, 1)$. Notice that the larger the individual, the lower the reserve density can be to continue reproduction. We have no growth, $\frac{d}{dt}L^3 = 0$, if $\kappa \dot{S}_C = \kappa L^3/L_m$, which happens for

$$f_G^0 = L/L_m \tag{4.104}$$

Notice that the larger the individual, the higher the reserve density must be to fulfil the somatic maintenance costs.

Maximum reproduction is given in (2.58). The maximum number of eggs accumulated over a time interval t_R is $N_m = t_R \dot{R}_m$, so if N_m represents the maximum observed number we have $t_R \geq N_m/\dot{R}_m$. If t_R does not meet this constraint we can obtain estimates of f that exceed the value 1.

The range of lengths of individuals with eggs is (1.12, 2.36) mm, which translates in estimates $L_p = 1$ mm and $L_m = 2.75$ mm. The latter value is well above the maximum observed length because maximum length can only be reached after prolonged exposure to abundant food, which is not likely in natural situations. The range of egg lengths is (0.137, 0.488) mm, which translates in an estimate $L_b = 0.48$ mm. The range of egg volumes is (0.0006, 0.1) mm^3; this covers a range of a factor of 16. In view of the finding that around half of the initial reserve is still present at birth in *D. magna* [661], this factor is much too large to be explained by differences in initial reserve. I conclude that during the incubation period, the volume of the egg must grow due to the uptake of water.

The values $\kappa = 0.8$, $\dot{v} = 3.24$ mm d^{-1}, $\dot{k}_J = \dot{k}_M = 1.7$ d^{-1} are chosen from *D. magna* [661] for a reference temperature of 20 °C, while $g = 0.69$ was corrected for differences in maximum body length. This leads to $U_H^b = 0.0046$ d mm^2 and $U_H^p = 0.042$ d mm^2 to arrive at the mentioned values for L_b and L_p. The implications are age at birth $a_b = 0.51$ d and von Bertalanffy growth rate $\dot{r}_B = 0.23$ d^{-1} at $f = 1$. About half of the initial reserve is used during the embryonic stage at $f = 1$ with these parameter settings.

The maximum number of eggs in the brood pouch is 41 in an individual of length 2.24 mm. To accommodate all these eggs with the above-mentioned parameter values, we need an intermoult period of $t_R \kappa_R = 4.8$ d, which seems somewhat long for $\kappa_R = 0.95$. If data on the real period were available, this could be used to adjust κ or g, which both have a large effect on the minimum period that is required. Too large an observed number of eggs depresses the reconstructed scaled food density considerably.

The conversion from initial scaled reserve U_E^0 to initial volume was obtained by regression, like the scale functional responses. Notice that all parameters with length in their units refer to physical length, not volumetric length. The shape coefficient for *D. hyalina* is probably close to $\delta_{\mathcal{M}} = 0.54$.

The estimates can be improved by including ecophysiological information on DEB parameters of *D. hyalina*.

Both scenarios produced similar $f(t)$ reconstructions, see Figure 4.29. The four experiments showed a very similar profile, but the peak in experiment 1 and 2 is before that of 3 and 4. A major difference is that in the scenario 1, some individuals have such a large number of eggs that the DEB parameters are forced to values such that the mean scaled functional response is rather low. If only scenario 2 had been tried, a wider choice

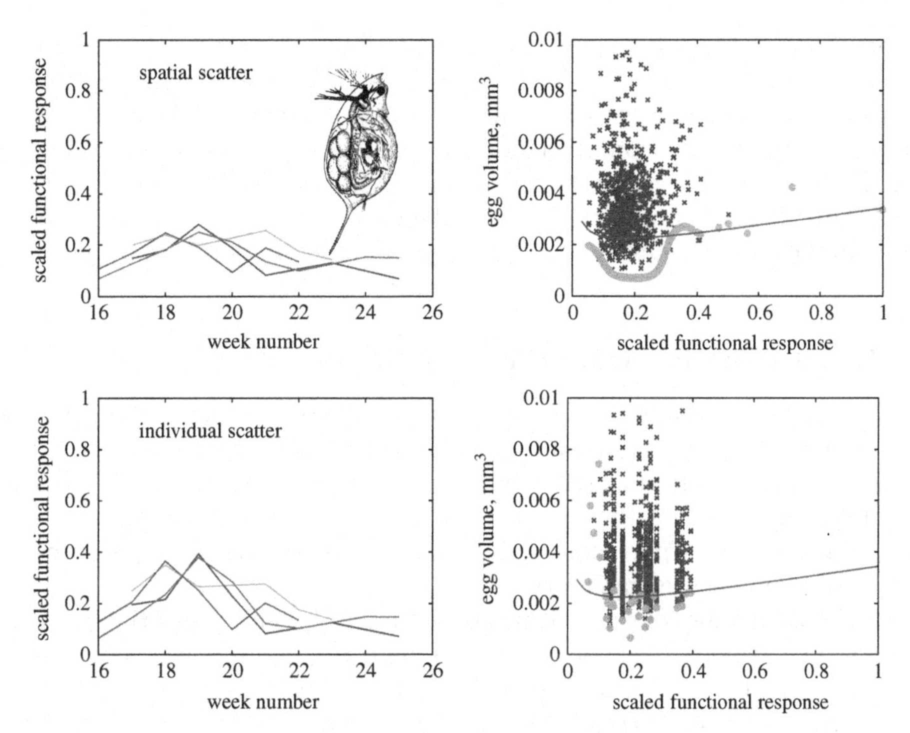

Figure 4.29 The reconstructed scaled functional response for *D. hyalina* as function of the week number (left), and the egg volume as function of the scaled functional response (right), for scenario 1 (top) and 2 (bottom). Data by Stella Berger, who used four parallel experiments. The crosses represent measured egg volumes, the points the estimated initial egg volumes and the curve the expected initial egg volume.

of DEB parameters would have been possible, such that the reconstructed mean scaled functional response fluctuates over a wider range of values. The large number of eggs in few individuals is then explained by deviating parameter values for those individuals. Scenario 2 involves relationships between number of eggs in the brood pouch and body length. Given the scatter, these relationships generally applied well.

The maternal effect is supported weakly only, see Figure 4.29, but reports in the literature on the contrary, i.e. that large eggs are produced at low food density (e.g. [413]), are not confirmed; I did not check the empirical basis of their claims. Apart from the problem of an increase in egg volume during development, another source of scatter in egg volume is that some individuals are likely to be in the stage of converting the reproduction buffer to eggs in the brood pouch. The number of eggs in the brood pouch might be small at the moment of sampling, but much larger a few moments later. Notice that the expected initial egg size is a U-shaped function of the functional response. The left branch has no ecological relevance because at the minimum of the function we have $f = L_b/L_m = f_G^0$, so no growth at birth. For $\dot{k}_J = \dot{k}_M$ this also means no maturation, so no birth. This calls for a revision of the parameter values, so for more information on the energetics of *D. hyalina*.

By decomposing observed egg volume into contributions from reserve and structure, they can also be used to study to what extend synchronisation of moulting cycles occur among individuals. The dry weights can be used to further test ideas on reserve, in combination with reproduction. Weights have contributions from structure, reserve, reproduction buffer and eggs. By adding assumptions about the relationship between number of individuals in a haul and that in the enclosure, these data can also be used to study population dynamics and the effect of sampling on population dynamics.

4.11.4 Reconstruction from otolith data

Collaborative work with Laure Pecquerie [877, 876] allowed reconstruction of the scaled food density $x(t)$ for $t \in (t_b, t_\dagger)$ from otolith data (from anchovy), where t_b is the time at mouth opening; first feeding often produces a specific mark on the otolith. The reconstruction supposes that the otolith's opacity O as a function of otolith length L_O is known from data for a particular individual fish as well as all required parameter values and the temperature trajectory.

We here make a number of simplifications, but none of them is essential, however, and all of them can be avoided. We assume that the maturity and somatic maintenance rate coefficients are equal, $\dot{k}_J = \dot{k}_M$ and so we have the scaled maturity $U_H = V(1-\kappa)g/\dot{v}$ and $\frac{d}{dt}U_H = 3(1-\kappa)\frac{L^2}{\dot{k}_M L_m}\frac{d}{dt}L$ and $S_J = \frac{1-\kappa}{\kappa}S_M$, $S_R = \frac{1-\kappa}{\kappa}S_G$ for $s_* = \dot{p}_*/\{\dot{p}_{Am}\}$. Temperature affects $\dot{k}_M$ and $\dot{v}$ given at reference temperature T_{ref} via a temperature correction factor $c_T = \exp\left(\frac{T_A}{T_{\text{ref}}} - \frac{T_A}{T}\right)$. Since the temperature effects on these two rates cancel in $L_m = \frac{\dot{v}}{\dot{k}_M g}$, we only have to take those on $\dot{v}$ into account. We can use elaborate methods to relate metabolic rates to temperatures that take deviations from the Arrhenius relationship into account at the high and low boundaries of the environmental temperature range, but here we only use the Arrhenius correction. We further assume that shrinking does not occur. Shrinking typically can happen at extreme starvation when the individual is relatively large, and we assume that the reproduction buffer of such individuals is large enough to cover maintenance costs.

Data suggested that the contribution of assimilation to otolith growth and opacity could be neglected. By setting the contribution from growth to opacity equal to zero, and that from dissipation equal to 1, the opacity as function of time is given by $O(t) = y^G_{\oslash E}\dot{J}_{EG}(y^D_{\oslash E}\dot{J}_{ED} + y^G_{\oslash E}\dot{J}_{EG})^{-1}$. It takes values between zero (in absence of growth) and one.

Scaled food density relates to scaled functional response as $x = \frac{f}{1-f}$, which translates the problem of finding $x(t)$ to that of finding $f(t)$. Given the states at the moment of observation (opacity, otolith length, fish length, scaled reserve density, scaled functional response, temperature), we might try to find $f(t_0)$, $c_T(t_0)$ and work our way backwards in time. This scheme, however, turns out to be hopelessly unstable, to the extent that it is useless. A stable scheme is to start from birth and integrate over otolith length, not time. This is possible because otolith length increases strictly monotonously in time (contrary to body length). Feeding starts at birth, so opacity at birth has no information about the food level. So we have to assume that between the first and the second data point food density is constant, and changes linearly in time after that at rates that we reconstruct from opacity data.

A continuation method for this change from one data point to the next one turns out to be satisfactory, except when growth is resumed after starvation. For these points we need a more robust method.

Figure 4.30 illustrates the reconstruction using parameters that are appropriate for anchovy. The first reconstruction uses the 'true' trajectory of the correction factor for temperature and reconstructs the otolith and body length trajectories perfectly. The scaled functional response and the reserve density trajectories are also perfectly reconstructed, except if the reserve density no longer supports growth. The second reconstruction assumes a constant temperature correction factor of 1, still leading to a very good reconstruction.

The reconstruction of $f(t)$ from $O(L_O)$ data is coded in routine `o2f` in toolbox 'animal' of software package DEBtool. The inverse routine, to construct $O(L_O)$ from $f(t)$ data, as done in routine `f2o` can be useful for checking the method. The comparison of the reconstructed body length at otolith collection with the measured one is another very useful check for consistency of the reconstruction method.

A weak component of our reconstruction method is the required knowledge about the temperature trajectory during the lifetime of the fish. It turns out, however, that the (unrealistic) assumption that the temperature was constant, despite that fact that

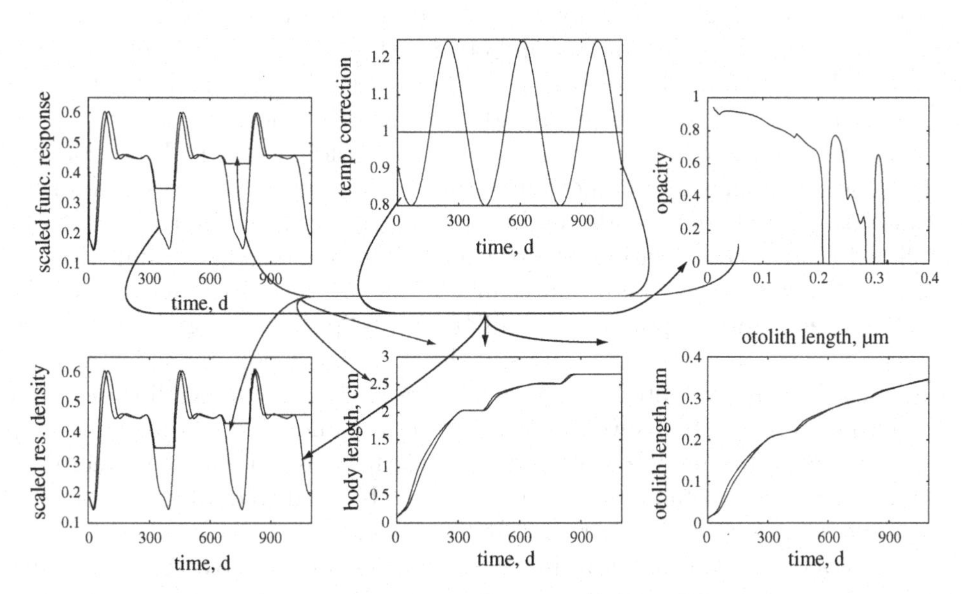

Figure 4.30 The construction of the opacity profile from the functional response trajectory and reconstruction of the functional response trajectory from the opacity profile. The first reconstruction uses the 'true' trajectory of the correction factor for temperature, the second reconstruction assumes a constant temperature correction factor. The match of the first reconstruction with the construction is almost perfect, so they coincide. Parameters: $L_b = 1\,\mathrm{cm}$, $L_p = 1.5\,\mathrm{cm}$, $\dot{v} = 0.526\,\mathrm{cm\,d^{-1}}$, $\dot{v}_{OD} = 1.186 \times 10^{-5}\,\mathrm{cm\,d^{-1}}$, $\dot{v}_{OG} = 1.1 \times 10^{-4}\,\mathrm{cm\,d^{-1}}$, $\dot{k}_M = 0.015\,\mathrm{d^{-1}}$, $g = 6$, $\kappa = 0.65$, $\kappa_R = 0.95$, $\delta_S = 1/20$.

it changed in reality, hardly affected the reconstructed food history in our simulations. The second reconstruction in Figure 4.30 illustrates this.

Modifications of this reconstruction can make use of other types of data and/or information, for instance that temperature extremes should match known points on the yearly cycle. Such calibrations transform an 'exact' reconstruction problem into a minimisation of deviations between predictions and measurements, but doubtlessly will improve the quality of the reconstruction.

Similar reconstruction methods can be applied to opacity variations in earplugs of whales, to rings in tree trunks or to ripples on bivalve shells, for instance.

4.12 Summary

Reactions to variations in food levels depend on the timescale of starvation; allocation rules are affected first, then follow reserve dynamics, and dormancy. During extreme starvation shrinking can occur; its dynamics can involve up to four extra parameters, but a first approximation does not require extra parameters.

Changes of shape are important if they affect the relationship between volumes and surface areas that are involved in food uptake. These changes can be implemented via the shape correction function, which quantifies the surface area relative to that of an isomorph. An important special case results if surface area is proportional to volume to the power one, V1-morphy, which is a good and simple approximation for dividing micro-organisms. The distinction between the individual and population levels disappears for V1-morphs. The popular models by Monod, Marr–Pirt and Droop turn out to be special cases of the univariate DEB model for V1-morphs.

The standard DEB model, as specified by the assumptions listed in Table 2.4 at {77}, fully determines the fluxes of organic compounds (food, faeces, reserves and structural mass); those of mineral compounds (carbon dioxide, dioxygen, water and nitrogenous waste) follow from the conservation law for chemical elements. The assumptions, therefore, specify *all* mass fluxes. These mass fluxes can all be written as weighted sums of three basic energy fluxes (powers): assimilation, dissipation and growth. Dissipating heat can also be written as weighted sums of the three basic powers, which means that dissipating heat is a also weighted sum of three mineral fluxes (carbon dioxide, dioxygen and nitrogenous waste). This is well known in empery, and used in the widely applied method of indirect calorimetry to obtain dissipating heat from the three mineral fluxes. Growth-related changes in biomass composition can be used to obtain the composition of reserves and structure, as is illustrated by examples.

Respiration is one of two mineral fluxes, carbon dioxide or dioxygen, or dissipating heat. The fluxes are proportional to each other, given certain constraints on the composition of reserves, relative to structural mass. Respiration that is not associated with assimilation is then proportional to the mobilisation rate of reserve. The theory also quantifies the respiration that is associated with assimilation, known as the Specific Dynamic Action; its nature is still considered to be enigmatic, but now explained in first principles.

Similar to other mineral fluxes, nitrogenous waste not only originates from assimilation directly, but also from maintenance (dissipation) and growth. Although this might not seem surprising, it differs from its treatment in Static Energy Budgets, see {433}, and turns out to be most useful in the analysis of trophic interactions, {337}.

Products can be included in just one single way, without changing the assumptions of Table 2.4; they, too, must be weighted sums of the three basic energy fluxes, the three weight coefficients per product are free parameters. In this way, products are included in the overhead costs of the three powers. Consequently, fermentation gives three constraints, which fully determine the three weight coefficients of a single product, or partly determine those of more products.

The drinking of water by terrestrial organisms and plants, to balance the metabolic turnover of water, can be quantified on the basis of two supplementary assumptions about water loss:

---

1 water evaporates in proportion to the surface area at a rate that depends on environmental conditions (temperature, humidity, wind speed)

2 water evaporates in proportion to respiration

These assumptions apply to animals as well as plants. Drinking by plants has complex interactions with nutrient uptake and is shown to affect the saturation constant. Water balance has intimate relationships with thermal balance, and so with the energetics of endotherms. These routes have been explored briefly.

There is just one way to include isotope dynamics in DEB theory, and this follows naturally from the isotope balance and has a mixing and a fractionation aspect. Fractionation can be from pools (nutrients, including dioxygen), but typically occurs from fluxes, by separating the anabolic and catabolic aspects in assimilation, dissipation and growth. This flux-based theory links up with that of Synthesising Units.

Simple measurements on amounts and composition of biomass can be used to access the primary DEB parameters from some compound DEB parameters and to separate reserve from structure, both in amounts and in composition. Methods are discussed to find chemical proxies for reserve (rRNA) and structure (DNA) to access reserve and structure more directly.

The entropy and chemical potentials of reserve and structure follow from the strong homeostasis assumption. They can be accessed indirectly from input–output relationships of the individual, or of the microbial population, but this method makes full use of mass balances.

Observations on growth can be used to reconstruct the trajectory of body temperature. Observations on growth, or reproduction, or opacity in (fish) otoliths, can be used to reconstruct the trajectory of (scaled) food density. In the latter two cases that can even be done from a single (dead) individual. In combination with isotope data, trajectories of temperature can also be reconstructed.

5

Multivariate DEB models

As long as all required nutrients and energy are available to the organism in fixed relative amounts, it can buffer temporal variations in abundance using a single reserve. This situation is approximated in organisms that eat other organisms, as discussed in the previous chapters. If energy and various nutrients are taken up independently, however, a reserve is required for each of them to buffer variations in abundance. The surface layers of seas are poor in nutrients and rich in light, while the reverse holds for the bottom of the photozone. Algal cells, which commute between these two environments on the wind-induced currents, can barely grow and survive, unless they use intracellular energy and nutrient reserves.

The purpose of this chapter is to show how the univariate DEB models can be extended to include several substrates, reserves and structural masses, in a way that reduces to the one-reserve, one-structure case if just one nutrient (or light) is limiting, or if nutrient abundances covary, and the reserve turnover times are identical. The concept of the Synthesising Unit, see {101}, will be used to show that a nutrient becomes almost non-limiting as soon as its availability exceeds that of the limiting nutrient, only by a small amount, relative to its needs. Simultaneous limitations of growth by nutrients and light only occur incidentally, and usually during a short period. This is why the simple one-reserve DEB theory can be applied so widely.

Each reserve requires specifications of its assimilation process and of its contribution to maintenance costs. Together with a single structural mass, and so a single growth process, $2n+1$ powers have to be specified to delineate n reserves. Each of these powers contributes to the dissipation of heat; the fixed weight coefficients directly follow from the conservation law for energy. Product formation is directly associated with these powers, and generally requires $2n+1$ coupling parameters per product for quantification; excretion is basic to multiple reserve systems and follows a deviating dynamics. Fluxes of non-limiting nutrients are also directly associated with the powers, and the $2n + 1$ coupling parameters follow from the conservation law for mass.

To structure the model appropriately, fast processes are separated from slow ones, and many transport processes are only included implicitly at the whole-individual level. Transport of metabolites through phloem in plants, for instance, shares important system properties with blood in animals: a small capacity is combined with a high turnover, which means that material in phloem should not play an explicit role at the whole-individual level. The transformation from nutrients and light to reserves is taken as a single step, while in fact many intermediary metabolites are formed.

5.1 Several substrates

Several extensions are possible from one to more types of food (or substrate). Details of growth and reproduction patterns can only be understood in relation to selection of food items and choice of diet. The reverse relationship holds as well, especially for 'demand' systems. I will, therefore, mention some aspects briefly.

5.1.1 Diet and preference

Many species change their diet during development in relation to their shifting needs with an emphasis on protein synthesis during the juvenile period and on maintenance during the adult one. Many juvenile holometabolic insects live on different types of food compared with adults. Most wasps and butterflies, for instance, feed on nectar as adults, but on animals and leaves, respectively, as juveniles. Stickleback fish change from being carnivorous to being herbivorous at some stage during development [267]. Plant-eating ducks live on insects during the first period after hatching. The male emperor penguin *Aptenodytes* and mouth-brooding frog *Rhinoderma darwinii* provide their young initially with secretions from the stomach. Mammals live on milk during the baby stage, see {6}.

The first hatching tadpoles of the alpine salamander *Salamandra atra* live on their siblings inside the mother, where they are also supported by blood from her reproductive organs, and the one to four winners leave the mother when fully developed. The same type of prenatal cannibalism seems to occur in the coelacanth *Latimeria* [1152], and several sharks (sand tiger sharks Odontaspidae, mackerel sharks Lamnidae, thresher sharks Alopiidae [936]), and the sea star *Patiriella* [178]. Some species of poison dart frog *Dendrobatus* feed their offspring with unfertilised eggs in the water-filled leaf axils of bromeliads, high up in the trees [299, 300].

Shifts in food selection that relate to shifts in nutritional requirements can be modelled using at least two reserves, e.g. carbohydrates plus lipids and proteins, which differ in their contributions to maintenance costs, and in the requirements for growth. Changes in behavioural aspects, such as food selection, can then be based on efficiency arguments.

Some species select for different food items in different seasons for reasons other than changes in the relative abundance of the different food sources. This is because of the tight coupling between feeding and digestion. The bearded tit *Panurus biarmicus* is a spectacular example; it lives on the seed of bulrush *Typha* and reed *Phragmites* from September to March and on insects in summer [1091, 1225]. This change in diet comes with an adaptation of the stomach which is much more muscular in winter when it contains stones to grind the seeds. Once converted to summer conditions, the bearded tit is unable to survive on seeds. The example is remarkable because the bearded tit stays in the same habitat all year round. Many temperate birds change habitats over the seasons. Divers, for instance, inhabit freshwater tundra lakes during the breeding season and the open ocean during winter. Such species also change prey, of course, but the change is usually not as drastic as the one from insects to seeds.

The relationship between feeding rates and diet composition gives a clue as to what actually sets the upper limit to the ingestion rate. An indication that the maximum ingestion rate is determined by the digestion rate comes from the observation that the maximum ingestion rate of copepods feeding on diatoms expressed as the amount of carbon is independent of the size of the diatom cells, provided that the chemical composition of the cells is similar [373]. The maximum ingestion rate is inversely related to protein, nitrogen and carbon contents fed to the copepod *Acartia tonsa* [530]. The observation that the maximum ingestion rate is independent of cell size on the basis of ingested volume [393] points to the capacity of gut volume being the limiting factor.

These examples should make clear that the quantitative details of the feeding process cannot be understood without some understanding of the fate of the food. This involves the digestion process in the first place, but a whole sequence of other processes follow.

Prokaryotes show a diversity and adaptability of metabolic pathways that is huge in comparison to that of eukaryotes. Many bacteria, for example, are able to synthesise all the amino acids they require, but will only do so if these are not available from the environment. The fungus *Aspergillus niger* only feeds on cellulose if no compounds are available that are easier to decompose. The relationship between food quality and physiological performance is discussed again in the treatment of food intake reconstructions {170}, dissipating heat {160} and adaptation {291}.

The decomposition of biomass into a structural component and a reserve component implies that a predator feeds on a mixture of two compounds, rather than just a single one, even if it specialises on a single species of prey. The significance of the contribution of prey reserves to predator nutrition is obvious in the example of waterfleas feeding on algae. Most of the organic carbon of algae consists of cellulose in the cell wall, and of chlorophyll. However, the waterflea cannot digest both compounds of the structural biomass, and mainly feeds on starch and lipids. The quantitative aspects of feeding on prey differ from the general case of sequentially processed substitutable substrates by the tight coupling of the abundances structural mass and reserves. The reserves of the prey can be treated as a kind of nutritional quality of prey biomass.

Suppose that the prey's reserves do not extend the predator's handling time. If the prey does not have an energy buffer allocated to its reproduction, the assimilation power of the predator amounts to $\dot{p}_A = (\mu_{AV} + \mu_{AE} m_E^\circ)\dot{J}_{XA}$, where μ_{AV} stands for the conversion of prey structural mass into predator assimilative power, μ_{AE} for the conversion of prey reserves into predator assimilative power, $m_E^\circ = M_E^\circ / M_V^\circ = e^\circ m_{Em}^\circ$ for the ratio of the reserve to the structural mass of the prey, and the feeding rate $\dot{J}_{XA}$ for the molar flux of prey structural biomass. Parameters and variables that relate to the prey are indicated with $^\circ$ to distinguish them from those of the predator.

Let $\mu_{AX} = \mu_{AV} + \mu_{AE} m_{Em}^\circ$ denote the conversion of well-fed prey biomass into assimilation power, and $\kappa_A = \left(1 + \frac{1}{m_{Em}^\circ} \frac{\mu_{AV}}{\mu_{AE}}\right)^{-1} = \left(1 + \kappa^\circ g^\circ \frac{\overline{\mu}_E^\circ}{\mu_{GV}^\circ} \frac{\mu_{AV}}{\mu_{AE}}\right)^{-1}$ the fraction of the assimilative power of the predator that originates from the digestion of prey reserves, when feeding on well-fed prey. The assimilative power can then be represented as $\dot{p}_A = (1 - \kappa_A + \kappa_A e^\circ)\mu_{AX} \dot{J}_{XA}$, so that the maximum assimilative power is $\dot{p}_{Am} = \mu_{AX} \dot{J}_{XAm}$, where $\dot{J}_{XAm}$ denotes the maximum ingestion rate in terms of structural biomass. This can be summarised as $\dot{p}_A = (1 - \kappa_A + \kappa_A e^\circ) f \dot{p}_{Am}$, since $\dot{J}_{XA} = f \dot{J}_{XAm}$. The dynamics of the scaled reserve density of a V1-morph predator becomes

$$\frac{d}{dt} e = \dot{k}_E (f - \kappa_A f + \kappa_A e^\circ f - e) \tag{5.1}$$

Energy extracted from reserves through digestion cannot exceed the energy invested in reserves, $\mu_{AE} < \overline{\mu}_E^\circ$, and energy extracted from structural biomass through digestion cannot exceed energy contained in this mass, which itself cannot exceed energy invested in the synthesis of this mass, $\mu_{AV} < \overline{\mu}_V^\circ < \mu_{GV}^\circ$. Therefore $\mu_{AX} < \mu_{GV}^\circ (1 + \frac{1}{\kappa^\circ g^\circ})$, and κ_A is probably, but not necessarily, larger than $(1 + \kappa^\circ g^\circ)^{-1}$.

If the prey has a reproduction buffer, it is possible that the assimilative power exceeds $\dot{p}_{Am}$, in this scaling, which indicates that the scaled reserve density of the predator can exceed 1, in principle. The quantitative description of feeding on prey can be further detailed by accounting for the selection of prey by the predator, based on the structural biomass and reserves of the prey, and/or by allowing the handling time to depend on these state variables. In this way, the saturation constant becomes dependent on the state variables of the prey as well. Although this might be realistic in particular applications, these mechanisms are not worked out here.

The four basic types of the uptake of substrates are discussed at {104}. The situation for bacteria that feed on glucose and fructose, for instance, is different because the carriers for glucose in the outer membrane of the bacterial cell cannot handle fructose, see {291}. These substrates, therefore, do not compete for access to the same carriers, and the uptake processes just add, not interacting in the transformation to reserve.

Data on the aerobic production of the yeasts *Saccharomyces cerevisiae* and *Kluyveromyces fragilis* strongly suggest the existence of two different uptake routes for glucose [457], see Figure 5.1. A low-affinity high-capacity carrier is active under anaerobic and aerobic conditions, and ethanol and acetaldehyde are produced in association with this assimilation process. A high-affinity low-capacity carrier is active under aerobic conditions only, and no products are produced in association with this assimilation process. Some strains, however, produce glycerol in association with the latter assimilation. When the process of glucose uptake and product formation is studied for increasing chemostat throughput rates under aerobic conditions, the quantitative dominance of the two carriers switches at a throughput rate of 0.2 h^{-1}, but no metabolic switches are required to capture this behaviour.

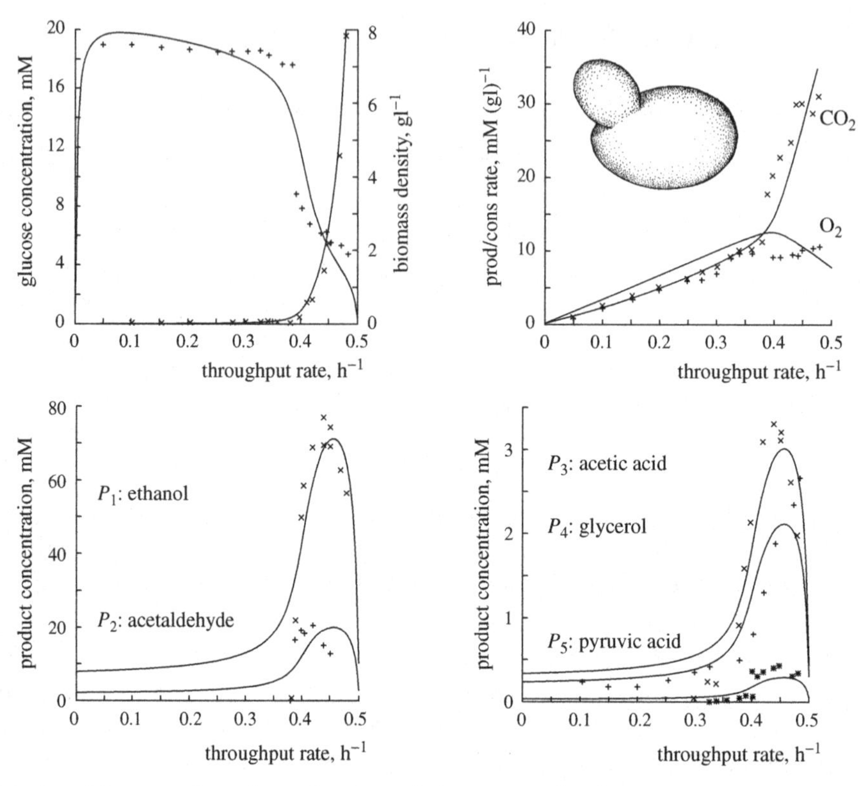

Figure 5.1 Aerobic growth and production of the yeast *Saccharomyces cerevisiae* at 30 °C in a chemostat. Data from Postma *et al.* [916, 917] and Verduyn [1188]. The data fits, modified from Hanegraaf [457], assume two assimilation processes for glucose, and product formation coupled to one assimilation process, which reduces the energy gain from glucose for metabolism by a factor $\kappa_A = 0.187$. The glucose concentration in the feed is 83.3 mM; the maximum throughput rate is $\dot{h}_m = 0.5\,\mathrm{h}^{-1}$; a measurement error on acetaldehyde is estimated to be 0.7 [457]. The compositions of structure and reserves have been set at $n_{HV} = 1.75$, $n_{OV} = 0.61$, $n_{NV} = 0.14$, $n_{HE} = 1.7$, $n_{OE} = 0.62$, $n_{NE} = 0.23$. Parameters: $j_{XAm1} = 2.16\,\mathrm{mM/M\,h}$, $j_{XAm2} = 81\,\mathrm{mM/M\,h}$, $X_{K1} = 0.1\,\mathrm{mM}$, $X_{K2} = 40\,\mathrm{mM}$, $\dot{k}_E = 0.54\,\mathrm{h}^{-1}$, $\dot{k}_M = 0.003\,\mathrm{h}^{-1}$, $g = 0.050$, $\zeta_{P_1A_2} = 55$, $\zeta_{P_2A_2} = 43$, $\zeta_{P_3A_2} = 2.35$, $\zeta_{P_4A_2} = 2.47$, $\zeta_{P_5A_2} = 0.34$, $y_{EX} = 0.51$, $y_{EV} = 0.78$. The curves follow from $\dot{h} = \frac{e\dot{k}_E - g\dot{k}_M}{e+g}$; $e = \frac{j_{XAm1}f_1 + \kappa_A j_{XAm2}f_2}{j_{XAm1} + \kappa_A j_{XAm2}}$; $\kappa_A = \frac{\mu_{A_2X}}{\mu_{A_1X}}$; $W = (w_V + e w_E y_{EV}/g)\frac{\dot{h}(X_r - X)}{j_{XAm1}f_1 + j_{XAm2}f_2}$; $f_i = \frac{X}{X + X_{Ki}}$; $X_{Pi} = \frac{\zeta_{P_iA_2}\dot{k}_E f_2(X_r - X)}{j_{XAm1}f_1 + j_{XAm2}f_2}$; $j_C = j_{XAm1}f_1(1 - y_{EX}) + j_{XAm2}f_2(1 - \kappa_A y_{EX}) + \dot{k}_M y_{EV} + \dot{h}(y_{EV} - 1) - \dot{k}_E f_2 n_C$; $j_O = j_{XAm1}f_1(y_{EX}n - 1) + j_{XAm2}f_2(\kappa_A y_{EX} n - 1) + \dot{k}_E f_2(n_C + n_H/4 - n_O/2) - \dot{k}_M y_{EV} n + \dot{h}(1 - n_{NV}/2 - n_{OV}/2 - n y_{EV})$; $n_C = \sum_i \zeta_{P_iA_2}$; $n_H = \sum_i n_{HP_i}\zeta_{P_iA_2}$; $n_O = \sum n_{OP_i}\zeta_{P_iA_2}$; $n = 1 + \frac{1}{4}n_{HE} - \frac{1}{2}n_{OE} - \frac{3}{4}n_{NE}$.

5.1.2 Pseudo-faeces and variations in half-saturation coefficients

Apart from faeces, bivalves produce pseudo-faeces: material, typically silt, that has been filtered from the water, but separated from food, which leaves the body before it would enter the gut. The production of pseudo-faeces can be quantified by considering silt as a second substrate with zero conversion efficiency to reserve [644]; it is processed sequentially with food by the filtering apparatus. Because silt competes with food for access to the filtering apparatus, the silt density in the water modifies food uptake. Its quantitative effect is increasing the half-saturation coefficient linearly in the density. The apparent half-saturation coefficient amounts to

$$K'(Y) = K(1 + Y/K_Y) \tag{5.2}$$

where Y is the silt density, K the half-saturation coefficient for food and K_Y that for silt. For varying silt densities, this means that the half-saturation coefficient for food becomes time-dependent as well. Given the half-saturation coefficient for food in absence of silt, estimates for the apparent half-saturation coefficient relationship (5.2) can be used to estimate mean silt densities.

5.1.3 Oxygenic photosynthesis

A detailed discussion of photosynthesis is beyond the scope of this book, see e.g. [333]; I here focus on the links with DEB theory. Photosynthesis concerns the process of carbon fixation, so the use of photons and dissolved inorganic carbon (DIC) to synthesise carbohydrates (starch or lipids in some taxa). This is just one of the assimilation processes in DEB theory, which forms a single reserve of a multiple-reserve system. Other reserves also contribute to the synthesis of structure (growth), so photosynthesis should not be identified with growth. Since phototrophy hardly depends on temperature, and nutrient uptake does, the composition of the biomass of phototrophs depends on temperature, with consequences for the grazers of this biomass. This means that photosynthesis can best be incorporated using several reserves. Moreover, most phototrophs have substantial heterotrophic activity, see {111} and Chapter 10, and should be classified as mixotrophs.

Oxygenic photosynthesis can be summarised as

$$CO_2 + H_2O + \text{light} \rightarrow CH_2O + O_2 \qquad \left\{ \begin{array}{l} 2\,H_2O + 4\,h\nu \rightarrow O_2 + 4\,H^+ + 4\,e^- \\ CO_2 + 4\,H^+ + 4\,e^- \rightarrow CH_2O + H_2O \end{array} \right.$$

Since all oxygen in dioxygen comes from water, not from carbon dioxide, we need an extra water molecule as substrate as well as product to follow oxygen isotopes, see {96}. Carbohydrate CH_2O has the role of carbon as well as energy source for the synthesis of structure.

Figure 5.2 gives a simplified schedule for the photosynthetic process. The photopigment system of cyanobacteria, photoautotrophic protoctists and plants consists of two Photo Systems (PSs). When a photon is captured by the antenna and transferred to an unexcited PS II, it switches to the excited state, transfers an electron from water to PS I, and switches back to the unexcited state. PS I can likewise accept a photon from its antenna, and also accepts an electron from PS II, which allows it to pass an electron via NADPH to the carbon-fixation cycle (Calvin–Benson cycle). The enzyme Rubisco

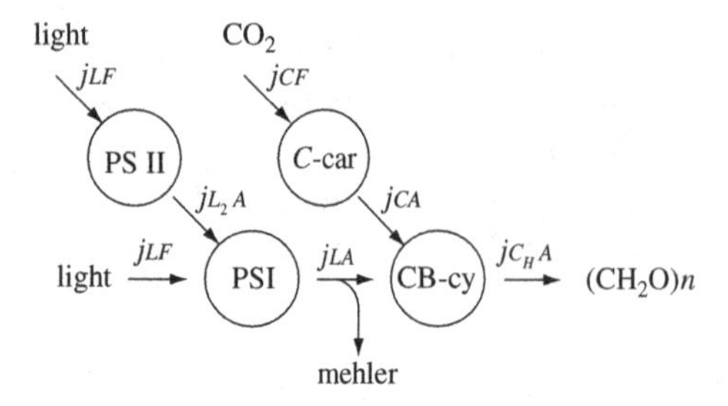

Figure 5.2 Diagram of the simplified carbon fixation, where light L and carbon dioxide C are converted into carbohydrates by Synthesising Units (circles, see text). Photorespiration and photoinhibition modify the synthesis of carbohydrates.

partakes in this cycle, and accepts the electron and a carbon dioxide molecule from its carrier, and reduces the latter to carbohydrate. Part of the carbohydrate is stored as such or excreted, part is delivered to the synthesis machinery. All units behave as 1,1-SUs {104}.

The 'binding' probability of a photon depends on its wavelength and on the photopigment, which differs between the various phototrophic taxa. In the water, the light extinction rate is constant according to the Beer–Lampert law, which means that light intensity decays exponentially with depth at rates that might depend on the chemical composition of the water (e.g. the presence of humic acids) and particles, such as 'detritus' and algal cells (self-shading). The decay rates might depend on the wavelength, which rapidly makes the model more complex.

Pigment systems

For (negative) photon flux $\dot{j}_{LF}$ and large values for the flux ratios z_{L_1} and z_{L_2}, assimilated light quantifies as

$$\dot{j}_{L_2A} = \dot{j}_{L_2Am}\left(1 + \frac{\dot{j}_{L_2FK}}{-\dot{j}_{LF}}\right)^{-1} \simeq -z_{L_2}\dot{j}_{LF} \quad \text{with } z_{L_2} = \frac{\dot{j}_{L_2Am}}{\dot{j}_{L_2FK}} \tag{5.3}$$

$$\dot{j}_{LA} = \dot{j}_{L_1Am}\left(1 + \frac{\dot{j}_{L_1FK}}{-\dot{j}_{LF}} + \frac{\dot{j}_{L_2AK}}{\dot{j}_{L_2A}} - \left(\frac{-\dot{j}_{LF}}{\dot{j}_{L_1FK}} + \frac{\dot{j}_{L_2A}}{\dot{j}_{L_2AK}}\right)^{-1}\right)^{-1} \tag{5.4}$$

$$\simeq \dot{j}_{L_1Am}\left(1 + \frac{\dot{j}_{L_1AK}}{-\dot{j}_{LF}}\right)^{-1} \quad \text{with}$$

$$\dot{j}_{L_1AK} = \dot{j}_{L_1FK} + \frac{\dot{j}_{L_2AK}}{z_{L_2}} - \left(\dot{j}_{L_1FK}^{-1} + \frac{z_{L_2}}{\dot{j}_{L_2AK}}\right)^{-1} \tag{5.5}$$

$$\dot{j}_{LA} \simeq -z_{L_1}\dot{j}_{LF} \quad \text{with } z_{L_1} = \dot{j}_{L_1Am}/\dot{j}_{L_1AK} \tag{5.6}$$

where $\dot{j}_{L_iFK}$ and $\dot{j}_{L_iAK}$ are specific half-saturation fluxes, i.e. parameters that are associated with the behaviour of SUs, and $\dot{j}_{L_iAm}$ are the maximum specific assimilation rates for photons for pigment system $i = 1, 2$. Although the electron input to the carbon-fixation cycle is (approximately) proportional to the light intensity, this does not mean that there is no upper limit to the light intensity that can be used, because the electrons experience increasing resistance to their use in the process of carbon fixation. Electrons that are not used in carbon fixation or photorespiration 'leak' away via the Mehler reaction [905], also known as pseudocyclic electron transport, which involves dioxygen uptake, and dioxygen production of equal size [333]. The interception of light barely depends on temperature, while other metabolic processes do, which explains the need to handle spoiled electrons.

Green, purple and heliobacteria photosynthesise under anaerobic conditions, using bacteriochlorophylls and a single pigment system (PS II in purple and green non-sulphur bacteria, and PS I in green sulphur bacteria and heliobacteria). They must have an equivalent of the Mehler reaction to get rid of the excess electrons.

Carbon fixation

The output from the carbon-fixation cycle can be derived according to a similar reasoning as applied for electron production. For $x_C = X_C/X_{KC}$ we have with substitution of (5.6)

$$\dot{j}_{CA} = \dot{j}_{CAm}(1 + x_C^{-1})^{-1} = \dot{j}_{CAm} f_C \tag{5.7}$$

$$\dot{j}_{C_HA} = \dot{j}_{C_HAm}(1 + z_C^{-1}) \left(1 + \frac{\dot{j}_{CAK}}{\dot{j}_{CA}} + \frac{\dot{j}_{LAK}}{\dot{j}_{LA}} - \left(\frac{\dot{j}_{CA}}{\dot{j}_{CAK}} + \frac{\dot{j}_{LA}}{\dot{j}_{LAK}}\right)^{-1}\right)^{-1}$$

$$\dot{j}_{C_HA} = \frac{\dot{j}_{C_HAm}(1 + z_C^{-1})}{1 + z_C^{-1} f_C^{-1} + \frac{\dot{j}_{LFK}}{-\dot{j}_{LF}} - \left(z_C f_C + \frac{-\dot{j}_{LF}}{\dot{j}_{LFK}}\right)^{-1}} = \dot{j}_{C_HAm} f_{C_H} \tag{5.8}$$

with $z_C = \dot{j}_{CAm}/\dot{j}_{CAK}$, $\dot{j}_{LFK} = \dot{j}_{LAK} z_{L_1}^{-1}$, $\dot{j}_{CAK}$ the specific half-saturation flux for carbon dioxide, $\dot{j}_{CAm}$ the maximum specific carbon dioxide assimilation rate, $\dot{j}_{C_HAm}$ the maximum specific carbohydrate assimilation rate.

Photorespiration

Rubisco is the most abundant enzyme on Earth: it constitutes 5–50 % of the soluble protein in algal cells [333], and is involved in the fixation of carbon dioxide. Rubisco can operate in two modes on the substrate ribulose-1,5-biphosphate (RuP_2)

Carboxylase activity: $RuP_2 + CO_2 + H_2O \rightarrow 2$[3P-glycerate]
Oxygenase activity: $RuP_2 + O_2 \rightarrow 1$[3P-glycerate] + 1[2P-glycolate]

The second reaction is known as photorespiration. The net effect is that the binding of carbon dioxide or dioxygen leads to the synthesis or degradation of carbohydrates. The binding is competitive, with widely varying relative strength among algal classes. The counterproductive effects of dioxygen might be a historical accident, since Rubisco evolved in a period that was essentially free of dioxygen [940]. C_4 plants, which bind carbon dioxide to an organic compound with four C-atoms in a micro-environment that is poor in dioxygen, avoid photorespiration almost completely. They do not use Rubisco, but phosphoenolpyruvate (PEP) carboxylase for the binding of carbon dioxide. Different species in the same genus can have C_3 and C_4 metabolism, and orache *Atriplex*

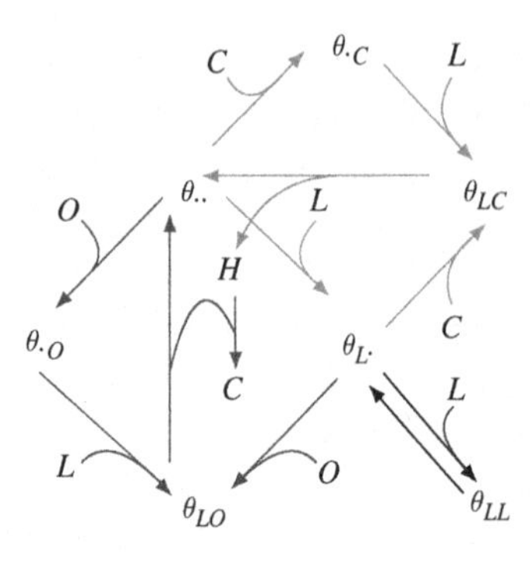

Figure 5.3 The coupled photosynthesis – photorespiration – photoinhibition in the transformations from carbon dioxide C plus photons (light) L (plus water) to hydrocarbon H (plus dioxygen), $C + L \to H$, and from dioxygen O plus photons plus hydrocarbon to carbon dioxide (plus water), $O + L + H \to C$. The state θ_{LL} is inactive, and accommodates photoinhibition.

prostrata, for instance, has both C_3 and C_4 metabolism. The dioxygen use that is associated with primary carboxylation only occurs in light, and is called photorespiration. This can be modelled as follows.

Figure 5.3 illustrates the scheme of photosynthesis and photorespiration. Ignoring photoinhibition, this scheme translates into the dynamics of the variously bounded fractions of SUs as follows. Let $\boldsymbol{\theta} = \theta_{..}, \theta_{.O}, \theta_{.C}, \theta_{L.}, \theta_{LO}, \theta_{LC}$ denote the fractions of the photosynthetic system (RuP_2 plus PSs) that are in complex with nothing, dioxygen, carbon dioxide, photon, photon and dioxygen, or photon and carbon dioxide, respectively. The changes in the fractions are given by

$$\begin{aligned} \tfrac{d}{dt}\theta_{..} &= \dot{k}_O\theta_{LO} + \dot{k}_C\theta_{LC} - (j'_L + j'_O + j'_C)\theta_{..} & \tfrac{d}{dt}\theta_{L.} &= j'_L\theta_{..} - (j'_O + j'_C)\theta_{L.} \\ \tfrac{d}{dt}\theta_{.O} &= j'_O\theta_{..} - j'_L\theta_{.O} & \tfrac{d}{dt}\theta_{LO} &= j'_L\theta_{.O} + j'_O\theta_{L.} - \dot{k}_O\theta_{LO} \\ \tfrac{d}{dt}\theta_{.C} &= j'_C\theta_{..} - j'_L\theta_{.C} & \tfrac{d}{dt}\theta_{LC} &= j'_L\theta_{.C} + j'_C\theta_{L.} - \dot{k}_C\theta_{LC} \end{aligned} \tag{5.9}$$

where $j'_* = \rho_* y_{C_H*} j_*$ denotes the arrival flux j_* times the binding probability ρ_*, and the coefficient y_{C_H*} couples $*$ to C_H; $\dot{k}_O$ and $\dot{k}_C$ stand for the dissociation rates of oxygenase and carboxylase products. The net flux of carbohydrate is found by equating the changes in fractions to zero and solving for $\boldsymbol{\theta}$. The result is

$$j_{C_H A} = \theta_{LC}\dot{k}_C - \theta_{LO}\dot{k}_O = \frac{j'_C - j'_O}{1 + \frac{j'_C}{\dot{k}_C} + \frac{j'_O}{\dot{k}_O} + \frac{j'_C + j'_O}{j'_L} - \frac{j'_C + j'_O}{j'_L + j'_C + j'_O}} \tag{5.10}$$

For $j'_O = 0$ this reduces to $j_{C_H A} = \left(\dot{k}_C^{-1} + {j'_L}^{-1} + {j'_C}^{-1} - (j'_L + j'_C)^{-1}\right)^{-1}$, which is identical to (5.8). At the compensation point $j'_O = j'_C$, no net synthesis of carbohydrate occurs.

Photoinhibition and photoadaptation

At high irradiance, photoinhibition can occur, see e.g. [1293]. This can be incorporated naturally into the carbon-fixation model using SUs, see {106}. To this end, we need to introduce a new state of the SU, where a photon can 'bind' to an SU that is already 'bound' to a photon, and send it into an inactive state, from which is can recover at a constant probability rate, see Figure 5.3. Likewise we can delineate two further inactive states θ_{LLC} and θ_{LLO}, which can be entered from the states θ_{LC} and θ_{LO}, respectively. Such further extensions should only be considered if necessary.

The light spectrum that is of relevance for phototrophy might in principle differ from that for photoinhibition. Accounting for such difference would obviously complicate the model considerably.

The amount of chlorophyll per cell turns out to be rather variable, see e.g. [1292]. Generally the relative amount is high at low mean levels of irradiance and the chlorophyll density acts as if it is compensation. Part of the variation of chlorophyll per cell weight can be explained by the contribution of the various reserves to the cell weight. A more detailed analysis might involve a workload allocation model, see {204}, for chlorophyll. This more detailed link between chlorophyll and primary production is of importance for the interpretation of remote sensing data.

5.1.4 Calcification

Bicarbonate is by far the dominant form of inorganic carbon in seawater. At the typical pH of about 8.3, 98% of the inorganic carbon is in this form. Few organisms can use this source; one problem is to deal with the electrical charge. Coccolithophorans, such as *Emiliania huxleyi* (right), have mastered this art, by using calcium in the transformation $Ca^{2+} + 2\,HCO_3^- \rightarrow Ca\,CO_3 + CO_2 + H_2O$, where the calcium carbonate is exported by the Golgi apparatus in the form of beautifully shaped extra-cellular coccoliths, and the carbon dioxide is used as carbon substrate for the synthesis of carbohydrates and lipids (for which they obviously need water and light as well). The coccoliths accumulate in a polysaccharide layer, and are shed at cell death. *Emiliania* is so abundant that the coccoliths can easily be seen on satellite images in huge areas in the northern Atlantic and Pacific Ocean where they bloom regularly. A substantial fraction of carbonates in rocks originates from coccoliths, and coccolithophorans may play a key role in the carbon metabolism of the Earth [1240].

Since carbon dioxide is relatively rare, and the transformation of carbonate and bicarbonate to carbon dioxide is slow, and the water that envelopes the cell is stagnant, see {266}, cells in the sea can become limited by carbon under otherwise optimal growth conditions [1269]. This points to the gain of using bicarbonate as an additional carbon source, with an inherent gradient in the CO_2/HCO_3^- ratio in the diffusive boundary layer [1270]. The process of calcification can be modelled in the context of the DEB theory by treating carbon dioxide and bicarbonate as substitutable substrates, with light as a supplementary 'substrate', for the synthesis of lipids as reserve, while calcium carbonate is formed as a product in this assimilation process. This implementation ties calcification to photosynthesis.

As long as calcium is not rate limiting, and the environment is homogeneous, the carbohydrate production amounts to

$$j_{C_H} = \left(\dot{k}_C^{-1} + (j'_C + j'_{C-})^{-1} + j'^{-1}_L - (j'_C + j'_{C-} + j'_L)^{-1}\right)^{-1} \tag{5.11}$$

where $j'_C = \rho_C j_C$ and $j'_{C-} = \rho_{C-} j_{C-}/2$ are the effective arrival rates of carbon dioxide and bicarbonate; and the factor 0.5 in j'_{C-} relates to the stoichiometry of the calcification process. The calcification rate now becomes

$$j_{Ca} = \frac{j_{C_H} j'_{C-}}{j'_C + j'_{C-}} \tag{5.12}$$

Calcification is also reported to occur in the dark, to some extent. This might relate to heterotrophic activity to acquire the energy for carbon fixation.

5.2 Several reserves

The number of reserves equals the number of nutrients and/or (generalised) substrates that are taken up independently. The case of two reserves and two nutrients serves as an example, see Figure 5.4; the model extends to more reserves and possibly limiting nutrients without causing additional problems. This will be worked out in some detail for V1-morphs in this section.

The general idea is to apply the rules for SUs to quantify the transformation of nutrients to each of the k reserves; each reserve is mobilised independently, first allocated to a maintenance SU for each reserve, which also receives input from mobilised structure. The remaining reserve fluxes are allocated to a (single) SU for growth. This takes $2k + 1$ SUs. The rejected nutrient and substrate fluxes do not pose any problem in the case of the assimilation SUs, because they are fed back into the environment. The rejected reserve fluxes, however, require special consideration, which is why I start with the specification of growth, given the reserve densities, and then consider reserve dynamics.

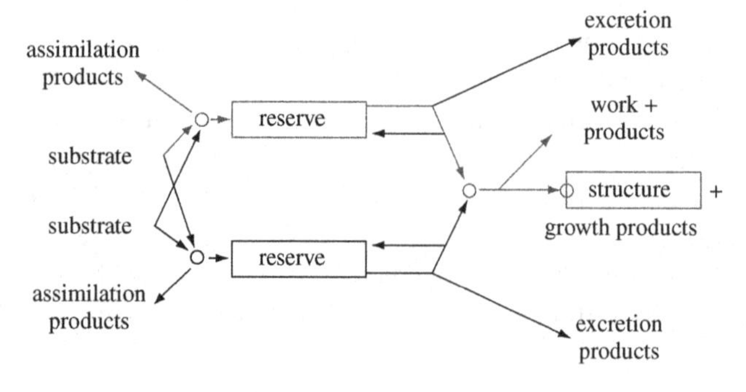

Figure 5.4 A diagram of the structure of a two-substrates, two-reserves DEB model. The circles indicate SUs. The diagram for a single-substrate, single-reserve model is in grey. Excretion products are then included in the products linked to growth and maintenance (work). The dynamics of other reserves interact with this process.

For a k-reserve system, we have $2k$ maintenance parameters: $j_{E_iM_i}$ and j_{VM_i} for $i = 1, \ldots, k$, where $j_{VM_i} \geq y_{VE_i} j_{E_iM_i}$. The actual fluxes of reserve and structure allocated to maintenance are $j_{E_i}^{M_i}$ and $j_V^{M_i}$, respectively, and might vary in time, where $j_{E_i}^{M_i} \leq j_{E_iM_i}$ and $j_{VM_i} \leq j_{VM_i}$. The decision to allocate structure to maintenance is made for each reserve separately, so $j_V^M = \sum_i j_V^{M_i}$, where $j_V^{M_i}$ is specified in (4.17) for each reserve.

5.2.1 Growth

If no structure is used to pay (somatic) maintenance costs the specific gross growth rate j_{VG} equals the specific net growth rate $\dot{r}$, but if (some) maintenance is paid from structure at rate $j_V^M < \sum_i j_{VM_i}$, then the growth rates relate to each other as $\dot{r} = j_{VG} - j_V^M$.

Just like the univariate case, reserve densities m_{E_i}, $i = 1, 2, \ldots$, follow first-order kinetics, which means that for a reserve mass M_{E_i}, the specific mobilised from the ith reserve, equals

$$j_{E_iC} = m_{E_i}(\dot{k}_{E_i} - \dot{r}) \tag{5.13}$$

where $\dot{k}_{E_i}$ denotes the turnover rate of the i-th reserve, and the net specific growth rate $\dot{r} = \frac{d}{dt} \ln M_V$ relates to the dilution by growth.

This means the i-th reserve sends a specific flux $j_{E_iG} = j_{E_iC} - j_{E_i}^{M_i}$ to the SU for growth of structural biomass which leads for $j_V^M = \sum_i j_V^{M_i}$ in the bivariate case $(i = 1, 2)$ to

$$j_{VG} = \dot{r} + j_V^M = \left(\sum_i \left(\frac{j_{E_iG}}{y_{E_iV}} \right)^{-1} - \left(\sum_i \frac{j_{E_iG}}{y_{E_iV}} \right)^{-1} \right)^{-1} \tag{5.14}$$

This equation can be solved numerically for the net specific growth rate $\dot{r}$ and readily extended to more than two reserves.

Figure 5.5 illustrates that this model for simultaneous growth limitation by reserves is realistic. Note that cell content of phosphorus and vitamin B_{12} have been measured, rather than reserves. In view of the very small values, the reserves hardly contribute to total biomass, which can then be conceived as structural biomass. The overhead costs in the synthesis of structural mass and the maintenance costs for these nutrients have been neglected.

5.2.2 Reserve dynamics and excretion

The growth SU rejects the reserve fluxes at specific rates

$$j_{E_iR} = (\dot{k}_{E_i} - \dot{r})m_{E_i} - j_{E_i}^{M_i} - y_{E_iV} j_{VG} \tag{5.15}$$

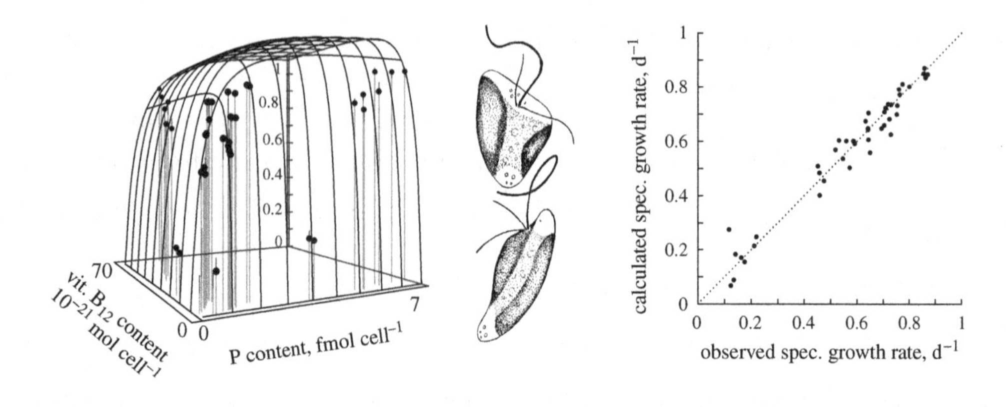

Figure 5.5 The specific growth rate $\dot{r}$ of the haptophyte *Pavlova lutheri* as a function of the intracellular reserves of phosphorus (reserve 1) and vitamin B_{12} (reserve 2) at 20 °C (left), and the relationship between the observed growth rate and the calculated one (right). Data from Droop [295]. The parameters are given in Figure 5.6.

where $j_{E_i}^{M_i} \leq j_{E_iM_i}$ is the specific flux of the i-th reserve spend on somatic maintenance. Each rejected reserve 'molecule' is excreted with probability $(1 - \kappa_{E_i})$ in one form or another, and fed back to the reserves with probability κ_{E_i}, so the balance equation for reserve densities $m_{E_i} = M_{E_i}/M_V$ becomes

$$\frac{d}{dt}m_{E_i} = j_{E_iA} - j_{E_iC} + \kappa_{E_i}j_{E_iR} - \dot{r}m_{E_i} \tag{5.16}$$

$$= j_{E_iA} - (1 - \kappa_{E_i})(\dot{k}_{E_i} - \dot{r})m_{E_i} - \kappa_{E_i}(j_{E_i}^{M_i} + y_{E_iV}j_{VG}) - \dot{r}m_{E_i} \tag{5.17}$$

5.2.3 Simultaneous nutrient limitation

Data from an experiment with the chemostat in steady state are used to test the simultaneous limitation model for realism. The balance equations for the nutrients in the medium of a chemostat with throughput rate $\dot{h}$ are

$$\frac{d}{dt}X_j = (X_{rj} - X_j)\dot{h} - \sum_i y_{jE_i}\dot{J}_{E_iA} \tag{5.18}$$

$$\frac{d}{dt}X_j^* = \sum_i (1 - \kappa_{E_i})y_{jE_i}\dot{J}_{E_iR} + \sum_i y_{jE_i}\dot{J}_{E_iM_i} - X_j^*\dot{h} \tag{5.19}$$

where X_j is the concentration of nutrient j, X_j^* the nutrient content of excretions due to reserves that are mobilised but rejected by the growth SU and not fed back to the reserves, and (the second term) nutrients involved in maintenance losses. X_{rj} denotes the concentration of substrate j in the feed. The summation is over all reserves E_i. I suppose that the excreted nutrients are metabolically changed such that they cannot be reused immediately.

At steady state, the substrate concentrations X_j in the chemostat do not change, so $\frac{d}{dt}X_j = 0$, $j = 1, 2$, and

$$(X_{rj} - X_j)\dot{h} = M_V \sum_i y_{jE_i} j_{E_iA} \quad (5.20)$$

$$X_j^*\dot{h} = M_V \sum_i y_{jE_i} \left((1 - \kappa_{E_i})\left(\dot{k}_{E_i} m_{E_i} - (m_{E_i} + y_{E_iV})\dot{r}\right) + (.\kappa_{E_i} j_{E_iM_i})\right) \quad (5.21)$$

The biomass density in the chemostat follows from the fact that the specific growth rate $\dot{r} = \dot{h}$ is known. The equations (5.14), (5.17) and (5.20) together define the biomass density M_V, the nutrient concentrations X_j and the reserve densities at steady state m_{E_i}, given the throughput rate $\dot{h}$ and the nutrient concentrations in the feed X_{rj}. Although the system consists of five coupled equations, it can be reduced to a single one in X_1 for uncoupled assimilation fluxes ($y_{1E_1} = 1$, $y_{1E_2} = 0$, $y_{2E_1} = 0$, $y_{2E_2} = 1$), while the range of X_j is given by $(\delta_j^{-1} - 1)^{-1} < X_j/X_{Kj} < X_{rj}/X_{Kj}$, with $\delta_j = \kappa_{E_j}(j_{E_jM_j} + y_{E_jV}\dot{r})/j_{E_jAm}$. It can be shown that the resulting equation in X_1 has one or three roots, while only one root satisfies the range restriction for X_2. A bisection method can be used to arrive at a high-quality initial estimate for the proper root, followed by a Newton–Raphson method to obtain that root accurately.

The details of the measurement method determine whether or not the excretions are included in the medium concentrations. In the data presented and analysed in Figure 5.6, phosphorus and cobalt (in vitamin B_{12}) were measured using isotopes. As a consequence, the measured medium concentrations include the excreted labelled phosphorus and cobalt, and correspond to $X_j + X_j^*$. The cellular contents correspond to $\sum_i y_{jE_i}(y_{E_iV} + m_{E_i})$. If the assimilation fluxes for phosphorus and vitamin B_{12} are not coupled, the cellular content reduces to $y_{E_jV} + m_{E_j}$. This simplification reduces the total number of parameters to be estimated to 10 for 20 data sets, or 220 data points. The balance equation for nutrient j in the medium plus that in the cells at steady state reads

$$X_{rj} = X_j + X_j^* + \sum_i y_{jE_i}(y_{E_iV} M_V + M_{E_i}) = X_j + X_j^* + M_V \sum_i y_{jE_i}(y_{E_iV} + m_{E_i}) \quad (5.22)$$

These balance equations have been checked for the model fits in Figure 5.6, but they apply only approximately to the data, because of measurement errors. Since tiny deviations in the amount of biomass and cellular content substantially change medium concentrations, the latter have been given a low weight in the simultaneous regressions of the 20 curves in Figure 5.6.

5.2.4 Non-limiting reserves can dam up

The significance of the excretion is in avoiding the possible occurrence of 'explosion'; if a cell cannot grow because of the absence of an essential nutrient, and it

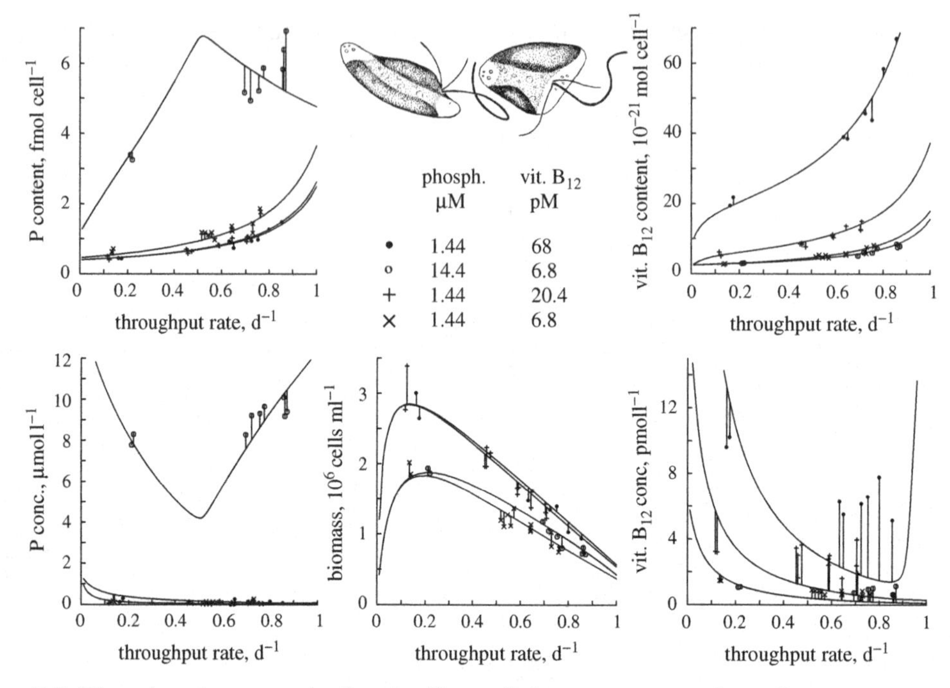

Figure 5.6 The phosphorus and vitamin B_{12} cellular contents and medium concentrations, and the biomass density, as functions of throughput rate $\dot{h}$ of the haptophyte *Pavlova lutheri* at four levels of these nutrients in the feed. Data from Droop [295]. The parameters are the reserve turnover rates $\dot{k}_{E_1} = 1.19\,\mathrm{d}^{-1}$, $\dot{k}_{E_2} = 1.22\,\mathrm{d}^{-1}$, stoichiometric requirements $y_{E_1V} = 0.39\,\mathrm{fmol\,cell}^{-1}$, $y_{E_2V} = 2.35\,10^{-21}\,\mathrm{mol\,cell}^{-1}$, maximum specific assimilation rates $j_{E_1Am} = 4.91\,\mathrm{fmol\,(cell\,d)}^{-1}$, $j_{E_2Am} = 76.6\,10^{-21}\,\mathrm{mol\,(cell\,d)}^{-1}$, recovery fractions $\kappa_{E_1} = 0.69$, $\kappa_{E_2} = 0.96$, maintenance rates $\dot{k}_{M_1} = 0.0079\,\mathrm{d}^{-1}$, $\dot{k}_{M2} = 0.135\,\mathrm{d}^{-1}$, given the saturation constants $X_{K_1} = 0.017\,\mu\mathrm{M}$, $X_{K_2} = 0.12\,\mathrm{pM}$ by Droop. The simultaneously fitted curves obey mass balances, and reveal measurement errors in the vitamin concentrations.

would continue to take up other nutrients, the accumulation of those nutrients would be unbounded without excretion. The combination of a first-order dynamics of reserve densities and $0 \leq \kappa_{E_i} < 1$ implies the existence of an upper boundary for reserve densities if upper boundaries for the assimilation rates exist. The steady-state reserve density m_{E2} is maximal if assimilation is maximal, $j_{E_2A} = j_{E_2Am}$, while expenditure is minimal, which occurs when growth is zero, $\dot{J}_{VG} = 0$, i.e. when $\dot{J}_{E_1C} = \dot{J}_{E_1M_1} = \dot{J}_{E_1A}$ or $m_{E_1} = j_{E_1M_1}/\dot{k}_{E_1}$. The maximum reserve density is found from (5.17) to be $m_{E_2m} = \frac{j_{E_2Am} - \kappa_{E_2} j_{E_2M_2}}{(1-\kappa_{E_2})\dot{k}_{E_2}}$. This illustrates the point that excretion is essential: $m_{E_im} \to \infty$ for $\kappa_{E_i} \to 1$. I will call the fractions κ_{E_i} recovery fractions. The density of the reserve that fully arrests growth is at minimum, and has the value $\frac{j_{EM}}{\dot{k}_E}$. Excretion is a common feature; extracellular release of organic carbon in phytoplankton has been reported to be as high as 75% of the totally fixed carbon [733].

The density of the limiting reserve *in*creases (hyperbolically) with the growth rate, while the non-limiting reserves can *de*crease with the growth rate. This very much depends on the recovery fraction κ_E. The reserve density of the non-limiting nutrients can build up to spectacular levels, which easily leads to the wrong conclusion that (all) reserve densities decrease with the growth rate.

If (traces of) all essential nutrients are required for the assimilation of each reserve, rare nutrients reduce the uptake of abundant ones and 'explosion' is avoided in almost all cases of practical interest, even if $\kappa_{Ei} = 1$; 'explosion' can still occur theoretically, in the absence of maintenance costs ($j_{E_iM_i} = 0$). The DEB model accommodates, therefore, two controls on reserve accumulation: via assimilation of nutrients and via recovery.

Biological phosphate removal

The accumulation of reserves that are synthesised from non-limiting nutrients is exploited technically in the process of phosphate removal in sewage treatment plants, using *Acinetobacter calcoaceticus*. These remarkable bacteria cannot use hexoses as carbon and energy source [1098]. Sewage water typically contains 10–30 mg l^{-1} phosphorus. Under aerobic conditions, actinobacters decompose carbohydrates, such that they extract energy but little carbon. The energy is fixed in polyphosphates, by taking up phosphate. Under anaerobic conditions, energy is limiting and volatile fatty acids, such as acetates, are taken up and converted into poly-2-hydroxybutyrate (PHB), while stored polyphosphates are used for energy supply in this transformation [526]. The quantitative details of this coupling are not quite clear yet; one possibility is that the rejected polyphosphate flux is used for the assimilation of PHB. The excreted phosphate is technically precipitated with calcium carbonate. This gives the scope for phosphate removal by alternating between aerobic and anaerobic conditions; the specific maintenance requirement j_{EM} for phosphate is probably very small.

5.2.5 Dioxygen flux

The physiological literature frequently presents Photosynthesis–Irradiance (PI) curves, where photosynthesis is usually measured via dioxygen production. The rate of photosynthesis is in practice frequently measured by the rate of dioxygen production, but the relationship is, however, rather indirect; dioxygen is produced in association with assimilation, but consumed in association with maintenance and growth. Almost all phototrophs have heterotrophic capabilities which makes the presence of a generalised reserve likely; this requires dioxygen in its assimilation.

The interpretation of experimental data is further hampered by the common practice of presenting dioxygen fluxes relative to chlorophyll, usually chlorophyll *a*; this is practical, because chlorophyll is rather easy to measure. This compound represents, just like all other compounds in the body, a weighted sum of the generalised reserves and the structural mass: $M_{Chl} = y_{Chl\,E} M_E + y_{Chl\,V} M_V$, or $m_{Chl} = y_{Chl\,E} m_E + y_{Chl\,V}$. The chlorophyll-specific dioxygen flux, therefore, amounts to j_O/m_{Chl}, which can be related to environmental and growth conditions, but involves many aspects of physiology, not just photosynthesis.

5.2.6 Ammonia–nitrate interactions

Many organisms can use several nitrogen substrates for assimilation, including ammonia, nitrite, nitrate, urea, amines and amino acids. Plants have access to nitrogen in organic compounds via mycorrhizae. Ammonia is rather toxic, so it does not accumulate as such; it is directly assimilated into amino acids, such as glutamate and glutamine. Nitrite is also rather toxic, and has mutagenic properties, see {247}; nevertheless, it is stored by some organisms. Nitrate is first reduced to nitrite, and then to ammonia, before further use [358]. These reductions require substantial energy, which is probably the reason why ammonia is usually strongly preferred as a substrate. It is even generally believed that ammonia inhibits nitrate uptake, but this does not seem to hold true [284]. Organisms vary in their properties with respect to nitrogen uptake. The intensively studied yeast *Saccharomyces cerevisiae* cannot assimilate nitrogen oxides [1213]. Some yeasts and bacteria nitrify ammonia to nitrate. Selective preferences for ammonia and nitrate can explain main patterns in plant associations [105]. Soil types differ substantially in ammonia and nitrate availability for plants [332], and their ratio strongly influences the occurrence of plant species, even at a very small spatial scale, such as the shifting mosaic of gap and understorey conditions in a forest [226, 670]. Probably because of its toxicity, ammonium assimilation occurs in the roots and not in the shoots of plants.

Figure 5.7 indicates how an alternative nitrogen substrate for ammonia can be implemented in a DEB framework. Ammonia is stored before use, just like nitrate, but the maximum storage capacity is very low, and the turnover rate very high. Homeostasis of structural mass requires that the product of the synthesis of ammonia and carbohydrate is identical to the generalised reserve, which means that the synthesis occurs twice: just after assimilation (prior to storage) from assimilates and after storage, just prior to synthesis of structural mass from catabolised products. The rules for sequential processing of substitutable substrates can be used to quantify the fluxes, see {104}. The extra requirement of energy in the processing of nitrate can be taken into account by the stoichiometric coupling with carbohydrates, which can depend on the substrate

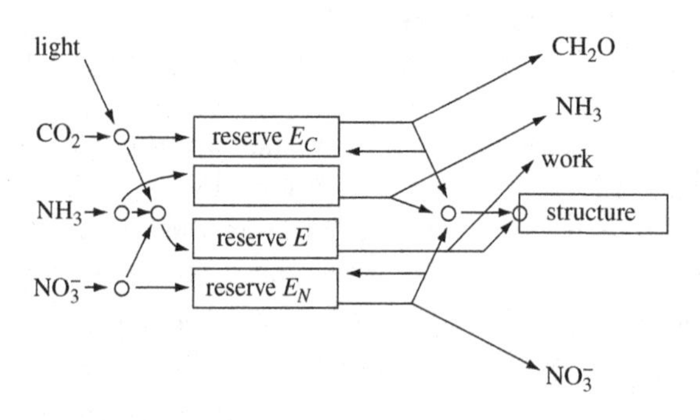

Figure 5.7 Diagram of nitrogen assimilation, where light and carbon dioxide are converted into carbohydrate reserves E_C, and nitrate into nitrate reserves E_N, which are used to synthesise structure. Ammonia can be used as an alternative nitrogen source, but barely accumulates. The circles indicate Synthesising Units.

that is used. Many applications allow a reduction of this redundancy, and a description without generalised reserves will be adequate. Ammonia is not only taken up, but is also excreted in association with growth and maintenance.

The assimilation of ammonia, nitrate and carbohydrates is given by (5.10) and (4.12). Treating ammonia and nitrogen as substitutable substrates, and complementary to carbohydrates, the specific assimilation of generalised reserves is

$$j_{EA} = \left(j_{EAm}^{-1} + (j'_{N_HA} + j'_{N_OA})^{-1} + j'^{-1}_{C_HA} - (j'_{N_HA} + j'_{N_OA} + j'_{C_HA})^{-1}\right)^{-1} \quad (5.23)$$

where $j'_{*A} = \rho_* y_{*E} j_{*A}$, and $y_{C_HE} = \theta^A_{N_H} y^{N_H}_{C_HE} + \theta^A_{N_O} y^{N_O}_{C_HE}$, where $\theta^A_{N_H} + \theta^A_{N_O} = 1$, and $\theta^A_{N_H} = j'_{N_HA}(j'_{N_HA} + j'_{N_OA})^{-1}$. The maximum of j_{EA} is not necessarily constant: $j_{EAm} = \theta^A_{N_H} j^{N_H}_{EAm} + \theta^A_{N_O} j^{N_O}_{EAm}$. Since the reduction of nitrate is rather energy consuming, and extracted from the oxidation of carbohydrates, the relationship $y^{N_H}_{C_HE} < y^{N_O}_{C_HE}$ holds. The requirement for carbohydrates can vary in time, and depends on the nitrogen source.

The specific mobilisation rates of the four reserves are $j_{*C} = (\dot{k}_* - j_{VG})m_*$. The specific mobilisation rate of the reserves E is $j_{EC_1} = (\dot{k}_E - j_{VG})m_E$. The synthesis of a compound identical to generalised reserves from catabolic products for metabolic use (maintenance and growth), j_{EC_2} is similar to that from assimilation products (5.23), with j_{*C} replacing j_{*A}. The growth SU is assumed to be fast enough to avoid spoiling of reserves, so $j_{VG} = y_{VE}(j_{EC_1} + j_{EC_2} - j_{EM})$.

Ammonia is hardly stored, which means that rejected ammonia is not fed back to the reserves ($\kappa_{E_{NH}} = 0$), but excreted. The turnover rate $\dot{k}_{E_{NH}}$ is large; this gives an extremely low ammonia reserve, $m_{E_{NH}} \simeq 0$, and the catabolic rate equals $j_{E_{NH}C} = j_{N_HA} - \theta^A_{N_H} y_{NE} j_{EA}$. The rejected ammonia flux is $j_{E_{NH}R} = j_{E_{NH}C} - \theta^C_{N_H} y_{NE} j_{EC_2}$, with $\theta^C_{N_H} + \theta^C_{N_O} = 1$ and $\theta^C_{N_O} = j'_{N_OC}(j'_{N_HA} - \theta^A_{N_H}\rho_{N_H} j_{EA} + j'_{N_OC})^{-1}$. The rejected fluxes of nitrate and carbohydrate reserves are $j_{E_{NO}R} = j_{E_{NO}C} - \theta^C_{N_O} y_{NE} j_{EC_2}$ and $j_{E_CR} = j_{E_CC} - j_{EC_2}$, from which fractions $\kappa_{E_{NO}}$ and κ_{E_C} are fed back to the reserves, the rest being excreted. The dynamics of the reserve densities m_{E_C} and $m_{E_{NO}}$ is given by (5.17). The specific rate of ammonium excretion amounts to $j_{N_HE} = j_{E_{NH}R} + y_{NE} j_{EM} + (n_{NE} y_{EV} - n_{NV}) j_{VG}$. The middle term relates to maintenance, the third one to growth overheads.

When nitrogen is limiting, the assimilation of generalised reserves (5.23) reduces to $j_{EA} = (j^{-1}_{EAm} + (j'_{N_HA} + j'_{N_OA})^{-1})^{-1}$. The carbohydrate reserve no longer limits growth and $j_{EC_2} = (j^{-1}_{EAm} + (y_{EN}\rho_{N_H}(j_{N_HA} - \theta^A_{N_H} y_{NE} j_{EA}) + j'_{N_OC})^{-1})^{-1} = (j^{-1}_{EAm} + (j'_{N_HA} - \theta^A_{N_H}\rho_{N_H} j_{EA} + j'_{N_OC})^{-1})^{-1}$.

The nitrogen in biomass can be decomposed into contributions from structural mass and the reserves, $n_{NW} = n_{NV} + n_{NE} m_E + m_{N_O}$. The specific nitrogen content is not constant during transient phases, but will become constant during the cell cycle in constant environments. This is an implication of the weak homeostasis assumption that is basic to the dynamics of reserves.

5.3 Several structural masses

The bill of the guillemot *Uriaaalge* is just one example of non-isomorphic growth. Although of little energetic significance, the κ-rule provides the structure to describe such deviations.

The assumption of isomorphy implies that any tissue is a fixed fraction of the somatic tissue, conceived as a lumped sum. Static and dynamic generalisations of the κ-rule for allocation, see at {42}, imply a particular type of growth regulation and reveals the intimate connection between the κ-rule and allometric growth.

5.3.1 Static generalisation of the κ-rule

If structural volume V can be decomposed into that of some organ (e.g. the heart) V_H and of the rest V_R, so $V = V_R + V_H$, the static multivariate extension of the κ-rule amounts to

$$\kappa\kappa_H\dot{p}_C = [E_{GH}]\frac{d}{dt}V_H + [\dot{p}_{MH}]V_H \tag{5.24}$$

$$\kappa(1-\kappa_H)\dot{p}_C = [E_G]\frac{d}{dt}V_R + [\dot{p}_M]V_R(1 + L_T/L_R) \tag{5.25}$$

where $[E_{GH}]$ and $[\dot{p}_{MH}]$ are the specific costs for synthesis and maintenance of organ H. The mobilisation power $\dot{p}_C$ is given in (2.20), so these equations fully specify the growth of V_R and V_H. This dynamics still has full isomorphy as special case, and can show near-allometric relationships between organ and whole body weight; see Figure 5.8. The mechanism behind allometric growth of body parts is intimately connected to the κ-rule. If $\dot{k}_{MH} = \dot{k}_M$, with $\dot{k}_{MH} = [\dot{p}_{MH}]/[E_{GH}]$, and the heating length L_T is small, we have $\frac{d}{d\tau}V_R + V_R \propto \frac{d}{d\tau}V_H + V_H$ for scaled time $\tau = \dot{k}_M^{-1}$. The somatic maintenance rate coefficients for the heart and the rest of the body for the ducks of Figure 5.8 differ by a factor 10 in four cases, but by a factor 0.2 in the case of the NF 20 Pekin. This big difference needs further explanation.

Allometric growth of a body part occurs if the contribution of part i to total body volume is insignificant, because $V_+ \neq \sum_i \alpha_i V^{\beta_i}$ if $\beta_i \neq 1$ for some i, whatever the values of positive α_i's. Huxley [540] described how certain parts of the body can change in size

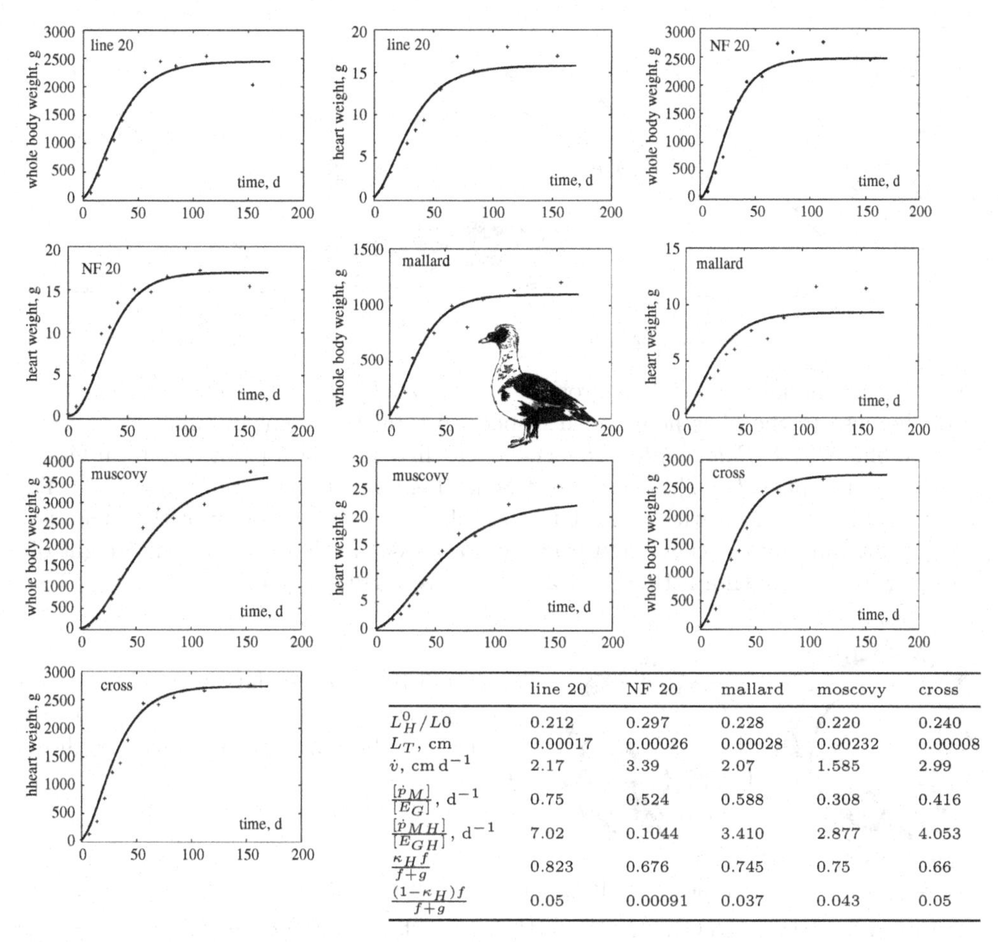

	line 20	NF 20	mallard	moscovy	cross
$L_H^0/L0$	0.212	0.297	0.228	0.220	0.240
L_T, cm	0.00017	0.00026	0.00028	0.00232	0.00008
$\dot{v}$, cm d^{-1}	2.17	3.39	2.07	1.585	2.99
$\frac{[\dot{p}_M]}{[E_G]}$, d^{-1}	0.75	0.524	0.588	0.308	0.416
$\frac{[\dot{p}_{MH}]}{[E_{GH}]}$, d^{-1}	7.02	0.1044	3.410	2.877	4.053
$\frac{\kappa_H f}{f+g}$	0.823	0.676	0.745	0.75	0.66
$\frac{(1-\kappa_H)f}{f+g}$	0.05	0.00091	0.037	0.043	0.05

Figure 5.8 Whole body weight and heart weight as function of time since birth in duck species: mallards, two lines of white Pekins (*Anas platyrhynchos*), muscovys (*Cairina moschata*), and a muscovy × white Pekin cross. Data from Gille and Salomon [408]. The static generalisation of the κ-rule can capture the decreasing relative size of the heart. This suggests that the relative workload of the heart remains rather constant.

relative to the whole body using allometric functions and highlighted the problem that if some parts change in an allometric way, other parts cannot. Absolute growth requires specification of how feeding and digestion (and heating for endotherms) depend on the volume and shape of the different tissues. It is likely to become complex. Allometric growth of extremities and skeletal elements frequently occurs, as illustrated in Figure 5.9. Houck *et al.* [529] used this growth as a criterion to delineate taxa in fossil bird *Archaeopteryx*. It is improbable that whole-organism energetics is seriously affected by these relative changes.

Isomorphs thus require growth regulation over the different body parts. Without control, allometric growth results. For isomorphs $[V_i] \equiv V_i/V_+$ must remain fixed, so that

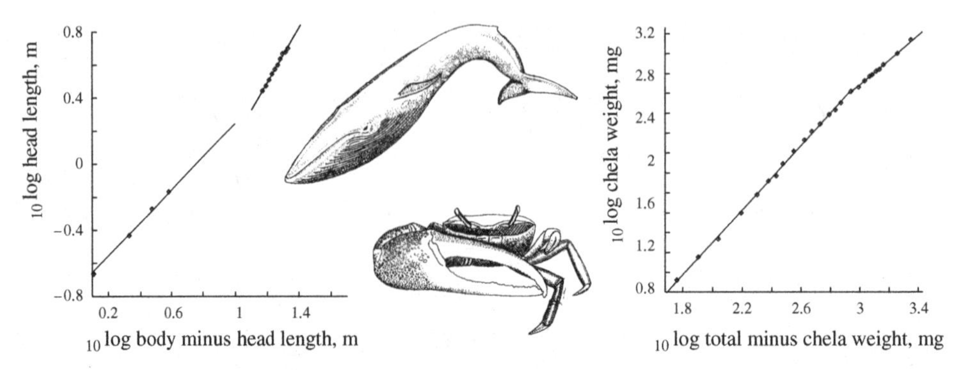

Figure 5.9 Examples of allometric growth: $\log y = a + b \log x$. Left: The head length (from the tip of the nose to the blow hole), with respect to total body length minus the head length in the male blue whale, *Balaenoptera musculus*. The first four data points are from fetuses, where growth is isomorphic ($b = 1$). Thereafter the head extends more rapidly ($b = 1.65$). Right: The weight of the large chela with respect to that of the rest of the body in the male fiddler crab *Uca pugnax*. Initially the chela grows rapidly ($b = 1.63$) until a rest of body weight of 850 mg, thereafter it slows down a little ($b = 1.23$). Data from Huxley [540].

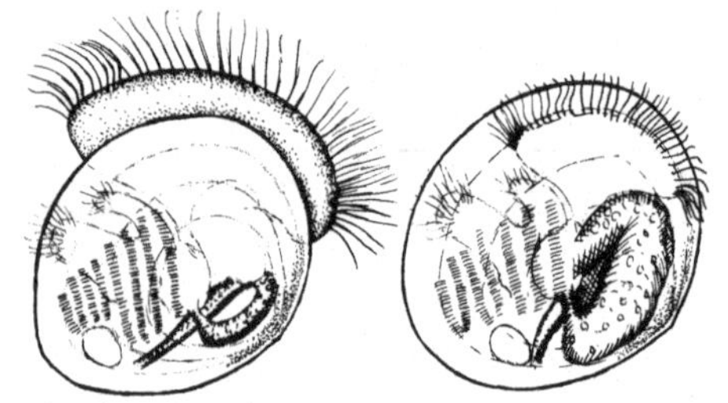

Figure 5.10 *Macoma* larvae develop a large velum (a filtering organ) and a small gut at low food levels (left), and the other way around at high food levels (right).

$\frac{d}{dt}V_i = [V_i]\frac{d}{dt}V_+$ must hold. For the DEB model this implies that the organism must accelerate or retard the growth of organ/tissue/part i by a factor $[V_i]\frac{dV_+}{dV_i} \simeq g_i \sum_j [V_j]/g_j$, with $g_i \equiv \frac{[E_{Gi}]}{\kappa_i [E_m]}$. (The approximation holds for $\dot{p}_{Mi} << \kappa_i \dot{p}_C$.) The mechanism of control may be via the density of carriers that transfer resources from the blood to the tissue. The carrier density in membranes of large tissues/parts should be less than that in small tissues/parts for a particular value.

The acceleration/retardation factor demonstrates that the carrier density does not have to change during growth. Other types of growth regulation are also possible. This discussion is only about the effects of regulation, rather than about its mechanism.

5.3.2 Dynamic generalisation of the κ-rule

The fraction κ_i of the mobilised reserve that is allocated to a particular organ i can change when the allocation is linked to the relative workload of that organ; this extension of the κ-rule is called the workload model and developed in collaboration with Ingeborg van Leeuwen [691]. Such dynamic extensions are necessary to capture e.g. the differential growth of velum and gut in bivalve larvae in response to changes in food levels [644], see Figure 5.10. Although the workload allocation is really general, I will specify it for the velum and the gut, assuming that the filtering rate is fully controlled by the size of filtering organ (velum) of volume V_F, and the digestion by the food-processing organ (gut) of volume V_X. The total structural volume thus amounts to $V = V_F + V_X + V_G$, where V_G is the general, i.e. non-assimilatory, part of the body.

Isomorphy implies that $V_F = \theta_F V$, for constant fraction θ_F, while $\dot{F} = \{\dot{F}\}(V_F/\theta_F)^{2/3}$, where $\{\dot{F}\}$ does not depend on the size of structure. The same applies for V_X, and $\dot{J}_{XA} = \{\dot{J}_{XA}\}(V_X/\theta_X)^{2/3}$. This couples organ size and function.

The arrival rate of food particles in density X at the individual that filters at rate $\dot{F}$ equals $\dot{F}X$. We assume a parsimonious design, so the filtering rate is such that $\dot{F}X = \dot{J}_{XA}$ and the amount of rejected particles is negligibly small. This means that the filtering rate equals $\dot{F} = (\dot{F}_m^{-1} + X\dot{J}_{XAm}^{-1})^{-1}$, and the half-saturation constant equals

$$K = \frac{\dot{J}_{XAm}}{\dot{F}_m} = \frac{\{\dot{J}_{XAm}\}}{\{\dot{F}_m\}}\left(\frac{V_X/\theta_X}{V_F/\theta_F}\right)^{2/3} = \frac{\{\dot{J}'_{XAm}\}}{\{\dot{F}'_m\}}\left(\frac{V_X}{V_F}\right)^{2/3} = K'\left(\frac{V_X}{V_F}\right)^{2/3} \tag{5.26}$$

with $\{\dot{J}'_{XAm}\} = \{\dot{J}_{XAm}\}\theta_X^{-2/3}$ and $\{\dot{F}'_m\} = \{\dot{F}_m\}\theta_F^{-2/3}$ and $K' = \{\dot{J}'_{XAm}\}/\{\dot{F}'_m\}$. The feeding rate amounts to $\dot{J}_{XA} = f\{\dot{J}'_{XAm}\}V_X^{2/3}$ with scaled functional response $f = \frac{X}{K+X}$.

Notice that this expression for the half-saturation constant is identical with that for nutrient uptake by plant roots, see {137}, where this uptake depends on the transport of water in the soil, and so on the evaporation by the shoot, thus on the surface area of the shoot. This resemblance of saturation constants is more than superficial if we look beyond morphology to functions of organs, where shoot and velum or root and gut have functional properties in common.

The workload of the filtering and digestion organs can be defined as

$$f_F = \dot{F}/\dot{F}_m = 1 - f_X \quad \text{and} \quad f_X = \dot{J}_{XA}/\dot{J}_{XAm} = f \tag{5.27}$$

The elasticities of food uptake for the volumes of velum and gut are complementary

$$\frac{V_F}{\dot{J}_{XA}}\frac{d\dot{J}_{XA}}{dV_F} = \frac{2}{3}(1-f) \quad \text{and} \quad \frac{V_V}{\dot{J}_{XA}}\frac{d\dot{J}_{XA}}{dV_X} = \frac{2}{3}f \tag{5.28}$$

Assuming that the costs of structure and its somatic maintenance are independent of the type of structure, the static generalisation of the κ-rule amounts for $\kappa_F = 1 - \kappa_X$ to

$$\kappa\kappa_A\kappa_F\dot{p}_C = [E_G]\frac{d}{dt}V_F + [\dot{p}_M]V_F \tag{5.29}$$

$$\kappa\kappa_A\kappa_X\dot{p}_C = [E_G]\frac{d}{dt}V_X + [\dot{p}_M]V_X \tag{5.30}$$

$$\kappa(1-\kappa_A)\dot{p}_C = [E_G]\frac{d}{dt}V_G + [\dot{p}_M]V_G \tag{5.31}$$

where κ_A is the fraction of the mobilised reserve that is allocated to the assimilation machinery, i.e. to $V_A = V_F + V_X$ and $1 - \kappa$ to maturity maintenance plus maturation (or reproduction in adults). The workload model now simply states that κ_F and κ_X are not constant but $\kappa_F = f_F$ and $\kappa_X = f_X = 1 - \kappa_F$, assuming that the coupling between organ size and function does not change with the relative size of the organ. Given the mobilisation power $\dot{p}_C$ by (2.20), these equations fully determine growth of body parts.

A nice property of the workload model is that it is weakly homeostatic: a sudden change in food density is followed by a rather rapid adaptation of the relative size of the organ, but after this adaptation the relative size remains constant, while growth continues. Another nice property is that the functional response after adaptation is found to be $f(X) = (1 + (K'/X)^{1/2})^{-1}$, which is a special case of the Hill's functional response $f(X) = (1 + (K/X)^n)^{-1}$; this model has an origin in biochemistry [508], and its application in ecology was empirical only; the workload model offers a mechanistic explanation.

This type of adaptation enhances growth, compared to the standard DEB model, both at low and high food densities. This is revealed by the steady-state value of the scaled reserve density, which is somewhat higher.

This line of reasoning can also be applied to kidneys, for instance, which remove nitrogenous waste from the body, and the DEB theory predicts how the nitrogen excretion rate should depend on body size. The only extra modelling step that is required to specify the detailed growth of kidneys is how the function of kidneys relates to their size. From a more abstract point of view, the interactive growth of the kidney and its host has much in common with the material discussed in the section on syntrophy, {337}, where the body acts as a donor of nitrogenous waste, and the kidney as receiver.

Like kidneys, lungs have a workload that can be quantified by DEB theory in a straightforward way, since their function is in the transport of dioxygen and carbon dioxide, and the theory fully specifies these fluxes. The static allocation model for the heart of the previous subsection can be reformulated dynamically as a workload model after specification of how the size of the heart relates to its function in transport. The fact that the static model gives a good description suggests that the relative workload of the heart remains rather constant, which indirectly specifies how heart size relates to transport function. Some more steps are required to specify the workload of the brain in DEB theory. Most of its mass is involved in computations on muscle coordination and information processing from the sensors, while some mass is linked to metabolic regulation and the archiving of knowledge. Although the details still have to be worked out, most of these activities must occur in an early phase in the development of an individual, and these activities don't increase a lot during further growth. Qualitatively it is easy to understand that a workload model will result in a relatively large brain at young age, which is consistent with observations on vertebrates.

Another interesting application of the workload model is in the specification of how the growth of tumours depends on the nutritional status of the host [693]. The 'work' by the tumour can be defined as the consumption of somatic maintenance relative to that of the tumour plus host. The model specifies the effects of caloric restriction as therapy to reduce the growth of tumours. The model also helps in understanding why tumours grow much faster in young hosts, relative to older ones. Rather than delineating just host and tumour as done in [693], the theory can also be extended to the growth of tumours in specific organs (or tissues), if the workload of the organ can be quantified by DEB theory as function of the size of the organ. This would be helpful for evaluating the effects of changes in diet on tumour growth, for instance.

The workload model links up nicely with the adaptation model [138] which was found to be adequate to capture diauxic growth of micro-organisms, see {291}. This model for the regulation of the relative abundance of carriers for the uptake of various substrates also uses the workload of the carriers as key for the production of the various substrate-specific carriers.

5.3.3 Roots and shoots: translocation

The delineation of (at least) two types of structural mass is essential if we are to understand the development and growth of plants that use roots for the uptake (and excretion) of nutrients, and shoots for light uptake, gas exchange (carbon dioxide and dioxygen), the evaporation of water (necessary for nutrient uptake by the roots) and reproduction. Bijlsma [105] makes a distinction between primary and secondary structures for both roots and shoots. The argument for such a refinement is to incorporate mechanical arguments to model stiffness versus transport. A simpler alternative is to use a single-state variable and change stiffness via products (cellulose, lignin) that accumulate in the plant. Environmental factors can affect this production. Plants seem to follow the obvious strategy of investing relatively more in roots when water or nitrogen is limiting, and in shoots if light is limiting [162].

The interactions between the roots and shoots of plants seem to be a mixture between a two-structure organism and a symbiosis, which gives them a substantial relative flexibility in growth. Table 5.1 presents a summary of the fluxes, Figure 5.11 gives a diagram of fluxes, while Table 5.2 specifies the fluxes, as follows from the DEB theory in its simplest form. Figure 5.12 gives an example of a plant growth curve for a single choice of parameter values, and illustrates the effect of light restriction. Many extensions of the model are conceivable, such as limitations by other nutrients.

The proposed model has eight state variables (structure, and three reserves for root and shoot). We need generalised reserves to accommodate all micro-nutrients that are required to generate structure, and carbon and nitrogen reserves to allow nitrogen uptake during darkness. The carbon and nitrogen reserves do not necessarily consist of pure carbohydrates and nitrates, respectively; they can be thought of as generalised reserves that are enriched in these compounds. Limitation by micro-nutrients is not discussed here, so they are assumed to be non-limiting.

All reserves are initially zero except the root's generalised reserve M_{ER}, which represents the initial mass of the seed. Due to the translocation mechanism, the generalised root reserve soon partitions itself across those of shoots and roots, at a rate that depends

Table 5.1 The chemical compounds of the plant and their transformations and indices. The + sign means appearance, the − sign disappearance. The signs of the mineral fluxes depend on the chemical indices and parameter values. The labels on rows and columns serve as indices to denote mass fluxes and powers. The table shows the flux matrix $\dot{\boldsymbol{J}}^T$, rather than $\dot{\boldsymbol{J}}$, if the signs are replaced by quantitative expressions presented in Table 5.2.

	compounds →		minerals						shoot					root				
↓ transformations			light	carbon dioxide	water	dioxygen	ammonia	nitrate	product	structure	carbon reserve	nitrogen reserve	reserves	product	structure	carbon reserve	nitrogen reserve	reserves
			L	C	H	O	N_H	N_O	PS	VS	E_CS	E_NS	ES	PR	VR	E_CR	E_NR	ER
shoot	assimilation	AS	−	−	−	+					+	+	+				−	
shoot	growth	GS		+	+	−	+	+	+	+	−	−	−					
shoot	dissipation	DS		+	+	−	+	+	+		−	−	−					
shoot	reproduction	R		+	+	−	+	+			−	−	−					
	translocation	T		+	+	−	+	+			−	−	±			−	−	∓
root	assimilation	AR		+	±	−	−	−			−					+	+	+
root	growth	GR		+	+	−	+	+						+	+	−	−	−
root	dissipation	DR		+	+	−	+	+						+		−	−	−

on the values $\dot{k}_{ES}$ and $\dot{k}_{ER}$. The initial structural masses of roots and shoots are infinitesimally small, just like those of animal embryos. Flowering plants first develop one or two cotyledons, leaf-like structures that differ morphologically from normal leaves, and are usually rather thick, because of the shoot's relatively high generalised reserves, M_{ES}. When the shoot develops further, these reserves are reorganised over stem and leaves.

The generalised reserves are actively translocated between roots and shoots, as proposed by Bijlsma [105]. The translocation from one reserve to another is discussed in the section on foetal development {63}. If the reserve turnover rates $\mathcal{A}_*\dot{k}_{E*}$ are identical, and the translocation fast, the κ-rule emerges, as has been discussed in the previous section. Generally, however, these turnover rates differ because they involve surface area/volume relationships, and so the shape correction function, as discussed on {137}. The nitrogen (nitrate) and carbon (carbohydrate) reserves are used independently by roots and shoots; only the 'spoils' are translocated, in a way similar to symbiotic partners, see {340}. The translocated fluxes partake in the assimilation of the receiver.

The synthesis of generalised reserves, as a chemical compound, occurs twice in the root and in the shoot, as described on {200}:

- The assimilation process (AS and AR). In the root, ammonia and nitrate are taken up from the soil, and carbohydrate is received from the shoot. In the shoot, nitrate is received from the root, and carbohydrate is photosynthesised. The resulting

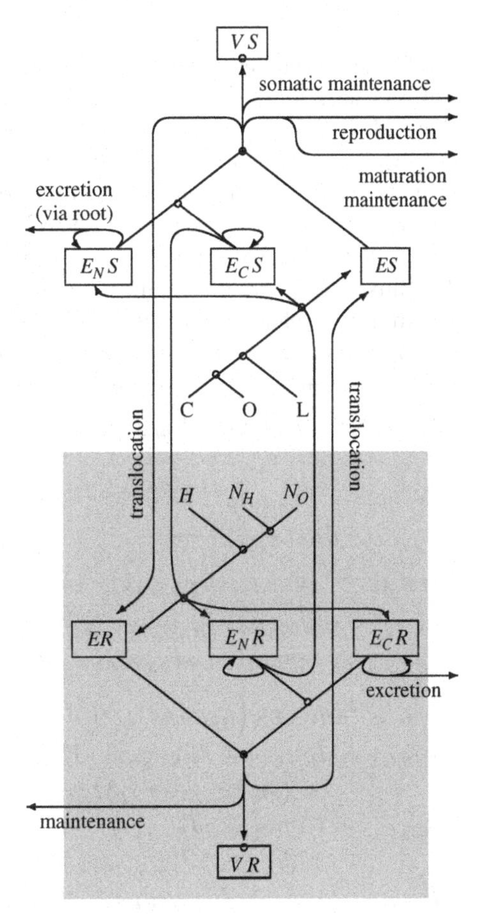

Figure 5.11 The diagram of a DEB model for the interactions between the root and the shoot of a plant. A seed has initially only an amount of generalised root reserve ER, all other reserves and the structural masses of the root, VR, and the shoot, VS, are negligibly small; translocation fills the generalised reserves of the shoot, ES, during the embryonic stage. Assimilation of ammonia, N_H, nitrate, N_O, carbon dioxide C and light, L, is switched on at birth. Water, H, interferes with the uptake of nutrients from the soil; dioxygen, O, interferes with the assimilation of carbon dioxide. Besides generalised reserves, carbohydrate reserves $E_C R$ and $E_C S$ and nitrogen reserves $E_N R$ and $E_N S$ are filled (and used) during the juvenile and adult stages. A fixed fraction of the rejected carbohydrate and nitrogen reserves translocated, and enters via the assimilation systems. The root remains in the juvenile stage; the allocation to maturity maintenance can be combined with that to somatic maintenance, and the allocation to maturation can be combined with that to growth. The shoot generally enters the adult stage, and requires explicit treatment of these fluxes. Maturation converts to reproduction at puberty. Circles indicate SUs. See Figure 10.4 for an evolutionary setting of this diagram.

compound is stored in the reserves ER and ES, respectively. The nitrogen and carbon that cannot be transformed into generalised reserves are stored in the specialised reserves.

- The mobilisation processes (C_2S and C_2R). Nitrate and carbohydrate are mobilised from the reserves; the resulting flux is merged with the mobilised generalised reserves and used for development and/or reproduction, and growth plus somatic maintenance. A fixed fraction of the nitrate and carbohydrate that is rejected by the Synthesising Units that produce generalised reserves is fed back to the reserves, the remaining fraction is translocated.

The binding probability ρ_{N_O} regulates the priority of ammonia relative to nitrate in the assimilation of reserves, as is discussed on {200}. The assimilated nitrate that is not used in this pathway is stored, but ammonia that is not used is excreted ($\rho_{N_H} = 1$ is taken here). The costs of synthesis of reserves ER from ammonia or nitrate are accommodated in the conversion coefficients $y^{N_O}_{C_H E}$ and $y^{N_H}_{C_H E}$; nitrate has to be reduced prior to this synthesis, and the costs are covered by the oxidation of carbohydrates, which gives $y^{N_O}_{C_H E} > y^{N_H}_{C_H E}$.

Table 5.2 The fluxes in and between the shoot S and the root R of a plant that experiences the forcing variables: light $\dot{J}_{LF}$ and concentrations of carbon dioxide X_C and dioxygen X_O (in the air), ammonia X_{N_H}, nitrate X_{N_O} and water X_H (in the soil). The dimensionless quantities $\mathcal{A}_S = (V_S/V_{dS})^{-1/3}\mathcal{M}_S(V_S) = (V_S/V_{dS})^{-V_S/V_{mS}}$ and $\mathcal{A}_R = (V_R/V_{dR})^{-1/3}\mathcal{M}_R(V_R) = (V_R/V_{dR})^{-V_R/V_{mR}}$ are introduced to simplify the notation, where V_{d*} are reference volumes that occur in the surface area/volume relationship, and $\mathcal{M}$ is the shape correction function, and V_{m*} parameters, see {137}. The relations $1 = \kappa_{SS}+\kappa_{RS}+\kappa_{TS}$ and $1 = \kappa_{SR}+\kappa_{TR}$ hold (the first index refers to soma, reproduction and translocation as destinations of catabolic fluxes). The fluxes $\dot{J}'_{*_1*_2}$ are gross fluxes, i.e. help fluxes for specifying the net fluxes $\dot{J}_{*_1*_2}$. Flux indices RS and RR refer to rejection, C_1S and C_1R to catabolism of reserves $E*$, C_2S and C_2R to catabolism of reserves E_N* plus E_C*; the other indices are listed in Table 5.1.

$$\dot{J}_{E_CS,AS} = \dot{J}'_{E_CS,AS} - y^{N_O}_{C_H,ES}\dot{J}_{ES,AS}; \quad \dot{J}'_{E_CS,AS} = \frac{(\dot{j}'_C - \dot{j}'_O)M_{VS}\mathcal{A}_S}{1+\frac{\dot{j}'_C}{k_C}+\frac{\dot{j}'_O}{k_O}+\frac{\dot{j}'_C+\dot{j}'_O}{\dot{j}'_L} - \frac{\dot{j}'_C+\dot{j}'_O}{\dot{j}'_L+\dot{j}'_C+\dot{j}'_O}}$$

$$\dot{J}_{E_NS,AS} = \dot{J}'_{E_NS,AS} - y_{E_NS,ES}\dot{J}_{ES,AS}; \quad \dot{j}'_L = \frac{\dot{j}_{L,Am}}{1+\dot{J}_{L,AK}/\dot{J}_{L,F}}; \quad \dot{j}'_C = \frac{\dot{j}_{C,Am}}{1+X_{KC}/X_C};$$
$$\dot{j}'_O = \frac{\dot{j}_{O,Am}}{1+X_{KO}/X_O}$$

$$\dot{J}_{ES,AS} = \left((y_{ES,E_NS}\dot{J}'_{E_NS,AS})^{-1} + (y^{N_O}_{ES,C_H}\dot{J}'_{E_CS,AS})^{-1} - (y_{ES,E_NS}\dot{J}'_{E_NS,AS} + y^{N_O}_{ES,C_H}\dot{J}'_{E_CS,AS})^{-1}\right)^{-1}$$

$$\dot{J}'_{E_NS,AS} = -y_{E_NS,E_NR}\dot{J}_{E_NR,AS}; \quad \dot{J}_{E_NR,AS} = -(1-\kappa_{E_NR})\dot{J}_{E_NR,RR}$$

$$\dot{J}_{VS,GS} = y_{VS,ES}\left(\kappa_{SS}\dot{J}_{ES,CS} + \dot{J}_{ES,MS}\right) \equiv \dot{r}_S M_{VS}; \quad \dot{J}_{ES,C_1S} = (\mathcal{A}_S\dot{k}_{ES} - \dot{r}_S)M_{ES}$$

$$\dot{J}_{ES,CS} = \dot{J}_{ES,C_1S} + \dot{J}_{ES,C_2S}; \quad \dot{J}'_{E_CS,CS} = (\mathcal{A}_S\dot{k}_{E_CS} - \dot{r}_S)M_{E_CS}; \quad \dot{J}'_{E_NS,CS} = (\mathcal{A}_S\dot{k}_{E_NS} - \dot{r}_S)M_{E_NS}$$

$$\dot{J}_{ES,C_2S} = \left((y_{ES,E_NS}\dot{J}'_{E_NS,CS})^{-1} + (y^{N_O}_{ES,C_H}\dot{J}'_{E_CS,CS})^{-1} - (y_{ES,E_NS}\dot{J}'_{E_NS,CS} + y^{N_O}_{ES,C_H}\dot{J}'_{E_CS,CS})^{-1}\right)^{-1}$$

$$\dot{J}_{ES,GS} = -y_{ES,VS}\dot{J}_{VS,GS}\theta_{ES}; \quad \dot{J}_{E_CS,GS} = y^{N_O}_{C_H,ES}\dot{J}_{ES,GS}(\theta^{-1}_{ES}-1); \quad \dot{J}_{E_NS,GS} = y_{E_NS,ES}\dot{J}_{ES,GS}(\theta^{-1}_{ES}-1)$$

$$\dot{J}_{E_CS,RS} = \dot{J}'_{E_CS,CS} - y^{N_O}_{C_H,ES}\dot{J}_{ES,C_2S}; \quad \dot{J}_{E_NS,RS} = \dot{J}'_{E_NS,CS} - y_{E_NS,ES}\dot{J}_{ES,C_2S}$$

$$\dot{J}_{ES,MS} = -\dot{j}_{ES,MS}M_{VS}; \quad \dot{J}_{ES,JS} = -\dot{j}_{ES,JS}\min\{M_{VS}, M_{V_pS}\}; \quad \theta_{ES} = \dot{J}_{ES,C_1S}\dot{J}^{-1}_{ES,CS}$$

$$\dot{J}_{E_CS,DS} = y^{N_O}_{C_H,ES}\dot{J}_{ES,DS}(\theta^{-1}_{ES}-1); \quad \dot{J}_{E_NS,DS} = y_{E_NS,ES}\dot{J}_{ES,DS}(\theta^{-1}_{ES}-1) - (1-\kappa_{E_NS})\dot{J}_{E_NS,RS}$$

$$\dot{J}_{ES,DS} = \theta_{ES}(\dot{J}_{ES,MS} + \dot{J}_{ES,JS}); \quad \dot{J}_{ES,R} = -\theta_{ES}(\kappa_{RS}\dot{J}_{ES,CS} + \dot{J}_{ES,JS})$$

$$\dot{J}_{E_CS,R} = y^{N_O}_{C_H,ES}\dot{J}_{ES,R}(\theta^{-1}_{ES}-1); \quad \dot{J}_{E_NS,R} = y_{E_NS,ES}\dot{J}_{ES,R}(\theta^{-1}_{ES}-1)$$

$$\dot{J}_{E_CS,T} = -\kappa_{TS}y^{N_O}_{C_H,ES}\dot{J}_{ES,C_2S}; \quad \dot{J}_{E_NS,T} = -\kappa_{TS}y_{E_NS,ES}\dot{J}_{ES,C_2S}$$

$$\dot{J}_{ES,T} = -\kappa_{TS}\dot{J}_{ES,C_1S} + y_{ES,ER}\kappa_{TR}\dot{J}_{ER,CR}; \quad \dot{J}_{ER,T} = -\kappa_{TR}\dot{J}_{ER,C_1R} + y'_{ER,ES}\kappa_{TS}\dot{J}_{ES,CS}$$

$$\dot{J}_{E_CR,T} = -\kappa_{TR}y^{N_O}_{C_H,ER}\dot{J}_{ER,C_2R}; \quad \dot{J}_{E_NR,T} = -\kappa_{TR}y_{E_NR,ER}\dot{J}_{ER,C_2R}$$

$$\dot{J}'_{N_H,AR} = \frac{M_{VR}\mathcal{A}_R\dot{j}_{N_H,Am}}{1+X_{KN_H}/X_{N_H}}; \quad \dot{J}_{N_O,AR} = \frac{M_{VR}\mathcal{A}_R\dot{j}_{N_O,Am}}{1+X_{KN_O}/X_{N_O}}; \quad X_{KN_H} = \frac{X'_{KN_H}}{1+\frac{X_H}{X_{KH}}\frac{\mathcal{A}_S}{\mathcal{A}_R}};$$
$$X_{KN_O} = \frac{X'_{KN_O}}{1+\frac{X_H}{X_{KH}}\frac{\mathcal{A}_S}{\mathcal{A}_R}}$$

$$\dot{J}_{E_CR,AR} = \dot{J}'_{E_CR,AR} - y_{C_H,ER}\dot{J}_{ER,AR}; \quad \dot{J}'_{E_CR,AR} = (1-\kappa_{E_CS})y_{E_CR,E_CS}\dot{J}_{E_CS,RS}$$

Table 5.2 (cont.)

$$\dot J_{E_N R,AR} = n^{-1}_{N,E_N R}(\dot J_{N_O,AR} - n_{N,ER}\theta_{N_O}\dot J_{ER,AR}); \quad \dot J_{N,AR} = \dot J'_{N_H,AR} + \rho_{N_O}\dot J_{N_O,AR}$$

$$\dot J_{ER,AR} = \left(n_{N,ER}\dot J^{-1}_{N,AR} + (y_{ER,C_H}\dot J'_{E_C R,AR})^{-1} - (n^{-1}_{N,ER}\dot J_{N,AR} + y_{ER,C_H}\dot J'_{E_C R,AR})^{-1}\right)^{-1}$$

$$y_{ER,C_H} = \theta_{N_H} y^{N_H}_{ER,C_H} + \theta_{N_O} y^{N_O}_{ER,C_H}; \quad \theta_{N_H} = 1 - \theta_{N_O} = J'_{N_H,AR}(J'_{N_H,AR} + \rho_{N_O} J_{N_O,AR})^{-1}$$

$$\dot J_{VR,GR} = y_{VR,ER}\left(\kappa_{SR}\dot J_{ER,CR} + \dot J_{ER,MR}\right) \equiv \dot r_R M_{VR}; \quad \dot J_{ER,C_1R} = (\mathcal{A}_R \dot k_{ER} - \dot r_R)M_{ER}$$

$$\dot J_{ER,CR} = \dot J_{ER,C_1R} + \dot J_{ER,C_2R}; \quad \dot J'_{E_C R,CR} = (\mathcal{A}_R \dot k_{E_C R} - \dot r_R)M_{E_C R};$$

$$\dot J'_{E_N R,CR} = (\mathcal{A}_R \dot k_{E_N R} - \dot r_R)M_{E_N R}$$

$$\dot J_{ER,C_2R} = \left(\,(y_{ER,E_N R}\dot J'_{E_N R,CR})^{-1} + (y^{N_O}_{ER,C_H}\dot J'_{E_C R,CR})^{-1}\right.$$
$$\left. -(y_{ER,E_N R}\dot J'_{E_N R,CR} + y^{N_O}_{ER,C_H}\dot J'_{E_C R,CR})^{-1}\,\right)^{-1}$$

$$\dot J_{ER,GR} = -y_{ER,VR}\dot J_{VR,GR}\theta_{ER}; \quad \dot J_{E_C R,GR} = y^{N_O}_{C_H,ER}\dot J_{ER,GR}(\theta^{-1}_{ER} - 1);$$

$$\dot J_{E_N R,GR} = y_{E_N R,ER}\dot J_{ER,GR}(\theta^{-1}_{ER} - 1)$$

$$\dot J_{E_C R,RR} = \dot J'_{E_C R,CR} - y^{N_O}_{C_H,ER}\dot J_{ER,C_2R}; \quad \dot J_{E_N R,RR} = \dot J'_{E_N R,CR} - y_{E_N R,ER}\dot J_{ER,C_2R}$$

$$\dot J_{E_C R,DR} = y^{N_O}_{C_H,ER}(1 - \theta_{ER})\dot J_{ER,MR} - (1 - \kappa_{E_C R})\dot J_{E_C R,RR};$$

$$\dot J_{E_N R,DR} = y_{E_N R,ER}(1 - \theta_{ER})\dot J_{ER,MR}$$

$$\dot J_{ER,MR} = -\dot j_{ER,MR}M_{VR}; \quad \dot J_{ER,DR} = \theta_{ER}\dot J_{ER,MR}; \quad \theta_{ER} = \dot J_{ER,C_1R}\dot J^{-1}_{ER,CR}$$

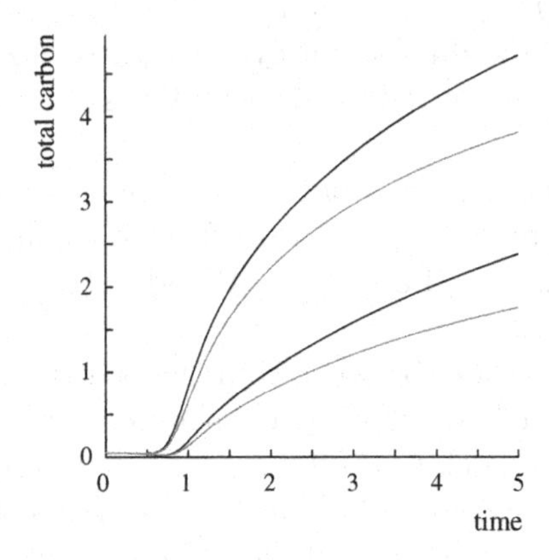

Figure 5.12 Examples of plant growth curves that result from the DEB model. The lower curves refer to root carbon, the upper ones to total plant carbon (root plus shoot). The grey curves refer to light restriction, and show that this affects the root more than the shoot. This realistic trait naturally results from the mechanism of exchange of carbohydrates and nitrogen; no optimisation argument is involved. The model has a rich repertoire of growth curves, of root/shoot mass ratio ontogenies, and of responses to changes in environmental factors, depending on parameter values.

The balance equations for the catabolic processes are

$$\begin{aligned}
&\dot{J}'_{E_CS,CS} - \kappa_{E_CS}\dot{J}_{E_CS,RS} = (1-\kappa_{E_CS})\dot{J}_{E_CS,RS} + \dot{J}_{E_CS,T} + \dot{J}_{E_CS,GS} + \dot{J}_{E_CS,DS} + \dot{J}_{E_CS,R} \\
&\dot{J}'_{E_NS,CS} - \kappa_{E_NS}\dot{J}_{E_NS,RS} = \dot{J}_{E_NS,T} + \dot{J}_{E_NS,GS} + \dot{J}_{E_NS,DS} + \dot{J}_{E_NS,R} \\
&\dot{J}_{ES,C_1S} = \dot{J}_{ES,T} + \dot{J}_{ES,GS} + \dot{J}_{ES,DS} + \dot{J}_{ES,R}; \quad \dot{J}_{ER,C_1R} = \dot{J}_{ER,T} + \dot{J}_{ER,GR} + \dot{J}_{ER,DR} \\
&\dot{J}'_{E_NR,CR} - \kappa_{E_NR}\dot{J}_{E_NR,RR} = (1-\kappa_{E_NR})\dot{J}_{E_NR,RR} + \dot{J}_{E_NR,T} + \dot{J}_{E_NR,GR} + \dot{J}_{E_NR,DR} \\
&\dot{J}'_{E_CR,CR} - \kappa_{E_CR}\dot{J}_{E_CR,RR} = \dot{J}_{E_CR,T} + \dot{J}_{E_CR,GR} + \dot{J}_{E_CR,DR}
\end{aligned} \tag{5.32}$$

where all fluxes are here taken to be positive. The left-hand sides specify what is leaking away from the reserves, and the right-hand sides specify the various destinations. The fluxes RS and RR on the left-hand sides specify the return fluxes of what cannot be used by the SUs. The 'spoil' fluxes RS and RR on the right-hand sides appear in the assimilation fluxes of the partner (root and shoot, respectively).

Assimilation in Table 5.2 should be set to zero for embryos. The root remains in the juvenile stage, the shoot is adult if $M_{VS} > M_{V_pS}$. The fluxes to reproduction (or maturation in embryos and juveniles) as specified in Table 5.2 represent outgoing fluxes from the reserves and includes overheads; a fraction κ_R of this flux is fixed in seeds, and the flux should be divided by the initial root reserve to arrive at a reproduction rate in terms of seeds per time.

If a plant lives for many years, and the resolution of the growth process is limited, the yearly shedding of leaves can be accommodated (approximately) in the constant specific maintenance costs $j_{ES,MS}$. For short-lived species, this will be less satisfactory and the maintenance costs should show a cyclical pattern explicitly. Ageing in plants does not follow the pattern of animals, because plants can replace cells that are hit by the ageing process.

The formulation in Table 5.2 does not account for water reserves. Water controls the saturation constant of the nitrogen uptake from the soil, see {154}. Photosynthesis and photorespiration are discussed on {189}. Wood production can be associated with growth and maintenance.

The weight of a plant has contributions from the structure, and all reserves, but also from the accumulation products, which can have supporting functions for the structures. Their production is associated with somatic maintenance, and continues even when growth of the structure ceases. In that case, weights do not have an asymptote. Chlorophyll is part of the structure and the generalised reserves of the shoot. The prime in the conversion coefficient $y'_{ER,ES}$ from root to shoot reserves indicates that $y'_{ER,AS} \neq y'_{ES,AR}$; root reserves have no chlorophyll and other differences in composition exist as well.

5.4 Summary

The DEB model can be extended in a straightforward way to deal with several substrates (nutrients), reserves and structural masses, using the rules for the behaviour of Synthesising Units (SUs) on the basis of the following supplementary assumptions:

---

1 Each reserve has an assimilation SU, and each structural mass a growth SU for their synthesis from substrates.

2 A fixed part of any catabolic flux that is rejected by a growth SU adds to the originating reserve, the rest is excreted.

3 Each structural mass requires a fixed mass-specific maintenance and each reserve contributes to this flux with a constant structure-specific rate. The maintenance requirements, therefore, have a fixed stoichiometry.

4 A single reserve can fuel the synthesis of several structural masses by partitioning the catabolic flux; a straightforward generalisation of the κ-rule. Maintenance competes with growth for each structure (tissue, organ).

The resulting dynamics allows a rich repertoire, because of the flexible behaviour of SUs. Two routes exist for the convergence to a single-substrate, single-reserve DEB model: a single substrate and reserve is limiting growth, or the abundances of the various substrates covary and all rejected catabolic fluxes are excreted. The transition of limiting to (almost) non-limiting behaviour of substrate and/or reserve is rather sharp, as results from the rules for the behaviour of SUs.

The limiting reserve *in*creases with the growth rate, but the non-limiting reserves can *de*crease with the growth rate as a consequence of the damming, which depends on the fraction of reserves that is excreted.

Carbon fixation by photosynthesis is modified by photorespiration and photoinhibition in ways that directly follow from the dynamics of SUs. The production and use of dioxygen is shown to have a rather indirect relationship with photosynthesis, although it is widely used as a quantifier.

Calcification is described, where carbon dioxide and bicarbonate are sequentially processed, substitutable substrates for photosynthesis, and calcification is stoichiometrically coupled with bicarbonate uptake.

The interaction between ammonia and nitrogen has been worked out, because of its ecological importance. The nitrogen compounds are treated as substitutable substrates in the synthesis of generalised reserves. The latter compound is synthesised from ammonia, nitrate and carbohydrate reserves prior to storage from assimilates, and prior to growth from catabolic fluxes. The ammonia reserves have an extremely low capacity.

The growth of body parts can be very close to the widely applied allometric growth on the basis of the static multivariate extension of the κ-rule. Dynamic extensions of the κ-rule can capture adaptations linked to the workload of organs; the growth of tumours, as modified by age of the host and caloric restriction, serves as an example.

A model for plant growth is proposed, which represents a mixture between a bivariate structured individual and a symbiosis between root and shoot; the generalised reserves are actively translocated, while each has its own nitrogen and carbohydrate reserve, where the partner receives the 'spoils', as in a symbiotic relationship.

6

Effects of compounds on budgets

The uptake rate of food and/or limiting nutrients depends on their availability and the processing capacity of the individual. The uptake rate of non-limiting nutrients can follow the same pattern in multiple reserve systems, or can be linked to the uptake of limiting compounds in single-reserve systems; transitions between these modes will be discussed later. The uptake of non-essential compounds differs from that of essential ones by the absence of regulated use.

The first purpose of this chapter is to show how energetics interferes with many aspects of the uptake of non-essential compounds and how compounds affect metabolism. This chapter starts with the effects of reactive oxygen species (ROS), a process known as ageing, and its links with energetics. After a brief introduction to other toxins and toxicants, the kinetics of non-essential compounds is discussed in terms of variations on the core model: a one-compartment model. Then follows a discussion of the inverse relationship, i.e. how compounds affect the energetics of individuals and the consequences for populations.

Effects of different compounds can be compared on the basis of parameter values. The covariation of parameter values is discussed in Chapter 8.

6.1 Ageing: effects of ROS

The frequently observed correlation between life span and the inverse volume-specific metabolic rate for different species (see, e.g. [1022]) has guided a lot of research, see [350]. Animals tend to live longer at low food levels than at high ones, which couples ageing to energetics. The experimental evidence, however, is rather conflicting on this point. For example, Ingle *et al.* [544] found such a negative relationship, while McCauley *et al.* [765] found a positive one for daphnids. This is doubtless due to the fundamental problem that death can occur for many reasons, such as food-related poisoning, that are not directly related to ageing.

Some species such as salmon, octopus, *Oikopleura* die after (first) reproduction, see {289}. They are said to be semelparous species, while species that reproduce more than once are called iteroparous. The sessile colonial sea squirt *Botryllus schlosseri* follows both genetically determined strategies within one population [437]. The semelparous colonies numerically dominate the population through midsummer, while the iteroparous ones do so in late summer. Death after first reproduction, like many other causes of death, does not relate to ageing.

The residence time of compounds in the reserve, $E/\dot{p}_C$, increases with the size of an isomorph. Some reserve compounds, such as proteins, can lose their metabolic activity. This implies that the fraction of inactive reserve compounds increases with size. This can be an important mechanism behind the gradual changes of metabolism during the life cycle.

On approaching the end of the life span, the organism usually becomes very vulnerable, which complicates the interpretation of the life span of a particular individual in terms of ageing. Experiments usually last a long time, which makes it hard to keep food densities at a fixed level and to prevent disturbances.

In a first naïve attempt to model the process of ageing, it might seem attractive to conceive the senile state, followed by death as the next step in the sequence embryo, juvenile, adult, and then tie it to energy investment in development just as has been done for the transitions to the juvenile and adult stages. This is not an option in the context of DEB theory, since at sufficiently low food densities the adult state is never entered, even if the animal survives for nutritional reasons. This means that it would live for ever, as far as ageing is concerned. Although species exist with very long life spans (excluding external causes of death [350]), this does not seem acceptable.

Not all species age in the way most animals do; ageing seems to be restricted to species that sport irreversible cell differentiation. This excludes coelenterates like *Hydra* and plants for instance, which does not imply that they have infinite life, due to e.g. mechanical wear and tear.

Free radicals, or related ROS formed as a spin-off of respiration, are thought to cause irreparable damage to the (nuclear and mitochondrial) DNA in organisms with irreversible cell differentiation and have a direct relationship with ageing [455, 460, 461, 1100, 1157]; see Figure 7.22. The damage by ROS depends on the specific activity of antioxidants and correlates with life span within mammals [33, 350, 1249], but also holds more generally [350]; the structure of the antioxidant enzyme manganese superoxide dismutase has been solved [1004]. Most use of dioxygen occurs in mitochondria, and damaged mitochondria might produce even more ROS [865, 692, 691], which explains the amplification ageing process, which is mainly observed in demand systems (birds and mammals), but much less so in supply systems.

The strategy to model the hazard rate for effects of ageing and toxic compounds on survival turned out to be effective. The hazard rate relates to the survival probability according to the differential equation $\frac{d}{dt}\Pr\{\underline{a}_\dagger > t\} = -\Pr\{\underline{a}_\dagger > t\}\dot{h}(t)$ or $\dot{h}(t) = -\frac{d}{dt}\ln\Pr\{\underline{a}_\dagger > t\}$. The mean life span equals $\mathcal{E}\underline{a}_\dagger = \int_0^\infty \Pr\{\underline{a}_\dagger > t\}\,dt = \int_0^\infty \exp(-\int_0^t \dot{h}(t_1)\,dt_1)\,dt$.

The effects of ageing on survival, and its interaction with energetics, can be captured efficiently on the basis of the idea that damage-inducing compounds Q (changed genes, affected mitochondria), accumulate at a rate that is proportional to the mobilisation rate $\dot{p}_C$ (so approximately proportional to the use of dioxygen excluding contributions by assimilation, see {149}), each of its molecules is copied at a rate proportional to the mobilisation rate [692], and damage-inducing compounds produce damage D ('wrong' proteins [1289] and other metabolic products) at a constant rate. The hazard rate $\dot{h}$ due to ageing is just proportional to the damage density $m_D = M_D/M_V$. Because of the uncertainty in coupling with molecular processes, I prefer to talk about damage and damage-inducing compounds. Kowald and Kirkwood [664, 665] followed a very similar line of reasoning, and incorporated much more detail.

For the mobilisation rate $\dot{p}_C$ and specific growth rate $\dot{r}$ given in (2.12) and (2.21), respectively, the changes in the densities of damage-inducing compounds Q and damage compounds D are

$$\frac{d}{dt}m_Q = \eta_{QC}\frac{\dot{p}_C}{M_V} + \frac{s_G \dot{p}_C}{[E_m]L_m^3}m_Q - \dot{r}m_Q; \quad \frac{d}{dt}m_D = \dot{k}_W y_{DQ} m_Q - \dot{r}m_D \tag{6.1}$$

where $m_Q(0) = 0$ and $m_D(0) = 0$. These two equations can be rewritten in equations for the change in acceleration $\ddot{q} = \ddot{h}_a \dot{k}_W y_{DQ} m_Q / m_D^{\text{ref}}$ and in hazard $\dot{h} = \ddot{h}_a m_D / m_D^{\text{ref}}$ as

$$\frac{d}{dt}\ddot{q} = (\ddot{q}\frac{L^3}{L_m^3}s_G + \ddot{h}_a)e(\frac{\dot{v}}{L} - \dot{r}) - \dot{r}\ddot{q}; \quad \frac{d}{dt}\dot{h} = \ddot{q} - \dot{r}\dot{h} \tag{6.2}$$

where $\ddot{q}(0) = 0$ and $\dot{h}(0) = 0$. The variables m_Q and m_D, or $\ddot{q}$ and $\dot{h}$, serve as state variables, supplementing reserve, structure and maturity. From these equations it can be seen that the model has two ageing parameters: Weibull ageing acceleration $\ddot{h}_a = \dot{h}_a \dot{k}_W y_{DQ} \frac{\eta_{QC}}{m_D^{\text{ref}}} \frac{[E_m]}{[M_V]}$ and Gompertz stress coefficient s_G. The model represents an extension of the single-parameter DEB model for ageing [641] into the direction of the three-parameter model by van Leeuwen *et al.* [692].

The strength of the model is in revealing the role of energetics in ageing; energy budget parameters show up in the survival process, which can also be obtained from data that do not relate to survival; differences in energetics affect survival probabilities. Male daphnids remain quite a bit smaller than females and this is the only difference in the fitted survival propabilities in Figure 6.1.

Figure 6.2 shows that the survival of mice is substantially affected by the feeding level, which mainly occurs via the amplification process quantified by s_G. The values of the energy conductance $\dot{v}$ and the somatic maintenance rate coefficient are similar to those of the rat, see Figure 4.7.

Figure 6.3 illustrates the effects of a temporary caloric restriction on growth and ageing in the guppy. The fit is remarkably good, and zero Gompertz stress also fits reasonably well. The present formulation allows for a separation of the ageing- and energy-based parameters. The estimation of the 'pure' ageing parameter in different situations and for different species will hopefully reveal patterns that can guide the search for more detailed molecular mechanisms; however, many factors may be involved,

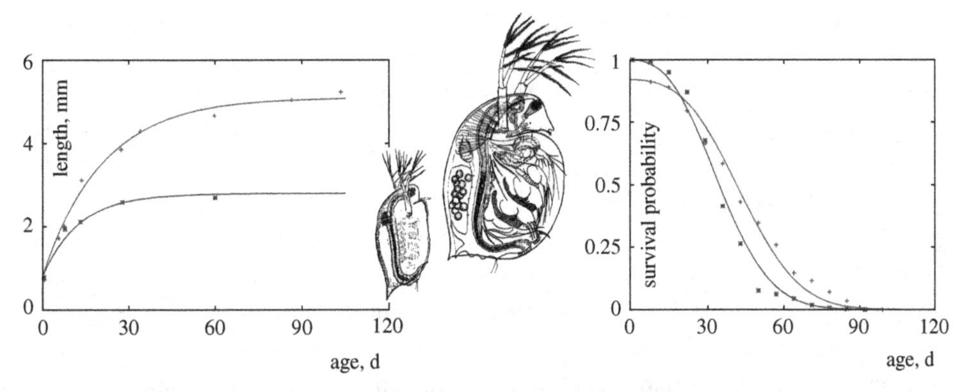

Figure 6.1 The growth curves and survival probabilities of female (+) and male (∗) *Daphnia magna* at 18 °C. Data from MacArthur and Baillie [724]. The growth curves are of the von Bertalanffy type with common length at birth. The survival probabilites have a common ageing acceleration $\ddot{h}_s$ and Gompertz stress s_G; the difference in survival is due to the difference in specific assimilation. The 8% initial female death is assumed to be accidental. Parameters: $L_b = 0.82\,\text{mm}$, $\dot{k}_M = 1\,\text{d}^{-1}$, $\dot{v} = 0.862\,\text{mm}\,\text{d}^{-1}$, $g_{\text{female}} = 0.169$, $g_{\text{male}} = 0.308$, $\ddot{h}_a = 1.25\,10^{-3}\,\text{d}^{-2}$, $s_G = -0.5$.

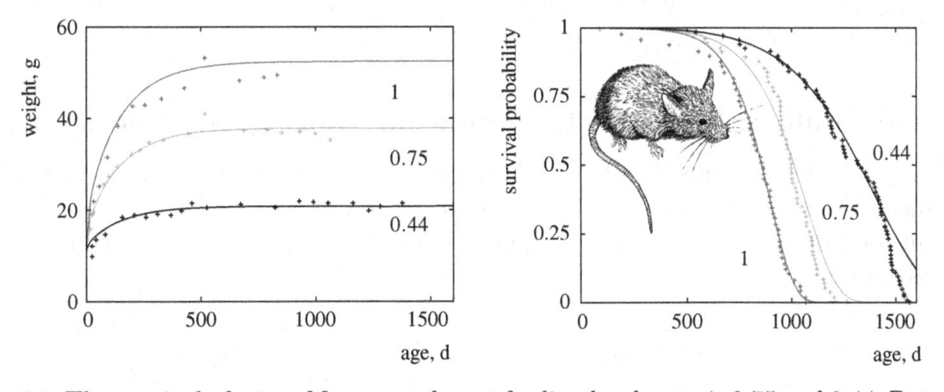

Figure 6.2 The survival of mice, *Mus musculus*, at feeding levels $\rho = 1$, 0.75 and 0.44. Data from Weindruch *et al.* [1231]. The feeding rate was experimentally kept constant, as in Figure 4.7, not the food density. The six curves, and that of Figure 2.15, were fitted simultaneously with ageing parameters: $\ddot{h}_a = 3\,10^{-9}\,\text{d}^{-2}$ and $s_G = 0.1$; DEB parameters $\dot{v} = 0.28\,\text{cm}\,\text{d}^{-1}$, $\dot{k}_M = 0.0065\,\text{d}^{-1}$, $L_T = 0\,\text{cm}$, $g = 13.6$; auxiliary parameters: $V(0) = 4\,\text{cm}^3$, $e(0) = 1$, $\omega_w = 0.681$, see (3.2). For foetal growth I assumed $\rho = 0.8$ for the mother and a start of development at 8.2 d.

see {247}. It has been suggested in the literature that the neural system may be involved in setting the ageing rate. The fact that brain weight in mammals correlates very well with respiration rate [520] makes it difficult to identify factors that determine life span in more detail. The mechanism may be again via the neutralisation of ROS.

The energy parameters can be linked to the accumulated damage to account for the well-known phenomenon that older individuals eat less and reproduce less than younger ones with the same body volume. Senescence can be modelled this way. It is a special

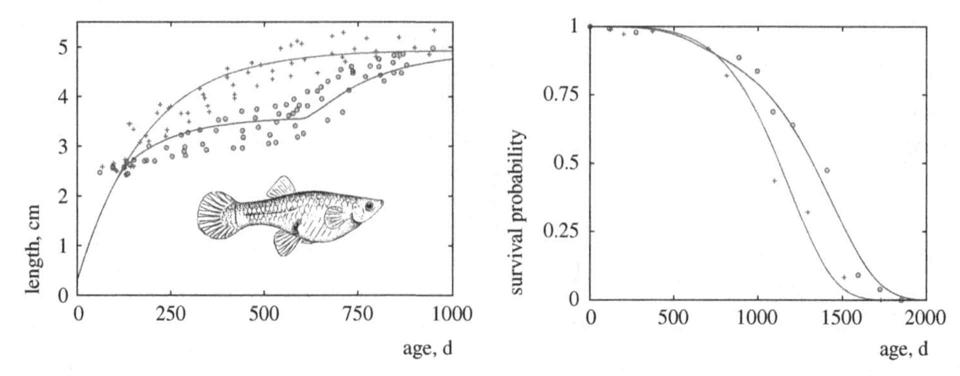

Figure 6.3 The length-at-age (left) and survival probability (right) of guppies (*Poecilia reticulata*), where one cohort was fed *ad libitum* (upper growth curve, lower survival curve), and another cohort was calorie-restricted between 100 and 600 days, but otherwise also fed *ad libitum*. Data from Comfort [216], after [350]. Parameters: $L_b = 0.3\,\text{cm}$, $g = 0.498$, $\dot{k}_M = 0.05\,\text{d}^{-1}$ (fixed), $\dot{v} = 0.123\,\text{cm}\,\text{d}^{-1}$ (physical length), $\ddot{h}_a = 0.1137\,10^{-6}\,\text{d}^{-2}$, $s_G = 0.129$. The scaled functional response during caloric restriction is estimated at $f = 0.726$.

case of a more general principle, that non-essential compounds can affect parameter values, {234}. The occurence of tumours is another typical sublethal effect of ageing [693]; the growth of tumours is discussed at {136, 206}.

6.1.1 Weibull and Gompertz models for short growth periods

Humans grow for 18 years, while their life span is some 80 years, as far as ageing is concerned. For short growth periods, relative to life span ($\dot{r} \simeq 0$; $L \simeq eL_m - L_T$), constant food density ($e = f$), and negligible effects of ageing during the embryo stage, (6.2) reduces to

$$\ddot{q}(t) = \frac{6\dot{h}_W^3}{\dot{h}_G}(\exp(\dot{h}_G t) - 1); \quad \dot{h}_W^3 = \frac{\ddot{h}_a e\dot{v}}{6L}; \quad \dot{h}_G = \frac{s_G e\dot{v}L^2}{L_m^3} \tag{6.3}$$

$$\dot{h}(t) = \frac{6\dot{h}_W^3}{\dot{h}_G^2}(\exp(\dot{h}_G t) - 1 - \dot{h}_G t) \tag{6.4}$$

$$\Pr\{\underline{a}_\dagger > t\} = \exp\left(\frac{6\dot{h}_W^3}{\dot{h}_G^3}\left(1 - \exp(\dot{h}_G t) + \dot{h}_G t + \frac{\dot{h}_G^2 t^2}{2}\right)\right) \tag{6.5}$$

For reasons that become clear in the next subsections, I call $\dot{h}_W$ the Weibull ageing rate, and $\dot{h}_G$ the Gompertz ageing rate, referring to two famous models that typically fit experimental data very well [966].

Weibull model for small Gompertz ageing rates

For small s_G, so small $\dot{h}_G$, acceleration, hazard rate and survival probability, (6.3) – (6.5), reduce to

$$\ddot{q}(t) = 6\dot{h}_W^3 t; \quad \dot{h}(t) = 3\dot{h}_W^3 t^2; \quad \Pr\{\underline{a}_\dagger > t\} = \exp(-(\dot{h}_W t)^3) \tag{6.6}$$

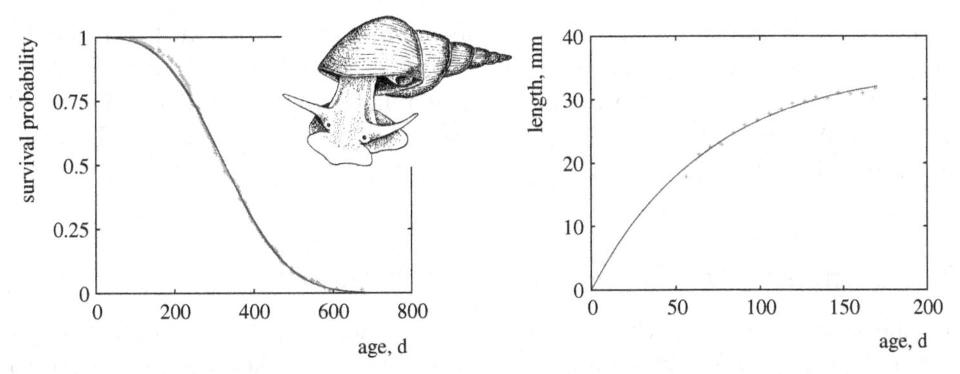

Figure 6.4 The survival probability and the growth curve of the pond snail *Lymnaea stagnalis* at 20 °C. Data from Slob and Janse [1072] and Bohlken and Joosse [124, 1294]. The fitted growth curve is the von Bertalanffy one, and the survival curve is on the basis of (6.2). The Gompertz stress s_G is close to zero; and the survival curve is indistinguishable from the Weibull one with rate $\dot{h}_W = 0.00269\,\mathrm{d}^{-1}$. Parameters: length at birth $L_b = 0.5\,\mathrm{mm}$, energy investment ratio $g = 1.43$, somatic maintenance rate coefficient $\dot{k}_M = 0.073\,\mathrm{d}^{-1}$, energy conductave $\dot{v} = 3.68\,\mathrm{mm\,d}^{-1}$, ageing acceleration $\ddot{h}_a = 3.56\,10^{-6}\,\mathrm{d}^{-1}$, Gompertz stress $s_G = -0.0652$.

The mean age at death equals $\Gamma(\frac{4}{3})/\dot{h}_W$, where Γ stands for the gamma function $\Gamma(x) \equiv \int_0^\infty t^{x-1}\exp(-t)\,dt$.

This survival propability is a special case of $\exp(-(\dot{h}_W t)^\beta)$ with $\beta = 3$, which was first proposed by Fisher and Tippitt [353] in 1928 as a limiting distribution of extreme values, and Weibull [1229] has used it to model the failure of a mechanical device composed of several parts of varying strength, according to Elandt-Johnson and Johnson [316].

In absence of growth, this survival probability fits well, see Figure 7.21, and even if growth occurs during a considerable part of the life span, see Figure 6.4. The slightly negative value of the Gompertz stress, see also Figure 6.1, suggests some decay of the damage-inducing compounds, but this needs further confirmation, because $s_G = 0$ also fits well.

Gompertz model for small Weibull ageing rates

If, on the other hand, $\dot{h}_W \to 0$ is small, but $m_Q(0)$, and so $\ddot{q}(0)$, slightly larger than zero, acceleration, hazard rate and survival probability, (6.3) – (6.5), reduce for $\beta = \ddot{q}(0)/\dot{h}_G^2$ to

$$\ddot{q}(t) = \dot{h}_G^2 \beta \exp(\dot{h}_G t); \quad \dot{h}(t) = \dot{h}_G \beta(\exp(\dot{h}_G t) - 1)$$
$$\Pr\{\underline{a}_\dagger > t\} = \exp(\beta(1 - \exp(\dot{h}_G t) + \dot{h}_G t)) \tag{6.7}$$

For small β, the survival probability further reduces to the Gompertz model $\exp(\beta(1 - \exp(\dot{h}_G t)))$, see [420], which is also frequently used as a model for ageing. Finch [350] favours it because of its property of a constant mortality rate doubling time, $\dot{h}_G^{-1} \ln 2$, which provides a simple basis for comparison of taxa. It can be mechanistically underpinned by a constant and independent failure rate for a fixed number of hypothetical critical elements. Death strikes if all critical elements cease functioning. The curvature of the survival probability then relates to the number of critical elements, which Witten [1264] found to be somewhere between 5 and 15. Their nature still remains unknown.

6.1.2 Ageing in unicellulars: stringent response

One of the many questions that remain to be answered is how ageing proceeds in animals that propagate by division rather than by eggs. Unlike eggs, they have to face the problem of initial damage. It might be that such animals have (relatively few) undifferentiated cells that can divide and replace the damaged (differentiated) ones. A consequence of this point of view is that the option to propagate by division is only open to organisms whose differentiation of specialised cells is not pushed to the extreme. If ageing affects all cells at the same rate, it becomes hard to explain the existence of dividing organisms. This is perhaps the best support for the damage interpretation of the ageing process. Theories that relate ageing, for instance, to the accumulation of compounds as an intrinsic property of cellular metabolism should address this problem. The same applies to unicellulars. If accumulated damage carries over to the daughter cells, it becomes hard to explain the existence of this life style. The assumption of the existence of cells with and without damage seems unavoidable. Organisms that live in anaerobic environments cannot escape ageing, because other radicals will occur that have the same effect as dioxygen. Note that if one follows the fate of each of the daughter cells, this theory predicts a limited number of divisions until death occurs, so that this event itself gives no support for ageing theories built on cellular programming. Only the variation in this number can to some extent be used to choose between both approaches. The present theory can be worked out quantitatively for unicellulars as follows.

Since unicellulars cannot dilute changed DNA with unchanged DNA and cannot compensate for its effect, the hazard rate for unicellulars must equal $\dot{h}(t) = \eta_{QC}\dot{p}_C/M_V$, where η_{QC} couples the generation of damage-inducing compounds to catabolic power. The hazard rate for V1-morphs and isomorphs for $s_G = 0$ amounts to

$$\dot{h}(e) = \dot{h}_a e \frac{1+g}{e+g}; \quad \dot{h}(e,l) = \dot{h}_a e \frac{1+g}{e+g}\frac{1+g/l}{1+g/l_d} \tag{6.8}$$

where $\dot{h}_a$ represents the maximum ageing rate. At constant substrate densities, the scaled energy reserve density, e, equals the scaled functional response, f, so the hazard rate is constant for V1-morphs, and decreases during the cell cycle for isomorphs.

If DNA is changed, the cell will cease functioning. This gives a lower boundary for the (population) growth rate because the population will become extinct if the division interval becomes too long. To prevent extinction (in the long run) the survival probability to the next division should be at least 0.5, so the lower boundary for substrate density can be found from $\Pr\{\underline{a}_\dagger > t_d\} = \exp(-\int_0^{t_d} \dot{h}(t)\, dt) = 0.5$. The lower boundary for the substrate density for rods must be found numerically. It is tempting to relate this ageing mechanism, which becomes apparent at low substrate densities only, to the occurrence of stringent responses in bacteria, as described by, for example, Cashel and Rudd [190]. This is discussed further when populations are considered, {352}.

6.1.3 Functionality of ageing

The observation that the hazard rate typically increases sharply at the onset of reproduction suggests that organisms use ROS to change their DNA. Although most changes are lethal or adverse, some can be beneficial to the organism. Using a selection process, the species can exploit ROS for adaptation to changing environments. By increasing the specific activity of antioxidants, a species can prolong the life span of individuals in non-hostile environments, but it reduces its adaptation potential as a species if the environment changes. This trait defines an optimal specific activity for antioxidants that depends on the life history of the organism and the environment. Large body size, which goes with a long juvenile period, as is discussed on {322}, requires efficient antioxidants to ensure survival to the adult state. It implies that large-bodied species have little adaptation potential, which is further reduced by the long generation time and low reproduction rate; this makes them vulnerable from an evolutionary perspective.

6.2 Toxins and toxicants

The appearance of dioxygen as a by-product of photosynthesis in the atmosphere was probably fatal for most pre-Cambrian organisms. Botulin, one of the most potent toxins known, is produced by the bacterium *Clostridium botulinum* and causes frequent casualties among fish and birds in fresh waters. The soil bacterium *Bacillus thuringiensis* produces a toxin that kills insects effectively [524]. The bacterium *Vibrio alginolyticus* excretes tetrodotoxin, which is a potent toxin that several unrelated organisms use for various purposes. The dinoflagellate *Pfiesteria piscicida* excretes toxins that kill significant numbers of fish in the coastal areas of the West Atlantic.

Some bacteria quickly transform sugar into acetate for later consumption, while acetate suppresses the growth of competitors. *Sphagnum*, a genus of mosses that dominate in peat land, suppress other plant growth by lowering the pH [235]. Natural growth-suppressing compounds that are produced by fungi, such as penicillin, are intensively applied in medicine.

The production of cyanides, alkaloids and other secondary metabolic products by plants obviously functions to deter herbivores. This is not always fully successful, and herbivores sometimes use these toxins to deter predators. Heliconid caterpillars accumulate toxins from passion flowers, and advertise this with bright warning colours. The protection from predation is sometimes so effective that similar species, that cannot handle the toxins metabolically, mimic the colour pattern to acquire the same protection. This is Batesian mimicry, well known in the case of the monarch butterfly *Danaus plexippus* (which accumulates toxins from euphorbids); its colours are adopted by the viceroy *Limenitus archippus* [1179]. The male rattlebox moth *Utetheisa ornatrix* has another striking use of toxins; he supplies his mate with plant-derived pyrrolizidine alkaloids, together with his sperm, which will protect her against predation for several hours.

The reason why the Australian brushtail possum *Trichosurus vulpecula* turned into a pest in New Zealand, but not in Australia, is probably because it does not feed on *Eucalyptus* leaves there, but on trees that lack the cyanides that restrict its reproduction in Australia. The tannins of acorns effectively block digestion by the European red squirrel *Sciurus vulgaris*, for instance, but the American grey squirrel *Neosciurus carolinensis* found a way to deal with this defence of the oak and so managed to outcompete the red squirrel in parts of Europe [726].

Like plants, many species of animal use toxins to *protect* themselves against predation; most nudibranchs (snails of the subclass Opisthochranchia) accumulate nematocysts from their cnidarian prey for protection, while their prey use these formic acid harpoons to collect food; termites [920], arrow frogs (Dendrobatidae), and some birds (the hooded pitohui *Pitohui dichrous* [322]) produce and accumulate chemicals to protect themselves against predators. The tetraodontid fish *Fugu vermicularis* and the starfish *Astropecten polyacanthus* use tetrodotoxin for this purpose [1156]. Cephalopods excrete a mixture of ink and toxins to confuse and paralyse an approaching predator; *Peripatus* emits some glue when offended.

Examples of chemical offence are easy to find. Snakes, wasps, spiders, centipedes, cone shells and many other organisms use venoms to *kill* offensively. Tetrodotoxin is used by chaetognaths and the blue-ringed octopus *Hapalochlaena maculata* to capture prey via sodium channel blocking. Mosquitoes and leeches use chemicals to prevent blood from coagulating.

The ability of the parasitic bacterium *Wolbachia* to *induce* parthenogenesis in normally sexually reproducing species (doubtlessly via chemical interference) has recently attracted a lot of attention [805]. The parasitic cirripede *Sacculina* converts a male crab *Carcinus* into a female, with all secondary sex characteristics, but the allocation to reproduction is converted to the parasite. Many parasites use endocrine disrupters to interfere in the host's allocation of resources.

Biology is full of examples of chemical warfare with sometimes striking responses and defence systems [6]. This collection of random examples serves to illustrate the wide occurrence of chemicals that affect organisms, the function of their production being frequently rather obscure.

Chemical pollution of the environment that is linked to economic activity represents a substantial upscaling of the occurrence of chemicals that affect organisms and comes with the need to study the transport, transformation and fate of chemical pollutants in combination with their effects. The quantification of effects of compounds as provided by DEB theory is especially useful in the context of environmental risk assessment, and this application has been the initial motivation to develop DEB theory.

6.3 One-compartment kinetics is the standard

Dioxygen uptake is directly linked to energetics; most compounds enter in ways less directly linked to energetics. The simplest model for toxicokinetics is the one-compartment uptake/elimination model [1251] in a variable environment: the uptake rate is proportional to the concentration in the environment and the elimination rate is proportional to the concentration in the compartment (= organism), see [1149] for a lucid introduction. The concentration in the environment, $c_d(t)$, is considered to be a specified function of time and both the compartment and the environment is supposed to be well mixed. If the environment (e.g. a small tank) is also treated as a compartment, and the concentrations in the compartment and the environment change interactively, the model is called a 1,1-compartment model.

The uptake kinetics is the same as in the Lotka–Volterra model for the uptake of food, {350}, and can be considered as a linear approximation of the hyperbolic functional response for low concentrations. The change in the mean concentration in the body, $[M_Q](t)$, i.e. the ratio of the amount of the compound in the body to the body volume V, amounts to

$$\frac{d}{dt}[M_Q] = i_Q c_d(t) - \dot{k}_e[M_Q] = \dot{k}_e(P_{Vd}c_d(t) - [M_Q]) \quad \text{for } P_{Vd} = i_Q/\dot{k}_e \tag{6.9}$$

$$\frac{d}{dt}c_V = \dot{k}_e(c_d(t) - c_V) \quad \text{for } c_V = [M_Q]/P_{Vd} \tag{6.10}$$

where $\dot{k}_e$ is the elimination rate (dimension time^{-1}) and i_Q is the uptake rate (dimension $\frac{\text{volume of environment}}{\text{volume of tissue} \times \text{time}}$). The product $\dot{k}_e[M_Q]V$ is interpreted as elimination flux (dimension mass time^{-1}) and the product $i_Q c_d(t)V$ is the uptake flux (dimension mass time^{-1}). Index V refers to the structural body volume, and index d to the dissolved fraction in the environment; both are preparations for more complex situations that are discussed later. P_{Vd} is the partition coefficient: the ultimate ratio of the concentrations in the tissue to that in the environment, also known as the bioconcentration coefficient. It is treated as a constant, which can be less than 1. The interpretation of this partition coefficient refers to the steady-state situation. A better definition for P_{Vd}, which I use here, is the ratio of the uptake to the elimination rate. Both definitions are equivalent for simple one-compartment models, but not for more elaborate ones. Although many texts treat the bioconcentration coefficient as a dimensionless one, it actually has dimension $\frac{\text{volume of environment}}{\text{volume of tissue}}$ because the sum of both types of volume does not have a useful role to play. Most texts in fact use $\frac{\text{environmental volume}}{\text{body dry weight}}$, or for soils $\frac{\text{environmental dry weight}}{\text{body dry weight}}$.

The concentration $c_V \equiv [M_Q]P_{dV}$, with $P_{dV} = P_{Vd}^{-1}$ is proportional to the tissue concentration, but has the dimensions of an environment concentration. It has a very useful role in practical applications, because the tissue concentration frequently plays the role of a hidden variable, because it is not measured.

The explicit expression of $[M_Q](t)$ in terms of $c_d(t)$ is found from (6.9) to be

$$[M_Q](t) = [M_Q](0)\exp(-t\dot{k}_e) + \dot{k}_e P_{Vd}\int_0^t \exp(-(t-t_1)\dot{k}_e)c_d(t_1)\,dt_1 \tag{6.11}$$

If $c_d(t)$ is actually constant, (6.11) reduces to

$$[M_Q](t) = [M_Q](0)\exp(-t\dot{k}_e) + \Big(1 - \exp(-t\dot{k}_e)\Big) P_{Vd} c_d \quad \text{or} \tag{6.12}$$

$$c_V(t) = c_V(0)\exp(-t\dot{k}_e) + \Big(1 - \exp(-t\dot{k}_e)\Big) c_d \tag{6.13}$$

which is known as the accumulation curve.

6.3.1 Ionisation affects kinetics

$$\begin{array}{ccc} N_1 & \overset{\dot{k}^1_{-.}}{\underset{\dot{k}^1_{.-}}{\rightleftarrows}} & N_1^- \\ \dot{k}_{01} \uparrow\downarrow \dot{k}_{10} & & \dot{k}^-_{01} \uparrow\downarrow \dot{k}^-_{10} \\ N_0 & \overset{\dot{k}^0_{-.}}{\underset{\dot{k}^0_{.-}}{\rightleftarrows}} & N_0^- \end{array}$$

The situation is a bit more complex when the compound can be present in molecular as well as in ionic form. Let N_i^- denote the number of ions in matrix i, and $\dot{k}^i_{.-}$ and $\dot{k}^i_{-.}$ the ionisation and de-ionisation rates in matrix i, and $\dot{k}^-_{01}$ the transport rate of the ionic form from matrix 0 to matrix 1. The dynamics now becomes

$$\frac{d}{dt}\begin{pmatrix} N_0 \\ N_0^- \\ N_1 \\ N_1^- \end{pmatrix} = \begin{pmatrix} -\dot{k}^0_{.-} - \dot{k}_{01} & \dot{k}^0_{-.} & \dot{k}_{10} & 0 \\ \dot{k}^0_{.-} & -\dot{k}^0_{-.} - \dot{k}^-_{01} & 0 & \dot{k}^-_{10} \\ \dot{k}_{01} & 0 & -\dot{k}^1_{.-} - \dot{k}_{10} & \dot{k}^1_{-.} \\ 0 & \dot{k}^-_{01} & \dot{k}^1_{.-} & -\dot{k}^1_{-.} - \dot{k}^-_{10} \end{pmatrix} \begin{pmatrix} N_0 \\ N_0^- \\ N_1 \\ N_1^- \end{pmatrix} \tag{6.14}$$

When a compound can ionise, transport becomes dependent on the acidity (pH) and deviations occur from 1,1-compartment kinetics. The definition of the ionisation constant in matrix i is $10^{-\mathrm{pK}_i} = 10^{-\mathrm{pH}_i} N_i^{-*}/N_i^*$, where pH_i stands for the pH in matrix i. Suppose that the processes of ionisation and de-ionisation are fast with respect to the transport processes, i.e. $N_i^-(t)/N_i(t) = \dot{k}^i_{.-}/\dot{k}^i_{-.} = 10^{\mathrm{pH}_i - \mathrm{pK}_i} = P^i_{-.}$. This seems to be acceptable because ionisation and de-ionisation do not require macro-scale movements of molecules. This implies that $N_i(t) + N_i^-(t) = (1 + P^i_{-.})N_i(t)$ and $\frac{d}{dt}(N_i(t) + N_i^-(t)) = (1 + P^i_{-.})\frac{d}{dt}N_i(t)$. Suppose also that the ratio of the binding forces of the ionic forms to the molecules of both matrices equals that of the molecular forms, so $\dot{k}^-_{01} = \dot{k}^-\rho_1/\rho_0$ and $\dot{k}^-_{10} = \dot{k}^-\rho_0/\rho_1$. Substitution into (6.14) gives

$$\frac{d}{dt}\begin{pmatrix} N_0 + N_0^- \\ N_1 + N_1^- \end{pmatrix} = \begin{pmatrix} -\dot{k}'_{01} & \dot{k}'_{10} \\ \dot{k}'_{01} & -\dot{k}'_{10} \end{pmatrix}\begin{pmatrix} N_0 + N_0^- \\ N_1 + N_1^- \end{pmatrix} \tag{6.15}$$

with $\dot{k}'_{01} = \frac{\dot{k} + \dot{k}^- P^0_{-.}}{1 + P^0_{-.}}\frac{\rho_1}{\rho_0}$ and $\dot{k}'_{10} = \frac{\dot{k} + \dot{k}^- P^1_{-.}}{1 + P^1_{-.}}\frac{\rho_0}{\rho_1}$ denoting the overall specific exchange rates of molecules plus ions between the matrices. The partition coefficient being defined as $P_{01} = \frac{N_0(\infty) + N_0^-(\infty)}{N_1(\infty) + N_1^-(\infty)}$, $P_{01} = \left(\frac{\rho_0}{\rho_1}\right)^2$ only holds if $\dot{k}^- = \dot{k}$. This is very unlikely, however. Generally we have

$$P_{01} = \frac{\dot{k}'_{10}}{\dot{k}'_{01}} = \frac{1 + P^0_{-.}}{\dot{k} + \dot{k}^- P^0_{-.}} \frac{\dot{k} + \dot{k}^- P^1_{-.}}{1 + P^1_{-.}} \frac{\rho_0^2}{\rho_1^2} \tag{6.16}$$

Suppose now that we change the pH in matrix 1 (e.g. the environment), while the pH in matrix 0 (e.g. the organism) is kept fixed. If the pH in matrix 1 is extremely low, say $\text{pH}_1 = -\infty$, or $P^1_{-.} = 10^{\text{pH}_1-\text{pK}_1} = 0$, and all of the compound is present in molecular form, (6.16) reduces to $P_{01} = \frac{1+P^0_{-.}}{\dot{k}+\dot{k}^- P^0_{-.}} \dot{k} \frac{\rho_0^2}{\rho_1^2}$. If, on the other hand, the pH in matrix 1 is extremely high, say $\text{pH}_1 = \infty$, or $P^1_{-.} = \infty$, and all of the compound is present in ionic form, (6.16) reduces to $P_{01} = \frac{1+P^0_{-.}}{\dot{k}+\dot{k}^- P^0_{-.}} \dot{k}^- \frac{\rho_0^2}{\rho_1^2}$. It directly follows that (6.16) can be rewritten as

$$P_{01}(\text{pH}_1) = \frac{P_{01}(-\infty) + P_{01}(\infty)10^{\text{pH}_1-\text{pK}_1}}{1 + 10^{\text{pH}_1-\text{pK}_1}} \tag{6.17}$$

which shows how the partition coefficient depends on the pH in matrix 1 (environment), where $P_{01}(-\infty)$ and $P_{01}(\infty)$ play the role of parameters, on the assumption that the pH in matrix 0 (organism) is independent of the pH in matrix 1.

Substitution into the expressions for $\dot{k}'_{10}$ and $\dot{k}'_{01}$, with $\dot{k}^-/\dot{k} = \dot{k}^-_{10}/\dot{k}_{10} = \dot{k}^-_{01}/\dot{k}_{01}$, results in

$$\dot{k}'_{10} = \dot{k}\sqrt{\frac{1+\frac{\dot{k}^-_{10}}{\dot{k}_{10}}P^1_{-.}}{1+P^1_{-.}} \frac{1+\frac{\dot{k}^-_{10}}{\dot{k}_{10}}P^0_{-.}}{1+P^0_{-.}} P_{01}} \quad \text{and} \quad \dot{k}'_{01} = \dot{k}\sqrt{\frac{1+\frac{\dot{k}^-_{01}}{\dot{k}_{01}}P^0_{-.}}{1+P^0_{-.}} \frac{1+\frac{\dot{k}^-_{01}}{\dot{k}_{01}}P^1_{-.}}{1+P^1_{-.}} P_{10}} \tag{6.18}$$

It directly follows that

$$\dot{k}'_{01}(\text{pH}_1) = \sqrt{\frac{\dot{k}'^2_{01}(-\infty) + \dot{k}'^2_{01}(\infty)10^{\text{pH}_1-\text{pK}_1}}{1 + 10^{\text{pH}_1-\text{pK}_1}}} \tag{6.19}$$

where $\dot{k}'_{01}(-\infty) = \dot{k}\sqrt{\frac{1+\frac{\dot{k}^-_{01}}{\dot{k}_{01}}P^0_{-.}}{1+P^0_{-.}} P_{10}}$ and $\dot{k}'_{01}(\infty) = \dot{k}\sqrt{\frac{1+\frac{\dot{k}^-_{01}}{\dot{k}_{01}}P^0_{-.}}{1+P^0_{-.}} \frac{\dot{k}^-_{01}}{\dot{k}_{01}} P_{10}}$ denote the exchange rates if all the compound is present in, respectively, the molecular and the ionised form in matrix 1, and the pH in matrix 0 is fixed.

When applied to toxicokinetics, one matrix corresponds to animal tissue, and one to fresh or sea water. Ionised and un-ionised (molecular) forms of a compound are taken up at different rates, while the pH affects their relative abundance and so the toxicokinetics [618]. If ions hardly exchange, so $\dot{k}^-_{10} = \dot{k}^-_{01} = 0$, knowledge about $\dot{k}_{10}$ or $\dot{k}_{01}$ is then no longer required; knowledge about pH_i, pK_i and $P_{ow} = P^{-1}_{wo}$ can be used to relate elimination rates of different compounds to each other, where octanol serves as a chemical model for animal tissue. Octanol might be a good model compound for studying lipophilicity, but a poor model for studying the ionisation tendency. The derivation above shows that compounds that can ionise must be compared with care; an increase in lipophilicity frequently comes with a decrease in ionisation tendency. It also shows how the pH affects the elimination rate and the partition coefficient via $P^i_{-.} = 10^{\text{pH}_i-\text{pK}_i}$. This can be useful for comparing the toxicokinetics of a single compound under different environmental conditions. Homeostasis ensures that the pK and pH in animal tissue hardly depend on the environmental conditions.

6.3.2 Resistance at interfaces: film models

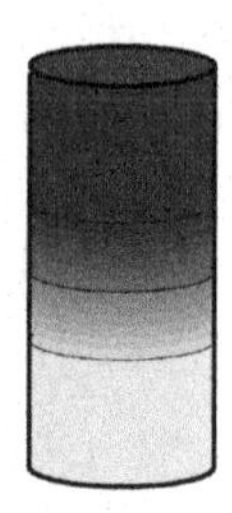

Film models are variations on the 1,1-compartment model where each well-mixed compartment has a non-mixed layer of thickness L_i, say, at each side of the interface between the compartments where the transport is limited by diffusion, see left. They are frequently used to model transport of compounds through the various environmental compartments, but might also be applicable to transport to and from organisms. We will use the notation that the depth of the layer $L = 0$ at the mixed bulk for both media, and $L = L_i$ at the interface. The volume between lengths L_a and L_b is given $V(L_a, L_b) = (L_b - L_a)S$, in both media. The density n of the compound in layer i relates to the number of molecules as $N_i(L_a, L_b, t) = \int_{L_a}^{L_b} n_i(L, t)\, dL$; we have concentration $c_i(L_a, L_b, t) = N_i(L_a, L_b, t)/V(L_a, L_b)$. If the bulk has depth $\mathcal{L}_i$, there are $N_i = n_i(0)\mathcal{L}_i$ molecules in the bulk, which means that the total amount of molecules in medium i is $N_i^+ = n_i(0)\mathcal{L}_i + N_i(0, L_i)$. The volume of the well-mixed medium is $\mathcal{V}_i = \mathcal{L}_i S$, and of the total medium is $V_i^+ = (\mathcal{L}_i + L_i)S$, so the (mean) concentration is $c_i^+ = N_i^+/V_i^+$. The concentration in the bulk is $c_i = c_i(0) = n_i(0)/S$.

Assuming that the initial densities $n_i(L, 0)$ are given such that the boundary conditions in (6.22), the dynamics for the densities is given for $i = 1 - j$ by PDEs

$$0 = \frac{\partial}{\partial t} n_i(L, t) - \dot{D}_i \frac{\partial^2}{\partial L^2} n_i(L, t) \quad \text{for } L \in (0, L_i) \tag{6.20}$$

with boundary conditions at $L = 0$ for $\dot{v}_i = \dot{D}_i / L_i$

$$0 = \frac{\partial}{\partial t} n_i(0, t) - \dot{v}_i \frac{\partial}{\partial L} n_i(0, t) \tag{6.21}$$

and boundary conditions at $L = L_i$

$$0 = \dot{v}_{ji} n_j(L_j, t) - \dot{v}_{ij} n_i(L_i, t) + \dot{D}_i \frac{\partial}{\partial L} n_i(L_i, t) \tag{6.22}$$

For increasing diffusivities $\dot{D}_i$, and/or decreasing thickness of the non-mixed layers L_i, this two-film model reduces to the 1,1-compartment model.

Steady-flux approximation

Suppose now that transport in the films is steady, i.e. the density profiles do not change in time, so $\frac{\partial}{\partial t} n_i(L, t) = 0$. Suppressing argument t, we then have according to (6.20) that

$$0 = \frac{d^2}{dL^2} n_i(L) \text{ for } L \in (0, L_i) \tag{6.23}$$

The density profiles in the films are thus linear:

$$\frac{d}{dL}n_i(L) = (n_i(L_i) - n_i(0))\,/L_i \tag{6.24}$$

The mass balance across the bi-film gives $L_i \frac{d}{dt} n_i(0) = -L_j \frac{d}{dt} n_j(0)$, which leads via (6.21) to $(n_j(L_j) - n_j(0))\dot{v}_j = -(n_i(L_i) - n_i(0))\dot{v}_i$. Substitution of this result in (6.22) gives $n_i(L_i)$ as a weighted sum of $n_i(0)$ and $n_j(0)$. Back-substitution in (6.21) finally leads to the first order kinetics for the bulk densities $\frac{d}{dt} n_i(0) = \dot{k}_e(P_{ij}n_j(0) - n_i(0))$ with elimination rate

$$\dot{k}_e = \dot{k}_i(1 + P_{ij}\dot{v}_i/\dot{v}_j - \dot{v}_i/\dot{v}_{ij})^{-1} \tag{6.25}$$

for $\dot{v}_i\dot{v}_j < \dot{v}_{ij}\dot{v}_j + \dot{v}_{ji}\dot{v}_i$. The restriction of this approximation is that the change in bulk densities n_i is sufficiently small to allow the transport flux in the bi-film to be steady and that transport within the film is strictly limiting; this is not necessarily true. If the transport within the bi-film is very slow, relative to the exchange velocities across the interface, $\dot{v}_i\dot{v}_j \ll \dot{v}_{ij}\dot{v}_j + \dot{v}_{ji}\dot{v}_i$, the elimination rate simplifies to $\dot{k}_e \simeq \dot{k}_i(1 + P_{ij}\dot{v}_i/\dot{v}_j)^{-1}$, while $n_i(L_i) \simeq P_{ij}n_j(L_j)$. Boundary condition (6.22) shows that this can only be a crude approximation indeed, because a gradient is required to drive diffusive transport, and the concentration jump across the interface only equals the partition coefficient in absence of a gradient in the films.

At {327} we will see that the expected relationship between elimination rates and partition coefficients for film models deviates substantially from that for 1,1-compartment models.

6.4 Energetics affects toxicokinetics

The one-compartment model does not always give a satisfactory fit with experimental data. For this reason more-compartment models have been proposed [236, 415, 548, 1008]; because of their larger number of parameters, the fit is better, but an acceptable physical identification of the compartments is usually not possible. These models, therefore, contribute little to our understanding of kinetics as a process. A more direct link with the physiological properties of the organism and with the lipophilicity of the compound seems an attractive alternative, which does not, however, exclude more-compartment models. As usual, the problem is not so much in the formulation of those complex models but in the useful application. Too many parameters can easily become a nuisance if few, scattered, data are available.

Frequent reasons for deviations from one-compartment models are the following. A chemical compound is usually present in the environment in several, and sometimes many, chemical species. Molecules of many compounds can dissociate into ions, which easily bind to ligands that are usually abundantly present, and can transform into other compounds. These species differ in their availability to the organism, which makes the subject of toxicokinetics in natural environments a rather complex one. The compound can enter the organism via different routes: directly from the environment across the skin, via specialised surfaces that play a role in gas exchange, via food, etc. In the aquatic environment uptake directly from water is especially important for hydrophilic organic compounds [165], and metals [128, 129, 971]. In aquatic animals that are chemically isolated from their environment, such as aquatic insects, birds and mammals, the common uptake route is via food. Walker [1211] gives a discussion of uptake routes. The compound can leave the organism using the uptake routes in reverse direction, or via reproductive output and/or products (e.g. moults in arthropods, milk in mammals). Several taxon-specific mechanisms occur. Collembola, for instance, can accumulate metal in mid-gut epithelium and excrete this tissue periodically as part of the moult [914, 915]. This epithelium contains granules, probably filled with calcium phosphate, which may be excreted into the gut lumen. These granules probably play a role in the excretion of an overload of lead in the food [565, 1121].

Apart from these species- and compound-specific reasons for deviation from one-compartment models, the individual can grow, change in chemical composition and metabolise the compound. I will discuss in more detail the more general deviations that demonstrate the relationships with DEB theory.

6.4.1 Dilution by growth

Body growth affects the toxicokinetics even at very low values, as Figure 6.5 illustrates. The physics of the transport processes strongly suggests that uptake and elimination are proportional to the surface area of the organism; it thus links up beautifully with the structure of the standard DEB model. Since the elimination rate is also proportional to the tissue concentration, thus to the amount per volume, it is proportional to the ratio of the surface area to the volume, thus inversely proportional to the volumetric length. This is why the elimination rate must be divided by a scaled length if the body size changes, as has been experimentally verified [1058, 1061]. The change in scaled tissue concentration c_V is given by

$$\frac{d}{dt}c_V = \frac{\dot{k}_e}{l}(c_d - c_V) - c_V \frac{d}{dt}\ln l^3 \tag{6.26}$$

where the term $c_V \frac{d}{dt} \ln l^3$ accounts for the dilution by growth. If food density is constant, the DEB model reduces to $\frac{d}{dt}l = (f - l)\dot{r}_B$, where $\dot{r}_B$ is the von Bertalanffy growth rate, so $\frac{d}{dt}\ln l^3 = 3\dot{r}_B \frac{f-l}{l}$. This model still classifies as a one-compartment kinetics model with time-varying coefficients.

Newman and Mitz [834] found that the elimination rate of zinc in guppies was about proportional to weight$^{-0.42}$ (which is consistent with the expected proportionality with length^{-1}, in view of the scatter), but the zinc uptake rate was about proportional to weight$^{-0.9}$. This has the unexpected consequence that the bioconcentration coefficient is proportional to weight$^{-0.48}$. The elimination rate of mercury did not seem to depend on the size of the mosquitofish, while the mercury uptake rate tended to decrease with size, so that the bioconcentration coefficient also decreased with size [833]. Boyden [135] also found negative correlations between body size and concentrations of cadmium,

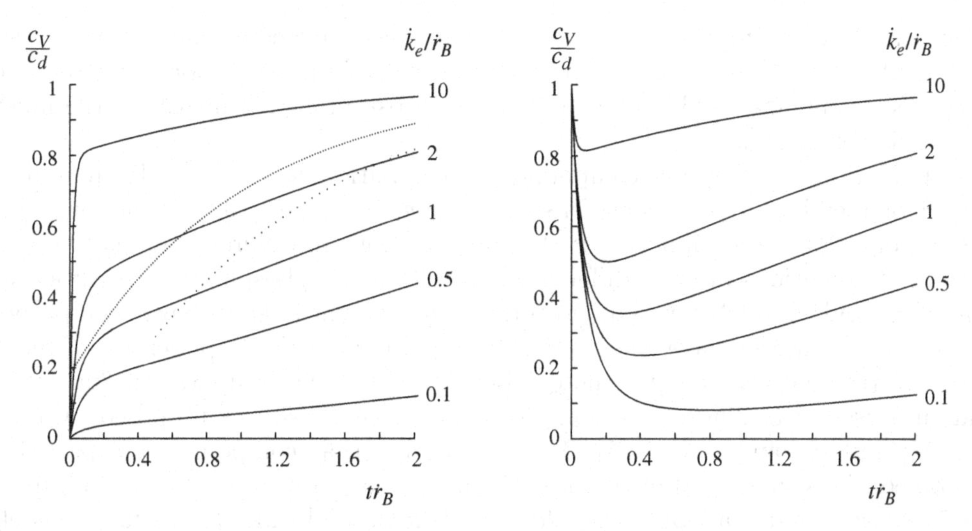

Figure 6.5 Uptake and elimination during growth. The scaled tissue concentrations start from $c_V(0) = 0$ (left), or $c_V(0) = c_d$ (right), where c_d stands for the concentration in the environment. The different curves represent different choices for the value of the elimination rate $\dot{k}_e$, relative to the von Bertalanffy growth rate $\dot{r}_B$. The finely dotted curve represents (scaled) body length and the coarsely dotted curve the (scaled) reproduction rate. The (scaled) lengths at the start of exposure and reproduction are realistic for the waterflea *Daphnia magna* and the value $t\dot{r}_B = 2$ corresponds with 21 d for *D. magna* at 20 °C. All curves in both graphs have an asymptote at the value 1. If the product of the von Bertalanffy growth rate and the exposure time $t\dot{r}_B > 0.4$, the curves in the left and right panels are almost identical, i.e. independent of the initial tissue concentration. The deviations from $c_V = c_d$ can therefore be attributed to 'dilution by growth'.

copper, iron, lead and zinc in some species of mollusc, but no correlations for cadmium, iron, nickel, lead and zinc in other species of mollusc and a positive correlation for cadmium in *Patella vulgata*.

The kinetics of these metals seems to interfere with the metabolism in a more complex way. The substantial scatter in the data hampers firm conclusions from being drawn. When the experimental protocol involves a shift up and thus a transition from low to high concentrations of contaminant, negative correlations between body size and concentrations of contaminant can be expected if elimination and uptake rates decrease with body size: it takes longer for big bodies to reach equilibrium. This mechanism can at best only explain part of the observations.

6.4.2 Changes in lipid content

Changes in lipid content, and thus in reserve density, affect toxicokinetics. Since energy kinetics has a direct link with food uptake, and uptake of a compound from food can be substantial, the link between toxicokinetics with food uptake and reserves kinetics is here discussed in the context of the DEB model. Changes in lipid content frequently occur in uptake experiments; it is practically impossible to feed a cohort of blue mussels in a 2-month uptake/elimination experiment adequately in the laboratory; at the end of the experiment, the lipid content is reduced substantially. This affects the kinetics of lipophilic compounds.

Accumulation of lipophilic compounds and partitioning between different organs can be explained by the occurrence of stored lipids. Schneider [1023] found large differences of polychlorinated biphenyl (PCB) concentrations in different organs of the cod, but the concentrations did not differ when based on the phospholipid-free fraction of extractable lipids. Models for feeding-condition-dependent kinetics have been proposed [451, 454, 685], but they have a large number of parameters. The application of the DEB model involves relatively few parameters, because of the one-compartment kinetics and instantaneous partitioning of the compound in the organism, as proposed by Barber *et al.* [58] and Hallam *et al.* [451]. The assumption that compounds are partitioned instantaneously is supported by a study of the elimination rate of 4,4′-dichlorobiphenyl (PCB15) in the pond snail *Lymnaea stagnalis* [1254]; Wilbrink *et al.* found that elimination rates are equal for different organs, implying that ratios of concentrations in different organs do not change. The fact that structural biomass consists of organs that have different partition coefficients for the xenobiotic is covered by the assumptions of isomorphism, homeostasis and instantaneous partitioning. The combination of these three assumptions implies that the concentration–time curve in one organ can be obtained from that in another organ by applying a fixed multiplication factor.

The amount of compound in the body can be partitioned as $M_Q = M_{QV} + M_{QE} + M_{QR}$ in contributions from structural body volume, M_{QV}, reserves, M_{QE}, and the reproduction buffer, M_{QR}. The latter contribution can be substantial in species like the blue mussel, which reproduces once a year and discharges half its body mass at spawning. Again, we assume instantaneous partitioning of the compound over these compartments, and introduce partition coefficients based on moles of compound per C-mole of body compartment. Since the reproduction buffer has the same chemical composition as the reserves, the amount of compound can be written as

$$M_Q = M_{QV}\left(1 + \frac{[M_{Em}]}{[M_V]}P_{EV}(e + e_R)\right) = M_{QV}P_{WV} \tag{6.27}$$

where P_{EV} denotes the partition coefficient between reserves and structural biomass on the basis of C-moles. The factor P_{WV}, which depends on the (changing) reserve density, can formally be considered as a partition coefficient between the total body mass and the structural body mass.

According to the DEB model, ingestion of food occurs at rate $\dot{k}_X = fl^2\dot{k}_{Xm}$. Suppose that the compound is present in food at concentration c_X (mole per C-mole). Egestion of faeces occurs at rate $\dot{k}_P = fl^2\dot{k}_{Pm}$, with $\dot{k}_{Pm}$ the maximum specific egestion rate. Suppose that the compound is present at concentration c_P in the fresh faeces, and that $\frac{c_X}{c_P} = P_{XP}$ is constant. The partition coefficient P_{XP} can be conceived as a measure of the extraction efficiency of the compound from food. The uptake flux via food amounts to $c_X\dot{k}_X - c_P\dot{k}_P = c_X\dot{k}_X - c_X\dot{k}_X\frac{\dot{k}_{Pm}}{\dot{k}_{Xm}}P_{PX} = c_Xfl^2(\dot{k}_{Xm} - \dot{k}_{Pm}P_{PX}) = c_Xfl^2\dot{k}_eP_{VX}$, where $\dot{k}_e$ denotes the elimination rate from the body, and $P_{VX} \equiv (\dot{k}_{Xm} - \dot{k}_{Pm} P_{PX})\dot{k}_e^{-1}$ is introduced to simplify the notation.

Suppose that the compound is present in the dissolved form at concentration c_d (mole per volume), while the exchange rates between water and body are again taken to be proportional to surface area. The nature of the uptake can be passive or active, but the rate is taken to be proportional to the concentration in the environment and/or to food uptake. Allowing for these two uptake routes, and for dilution by growth, the kinetics amounts to

$$\frac{d}{dt}[M_Q] = \frac{\dot{k}_e}{l} P_{Vd} c_d + \frac{\dot{k}_e}{l} P_{VX} f c_X - [M_Q]\left(\frac{\dot{k}_e}{l} P_{VW} + \frac{d}{dt}\ln l^3\right) \quad (6.28)$$

$$\frac{d}{dt} c_V = \frac{\dot{k}_e}{l}(c_d + P_{dX} f c_X - P_{VW} c_V) - c_V \frac{d}{dt}\ln l^3 \quad (6.29)$$

where the partition coefficient $P_{VW} = P_{WV}^{-1}$ is given in (6.27), $\frac{\dot{k}_e}{l} P_{Vd}$ is the uptake rate from the water, $\frac{\dot{k}_e}{l} P_{VX} f$ is the uptake rate from the food, $\frac{\dot{k}_e}{l}$ is the elimination rate from the body, $P_{dX} = P_{VX}/P_{Vd}$, and $c_V = [M_Q] P_{dV}$, as before. The definition of the partition coefficient P_{Vd} is the ratio of the uptake rate from water to the elimination rate; it is no longer interpreted as the ultimate ratio of the concentration in the body to that in the water. Likewise, P_{dX} is not interpreted as the ultimate ratio of the concentration in the food to that in the water. For $P_{VW} = 1$ and $P_{dX} = 0$, (6.29) reduces to (6.26), and for $l = f$ it further reduces to (6.10). The model still classifies as a one-compartment model with time-varying coefficients.

Since most measurements are done on the basis of weights, the kinetics of the variable $\langle M_Q \rangle_w = [M_Q]/[W_w]$ is of practical interest; it represents the number of moles per unit of wet weight. Like the total amount of compound, wet weight can be decomposed into the contributions made by the structural body volume, the reserves and the reproduction buffer, as done in (3.2). The change in concentration on the basis of weights is

$$\frac{d}{dt}\langle M_Q \rangle_w = \frac{1}{[W_w]}\frac{d}{dt}[M_Q] - \langle M_Q \rangle_w w_E \frac{[M_{Em}]}{[W_w]}\left(\frac{d}{dt}e + \frac{d}{dt}e_R\right) \quad (6.30)$$

where the second term relates to the change in weight, as implied by $[W_w] = d_V + w_E[M_{Em}](e+e_R)$, cf. (3.2). Apart from the initial conditions, this specifies the dynamics in the period between the moments of spawning or reproduction. At such moments, (wet) weight as well as the amount of xenobiotic compounds are discontinuous, because the buffer of energy allocated to reproduction is emptied, possibly together with its load of xenobiotic compound. The most simple assumption is to let the compound in that buffer transfer to the egg. If reproduction occurs at time t_R, and if t_R^- denotes a moment just before t_R, and t_R^+ just after, the ratio of the concentrations of compound equals

$$\frac{\langle M_Q \rangle_w(t_R^+)}{\langle M_Q \rangle_w(t_R^-)} = \frac{d_V + w_E[M_{Em}]P_{EV}(e + e_R)}{d_V + w_E[M_{Em}]P_{EV} e} \, \frac{d_V + w_E[M_{Em}]e}{d_V + w_E[M_{Em}](e + e_R)} \quad (6.31)$$

The first factor corresponds to the ratio of xenobiotic masses in moles, the second factor to the ratio of body weights. This result can be larger or smaller than 1, depending primarily on the partition coefficient P_{EV}. If the moments of reproduction are frequent enough to neglect the contribution of e_R to wet weight and compound load, $\frac{d}{dt}e_R$ can be replaced by $e_0\dot{R}$, which can be left out if the reproductive output is negligibly small.

The elimination route via reproduction can be very important for rapidly reproducing species, such as daphnids. Even in guppies it can be noticeable [1060]. It is also possible that no compound is transduced through the reproduction process, as has been found for 4,4′-DCB in *Lymnaea* [1254]. This implies a (sudden) increase of the concentration at reproduction.

The change of concentration at reproduction has, of course, an intimate relationship with the initial conditions for the offspring, which depend on the feeding conditions and the loading of the mother. Experience with chronic toxicity tests shows that most effects occur at hatching, which means that an egg must be considered to be rather isolated, chemically, from its environment apart from gas exchange. An extreme consequence is that the amount of compound at egg formation is the same as that at hatching. This means that the concentration at hatching relates to that of the mother just after reproduction as

$$\langle M_Q \rangle_w(a_b) = \langle M_Q \rangle_w(t_R^+) \frac{P_{EV} V_m}{P_{WV} V_b} e_0 \tag{6.32}$$

where the ratio P_{WV} is given in (6.27) and should now be evaluated at $e_R = 0$.

The parameters that relate to the kinetics of the compound are the elimination rate $\dot{k}_e$, and the partition coefficients P_{Vd}, P_{VX} and P_{EV}. In addition, there are a number of parameters that relate volumes to weights. The third class of parameters is from the DEB model via the expressions for $\frac{d}{dt}l$, $\frac{d}{dt}e$ and $\frac{d}{dt}e_R$. Not all parameters are required to fit the model to experimental data. If food density and c_d/c_X do not change, for instance, and the reproduction buffer plays a minor role, P_{WV} is constant, and the four toxicokinetic parameters combine in just two compound parameters $(P_{Vd} + P_{VX}c_X/c_d)/P_{VW}$, and $\dot{k}_e P_{VW}$. It is obvious that additional physiological knowledge will help us to interpret experimental results, especially if the physiological condition changes during the experiment. Although some of the physiological parameters can be estimated from uptake/elimination curves in principle, an independent and more direct estimation is preferable.

Figures 6.6 and 6.7 illustrate the performance of the model to describe the uptake/elimination behaviour of the compounds hexachlorobenzene (octanol/water partition coefficient $\log P_{ow} = 5.45$ [1000]) and 2-monochloronaphthalene ($\log P_{ow} = 3.90$ [853]). The mussels and fish were not fed during the experiment, which implies that their energy reserves decreased during this time. The fish depleted its energy reserves faster, because it was smaller than the mussel and its temperature was higher. As a result of the decrease in reserves, the fish started to eliminate the compound during the accumulation phase of the experiment. The model successfully describes this phenomenon. The experiments were short enough to assume that the size of the test animals did not change and that the energy allocation to reproduction was negligibly small during the experiment. The concentration of xenobiotic compounds in

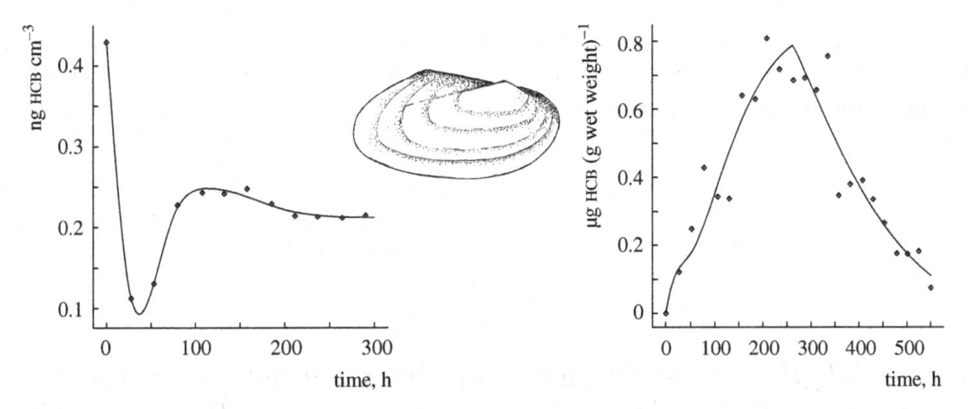

Figure 6.6 Measured concentration of hexachlorobenzene (HCB) in water and in a starving $6.03\,\mathrm{cm}^3$ freshwater mussel *Elliptio complanata* at 20 °C during a 264-h uptake/elimination experiment. Data from Russel and Gobas [1000]. The least-squares-fitted curves are the cubic spline function for concentrations in the water and the model-based expectation for the concentration in the wet weight. From Kooijman and van Haren [656].

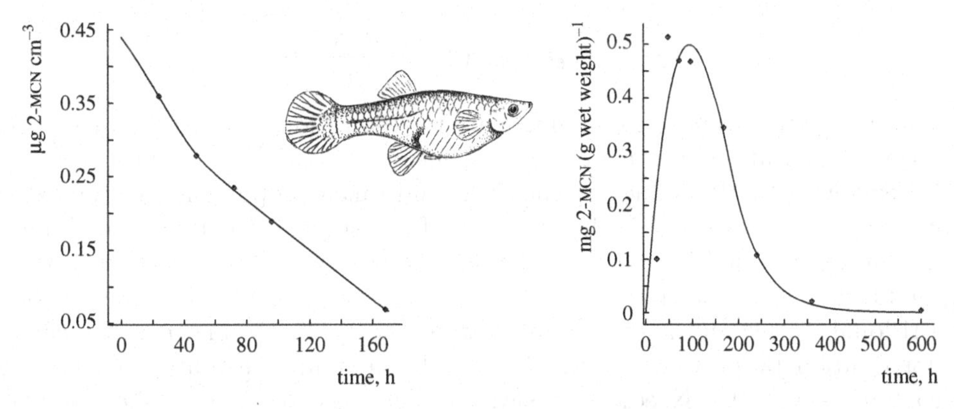

Figure 6.7 Measured concentration of 2-monochloronaphthalene (2-MCN) in water and in a starving $0.22\,\mathrm{cm}^3$ female guppy *Poecilia reticulata* at 22 °C during a 168-h uptake/elimination experiment. Data from Opperhuizen [853]. The least-squares-fitted curves are the cubic spline function for the concentrations in the water and the model-based expectation for the concentration in the wet weight. From Kooijman and van Haren [656].

the water changed during accumulation. A cubic spline was, therefore, fitted to these concentrations and used to obtain the concentrations in the wet weight.

6.4.3 Metabolic transformations

If compounds are metabolised, the usual effect is that the products are less lipophilic than the original compound, so P_{EV} is reduced. In this way, the product will be eliminated at a higher rate. If the metabolic transformation behaves as a first-order process, this only affects the value of the elimination rate, and not the model structure. It has long been recognised, however, that elimination frequently involves a metabolic activity that can be satiated [1208, 1209, 1210]. Many compounds, such as salicylurate [700, 701], are found to have a capacity-limited elimination route. Wagner [1207] used Michaelis–Menten (MM) kinetics to describe the elimination of ethanol from human serum, i.e.

$$\frac{d}{dt}[M_Q] = \dot{k}_e P_{Vd} c - \frac{\dot{k}_e [M_Q]}{1 + [M_Q]/[M_Q]_M} \quad \text{or} \tag{6.33}$$

$$\frac{d}{dt} c_V = \dot{k}_e c - \frac{\dot{k}_e c_V}{1 + c_V/c_M} \tag{6.34}$$

with $c_V = [M_Q]P_{Vd}$, $c_M = [M_Q]_M P_{Vd}$, P_{Vd} being the ratio of the uptake to the elimination rate, and $\dot{k}_e$ and $[M_Q]_M$ or c_M are the parameters of the MM-elimination. The concentration c_M has the interpretation of the maximum sustained concentration in the environment that can be 'handled' by the organism. If the concentration exceeds this value, the concentration in the organism will build up continuously.

The MM elimination route can supplement a first-order elimination route, which gives

$$\frac{d}{dt} c_V = \dot{k}_e c - \dot{k}_l c_V - \frac{\dot{k}_e c_V}{1 + c_V/c_M} \tag{6.35}$$

The first-order elimination route might relate to respiration, and can be taken proportional to the respiration rate $\dot{k}_C$ of the organism. The MM elimination route might relate to excretion by the kidney or the liver, and taken proportional to the excretion rate of nitrogenous waste $\dot{k}_N$ of the organism, for instance. A coupling of elimination with exudate excretion in algae has been suggested [1057]. These couplings with the energy budget reveal how these parameters change with size during growth, or with the nutritional status, and how they differ from one species to another. We can again allow for dilution by growth, see (6.26), and different uptake routes and changes in lipid content, see (6.29). Needless to say, we then need a rather elaborate series of measurements, because of the six parameters that have to be estimated.

Note that (6.35) collapses to first-order kinetics if $c_V \ll c_M$, and if $c_V \gg c_M$, with elimination rate $\dot{k}_e + \dot{k}_l$ or uptake rate $\dot{k}_e c - \dot{k}_e c_M = \dot{k}_e(c - c_M)$, respectively. If c_V varies in a rather small window around c_M, (6.35) approximates first-order kinetics, with elimination rate $\dot{k}_l + \dot{k}_e/4$. In all those cases, c_M cannot be estimated; we need a rather big window for measurements of c_V around c_M for that purpose. Problems disappear if the elimination flux via the MM route can be measured directly, by measuring the compound or its products, in the urine, for instance.

6.5 Toxicants affect energetics

Toxic effects should be linked to internal concentrations [500, 550], and the one-compartment model links internal concentrations to external ones in the simplest way; effects disappear as soon as internal concentrations are below a threshold value and reappear if they are above this value [883]. Theory for effects during variations of external concentrations is given in [882].

Organisms have evolved in a chemically varying environment; consequently they can cope with varying concentrations of any particular compound, as long as the variations are within a certain range. In general, three ranges of internal concentrations can be distinguished for any compound: too little, enough and too much. The definition of the enough-range is that changes of internal concentrations within this range hardly affect metabolism. This does not mean that some molecules have effect and others don't; it means that the metabolic system can compensate effects at the individual level to a limited extent. Take one kidney out of a healthy person, and the other kidney will do all that is necessary as long as the metabolic capabilities are not tested to the extreme.

Only effects on survival and reproduction are of primary, ecotoxicological, interest; these effects determine population dynamics, and thus production and existence. Due to the coupling between the various processes of energy uptake and use, many other effects of compounds have an indirect effect on reproduction. DEB models describe the various modes of action translated into an effect on reproduction; allocation to reproduction depends on reserve density, which depends on feeding rate, which depends on body size, which depends on growth. Maintenance competes with growth for allocation, so effects on maintenance can be translated into effects on growth, and thus into effects on reproduction. Small individuals eat less than large ones, so less energy is available for reproduction. Effects on feeding, growth and maintenance indirectly affect reproduction on the basis of the DEB model. These types of effects relate directly to energetics. Their consequences can be evaluated by changing one or more parameter values of the DEB model. Such a study is not very different from a more general one on the evolutionary implications of parameter settings.

The environmental relevance of mutagenic effects is still in debate. A frequently heard opinion from some industrialists is that mutagenic effects have no environmental impact at all, stating that the direct effect on survival is negligibly small and the loss of gametes does not count from an ecological point of view. The way ageing is treated within the DEB model closely links up with mutagenic effects, particularly if the ROS mechanism is correct. Mutagenic compounds have about the same effect on organisms as ROS. As a consequence, mutagenic effects can be studied by changing ageing acceleration (in the case of metazoans). The DEB model offers the possibility of evaluating the consequences of mutagenic effects along the same lines as the effects on energy fluxes. I have already mentioned the setting of ageing acceleration as a compromise between the life span of individuals and the evolutionary flexibility of the genome. The effects of changes in ageing acceleration must then be found over a timescale of many generations and involve inter-species relationships. This makes such effects extremely hard to study, both experimentally and theoretically. The lack of reliable models for this timescale makes it difficult to draw firm conclusions. The fact that mutagenic compounds tend to be rather reactive and, therefore, generally have a short life in the environment is part of the problem, which perhaps makes them less relevant to the problem of environmental pollution if emissions are only incidental.

The significance of mutagenic effect on human health is widely recognised, particularly in relation to the occurrence of tumours and cancer. The Ames test is frequently applied to test compounds for mutagenic effects. The DEB model offers a framework for interpreting the sometimes unexpected results from these tests, see {245}. The environmental significance of teratogenic effects, i.e. effects on the development of organisms, is even less well recognised than the significance of mutagenic effects. Fortunately, only a few compounds seem to have a teratogenic effect as their primary one, and these fall outside the scope of this book.

Basic to the description of small effects of toxicants is the notion that each molecule that exceeds the tolerance range contributes to the same extent to the effect. Interactions between the molecules only occur at higher tissue concentrations. Hence, the effect size is, as a first approximation, a linear function of the tissue concentration. This point of view relates to the Taylor approximation for non-linear functions that describe how effect size relates to tissue concentrations: we use only the first term of the Taylor approximation at the upper boundary of the tolerance range. The theorem by Taylor states that we can describe any non-linear function in a given interval arbitrarily well with an appropriate polynomial function if we include enough higher-order terms. So when we want to improve the description of effects, if they happen to deviate from a linear relationship with tissue concentrations, we simply include the squared term, the cubed term, etc. Such improvements will rapidly become counter-productive because we increase the number of parameters that must be estimated and because higher tissue concentrations will affect more metabolic processes. So we are increasing precision at the wrong points. Practice teaches that very good descriptions can be obtained by just taking effect size to be linear in the tissue concentration, even at rather high effect sizes, provided that we focus on the correct physiological process.

6.5.1 No effects

The upper boundary for the enough-range, i.e. the internal no-effect concentration (NEC), might be zero for particular compounds. Each molecule of such compounds induces effects with a certain probability, but for most compounds, the upper boundary is positive. The lower boundary of the enough-range is zero for most compounds, because they are not necessary for life. Elements such as copper are required, so the lower boundary for copper is positive. NECs are important for environmental risk assessment, but also for ecology. An example is the zinc-resistant *Viola calaminaria*, where zinc, in concentrations between its NEC and that of other plants, reduces competition, and so promotes its abundance.

Effects of a shortage of a compound resemble those of an overdose in their kinetics. The founder of ecotoxicology, Sprague [1093] studied the effects of toxicants in bioassays, using dioxygen shortage as an example. Although many interrelationships exist between nutrition and toxic effects, the upper boundary of the tolerance range attracted most attention in ecotoxicology, because of its application in risk assessment studies, while ecology focused on the lower boundary (see [1244]).

NEC estimates from routine toxicity data turn out to be insensitive for a small number of tested concentrations or a small number of test organisms or even for differences in NEC values among individuals [23, 44]. The confidence interval can best be accessed via the profile likelihood function. The NEC concept can be extended to for applications to mixtures of compounds [43].

6.5.2 Hormesis

Hormesis is the phenomenon that low concentrations of a toxicant seem to have a stimulating rather than an inhibiting effect on some endpoint, especially on reproduction. It can result from interactions of the compound with a secondary stress, such as resulting from very high levels of food availability. I found this for the daphnia reproduction test where large concentrations of *Chlorella* cells are used; when I repeated the test after having observed hormesis with lower food levels, reproduction was less (of course), but the hormesis effect disappeared. This seemed independent of the type of toxicant.

Cadmium has been found to elongate the life span of the nematode *Acrobeloides nanus* [12], which can be captured by a decrease of the Gompertz stress s_G, see {219}, since the amplification of ROS occurrence is a metabolic process. Cadmium also decreased growth, so this can hardly be considered as 'beneficial'.

Hormesis might have many causes. If a compound decreases the yield of structure on reserve, y_{VE}, it reduces growth and delays birth (if an embryo is exposed) and puberty (in the case of juveniles), but also reduces the size at birth. A reduction of growth indirectly reduces reproduction, because food uptake is linked to size. Since it also reduces size at birth, the overall effect can be a hormesis effect on reproduction (in terms of number of offspring per time) [645]. The reduction in the size of offspring can be small and difficult to observe, and still be of importance to explain hormesis. This explanation for hormesis shows that reproduction can be stimulated, but the effect is hardly beneficial.

6.5.3 Effects on survival

The standard survival model in ecotoxicity assumes that the hazard rate jumps from zero to infinity as soon as the internal concentration exceeds some threshold value. The reason why not all individuals die at the same moment is because the threshold values are supposed to be individual-specific; their values are log-normally or log-logistically distributed [117]. This model suffers from several shortcomings [639] and cannot capture, for instance, the generally observed pattern that the slope of the concentration–response curve increases during exposure. A much better alternative is to take the hazard rate linear in the internal concentration like

$$\dot{h}_c \propto ([M_Q^{0,l}] - [M_Q])_+ \quad \text{and/or} \quad \dot{h}_c \propto ([M_Q] - [M_Q^{0,u}])_+ \tag{6.36}$$

where $[M_Q^{0,l}]$ and $[M_Q^{0,u}]$ stand for the lower and the upper boundary of the concentrations of compound that do not affect survival.

The proportionality constant that describes the effect on the hazard rate probably differs for shortages and excesses. This relates to differences in mechanisms. If the concentration exceeds the tolerance range substantially, it is likely that death will strike via other mechanisms than for small excesses. This restricts the applicability of the model to relatively small ranges of concentration. In practice, however, very wide concentration ranges are frequently used, as in range-finding tests on a routine basis.

In the rest of this subsection, I assume that $[M_Q^{0,l}] = 0$ for simplicity's sake, and reduce the notation $[M_Q^{0,u}]$ to $[M_Q^0]$. The idea for hazard modelling can be worked out quantitatively as follows for a constant concentration in the environment.

Because of the general lack of knowledge about relevant concentrations in tissue, those in the environment will be used to specify the hazard rate. If the initial concentration in the tissue is negligibly small and if the concentration of compound in the tissue follows simple first-order (i.e. one-compartment) kinetics (6.10), the hazard rate at constant concentration c in the environment is

$$\dot{h}_c = \dot{b}_\dagger c_e = \dot{b}_\dagger((1 - \exp(-t\dot{k}_e))c - c_0)_+ \tag{6.37}$$

where $c_e = (c_V - c_0)_+$ is the concentration above the NEC; the NEC $c_0 \equiv [M_Q^0]P_{dV}$ is the highest concentration in the environment that will never result in an effect if the concentration in the environment is constant (short peaks do not necessarily give effects). The proportionality constant $\dot{b}_\dagger$ is the killing rate with dimension (environment concentration $\times$ time)$^{-1}$; it is a measure of the toxicity of the compound with respect to survival. If $c > c_0$, but constant, and if the initial concentration in the tissue is 0, effects start to show at $t_0 = -\dot{k}_e^{-1}\ln\{1 - c_0/c\}$, the moment at which the concentration in the tissue exceeds the NEC. In the absence of 'natural' mortality, the survival probability q for $c > c_0$ and $t > t_0$ is

$$q(c,t) = \exp\left(-\int_0^t \dot{h}_c(t_1)\,dt_1\right) \tag{6.38}$$

$$= \exp\left(\dot{b}_\dagger\dot{k}_e^{-1}c(\exp(-t_0\dot{k}_e) - \exp(-t\dot{k}_e)) - \dot{b}_\dagger(c - c_0)_+(t - t_0)\right) \tag{6.39}$$

This equation has three parameters, which are of all of practical interest: the NEC c_0, the killing rate $\dot{b}_\dagger$ and the elimination rate $\dot{k}_e$. The more elaborate description of the DEB-based kinetics could be used to describe survival patterns in more detail. Practical limitations are likely to ruin such an attempt if no measurements for the concentration in the tissue are available. An appropriate experimental design can usually avoid such complications.

Figure 6.8 illustrates the application of (6.39) to the results of some standard toxicity tests. Note that this formulation implies that the concentration–response relationships become steeper for longer exposure periods.

An interesting special case concerns extremely small elimination rates, so $\dot{k}_e \to 0$, and $P_{Vd} \to \infty$, such that the uptake rate $\dot{k}_e P_{Vd} = \dot{k}_{dV}$ remains fixed. The accumulation process reduces to $\frac{d}{dt}[M_Q] = \dot{k}_{dV}c$, so that $[M_Q](t) = \dot{k}_{dV}ct$ if the initial concentration in the tissue is negligibly small. The NEC (in the environment) is now 0, because a very

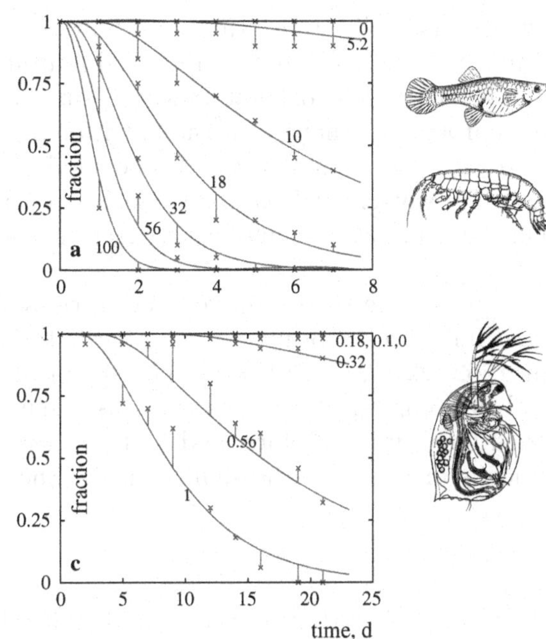

Figure 6.8 The expected fraction of surviving individuals as a function of exposure time at constant concentrations (indicated) when the hazard rate is linear in the internal concentration that follows first-order kinetics. Unpublished data, kindly provided by Thea Adema. Parameters:

no.	species	N	compound	unit	$\dot{b}_\dagger$, (unit d)$^{-1}$	$\dot{k}_e$, d^{-1}	c_0, unit
a	*Poecilia reticulata*	20	dieldrin	µg l^{-1}	0.038	0.712	4.49
b	*Chaetogammarus marinus*	50	3,4-dichloroanil.	mg l^{-1}	0.40	0.335	1.41
c	*Daphnia magna*	50	$K_2Cr_2O_7$	mg l^{-1}	0.40	0.125	0.26

small concentration in the environment will result ultimately in a very high concentration in the tissue. A NEC in the tissue, i.e. the upper boundary of the tolerance range, still exists, of course, and is exceeded at $t_0 = [M_Q^0](\dot{k}_{dV}c)^{-1}$. The hazard rate amounts to $\dot{h}_c = \ddot{b}_\dagger c(t - t_0)_+$. The relationship between the killing acceleration $\ddot{b}_\dagger$ and the killing rate $\dot{b}_\dagger$, in the case that $\dot{k}_e \neq 0$, is $\ddot{b}_\dagger = \dot{b}_\dagger \dot{k}_e$. The survival probability is

$$q(c, t) = \exp(-\ddot{b}_\dagger c(t - t_0)^2/2) \tag{6.40}$$

For small NECs in the tissue, so $t_0 \to 0$, this represents a Weibull distribution with shape parameter 2. The only difference with the survival probability related to ageing, see {218}, is the extra accumulation step of products made by affected DNA, which results in a Weibull distribution with shape parameter 3.

In this special case, the full response surface in the concentration–exposure time plane is described by just one parameter, the killing acceleration $\ddot{b}_\dagger$. One step towards more elaborate models is the introduction of the upper boundary of the tolerance range, via $[M_Q^0]/\dot{k}_{da}$ in t_0. Next comes the introduction of the elimination rate $\dot{k}_e$, which allows a new parameter basis: $\dot{b}_\dagger$, $\dot{k}_e$ and c_0. Then follow changes in the chemical composition (and size) of the animal by introduction of the partition coefficients P_{EV} and P_{PX}, and/or a separation of uptake routes via the dissolved fraction $\{\dot{k}_{dV}\}$ or via food $\{\dot{k}_{xV}\}$. Finally, we should allow for metabolic transformations. So the level of the model's complexity can be fully trimmed to the need and/or practical limitations. The more complex the model is, the more one needs to know (and measure) about the behaviour of the compound in the environment, changes in the nutritional status of the animals, growth, reproduction, etc. If experimental research and model-based analysis of results are combined in the proper way, one will probably feel an increasing need to define precise experimental conditions and avoid complicating factors, such as uncontrolled changes in exposure.

This description of effects on survival makes the theory on competing risks available for direct application to toxicity and links up smoothly with standard statistical analyses of hazard rates; see for instance [62, 225, 247, 572, 766]. The significance of a toxic stress for a particular individual depends on other risks, such as ageing and starvation. If $\dot{h}$ stands for the hazard tied to ageing as before, and $\dot{h}_p$ for other risks, such as predation, an obvious instantaneous measure for the significance of the toxic stress is $\dot{h}_c(\dot{h} + \dot{h}_c + \dot{h}_p)^{-1}$.

Ionisation and P_{ow}

Since effect is linear in the number of molecules of toxicant in the organism, it follows directly from (6.17) that the killing rate and the NEC depend on the pH and the pK as

$$\dot{b}_\dagger(\text{pH}) = \frac{\dot{b}_\dagger(-\infty) + \dot{b}_\dagger(\infty)10^{\text{pH}-\text{pK}}}{1 + 10^{\text{pH}-\text{pK}}}; \quad c_0^{-1}(\text{pH}) = \frac{c_0^{-1}(-\infty) + c_0^{-1}(\infty)10^{\text{pH}-\text{pK}}}{1 + 10^{\text{pH}-\text{pK}}} \tag{6.41}$$

where $\dot{b}_\dagger(-\infty)$ and $\dot{b}_\dagger(\infty)$ stand for the killing rate if all of the compound were present in, respectively, the molecular and the ionised form and pK is the dissociation coefficient. A similar relationship has been proposed by Könemann [618] for LC50^{-1}, where the LC50 is defined as the concentration $c_{L50}(t)$ for which $q(c_{L50}(t), t) = 0.5$ holds; it is frequently used as a quantifier for lethal effects.

6.5.4 Effects on growth and reproduction

Toxic effects of chemicals change the allocation via the parameter values. Since the processes of assimilation (i.e. the combination of feeding and digestion), growth, maintenance and reproduction are intimately interlinked, changes in any of these processes will result in changes in reproduction [651]. Two classes for the mode of action of compounds will be distinguished: direct and indirect effects on reproduction.

When reproduction is affected directly, assimilation, growth and maintenance are not affected. There are two closely related routes within the DEB framework to affect reproduction directly. One is via survival of each ovum, and the other is via the energy costs of each egg.

Direct effects on reproduction

The survival probability of each ovum is affected as discussed in the previous section on effects on survival, except that the sensitive period is taken to be relatively short and fixed rather than the whole life span. (Age zero refers to the moment at which the ovum starts to develop, rather than the moment of hatching or birth.) The combination of an effect on the hazard rate of the ovum and a fixed sensitive period results in a survival probability that depends on the local environment of the ovum. This leads to another important difference with the previous section: the local environment of the ovum is the tissue of the mother rather than the environment concentration. The relevant concentration, therefore, changes in time even if the environment concentration is constant. The toxicity parameters that appear in the survival probability of an ovum are the NEC, as before, and the tolerance concentration, which is inversely related to the product of the killing rate and the length of the sensitive period. The elimination rate defines how the effect builds up during exposure.

In terms of number of eggs per time, the reproduction rate equals the ratio of the energy allocated to reproduction and the energy costs of an egg. If the compound affects the latter, it can be modelled by making the reproduction overhead $1 - \kappa_R$ a (linear) function of the tissue concentration. The model is mathematically different from the hazard model but behaves quantitatively rather similarly, as is illustrated in Figure 6.9.

Indirect effects on reproduction

Allocation to reproduction starts as soon as the cumulative investment in the increase of the state of maturity exceeds some threshold value. Since direct effects on reproduction only affect the translation from energy allocated to reproduction into number of

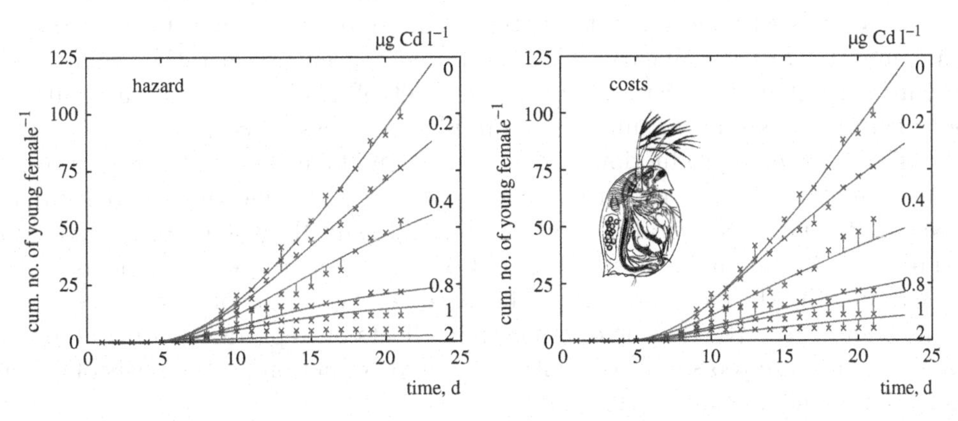

Figure 6.9 Direct effects of cadmium on *Daphnia* reproduction. The mean cumulated number of young per female daphnid as a function of the exposure time to several concentrations of cadmium. The fitted curves represent least-squares fits of the hazard (left) and the cost (right) model for effects on reproduction to the same data. The DEB parameters of Figure 2.10 are used: $\kappa = 0.8$, $\kappa_R = 0.95$, $g = 0.15$, $\dot{k}_J = 3.6\,\text{d}^{-1}$, $\dot{k}_M = 4.1\,\text{d}^{-1}$, $\dot{v} = 1.62\,\text{mm}\,\text{d}^{-1}$ and $L(0) = 0.8\,\text{mm}$. The estimated parameters are

	c_0, $\mu\text{g}\,\text{l}^{-1}$	c_*, $\mu\text{g}\,\text{l}^{-1}$	$\dot{k}_e$, d^{-1}	U_H^b, mm^2d	U_H^b, mm^2d
hazard	0.0173	0.1184	0.0317	0.0023	0.1215
cost	0.0172	0.0233	0.0138	0.0022	0.1262

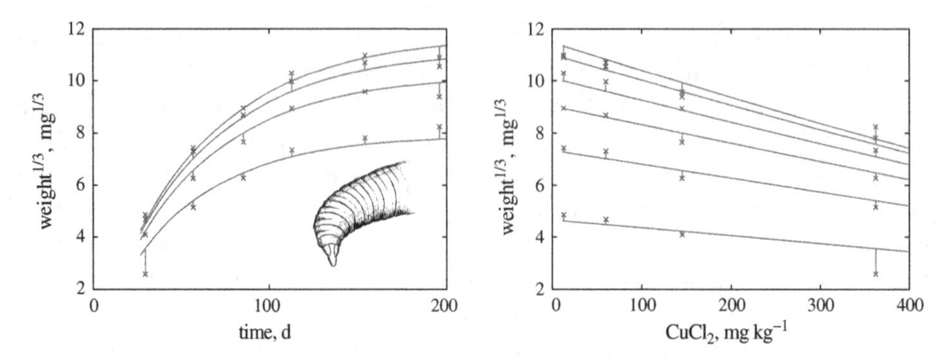

Figure 6.10 The effect of $CuCl_2$ on the assimilation of earthworm *Lumbricus rubellus*. Data kindly provided by Chris Klok [606] and fits by Jacques Bedaux. The newly hatched worms were exposed in sandy loam soil and fed *ad libitum* with *Alnus* leaves at 15 °C and 90% relative humidity. Parameter values: $W_b = 0\,\text{mg}$, $\dot{v} = 1.18\,\text{mg}^{1/3}\text{d}^{-1}$, $\dot{k}_M = 0.099\,\text{d}^{-1}$, $g = 1$, $c_0 = 4.45\,\text{mg kg}^{-1}$, $c_A = 1946\,\text{mg kg}^{-1}$, $\dot{k}_e = \infty\,\text{d}^{-1}$.

offspring, these modes of action do not affect the time of onset of reproduction. Indirect effects on reproduction via assimilation, maintenance and growth do delay the onset of reproduction. The occurrence of such delays is the best criterion for distinguishing direct from indirect effects.

Indirect effects on reproduction all follow the same basic rules: the relevant parameter (surface-specific assimilation rate, volume-specific maintenance costs or volume-specific costs of structure) is taken to be a linear function of the tissue concentration. Since the assimilation rate represents a source of income rather than costs, it is assumed to *decrease* linearly with the tissue concentration rather than increase, see Figure 6.10. This is consistent with the effect of dioxygen on the assimilation of autotrophs: photorespiration subtracts from photosynthesis, see {191}.

The effects on the reproduction rate as a function of environment concentration and exposure time all work out rather similarly and have the same three toxicity parameters: NEC, tolerance concentration and elimination rate. If growth is measured during exposure, or if the animal's size at the end of the exposure period is measured, it is possible to identify the mode of action. The increase of the costs of structure comes with a hormesis effect, see {237}. The differences in effects on reproduction via assimilation and maintenance are too small to identify the mode of action on the basis of effects on reproduction alone in Figure 6.11.

The quantitative aspects of the various modes of action can be captured as follows. The compound affects a single target parameter $*$ at low concentrations via the dimensionless stress value

$$s = \frac{c_e}{c_*} = \frac{(c_V - c_0)_+}{c_*} = (c_V/c_* - s_0)_+ \tag{6.42}$$

where c_V is the scaled tissue concentration (that has the dimensions of an environment concentration), c_0 the NEC and c_* the tolerance concentration. The tolerance

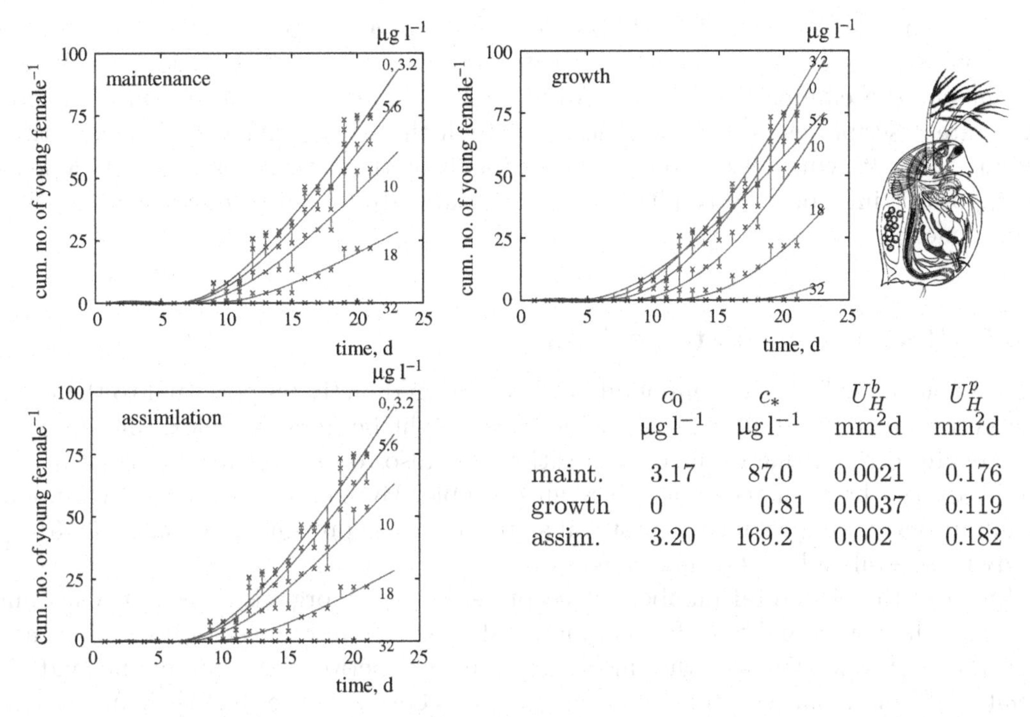

	c_0 μg l^{-1}	c_* μg l^{-1}	U_H^b mm^{2}d	U_H^p mm^{2}d
maint.	3.17	87.0	0.0021	0.176
growth	0	0.81	0.0037	0.119
assim.	3.20	169.2	0.002	0.182

Figure 6.11 Indirect effects of 3,4-dichloroaniline on *Daphnia* reproduction. The mean cumulative number of young per female daphnid as a function of the exposure time to several concentrations of 3,4-dichloroaniline. The fitted curves represent least-squares fits of the model for effects on reproduction via maintenance, growth and assimilation to the same data for large elimination rates. The estimated values for the NEC c_0, the tolerance concentration c_*, the scaled muturity at birth and puberty are given in the table of parameters, assuming $\kappa = 0.8$, $\kappa_R = 0.95$, $g = 0.15$, $\dot{k}_J = 3.6\,\mathrm{d}^{-1}$, $\dot{k}_M = 4.1\,\mathrm{d}^{-1}$.

concentration is a parameter that has the dimension of an environment concentration, and it belongs to a specific physiological target parameter; its name refers to the fact that the value decreases for increasing toxicity of the compound. The parameter s_0 has the interpretation of the stress value with which the individual can cope without showing effects.

Typical examples of target parameters are the maximum specific assimilation rate $\{\dot{p}_A\}$, the specific maintenance rate $[\dot{p}_M]$, the costs of structure $[E_G]$, the costs of reproduction $1 - \kappa_R$, and the hazard of the ovum $\dot{h}$ (during a short period). It is conceivable that other parameters can be affected as well, such as the maturity thresholds E_H^b or E_H^p, or in the case of endrocrine disrupters, the partition coefficient κ. These target parameters can be multiplied by a factor $(1 + s)$ if the compound increases the value of the parameter, e.g. in the case of $[E_G]$ or $[\dot{p}_M]$, or by a factor $(1 - s)$ if the compound decreases the value, e.g. in the case of $\{\dot{p}_A\}$.

The dynamics of sublethal effects are thus characterised by just two parameters, the NEC c_0 and the tolerance concentration c_*, and at least one toxicokinetic parameter, the elimination rate $\dot{k}_e$. Additional parameters can be included in more elaborate descriptions of toxicokinetics. Although the stress value can change in time, because of a varying tissue concentration, none of these three parameters depends on exposure time, but the resulting effects can already be quite complex in transient environments.

6.5.5 Receptor-mediated effects

Up till now, the effect of a compound has been taken directly proportional to the tissue concentration. In a number of cases, the effect might be more complex, and does not only relate to the actual tissue concentration, but also to its (recent) history. A simple model on the basis of receptors gives an example. By considering several endpoints simultaneously, we have found empirical support for receptor-mediated effects [551] by studying several endpoints simultaneously.

Suppose that the total number of receptors N_+ in an organism remains constant, and that the compound transforms functional receptors into non-functional ones at a rate that is proportional to the 'meeting frequency' between the compound and the number of functional receptors. Non-functional receptors can resume their functioning at a given probability rate, or the organism can produce new functional receptors at a rate that is proportional to the number of non-functional receptors. Let $N_n(t)$ denote the number of non-functional receptors, and $N_f(t)$, the number of functional ones, while $N_n(t) + N_f(t) = N_+$. The change in the number of non-functional receptors then amounts to

$$\frac{d}{dt}N_n = \dot{b}_{fn}c_V N_f - \dot{r}_{nf}N_n = \dot{b}_{fn}c_V N_+ - (\dot{r}_{nf} + \dot{b}_{fn}c_V)N_n \tag{6.43}$$

with $\dot{r}_{nf}$ the specific recovery rate, $\dot{b}_{fn}$ the knock-out rate, and c_V the scaled tissue concentration. The stress value can be taken linear in the number of non-functional receptors, $s = N_s^{-1}(N_n - N_0)_+ = (N_n/N_s - s_0)_+$, where the parameter N_s scales the number of non-functional receptors to the stress, and N_0 is the number of non-functional receptors that does not result in an effect on the stress. If we start with unexposed individuals, we have $N_n(0) = 0$, $N_f(0) = N_+$ and $c_V(0) = 0$. This formulation can be combined with a simple first-order kinetics for the tissue concentration, if the amount of compound involved in the binding process is negligibly small.

The model has the interesting property that the amount of memory of the effect is tunable. For large values of $\dot{b}_{fn}$ and $\dot{r}_{nf}$, the number of non-functional receptors is in pseudo steady state, the amount of memory is negligibly small, and the stress is a Michaelis–Menten function of the tissue concentration, rather than a linear one, since $N_n \simeq N_+ \left(1 + \frac{\dot{r}_{nf}}{\dot{b}_{fn}c_V}\right)^{-1}$. If $\dot{r}_{nf} \gg \dot{b}_{fn}$ and N_s is small, or the concentration is small, the model converges to the earlier one, where the effect depends linearly on the actual tissue concentration.

Receptor-mediated effects on survival can be modelled by simply taking the hazard rate as being proportional to the stress, which amounts to the coupled differential equations for the scaled number of non-functional receptors $n_n = N_n/N_+$ and the survival probability q

$$\frac{d}{dt}n_n = \dot{b}_{fn}c_V - (\dot{r}_{nf} + \dot{b}_{fn}c_V)n_n; \quad \frac{d}{dt}q = -q(\dot{k}_\dagger n_n - \dot{h}_0)_+ \tag{6.44}$$

on the assumption that all of the compound in tissues contributes to knocking out receptors, but that the individual can handle a threshold level of non-functioning receptors.

6.5.6 Mutagenic effects

Ames test

The *Salmonella* test, also known as the Ames test, is a popular test for the mutagenic properties of a compound [21, 745]. It is discussed here because the results of the test can sometimes only be understood if energy side-effects are taken into account, for which the DEB model gives a useful framework [518].

The test is carried out as follows. Bacteria (mutants of *Salmonella typhimurium*) that cannot produce the amino acid histidine are grown on an agar plate with a small amount of histidine but otherwise large amounts of all sorts of nutrients. When the histidine becomes depleted, these histidine auxotrophs stop growing at a colony size of typically 8–32 cells. Histidine auxotrophic bacteria can undergo a mutation enabling them to synthesise the necessary histidine themselves, as can the wild strain. They become histidine-prototrophic and continue to grow, even if the histidine on the plate is depleted. (They only synthesise histidine if it is not available in the environment.) Colonies that contain histidine-prototrophs are called revertant colonies and can eventually be observed with the naked eye when the colony size is thousands of cells. The number of revertant colonies relates to the concentration of the compound that has been added to the agar plate and its mutagenic capacity.

Liver homogenate of metabolically stimulated rats is sometimes added to simulate mutagenicity for vertebrates. The primary interest in mutagenicity is because of human health problems, as explained. Vertebrates have many metabolic pathways that prokaryotes do not have. Enzymes in this homogenate sometimes transform non-mutagenic compounds into mutagenic ones, sometimes they do the opposite or have no effect at all.

Some initial histidine is necessary, because bacteria that do not grow and divide do not seem to mutate, or, at least, the mutation is not expressed. This ties mutation frequency to energetics. It is a most remarkable observation, with many consequences. Since maintenance processes also involve some protein synthesis, one would think that mutations should also be expressed if growth ceases, but observation teaches otherwise. If a compound is both mutagenic and reduces growth, the moment of histidine depletion is postponed, so that effective exposure time to the mutagenic compound is increased. Some brands of agar contain small amounts of compounds that become (slightly) mutagenic after autoclaving. This gives a small mutagenic response in the blank. If a test compound only affects growth and is not mutagenic at all, the number of revertant colonies will increase with the concentration of test compound. Such responses make it necessary to model the combined mutation/growth process for the interpretation of the test results.

The rest of this section gives a simple account, appropriate for DEB V1-morphs from a culture that resembles the (initial) growth conditions on the agar plate. For a more detailed account; see [518].

Suppose that the initial amount of histidine on a plate is just enough for the synthesis of N_h cells. Figure 9.7 shows that the histidine reserves are small enough to be neglected. If the inoculum size on the plate is N_0, the number of cells develops initially as $N(t) = N_0 \exp(\dot{r}t)$. Histidine thus becomes depleted at time $t_h = \dot{r}^{-1} \ln\{1 + N_h/N_0\}$. If the mutation rate per unit of DNA is constant, say at value $\dot{h}_M$, the probability of at least one mutation occurring in the descendants of one auxotrophic cell becomes for low mutation rates

$$1 - \exp\left(-\dot{h}_M \int_0^{t_h} (N(t)/N_0)\, dt\right) = 1 - \exp\left(-\frac{\dot{h}_M N_h}{\dot{r} N_0}\right) \simeq \frac{\dot{h}_M N_h}{\dot{r} N_0} \qquad (6.45)$$

The probability of back-mutation is small enough to be neglected. The expected number of revertant colonies is N_0 times (6.45), so that the number of revertant colonies is hardly affected by the inoculum size. The effect of an increase in the number of micro-colonies on the plate is cancelled by the resulting reduction of exposure time.

A consequence of the assumption that the mutation frequency per unit of DNA is constant is that the mutations are independent of each other. This means that the number of revertant colonies on a plate follows a binomial distribution, which is well approximated by the Poisson distribution for low mutation rates. (There are typically fewer than 100 revertant colonies with a typical inoculum size of 10^8 per plate.)

The significance of this expression is that the effect of inoculum size and the amount of histidine become explicit. Variations in these variables, which are under experimental control, translate directly into extra variations in the response. If a compound affects the population growth rate, it also affects the expected number of revertant colonies. I refer to the subsection on population growth rates, {250}, for a discussion of how individual performance (substrate uptake, maintenance, growth) relates to population growth rates. This defines how effects on individual performance translate into effects on population growth rates. This remark not only applies to effects of the test compound, but also to the nutritional quality of the agar.

The mutation rate is usually found to be proportional to the concentration of test compound. This means that each molecule has a certain probability of causing a mutation. Deviations from this relationship can usually be related to changes in the stability of the compound on the plate. Many mutagenic compounds are rather reactive, so the concentration usually decreases substantially before t_h. Others, such as nitrite, diffuse to the deeper layers of the agar plate and become less available to the bacteria in the upper layer. It is easy to circumvent this problem by adding the compound to the

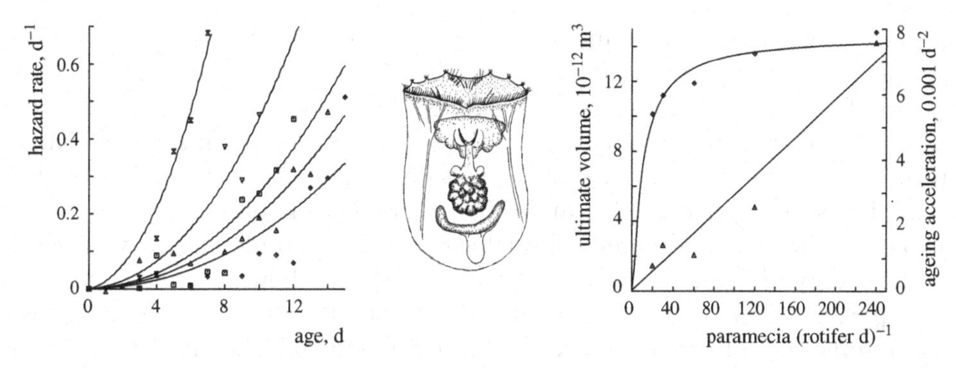

Figure 6.12 The hazard rates for the rotifer *Asplanchna girodi* for different food levels: 20 (◇) 30 (△) 60 (□) 120 (▽) and 240 (⋈) paramecia rotifer^{-1} d^{-1} at 20 °C. Data from Robertson and Salt [978]. The one-parameter hazard curves are based on the scaled food densities as estimated from the ultimate volumes (◇, right), which give f = 0.877, 0.915, 0.955, 0.977, 0.988. The resulting five ageing accelerations are plotted in the right figure (△). They proved to depend linearly on food density, with an intercept that is consistent with the ageing acceleration found for daphnids.

(thick) nutritive bottom layer when it is still liquid, rather than to the (thin) top layer. However, this would increase the financial costs of the test. If metabolic activation is applied, the concentrations of the original compound and the products are likely to become complex compound-specific functions of time. One strategy for interpreting the test results is to analyse and model the time stability of compounds in the Ames test. A better strategy would be to change the experimental procedure in such a way that these complexities do not occur.

Food-induced ageing acceleration

Some data sets, such as that of Robertson and Salt [978] on the rotifer *Asplanchna girodi* feeding on the ciliate *Paramecium tetraurelia*, indicate that the hazard rate increases sharply with food density, see Figures 6.2 and 6.12, which is explained by the effect of the metabolic activity (mobilisation rate) on the emplification of damage inducing compounds in (6.2). This particular data set shows that ageing acceleration is linear in the food density, which suggests that something that is proportional to food density affects the build-up of damage-inducing compounds or the transformation of these compounds into damage. One possibility is nitrite derived from the lettuce used to culture the ciliates; nitrite is known for its mutagenic capacity [518].

6.5.7 Effects of mixtures

The toxicity of mixtures of compounds is of substantial practical interest, which explains the wide interest in the subject. A compound that can be present in molecular and ionic forms can be thought of as a mixture. Many economically important compounds, such as polychlorobiphenyls (PCBs) and polycyclic aromatic hydrocarbons (PAHs), naturally occur as mixtures of many compounds. The toxicity of these mixtures can be understood in terms of that of the participating compounds [45, 45], sometimes using ideas on the covariation of parameter values, see {327}. Since the stress value is assumed to depend linearly on the tissue concentration, the evaluation of effects of mixtures of compounds within the DEB context is relatively straightforward. The discussion below is given for binary mixtures, but generalises to an arbitrary number of compounds.

If two compounds have different physiological target parameters, they always interact via the energy budget; the DEB theory aims to specify how. If two compounds have the same physiological target parameter, and do not interact via the energy butget, they can interact directly in complex ways in their toxic effects. Hewlett and Plackett [503] found that the insecticide thanite intensifies biochemically the toxicity of aprocarb, but that the inverse was not the case. Excluding this type of complex interaction, we can think of the stress value as some non-linear function of the two tissue concentrations, where we are interested in small stress values only. Where a two-term Taylor approximation of a univariate function amounts to a linear function, and was used for effects of one compound on parameter values, a three-term Taylor approximation of a multivariate function involves an interaction term, as is used in the analysis of variance. For two compounds A and B that affect target parameter $*$ this amounts to

$$s = \frac{c_e^A}{c_*^A} + \frac{c_e^B}{c_*^B} + B_*^{AB} c_e^A c_e^B \tag{6.46}$$

where c_e^A and c_e^B are the scaled tissue concentrations above the NEC (see next subsection), c_*^A and c_*^B are the tolerance concentrations, and the interaction parameter B_*^{AB} (dimension: (environment concentration)$^{-2}$) can be positive, in the case of synergism, and negative, in the case of antagonism. Like in the analysis of variance, this idea can readily be extended to an arbitrary number of compounds.

For effects on the hazard rate, this translates to

$$\dot{h}_c = \dot{b}_\dagger^A c_e^A + \dot{b}_\dagger^b c_e^B + \dot{B}_\dagger^{AB} c_e^A c_e^B \tag{6.47}$$

where $\dot{B}_\dagger^{AB}$ is the interaction parameter, see [43]. Figure 6.13 gives an example of application to mixtures of cadmium and copper for $\dot{h} = \dot{h}_0 + \dot{h}_c$, where $\dot{h}_0$ is the constant hazard rate in the blank.

Compounds A and B do not interact if $B_*^{AB} = 0$ or $\dot{B}_\dagger^{AB} = 0$. This situation seems to be called 'concentration addition' or 'independent action', which are two words for the same concept in this context.

NECs of mixtures

From a conceptual point of view, the simplest form of competition of compounds A and B for capacity to cancel effects is that no effects occur if

$$1 > [M_A]/[M_A^0] + [M_B]/[M_B^0] \tag{6.48}$$

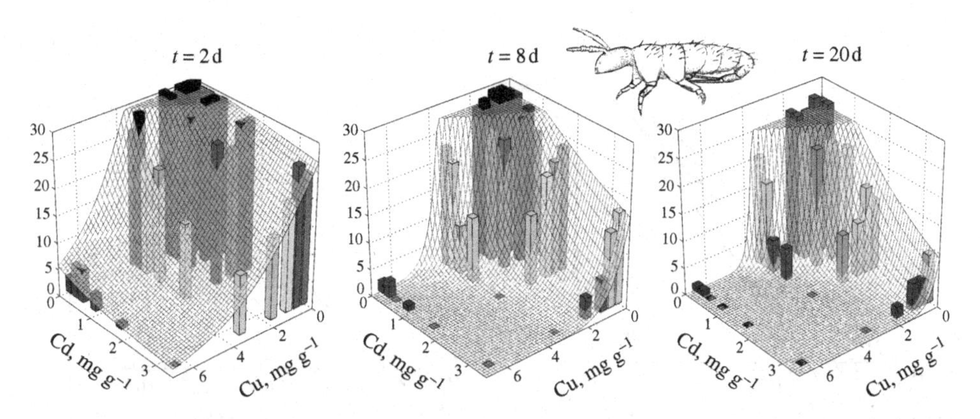

Figure 6.13 The effect of mixtures of cadmium and copper on the survival of the springtail *Folsomia candida* at 18 °C after 2, 8 and 20 days of exposure. Data from Baas *et al.* [43]. Parameters: $\dot{h}_0 = 0.016\,\mathrm{d}^{-1}$, $c_0^{Cd} = 3.60\,\mathrm{mg\,g}^{-1}$, $c_0^{Cu} = 1.13\,\mathrm{mg\,g}^{-1}$, $\dot{b}_\dagger^{Cd} = 0.294\,\mathrm{mg}^{-1}\,\mathrm{g\,d}^{-1}$, $\dot{b}_\dagger^{Cu} = 0.024\,\mathrm{mg}^{-1}\,\mathrm{g\,d}^{-1}$, $\dot{k}_e^{Cd} = 5\,\mathrm{d}^{-1}$, $\dot{k}_e^{Cu} = 1.7\,\mathrm{d}^{-1}$, $\dot{B}_\dagger^{Cd,Cu} = 0.012\,\mathrm{mg}^{-2}\,\mathrm{g}^2\,\mathrm{d}^{-1}$. The interaction parameter is not significantly different from zero.

where $[M_A^0]$ and $[M_B^0]$ are the internal NECs [43] for the compounds separately. If this condition is not fulfilled, compounds A an B take fractions

$$w_A = \frac{[M_A]}{[M_A^0]}\left(\frac{[M_A]}{[M_A^0]} + \frac{[M_B]}{[M_B^0]}\right)^{-1}; \quad w_B = \frac{[M_B]}{[M_B]^0}\left(\frac{[M_A]}{[M_A^0]} + \frac{[M_B]}{[M_B^0]}\right)^{-1} \tag{6.49}$$

of the effect cancel capacity. The internal concentrations of A and B that cause effect are

$$[M_A^e] = \max(0, [M_A] - w_A[M_A^0]); \quad [M_B^e] = \max(0, [M_B] - w_B[M_B^0]) \tag{6.50}$$

Internal concentrations are rarely measured, however; we can replace them by scaled external concentrations by substituting $c_V^A = [M_A]/P_{Ad}$, $c_V^B = [M_B]/P_{Bd}$, $c_0^A = [M_A^0]/P_{Ad}$, $c_0^b = [M_B^0]/P_{Bd}$. If the concentrations in the environment are constant, we have

$$c_V^A(t) = c_A(1 - \exp(-t\dot{k}_e^A)); \quad c_V^B(t) = c_B(1 - \exp(-t\dot{k}_e^B)) \tag{6.51}$$

$$w_A(t) = \frac{c_V^A(t)}{c_0^A}\left(\frac{c_V^A(t)}{c_0^A} + \frac{c_V^B(t)}{c_0^B}\right)^{-1}; \quad w_B(t) = \frac{c_V^B(t)}{c_0^B}\left(\frac{c_V^A(t)}{c_0^A} + \frac{c_V^B(t)}{c_0^B}\right)^{-1} \tag{6.52}$$

$$c_e^A(t) = \max(0, c_V^A(t) - w_A(t)c_0^A); \quad c_e^B(t) = \max(0, c_V^B(t) - w_B(t)c_0^B) \tag{6.53}$$

$$\dot{h}_c(t) = \dot{b}_\dagger^A c_e^A(t) + \dot{b}_\dagger^B c_e^B(t) + \dot{B}_\dagger^{AB} c_e^A(t) c_e^B(t) \tag{6.54}$$

The complete hazard rate is given by $\dot{h}(t) = \dot{h}_0 + \dot{h}_c(t)$, where $\dot{h}_0$ is the hazard rate in the blank. Effects on other target parameters can be worked out in a similar way.

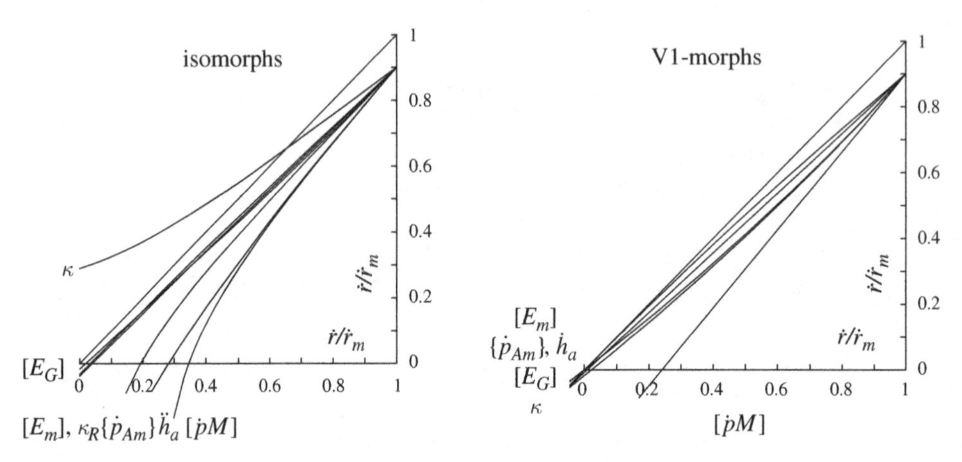

Figure 6.14 Population growth rate in a stressed situation is plotted against that in a blank situation, when only one energy parameter is affected at the same time for reproducing isomorphs (left) and dividing V1-morphs (right). The effect of compounds with different modes of action is standardised such that the maximum population growth rate is 0.9 times that in the blank. Food density is assumed to be constant. Relative effects in isomorphs on structure costs $[E_G]$, reserve capacity $[E_m]$ and reproduction κ_R are almost independent of the feeding conditions, while those on assimilation $\{\dot{p}_{Am}\}$, maintenance $[\dot{p}_M]$ and survival $\ddot{h}_a$ are much stronger under poor feeding conditions. The effect on the partitioning fraction κ is different from the rest and probably does not correspond to an effect of a toxic compound. The relative effects on filaments are largely comparable to those on isomorphs for growth and maintenance. Effects on assimilation $[\dot{p}_{Am}]$ coincide with effects on survival $\ddot{h}_a$.

Effects occur at finite time t_0 if

$$1 < c_A/c_0^A + c_B/c_0^B \tag{6.55}$$

A consequence of this competition model for cancel capacity is that $c_e^A > 0$ if $c_e^B > 0$, and vice versa. This occurs at time t_0, where

$$1 = \frac{c_V^A(t_0)}{c_0^A} + \frac{c_V^B(t_0)}{c_0^B} = (1 - \exp(-t_0\dot{k}_e^A))\frac{c_A}{c_0^A} + (1 - \exp(-t_0\dot{k}_e^B))\frac{c_B}{c_0^B} \tag{6.56}$$

This time point t_0 must be obtained numerically, but with Octave's *fsolve* convergence is fast from the initial choice $t_0 = 0$.

This formulation allows changes in the use of the cancel capacity after the moment effects show up. Computationally simpler is when the use of this capacity is frozen at the moment effects show up. Experience so far indicates that this variant resembles the dynamic one very much, quantitatively.

6.5.8 Population consequences of effects

The general theory to evaluate properties of individuals in terms of dynamics of populations is discussed on {336ff}. Here I only remark that different modes of

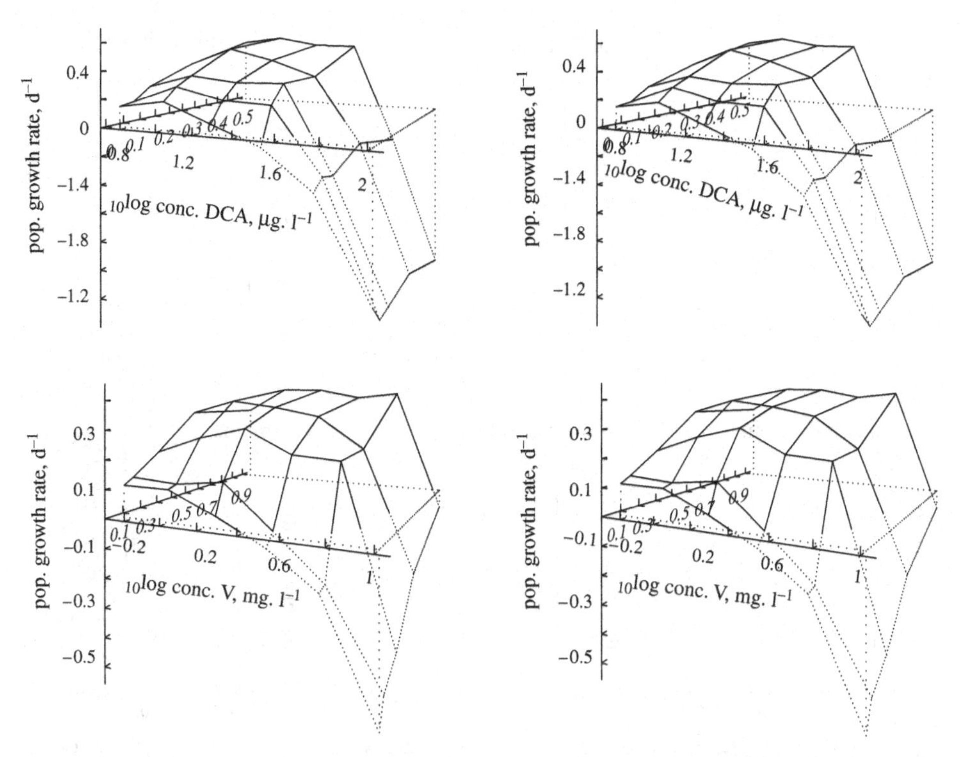

Figure 6.15 Stereo view of the population growth rate of the rotifer *Brachionus rubens* (z-axis) as a function of food density (y-axis) and concentration of toxic compound (x-axis): 3,4-dichloroaniline (above) and potassium metavanadate (below). Food density is in 1.36×10^9 cells *Chlorella pyrenoidosa* per litre, temperature is 20 °C. The difference in shape of the response surfaces is due to differences in the mode of action of the compounds, as predicted by the DEB theory.

action translate differently to consequences for the population, which can be understood intuitively as follows. (See [653] for a more detailed discussion.) If the population is at its carrying capacity, $\dot{r} = 0$, and reproduction and loss rates are both very low, food availability completely governs the reproduction rate. All resources are used for maintenance. Effects on maintenance, therefore, show up directly in this situation, but effects on growth and reproduction remain hidden, unless the effect is so strong that replacement is impossible. If the population is growing at a high rate, energy allocation to maintenance is just a small fraction of available energy. Even considerable changes in this small fraction will, therefore, remain hidden, but effects on production rates are now revealed. These principles are illustrated in Figure 6.14. They imply that at a constant concentration of compound in the environment, the effect at the population level depends on food availability and thus is of a dynamic nature. This reasoning does not yet use the more subtle effects of uptake via food as opposed to those via the environment directly.

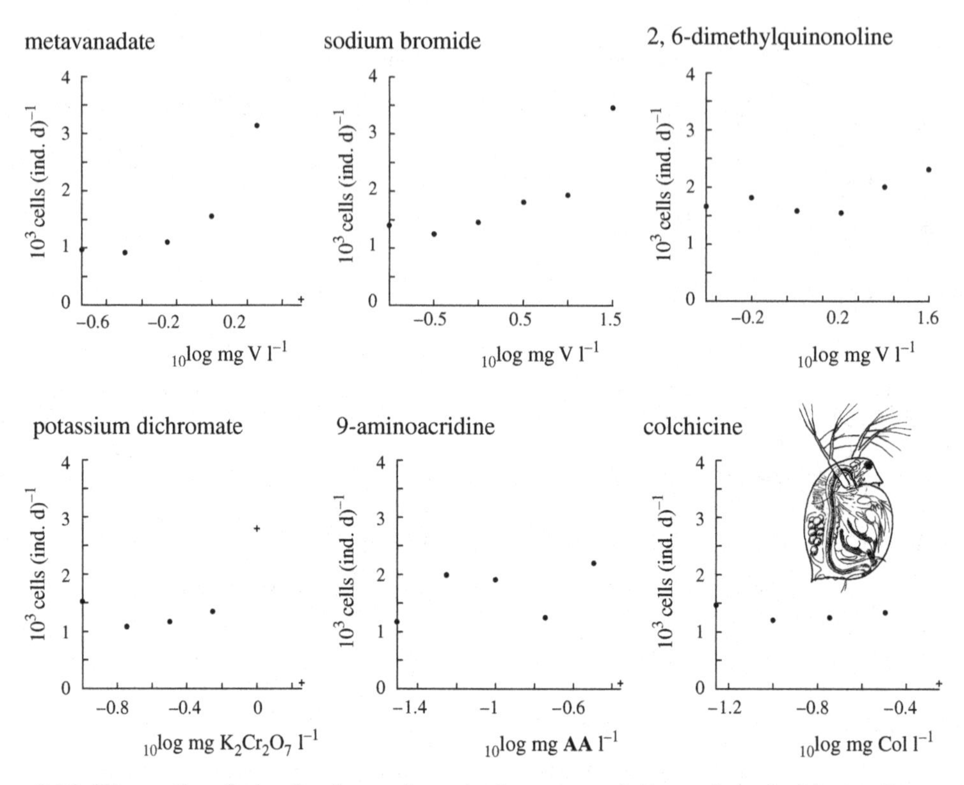

Figure 6.16 The ratio of the food supply rate for a population of daphnids to the number of individuals at carrying capacity in fed-batch cultures as a function of the concentration of compound at 20 °C. The crosses, +, refer to the occurrence of mortality. Only compounds that affect maintenance give a positive response.

The effects at low population growth rates can be studied if the population is at its carrying capacity. If food supply to a fed-batch culture is constant, the number of individuals at carrying capacity is proportional to the food supply rate, see Figure 9.13. If the loss rate, and so the reproduction rate, is small, the ratio of the food supply rate to the number of individuals is a good measure of the maintenance costs. Figure 6.16 illustrates that some compounds, such as vanadium and bromide, affect these maintenance costs, while others do not and 'only' cause death in this situation. It also shows that the effect is almost linear in the concentration, as are the effects on survival, ageing and mutagenicity.

6.6 Summary

Ageing is thought to result as a by-product of respiration via ROS. The DEB theory specifies the quantitative aspects for multicellulars with differentiated cells on the basis of the following supplementary assumptions

---

1 damage-inducing compounds (modified nuclear and mitochondrial DNA) are generated at a rate that is proportional to the mobilisation rate

2 damage-inducing compounds induce themselves at a rate that is proportional to the mobilisation rate

3 damage-inducing compounds generate damage compounds ('wrong' proteins) at constant rate, which cumulate in the body

4 the hazard rate is proportional to the density of damage compounds

---

This results in a module for ageing with two parameters: the (Weibull) ageing acceleration and the Gompertz stress coefficient; the latter is positive for demand systems. Unicellulars do not age gradually, but instantaneously; the parameter is the ageing rate. Interaction of ageing and energetics is via the mobilisation rate as quantifier for metabolic activity. Mutagenic compounds, such as nitrite, have effects very similar to those of ROS, and accelerate ageing.

Non-essential compounds are taken up in a similar way to essential ones, the difference is in their use: non-essential ones are not used, but eliminated. The concentrations in the environment are usually small enough to let the uptake rate be proportional to the concentration, and densities in the body are usually small enough to let the elimination rate be proportional to the density in the body. The one-compartment model results and is basic to all toxicokinetic models. Deviations from this standard model are discussed, which lead to a family of related more realistic and more complex models.

Energetics modifies the kinetics in a number of ways: dilution by growth, changes in the body's lipid content, the existence of several uptake and elimination routes, and metabolic transformation. Since the exchange rate with the environment is proportional to the surface area of the body, these various modifications link up beautifully with the structure of the DEB model, and are evaluated in this chapter.

Non-essential compounds can modify energetics in a number of ways, by changing one or more parameters of the DEB model. Small changes in the parameter values can be taken to be linear in the density of non-essential compounds in the body, on the basis of the Taylor approximation. The changes can be effectuated by multiplying the appropriate parameter(s) with a time-varying stress factor. This quantifies the direct and indirect effects of compounds on energetics dynamically as a function of the concentration in the environment and exposure time, and provides the basis of the estimation of NECs of non-essential compounds.

Effects of mixtures of compounds can be understood from those of single compounds. If the various compounds affect different parameters, interaction always occurs, as quantified by DEB theory. If they affect the same target parameter, interaction can occur in a way that is well captured by the interaction parameter as used in the analysis of variance. The effects on the NECs of mixtures are discussed in some detail. This is of substantial value for Environmental Risk Assessment for toxicants produced by humans.

The description of the effects of non-essential compounds in terms of changes in the parameter values allows the effects of compounds on individuals to be translated into those on populations. Effects at the molecular level have an NEC of zero, because each molecule can react. At the individual level, it is generally larger than zero, because individuals can handle small physiological handicaps. At the population level, effects can vary with food levels even if toxicokinetics is in full steady state; this depends on the mode of action of the compound.

7

Extensions of DEB models

So far, the uni- and multivariate DEB models have been kept as parameter-sparse as possible, with a strong focus on the slow processes that matter for the life cycle. For particular applications is it essential to include more detail, especially if shorter space and timescales need to be included. The purpose of this chapter is to discuss some of these extensions. Each section can be read independently, and deals with a problem that may have taxon-specific elements. Although the sections cover a range of topics, many important ones are painfully lacking, which only reflects that the theory is still in a stage of development. My hope is that it is possible to reduce the dazzling amount of seemingly complex eco-physiological phenomena to a small set of simple underlying principles that can be based on lower levels of organisation.

7.1 Handshaking protocols for SUs

If SUs are physically close to each other, they can interact. I discuss two types of interaction: that between an SU and a carrier and between SUs in a chain. This extension of the behaviour of SUS is applied later for the behaviour of metabolons in mitochondria, which perform the TCA cycle, see {282}.

7.1.1 Handshaking protocols for carriers

Suppose that a substrate X is taken up from the environment by a Carrier (C), which passes its product Y to a Synthesising Unit (SU), which delivers its product Z to the rest of the metabolism of the cell, see Figure 7.1. One molecule of substrate converts into y_{YX} molecules of product Y or y_{ZX} molecules of product Z. I will evaluate the dynamics of the Carrier–Synthesising Unit (CSU) complex, under various assumptions about the exchange of compounds between the two components, given c Carriers and s SUs per unit of biomass.

Three processes should be delineated: feeding F, rejection R, and production P. Appearing fluxes are taken positive (R and P), disappearing ones negative (F). Fluxes

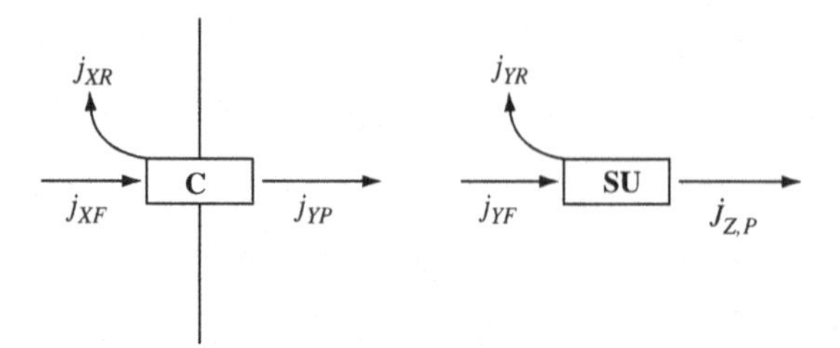

Figure 7.1 The Carrier–Synthesising Unit complex binds substrate X in the environment reversibly and delivers product Z to the cellular metabolism. The inherent rejected fluxes of substrate X and intermediary metabolite Y are indicated, and quantified in the text.

are denoted by two indices: one represents the compound, the other the process. The feeding flux is the flux of substrate molecules that arrives in the catching area of the c Carriers or s SUs.

The mass balances for the Carriers, the SUs and the CSU complex are

$$0 = j_{XF} + j_{XR} + y_{XY} j_{YP} \tag{7.1}$$
$$0 = j_{YF} + j_{YR} + y_{YZ} j_{ZP} \tag{7.2}$$
$$0 = j_{XF} + j_{XR} + y_{XY} j_{YR} + y_{XZ} j_{ZP} \tag{7.3}$$

for $y_{XZ} = y_{XY} y_{YZ}$ and $j_{YP} = -j_{YF}$. The problem now is to write all these fluxes as functions of the feeding flux j_{XF} given a specification of the interaction between the Carriers and the SUs.

The behaviour of the CSU complex depends on the handshaking protocol between the Carrier and the SU. Two extremes are evaluated. In the 'closed' protocol, the Carrier only passes its product to the SU if the SU is in the unbounded state. In the 'open' protocol, the Carrier releases its product irrespective of the state of the SU. The derivation of the behaviour of the CSU complex under both handshaking protocols starts with the changes in the binding fractions, θ_c and θ_s, among the c Carriers and the s SUs, followed by a pseudo-steady-state assumption.

Closed protocol

The changes in the binding fractions amount to

$$\frac{d}{dt}\theta_c = (\dot{k}_X + \dot{k}_Y \theta_s)(1 - \theta_c) + \rho_X j_{XF} \theta_c / c \tag{7.4}$$
$$\frac{d}{dt}\theta_s = \dot{k}_Z (1 - \theta_s) - \rho_Y \dot{k}_Y \theta_s (1 - \theta_c) y_{YX} c / s \tag{7.5}$$

where ρ_X denotes the binding probability of substrate X to the Carrier, ρ_Y the binding probability of assimilated substrate Y (i.e. Carrier product) to the SU, and $\dot{k}_*$ the dissociation rates.

The fluxes can now be quantified as

$$j_{XR} = \dot{k}_X (1 - \theta_c) c - j_{XF} (1 - \rho_X \theta_c) \qquad j_{YR} = (1 - \rho_Y) y_{YX} \dot{k}_Y (1 - \theta_c) \theta_s c$$
$$j_{YP} = y_{YX} \dot{k}_Y \theta_s (1 - \theta_c) c \qquad j_{ZP} = y_{ZY} \dot{k}_Z (1 - \theta_s) s \tag{7.6}$$

Suppose now that the binding fractions are in steady state, i.e. $\frac{d}{dt}\theta_* = 0$. The binding fractions and all fluxes can then be written as functions of the feeding flux j_{XF}. The result is

$$\theta_c = \frac{2c^2\dot{k}_X\dot{k}'_Y + 2cs\dot{k}_Y\dot{k}_Z - \ddot{k}_1 - \ddot{k}}{-2c\dot{k}'_Y j'_{XF}}; \quad \theta_s = \frac{\ddot{k}_1 + \ddot{k}}{2cs\dot{k}_Y\dot{k}_Z} \tag{7.7}$$

$$\dot{k}'_Y = \rho_Y\dot{k}_Y y_{YX}; \quad j'_{XF} = \rho_X j_{XF} - c\dot{k}_X \tag{7.8}$$

$$\ddot{k} = \sqrt{\ddot{k}_1^2 - 4cs^2\dot{k}_Y\dot{k}_Z^2 j'_{XF}}; \quad \ddot{k}_1 = \rho_X j_{XF}(s\dot{k}_Z + c\dot{k}'_Y) + cs(\dot{k}_Y - \dot{k}_X)\dot{k}_Z \tag{7.9}$$

For $\rho_Y = 1$, no products Y are produced and all assimilated X is transformed into Z. For $\rho_Y = 0$, all assimilated X is transformed into Y. The binding probability ρ_Y can be tuned by inhibitors, allowing the CSU complex to branch flux X into fluxes Y and Z.

Open protocol

The changes in the binding fractions amount to

$$\frac{d}{dt}\theta_c = (\dot{k}_X + \dot{k}_Y)(1 - \theta_c) + \rho_X j_{XF}\theta_c/c \tag{7.10}$$

$$\frac{d}{dt}\theta_s = \dot{k}_Z(1 - \theta_s) - \rho_Y\dot{k}_Y\theta_s(1 - \theta_c)y_{YX}c/s \tag{7.11}$$

where the ρ_* denote the binding probabilities and $\dot{k}_*$ the dissociation rates. The only difference with the closed protocol is the absence of θ_s in the change of θ_c.

Assuming a steady state for the binding fractions, the fluxes can be quantified as

$$\begin{aligned} j_{XR} &= \dot{k}_X(1 - \theta_c)c - j_{XF}(1 - \rho_X\theta_c) & j_{YR} &= (1 - \rho_Y\theta_s)y_{YX}\dot{k}_Y(1 - \theta_c)c \\ j_{YP} &= y_{YX}\dot{k}_Y(1 - \theta_c)c & j_{ZP} &= y_{ZY}\dot{k}_Z(1 - \theta_s)s \end{aligned} \tag{7.12}$$

Assuming a steady state again, the binding fractions can be solved through $\frac{d}{dt}\theta_* = 0$, giving all fluxes as functions of the feeding flux j_{XF}.

The solutions amount to

$$\theta_c = \frac{c(\dot{k}_X + \dot{k}_Y)}{c(\dot{k}_X + \dot{k}_Y) - \rho_X j_{XF}}; \quad \theta_s = \frac{(c(\dot{k}_X + \dot{k}_Y) - \rho_X j_{XF})s\dot{k}_Z}{(c(\dot{k}_X + \dot{k}_Y) - \rho_X j_{XF})s\dot{k}_Z - \rho_X j_{XF}c\rho_Y\dot{k}_Y y_{YX}} \tag{7.13}$$

The production of Y relative to Z can be modified by the binding probability ρ_Y, as in the closed protocol, but even when $\rho_Y = 1$ the CSU complex still produces Y.

Comparison

Figure 7.2 compares the performances of the CSU complex using a closed and an open handshaking protocol between the Carrier and the SU. The closed protocol allows a slightly greater production rate of product Z, but less of precursor plus product, $Y + Z$.

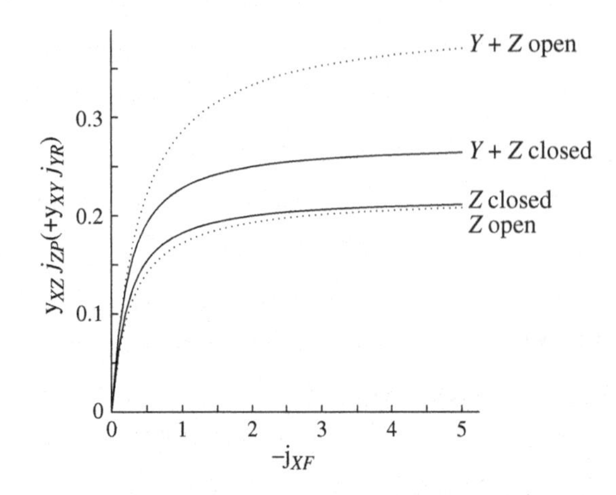

Figure 7.2 The production of product, Z, and of precursor plus product, $Y + Z$, from substrate X of a CSU complex, as functions of the substrate arrival flux, using the closed (solid curves) or the open (dotted curves) handshaking protocol. The open protocol leads to hyperbolic production curves. The parameters are $\dot{k}_X = 0\,\mathrm{s}^{-1}$, $\dot{k}_Y = 0.4\,\mathrm{s}^{-1}$, $\dot{k}_Z = 0.7\,\mathrm{s}^{-1}$, $\rho_X = 1$, $\rho_Y = 0.8$, $y_{XY} = 1$, $y_{XZ} = 1$, $c = s = 1$.

This is because the Carrier waits to dissociate from its substrate until the SU is ready for acceptance, so no precursor is 'spoiled', but the carrier is busy for longer time intervals.

The SU can be thought of as a resistance that leads to deviation of the hyperbolic production curve as a function of substrate density, making it somewhat steeper. The closed protocol is optimal for regulation of flux Y versus flux Z, while the open protocol maximises substrate uptake, with inherent production of Y, and a slightly reduced production of Z. The closed protocol requires compact spatial organisation to allow information exchange between the Carriers and the SUs with respect to the binding state of the SUs, which is not required for the open protocol.

7.1.2 Handshaking protocols for chains

SUs can be organised in a metabolic chain or network; sometimes they are spatially organised in a metabolon and pass intermediate metabolites to each other by channelling. This occurs, for instance, in the TCA cycle [660], where a single copy of each of the nine enzymes is organised in a metabolon in the correct sequence, and two of the enzymes are anchored to the membrane, in most eukaryotes the inner membrane of mitochondria, see {282}. The enzymes might use the open handshaking protocol for dissociation, meaning that the dissociation process is independent of the binding state of the neighbouring SUs, the closed handshaking protocol, meaning that dissociation only occurs if the neighbouring SUs are in the unbounded state, or a mixture of both protocols. Closed handshaking involves communication, and typically physical contact (so spatial structure). If handshaking is fully closed, the whole metabolon acts as if it is a single SU.

Consider the transformation $X_{i-1} \rightarrow y_{X_iX_{i-1}}X_i$ for $i = 1, \ldots, n$, see [660]. We take $y_{X_iX_{i-1}} = 1$ for simplicity. After an introduction of the behaviour of a single SU, we discuss closed and open handshaking, followed by mixtures of these two extremes. Finally we discuss synthesis from two substrates, to model cyclic pathways.

Chain of length $n = 1$

For a given flux $\dot{J}_{X_0,F}$ of substrate to the M_{S_1} SUs, we have

$$\text{Change in unbound fraction:} \quad \frac{d}{dt}\theta_1 = (1-\theta_1)\dot{k}_1 - \theta_1\rho_1\dot{J}_{X_0,F}/M_{S_1} \tag{7.14}$$

$$\text{Steady state unbound fraction:} \quad \theta_1^* = \left(1+\rho_1\dot{J}_{X_0,F}(\dot{k}_1 M_{S_1})^{-1}\right)^{-1} \tag{7.15}$$

$$\text{Production flux:} \quad \dot{J}_{X_1,P} = \dot{k}_1 M_{S_1}(1-\theta_1^*) = \frac{\rho_1\dot{J}_{X_0,F}}{1+\rho_1\dot{J}_{X_0,F}(\dot{k}_1 M_{S_1})^{-1}} \tag{7.16}$$

Closed handshaking at all nodes

Closed handshaking is defined as an interaction between subsequent SUs in a linear pathway such that, given perfect binding, the product of SU i is directly piped to the SU $i+1$ for further processing. To find appropriate expressions for the dynamics of SUs that have this property, we introduce (yet) unknown functions $\dot{b}_i$ of the θ_i's that specify the appearance of unbound fractions. The dynamics of the last SU is simple, because the release of product does not depend on the binding state of any other SU. The release rate of product from the last SU–product complex is proportional to the bound fraction, so $\dot{b}_n = (1-\theta_n)\dot{k}_n$. We now have

$$\frac{d}{dt}\theta_1 = \dot{b}_1 - \theta_1\rho_1\dot{J}_{X_0,F}/M_{S_1} \tag{7.17}$$

$$\frac{d}{dt}\theta_i = \dot{b}_i - \dot{b}_{i-1}M_{S_{i-1}}/M_{S_i} \quad \text{for } i = 2, \ldots, n-1 \tag{7.18}$$

$$\frac{d}{dt}\theta_n = (1-\theta_n)\dot{k}_n - \dot{b}_{n-1}M_{S_{n-1}}/M_{S_n} \tag{7.19}$$

$$\dot{J}_{X_n,P} = \dot{k}_n M_{S_n}(1-\theta_n^*) = \frac{\rho_1\dot{J}_{X_0,F}}{1+\rho_1\dot{J}_{X_0,F}\sum_j(\dot{k}_j M_{S_j})^{-1}} \tag{7.20}$$

The latter follows from the idea that $(\dot{k}_i M_{S_i})^{-1}$ acts as a resistance, and that the n-chain should operate as if it is a single SU; compare (7.16) with (7.20). At steady state, we have $\dot{b}_i^* M_{S_i} = \dot{b}_{i-1}^* M_{S_{i-1}}$, so $\dot{b}_1^* M_{S_1} = \dot{b}_{n-1}^* M_{S_{n-1}}$. From (7.20) follows

$$\dot{b}_i^* = \frac{\rho_1\dot{J}_{X_0,F}/M_{S_i}}{1+\rho_1\dot{J}_{X_0,F}\sum_{j=1}^{n}(\dot{k}_j M_{S_j})^{-1}} \tag{7.21}$$

$$\theta_1^* = \frac{1}{1+\rho_1\dot{J}_{X_0,F}\sum_i(\dot{k}_i M_{S_i})^{-1}} \tag{7.22}$$

$$\theta_n^* = \frac{1+\rho_1\dot{J}_{X_0,F}\sum_{j=1}^{n-1}(\dot{k}_j M_{S_j})^{-1}}{1+\rho_1\dot{J}_{X_0,F}\sum_{j=1}^{n}(\dot{k}_j M_{S_j})^{-1}} \tag{7.23}$$

We see that

$$\theta_{i+1}^* - \theta_i^* = \frac{\rho_1 \dot{J}_{X_0,F} (\dot{k}_i M_{S_i})^{-1}}{1 + \rho_1 \dot{J}_{X_0,F} \sum_{j=1}^n (\dot{k}_j M_{S_j})^{-1}} \tag{7.24}$$

which suggests

$$\dot{b}_i = (\theta_{i+1} - \theta_i)\dot{k}_i \tag{7.25}$$

Substitution into (7.17)–(7.19) gives

$$\frac{d}{dt}\theta_1 = (\theta_2 - \theta_1)\dot{k}_1 - \theta_1 \rho_1 \dot{J}_{X_0,F}/M_{S_1} \tag{7.26}$$

$$\frac{d}{dt}\theta_i = (\theta_{i+1} - \theta_i)\dot{k}_i - (\theta_i - \theta_{i-1})\dot{k}_{i-1} M_{S_{i-1}}/M_{S_i} \quad \text{for } i = 2, \ldots, n-1 \tag{7.27}$$

$$\frac{d}{dt}\theta_n = (1 - \theta_n)\dot{k}_n - (\theta_n - \theta_{n-1})\dot{k}_{n-1} M_{S_{n-1}}/M_{S_n} \tag{7.28}$$

For the purpose of mixing this dynamics with open handshaking, we substitute the feeding fluxes $\dot{J}_{X_{i-1},F} = \dot{J}_{X_{i-1},P} = (\theta_i - \theta_{i-1})\dot{k}_{i-1} M_{S_{i-1}}$ and allow for non-perfect binding ($0 \leq \rho_i \leq 1$). Moreover, we remove θ_1 in front of the flux of X_0 that arrives to the pathway to avoid leaks of X_0. (This can be done because the open handshaking already has this factor.) The result is

$$\frac{d}{dt}\theta_1 = (\theta_2 - \theta_1)\dot{k}_1 - \rho_1 \dot{J}_{X_0,F}/M_{S_1} \tag{7.29}$$

$$\frac{d}{dt}\theta_i = (\theta_{i+1} - \theta_i)\dot{k}_i - \rho_i \dot{J}_{X_{i-1},F}/M_{S_i} \quad \text{for } i = 2, \ldots, n-1 \tag{7.30}$$

$$\frac{d}{dt}\theta_n = (1 - \theta_n)\dot{k}_n - \rho_n \dot{J}_{X_{n-1},F}/M_{S_n} \tag{7.31}$$

The steady-state unbound fractions are

$$\theta_i^* = 1 - \dot{J}_{X_0,F} \sum_{j=n+1-i}^{n} (\dot{k}_j M_{S_j})^{-1} \Pi_{k=1}^j \rho_k = 1 - \dot{j}_{X_0F} \sum_{j=n+1-i}^{n} (\dot{k}_j m_{S_j})^{-1} \Pi_{k=1}^j \rho_k \tag{7.32}$$

with $m_{S_j} = M_{S_j}/M_V$ and $\dot{j}_{X_0F} = \dot{J}_{X_0,F}/M_V$. The production fluxes are

$$\dot{J}_{X_i,P} = \dot{k}_i M_{S_i}(\theta_{i+1}^* - \theta_i^*) = \dot{J}_{X_0,F} \Pi_{k=1}^i \rho_k \tag{7.33}$$

Open handshaking at all nodes

Open handshaking is defined as the lack of any interaction between subsequent SUs in a linear pathway. We here simply have (compare with (7.14))

$$\frac{d}{dt}\theta_i = (1 - \theta_i)\dot{k}_i - \theta_i \rho_i \dot{J}_{X_{i-1},F}/M_{S_i} \quad \text{for } i = 1, \ldots, n \tag{7.34}$$

$$\theta_i^* = \left(1 + \rho_i \dot{J}_{X_{i-1},F} (\dot{k}_i M_{S_i})^{-1}\right)^{-1} \tag{7.35}$$

$$\dot{J}_{X_i,P} = \dot{k}_i M_{S_i}(1 - \theta_i^*) = \frac{\rho_i \dot{J}_{X_{i-1},F}}{1 + \rho_i \dot{J}_{X_{i-1},F} (\dot{k}_i M_{S_i})^{-1}} \tag{7.36}$$

General handshaking

We now combine the dynamics of (7.29)–(7.31) and (7.34) linearly with n handshaking parameters α_i and arrive for $i = 1, \ldots, n-1$ at

$$\frac{d}{dt}\theta_i = (1 - \alpha_i(1-\theta_{i+1}) - \theta_i)\,\dot{k}_i - (\theta_i + \alpha_{i-1}(1-\theta_i))\,\rho_i \dot{J}_{X_{i-1},F}/M_{S_i} \quad (7.37)$$

$$\frac{d}{dt}\theta_n = (1-\theta_n)\dot{k}_n - (\theta_n + \alpha_{n-1}(1-\theta_n))\,\rho_n \dot{J}_{X_{n-1},F}/M_{S_n} \quad (7.38)$$

We can check that the system reduces to closed handshaking for $\alpha_i = 1$ and open handshaking for $\alpha_i = 0$ for $i = 1, \ldots, n-1$. The motivation for the linear combination of the two handshaking protocols is that a fraction α_i of the SUs is following the open handshaking protocol, and a fraction $1-\alpha_i$ the closed one. Notice that α_0 controls the flux from the cell to the pathway, while the other α_i's only deal with the metabolite traffic between SU i and $i+1$.

7.2 Feeding

7.2.1 Food deposits and claims

Quite a few animal species stock food and claim resources via defending a territory as a kind of 'external reserve', which differs from internal reserves by not having active metabolic functions. Many food deposits relate to survival during winter, frequently in combination with dormancy, see {122}. The hamster is famous for the huge piles of maize it stocks in autumn. In the German, Dutch and Scandinavian languages, the word 'hamster' is the stem of a verb meaning to stock food in preparation for adverse conditions. The English language has selected the squirrel for this purpose. This type of behaviour is much more widespread, for example in jays and shrikes, see Figure 7.3. Bees produce honey, and many other adult hymenopterans catch prey for their juveniles. Dung beetles also sport comparable parental care.

Many animal species defend territories just prior to and during the reproductive season. Birds do it most loudly. The size of the territories depends on bird as well as food density. One of the obvious functions of this behaviour is to claim a sufficient amount of food to fulfil the peak demand when the young grow up. The behaviour of stocking and reclaiming food typically fits 'demand' systems and is less likely to be found in 'supply' systems.

The importance of food storing and claiming behaviour is at the population level, where the effect is strongly stabilising for two reasons. The first is that the predator lives on deposits if prey is rare, which lifts the pressure on the prey population under those conditions. The second one is that high prey densities in the good season do not directly result in an increase in predator density. This also reduces the predation pressure during the meagre seasons. Although the quantitative details are not worked out here because of species specificity, I want to highlight this behaviour as an introduction to other smoothing phenomena that are covered.

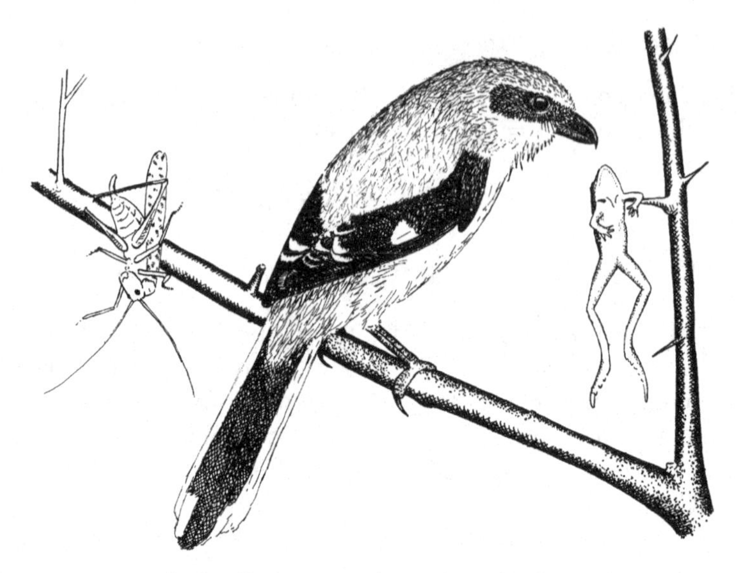

Figure 7.3 The great grey shrike *Lanius excubitor* hoards throughout the year, possibly to guard against bad luck when hunting. Many other shrikes do this as well.

7.2.2 Fast food intake after starvation: hyperphagia

A phenomenon shared by many taxa is that food (substrate) intake after a period of starvation is substantially higher during a short period. Variations of food availability can stimulate growth [445]. Morel [806] modelled a fast short-term uptake (at maximum specific rate j^h_{XAm}) in combination with a much lower longer-term uptake (at maximum specific rate j^l_{XAm}) in algae by assuming empirically that nutrient uptake decreases linearly with the reserve density. A problem with his empirical extension to include fast short-term uptake is that it modifies the well-tested long-term uptake. A variant of this idea that leaves the long-term uptake unaltered is

$$j_{XA}(Xm_E) = fj^h_{XAm} - (j^h_{XAm}/j^l_{XAm} - 1)m_E\dot{k}_E/y_{EX} \quad \text{with} \quad f = \frac{X}{K+X} \tag{7.39}$$

$$\frac{d}{dt}m_E = y_{EX}fj^h_{XAm} - m_E\dot{k}_Ej^h_{XAm}/j^l_{XAm} \quad \text{with} \quad m^*_E = y_{EX}fj^l_{XAm}/\dot{k}_E \tag{7.40}$$

To avoid negative uptake rates, we must have

$$\frac{1}{j^l_{XAm}} - \frac{1}{j^h_{XAm}} > \frac{fy_{EX}}{\dot{k}_Em_E} \tag{7.41}$$

In animals very short-term food uptake after starvation is typically even higher due to filling of the digestive system (stomach plus gut). This can be modelled similarly and linked to a more detailed module for digestion.

7.2.3 Digestion parallel to food searching: satiation

The standard module for food uptake, as presented by the SU scheme in Figure 3.6, classifies behaviour in food searching and food handling. Food handling can be partitioned in mechanical and metabolic handling, where metabolic handling follows mechanical handling, see Figure 7.4. This itself does not affect food uptake. This can be checked as follows

$$1 = \theta_{.} + \theta_X + \theta_E \tag{7.42}$$

$$\frac{d}{dt}\theta_{.} = \dot{k}_E\theta_E - X\dot{b}_X\theta_{.} \quad \text{and} \quad \frac{d}{dt}\theta_E = \dot{k}_X\theta_X - \dot{k}_E\theta_E \tag{7.43}$$

and leads to $\dot{J}_{EA} = y_{EX}\dot{k}_E\theta_E^* = \frac{y_{EX}\dot{b}_X X}{1+\dot{b}_X X(\dot{k}_X^{-1}+\dot{k}_E^{-1})}$, where $(\dot{k}_X^{-1} + \dot{k}_E^{-1})$ serves as a single parameter.

Suppose now that food searching can be parallel to metabolic handling. This extension leads to

$$1 = \theta_{..} + \theta_{.X} + \theta_{E.} + \theta_{EX}; \quad \text{and} \quad \frac{d}{dt}\theta_{..} = \dot{k}_E\theta_{E.} - X\dot{b}_X\theta_{..} \quad \text{and} \tag{7.44}$$

$$\frac{d}{dt}\theta_{EX} = X\dot{b}_X\theta_{.X} - \dot{k}_E\theta_{EX} \quad \text{and} \quad \frac{d}{dt}\theta_{.EX} = X\dot{b}_X\theta_{..} - (\dot{k}_X + X\dot{b}_x)\theta_{.X} \tag{7.45}$$

and $\dot{J}_{EA} = y_{EX}\dot{k}_E(\theta_{EX}^* + \theta_{E.}^*)$, where $\theta_{E.}^* = (1 + \frac{\dot{k}_E}{x\dot{b}_X} + \frac{\dot{k}_S + x\dot{b}_X}{\dot{k}_X + x\dot{b}_X})^{-1}$ and $\theta_{EX}^* = \theta_{E.}^* \frac{x\dot{b}_X}{\dot{k}_X + x\dot{b}_X}$. This model has one parameter extra, and has been fitted successfully to feeding data for sea bream *Sparus aurata* larvae [707].

This idea can be further extended into many directions. One is that the searching rate $\dot{b}_X$ during metabolic handling differs from that after completion of metabolic handling, to introduce the notion of satiation (acquisition homeostasis). A next step is to partition metabolic handling further in steps 1, 2 to n, and let the searching rate increase with the metabolic steps. This way of modelling satiation does not require new state variables. Satiation can also be linked to gut filling, see {273}, and/or to concentrations of metabolites in the blood. Such extensions typically involve quite a few extra parameters.

Likewise more behavioural traits can be introduced, such as sleeping, which can be (partially) parallel to other traits. Behaviour is most realistically described by stochastic models. The dynamics of SUs is intrinsically Markovian, see {104}, which gives access to powerful statistical techniques [447], while respecting the metabolic functions of behaviour.

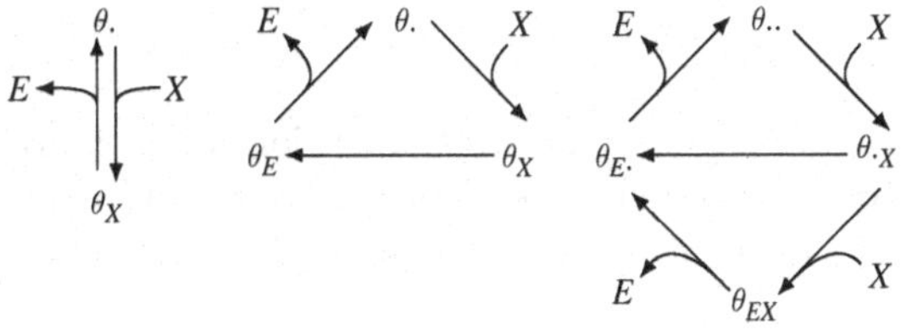

Figure 7.4 The extension of the simple two-state scheme (left, cf. Figure 3.6) to the three-state scheme (middle) does not affect food uptake, but the extension to the four-state (right) does.

7.2.4 Social interaction

Especially among animals at the demand end of the supply–demand spectrum social interaction is an important feature. It can be seen as an association between two individuals that dissociates without transformation; the effect on the feeding rate is via loss of time that depends in a particular way on the population density; the process is formally equivalent to an inhibition process of a special type. Figure 7.5 shows schemes for the cases that socialisation can be initiated during food processing (parallel case) or can not (sequential case), while searching for food cannot be initiated during socialisation. Socialisation can be intra- and/or inter-specific.

For species Y that interacts intra-specifically only and feeds on food X, the possible 'binding' fractions are $1 = \theta_{..} + \theta_{X\cdot} + \theta_{\cdot Y} + \theta_{XY}$. The changes in the 'binding' fractions for the parallel case are

$$\frac{d}{dt}\theta_{..} = \dot{k}_X\theta_{X\cdot} + \dot{k}_Y\theta_{\cdot Y} - (\dot{b}_X X + \dot{b}_Y Y)\theta_{..} \tag{7.46}$$

$$\frac{d}{dt}\theta_{X\cdot} = \dot{b}_X X\theta_{..} + \dot{k}_Y\theta_{XY} - \dot{b}_Y Y\theta_{X\cdot} \tag{7.47}$$

$$\frac{d}{dt}\theta_{\cdot Y} = \dot{b}_Y Y\theta_{..} + \dot{k}_X\theta_{XY} - \dot{b}_X X\theta_{\cdot Y} \tag{7.48}$$

where $\dot{b}_*$ are the affinities and $\dot{k}_*$ dissociation rates. For the sequential case, we exclude all double binding.

The scaled functional response equals $f = \theta_{\cdot}^* x$ with

$$\theta_{\cdot}^* = (1 + x + y)^{-1} \quad \text{sequential case} \tag{7.49}$$

$$= \left(1 + x + y + \frac{xy}{1 + w' + w'y}\right)^{-1} \quad \text{parallel case} \tag{7.50}$$

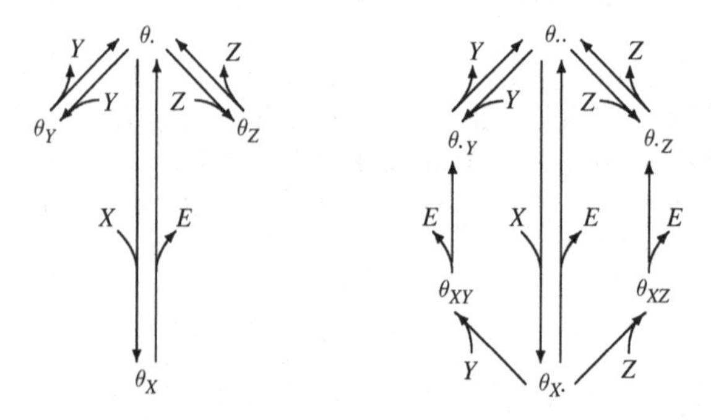

Figure 7.5 The various associations of an individual of species Y with substrate X, which leads to a conversion $X \rightarrow E$, or with other individuals of species Y or Z. The left scheme treats food processing as a process sequential to socialisation, the right one as parallel. From Kooijman and Trost [662].

where scaled food density $x = X/K_X$ and scaled population density $y = Y/K_Y$ are scaled with saturation constants $K_X = \dot{k}_X/\dot{b}_X$ and $K_Y = \dot{k}_Y/\dot{b}_Y$, i.e. ratios of the dissociation rates and the affinities. The socialisation parameter $w' = \dot{k}_X/\dot{k}_Y$ is the ratio of the dissociation rates for food and social interaction and plays the role of an inhibition parameter.

If food X is supplied to a population of socially interacting consumers Y in a chemostat run at throughput rate $\dot{h}$, the changes in food and population densities are given for V1-morphs by

$$\frac{d}{dt}X = \dot{h}(X_r - X) - f\dot{j}_{XAm}Y, \tag{7.51}$$

$$\frac{d}{dt}Y = (\dot{r} - \dot{h})Y, \tag{7.52}$$

with specific growth rate $\dot{r} = \frac{\dot{k}_E f - \dot{k}_M g}{f+g}$, where $\dot{k}_M$ is the maintenance rate coefficient, $\dot{k}_E$ the reserve turnover rate and g the energy investment ratio. At steady state we have $\dot{h} = \dot{r}$.

Figure 7.6 illustrates the effects of socialisation in a single-species situation. After finishing a food-processing session, a sequentially interacting individual starts food searching, but one interacting in parallel first has to complete any social interaction that started during food processing. If social interaction is parallel, it can always be initiated; if sequential, it can only be initiated during searching. This explains the substantial difference between both models; sequential socialisation has relatively little impact because low growth rates accompany low densities (because of maintenance), and so rare social encounters, whereas high growth rates accompany high food levels, so most time is spend on food processing and not on social interaction. The models are more similar for higher values of K and/or w'. While the sequential model is well known [73, 253], the parallel model is not.

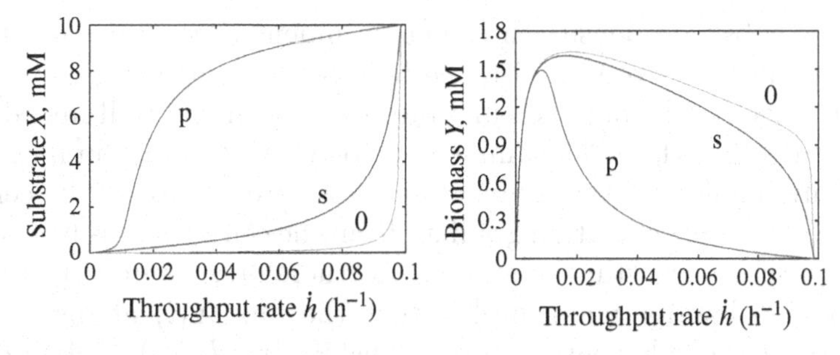

Figure 7.6 No socialisation (0), and sequential (s) and parallel (p) socialisation in a single-species population in a chemostat. Parameter values: substrate concentration in the feed X_r = 10 mM, maximum specific substrate uptake rate $\dot{j}_{XAm} = 1$ h^{-1}, energy investment ratio $g = 1$, maintenance rate coefficient $\dot{k}_M = 0.002\,\text{h}^{-1}$, reserve turnover rate $\dot{k}_E = 0.2\,\text{h}^{-1}$, half-saturation coefficients $K_x = 0.1\,\text{mM}$ and $K_y = 0.1\,\text{mM}$, socialisation $w' = 0.01$. The latter parameter only occurs in the parallel case.

The significance of this deviation from the standard Holling type II formulation is that the feeding rate is no longer a function of food density only, but also of the population density. This has the effect that outcompetition is now much more rare, and species diversity is more easy to maintain in the model system.

7.2.5 Diffusion limitation

The purpose of this subsection is to show why small deviations from the hyperbolic functional response can be expected under certain circumstances, and how the functional response should be corrected. Diffusion limitation as discussed here is key to the understanding of the flocculated growth of micro-organisms, as typically occurs in sewage treatment plants, see {136}.

Any submerged body in free suspension has a stagnant water mantle of a thickness that depends on the roughness of its surface, its electrical properties and on the turbulence in the water. The uptake of nutrients by cells that are as small as those of a bacterium can be limited by the diffusion process through this mantle [612]. Logan [712, 713] related this limitation to the flocculation behaviour of bacteria at low food densities. The existence of a diffusion-limited boundary layer is structural in Gram-negative bacteria such as *Escherichia* [610], which have a periplasmic space between an inner and outer membrane. The rate of photosynthesis of aquatic plants [1078, 1242] and algae [969] can also be limited by diffusion of CO_2 and HCO_3^- through the stagnant water mantle that surrounds them. Coccolithophores, such as *Emiliania*, have a layer of polysaccharides with coccoliths (i.e. calcium carbonate platelets), which might limit diffusion. Since diffusion limitation affects the functional response, it is illustrative to analyse the deviations a bit more in detail. For this purpose I reformulate some results that originate from Best [99] and Hill and Whittingham [509] in 1955.

Homogeneous mantle

Suppose that the substrate density in the environment is constant and that it can be considered as well mixed beyond a distance l_1 from the centre of gravity of a spherical cell of radius l_0. Let X_1 denote the substrate density in the well-mixed environment and X_0 that at the cell surface. The aim is now to evaluate uptake in terms of substrate density in the environment, given a model for substrate uptake at the cell surface.

The build-up of the concentration gradient from the cell surface is fast compared with other processes, such as growth; the gradient is, therefore, assumed to be stationary.

The conservation law for mass implies that the flux $\dot{X}(l)$ at distance l from the centre of gravity of the cell obeys the relationship $4\pi l_1^2 \dot{J}_X(l_1) = 4\pi l_2^2 \dot{J}_X(l_2)$ for any two choices of distances l_1 and l_2. From the choice $l_1 = l$ and $l_2 = l + dl$ follows $l^2 \frac{d}{dl} \dot{J}_X + 2l \dot{J}_X = 0$. According to Fick's diffusion law, the mass flux over a sphere with radius l is proportional to the substrate density difference in the adjacent inner and outer imaginary tunics (i.e. three-dimensional annulus), so $\dot{J}_X \propto -\frac{d}{dl} X$. This leads to the relationship $l^2 \frac{d^2}{dl^2} X + 2l \frac{d}{dl} X = 0$ or $\frac{d}{dl} \left(l^2 \frac{d}{dl} X \right) = 0$, which is known as the Laplace equation. The boundary conditions $X(l_0) = X_0$ and $X(l_1) = X_1$ determine the solution $X(l) = X_1 - (X_1 - X_0)\frac{1-l_1/l}{1-l_1/l_0}$.

The mass flux at l_0 is, according to Fick's law, $4\pi l_0^2 \dot{D} \frac{d}{dl} X(l_0)$, where $\dot{D}$ is the diffusivity. It must be equal to the uptake rate $\dot{J}_X = \dot{J}_{Xm} X_0/(K + X_0)$. This gives the relationship between the density at the cell surface and the density in the environment as a function of the thickness of the mantle

$$X_0 = g(X_1|K, K_1) = \frac{1}{2}X_c + \frac{1}{2}\sqrt{X_c^2 + 4X_1K} \tag{7.53}$$

with $X_c \equiv X_1 - K - K_1$ and $K_1 = \frac{\dot{J}_{Xm}}{4\pi \dot{D} l_0}\left(1 - \frac{l_0}{l_1}\right)$. Since the cell can only 'observe' the substrate density in its immediate surroundings, X_0 must be taken as the argument for the hyperbolic functional response and not X_1. Measurements of substrate density, however, refer to X_1, which invites one to write the functional response as a function of X_1, rather than X_0, so $\dot{J}_X(X_1) = \dot{J}_{Xm}\frac{g(X_1|K, K_1)}{K + g(X_1|K, K_1)}$.

The extent to which a stagnant water mantle changes the uptake rate and the shape of the functional response depends on the value of the mantle saturation coefficient K_1, and therefore on the thickness of the mantle relative to the size of the individual and the diffusivity relative to the maximum uptake rate. If the mantle saturation coefficient is small, the mantle has hardly any effect, i.e. $X_0 \to X_1$ for $K_1 \to 0$, and the functional response is of the hyperbolic type. If it is large, however, the functional response approaches Holling's type I [523], also known as Blackman's response [111], where the ingestion rate is just proportional to food density up to some maximum; see Figures 7.7 and 7.8. This exercise thus shows that the two types of Holling's functional response are related and mixtures are likely to be encountered. This response is at the root of the concept of limiting factors, which still plays an important role in ecophysiology.

The uptake rate depends on the size of the individual in a rather complex way if diffusion is rate limiting. Figure 7.9 illustrates that irregular surfaces are smoothed out. For relatively thick water mantles and at low substrate densities, especially, it is not important that the cell is spherical. The approximate relationship $V \simeq l_0^3 \pi 4/3$ will be appropriate for most rods. The rod then behaves as a V0-morph, since the boundary of the mantle is limiting the uptake and hardly changes during growth of the cell.

Increasing water turbulence and active motion by flagella will reduce the thickness of the water mantle. Its effect on mass transfer is usually expressed by the Sherwood number, which is defined as the ratio of mass fluxes with to those without turbulence. If $X_1 \ll K$, the Sherwood number is independent of substrate density, and amounts to $\left(1 + \frac{K_1/K}{1-l_0/l_1}\right)(1 + K_1/K)^{-1}$. For larger values of X_1, the Sherwood number becomes dependent on substrate density and increasing turbulence will less easily increase mass transfer, because uptake will be rate limiting; see Figure 7.10. This probably defines the conditions for producing sticky polysaccharides which result in the development of films of bacteria on hard substrates or of flocs. If a cell attaches itself, it loses potentially useful surface area for uptake, but increases mass transfer via convection.

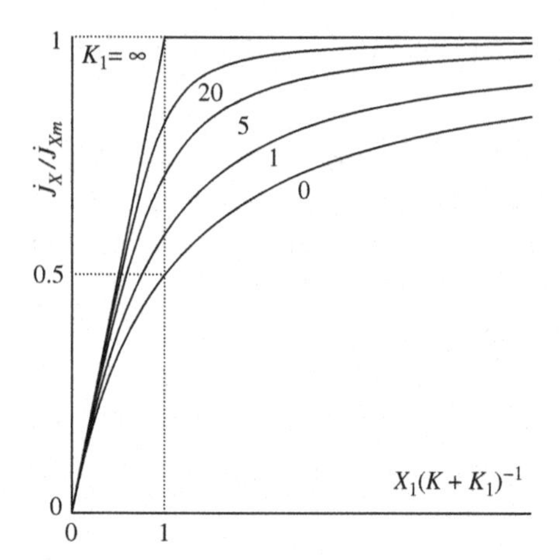

Figure 7.7 The shape of the functional response depends on the value of the mantle saturation coefficient; it can vary from a Holling type II for small values of the mantle saturation coefficient, to Holling type I for large values.

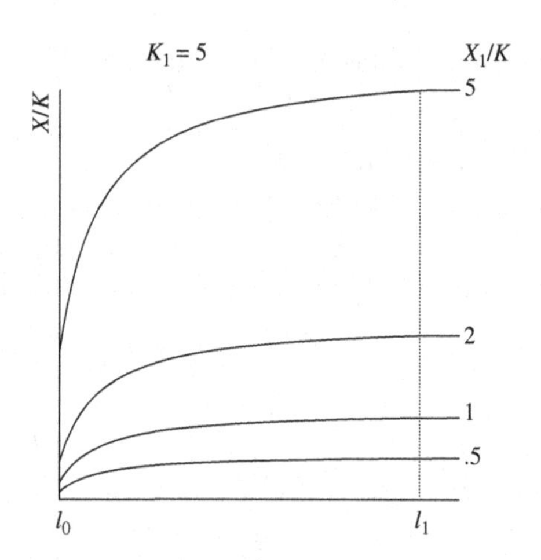

Figure 7.8 Substrate density as a function of the distance from the cell centre in the case of a homogeneous water mantle, for different choices for the substrate densities X_1 in the well-mixed medium.

Although the quantitative details for the optimisation of uptake can be rather complex, the qualitative implication that cells usually occur in free suspension when substrate densities are high, and in flocs when they are low can be understood from Sherwood numbers.

Since diffusivity is proportional to (absolute) temperature, see e.g. [38], and uptake rates tend to follow the Arrhenius relationship, {17}, the temperature dependence of diffusion-limited uptake is likely to depend on temperature in a more complex way.

It is conceivable that slowly moving or sessile animals exhaust their immediate surroundings in a similar way to that described here for bacteria in suspension, if the transport of food in the environment is sufficiently slow. Trapping devices suffer from this problem too [553]. Patterson [872] showed by changing the flow rate that the physical state of the boundary layer surrounding the symbiosis of coral and algae directly affects nutrient transfer. The shape, size and polyp-wall thickness of scleractinian corals could be related to diffusion limitation of nutrients. Some processes of transport can be described accurately by diffusion equations, although the physical mechanism may be different [847, 1070].

Mantle with barrier

For Gram-negative bacteria, which have an inactive outer membrane with a limited permeability for substrate transport, the relationship between the substrate density at the active inner membrane and that in the well-mixed environment is a bit more complicated. On the assumption that the substrate flux through the outer membrane is proportional to the difference of substrate densities on either side of the outer membrane, the permeability affects the mantle saturation coefficient K_1, i.e.

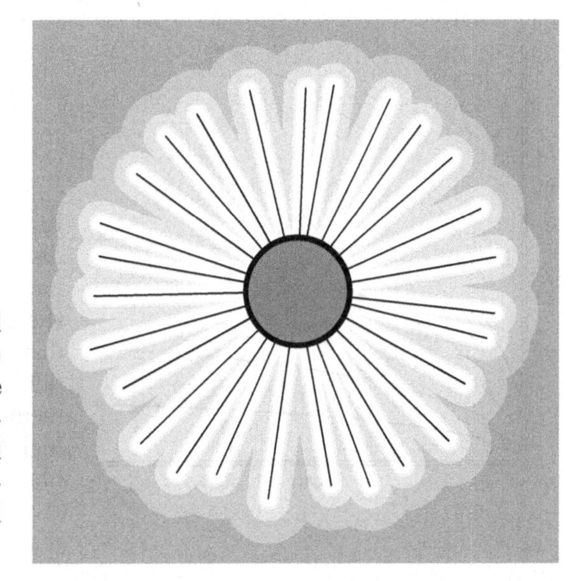

Figure 7.9 Irregular surfaces that catch food are smoothed out by a water mantle if (eddy) diffusion through this layer limits the uptake rate; the thicker the mantle, the more efficient the smoothing. This is here illustrated for a heliozoan, which has thin protoplasm-covered spines that help to catch small food particles (bacteria, algae, micro-organisms).

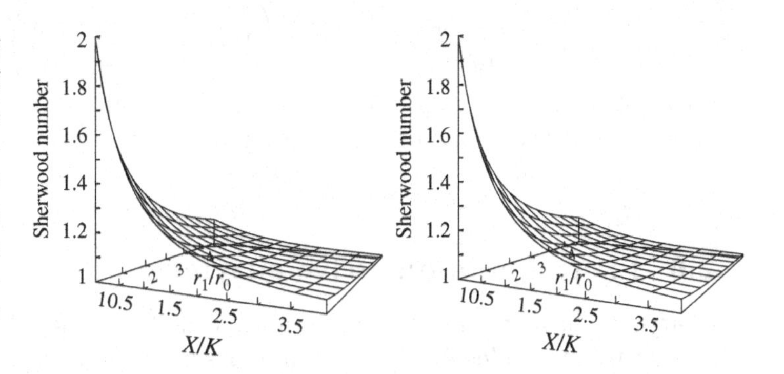

Figure 7.10 Stereo view of the substrate uptake rate of a cell in suspension relative to that in completely stagnant water, as a function of the substrate density in the medium (x-axis) and the thickness of the water mantle (y-axis). Parameter choice: $\dot{J}_{Xm} = 4\pi \dot{D} K l_0$

$K_1 = \frac{\dot{J}_{Xm}}{4\pi \dot{D} l_0}\left(1 - \frac{l_0}{l_1} + \frac{l_0 \dot{D}}{l_2^2 \dot{P}}\right)$, where l_2 is the radius at which the outer membrane occurs and $\dot{P}$ is the permeability of that membrane (dimension length.time^{-1}). The periplasmic space is typically some 20–40 % of the cell volume [829], so that $l_0/l_2 \simeq 0.9$. If $l_2\dot{P} \gg \dot{D}$, the resistance of the outer membrane for substrate transport is negligible. Figure 7.11 illustrates how substrate density decreases towards the inner membrane.

Non-homogeneous mantle

Suppose now that the cell has, besides a stagnant water mantle, also a layer of polysaccharides, where the diffusivity has value $\dot{D}_0$, while it has value $\dot{D}_2$ in the water mantle. Suppose that the boundary of the layer is at distance l_2 from the cell centre, so $l_0 < l_2 < l_1$. The substrate density at the boundary of the polysaccharide layer can now be solved by equating the uptake rate to the flux at the cell membrane. This value can be substituted when the flux at the layer boundary is set equal to the uptake rate. The

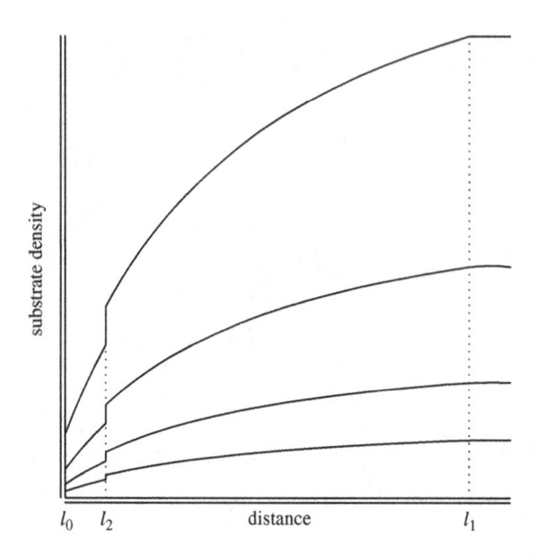

Figure 7.11 Substrate density as a function of the distance from the cell centre for a Gram-negative bacterium. The inner membrane is at distance l_0, the outer membrane at distance l_2, and beyond distance l_1 the medium is completely mixed. Four different choices for substrate densities X_1 in the medium have been made, to illustrate that the higher X_1 the more the substrate density at the inner membrane X_0 is reduced.

relationship between X_0 and X_1 is still given by (7.53), but the mantle saturation constant is now $K_1 = \frac{\dot{J}_{Xm}}{4\pi \dot{D}_2 l_2}\left(1 - \frac{l_2}{l_1}\right)\left(1 + \frac{\dot{D}_2 l_1}{\dot{D}_0 l_2}\frac{1 - l_2/l_0}{1 - l_1/l_2}\right)$. So, just like the barrier in the preceding section, inhomogeneities in the mantle only affect the mantle saturation coefficient, not the shape of the functional response.

7.2.6 Excretion of digestive enzymes

Prokaryotes have no phagocytosis and, therefore, they have to excrete enzymes to digest substrate molecules that cannot pass the membrane. These enzymes transform substrate into product (metabolites); the resulting metabolites can be taken up and used for metabolism. We here compare this digestion mode with endocellular digestion, assuming that the concentration of (solid) substrate is very large relative to the biomass (so the decrease of solid substrate is negligibly small) and the enzyme molecules have a limited active life span.

At lower substrate concentrations, extracellular feeding becomes rapidly even less efficient, because enzymes lose time in their unbound state.

Intracellular digestion

Suppose the digestive enzyme becomes inactive at constant specific rate $\dot{k}_P$, and the mean production time per product molecule is $\dot{k}_X^{-1}$. The maximum yield of product per enzyme molecule thus amounts to $y_{XP}^m = \dot{k}_X/\dot{k}_P$ and serves as a reference for extracellular digestion. Although no metabolites become lost, this mode of digestion comes with costs of phagocytosis, and processing of inactive enzymes. The latter might represent a cost or a further benefit.

Social digestion

Suppose now that bacteria are tightly packed in a one-cell-thick layer on a solid substrate, and they excrete enzyme molecules at specific rate $\{\dot{J}_{PA}\}$ (moles per surface area of cell per time). If the cells are spherical with radius L_R, they excrete enzymes at rate $\{\dot{J}_{PA}\}4\pi L_R^2$. One cell occupies surface πL_R^2 in the layer, so a unit surface area has $(\pi L_R^2)^{-1}$ cells. In surface area S of medium enzymes are excreted at rate $\dot{J}_{PA} = 4\{\dot{J}_{PA}\}S$ (mol/t). Assuming that the cells are half embedded in the medium and the maximum specific uptake rate $\{\dot{J}_{XAm}\}$ is large enough to ensure that the concentration $n_X(L_R)/S$ at the cell membrane is small, the uptake rate of a cell is $\{\dot{J}_{XAm}\}\pi L_R^2 \frac{n_X(L_R)}{n_{XK}}$, where n_{XK} is the half-saturation density. In surface area S of medium the uptake rate is $\dot{J}_{XA} = \{\dot{J}_{XAm}\}S\frac{n_X(L_R)}{n_{XK}}$ (mol/t). The yield of metabolite on enzyme equals $y_{XP} = \frac{\{\dot{J}_{XAm}\}}{\{\dot{J}_{PA}\}}\frac{n_X(L_R)}{4n_{XK}}$.

Choosing the origin of length L in the centre of a cell on the solid medium (for reasons that are obvious in the case of solitary feeding), the change in densities of enzyme and product concentrations is for diffusivities $\dot{D}_P$ and $\dot{D}_X$

$$0 = \dot{J}_{PA} + \dot{D}_P \frac{\partial}{\partial L} n_P(L_R, t) \tag{7.54}$$

$$0 = \frac{\partial}{\partial t} n_P(L,t) + \dot{k}_P n_P(L,t) - \dot{D}_P \frac{\partial^2}{\partial L^2} n_P(L,t) \tag{7.55}$$

$$0 = \dot{J}_{XA} - \dot{D}_X \frac{\partial}{\partial L} n_X(L_R, t) \tag{7.56}$$

$$0 = \frac{\partial}{\partial t} n_X(L,t) - \dot{k}_X n_P(L,t) - \dot{D}_X \frac{\partial^2}{\partial L^2} n_X(L,t) \tag{7.57}$$

The steady state profiles follow from the balance for enzyme molecules $\int_{L_R}^{\infty} n_P(L)\,dL = \dot{J}_{PA}/\dot{k}_P$, which have solution

$$n_P(L) = \frac{\dot{J}_{PA} L_P}{\dot{D}_P} \exp\left(\frac{L_R - L}{L_P}\right) \quad \text{for } L_P = \sqrt{\dot{D}_P/\dot{k}_P} \tag{7.58}$$

$$n_X(L) = \frac{\dot{k}_X}{\dot{k}_P}\frac{\dot{D}_P}{\dot{D}_X}\left(\frac{\dot{J}_{PA} L_P}{\dot{D}_P} - n_P(L)\right) \tag{7.59}$$

We have $\frac{d}{dL} n_X(L_R) = \frac{\dot{J}_{PA}\dot{k}_X}{\dot{D}_X \dot{k}_P}$. The uptake equals $\dot{J}_{XA}(t) = \dot{D}_X \frac{d}{dL} n_X(L_R, t)$, while $\dot{J}_{PA}\dot{k}_X/\dot{k}_P$ metabolites are produced when the extracellular enzyme buffer is full. The difference is lost in the environment. The yield coefficient at infinite time is $y_{XP} = \dot{J}_{XA}/\dot{J}_{PA}$. We define the relative efficiency to be $\theta = \frac{y_{XP}}{y_{XP}^m} = \frac{\dot{J}_{XA}\dot{k}_P}{\dot{J}_{PA}\dot{k}_X}$. Initially, when $n_P(L,0) = n_X(L,0) = 0$, we have $\theta = 0$; it takes a long time to build up to $\theta = 1$, when all of the medium (apart from the direct neighbourhood of the bacteria) has metabolite density $n_X(\infty)$.

Solitary digestion

Suppose now that a single spherical cell of radius L_R lives half-embedded on a homogeneous medium and excretes enzyme molecules at specific rate $\{\dot{J}_{PA}\}$ (moles per cell's surface area per time) or at rate $\dot{J}_{PA} = \{\dot{J}_{PA}\}4\pi L_R^2$ in total. The cell's uptake rate of metabolites is $\{\dot{J}_{XAm}\}2\pi L_R^2 \frac{n_X(L_R)}{n_{XK}}$, so the yield of metabolites on enzyme is $y_{PX} = \frac{\{\dot{J}_{XAm}\}}{\{\dot{J}_{PA}\}} \frac{n_X(L_R)}{2n_{XK}}$.

The change in densities of enzyme and product concentrations is for diffusivities $\dot{D}_P$ and $\dot{D}_X$

$$0 = \dot{J}_{PA} + \dot{D}_P \frac{\partial}{\partial L} n_P(L_R, t) \tag{7.60}$$

$$0 = \frac{\partial}{\partial t} n_P(L,t) + \dot{k}_P n_P(L,t) - \dot{D}_P \frac{\partial^2}{\partial L^2} n_P(L,t) - 2\frac{\dot{D}_P}{L}\frac{\partial}{\partial L} n_P(L,t) \tag{7.61}$$

$$0 = \dot{J}_{XA} - D_X \frac{\partial}{\partial L} n_X(L_R, t) \tag{7.62}$$

$$0 = \frac{\partial}{\partial t} n_X(L,t) - \dot{k}_X n_P(L,t) - \dot{D}_X \frac{\partial^2}{\partial L^2} n_X(L,t) - 2\frac{\dot{D}_X}{L}\frac{\partial}{\partial L} n_X(L,t) \tag{7.63}$$

The steady-state profile of the enzyme and metabolite is

$$n_P(L) = \frac{\dot{J}_{PA}}{\dot{D}_P} \frac{L_P L_R^2/L}{L_P + L_R} \exp\left(\frac{L_R - L}{L_P}\right) \tag{7.64}$$

$$n_P(L) = \frac{\dot{k}_X}{\dot{k}_P} \frac{\dot{D}_P}{\dot{D}_X} \left(\frac{\dot{J}_{PA}}{\dot{D}_P} \frac{L_P L_R}{L_P + L_R} - n_P(L)\right) \tag{7.65}$$

Figure 7.12 compares enzyme and metabolite profiles for the social and solitary digestion modes. Although the results depend on parameter values, quite a lot of metabolites

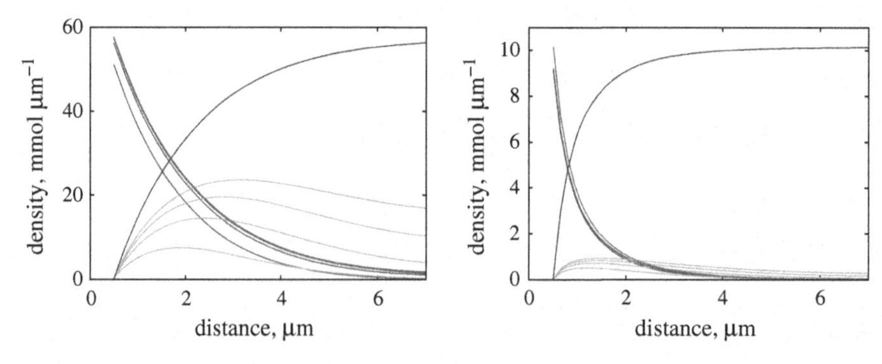

Figure 7.12 The enzyme (decreasing) and metabolite (hump-backed) profiles for social (top left) and solitary (top right) digestion for times 100, 200, ..., 500 h. The magenta and blue curves are the steady state profiles for enzyme and metabolite. Parameters: $\dot{J}_{PA} = 1\,\text{mmol}\,\text{h}^{-1}$, $\dot{D}_P = 0.03\,\mu\text{m}^2\,\text{h}^{-1}$, $\dot{D}_X = 0.03\,\mu\text{m}^2\,\text{h}^{-1}$, $\dot{k}_P = 0.01\,\text{h}^{-1}$, $\dot{k}_X = 0.01\,\text{h}^{-1}$, $\{\dot{J}_{XAm}\}\pi L_R/n_{XK} = 20\,\text{h}^{-1}$, $L_R = 0.5\,\mu\text{m}$. See Figure 7.13 for the yield of metabolite on enzyme as function of time.

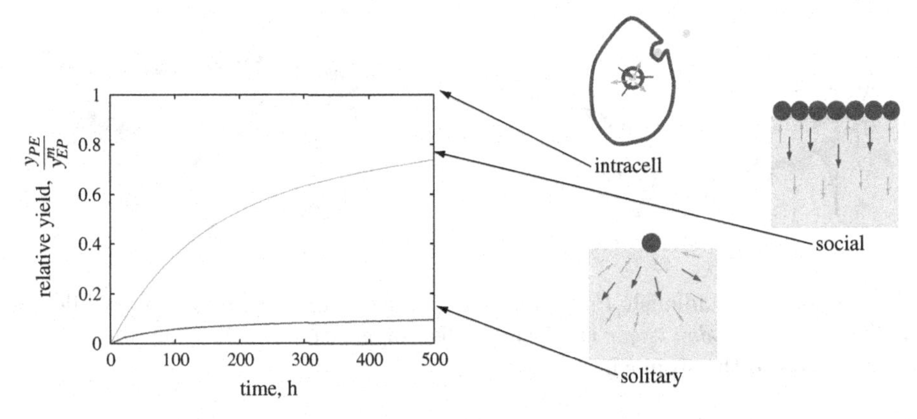

Figure 7.13 Relative to the yield of metabolite on enzyme for intracellular digestion, the yield for social extracellular digestion builds up slowly in time, while that for solitary digestion really takes a long time and builds up to a lower level. The bold arrows stand for enzyme flux, the lighter ones for metabolite flux.

are unavailable for the cell, and the problem is much worse for solitary cells. It also takes a long time to build up some yield, compared with intracellular digestion. The enzyme profile reaches its steady state much earlier than the metabolite profiles; the metabolites first must flush the whole medium before a steady-state profile can build up.

7.3 Digestion in guts

7.3.1 Smoothing and satiation

The capacity of the stomach/gut volume depends strongly on the type of food a species specialises on. Fish feeding on plankters, i.e. many small constantly available particles, have a low stomach capacity, while fish such as the swallower, which feed on rare big chunks of food (see Figure 7.14), have high stomach capacities. It may wait for weeks before a new chunk of food arrives. The stomach/gut volume, which is still 'environment' rather than animal, is used to smooth out fluctuations in nutritional input to the organism. Organisms attempt to run their metabolic processes under controlled and constant conditions. Food in the digestive tract and reserves inside the organism together make it possible for regulation mechanisms to ensure homeostasis. Growth, reproductive effort and the like do not depend directly on food availability but on the internal state of the organism. This even holds, to some extent, for those following the 'supply' strategy, where energy reserves are the key variable. These reserves rapidly follow the feeding conditions.

If the food in the stomach, M_s, follows a simple first-order process, the change of stomach contents is

$$\frac{d}{dt}M_s = \{\dot{J}_{XAm}\}V^{2/3}\left(f - \frac{M_s}{[M_{sm}]V}\right) \tag{7.66}$$

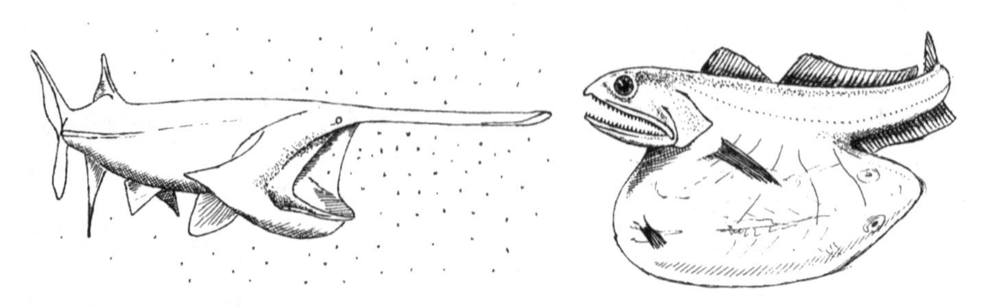

Figure 7.14 The 2-m paddlefish *Polyodon spathula* feeds on tiny plankters, while the 18-cm black swallower *Chiasmodon niger* can swallow fish bigger than itself. They illustrate extremes in buffer capacities of the stomach.

where $[M_{sm}]V$ is the maximum food capacity of the stomach. The derivation is as follows. A first-order process here means that the change in stomach contents can be written as $\frac{d}{dt}M_s = \dot{J}_{XA} - t_s^{-1}M_s$, where the proportionality constant t_s^{-1} is independent of the input, given by $\dot{J}_{XA} = \{\dot{J}_{XAm}\}f$, see (2.2). Since food density is the only variable in the input, t_s^{-1} must be independent of food density X, and thus of scaled functional response f. If food density is high, stomach content converges to its maximum capacity $\dot{J}_{XAm}t_s = \{\dot{J}_{XAm}\}V^{2/3}t_s$. The assumption of isomorphism implies that the maximum storage capacity of the stomach is proportional to the volume of the individual. This means that we can write it as $[M_{sm}]V$, where $[M_{sm}]$ is some constant, independent of food density and body volume. This allows one to express t_s^{-1} in terms of $[M_{sm}]$, which results in (7.66).

The mean residence time in the stomach is thus $t_s = V^{1/3}[M_{sm}]/\{\dot{J}_{XAm}\}$, and so it is proportional to length and independent of the ingestion rate. First-order dynamics implies complete mixing of food particles in the stomach, which is unlikely if fermentation occurs. This is because the residence time of each particle is then exponentially distributed, so a fraction $1 - \exp(-1) = 0.63$ of the particles spends less time in the stomach than the mean residence time, and a fraction $1 - \exp(-\frac{1}{2}) = 0.39$ less than half the mean residence time. This means incomplete, as well as over-complete, and thus wasteful, fermentation.

The extreme opposite of complete mixing is plug flow, where the variation in residence times between the particles is nil in the ideal case. Pure plug flow is not an option for a stomach, because this excludes smoothing. These conflicting demands probably separated the tasks of smoothing for the stomach and digestion for the gut to some extent. Most vertebrates do little more than create an acid environment in the stomach to promote protein fermentation, while actual uptake is via the gut. For a mass of food in the stomach of M_s, and in the gut of M_g, plug flow of food in the gut can be described by

$$\frac{d}{dt}M_g(t) = t_s^{-1}(M_s(t) - M_s(t - t_g)) \tag{7.67}$$

where t_g denotes the gut residence time and t_s the mean stomach residence time. This equation follows directly from the principle of plug flow. The first term, $t_s^{-1}M_s(t)$, stands for the influx from the stomach and follows from (7.66). The second one stands for the outflux, which equals the influx with a delay of t_g. Substitution of (7.66) and (2.2) gives $\frac{d}{dt}M_g(t) = \dot{J}_{XA}(t) - \dot{J}_{XA}(t - t_g) + \frac{d}{dt}M_s(t - t_g) - \frac{d}{dt}M_s(t)$. Since $0 \leq M_s \leq [M_{sm}]V$, $\frac{d}{dt}M_s \rightarrow 0$ if $[M_{sm}] \rightarrow 0$. So the dynamics of food in the gut reduces to $\frac{d}{dt}M_g(t) = \dot{J}_{XA}(t) - \dot{J}_{XA}(t - t_g)$ for animals without a stomach.

Some species feed in meals, rather than continuously, even if food is constantly available. They only feed when 'hungry' [285]. Stomach filling can be used to link feeding with satiation. From (7.66) it follows that the amount of food in the stomach tends to $M_s^* = f[M_{sm}]V$, if feeding is continuous and food density is constant. Suppose that feeding starts at a rate given by (2.2) as soon as food in the stomach is less than $\delta_{s0}M_s^*$, for some value of the dimensionless factor δ_{s0} between 0 and 1, and feeding ceases as soon as food in the stomach exceeds $\delta_{s1}M_s^*$, for some value of $\delta_{s1} > \delta_{s0}$. The mean ingestion rate is still of the type (2.2), where $\{\dot{J}_{XAm}\}$ now has the interpretation of the *mean* maximum surface-area-specific ingestion rate, not the one during feeding. A consequence of this on/off switching of the feeding behaviour is that the periods of feeding and fasting are proportional to a length measure. This matter links up with {114}.

7.3.2 Gut residence time

The volume of the digestive tract is proportional to the whole body volume in strict isomorphs. The fraction is ≃11% for ruminant and non-ruminant mammals [261] and ≃2.5% for daphnids if the whole space in the carapace is included [330]. If the animal keeps its gut filled to maximum capacity, $[M_{gm}]V$ say, and if the volume reduction due to digestion is not substantial, this gives a simple relationship between gut residence time of food particles t_g, ingestion rates $\dot{J}_{XA}$, and body volume V

$$t_g = [M_{gm}]V/\dot{J}_{XA} = \frac{L[M_{gm}]}{f\{\dot{J}_{XAm}\}} \tag{7.68}$$

This has indeed been found for daphnids [330], see Figure 7.15, and mussels [459]. Copepods [207] and carnivorous fish [558] seem to empty their gut at low food densities, which gives an upper boundary for the gut residence time. The gut residence time has a lower boundary of $L[M_{gm}]/\{\dot{J}_{XAm}\}$, which is reached when the throughput is at maximum rate.

Since ingestion rate, (2.2), is proportional to squared length, the gut residence time is proportional to length for isomorphs. For V1-morphs, which have a fixed diameter, ingestion rate is proportional to cubed length, (4.12), so gut residence time is independent of body volume.

Daphnids are translucent, which offers the possibility of studying the progress of digestion as a function of body length, see Figure 7.16.

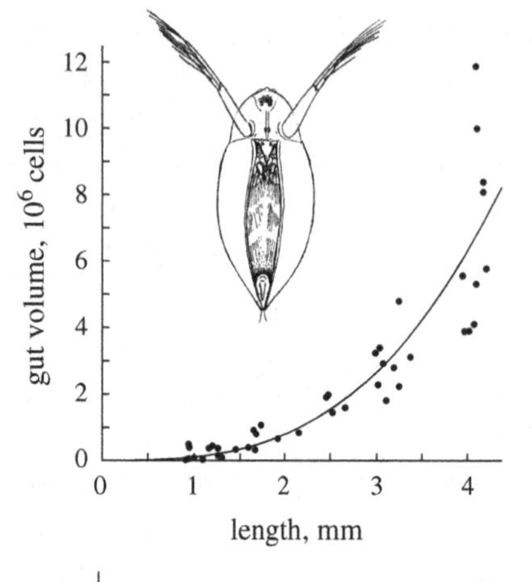

Figure 7.15 Gut volume is proportional to cubed length (right) and gut residence time is proportional to length (lower left), while the latter depends hyperbolically on food density (lower right), as illustrated for daphnids. The first two figures relate to *D. magna* feeding on the green alga *Scenedesmus* at 20 °C. Data from Evers and Kooijman [330]. The third one relates to a 2-mm *D. pulex* feeding on the diatom *Nitzschia actinastroides* at 15 °C. Data from Geller [393].

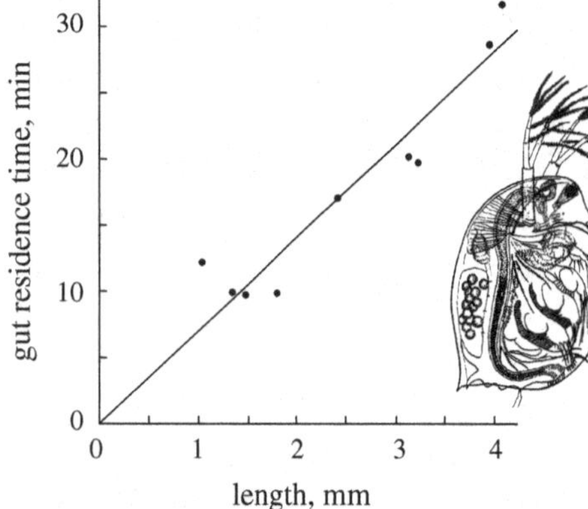

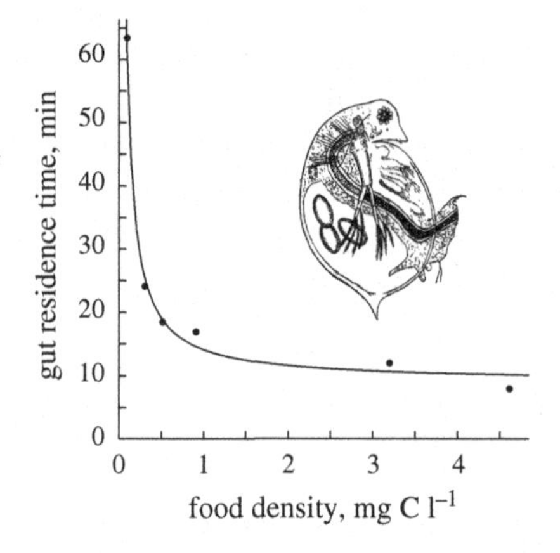

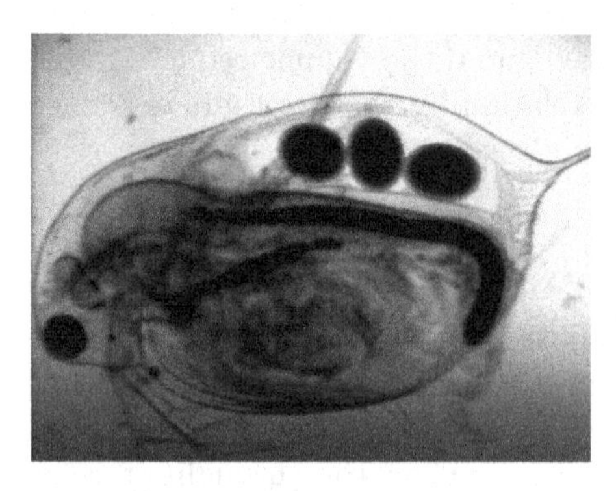

Figure 7.16 The photograph of *Daphnia magna* shows the sharp transition between the chlorophyll of the green algae and the brown-black digestion products, which is typical for high ingestion rates. The relative position of this transition point depends on the ingestion rate, but not on the body length. Even in this respect daphnids are isomorphic. At low ingestion rates, the gut looks brown from mouth to anus. The paired digestive caeca are clearly visible just behind the mouth.

7.3.3 Gut as a plug flow reactor

Microflora is likely to play an important role in the digestion process of all herbivores. It can provide additional nutrients by fermenting carbohydrates and by synthesising amino acids and essential vitamins. Daphnids are able to derive structural body components and lipids from the cellulose of algal cell walls [1024], though it is widely accepted that daphnids, like almost all other animals, are unable to produce cellulase. Endogenous cellulase production is only known to occur in some snails, wood-boring beetles, shipworms and thysanurans [719]. The leaf-cutting ant *Atta* specifically cultures fungi, probably to obtain cellulase [747]. Bacteria have been found in the guts of an increasing number of crustaceans [822], but not yet in daphnids [1024]. In view of the short gut residence times for daphnids, it is improbable that the growth of the daphnid's gut flora plays an important role. Digestion of cellulose is a slow process, and the digestive caecum is situated in the anterior part of the gut. Daphnids, therefore, probably produce enzymes that can pass through cell walls, because they do not have the mechanics to rupture them.

Many studies of energy transformations assume that the energy gain from a food item does not depend on the size of the individual or on the ingestion rate. The usefulness of this assumption in ecological studies is obvious, and the DEB model uses it as well. In view of the relationship of gut residence time to both size and ingestion rate, this assumption needs further study.

The nutritional gain from a food particle has been observed to depend on gut residence time [961, 1019]. These findings are suspect for two reasons, however. The first reason is that assimilation efficiencies are usually calculated per unit of dry weight of consumer, while the energy reserves contribute increasingly to dry weight with increasing food density, but do not affect digestion. The second reason is that, while the nutritional value of faecal pellets may decrease with increasing gut residence time, it is not obvious whether the animal or the gut microflora gains from the difference. I discuss here to what extent digestion is complete and the composition of faeces does not change if the composition of food does not change.

When animals such as daphnids are fed with artificial resin particles mixed through their algal food, the appearance of these particles in the faeces supports the plug flow type of model for the digestion process, as proposed by Penry and Jumars [330, 879, 880].

The shape of the digestive system also suggests plug flow. The basic idea is that materials enter and leave the system in the same sequence and that they are perfectly mixed radially. Mixing or diffusion along the flow path is assumed to be negligible. (This is at best a first approximation, because direct observation shows that particles sometimes flow in the opposite direction.)

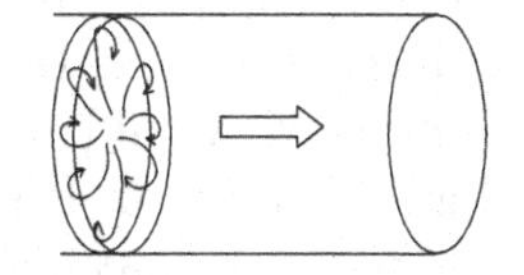

Suppose that a thin slice of gut contents can be followed during its travel along the tube-like digestive tract, under conditions of a constant ingestion rate. The small changes in the size of the slice during the digestion process are ignored. The gut content of a 4-mm *D. magna* is about 0.1 mm^3, while the capacity is about 6.3×10^5 cells of *Scenedesmus*, see Figure 7.15, of some 58 μm^3 per cell, which gives a total cell volume of 0.0367 mm^3. The cells occupy some 37% of the gut volume, which justifies the neglect of volume changes for the slice. The volume of the slice of thickness L_λ is $V_s = \pi L_\lambda L_\phi^2/4$, where L_ϕ is the diameter of the gut, and $\pi L_\lambda L_\phi$ is the surface area of contact between slice and gut.

Suppose that the gut wall secretes enzymes into the slice, which catalyse the transformation of food X into faeces and a product P, which can be absorbed through the gut wall. The rate of this transformation, called digestion, is taken proportional to the concentration of active enzymes that have been secreted. If the secretion of enzymes is constant and the deactivation follows a simple first-order process, the amount M_g of active enzyme in the slice will follow $\frac{d}{dt}M_g = \{\dot{J}_g\}\pi L_\lambda L_\phi - \dot{k}_g M_g$, where $\{\dot{J}_g\}$ is the (constant) secretion rate of enzyme per unit of gut wall surface area and $\dot{k}_g$ is the decay rate of enzyme activity. The equilibrium amount of enzyme is thus $M_g = \{\dot{J}_g\}\pi L_\lambda L_\phi / \dot{k}_g$ and I assume that this equilibrium is reached fast enough to neglect changes in the concentrations of active enzyme. So the enzyme concentration is larger in smaller individuals because of the more favourable surface area/volume ratio of the slice.

A simple Michaelis–Menten kinetics for the change in the amount of food gives $\frac{d}{dt}M_X = -\dot{k}_X y_{Xg} f_X M_g$, where $f_X = M_X/(M_K^X + M_X)$ is the scaled functional response for digestion. The compound parameter $\dot{k}_X y_{Xg}$ is a rate constant for digestion.

If the absorption of product through the gut wall again follows Michaelis–Menten kinetics, the change of the amount of product in the slice is given by $\frac{d}{dt}M_P = -y_{PX}\frac{d}{dt}M_X - \dot{k}_P y_{Pc} f_P M_c$ with $f_P \equiv M_P/(M_K^P + M_P)$ the scaled functional response for absorption and $M_c = \{M_c\}\pi L_\lambda L_\phi$ the amount of carriers in the gut wall with which the slice makes contact, while the surface-area-specific number of carriers $\{M_c\}$ is taken to be constant. The parameter $\dot{k}_P y_{Pc}$ is a rate constant for absorption. This two-step Michaelis–Menten kinetics for digestion with plug flow has been proposed independently by Dade *et al.* [240].

The digestion process in the slice ends at the gut residence time t_g, given in (7.68), which decreases for increasing ingestion rate and is minimal for the scaled functional response for feeding $f = 1$. The conservation law for mass can be used to deduce that the total amount of product taken up from the slice equals $M_{Pu}(t_g) = ((M_X(0) - M_X(t_g))y_{PX} - M_P(t_g))$, where $M_X(0)$ denotes the amount of food in the slice at ingestion. An ideal gut will digest food completely ($M_X(t_g) = 0$) and absorb all product ($M_P(t_g) = 0$).

To evaluate to what extent food density in the environment and the size of the organism affect digestion, via gut residence time and gut diameter, it is helpful to define the digestion and uptake efficiency $M_{Pu}(t_g)(y_{PX}M_X(0))^{-1}$. For isomorphs, where gut diameter L_ϕ is proportional to whole body length L, the energy uptake from food is independent of body size. A shorter gut residence time in small individuals is exactly compensated by a higher enzyme concentration. This is because the production of short-living enzymes is taken to be proportional to the surface area of the gut. An obvious alternative would be a long-lived enzyme that is secreted in the anterior part of the digestive system. If this part is a fixed proportion of the whole gut length the result of size independence is still valid.

Efficiency depends on food density as long as digestion is not complete. The undigested amount of food $M_X(t_g)$ can be solved implicitly and a relationship results between the rate of enzyme secretion and the ingestion rate of food items by imposing the constraint that the $M_X(t_g)$ must be small. So it relates ingestion rate to food quality.

If the saturation coefficient M_K^X of the digestion process is negligibly small, digestion becomes a zero-th order process, and the amount of food in the slice decreases linearly with time (and distance). This has been proposed by Hungate [537], who

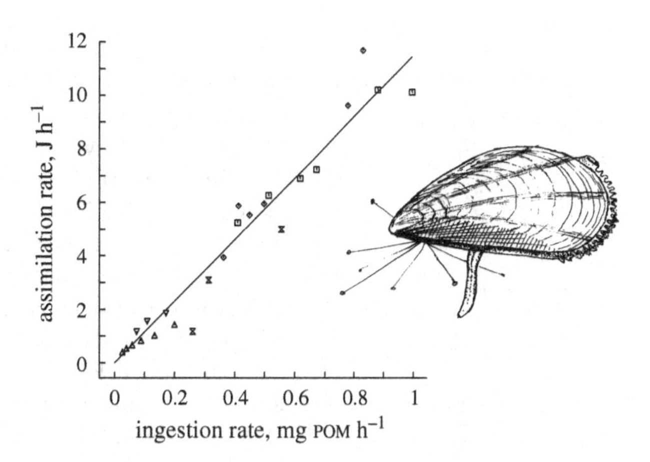

Figure 7.17 The assimilation rate as a function of ingestion rate for mussels *Mytilus edulis* ranging from 1.75 to 5.7 cm. Data from Bayne *et al.* [68, 69], Borchardt [129], Hawkins and Bayne [476] and Kiørboe *et al.* [600], figure from van Haren and Kooijman [459]. All rates are corrected to 15 °C. The fitted line is $\dot{p}_A = \mu_{AX}\dot{J}_{XA}$ with $\mu_{AX} = 11.5\,\text{J mg POM}^{-1}$.

modelled the 42-h digestion of alfalfa in ruminants. Digestion is complete if $t_g > M_X(0)(\dot{k}_X y_{Xg} M_g)^{-1}$, and so $\dot{k}_g \dot{J}_{XAm} < \dot{k}_X y_{Xg} \dot{J}_g$, where $\dot{J}_g$ denotes the total enzyme production by the individual; both $\dot{J}_{XAm}$ and $\dot{J}_g$ are proportional to L^2 for an isomorph.

The above model can be extended to cover a lot of different enzymes in different sections of the gut, without becoming much more complicated, as long as the additivity assumptions of their mode of action and their products hold. Food usually consists of many components that differ in digestibility. Digestion can only be complete for the animal in question if the most resistant component is digested.

The existence of a maximum ingestion rate implies a minimum gut residence time. With a simple model for digestion, it is possible to relate the digestive characteristics of food to the feeding process, on the assumption that the organism aims at complete digestion. The energy gain from ingested food is then directly proportional to the ingestion rate, if prolonged feeding at constant, different, food densities is considered, see Figure 7.17. Should temperature affect feeding in a different way than digestion, the close harmony between both processes would be disturbed, which would lead to incomplete digestion under some conditions.

7.4 Division

If propagation is by division, the situation is comparable to the juvenile stage of species that propagate via eggs. A cell divides as soon as the energy invested in the increase of the state of maturity exceeds a threshold value. If the maturity and somatic maintenance rate coefficients are equal, $\dot{k}_J = \dot{k}_M$, division also occurs at a fixed structural volume, say V_d. Donachie [281] pointed out that in fast-growing bacteria the initiation of DNA duplication occurs at a certain volume V_p, but it requires a fixed and non-negligible amount of time t_D for completion. This makes the volume at division, V_d, dependent on the growth rate, so indirectly on substrate density, because growth proceeds during this period.

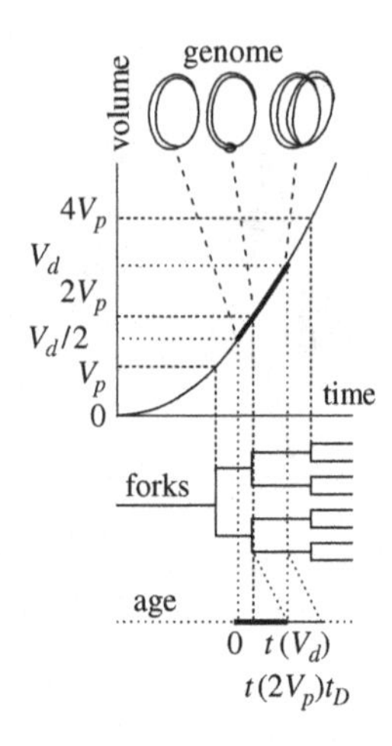

Figure 7.18 A schematic growth curve of a cell, where the fat part is used in steady state. This is the situation for $i = 2$, the number of forks switching between 1 and 3. If $V_d/V_p = 2^i$ (7.69) reduces to $t_D = it(2^i V_p) = it(V_d)$, with $t(2^{i-1}V_p) = 0$, which means that the time required to duplicate DNA is exactly i times the division interval. So, during each cell cycle, a fraction i^{-1} of the genome is duplicated, which implies that $2^i - 1$ DNA duplication forks must be visible during the cell cycle. At the moment that the number of forks jumps from $2^i - 1$ to $2^{i+1} - 1$, the cell divides and the number of forks resets to $2^i - 1$. This is obviously a somewhat simplified account, as cell division is not really instantaneous. If $V_d/V_p \neq 2^i$, the age of the cell at the appearance of the new set of duplication forks somewhere during the cell cycle is $t(2^{i-1}V_p)$, which thus has to be subtracted from $it(V_d)$ to arrive at the genome duplication time.

The mechanism (in eukaryotic somatic cells) of division at a certain size is via the accumulation of two mitotic inducers, cdc25 and cdc13, which are produced coupled to cell growth. (The name for the genes 'cdc' stands for cell division cycle.) If these inducers exceed a threshold level, protein kinase $p34^{cdc2}$ is activated and mitosis starts [807, 817]. During mitosis, the protein kinase is deactivated and the concentration of inducers resets to zero. This mechanism indicates that for shorter interdivision periods, the cell starts a new DNA duplication cycle when its volume exceeds $2V_p$, $4V_p$, $8V_p$, etc. The interdivision time for *Escherichia coli* can be as short as 20 min under optimal conditions, while it takes an hour to duplicate the DNA. The implementation of this trigger is not simple in a dynamic environment. At constant substrate densities, the scaled cell length at division, $l_d \equiv (V_d/V_m)^{1/3}$, and the division interval, $t(l_d) \equiv t_d$, can be obtained directly. When i is an integer such that $2^{i-1} < V_d/V_p \leq 2^i$, V_d can be solved from

$$t_D = it(V_d) - t(2^{i-1}V_p) \tag{7.69}$$

Figure 7.18 illustrates the derivation.

The volume at division V_d can be found numerically when (2.23), (4.20) or (4.27) is substituted for $t(V)$ in (7.69), for isomorphs, V1-morphs or rods, respectively. If the maturity and somatic maintenance rate coefficients are not equal, the size at division must be obtained from equating the cumulative investment in maturation to a threshold level, the size at division generally increases with substrate density, even apart from delays due to DNA duplication.

Many organisms that propagate vegetatively produce spores, and the mother cell dies upon release. The number usually varies between species and growth conditions, and frequently is a power of 2. In the green alga *Scenedesmus* it is usually 4 or 8, but in the waternet, it can be several thousand, see Figure 7.19.

Figure 7.19 The waternet *Hydrodictyon reticulatum* forms a cylindrical sac-like net; the largest recorded size is 114 cm long and 4–6 cm broad [187]. Several thousand spores in each cell grow into small cylindrical cells, which make contact, stick together and form a minute net. The mother cells disintegrate synchronously, each giving birth to a new net. This green alga recently arrived in New Zealand, where it causes water-quality problems in eutrophic fresh waters.

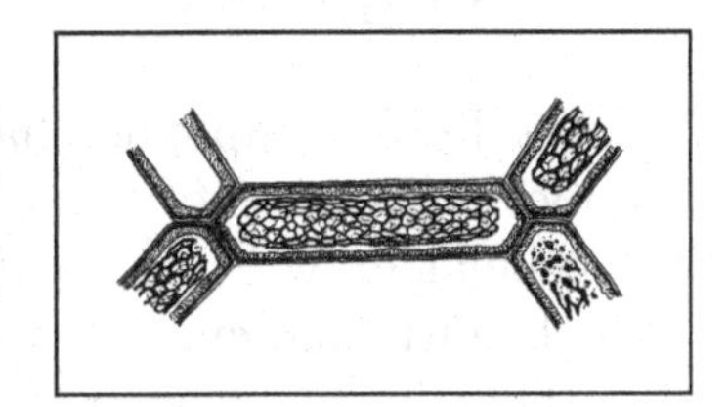

7.5 Cell wall and membrane synthesis

The cell has to synthesise extra cell wall material at the end of the cell cycle. Since the cell grows in length only, the growth of surface material is directly tied to that of cytoplasmic material. Straightforward geometry shows that the change in surface area A of a rod, see {132}, with aspect ratio δ and volume at division V_d, is given by $\frac{d}{dt}A = (16\pi\frac{1-\delta/3}{\delta V_d})^{1/3}\frac{d}{dt}V$. So the energy costs of structure can be partitioned as $[E_G] = [E_{GV}] + \{E_{GA}\}(16\pi\frac{1-\delta/3}{\delta V_d})^{1/3}$, where $\{E_{GA}\}$ denotes the energy costs of the material in a unit surface area of cell wall and $[E_{GV}]$ that for the material in a unit volume of cytoplasm. For reasons of symmetry, it is more elegant to work with $[E_{GA}] \equiv \{E_{GA}\}V_d^{-1/3}$ rather than $\{E_{GA}\}$. The dimensions of $[E_{GV}]$ and $[E_{GA}]$ are then the same: energy per volume. At the end of the cell cycle, when cell volume is twice the initial volume, the surface material should still increase from $A(V_d)$ to $2A(V_d/2) = (1+\delta/3)A(V_d)$. This takes time, of course. If all incoming energy not spent on maintenance is used for the synthesis of this material, the change in surface area is given by $\frac{d}{dt}A = \frac{\dot{k}_E}{g_A}(fA - V_d/V_m^{1/3})$, where $g_A \equiv [E_{GA}]/\kappa[E_m]$. So $A(t) = (A(0) - V_d/fV_m^{1/3})\exp(tf\dot{k}_E/g_A) + V_d/fV_m^{1/3}$. The time it takes for the surface area to reach $(1+\delta/3)V_d^{2/3}$, starting from $A(0) = V_d^{2/3}$, equals

$$t_A = \frac{g_A}{f\dot{k}_E}\left(\ln 2 + \ln\frac{V_\infty - V_d/2}{V_\infty - V_d}\right) \tag{7.70}$$

For the time interval between subsequent divisions, $t(V_d)$ must be added, giving

$$t_d = \frac{g_A}{f\dot{k}_E}\ln 2 + \left(\frac{g_A}{f\dot{k}_E} + \frac{(f+g)V_\infty}{f\dot{k}_E V_d\delta/3}\right)\ln\frac{V_\infty - V_d/2}{V_\infty - V_d} \tag{7.71}$$

The extra time for cell wall synthesis at the caps is not significant for filaments, as their caps are comparatively small. Neither does it play a significant role in unicellular eukaryotic isomorphs, because they do not have cell walls to begin with. The cell's volume is full of membranes in these organisms, so the amount of membrane at the end of the cell cycle does not need to increase as abruptly as in bacteria, where the outer membrane and cell wall (if present) are the only surfaces. Comparable delays occur in ciliates for instance, where the cell mouth does not function during and around cell division.

7.6 Organelle–cytosol interactions and dual functions of compounds

Many cellular compounds have a dual function, as source for energy as well as building blocks. Because of this duality, the fate of metabolites should generally depend on the cellular growth rate. This problem is studied in [660] for the interaction between mitochondria and cytosol. When the cytosol passes pyruvate and ADP to its mitochondria, how do the enzymes of the TCA cycle in the mitochondria 'know' the metabolic needs of the cell in terms of export of the correct mix of intermediary metabolites and ATP? Can these enzymes use the size of the pyruvate flux as the only source of information? The cell uses these mitochondrial products for maintenance and growth; these two processes differ in their metabolic needs and they vary in relative intensity. If the export of metabolites from the mitochondria did not match cellular needs, a growth rate dependent waste flux would result, and such a flux is not observed empirically.

The spatial organisation of the enzymes of the TCA cycle suggests that the answer to the question might involve handshaking protocols: the enzymes of the TCA cycle occur in metabolons, where a single copy of each enzyme makes physical contact with its neighbours in the correct sequence and the metabolons are linked to the (inner) mitochondrial membrane. The answer, detailed in [660], is that using two degrees of freedom, the size of the pyruvate flux to the mitochondria has indeed all the information that the enzymes require to match the export of metabolites from the mitochondria to the cytosol to the cellular needs in terms of maintenance and growth. This holds even at varying substrate levels in cell's environment. The first degree of freedom is that handshaking is a mix between the open and closed protocols, as discussed at {261}, with fixed weight coefficients. The second one is that the metabolons comprise fixed fractions of reserve and structure, in accordance with the strong homeostasis assumption.

The cytosol–mitochondria system can be seen as a complex syntrophic system that might serve as a more general template for interaction between the various modules of the central metabolism, see {111}, and more generally between trophic systems, see {340}.

7.7 Mother–foetus system

Temporarily elevated food intake can be observed in birds preparing for migration or reproduction, in mammals preparing for hibernation or in pregnant mammals [1181]. Vitellogenin in eggs and casein in milk have similar metabolic functions. The three vitellogenin-encoding genes were progressively lost in the Mammalia around 70–30 Ma ago, except in the prototherians, while casein-encoding genes already appeared in the mammalian ancestor some 310–200 Ma ago [143]. The transition from egg-laying to placental development was probably incremental. The prototherians still lay yolky eggs, the marsupial oocyte still has some yolk, that of the eutherians has not.

The foetus is ectothermic, so $\{\dot{J}_{ET}\} = 0$; in placentalians the mother keeps the foetus warm. The mother provides the foetus with a reserve flux $\dot{J}_{EA}^F = \{\dot{J}_{EA}^F\}L_F^2$ through the placenta, which is proportional to the squared length of the foetus, while $\{\dot{J}_{EA}^F\}$ is assumed to be constant, but might depend on the general nutritional status of the mother. This supply bypasses the assimilatory system of the foetus, which only becomes functional in the juvenile stage.

Foetal development is discussed at {63}, so I here focus on the energetics of the mother, as a result of collaboration with Tânia Sousa and Tiago Domingos. The assimilation of the mother is up-regulated during pregnancy, where the surface area of the placenta is added to that of the mother. The idea behind this construct is that for demand systems like most organisms that produce fetuses, food uptake capacity is proportional to the gut surface area, where not only the actual transport of metabolites across the gut surface limits uptake, but also the further processing of the metabolites derived from food to reserve. The transport across the placenta accelerates this process. The assimilation process of the mother of length L and a foetus of length L_F then amounts to

$$\dot{J}_{EA} = f\{\dot{J}_{EAm}\}(L^2 + \delta L_F^2) \tag{7.72}$$

At constant food levels the extra assimilation will match the foetal needs, so

$$f\{\dot{J}_{EAm}\}(L^2 + \delta L_F^2) = f\{\dot{J}_{EAm}\}L^2 + \{\dot{J}_{EA}^F\}L_F^2, \quad \text{so } \{\dot{J}_{EA}^F\} = f\{\dot{J}_{EAm}\}\delta \tag{7.73}$$

The export of reserve from the mother to the foetus is from the somatic branch of the catabolic flux and has priority over the somatic maintenance of the mother unless starvation conditions are so severe that spontaneous abortion occurs.

The parameters in the following specification of the changes in reserve M_E and structure M_V refer to that of the mother:

$$\frac{d}{dt}M_E = \dot{J}_{EA} - \dot{J}_{EC} \quad \text{with } \dot{J}_{EC} = \{\dot{J}_{EAm}\}L^2\frac{ge}{g+e}\left(1 + \frac{L_T + L}{gL_m}\right) \tag{7.74}$$

$$\frac{d}{dt}M_V = (\kappa\dot{J}_{EC} - \dot{J}_{EM} - \dot{J}_{ET} - \dot{J}_{EA}^F)y_{VE} \tag{7.75}$$

with $\dot{J}_{EM} = [\dot{J}_{EM}]L^3$, $\dot{J}_{ET} = \{\dot{J}_{ET}\}L^2$, $L_T = \{\dot{J}_{ET}\}/[\dot{J}_{EM}]$ and $L_m = \kappa\{\dot{J}_{EAm}\}/[\dot{J}_{EM}]$.

The foetus increases the assimilation of the mother, but not the mobilisation rate directly, only indirectly via the increase of the reserve of the mother that is the consequence of the actions of the foetus. This is qualitatively consistent with empirical observations.

Since allocation to the foetal system has priority over somatic maintenance, and so over growth, the foetus might reduce the growth of the mother. If the mother is already fully grown at pregnancy, somatic maintenance might be reduced, e.g. by reducing activity, which typically comprises some 5–10% of the somatic maintenance costs. This too is qualitatively consistent with observations. It is probably no coincidence that species that sport foetal reproduction frequently developed advanced social systems to avoid the translation of a reduction in activity into a reduction in food intake or an increase in hazard rate via an increased risk of being catched by a predator.

Suppose that food density, and so f, as well as M_V, and so L, are constant at the start of pregnancy, so $\kappa \dot{J}_{EC} = \dot{J}_{EM} + \dot{J}_{ET} + \{\dot{J}_{EA}^F\} L_F^2$ and $L = fL_m - L_T$. The mobilisation flux reduces to $\dot{J}_{EC} = \{\dot{J}_{EAm}\} L^2 \frac{f+g}{e+g} e$. We now study the reduction of $[\dot{J}_{EM}]$ that is required to cope with foetal development. The value relative to the pre-pregnancy period and the dynamics of scaled reserve density amounts for $y_{TA} = \{\dot{J}_{ET}\}/\{\dot{J}_{EAm}\}$ and $y_{FA} = \{\dot{J}_{EA}^F\}/\{\dot{J}_{EAm}\}$ to

$$\frac{[\dot{J}_{EM}](t)}{[\dot{J}_{EM}](0)} = \frac{\kappa e(t) \frac{f+g}{e(t)+g} - y_{TA} - y_{FA} L_F^2/L^2}{\kappa f - y_{TA}} \tag{7.76}$$

$$\frac{d}{dt} e = \left(f + f\delta L_F^2/L^2 - e\right) \dot{v}/L \tag{7.77}$$

This dynamics implies a maximum reduction of somatic maintenance costs, and might match the fraction that activity takes in the somatic maintenance costs. In this way foetal reproduction could evolve without substantial metabolic adaptations.

In many placentalians pregnancy is followed by a period of lactation. This product of the mother is also paid from the somatic branch of the mobilised reserve flux, and also has the effect that the assimilation capacity is up-regulated to match this drain of reserve. It is typically a demand-driven process where the flux of milk taken by the baby is proportional to its squared length. The consequence is that the reserve of the mother remains elevated above the normal level during this period. In the marsupials, foetal development is really short and the length at birth is very small, but the lactation period is relatively long.

7.8 Extra life-stages

7.8.1 Pupa and imago

Insects do not grow in the adult stage, called the imago. They are thus much less flexible in their allocation of energy. Holometabolic insects (butterflies, wasps, beetles, flies) have a pupal stage between the juvenile and the adult one, which has a development pattern that strongly resembles that of the embryo or, more specifically, the foetus, since the energy reserves at eclosion are usually quite substantial so that there is hardly any growth retardation due to reserve depletion. This resemblance to a development pattern is not a coincidence because the adult tissue develops from a few tiny imaginal disks, the structural biomass of the larva being first converted to reserves for the pupa. So the initial structural volume of the pupa is very small indeed. Since no energy input from the environment occurs until development is completed, pupal weight decreases, reflecting the use of energy. This can be worked out quantitatively as follows.

As discussed under foetal development, see {63}, growth is given by $\frac{d}{dt} V = \dot{v} L^2$, so that, if temperature is constant, $L(t) = L_0 + t\dot{v}/3$, where L_0 represents the structural length of the imaginal disks. The energy in the reserves decreases because of growth, maintenance and maturation, so that

$$E(t) = E_0 - \frac{[E_G]}{\kappa} V(t) - \frac{[\dot{p}_M]}{\kappa} \int_0^t V(t_1)\, dt_1 \tag{7.78}$$

$$= E_0 - \frac{[E_G]}{\kappa} (V_0^{1/3} + t\frac{\dot{v}}{3})^3 - \frac{[\dot{p}_M]}{4\kappa\dot{v}} (V_0^{1/3} + t\frac{\dot{v}}{3})^4 + \frac{[\dot{p}_M]}{4\kappa\dot{v}} V_0^{4/3} \tag{7.79}$$

Together with the contribution of the structural volume, this translates via (3.2) into the wet weight development

$$W_w(t) = w_E \frac{E_0}{\overline{\mu}_E} - (g w_E [M_{Em}] - d_V) \left(V_0^{1/3} + t\frac{\dot{v}}{3}\right)^3 - \frac{w_E [M_{Em}]}{4 V_m^{1/3}} \left(\left(V_0^{1/3} + t\frac{\dot{v}}{3}\right)^4 - V_0^{4/3}\right) \tag{7.80}$$

Tests against experimental data quickly show that the contribution of the third term, which relates to maintenance losses, is too small to be noticed. So the weight-at-time curve reduces to a three-parameter one. It fits the data excellently, see Figure 7.20. Just as in fetuses, the start of the development of the pupa can be delayed, in a period known as the diapause. The precise triggers that start development are largely unknown.

Imagos do not grow, so if the reserve dynamics (2.10) still applies, the catabolic rate reduces to $\dot{p}_C = \{\dot{p}_{Am}\} L^2 e$ and the survival probability due to ageing is given by (6.6).

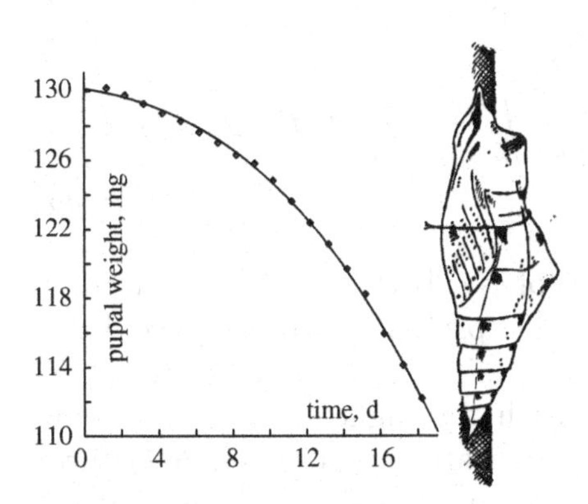

Figure 7.20 The wet weight development of the male pupa of the green-veined white butterfly *Pieris napi* at 17 °C until eclosion, after having spent 4 months at 4 °C. Data from Forsberg and Wiklund [360]. The fitted curve is $W_w(t) = 130.56 - (\frac{7.16+t}{9.61})^3$, with weight in mg and time in days, as is expected from the DEB theory.

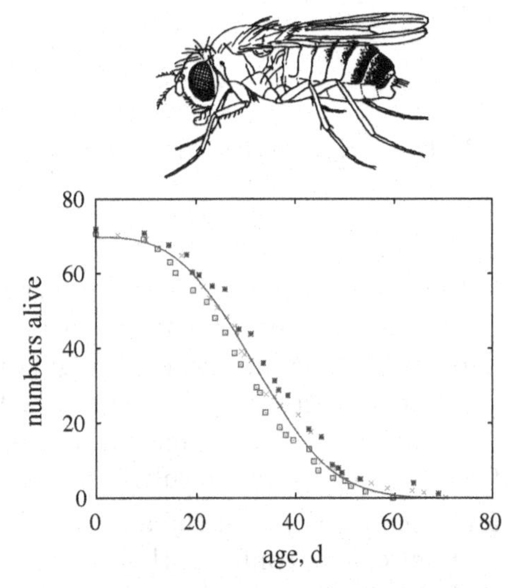

Figure 7.21 The survival curve of the female fruit fly *Drosophila melanogaster* at 25 °C and unlimited food. Data from Rose [990]. The fitted survival curve is $\exp(-(\dot{h}_W t)^3)$ with $\dot{h}_W = 0.0276\,\mathrm{d}^{-1}$.

Experimental results of Rose, Figure 7.21, suggest that this is realistic. He showed that longevity can be prolonged in female fruit flies by selecting offspring from increasingly older females for continued culture [990]. It cannot be ruled out, however, that this effect has a simple nutrient/energy basis with little support for evolutionary theory. Selection for digestive deficiency also results in a longer life span. Reproduction, feeding, respiration and, therefore, ageing rates must be coupled because of the conservation law for energy. This is beautifully illustrated with experimental results by Ernsting and Isaaks [326], who collected carabid beetles *Notiophilus biguttatus* from the field shortly after eclosion, kept them at a high and a low level of food supply (springtail *Orchesella cincta*) at 16 h 20 °C : 8 h 10 °C, and measured survival and egg production. A third cohort was kept at 10 °C at a high feeding level. They showed that the respiration rate of this 4–7-mg beetle is linear in the reproduction rate: $0.84 + 0.041\dot{R}$ in J d^{-1} at 20/10 °C and $0.57 + 0.051\dot{R}$ at 10 °C. This linear relationship is to be expected for imagos on the basis of the aforementioned interpretations. It allows a reconstruction of the respiration rate during the experiment from reproduction data and a detailed description of the ageing process. This is more complex than the Weibull model, because metabolic activity was not constant, despite standardised experimental conditions. The quantitative details are as follows.

The mobilisation rate is subdivided into the maintenance and reproduction costs as

$$\dot{p}_C = \dot{p}_M/\kappa + \dot{R}E_0/\kappa_R = (\dot{k}_M + \dot{R}e_0(g\kappa_R l^3)^{-1})[E_G]V/\kappa \tag{7.81}$$

where the scaled egg cost e_0 is given in (2.42). This gives the hazard rate and survival probability

$$\dot{h}(t) = \frac{1}{2}\ddot{h}_a\dot{k}_M t^2 + \frac{\ddot{h}_a e_0}{\kappa_R g l^3}\int_0^t\int_0^{t_1}\dot{R}(t_2)\,dt_2\,dt_1 \tag{7.82}$$

$$\Pr\{\underline{a}_\dagger > a_p + t|\underline{a}_\dagger > a_p\} = \exp\left(-\frac{1}{6}\ddot{h}_a\dot{k}_M t^3 - \frac{\ddot{h}_a e_0}{\kappa_R g l^3}\int_0^t\int_0^{t_1}\int_0^{t_2}\dot{R}(t_3)\,dt_3\,dt_2\,dt_1\right) \tag{7.83}$$

Although e_0 depends on the reserve energy density of the beetle, and so on feeding behaviour, variations will be negligibly small for the present purpose since food-dependent differences in egg weights have not been found. The low-temperature cohort produced slightly heavier eggs, which is consistent with the higher respiration increment per egg. The estimation procedure is now to integrate the observed $\dot{R}(t)$ three times and to use the result in the estimation of the two compound parameters $\frac{1}{6}\ddot{h}_a\dot{k}_M$ and $\ddot{h}_a e_0(\kappa_R g l^3)^{-1}$ of the survivor function from observations.

Figure 7.22 confirms this relationship between reproduction, and thus respiration, and ageing. The contribution of maintenance in respiration is very small and could not be estimated from the survival data. The mentioned linear regressions of respiration data against the reproduction rate indicate, however, that $\kappa_R g\dot{k}_M l^3/e_0 = 0.84/0.041 = 20.4$ d^{-1} at 20/10 °C or $0.57/0.051 = 11$ d^{-1} at 10 °C. This leaves just one parameter $\ddot{h}_a e_0(\kappa_R g l^3)^{-1}$ to be estimated from each survival curve. The beetles appear to age a bit faster per produced egg at high than at low food density. This might be caused by

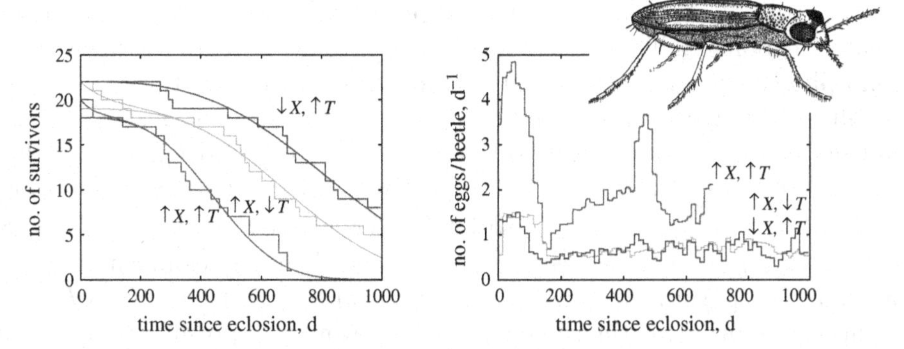

Figure 7.22 The reproduction rate (right) of the carabid beetle *Notiophilus biguttatus* feeding on a high density of springtails at 20/10 °C (↑ X, ↑ T) and at 10 °C (↑ X, ↓ T) and a lower density at 20/10 °C (↓ X, ↑ T). The survival probability of these cohorts since eclosion is given left. Data from Ger Ernsting, pers. comm. and Ernsting and Isaaks [326]. The survival probability functions (left) are based on the observed reproduction rates with estimated parameter $\ddot{h}_a e_0(\kappa_R g l^3)^{-1} = 0.63\,\mathrm{a}^{-2}$ for ↑ X, ↑ T, $0.374\,\mathrm{a}^{-2}$ for ↑ X, ↓ T, $0.547\,\mathrm{a}^{-2}$ for ↓ X, ↑ T. The contribution of maintenance costs to ageing is determined from respiration data. A small fraction of the individuals at the high food levels died randomly at the start of the experiments.

eggs being more costly at high food density, because of the higher reserves at hatching. Another aspect is that, at high food density, the springtails induced higher activity, and so higher respiration, by physical contact. Moreover, the substantial variation in reproduction rate at high food density suggests that the beetles had problems with converting the energy allocated to reproduction to eggs, which led to an increase in κ_R and a higher respiration per realised egg. Note that these variations in reproduction rate are hardly visible in the survival curve, which is due to triple integration. The transfer from the field to the laboratory seemed to induce early death for a few individuals at the high food levels. This is not related to the ageing process but, possibly, to the differences with field conditions.

The Weibull model for ageing with a fixed shape parameter of 3 should not only apply to holometabolic insects, but to all ectotherms with a short growth period relative to the life span. Gatto *et al.* [389] found, for instance, a perfect fit for the bdelloid rotifer *Philodina roseola* where the growth period is about 1/7th of the life span. Notice that constant temperature and food density are still necessary conditions for obtaining the Weibull model.

The presented tests on pupal growth and survival of the imago support the applicability of the DEB theory to holometabolic insects, if some elementary facts concerning their life history are taken into account. This suggests new interpretations for experimental results.

7.8.2 Metamorphosis in juvenile fish

In collaborative work with Laure Pecquerie [876], we extend the standard DEB model for anchovy *Engraulis encrasicolus* by splitting the juvenile stage, separated by metamorphosis, to accommodate the empirical observation that length increases approximately exponentially with age during the juvenile I stage. Pigmentation occurs at metamorphosis, which marks the metamorphosis event. In the embryo, juvenile II and adult stages, anchovy is isomorphic, but in the juvenile I state V1-morphic. Stage transitions occur at scaled maturity values U_H^b, U_H^j and U_H^p, where the structural volume has values V_b, V_j and V_p, respectively. The shape correction function is $\mathcal{M}(V) = (\min(\max(V, V_b), V_j)/V_b)^{1/3}$, so for $V < V_b$ we have $\mathcal{M}(V) = 1$, for $V > V_j$ we have $\mathcal{M}(V) = (V_j.V_b)^{1/3}$ and between V_b and V_j, $\mathcal{M}(V)$ increases linearly.

We assume that physical length L_w relates to volumetric length L as $L_w = L/\delta_\mathcal{M}$, for constant $\delta_\mathcal{M}$. In principle the value for $\delta_\mathcal{M}$ in the embryo, juvenile II and adult stage could be different, and many possibilities for the relationship exists for the juvenile I, since V1-morphy only concerns the relationship between surface area and structural volume.

For $U_H < U_H^p$ and $\dot{r} = \dot{v}^* \frac{e/L - (1+L_T/L)/L_m^*}{e+g}$ we arrive at:

$$\frac{d}{dt}e = (f - e)\frac{\dot{v}^*}{L}; \quad \frac{d}{dt}L = \frac{\dot{r}}{3}; \quad \frac{d}{dt}U_H = (1-\kappa)eL^2\frac{g^* + L/L_m}{g+e} - \dot{k}_J U_H \qquad (7.84)$$

with $\dot{v}^* = \dot{v}$ and $f = 0$ for the embryo but otherwise $\dot{v}^* = \dot{v}\mathcal{M}(V)$ and $g^* = g\mathcal{M}(V)$. Furthermore $L_m^* = \frac{\dot{v}^*}{\dot{k}_M g}$ and $L_m = \frac{\dot{v}}{\dot{k}_M g}$. Notice that $\dot{v}^*$ changes in a continuous (but not differentiable) way across stage transitions, as does the reserve turnover rate $\dot{v}^*/L$. For $U_H > U_H^p$ we have $\frac{d}{dt}U_H = 0$, and allocation to reproduction occurs. Notice that for $U_H^j = U_H^b$ we have no juvenile I stage; the individual then remains isomorphic during all stages, with $\dot{v}^* = \dot{v}$ and $\mathcal{M}(V) = 1$. The threshold U_H^j is the only parameter that we introduced for the change in shape.

We also have $L(0) = 0$, $U_H(0) = 0$ and $e(a_b) = f$, where age at birth a_b is given implicitly by $U_H(a_b) = U_H^b$. All rate parameters depend on temperature, including the parameter $\{\dot{J}_{EAm}\}$ or $\{\dot{p}_{Am}\}$ with which we scaled the cumulated investment into maturity; for the scaling, however, we use the constant value that applies to some reference temperature and avoid complex forms of dynamic scaling.

7.9 Changing parameter values

Parameter values are usually constant, by definition, but environmental and internal factors might make them vary in time; I discussed effects of temperature {17}, toxicants and parasites {234} and diurnal cycles {118} as external factors, and changes in shapes {124, 284, 287}, diet choices {185} and during pregnancy {282} as internal factors. Quite a few other causes exist at the interface between internal and external factors; starvation can induce changes energy allocation through κ, and prolonged starvation can invoke drastic qualitative changes {118}.

Although not worked out quantitatively, parameter values can be coupled to the ageing process, where maintenance, reproduction and feeding usually tend to decrease

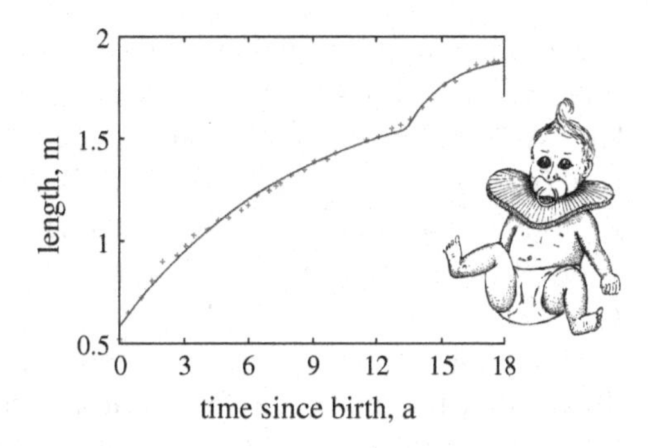

Figure 7.23 Length-at-age of man, de Montbeillard's son, in the years 1759–1777. Data from Cameron [184]. The curve is the von Bertalanffy one with an instantaneous change of the ultimate length from 177 cm to 190 cm and of the von Bertalanffy growth rate from 0.123 a^{-1} to 0.549 a^{-1} at 13.3 a since birth.

with age. Some other types of changes are briefly discussed in the next subsections, to reveal possible interpretations of data in the light of the DEB theory. They are arranged on an increasing timescale.

7.9.1 Changes at puberty

Growth curves suggest that some species, e.g. humans, change the partition coefficient κ and the maximum surface-area-specific assimilation rate $\{\dot{p}_{Am}\}$ at puberty in situations of food abundance; see Figure 7.23.

The reasoning presented at {68} resulted in $\dot{k}_M = 0,5\,\mathrm{a}^{-1}$ and $g = 2.79$ for humans, assuming absence of jumps in parameter values at birth. The scaled length 58/177 = 0.33 from Figure 7.23 is in good agreement with that estimated from weights. Assuming that $L_T = 0$ for western humans (because of clothing) and food unrestrictive ($f = 1$), the (juvenile) human energy conductance is $\dot{v} = L_m \dot{k}_M g = 2.57\,\mathrm{m\,a}^{-1}$ or $6.76\,\mathrm{mm\,d}^{-1}$ in the basis of physical length at 37 °C.

Figure 7.23 shows that at puberty, both the maximum length $L_m = \frac{\dot{v}}{\dot{k}_M g}$ and the von Bertalanffy growth rate $\dot{r}_B = \frac{\dot{k}_M g}{3(1+g)}$ make an upward jump, by a factor 190/177 = 1.1 and 0.549/0.123 = 4.46 respectively, while they are negatively correlated (these data do not determine the ultimate length very well).

Indicating the values for juveniles and adults with j and a, respectively, this translates to $\frac{1}{1.1}\frac{\dot{v}_a}{\dot{v}_j} = \frac{\dot{k}_M^a g_a}{\dot{k}_M^j g_j} = 4.46\frac{1+g_a}{1+g_j}$. Additional information is required to interpret these changes to those of primary parameters. Some possibilities can be excluded, however; if $\dot{k}_M^j = \dot{k}_M^a$ we would have $\frac{g_a}{1+g_a} = 4.46\frac{g_j}{1+g_j}$, which is inconsistent with the observation that $g_j = 2.79$. So $\dot{k}_M$ must be affected. Deviations from strict isomorphism may affect estimates.

7.9.2 Suicide reproduction

Like *Oikopleura*, salmon, eel and most cephalopods die soon after reproduction. The distribution of this type of behaviour follows an odd pattern in the animal kingdom. Tarantula males die after first reproduction, but the females reproduce frequently and can survive for 20 years. Death after first reproduction does not follow the Weibull-type ageing pattern and probably has a different mechanism. Because the (theoretical) asymptotic size is not approached in cephalopods, they also seem to follow a different growth pattern. I believe, however, that early death, not the energetics, makes them different from iteroparous animals. The arguments are as follows.

Starting not close to zero, the surface area in von Bertalanffy growth is almost linear in time across a fairly broad range of surface areas. This has led Berg and Ljunggren [84] to propose an exactly linear growth of the surface area for yeast until a certain threshold is reached; see Figure 2.14. Starting from an infinitesimally small size, however, which is realistic for most cephalopods, length is almost linear in time, so the volume increases with cubed time: $V(t) = (\frac{\dot{v}ft}{3(f+g)})^3$. Over a small trajectory of time, this closely resembles exponential growth, as has been fitted by Wells [1235], for instance.

Squids show a slight decrease in growth rate towards the end of their life (2 or perhaps 3 years [1130]), just enough to indicate the asymptotic size, which happens to be very different for female and male in *Loligo pealei*. It will be explained in the section on primary scaling relationships, {300}, that the costs of structure $[E_G]$ in the von Bertalanffy growth rate $\dot{r}_B = \dfrac{[\dot{p}_M]/3}{[E_G] + \kappa[E_m]f}$ hardly contribute in large-bodied species because they are independent of asymptotic length, while maximum energy density is linear therein. So $\dot{r}_B \simeq \dfrac{[\dot{p}_M]}{3\kappa[E_m]f}$. The product $\dot{r}_B V_\infty^{1/3} \simeq \dot{v}/3$ should then be independent of ultimate size. On the basis of data provided by Summers [1130], the product of ultimate length and the von Bertalanffy growth rate was estimated to be 0.76 and 0.77 $\mathrm{dm\,a^{-1}}$ for females and males respectively. The equality of these products supports the interpretation in terms of the DEB model. The fact that the squids die well before approaching the asymptotic size only complicates parameter estimation.

A large (theoretical) ultimate volume goes with a large maximum growth rate. If the maximum growth rates of different species are compared on the basis of size at death, the octopus *Octopus cyana* grows incredibly quickly, as argued by Wells [1235]. Assuming that the maximum growth rate is normal, however, a (theoretical) ultimate volume can be inferred by equating $\dot{r}_B V_\infty^{1/3}$ for the octopus to that for the squid, after correction of temperature differences. Summers did not indicate the temperature appropriate for the squid data, but on the assumption that it has oscillated between 4 and 17 °C and that $T_A = 12.5$ kK, the growth rate has to be multiplied by 9.3 to arrive at the temperature that Wells used, i.e. 25.6 °C. The data of Wells indicate a maximum growth rate of $\frac{4}{9}\dot{r}_B V_m = 25.5\ \mathrm{dm^3\,a^{-1}}$. The ultimate volume is thus $\left(\frac{9\times 25}{4\times 9.3\times 0.77}\right)^{3/2} =$ 22 $\mathrm{dm^3}$ for the octopus. This is three times the volume at death.

While emergency reproduction, see {123}, is typically a response to an unpredictable food shortage, suicide reproduction can sometimes be a response to predictable food shortage, and is integrated in the life history of the species.

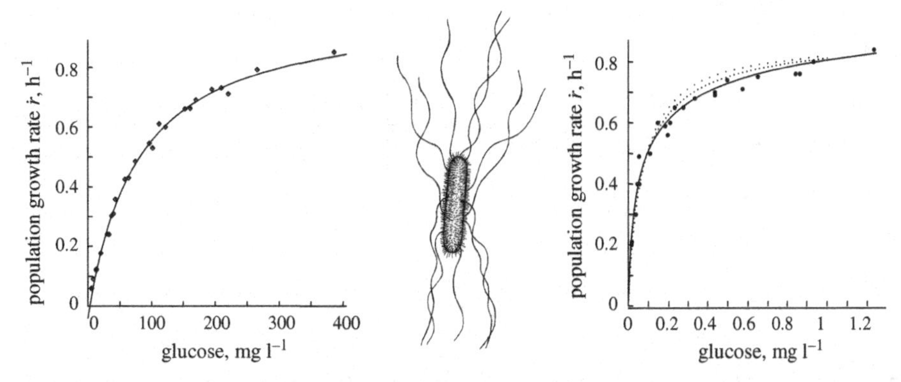

Figure 7.24 The population growth rate of *Escherichia coli* on glucose-limited media. Schulze and Lipe's culture [1034], left, had been exposed to glucose limitation just prior to the experiment, while that of Senn [1042], right, had been pre-adapted for a period of 3 months. The coarsely stippled curve in the right graph does not account for a time lapse between sampling from the continuous culture and measurement of the concentration of glucose [815]; the finely stippled one accounts for a time lapse of 0.001 h; the drawn one for a time lapse of 0.01 h.

7.9.3 Adaptation of uptake capacity

Figure 7.24 illustrates that prolonged exposure to limiting amounts of glucose eventually results in substantially improved uptake of glucose from the environment by bacteria. The difference in saturation constants between a 'wild' and an adapted population can amount to a factor of 1000. The outer membrane adapts to the specialised task of taking up a single type of substrate, which may jeopardise a rapid change to other substrates. This adaptation process takes many cell division cycles, as is obvious from the measurement of population growth rates, which itself takes quite a few division cycles.

When substrate is absent for a sufficiently long period of time, the metabolic machinery that deals with handling those substrates can be deleted from DNA. This is a route to speciation, which leads to permanent changes in parameter values.

7.9.4 Diauxic growth: inhibition and preference

Diauxic growth is the property of populations of micro-organisms to grow first more or less logistically to a certain level in a batch culture, using one substrate only, and then resume growth to a second higher level, using another substrate. So the use of the second substrate is delayed until the first substrate is exhausted. This behaviour is species as well as substrate-combination specific.

Carriers in the outer membrane typically only transport particular substrates from the environment into the cell. This comes with the requirement to regulate gene expression for carriers of substitutable substrates to match the substrate availability in the environment. Data strongly suggest that allocation to the assimilation machinery is a fixed fraction of the utilised reserve flux, and that the expression of one gene for a carrier

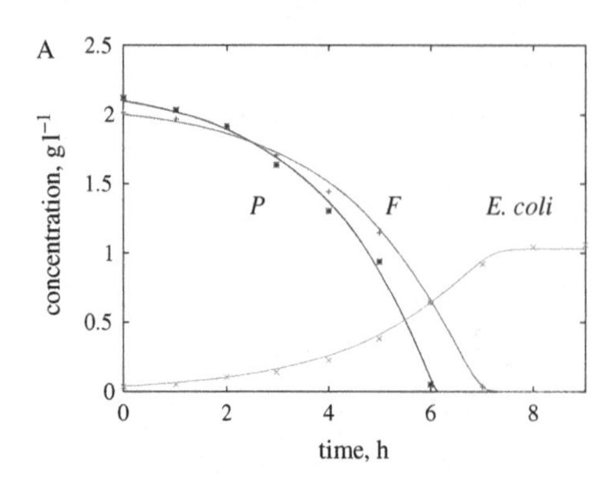

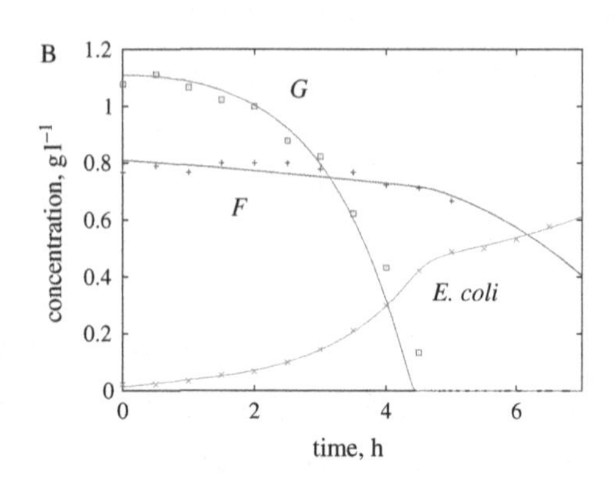

Figure 7.25 The uptake of fumarate (F) and pyruvate (P) (see Figure A), and of fumarate (F) and glucose (G) (see Figure B) by *E. coli* K12 in a batch culture. Data from Narang *et al.* [826]. Parameters: saturation coefficients (g l^{-1}) $K_F = 0.089$, $K_P = 0.012$, $K_G = 0.013$; yield coefficients (g g^{-1} dry weight) $y_{EF} = 0.577$, $y_{EP} = 0.015$, $y_{EG} = 0.446$, $y_{EV} = 1.2$ (fixed); max. specific uptake rates (g(h g dry weight)$^{-1}$), $j_{FAm} = 1.138$, $j_{PAm} = 40.15$, $j_{GAm} = 2.59$; reserve turnover rate (h^{-1}) $\dot{k}_E = 4.256$; maintenance rate coefficient (h^{-1}) $\dot{k}_M = 0$ (fixed); preference parameter (-) $w_P = 0.941 w_F$ for pyruvate *versus* fumarate; $w_G = 12.15 w_F$ for glucose versus fumarate; background expression (h^{-1}) $\dot{h} = 0$ (fixed). Initial conditions: (A) $F(0) = 2.0$ g l^{-1}, $P(0) = 2.1$ g l^{-1}, *E. coli*(0) $= 0.037$ g l^{-1}, $\kappa_F(0) = 0.96$, $m_E(0) = 0.288$ g g^{-1} dry weight; (B) $F(0) = 0.81$ g l^{-1}, $G(0) = 1.11$ g l^{-1}, *E. coli*(0) $= 0.013$ g l^{-1}, $\kappa_F(0) = 0.99$, $m_E(0) = 1.3$ g g^{-1} dry weight

inhibits in some cases the expression of another gene. Inhibition strength is linked to the workload of the carriers, see {204}. This regulation mechanism has similarities to that of differentiation.

Figure 7.25 illustrates this for two data sets on the uptake by *E. coli* K21 of fumarate and pyruvate and of fumarate and glucose (from [825]). Unlike pyruvate, glucose suppresses the uptake of fumarate. The background expression of carrier synthesis and the maintenance requirements were set to zero, because the data provide little information on this. The yield of structure on reserve was fixed (because the data give no information on biomass composition). The data were fitted simultaneously to ensure that the uptake parameters for fumarate and the reserve turnover rates are identical in the two data sets (so removing degrees of freedom). Apart from the initial conditions, 12 parameters were estimated for six trajectories. The fit is quite good, despite the constraint for the parameter values for fumarate to be identical. The data in Figure 7.25A clearly show continued growth after depletion of substrates, which requires reserves to capture; this cannot be done with e.g. a Monod model.

Although their derivation has been set up slightly differently, the supply formulation for inhibition is used in [138] to model substrate preference and diauxic growth in micro-organisms, while [1165] used a demand formulation, see Figure 3.8. The use of genes coding for substrate-specific carriers is here linked to the use of carriers; the expression of one gene inhibits the expression of the other. When embedded in a batch culture, the uptake rate of substrates S_1 and S_2 by biomass X (of V1-morphs) with reserve density m_E in a batch culture is given by

$$\frac{d}{dt}S_i = -j_{S_iA}X; \quad j_{S_iA} = \kappa_{S_i} j_{S_iAm} f_{S_i}; \quad f_{S_i} = \frac{S_i}{S_i + K_{S_i}} \quad \text{for } i = 1, 2 \tag{7.85}$$

$$\frac{d}{dt}X = \dot{r}X; \quad \dot{r} = \frac{\dot{k}_E m_E - \dot{k}_M}{m_E + y_{EV}} \tag{7.86}$$

$$\frac{d}{dt}m_E = y_{ES_1} j_{S_1A} + y_{ES_2} j_{S_2A} - m_E \dot{k}_E \tag{7.87}$$

$$\frac{d}{dt}\kappa_{S_1} = (\dot{r} + \dot{h})\left(\frac{w'_{S_1}\kappa_{S_1}f_{S_1}}{w'_{S_1}\kappa_{S_1}f_{S_1} + w'_{S_2}\kappa_{S_2}f_{S_2}} - \kappa_{S_1}\right); \quad \kappa_{S_2} = 1 - \kappa_{S_1} \tag{7.88}$$

where j_{*Am} is the maximum specific uptake flux of substrate $*$, f_* is the scaled functional response and K_* the half-saturation coefficient for substrate $*$. The coefficient y_{E*} is the yield of reserve E on substrate $*$, $\dot{k}_E$ the reserve turnover rate, $\dot{k}_M$ the maintenance rate coefficient and $\dot{r}$ the specific growth rate. The fraction κ_{S_1} between 0 and 1 quantifies the relative gene expression for the carrier of substrate S_1 and w'_{S_1} the inhibition of the expression of the gene for the carrier of substrate S_1 by the expression of the gene for the carrier of substrate S_2; without loss of generality we can assume that $1 = w'_{S_1} + w'_{S_2}$. Notice that a single substrate induces full gene expression ($\kappa_{S_1} \to 1$ if $f_{S_2} = 0$). The typically very low background expression rate $\dot{h}$ serves an antenna function for substrates that have been absent for a long time. This readily extends to an arbitrary number of substrates. See Figure 7.25 and [138] for an illustration of the application of this theory.

7.10 Summary

A variety of aspects are discussed to show how the univariate DEB models can be applied and extended to deal with details of energetics.

- Behavioural traits can be incorporated in a systematic way using SUs and affect the functional response due to conservation of time.
- Transport processes in the environment can also modify functional responses, and build up spatial structures.
- Constraints on digestion are discussed for an efficiency that is independent of body size and food density.
- Intra- and extracellular digestion are compared. Extracellular digestion is greatly improved by living together with other individuals.
- The synthesis of material that relates to surface area requires a waiting time that can be expressed in terms of energetic costs, and affects the population growth rate.

- The link between models for pathways and DEB models is discussed. Appropriate handshaking protocols and links of enzyme abundance to the amounts of reserve and structure are required to match both models.
- The number of life-stages can be extended to include changes in parameter values of a particular type. Pupae and imagos are discussed as examples of modifications in life-stage patterns, with the implied consequences for reproduction and ageing.
- Changes of parameter values are discussed that can occur at the various timescales; the evolutionary timescale involves changes that lead to speciation, as discussed in Chapter 10.
- Adaptation of the uptake of particular substrates is discussed, in relation to diauxic growth.

8

Covariation of parameter values

The range of body sizes is enormous. Prokaryotes span a huge cellular size range; the largest is the colourless sulphur bacterium *Thiomargarita namibiensis* with a cell volume of 2×10^{-10} m^3 [1033], the smallest is *Pelagibacter ubique* at 10^{-20} m^3. This small size has the remarkable implication that it has less than a single free proton in its cell if its internal pH is 7 as is typical for bacteria, see {430}. This has peculiar consequences for the molecular dynamics of metabolism. A typical bacterium with full physiological machinery has a volume of about 0.25×10^{-18} m^3. The blue whale has a volume of up to 135 m^3. A sequoia may even reach a volume of 2000 m^3, but one can argue that it is not all living matter. Ironically, the organism with the largest linear dimensions is usually classified as a 'micro-organism': the fungus *Armillaria bulbosa* is reported to occupy at least 15 hectares and exceeds 10 Mg or 10 m^3 [1080]. The factor between the volumes of bacterium and whale is 5.4×10^{20}, that between the volume a water molecule occupies in liquid water and that of a bacterium is 'only' 10^{10}. The interdivision interval of a bacterium can be as short as 20 min; the life span of whales may exceed a century [350], while some plants live for several millennia.

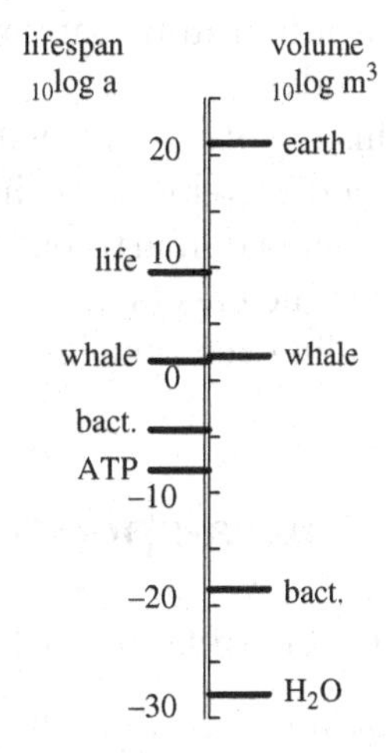

The maximum body length, $L_m = \kappa\{\dot{p}_{Am}\}/[\dot{p}_M]$, see {51}, the maximum reserve density $[E_m] = \{\dot{p}_{Am}\}/\dot{v}$, see {39}, and the bioconcentration coefficient $P_{Vd} = i_Q/\dot{k}_e$, see {223}, all share a common property: they are ratios of in- and outgoing fluxes. This property plays a key role in theory for the covariation of parameter values among different individuals or chemical compounds. The remarkable property of this theory is that it is fully implied already by the standard DEB model and the one-compartment model. No new assumptions are required. This is remarkable because the assumptions behind these models are about mechanisms, and these mechanisms turn out to imply rules for the covariation of parameter values.

This chapter deals with theory of covariation of parameter values, also known as body size scaling relationships and quantitative structure–activity relationships (QSARs).

Both terms don't catch the essence, however. Body size is not a cause but a consequence of underlying processes, and QSARs are about effects of chemical compounds, but the key is in transport. The covariation is also tested empirically in this chapter, and is found to be consistent with the predictions. This is perhaps the strongest empirical support for the assumptions behind DEB theory.

The ecological literature is full of references to what is known as r and K strategies, as introduced by MacArthur and Wilson [725]. The symbol r refers to the population growth rate and K to the carrying capacity; these two parameters occur in the logistic growth equation, which plays a central role in ecology. Under the influence of Pianka [891], organisms are classified relative to each other with respect to a number of coupled traits, the extremes being an 'r-strategist' and a 'K-strategist'. Many of these traits can now be recognised as direct results of body size scaling relationships for ecophysiological characteristics. The search for factors in the environment selecting for r or K strategies can, as a first approximation, be translated into that for factors selecting for a small or large body size.

This chapter starts with intra-species variations of parameter values of the standard DEB model, followed by inter-species variations. Then I discuss the parameter variation of the one-compartment model and its variations (e.g. the film models), followed by that for the effects of chemical compounds. Finally I discuss interactions between body size scaling relationships and QSARs.

8.1 Intra-specific parameter variations

8.1.1 Genetics and parameter variation

The parameter values undoubtedly have a genetically determined component, which can to some extent be modulated phenotypically. As I hopefully made clear, the processes of feeding, digestion, maintenance, growth and reproduction are intimately related. They involve the complete cellular machinery. Although mechanisms for growth which involve just one gene have been proposed [282], the many contributing processes make it likely that thousands are involved. This restricts the possibilities of population genetic theories dealing with auxiliary characters that do not have a direct link with energetics. (This is not meant to imply that such theories cannot be useful for other purposes.)

The parameter values for different individuals are likely to differ somewhat. Differences in ultimate volume at constant food density testify to this. To what extent this has a genetic basis is not clear, but the heredity of size in different races of dogs and transgenic mice and turkeys reveals the genetic basis of growth and size. Phenotypic factors and maternal effects exist as well. A statistical implication is that parameter estimates should be based on measurements on a single individual, rather than on means: the mean of von Bertalanffy curves with different parameters is not a von Bertalanffy curve. This problem obviously grows worse with increasing scatter.

8.1.2 Geographical size variations

The energy constraints on distribution, apart from physical barriers to migration, consist primarily of the availability of food in sufficient quantity and quality. The second determinant is the temperature, which should be in the tolerance range for the species for a long enough period. If it drops below the lower limit, the species must adopt adequate avoidance behaviour (migration, dormancy) to survive. By 'choosing' a convenient value for the lower edge of the temperature tolerance range, in combination with a high Arrhenius temperature at this edge, see {22}, a species can send itself into torpor during the cold season, which typically coincides with lack of food.

The minimum food density for survival relates to metabolic costs. If an individual is able to get rid of all other expenses, mean energy intake should not drop below $[\dot{p}_M]V + \{\dot{p}_T\}V^{2/3}$ for an individual of volume V, so the minimum ingestion rate, known as the maintenance ration, should be $\frac{\{\dot{J}_{Xm}\}}{\{\dot{p}_{Am}\}}([\dot{p}_M]V + \{\dot{p}_T\}V^{2/3})$. For a 3-mm daphnid at 20 °C this minimum ingestion rate is about six cells of *Chlorella* (diameter 4 μm) per second [632]. The minimum scaled food density X/K is $x_s = \frac{l_T + l}{1/\kappa - l_T - l}$.

This minimum applies to mere survival for an individual. For prolonged existence, reproduction is essential to compensate at least for losses due to ageing. The ultimate volumetric length, $fV_m^{1/3} - V_T^{1/3}$, should exceed that at puberty, $V_p^{1/3}$, which leads to the minimum scaled food density $x_R = \frac{l_T + l_p}{1/\kappa - l_T - l_p}$.

Several factors determine food density. It is one of the key issues of population dynamics. The fact that von Bertalanffy growth curves frequently fit data from animals in the field indicates that they live at relatively constant (mean) food densities. In the tropics, where climatic oscillations are at a minimum, many populations are close to their 'carrying capacity', i.e. the individuals produce a small number of offspring, just enough to compensate for losses. It also means that the amount of food per individual is small, which reduces them in ultimate size. Towards the poles, seasonal oscillations divide the year into good and bad seasons. In bad seasons, populations are thinned, so in the good seasons a lot of food is available per surviving individual. Breeding periods are synchronised with the good seasons, such that the growth period coincides with food abundance. So food availability in the growth season generally increases with latitude [678]. The effect is stronger towards the poles, which means that body size tends to increase towards the poles for individuals of one species. Figure 8.1 gives two examples. Other examples are known from, for instance, New Zealand including extinct species such as the moa *Dinornis* [173].

Geographical trends in body sizes can easily be distorted by regional differences in soils, rainfall or other environmental qualities affecting (primary) production. Many species or races differ sufficiently in diets to hamper a geographically based body size comparison. For example, the smallest stoats are found in the north and east of Eurasia, but in the south and west of North America [597]. The closely related weasels are largest in the south, both in Eurasia and in North America.

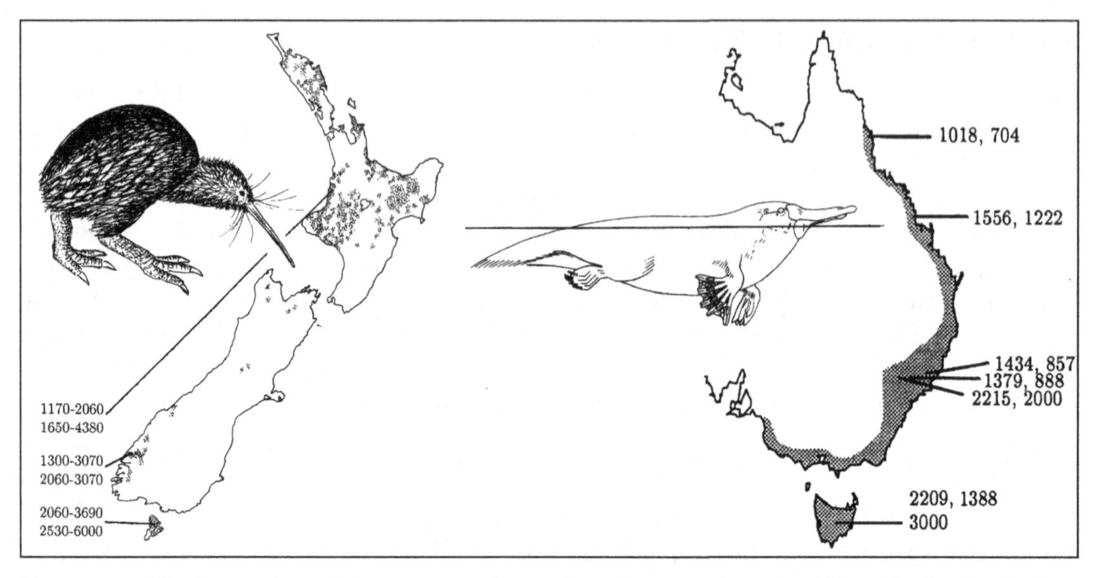

Figure 8.1 The brown kiwi *Apteryx australis* in the subtropical north of New Zealand is lighter than in the temperate south. The numbers give ranges of weights of male and female in grams, calculated from the length of the tarsus using a shape coefficient of $\delta_{\mathcal{M}} = 1.817 \text{ g}^{1/3} \text{cm}^{-1}$. Data from Fuller [381]. A similar gradient applies to the platypus *Ornithorhynchus anatinus* in Australia. The numbers give the mean weights of male and female in grams as given by Strahan [1123]. The DEB theory relates adult weights to food availability and so to the effect of seasons. This interpretation is supported by the observation that platypus weight increases with seasonal differences at the same latitude in New South Wales. The seasons at the three indicated sites are affected by the Great Dividing Range in combination with the easterly winds.

Patterns like these can only be understood after a careful analysis of the food relationships. Simpson and Boutin [1065] observed that muskrats *Ondatra zibethicus* of the northern population in Yukon Territory are smaller and have a lower reproduction rate than the southern population in Ontario. They could relate these differences to feeding conditions, which were better for the southern population, in this case.

Bergmann [85] observed the increase in body size towards the poles in 1847, but he explained it as an effect of temperature. Large body size goes with small surface area/volume ratios, which makes endotherms more efficient per unit body volume. This explanation has been criticised [774, 1025, 1037]. It is indeed hard to see how this argument applies in detail. Animals do not live on a unit-of-body-volume basis, but as a whole individual [774]. It is also hard to see why the argument applies within a species only, and why animals with body sizes as different as mice, foxes and bears can live together in the Arctic. The tendency to increase body size towards the poles also seems to occur in ectotherms, see e.g. [1185] for flatfish, which requires a different explanation. The DEB theory offers an alternative explanation for the phenomenon because of the relationship between food availability and ultimate body volume. Temperature alone works in the opposite direction within this context. If body temperature has to be maintained at some fixed level, individuals in the Arctic are expected to be smaller while living at the same food density, because they have to spend more energy on heating, which reduces their growth potential. The effect will, however, be small since insulation tends to be better towards the poles.

It is interesting to note that species with distribution areas large enough to cover climatic gradients generally tend to split up in isolated races or even subspecies. This can be seen as a form of adaptation, see {291}. The differences in ultimate size have usually become genetically fixed. This is typical for 'demand' systems where regulation mechanisms set fluxes at predefined values which are obtained through adaptation. Within the DEB theory this means that the parameter values are under genetic control and that the minimum food level at which survival is possible is well above the level required for maintenance. The matter is taken up again on {326}.

Organisms that cannot shrink ultimately reach a size that corresponds with the highest scaled reserve density in a periodically varying food environment [307] and this highest reserve density not only depends on the highest food density, but also on the frequency of food density oscillations relative to the energy conductance rate. As a result, a small organism in the tropics may experience its relatively constant environment as more volatile than a large organism living closer to one of the poles. If the abundant period is short relative to $L/\dot{v}$, smaller individuals can reach a substantially higher scaled reserve density than larger conspecifics, which may help the smaller ones to endure starvation better than the larger ones [307, Figs 3, 8].

Chapelle and Peck [198] explain the Bergmann rule by dioxygen availability; they reported a linear relationship of the upper 5-percentile of the length distribution for amphipods and the dioxygen concentration in the water. They compare inter-specifically, however, and did not consider food availability, which might well also covary with the dioxygen concentration.

8.2 Inter-specific parameter variations

The tendencies of covariation discussed in the next few sections can be inferred on the basis of general principles of physical and chemical design. On top of these tendencies, species-specific adaptations occur that cause deviations from the expected tendencies. The application of body size scaling relationships to marine fish and molluscs is discussed in [189, 1185, 1186]. A general problem in body size scaling relationships is that large-bodied species frequently differ from small-bodied species in a variety of ways, such as behaviour, diet, etc. These life styles require specific adaptations, which hamper simple inter-species comparison. McMahon [772] applied elasticity arguments to deduce allometric scaling relationships for the shape of skeletal elements. Godfrey *et al.* [416] demonstrated, for mammals, that deviations from a simple geometrical upscaling of skeletal elements are due to size-related differences in life styles.

Primary scaling relationships are about covariations of primary parameter values of the standard DEB model, secondary scaling relationships concern (simple) functions of

Table 8.1 The 12 primary parameters of the standard DEB model in a time–length–energy and a time–length–(dry)mass frame and typical values among species at 20 °C with maximum length $L_m = zL_m^{\text{ref}}$ for a dimensionless zoom factor z and $L_m^{\text{ref}} = 1$ cm. The two frames relate to each other via $\overline{\mu}_E = 550\,\text{kJ mol}^{-1}$ and $[M_V] = 4\,\text{mmol cm}^{-3}$. The typical value for the Arrhenius temperature $T_A = 8$ kK. See the text for a discussion of the values.

specific searching rate	$\{\dot{F}_m\}$	$6.5\,\text{l cm}^{-2}\,\text{d}^{-1}$	$\{\dot{F}_m\}$	$6.5\ \text{l cm}^{-2}\ \text{d}^{-1}$
assimilation efficiency	κ_X	0.8	y_{EX}	$0.8\,\text{mol mol}^{-1}$
max spec. assimilation rate	$\{\dot{p}_{Am}\}$	$22.5\,z\,\text{J cm}^{-2}\text{d}^{-1}$	$\{\dot{J}_{EAm}\}$	$0.041\,z\,\text{mmol cm}^{-2}\text{d}^{-1}$
energy conductance	$\dot{v}$	$0.02\,\text{cm d}^{-1}$	$\dot{v}$	$0.02\ \text{cm d}^{-1}$
allocation fraction to soma	κ	0.8	κ	0.8
reproduction efficiency	κ_R	0.95	κ_R	0.95
volume-spec. som. maint. cost	$[\dot{p}_M]$	$18\,\text{J cm}^{-3}\text{d}^{-1}$	$[\dot{J}_{EM}]$	$0.033\,\text{mmol cm}^{-3}\text{d}^{-1}$
surface-spec. som. maint. cost	$\{\dot{p}_T\}$	$0\,\text{J cm}^{-2}\text{d}^{-1}$	$\{\dot{J}_{ET}\}$	$0\,\text{mol cm}^{-2}\text{d}^{-1}$
maturity maint. rate coeff.	$\dot{k}_J$	$0.002\,\text{d}^{-1}$	$\dot{k}_J$	$0.002\ \text{d}^{-1}$
specific cost for structure	$[E_G]$	$2800\,\text{J cm}^{-3}$	y_{VE}	$0.8\,\text{mol mol}^{-1}$
maturity at birth	E_H^b	$275\,z^3\,\text{mJ}$	M_H^b	$500\,z^3\,\text{nmol}$
maturity at puberty	E_H^p	$166\,z^3\,\text{J}$	M_H^p	$0.3\,z^3\,\text{mmol}$

primary parameters, i.e. compound parameters, and tertiary scaling relationships deal with phenomena at larger scales in space and time, where environmental factors and population dynamics modify and distort the patterns.

8.2.1 Primary scaling relationships

The core of the argument for the covariation of primary parameter values is that parameters that relate to the physical design of the organism depend on the maximum size of the organism, while parameters that depend on the local (bio)chemical environment are size independent, both during ontogeny as well as between species. The latter parameters relate to molecular processes, which are thus essentially density based. The difference between physical design and density-based parameters relates to the difference between intensive and extensive quantities.

How physical design parameters relate to maximum size is fully prescribed by the standard DEB model. For this purpose I introduce a dimensionless zoom factor z, and consider two species, so two sets of parameters. Only three primary parameters are physical design parameters: $\{\dot{p}_{Am}^2\} = z\{\dot{p}_{Am}\}_1$, $E_{H2}^b = z^3 E_{H1}^b$ and $E_{H2}^p = z^3 E_{H1}^p$. All other primary parameters don't depend on maximum size, so $\kappa_X^2 = \kappa_X^1$, $\kappa_2 = \kappa_1$, $\kappa_R^2 = \kappa_R^1$, $\dot{v}_2 = \dot{v}_1$, $[E_G^1] = [E_G^1]$, $[\dot{p}_M^2] = [\dot{p}_M^1]$, $\{\dot{p}_T^2\} = \{\dot{p}_T^1\}$, $\dot{k}_J^2 = \dot{k}_J^1$, $\{\dot{F}_m^2\} = \{\dot{F}_m^1\}$, see Table 8.1.

The reason why the maximum surface-area-specific assimilation rate $\{\dot{p}_{Am}\}$ is a design parameter directly follows from the observation that $L_m = \kappa\{\dot{p}_{Am}\}/[\dot{p}_M]$ and κ and $[\dot{p}_M]$ are intensive parameters. This makes $\{\dot{p}_{Am}\}$ proportional to L_m and also provides a simple interpretation of the zoom factor z: it is the ratio of the maximum (structural) lengths of the compared species. The argument is so simple that it can easily be overlooked; yet it is very powerful.

The reason why E_H^b and E_H^p are design parameters is that the maturity densities E_H^b/V_m and E_H^p/V_m are intensive parameters, with $V_m = L_m^3$ as before.

The case for the scaling of the maximum specific searching rate $\{\dot{F}_m\}$ is weak because organisms of very different body size are likely to feed on very different types of prey, using very different searching methods. If comparisons are made at abundant food, the scaling of $\{\dot{F}_m\}$ becomes irrelevant, however.

The empirical support for these scaling relationships is given by the secondary scaling relationships.

8.2.2 Secondary scaling relationships

This section gives examples of the derivation of body size scaling relationships of a variety of ecophysiological quantities that can be written as compound parameters of the DEB model. The derivation follows the same path time and again and has the following structure. The quantity of interest is written as a function of scaled state variables (e and l) and primary parameters. Food density is taken to be high ($f = 1$) and the scaled reserve density is set at equilibrium ($e = 1$). I choose a (theoretical) reference species and apply the zoom factor $z = L_m/L_m^{\text{ref}}$ to arrive at the maximum structural volume of the species of interest. In many cases I compare the result with allometric functions because this is how they are typically presented, see e.g. [884].

Heating length

The heating length $L_T = \{\dot{p}_T\}/[\dot{p}_M] = L_T^{\text{ref}}$ does not depend on ultimate size, which means that the scaled heating length $l_T = L_T/L_m$ decreases with z. This parameter is likely to be rather variable, even for a single individual, because it depends on the temperature (for endotherms) or the chemical composition of the water in which they are (osmotic costs). Since the maximum ultimate length equals $L_m - L_T$, the significance of the heating length for energetics vanishes for large-bodied species.

Somatic maintenance rate coefficient

The somatic maintenance rate coefficient, see {45}, $\dot{k}_M = \frac{[\dot{p}_M]}{[E_G]} = \dot{k}_M^{\text{ref}}$, is independent of the zoom factor. Yet this parameter turns out to vary substantially among taxa.

The highest value so far is from *Daphnia*, $\dot{k}_M = 4\,\text{d}^{-1}$ at 20 °C, see Figure 2.10, and is probably typical for Ecdysozoa and possibly relates to the costs for frequent moulting. This is consistent with the high value $1.64\,\text{d}^{-1}$ at 20 °C for *Oikopleura*, see Figure 2.19, which possibly relates to the costs of house production. The high value for *D. magna* possibly links to the negative Gompertz stress, see Figure 6.1, which points to a high turnover rate for structure. Many prokaryotes vary from $1\,\text{d}^{-1}$ at 37 °C (*Aerobacter* {165}; *Streptococcus*, Figure 4.10), $0.5\,\text{d}^{-1}$ at 35 °C (*Klebsiella*, Figure 4.16) or $0.67\,\text{d}^{-1}$ at 30 °C (*E. coli* [746]). Yeast gave a value of $0.07\,\text{d}^{-1}$ at 30 °C in Figures 4.24, 5.1.

Table 2.2 gives a range from 0.03–2.3 d^{-1} for a variety of vertebrate eggs at 30 °C. Several species of bivalve gave a 0.013 d^{-1} at 20 °C [189, 188].

The conclusion is that the somatic maintenance rate coefficient varies considerably among taxa, demonstrating substantial evolutionary adaptation.

Reserve capacity and turnover: energy investment ratio

The reserve capacity $[E_m] = \frac{\{\dot{p}_{Am}\}}{\dot{v}} = z[E_m^{\text{ref}}]$ is proportional to the zoom factor z. The energy investment ratio, see {51}, $g = \frac{[E_G]}{\kappa[E_m]} = \frac{[E_G]}{\kappa z[E_m^{\text{ref}}]} = g_{\text{ref}}/z$, decreases with the zoom factor as a direct consequence of the scaling of the reserve capacity.

According to (2.20) for $e = 1$ and $L = L_m - L_T$, see (2.25), the minimum turnover time of reserve is given by

$$t_E = \frac{[E_m]}{[\dot{p}_C]} = \frac{1 - l_T}{\dot{k}_M g}\frac{1+g}{2+g} = \frac{z - l_T^{\text{ref}}}{\dot{k}_M g_{\text{ref}}}\frac{z + g_{\text{ref}}}{2z + g_{\text{ref}}} \tag{8.1}$$

so the minimum turnover time increases with the maximum body size.

The fraction of the body mass of an individual without a reproduction buffer at abundant food that consists of structure is, see Table 3.3

$$\theta_V = \frac{M_{Vm}}{M_{Vm} + M_{Em}} = \frac{\kappa g}{\kappa g + y_{EV}} = \frac{\kappa g_{\text{ref}}}{\kappa g_{\text{ref}} + z y_{EV}} = \frac{1}{1 + z[E_m^{\text{ref}}]y_{EV}/[E_G]} \tag{8.2}$$

so a large-bodied species not only has more reserve, but also as a fraction of its body mass. Working with grams rather than moles, (3.2) implies $\omega_w = \frac{w_E}{w_V}\frac{y_{EV}}{\kappa g}$ and $\theta_V = \frac{1}{1+\omega_w} = \frac{1}{1+z\omega_w^{\text{ref}}}$.

Body weight

Since the independent variable in body size scaling relationships is typically wet weight, it is of special interest. For $[E] = [E_m]$ and $L = L_m - L_T$, see (2.25), and without a reproduction buffer, $E_R = 0$, the maximum wet weight according to (3.2) equals

$$W_w = V_m(d_V + [E_m]w_E/\overline{\mu}_E)(1 - l_T)^3 = z^3 V_m^{\text{ref}}(d_V + z[E_m^{\text{ref}}]w_E/\overline{\mu}_E)(1 - z^{-3}l_T^{\text{ref}})^3 \tag{8.3}$$

It increases faster than z^3 because of the increasing contribution of reserve to (wet) weight, see (8.2); the maximum weights of rat and mouse differ by a factor of 17, but their maximum structural volumes by a factor of 13, see Figures 6.2 and 4.7. For large-bodied species the surface-area-linked maintenance costs (in the scaled heating length l_T) almost vanish, which further contributes to the increase of W_w/V_m with V_m.

Table 8.2 Body weight has contributions from reserve and structure; the parameters d_E and d_V stand for the specific density of reserve (E) and structure (V). The structure volume is proportional to cubed volumetric length, L^3. The heating length L_T is a constant for endotherms, and zero for most ectotherms. The length parameters L_g and L_s are constant (under certain conditions). The inter-species comparison is based on fully grown (adult) individuals; κ is large. From Kooijman *et al.* [650].

	intra-species	inter-species
maintenance	$\propto L_T L^2 + L^3$	$\propto L_T L^2 + L^3$
growth	$\propto L_g L^2 - L^3$	0
$\frac{\text{reserve}}{\text{structure}}$	$\propto L^0$	$\propto L$
$\frac{\text{respiration}}{\text{weight}}$	$\propto \frac{L_s L^2 + L^3}{d_V L^3 + d_E L^3}$	$\propto \frac{L_T L^2 + L^3}{d_V L^3 + d'_E L^4}$

Wet weights are sensitive to body composition. The structural body mass and in particular water content and type of reserve materials are different in unrelated species. This hampers comparisons that include species as different as jellyfish and elephants. If comparisons are restricted to related species, for example among mammals, the structural volume–weight conversion d_V will be independent of body volume, while the maximum reserve increases.

Respiration

The scaling of respiration rate with body size has captured the imagination since the formulation of Kleiber's law, see {303}. The scaling exponent has been found to be 0.66 for unicellulars, 0.88 for ectotherms and 0.69 for endotherms [890]. The exact value differs among authors taking their data from the literature. The variations are due, in part, to differences in the species included and in the experimental conditions under which respiration rates were measured. For crustaceans Vidal and Whitledge [1194] present values of 0.72 and 0.85, and Conover [219] gives 0.74. If species ranging from bacteria to elephants are included, the value 0.75 emerges. It has become an almost magic number in body size scaling relationships.

The predictions of intra- and inter-specific scaling of the respiration by DEB theory are presented in Table 8.2 for juveniles and adults at constant food. Although assimilation contributions are excluded, assimilation controls reserve density; this is why respiration increases, while weight decreases, during the embryo stage, see Figures 2.12 and 2.13. The fact that intra- and inter-specific scaling are numerically very similar, see Figures 2.10 and 4.21, while the explanation is very different, has caused a lot of confusion in the literature. Many alternative explanations have been proposed, but all basically failed [776]; none of them distinguishes between intra- and inter-specific comparisons. For increasing weight, the intra-specific scaling results from the reducing contribution of growth to respiration, and the inter-specific scaling from the increasing contribution of reserve to weight.

The carbon dioxide production of well-fed animals that is not associated with assimilation is given by (4.49). If we compare individuals with the same parameter set, this expression shows that the mineral fluxes, and so the dioxygen consumption rate and the carbon dioxide production rate, depend on structural body length via the powers $\dot{p}_D$ and $\dot{p}_G$, which are both weighted sums of surface area and volume, i.e. of l^2 and l^3. This is nothing new. If we compare species of different (maximum) body size, however, we keep the state variables constant, and vary the parameters that depends on the maximum structural body volume. In the respiration rate (4.49), L_m, l_T^{-1} and g^{-1} are proportional to structural body length; none of the other parameters depend on maximum structural body length (or volume). Generally, this again results in a scaling of respiration somewhere between a surface area and a volume, but it is rather critical which individuals are included. If we include only fully grown individuals of ectothermic species, the dissipating and growth powers no longer depend on the investment ratio g, and respiration is proportional to structural body volume. Even in this case, however, the weight-specific respiration decreases with body weight, because of the increasing contribution of reserves to body weight. Growth is asymptotic, however, and if individuals are selected with a structural length of some fixed fraction of the maximum possible one, the contribution of surface area will be more important.

In conclusion, the respiration rate will appear almost as a straight line in a double-log plot against body weight, the slope being somewhere between 0.66 and 1, depending on the species and the relative size of the individuals that have been included, see Figure 8.2. The scaling relationship for unicellulars is less informative, because assimilation is included and respiration depends sensitively on substrate composition. Surface-bound heating costs dominate in endotherms, so a plot that includes them

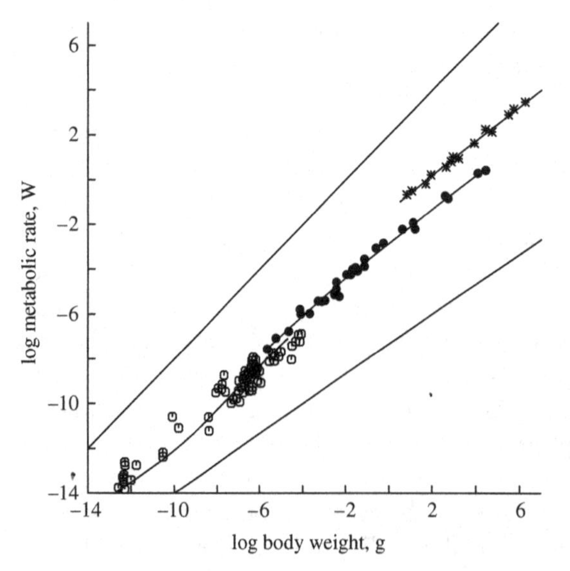

Figure 8.2 The metabolic rate of unicellulars (∘, at 20 °C), ectotherms (•, at 20 °C) and endotherms (∗, at 39 °C) as a function of body weight. Modified from Hemmingsen [310, 491]. The difference between this figure and the many others of the frequently reproduced data set is that the curves relate to DEB-based expectations, and are not allometric regressions. Nonetheless they appear almost as straight lines. The lower line has slope of 2/3, the upper one a slope of 1.

will be close to a line with slope 0.66. The slope for the Bathyergidae, a family of rodents that are practically ectothermic, see Figure 1.5, is found to be close to 1 [721], as expected. Another conclusion is that respiration has a complex interpretation and should play a less central role in ecophysiological research.

Length at birth and initial amount of reserve

Huge fishes can lay very small eggs and thus have small values for V_b. For example, the ocean sunfish *Mola mola* can reach a length of 4 m and can weigh more than 1500 kg, it can produce clutches of 3×10^{10} tiny eggs. The other extreme within the bony fishes is the ovoviviparous coelacanth *Latimeria chalumnae*, which can reach a length of 2 m and a weight of 100 kg. It produces eggs with a diameter of 9 cm in clutches of some 26. (If we include the cartilaginous fish, the whale shark *Rhincodon typus* wins with a 12–18 m length, more than 8165 kg weight and eggs of some 30 cm.) Egg size hardly depended systematically on ultimate size in flatfish [1185] and bivalves [188], which probably relates to adaptation to the deviating larval life history. This lack of systematic pattern includes the size at metamorphosis in flatfish [1185].

The scaled maturity at birth U_H^b scales with z^2. This is because the (unscaled) maturity at birth $M_H^b = U_H^b\{\dot{J}_{EAm}\}$ scales with z^3, and the surface-area-specific reserve assimilation rate $\{\dot{J}_{EAm}\}$ scales with z.

If $\dot{k}_J = \dot{k}_M$, the structural volume at birth is proportional to the maturity at birth, so length at birth scales with maximum length. If $\alpha_b >> B_{x_b}(\frac{4}{3}, 0)$, the initial reserve scales approximately with maximum length to the power 4 and age at birth with maximum length. These scalings are confirmed in the analysis presented in Figure 8.3. If scaled

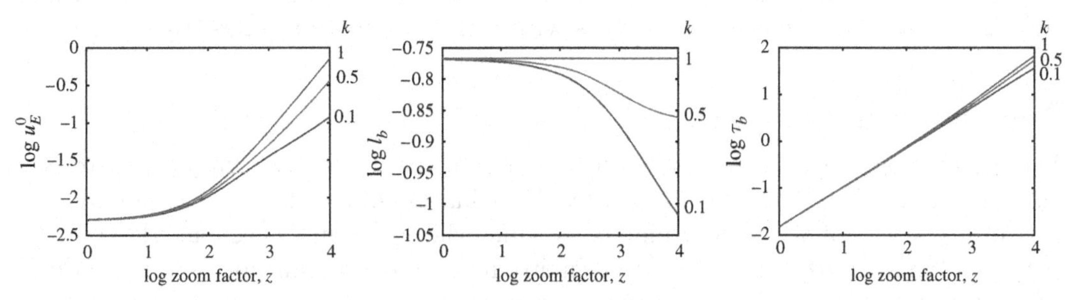

Figure 8.3 The scaled initial reserve u_E^0 (left), length at birth l_b (middle), and age at birth τ_b (right) as function of the zoom factor, log–log plotted (base 10). Each plot has three curves, corresponding to maintenance ratio $k = 0.1$ (lower), 0.5 (middle), 1 (upper). Parameters: $g = 80/z$, $u_H^b = 0.005$, $e_b = 1$. The curves are approximately allometric with slopes for large zoom factors.

maintenance ratio k	0.1	0.5	1.0
scaled initial reserve u_E^0	0.55	0.83	1.00
scaled length at birth l_b	−0.14	−0.04	0.00
scaled age at birth τ_b	0.85	0.89	0.93

initial reserve u_E^0 scales with z, U_E^0 scales with z^2 and initial reserve M_E^0 with z^4, which is consistent with the scaling of the incubation time (see below). If scaled length at birth l_b scales with z^0, length at birth $L_b = l_b L_m$ scales with z.

Figure 8.3 shows that if $\dot{k}_J \neq \dot{k}_M$, the scaling is more complex, especially for the length at birth and the initial reserve; I presented the approximate scaling exponents to comply with the traditional way to present these types of relationships. It is remarkable that taxa show a wide scatter in scaling relationships for specially these quantities, while age at birth shows much less scatter. This suggests that taxa might differ in the maintenance ratio. The increase in the maintenance ratio $k = \dot{k}_J/\dot{k}_M$ goes with an increase in the relative size at birth for any given value of the zoom factor z, but the effect is bigger for the large-bodied species. Since protein turnover is an important component of somatic maintenance costs, and activity typically a minor component, it is not likely that species differ a lot in the somatic maintenance costs. I expect that costs for defence (e.g. the immune system) vary more among species. It is tempting to speculate about the relatively small egg size of dinosaurs (indicating small maturity maintenance costs) versus the relatively large size at birth of mammals (indicating hight maturity maintenance costs).

European birds have egg weights approximately proportional to adult weights. Calder [182] and Rahn *et al.* [933] obtained egg weights proportional to adult weights$^{0.77}$; Birkhead [106] found that the egg weight of auks is proportional to adult weight$^{0.72}$.

The energy content, i.e. the chemical composition, may also show scaling relationships, which can cause deviations from simple predictions. The larger species also have to observe mechanical constraints, and small species can have problems with heating themselves during development.

In conclusion: the tendency of length at birth to be proportional to ultimate length only holds for related species at best, as within the squamate reptiles [1050].

Water loss from eggs

The use of energy (stored in lipids, etc.) relates to the water that will evaporate from bird eggs. Part of this water is formed by the oxidation of energy-rich compounds, and part of it consists of the watery matrix in which the compounds are embedded for the purpose of giving enzymes the correct environment and for transport of the products. The total loss of water during the incubation period, therefore, reflects the total use of energy $E_0 - E_b$. Since, like the energy investment in a single egg E_0, the amount of energy at birth $E_b = [E_m]V_b = z^4[E_m^{\text{ref}}]V_b^{\text{ref}}$, the loss of water must be a fixed proportion of egg volume. Rahn, Ar and Paganelli [31, 931] found that it is some 15% of the initial egg weight. If the use of energy relates to water loss directly, one would expect the initial loss rate to be small and build up gradually. The egg usually decreases linearly in weight, as Gaston [387] found for the ancient murrelet *Synthliboramphus antiqua*. This is to be expected on physical grounds, of course. The specific density of an egg can be used to determine the length of time it has been incubated. This process of water loss implies that the water content of the reserves changes during incubation, but its range is rather restricted. The functional and physical aspects of water loss from eggs thus coincide beautifully.

Minimum size for separation of embryonic cells

Suppose that the cells in the two-cell stage of an embryo are identical in terms of amounts of maturity, reserve and structure. If the cells are separated, the three amounts are halved. This is less obvious for maturity, but since its level is very small, compared to that at birth, this detail is numerically unimportant.

If one tried to separate cells in a species with the parameter values as found for *Daphnia* in Figure 2.10, the theory predicts that the initial reserve would not be enough to cover embryonic development. This result is remarkable because these parameters imply that a fraction of 0.8 of the initial amount of reserve is still left at birth at abundant food, see {62}. The explanation is that the mobilisation of reserve decreases with the reserve density and it still should cover the maintenance needs. It might be, of course, that maturity at birth is affected by cell separation, which can still allow this to occur successfully in small-bodied species. However, I am unaware of any empirical evidence for this.

Since the reserve density capacity $[E_m]$ scales with structural length, species with a larger ultimate body size tend to have a relatively larger reserve capacity. It turned out that for the combination of parameter values as found for *D. magna* we have to apply a zoom factor of at least $z = 1.87$ to arrive at a minimum maximum body size for which cell separation might be successful. The resulting parameter values are used in the figure; the scaling relations only affect the energy investment ratio gz^{-1}, while κ, k and u_H^b are independent of the zoom factor z.

Minimum embryonic period

Because the DEB model is maturity-structured rather than age-structured, the length of the various life-stages is closely tied to growth. The gestation time is proportional to volume$^{1/3}$, excluding any delay in implantation. Weasels and probably armadillos are examples of species that usually observe long delays, possibly to synchronise the juvenile period with favourable environmental conditions. Figure 8.4 illustrates that the expected scaling relationship is appropriate for 250 species of eutherian mammals. The mean energy conductance was found to be 2 mm d^{-1} at some 37 °C. This is less than half the mean-temperature-corrected value found from the von Bertalanffy growth rates of juveniles and adults, a difference that must be left unexplained at this moment.

Incubation time (2.39) depends on volume in a more complex way, but it is also scales approximately with z, so with egg volume$^{1/4}$; the scaled egg costs e_0 do not depend on the zoom factor, so that egg costs themselves $E_0 = e_0 E_m = z^4 e_0 E_m^{\text{ref}}$ scale with z^4, and $a_b \simeq z a_b^{\text{ref}} \propto E_0^{1/4}$. Figure 8.5 gives the log–log plot for the species that breed in Europe. These data are very similar to those of Rahn and Ar [930], who included species from all over the world. Although the scatter is considerable, the data are consistent with the expectation. Note that, within a species, large eggs hatch earlier than small

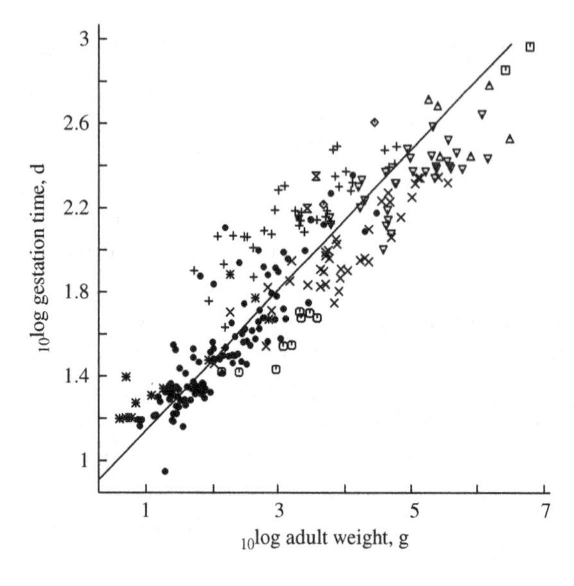

Figure 8.4 The gestation time of eutherian mammals tends to be proportional to volumetric length (line). Data from Millar [795]. The times have been corrected for differences in relative birth weight, i.e. birth weight as a fraction of adult weight, by multiplying by the ratio of the mean relative birth weight$^{1/3}$, 0.396, to the actual relative birth weight$^{1/3}$. The symbols refer to $*$ Insectivora, + Primates, ⋄ Edentata, ∘ Lagomorpha, • Rodentia, × Carnivora, □ Proboscidea, ⧖ Hyracoidea, △ Perissodactyla, ▽ Artiodactyla.

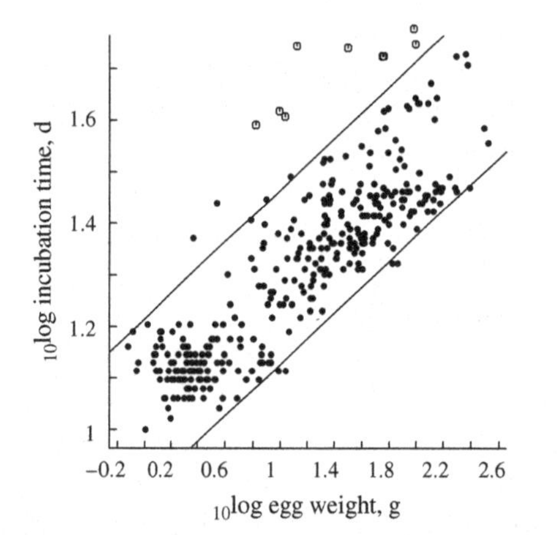

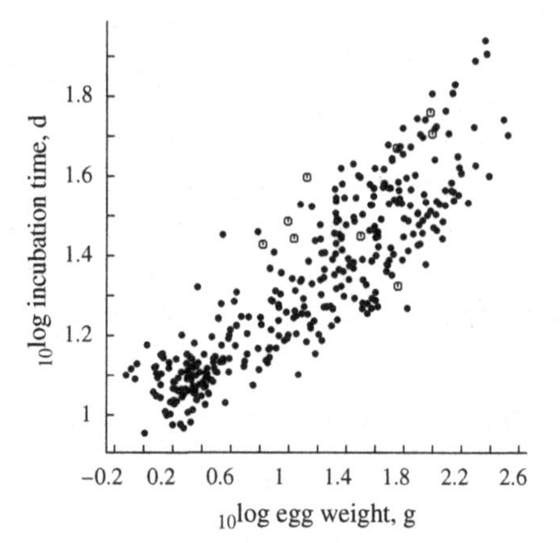

Figure 8.5 The incubation time for European breeding birds as a function of egg weight (left figure). Data from Harrison [464]. The lines have a slope of 0.25. The tube noses (∘) sport long incubation times. If corrected for a common relative volume at birth (right figure), this difference largely disappears.

ones, though one needs to look for species with egg dimorphism to find a large enough difference between egg weights.

The tube noses Procellariiformes incubate longer, while they also have relatively heavy eggs, and so relatively large chicks. If corrected for this large volume at birth, their incubation time falls within the range of other species. This correction has been done by calculating the egg weight first, from $([W_w]\pi/6)$(egg length)(egg breadth)^2. (Data from Harrison [464].) The weight at birth is about 0.57 times the initial egg weight [1199]. The scaled length at birth is about $(W_b/W_\infty)^{1/3}$. (This is not 'exact' because of the weight–volume conversion and the volume reduction due to heating.) Bergmann and Helb [86] give adults weights. The incubation time is then corrected for differences in scaled length at birth on the basis of (2.39) for small values of the investment ratio g and a common value for the scaled length at birth, 0.38.

The application of the DEB model has been useful in identifying the proper question, which is not why the incubation time of tube noses is so long, but why they lay so large an egg. The bird champion in this respect is the kiwi *Apteryx*, which produces eggs of 350–400 g, while the adult weight is only 2200 g. It has an incubation period of a respectable 78 days. The relatively low incubation temperature of 35.4 °C extends incubation in comparison to other birds, which usually incubate at 37.7 °C [181, 182]. This accounts for some 17–20 days extension with an Arrhenius temperature of 10–12.5 kK; however, most of this long incubation relates to the very large relative size of the egg. The relative size of the egg itself is a result of the energy uptake and use pattern.

If one or more primary parameters are known, the value of a certain compound parameter such as the (minimum) incubation time can be predicted with much more accuracy. On the basis of growth data for Cassin's auklet *Ptychoramphus aleuticus* during the juvenile phase, I predicted an incubation period of 40 days [635], not knowing that it had been measured and actually found to be 37–42 days [741].

The reptilian champion in incubation time is the tuatara *Sphenodon punctatus* where the 4-g hatchling leaves the egg after 15 months. The low temperature, 20–25 °C, contributes to this record.

The European cuckoo is a breeding parasite which parks each of its many eggs in the nest of a 'host', which has an adult body weight of only 10% of that of the cuckoo. The eggs of the host are one-half to three-quarters the size of that of the cuckoo. On the basis of egg size alone, therefore, the cuckoo egg should hatch later than the eggs of the host, while in fact it usually hatches earlier despite the later date of laying. If the relative size of the egg with respect to the adult is taken into account, the DEB theory correctly predicts the observed order of hatching. The essence of the reasoning is that, since the cuckoo is much larger than the host, the cuckoo uses the reserves at a higher rate (i.e. $\{\dot{p}_{Am}\}$ is larger), and, therefore, it grows faster in the absolute sense. Growth is so much faster that the difference in birth weight with the chicks of the host is more than compensated. In non-parasitic species of the cuckoo family, the eggs are much larger [1281], which indicates that the small egg size is an adaptation to the parasitic way of life. The extra bonus for the European cuckoo is that it can produce many small eggs (about 20–25), which helps it to overcome the high failure rate of this breeding strategy.

Maximum ingestion rate

The maximum specific ingestion rate for an individual is $\{\dot{J}_{XAm}\} = \{\dot{J}_{EAm}\}/y_{EX} = z\{\dot{J}^{\text{ref}}_{EAm}\}/y_{EX}$, in mass, or $\{\dot{p}_{Xm}\} = \{\dot{p}_{Am}\}/\kappa_X = z\{\dot{p}^{\text{ref}}_{Am}\}/\kappa_X$ in energy, so the ingestion is proportional to surface area intra-specifically, and to volume inter-specifically. Farlow [335] gives an empirical scaling exponent of 0.88, but value 1 also fits the data well. For endotherms especially, a scaling exponent of somewhat less than 1 is expected for weight as the independent variable, because of the increase in volume-specific weight, as explained. In a thorough study of scaling relationships, Calder [182] coupled the inter-specific ingestion rate directly to the respiration rate, without using an explicit model for energy uptake and use. The present DEB-based considerations force one to deviate from intuition.

Half-saturation coefficient

The (half) saturation coefficient $K = \{\dot{J}_{XAm}\}/\{\dot{F}_m\} = z\{\dot{J}^{\text{ref}}_{XAm}\}/\{\dot{F}_m\} = zK_{\text{ref}}$ is proportional to the zoom factor. The logic is less easy to see, because species differ so much in their feeding behaviour. At low food densities this constant can be interpreted as the ratio of the maximum ingestion to the filtering rates in a filter feeder such as *Daphnia*. If maximum beating rate is size independent, as has been observed, the filtering rate is proportional to surface area. Since the maximum specific ingestion rate $\{\dot{J}_{XAm}\}$ scales with a length measure, the saturation coefficient K should scale with a length measure as well. More detailed modelling of the beating rate would involve 'molecular' density-based formulations for the filtering process, which turns the saturation coefficient into a derived compound parameter. This is not attempted here, because the formulations would only apply to filtering, while many species do not filter.

Maximum filtering rate

The filtering rate is maximal at low food densities. If particle retention is complete, it is given by $\dot{F}_m = \dot{J}_{Xm}/K = L_m^2\{\dot{J}_{Xm}\}/K$. So, $\dot{F}_m = z^2 L_m^{\text{ref}}\{\dot{J}^{\text{ref}}_{Xm}\}/K^{\text{ref}} = z^2 \dot{F}^{\text{ref}}_m$. This was found by Brendelberger and Geller [147].

Gut capacity

Within a species, isomorphy implies a gut capacity that is a constant fraction of body volume. This must also hold for inter-species comparisons, as long as body design and diet are comparable and this has been found for birds and mammals [182]. The mean gut residence time of food particles is thus independent of body size as a consequence, because ingestion rate is proportional to body size, while it has been found to be proportional to a length measure for intra-species comparisons. This is of major ecological significance for herbivores, because it determines which type of food can be digested. The poorly digestible substrates can only be used successfully by animals with a large body size. The giant sauropods of the Jurassic fed mainly on cycads and conifers, which require long gut residence times for digestion. Giant carnivores probably evolved in response to giant herbivores; the explanation of their body size probably relates to the survival of meagre periods.

Speed

Since biomechanics is not part of the DEB theory, this is not the right place for a detailed discussion on Reynolds and Froude numbers, although interesting links are possible. Speed of movement has only a rather indirect relationship with feeding or other aspects that bear on energy budgets. A few remarks are, therefore, made here.

McMahon and Bonner [773] found that the speed of sustained swimming for species ranging from larval anchovy, via salmon, to blue whales scales with the square root of volumetric length; they underpinned this finding with mechanical arguments. Since the energy costs of swimming are proportional to squared speed and to surface area, see {33}, the total costs of movement would scale with cubed length, or V, for a common travelling time. This is consistent with the DEB theory, where the costs of travelling are taken to be a fixed fraction of the maintenance costs.

A similar result appears to hold for the speed of flying, but by a somewhat different argument. The cruising speed, where the power to fly is minimal, is proportional to the square root of the wing loading [1145]. If a rough type of isomorphy applies, comparing insects, bats and birds, wing loading, i.e. the ratio of body mass to wing area, scales with length, so that cruising speed scales with the square root of length [773].

Arguments for why the standard cruising rate for walking tends to be proportional to length are given on {27, 32}. If energy invested in movement is proportional to volume and taken to be part of the maintenance costs, the intra- and inter-species scaling relationships work out in the same way.

Maximum diving depth

Birkhead [106] found that the maximum diving depth for auks and penguins tends to be proportional to volumetric length. This can be understood if diving depth is proportional to the duration of the dive; the latter is proportional to length, see {27}, by the argument that the respiration rate of these endotherms is about proportional to surface area and dioxygen reserves to volume.

Minimum food density

The minimum food density at which an isomorph of structural length L can live for a long time is found from the condition that energy derived from ingested food just equals the maintenance costs, so $\dot{p}_A = \kappa_X \dot{p}_X = \dot{p}_M$, or $f_\dagger = \frac{X_\dagger}{K + X_\dagger} = \frac{[\dot{p}_M]}{\{\dot{p}_{Am}\}} L$, or $X_\dagger = K \frac{L[\dot{p}_M]/\{\dot{p}_{Am}\}}{1 - L[\dot{p}_M]/\{\dot{p}_{Am}\}}$. At this food density, the individual can only survive, not reproduce. For different species, we obtain the condition $X_\dagger = zX_\dagger^{\text{ref}}$. Minimum food density, also called the threshold food density, is thus proportional to volumetric length. An important ecological consequence is that, at a given low food density, small individuals can survive, while the large ones can not. This explains, for instance, why bacteria in oligotrophic seas are so small.

This result only applies to situations of constant food density. If it is fluctuating, storage capacity becomes important, which tends to increase with body size; see {118}. The possibility of surviving in dynamic environments then works out to be rather complex. Stemberger and Gilbert [1108] found that the threshold food density tends to increase with body size for rotifers, as expected, but Gliwicz [412] found the opposite for cladocerans. This result can be explained, however, by details of the experimental protocol. The threshold food density was obtained by plotting the growth rate against food density and selecting the value where growth is nil. Growth at the different food densities was measured from 2-day-old individuals exposed to a constant food density for 4 days. The reserves at the start of the growth experiment, which depend on culture conditions, will contribute substantially to the result.

Maximum growth

From (2.21) it can be seen that $\frac{d}{dt}V$ is at maximum for $e = 1$ and $L = \frac{2}{3}(L_m - L_T)$ and the maximum growth in cubed scaled length is $\frac{d}{dt}l^3 = \frac{4}{27}\frac{g\dot{k}_M}{1+g}(1 - l_T)^3$. The maximum growth rate for different species equals

$$\frac{4}{27}\frac{\dot{k}_M^{\text{ref}}g_{\text{ref}}}{(z + g_{\text{ref}})}(zL_m^{\text{ref}} - L_T^{\text{ref}})^3 \tag{8.4}$$

and thus scales approximately with z^2. This fits Calow and Townsend's data very well [183].

von Bertalanffy growth rate

The von Bertalanffy growth rate at high food density is

$$\dot{r}_B = \frac{\dot{k}_M^{\text{ref}}/3}{1 + z/g_{\text{ref}}} \tag{8.5}$$

for different species, see (2.24). It scales with z^{-1} for $z \gg g_{\text{ref}}$. This is consistent with empirical findings; see Figure 8.6. The parameters and data sources are listed in Table 8.3. This table is extensive because the fit with the von Bertalanffy growth curve supports the argument that it is possible to formulate a theory that is not species-specific. If one collects growth data from the literature, an amazingly large fraction fits the von Bertalanffy curve despite the fact that most data sets are from specimens collected in the field. Since it is hard to believe that food density has been constant during the growth period, this suggests that food has been abundant; this is relevant for population dynamics.

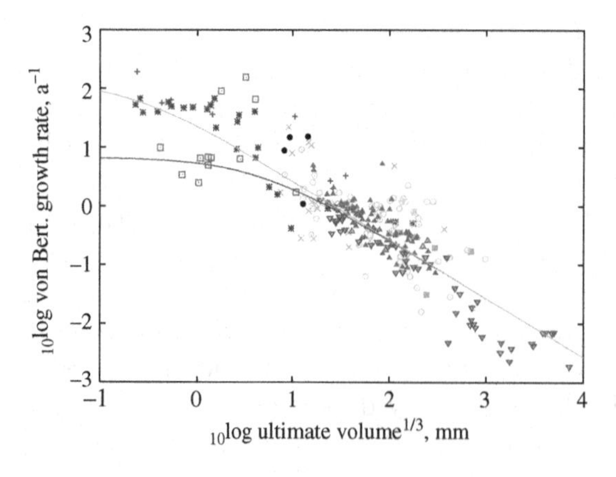

Figure 8.6 The von Bertalanffy growth rate as a function of maximum volumetric length corrected to a common body temperature of 20 °C using T_A = 8 kK. The markers refer to ▽ (open) Mammalia, △ (filled) Aves, △ (open) Reptilia, ○ (filled) Amphibia, ○ (open) Osteichthyes, □ (filled) Chondrichthyes, $*$ Crustacea, □ (open) Uniramia × Mollusca, + others. The curve presents the prediction by DEB theory, using $\dot{v}$ = 0.022 cm d^{-1} and $\dot{k}_M$ = 0.055 d^{-1} (lower curve) or 1.1 d^{-1} (upper curve).

Table 8.3 The von Bertalanffy parameters and their standard deviations as calculated by non-linear regression. The shape coefficient converts the size measure used to volumetric length. For shape coefficient 1, the data refer to wet weight, except for *Saccharomyces*, *Actinophrys* and *Asplanchna*, where volumes were measured directly. The data for *Mnemiopsis* and *Calanus* refer to dry weight. The other data are length measures, mostly total body length. Where the standard deviation is not given, the parameters from the authors are given. Temperatures in parentheses were inferred from the location on Earth. Where two temperatures are given, a sinusoidal fluctuation between these extremes is assumed. In the column 'sex': f = female, m = male, l = larva.

species	sex	length mm L_∞	sd mm	shape coeff $\delta_\mathcal{M}$	rate a^{-1} $\dot{r}_B$	sd a^{-1}	location NS	location EW	temp °C	source
Ascomyceta										
Saccharomyces carlsbergensis		4.59e-3	2.16e-5	0.806	11830	318	lab	lab	30	[84]
Heliozoa										
Actinophrys sp.		0.0043	2.2e-5	1	2891	368	lab	lab		[1134]
Rhizopoda										
Amoeba proteus		2.79	0.016	0.01	832.2	56.9	lab	lab	23	[919]
Ciliata										
Paramecium caudatum		2.969	0.062		1638	210	lab	lab	17	[1020]
Ctenophora										
Pleurobrachia pileus	fm	15.04	0.436	0.702	33.27	2.49	lab	lab	20	[432]
Mnemiopsis mccradyi	fm	8.851	0.927	3.90	11.61	1.88	lab	lab	26	[949]
Rotifera										
Asplanchna girodi	f	0.2400	7.32e-4	1	193.7	4.92	lab	lab	20	[978]
Annelida										
Dendrobeana veneta	fm	14.5	0.24	1	12.04	0.73	lab	lab	20	H. Bos, pc
Mollusca										
Aplysia californica	fm	112.2	6.05	1	4.840	0.871	lab	lab	18–20	[881]
Urosalpinx cinerea	fm	30.94	1.31	0.397	0.8116	0.11	31S	152E	-1–25	[368]
Achatina achatina	fm	106.5	2.45	0.543	1.121	0.0770	5N	0E	(25)	[515]
Helix aspera	fm	25.06	0.498	0.68	1.098	0.0960	lab	lab	(18–20)	[245]
Patella vulgata	fm	46.93	0.306	0.310	0.4296	7.91e-3	54N	4.40W	(4–17)	[1280]
Monodonta lineata	fm	21.92	0.130	0.716	0.6213	0.0171	52.25N	4.05W	(4–17)	[1258]
Biomphalaria pfeifferi	fm	7.538	0.0497	1	4.879	0.201	lab	lab	25	[788]
Lymnaea stagnalis	fm	15.37	0.0584	1	10.81	0.204	lab	lab	20	[1074]

Table 8.3 (cont.)

species	sex	length mm L_∞	sd mm	shape coeff $\delta_{\mathcal{M}}$	rate a^{-1} $\dot{r}_B$	sd a^{-1}	location NS	location EW	temp °C	source
Helicella virgata	fm	9.888	0.215	1	3.316	0.163	35S	139E	11–16	[906]
Macoma baltica		21.57	0.154	0.423	3.00	0.0869	41.31N	70.39W	10.56	[407]
Cerastoderma glaucum		29.24	1.86	0.558	2.221	0.380	40.50N	14.10E	13–30	[546]
Venus striatula		37.76	25.1	0.471	0.1961	0.210	55.50N	4.40W	6–13	[29]
Ensis directus		142.2		0.187	0.5830		54.35N	8.45E	4–17	[1133]
Mytilus edulis		95.92	2.02	0.394	0.1045	5.109e-3	53.36N	9.50W	7–17	[979]
Placopecten magellanicus		162.3	1.01	0.388	0.1671	2.842e-3	47.10N	53.36W	0–18	[727]
Perna canaliculus		191.2	10.6	0.394	0.3555	0.0342	36.55S	174.47E	17	[506]
Hyridella menziesi		74.62	2.05	0.400	0.1331	8.38e-3	36.55S	147.47E		[552]
Mya arenaria		91.31		0.407	0.1866		41.39N	70.42W	(4–17)	[160]
Loligo pealei	f	455.3	39.5	0.398	0.4201	0.0551	41.31N	70.39W	(4–17)	[1130]
Loligo pealei	m	918.2	111	0.398	0.2122	0.0315	41.31N	70.39W	(4–17)	[1130]
Brachiopoda										
Terebratalia transversa		48.39	1.09	0.640	0.3140	0.0163	47.30N	122.5W	(4–17)	[862]
Crustacea										
Daphnia pulex	f	2.366	0.0192	0.526	44.25	2.10	lab	lab	20	[961]
Daphnia longispina	f	2.951	0.0260	0.520	61.32	2.92	lab	lab	25	[544]
Daphnia magna	f	5.136	0.0970	0.526	35.04	1.83	lab	lab	20	[634]
Daphnia magna	m	2.813	0.0440	0.526	66.80	5.11	lab	lab	20	[634]
Daphnia cucullata	f	1.049	0.0214	0.480	58.25	9.71	lab	lab	20	[1195]
Daphnia hyalina	f	1.717	0.0399	0.520	47.52	5.93	lab	lab	20	[1195]
Ceriodaphnia pulchella	f	0.7503	0.0122	0.520	39.89	5.04	lab	lab	20	[1195]
Ceriodaphnia reticulata	f	1.038	0.0210	0.520	49.28	3.30	lab	lab	20	[634]
Chydorus sphaericus	f	0.4115	1.10e-3	0.560	52.63	0.969	lab	lab	20	[1195]
Diaphanosoma brachyurum	f	1.380	0.0198	0.520	46.50	3.72	lab	lab	20	[1195]
Leptodora kindtii	f	8.632	0.204	0.300	26.96	2.64	lab	lab	20	[1195]
Bosmina longirostris	f	0.5289	0.0215	0.520	38.73	6.50	lab	lab	20	[1195]
Bosmina coregoni	f	0.4938	0.0104	0.520	66.90	9.59	lab	lab	20	[1195]
Calanus pacificus		6.295	1.02	0.215	8.863	1.89	lab	lab	12	[857]
Dissodactylus primitivus	f	11.02	0.410	0.635	1.025	0.0732	lab	lab	(18)	[903]

Table 8.3 (cont.)

species	sex	length mm L_∞	sd mm	shape coeff $\delta_\mathcal{M}$	rate a^{-1} $\dot{r}_B$	sd a^{-1}	location NS	location EW	temp °C	source
Dissodactylus primitivus	m	9.013	0.212	0.635	1.362	0.0742	lab	lab	(18)	[903]
Euphasia pacifica		12.91	2.35	0.197	1.008	0.369	lab	lab	10	[757]
Homarus vulgaris		186.6	6.99	0.939	0.05543	3.36e-3	lab	lab	10	[502]
Cancer pagurus	f	9.707	0.385	1	0.2711	0.0122	50.30N	2.45W	(5–18)	[80]
Cancer pagurus	m	115.6	0.513	1	0.3513	0.0174	50.30N	2.45W	(5–18)	[80]
Dichelopandalus bonnieri		25.73	1.97	0.882	0.4795	0.0824	54N	4.40W	(4–17)	[9]
Gammarus pulex	m	4.355	0.0570	1	3.300	0.177	lab	lab	15	[1131]
Gammarus pulex	f	4.089	0.0554	1	2.218	0.123	lab	lab	15	[1131]
Calliopius laeviusculus		15.27	0.699	0.262	13.52	1.96	lab	lab	15	[241]
Uniramia										
Tomocerus minor		3.903	0.0848	0.351	6.600	0.379	lab	lab	20	[566]
Orchesella cincta		3.652	0.0858	0.351	4.948	0.351	lab	lab	20	[566]
Isotomata viridis		3.034	0.0751	0.351	6.52	0.469	lab	lab	20	[566]
Entomobrya nivalis		1.981	0.0830	0.351	3.416	0.418	lab	lab	20	[566]
Lepidocyrtus cyaneus		1.181	0.0666	0.351	9.840	2.17	lab	lab	20	[566]
Orchesella cincta		1.281	0.0151	1	6.817	0.354	lab	lab	20	[554]
Phaenopsectra coracina		1.745	0.147	1	2.388	0.779	63.14N	10.24E	4	[1]
Diura nanseni		2.782	0.0460		6.328	0.536	60.15N	6.15E	0–20	[46]
Capnia pygmaea		1.024	0.0967		2.493	0.663	60.15N	6.15E	1–20	[46]
Locusta migratoria		10.82	0.237	1	44.82	7.36	lab	lab	23–36	[718]
Chironomus plumosus	f	4.053	0.272	1	21.88	5.50	lab	lab	15	[543]
Chironomus plumosus	m	3.211	0.0415	1	52.74	4.77	lab	lab	15	[543]
Chaetognatha										
Sagitta hispida	fm	9.431	0.150	0.15	44.80	5.25	lab	lab	21	[948]
Echinodermata										
Lytechenus variegatus		46.10	0.147	0.70	3.913	0.199	18.26N	77.12W	26–29	[588]
Echinocardium cordatum		34.50	0.425	0.696	0.4590	0.0232	53.10N	4.15E	5–12	[301]
Echinocardium cordatum		36.70	0.375	0.696	0.5320	0.0259	53.40N	4.30E	5–14	[301]
Echinocardium cordatum		44.90	0.405	0.696	0.4960	0.0212	54.15N	4.30E	5–16	[301]

Table 8.3 (cont.)

species	sex	length mm L_∞	sd mm	shape coeff $\delta_{\mathcal{M}}$	rate a^{-1} $\dot{r}_B$	sd a^{-1}	location NS	location EW	temp °C	source
Tunicata										
Oikopleura longicauda	fm	0.829	0.049	0.520	56.56	6.62	lab	lab	20	[341]
Oikopleura dioica		0.952	0.327	0.560	63.97	37.3	lab	lab	20	[341]
Chondrichthyes										
Raja montaqui	fm	695.9	11.0	0.184	0.1874	0.0140	52–54N	3–7E	(4–17)	[521]
Raja brachyura		1589	213	0.184	0.1018	0.0261	52–54N	3–7E	(4–17)	[521]
Raja clavata	f	1303	107	0.184	0.09297	0.0163	52–54N	3–7E	(4–17)	[521]
Raja clavata	m	952.7	29.8	0.184	0.1557	0.0145	52–54N	3–7E	(4–17)	[521]
Raja erincea		542.9	32.6	0.184	0.2787	0.0542	41.05N	73.10W	1–19.1	[958]
Prionace glauca		4230		0.165	0.1100		48N	7W	(5–18)	[1111]
Osteichthyes										
Accipenser stellatus		2120	30.5	0.198	0.05396	1.46e-3	(45.10N)	(28.30E)	(4–23)	[94]
Clupea sprattus		157.0	0.557	0.200	0.5847	4.60e-3	52.30N	2E	(4–17)	[542]
Coregonus lavaretus		397.3	8.39	0.203	0.3295	0.0221	54.35N	2.50W	(5–15)	[48]
Salvelinus willughbii	f	385.4	72.9	0.225	0.2495	0.0973	54.20N	2.57W	(5–15)	[375]
Salvelinus willughbii	m	328.9	12.7	0.224	0.3545	0.0366	54.20N	2.57W	(5–15)	[375]
Salmo trutta		585.8	18.0	0.216	0.4769	0.0411	53.15N	4.30W	(4–17)	[538]
Salmo trutta		576.2	20.6	0.240	0.2921	0.0253	57.40N	5.10W	5–12.8	[185]
Salmo trutta		420.2	3.13	0.240	0.4157	0.0107	54.20N	2.57W	(5–15)	[228]
Oncorhynchus tschawytscha		155.2	11.9	1	0.9546	0.217	36S	147E	(11–16)	[179]
Thymallus thymallus		459.6	8.44	0.240	0.4656	0.0224	52.09N	2.41W	(5–15)	[489]
Esox lucius	f	948.7	88.3	0.209	0.2101	0.0718	50.17N	3.39W	(5–15)	[144]
Esox lucius	m	703.6	13.0	0.209	0.4016	0.0455	50.17N	3.39W	(5–15)	[144]
Esox masquinongy		2091	848	0.199	0.04503	0.0263	44N	79W	(5–15)	[813]
Rutilus rutilus		441.6	15.8	0.258	0.1661	0.0116	52.30N	0.30E	(5–15)	[227]
Leuciscus leuciscus		252.6	2.32	0.258	0.3329	0.0131	52.30N	0.30E	(5–15)	[227]
Barbus grypus		1036	25.2	0.206	0.1265	6.59e-3	35.75N	44.7E	(17–30)	[11]
Abramis brama		546.0		0.225	0.1142		53.15N	2.30W	(5–15)	[419]
Gambusia holbrookii	f	61.72	2.34	0.250	0.9366	0.216	38.40N	9.40W	(5–25)	[367]
Poecilia reticulata	f	50.58	1.14	0.252	1.667	0.0690	lab	lab	21	[1180]
Merluccius merluccius		1265	78.4	0.222	0.2075	0.0184	55.45N	5W	(8–12)	[47]
Lota lota		1009	60.3	0.193	0.09768	0.0103	53N	98W	(5–15)	[504]
Gadus merlangus	f	898.6	12.2	0.222	0.08626	2.07e-4	54N	4.40W	(8–12)	[133]
Gadus merlangus	m	772.8	9.03	0.222	0.08626	2.07e-4	54N	4.40W	(8–12)	[133]
Gadus morhua		1089	43.2	0.222	0.1308	9.26e-3	40N	60W	10	[615]

Table 8.3 (cont.)

species	sex	length mm L_∞	sd mm	shape coeff $\delta_{\mathcal{M}}$	rate a^{-1} $\dot{r}_B$	sd a^{-1}	location NS	location EW	temp °C	source
Gadus aeglefinus		106.5		1	0.2000		53–57N	0–7E	(4–17)	[101]
Atherina presbyter		124.0	3.20	0.238	1.091	0.109	51.55N	1.20W	(5–18)	[1174]
Gasterosteus aculeatus		52.41	2.62	0.250	1.019	0.249	52.20N	3W	(4–17)	[563]
Pungitius pungitius		41.28	1.03	0.200	1.777	0.468	52.20N	3W	(4–17)	[563]
Nemipterus marginatus		232.8	35.8	0.243	0.5047	0.227	6N	116E	(26–30)	[873]
Labrus bergylta		509.2	8.64	0.258	0.07170	3.30e-3	54N	4.40W	(4–17)	[272]
Ellerkeldia huntii		152.1	10.8	0.319	0.3350	0.0791	35.30S	174.40E	(12–22)	[561]
Lepomis gibbosus		61.86	9.04	1	0.1415	0.0342	45.40N	89.30W	(5–15)	[801]
Lepomis macrochirus		71.62	16.8	1	0.1292	0.0467	45.40N	89.30W	(5–15)	[801]
Perca fluviatilis		317.9	22.5	0.25	0.1615	0.0242	56.10N	4.45W	8–14	[1043]
Tilapia sp.		129.6	20.7	1	3.542	1.10	31.30N	35.30E	(37)	[722]
Liza vaigiensis		746.3	31.8	0.258	0.1758	0.0147	17S	145E	(18–27)	[427]
Mugil cephalus		595.0	27.2	0.258	0.3350	0.0370	17S	145E	(18–27)	[428]
Valamugil seheli		635.3	35.0	0.258	0.2725	0.0291	17S	145E	(18–27)	[428]
Seriola dorsalis		1373	30.7	0.231	0.1155	5.72e-3	33N	118W	(15–20)	[67]
Ammodytes tobianus		140.9	1.98	0.147	0.7305	0.0595	50.47N	1.02W	5–18	[945]
Thunnus albacares		2745	636	0.266	0.1481	0.0509	0–10N	165E	(26–30)	[845]
Thunnus thynnus		3689	448	0.266	0.06623	0.0144	53–57N	0–7E	(4–17)	[1163]
Coryphoblennius galerita		69.55	2.72	0.250	0.4011	0.0598	50.20N	4.10W	(5–18)	[797]
Pomatoschistus norvegicus		48.80	0.770	0.252	2.466	0.305	56.20N	5.45W	(8–14)	[405]
Gobio gobio		154.9	15.9	0.250	0.7519	0.495	51N	2.15W		[739]
Gobio gobio		174.8	3.84	0.250	0.4165	0.0321	51.50N	8.30W	(4–17)	[591]
Gobius cobitis		213.9	14.9	0.295	0.2082	0.0385	48.45N	4W	(5–18)	[402]
Gobius paganellus		79.89	1.94	0.200	0.4790	0.0463	54N	4.40W	(4–17)	[796]
Lesueurigobius friesii		65.82	0.623	0.252	0.5628	0.0349	55.45N	5W	8–12	[827]
Lesueurigobius friesii		63.72	0.409	0.252	0.6826	0.0322	56.20N	5.45W	(8–14)	[403]
Blennius pholis		150.5	3.36	0.250	0.2464	0.0176	50.20N	4.10W	(5–18)	[797]
Arnoglossus laterna		93.55	3.06	0.200	0.4544	0.0895	56.15N	5.40W	(8–14)	[404]
Hypoglossus hypoglossus		632.7	54.7	1	0.04797	6.04e-3	59N	152W	(3–14)	[1089]

Table 8.3 (cont.)

species	sex	length mm L_∞	sd mm	shape coeff $\delta_{\mathcal{M}}$	rate a^{-1} $\dot{r}_B$	sd a^{-1}	location NS	location EW	temp °C	source
Scophthalmus maximus	f	669.4	14.2	0.266	0.2165	0.0298	53–57N	0–7E	(3–14)	[560]
Scophthalmus maximus	m	495.3	6.93	0.272	0.3247	0.0222	53–57N	0–7E	(3–14)	[560]
Pleuronectes platessa		142.1		1	0.09500		53–57N	0–7E	(4–17)	[101]
Solea vulgaris		78.41		1	0.4200		53–57N	0–7E	(4–17)	[101]
Amphibia										
Rana tigrina	l	12.79	0.670	1	15.75	1.88	lab	lab	30–33	[246]
Rana sylvatica	l	8.201	0.154	1	30.97	6.64	36.05N	81.50W	21–26	[1255]
Triturus vulgaris	l	26.40		0.353	3.960		59.30N	10.30E	-5–14	[279]
Triturus cristatus	l	40.40		0.353	4.080		59.30N	10.30W	-5–14	[279]
Reptilia										
Emys orbicularis	f	182.1	1.98	0.500	0.2707	0.0124			(22)	[215]
Emys orbicularis	m	161.8	1.56	0.500	0.3453	0.0172			(22)	[215]
Vipera berus		539.0	33.0	0.075	0.3734	0.0657			(20)	[372]
Eunectes notaeus	f	3283	50.9	0.075	0.2552	0.0165	lab	lab	(20)	[889]
Eunectes notaeus	m	2946	94.5	0.075	0.2030	0.0251	lab	lab	(20)	[889]
Aves										
Eudyptula minor nov.		114.7	5.67	1	15.60	2.69			39.5	[599]
Pygoscelis papua		191.8	3.35	1	15.31	0.965			39.5	[1201]
Pygoscelis antarctica		163.6	5.29	1	16.88	2.12			39.5	[1201]
Pygoscelis adeliae		159.9	7.77	1	15.47	2.81			39.5	[1201]
Pygoscelis adeliae		188.7	3.47	1	14.32	0.698			39.5	[1143]
Aptenodytes patagonicus		250.0		1	8.508	0.164			39.5	[1117]
Pterodroma cahow		63.16	0.465	1	62.96	1.55			39.5	[1259]
Pterodroma phaeopygia		79.2	0.93	1	20.08	3.43			39.5	[463]
Puffinus puffinus		83.90	0.069	1	41.55	2.87			39	[157]
Diomedea exulans		229.1	1.02	1	5.541	0.176			39.5	[1158]
Oceanodroma leucorhoa		41.53	0.282	1	26.37	1.58			39.5	[968]
Oceanodroma furcata		44.73	0.339	1	23.28	1.16			39.5	[123]
Phalacrocorax auritus		149.5	6.31	1	18.18	1.81			39.5	[302]
Phaethon rubricaudata		101.1	1.45	1	13.03	0.923			39.5	[266]

Table 8.3 (cont.)

species	sex	length mm L_∞	sd mm	shape coeff $\delta_{\mathcal{M}}$	rate a^{-1} $\dot{r}_B$	sd a^{-1}	location NS	location EW	temp °C	source
Phaethon lepturus		72.79	1.12	1	18.77	2.03			39.5	[266]
Sula sula		80.01	1.18	1	11.82	1.53			39.5	[266]
Sula bassana		172.7	2.50	1	12.41	0.639			39.5	[831]
Cionia cionia		158.0	6.10	1	18.36	2.35			39.5	[229]
Phoeniconaias minor		116.8	3.01	1	11.31	1.30			39.5	[93]
Florida caerulea		68.19	1.16	1	42.63	3.61			39.5	[1238]
Anas platyrhynchos		117.3	0.330	1	17.75	0.410			39.5	[499]
Anas platyrhynchos		151.3	0.353	1	17.04	0.307			39.5	[499]
Anas platyrhynchos		145.5	1.94	1	10.26	0.680			39.5	[794]
Anas platyrhynchos		154.8	1.65	1	13.14	4.56			39.5	[996]
Anser anser		181.5	2.99	1	7.895	0.626			39.5	[794]
Buteo buteo	f	103.7	1.17	1	27.57	1.34			39.5	[909]
Buteo buteo	m	95.99	1.11	1	27.90	1.45			39.5	[909]
Falco subbuteo		66.16	0.689	1	46.77	3.57			39.5	[104]
Meleagris gallopavo		256.1	9.89	1	4.340	0.782			39.5	[212]
Meleagris gallopavo		296.2	26.2	1	3.657	1.18			39.5	[212]
Phasianus colchicus	f	100.3	1.86	1	6.610	0.738			39.5	[794]
Phasianus colchicus	m	118.8	4.25	1	5.004	0.746			39.5	[794]
Gallus domesticus	f	136.5	1.24	1	4.625	0.209			39.5	[867]
Gallus domesticus	m	153.5	2.22	1	4.522	0.305			39.5	[867]
Bonasia bonasia		85.17	2.68	1	7.807	0.740			39.5	[87]
Colinus virginianus		56.90	0.328	1	10.81	0.427			39.5	[991]
Coturnix coturnix		55.41	0.761	1	14.94	0.784			39.5	[150]
Rallus aquaticus		51.66	0.730	1	14.45	0.0882			39.5	[1055]
Gallinula chloropus		67.05	1.20	1	20.00	1.72			39.5	[324]
Philomachus pugnax	f	47.41	1.04	1	39.46	2.75			39.5	[1018]
Philomachus pugnax	m	59.94	2.18	1	29.09	2.97			39.5	[1018]
Haematopus moquini		103.4	5.69	1	10.63	1.40			39.5	[514]
Chlidonias leucopterus		42.76	0.502	1	66.39	4.08			39.5	[574]
Sterna fuscata		57.94	0.364	1	22.21	1.07			39.5	[164]
Sterna dougalli		50.15	1.12	1	33.97	3.77			39.5	[689]
Sterna hirundo		46.74	1.10	1	35.29	4.76			39.5	[689]
Rissa tridactyla		76.07	0.715	1	32.98	1.79			39.5	[758]
Larus argentatus		115.1	1.70	1	16.53	0.791			39.5	[1090]
Catharacta skua		131.3	4.64	1	17.42	2.37			39.5	[1116]
Catharacta skua		100.5	0.610	1	40.69	3.12			39	[384]
Catharacta maccormicki		104.8	0.310	1	60.29	3.18			39	[384]
Stercorarius longicaudus		83.90	0.069	1	41.55	2.87			30	[384]
Ptychoramphus aleuticus		59.66	0.373	1	23.73	0.913			39.5	[1189]
Cuculus canoris		45.49	0.884	1	49.29	4.00			39.5	[1281]
Cuculus canoris		50.26	1.45	1	38.56	3.68			39.5	[1281]
Cuculus canoris		52.02	7.20	1	42.11	2.12			39.5	[1281]

Table 8.3 (cont.)

species	sex	length mm L_∞	sd mm	shape coeff $\delta_{\mathcal{M}}$	rate a^{-1} $\dot{r}_B$	sd a^{-1}	location NS	location EW	temp °C	source
Cuculus canoris		52.44	1.40	1	39.91	3.60			39.5	[1281]
Glaucidium passerinum	f	42.36	0.309	1	46.98	2.02			39.5	[1028]
Glaucidium passerinum	m	41.86	0.484	1	41.57	2.51			39.5	[1028]
Asio otus		64.94	0.596	1	36.54	1.77			39.5	[1253]
Tyto alba		68.25	1.18	1	21.68	2.70			39.5	[436]
Strix nebulosa		98.26	0.960	1	16.43	0.730			39.5	[793]
Steatornis capensis		94.59	5.24	1	12.96	2.39			39.5	[1086]
Apus apus		37.44	0.274	1	45.55	2.88			39.5	[1234]
Selasphorus rufus		16.33	0.475	1	58.44	9.88			≤ 41	[220]
Amazilia fimbriata		16.12	0.110	1	69.86	3.54			≤ 41	[473]
Ramphastos dicolorus		70.11	1.89	1	28.52	4.01			39.5	[145]
Sturnus vulgaris		40.83	0.332	1	82.71	5.04			41	[1241]
Bombycilla cedrorum		34.16	0.392	1	73.37	4.31			41	[963]
Petrochelidon pyrrhonota		31.19	0.520	1	69.64	6.40			41	[963]
Toxostoma curvirostre		36.62	0.695	1	49.82	3.62			41	[963]
Tyrannus tyrannus		35.53	0.673	1	59.43	4.60			41	[816]
Sylvia atricapilla		25.59	0.142	1	108.2	11.7			41	[97]
Garrulus glandarius		52.34	2.85	1	39.82	8.52			41	[595]
Campylorhynchus brunneicap.		32.79	0.200	1	65.85	6.70			41	[964]
Emberiza schoeniclus		25.88	0.238	1	138.7	12.1			41	[118]
Troglodytes aedon		22.29		1	105.9				41	[34]
Phylloscopus trochilus		22.41	0.576	1	76.78	8.86			41	[1026]
Parus major		27.47	0.207	1	59.90	2.33			41	[53]
Parus ater		23.40	0.232	1	75.74	3.88			41	[714]
Motacilla flava		9.910	0.298	2.913	55.19	4.42			41	[274]
Agelaius phoeniceus	f	35.94	0.951	1	75.16	7.58			41	[232]
Agelaius phoeniceus	m	40.66	0.529	1	65.28	2.74			41	[232]
Gymnorhinus cyanocephalus		44.84	0.596	1	49.68	2.97			41	[63]
Eremophila alpestris		30.81	1.24	1	75.98	10.9			41	[72]
Mammalia										
Macropus parma		148.6	0.615	1	2.736	0.0942			35.5	[762]
Macropus fuliginosus		261.6	34.8	1	2.397	0.910			35.5	[908]
Trichosurus caninus		137.8	1.06	1	1.754	0.561			35.5	[532]
Trichosurus vulpecula		139.3	1.34	1	3.715	0.184			35.5	[723]
Perameles nasuta		100.5	0.967	0.961	4.743	0.175			35.5	[723]
Setonix brachyurus		116.6		1	1.728	0.117			35.5	[1176]
Suncus murinus	f	26.58	0.160	1	30.92	1.37			37	[297]
Suncus murinus	m	29.88	0.267	1	20.64	1.27			37	[297]
Sorex minutus		65.00		0.294	32.97	0.674			36	[539]
Desmodus rotundus		30.68	0.175	1	8.775	0.277			(35.5)	[1021]
Homo sapiens	m	1648	58.5	0.244	0.1490	0.0158			37	[184]

Table 8.3 (cont.)

species	sex	length mm L_∞	sd mm	shape coeff $\delta_{\mathcal{M}}$	rate a^{-1} $\dot{r}_B$	sd a^{-1}	location NS	location EW	temp °C	source
Lepus europaeus		148.3	1.60	1	5.034	0.530			37	[152]
Oryctolagus cuniculus		116.6	1.11	1	6.507	0.272			37	[1176]
Notomys mitchellii		27.09	0.412	1	21.54	1.64			38	[230]
Notomys cervinus		23.85	0.456	1	23.94	3.00			38	[230]
Notomys alexis		27.43	0.382	1	20.03	1.24			38	[230]
Pseudomys novaehollandiae		24.88	0.101	1	13.00	0.386			38	[589]
Castor canadensis		234.4	1.64	1	5.117	0.365			38	[13]
Mus musculus		34.24	0.474	1	15.09	0.924			38	[867]
Mus musculus	f	31.87	0.129	1	22.33	1.31			38	[867]
Mus musculus	m	33.98	0.118	1	26.66	1.28			38	[867]
Rattus fuscipes		171.5	4.08	0.280	9.333	0.843			38	[1140]
Rattus norvegicus		75.23	0.301	1	9.286	0.279			38	[867]
Tachyoryctes splendens		64.87	0.992	1	8.231	0.680			38	[929]
Balaenoptera musculus		37810	5420	0.188	0.05884	0.0208			37	[1075]
Balaenoptera musculus	f	26200		0.188	0.2240				37	[711]
Balaenoptera musculus	m	25000		0.188	0.2160				37	[711]
Balaenoptera physalus	f	22250		0.180	0.2220				37	[711]
Balaenoptera physalus	m	21000		0.180	0.2221				37	[711]
Balaenoptera borealis	f	15300		0.197	0.1337				37	[711]
Balaenoptera borealis	m	14800		0.197	0.1454				37	[711]
Delphinapterus leucas	f	3056	54.4	0.254	0.2700	0.0399			37	[400]
Delphinapterus leucas	m	3589	86.5	0.254	0.1876	0.0227			37	[400]
Canis domesticus		387.2	1.46	1	4.168	0.120			37	[867]
Lutra lutra	f	178.1	1.32	1	2.870	0.156			37	[1109]
Lutra lutra	m	197.7	1.38	1	2.692	0.143			37	[1109]
Pagaphilus groenlandicus		486.4	7.44	1	0.4787	0.0673			37	[686]
Mirounga leonina	m	5580	356	0.254	0.1492	0.0265			37	[687]
Mirounga leonina	f	2933	42.7	0.254	0.3094	0.0480			37	[687]
Mirounga leonina	m	1799	149	1	0.1185	0.0278			37	[169]
Mirounga leonina	f	704.0	20.4	1	0.3661	0.0982			37	[169]
Leptonychotes weddelli		685.4		1	0.3001	0.0184			37	[169]
Loxodonta a. africana	f	1392	14.5	1	0.1016	8.16e-3			37	[688]
Loxodonta a. africana	m	1723	45.4	1	0.07173	7.81e-3			37	[688]
Rangifer tarandus	f	470.2	1.84	1	1.263	0.0589			37	[767]
Rangifer tarandus	m	534.4	4.39	1	1.000	0.0617			37	[767]
Bos domesticus	f	815.4	3.66	1	0.9957	1.73e-3			38.5	[867]
Alces alces		712.6	12.7	1	0.5930	0.159			37	[495]

If the von Bertalanffy growth rate is plotted against maximum volumetric length, the scatter is so large that it obscures their relationship. This is largely due to differences in body temperature. A fish in the North Sea with a yearly temperature cycle between 3 and 14 °C grows much more slowly than a passerine bird with a body temperature of 41 °C. This is not due to fundamental energy differences in their physiology. If corrected to a common body temperature according to the Arrhenius relationship with an Arrhenius temperature of 8 kK, the expected relationship is revealed and the differences between fishes and birds disappear. Since temperature had not been measured in most cases, I had to estimate it in a rather crude way. For most molluscs and fish data I used general information on local climate and guessed water temperatures (which depend on the, frequently unknown, depth). The body temperatures of birds and mammals have also been guessed. Uncertainties about temperature doubtlessly contributed the most to the remaining scatter. The corrected rates are not meant as predictions of actual growth rates at this body temperature because most North Sea fish and birds would die almost instantaneously if the temperature was realised. The large-bodied species dominate the estimate for $\dot{v}$, and the small-bodied species that for $\dot{k}_M$. Most small-bodied species are Ecdysozoa, which might have a large $\dot{k}_M$ due to moulting. Ricklefs [965] also found that the von Bertalanffy growth rate is inversely proportional to length for birds.

Length at puberty

If $\dot{k}_J = \dot{k}_M$, length at first at puberty is proportional to maximum length and has been found for clupeoid fishes, see Figure 8.7, and for flatfish (Pleuronectiformes) [1185].

Minimum juvenile period

The juvenile period at high food density for different species, as given by (2.53) scales approximately with z for $z >> g_{\text{ref}}$. This relationship fits Bonner's data, as given in Pianka [126, 892] very well; however, this data set uses actual lengths, rather than the more appropriate volumetric ones.

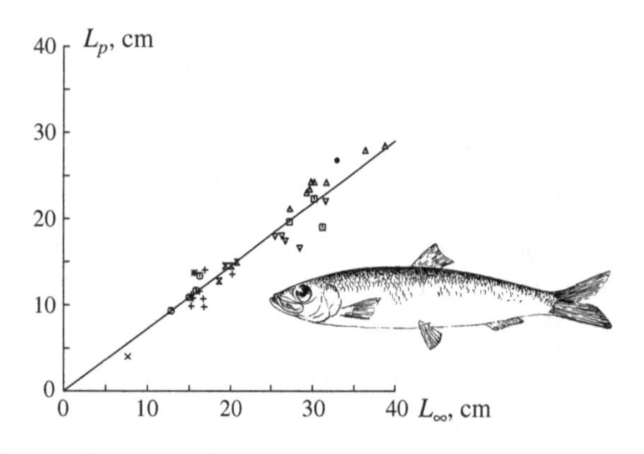

Figure 8.7 The length at first reproduction is proportional to the ultimate length in clupeoid fishes, Clupeoidei. Data from Blaxter and Hunter [115].

△	*Clupea*	▽	*Sardinella*
•	*Brevoortia*	+	*Engraulis*
○	*Sprattus*	⋆	*Centengraulis*
□	*Sardinops*	×	*Stolephorus*
⧖	*Sardina*		

Figure 8.8 The striped tenrec *Hemicentetes semispinosus* is a curious 'insectivore' of 110 g from the rain forests of Madagascar that feeds on arthropods and earthworms and finds its way about using sonar. Walking in the forest, you can spot it easily by its head shaking, not unlike that of an angry lizard. Its juvenile period of 35 days is the shortest among mammals. The gestation period is 58 days [315].

The *Guinness Book of World Records* mentions the striped tenrec *Hemicentetes semispinosus*, see Figure 8.8, as the mammal with the shortest juvenile period [683]. The cuis *Galea musteloides*, a 300- to 600-g South American hystricomorph rodent, usually ovulates at some 50 days, but sometimes does so within 11 days of birth [1052, 1212]. Many smaller mammals have a longer juvenile period, which points to the fact that body scaling relationships only give tendencies and not reliable predictions.

Maximum reproductive rate

The maximum reproductive rate, as given in (2.58), scales approximately with z^{-1}, since $l_T = z^{-3} l_T^{\text{ref}}$ hardly contributes for large zoom factors, v_H^p is independent of z and v_E^0 scales with z. This is a beautiful example showing that the size relationships within a species work out differently from those between species. Intra-species comparisons show that large individuals reproduce at a higher rate than small ones, while the reverse holds for inter-species comparisons. Like most of the other scaling relationships mentioned in this chapter, this only reflects tendencies that allow substantial deviations. The trade-off between a small number of large young and a large number of small young is obvious.

The allocation fraction κ does not depend on body size; thus, a small species spends the same fraction of energy that it utilises from its reserves on reproduction as a large species. (That is, if the energy required to maintain maturity is negligibly small.) If we express the energy spent on reproduction as a fraction of the energy taken up from the environment (at constant food density), as typically done in Static Energy Budgets, see {433}, this fraction decreases with increasing body volume. This is because ingestion rate increases with volume across species, see {309}, and utilised energy (respiration rate) with a weighted sum of surface area and volume. This illustrates once again the importance of explicit theories for the interpretation of data.

Time till death by starvation

The time till death by starvation for an individual with an initial scaled energy density of $e(0) = l$ was found to be $t_\dagger = \dot{v}^{-1} L \ln \kappa^{-1}$, see {118}, or $t_\dagger = \dot{v}^{-1} L \kappa^{-1}$ depending on its storage dynamics during starvation. In the first expression the individual does not change its storage dynamics, and in the second one it spends energy on maintenance only. In both cases, this time scales with z. Threlkeld [1155] found a scaling exponent of 3/4, but 1 also fits the data well.

Constant food densities thus select for small body volume, because small volume aids survival at lower food densities; fluctuating food densities select for large body volume, because a large body volume gives better survival over prolonged starvation [631]. Brooks and Dodson [158] observed that, in the absence of predators, the larger species of zooplankton dominate. The DEB theory suggests that the explanation does not lie in the size dependence of threshold food density (because this would operate the other way round), but in the length of periods for which no animal can find sufficient food. This has been confirmed experimentally by Goulden and Hornig [425].

Life span

Growth never stops in the most elementary formulation of the DEB model, but it is practical to consider the moment at which body volume exceeds $(1-\epsilon)^3 V_\infty$ as the end of the growth period, for some chosen small fraction $\epsilon = 0.05$, say. The length of the growth period at constant food density is given in (2.23) and amounts to $\dot{r}_B^{-1} \ln \epsilon(1 - l_b/f)$. It thus increases with volumetric length for different species, just as the juvenile period. Inspection of the ageing parameters as discussed at {214} reveals that $\ddot{h}_a = z\ddot{h}_a^{\text{ref}}$ and $s_G = s_G^{\text{ref}}$, so that $\dot{h}_W = \dot{h}_W^{\text{ref}}$ and $\dot{h}_G = \dot{h}_G^{\text{ref}}/z$. If the Weibull ageing pattern is followed (as most ectotherms do), the mean life span as far as ageing is concerned amounts to $\Gamma(\frac{4}{3})/\dot{h}_W$ and is thus independent of maximum size. Finch [350] concluded that the scanty data on life spans of ectotherms do not reveal clear-cut relationships with body volume. Large variations in life spans exist, both within and between taxa. The ratio of the growth period to the mean life span is $5.55\ddot{h}_a^{1/3}\dot{k}_M^{-2/3}(1 + f/g) \ln \epsilon(1 - l_b/f)$ and increases with volumetric length. If this ratio approaches 1, life span tends to increase with maximum body volume in a sigmoid manner.

If the Gompertz ageing pattern is followed (as most endotherms do), the median age at death is $\log(1 + \dot{h}_G^2 \log 2/\ddot{q}(0))/\dot{h}_G$. The incrementally small value of $\ddot{q}(0)$ relates to the small source of ROS, especially during early ages, and 'early' might depend on body size. The scaling of the median age is dominated by $\dot{h}_G$, however, so median age at death increases with volumetric length for endotherms. This is consistent with the conclusion of Finch [350] that the life span of mammals scales with weight$^{0.2}$, given that weight increases faster than structural body volume.

If ageing allows long life spans, individuals are likely to have effective means for dealing with a threatening environment, such as avoidance behaviour for dangerous situations (learning), physiological regulation to accommodate changes in diet, temperature and so on. This is likely to involve large brain size and thus an indirect coupling between brain size and life span. The brain may also be involved in the production of antioxidants or the regulation thereof, which makes the link between brain size and life span more direct. Birds have larger brain-to-body-weight ratios than mammals and live twice as long. The life spans of both mammals and birds tend to scale empirically with weight$^{0.2}$ [182, 350], which is close to volume$^{1/3}$. This is consistent with the

DEB-based expectation, because surface-bound heating costs dominate respiration, and thus ageing. Brain size is found, empirically, to be approximately proportional to surface area in birds and mammals [182]. Mammals tend to have higher volume-specific respiration rates than birds [1232], which contributes to the difference in mean life span and jeopardises easy explanations.

It must be stressed that these life span considerations relate to ageing, though it is doubtful that ageing is a major cause of death under field conditions. Suppose that size and age independent of death dominate under those conditions and that food web interactions work out such that the population remains at the same level while food is abundantly available. To simplify the argument, let us focus on species that have a size at first maturation close to the ultimate size. The death rate can then be found from the characteristic equation (9.20) for $\dot{r} = 0$ and $\Pr\{\underline{a}_\dagger > a\} \simeq \exp(-\dot{h}a)$ and $\dot{R}(a) \simeq (a > a_p)\dot{R}_m$. Substitution gives $\exp(-\dot{h}a_p) = \dot{h}/\dot{R}_m$. I have shown already that the age at first maturation a_p increases almost linearly with length, {322}, and the maximum reproduction rate $\dot{R}_m$ decreases with length, {323}. The death rate $\dot{h}$ must, therefore, decrease with length, so that the life span $\dot{h}^{-1}$ increases with length.

These considerations help to explain the results of Shine and Charnov [1050], who showed that the product of the von Bertalanffy growth rate and the life span is independent of body size for snakes and lizards. Charnov and Berrigan [202] argued that the ratio of the juvenile period to the life span is also independent of body size. They tried to understand this empirical result from evolutionary arguments. Since the juvenile period is approximately proportional to length as well, {322}, the ratio of this period to the life span is roughly independent of body size. The present derivation also specifies the conditions under which the result is likely to be found, without using evolutionary arguments.

8.2.3 Tertiary scaling relationships

Primary and secondary scaling relationships follow directly from the the separation of parameters in intensive and design parameters. The class of tertiary scaling relationships invokes indirect effects via the population level. The assumptions that lead to the DEB model, Table 2.4, must for tertiary scaling relationships be supplemented with assumptions about individual interactions. This makes tertiary scaling relationships a weaker type. Body size scaling relationships are usually much less obvious at the community level [238], because of a multitude of complicating factors. Nonetheless, they can be of interest for certain applications.

Abundance

Geographical distribution areas are frequently determined by temperature tolerance limits; see {17}. Temperature and food abundance also determine species abundance in more subtle ways.

Since both the maximum ingestion rate and maintenance costs are proportional to body volume across species, abundance is likely to scale with z^{-3}, so $N \propto z^{-3} N_{\rm ref}$. This has been found by Peters [884], but Damuth [243] gives a scaling of -0.76 with body volume. This relationship can only be an extremely crude one. Abundances depend on primary production levels, positions in the food web, etc. Nee *et al.* [828] point to the relationships between phylogenetic position, position in food webs and abundances in birds.

Distribution

High food densities go with large ultimate body sizes within a species. If different geographical regions which differ systematically in food availability are compared, geographical races can develop in which these size differences are genetically fixed. Since high food densities occur more frequently towards the poles and low food densities in the tropics, body sizes between these races follow a geographical pattern known as the Bergmann rule; see {297}.

It is tempting to extend this argument to different species feeding on comparable resources. This is possible to some extent, but another phenomenon complicates the result. Because of the yearly cycle of seasons, which are more pronounced towards the poles, food tends to be more abundant towards the poles in the good season, but at the same time the length of the good season tends to shorten. The time required to reach a certain size (for instance the one at which migration is possible) is proportional to volumetric length. This implies that maximum size should be expected at the polar side of the temperate regions, depending on parameter values, migratory behaviour, endothermism, etc. This probably holds for species such as geese, which migrate to avoid bad seasons. Geist [392] reported a maximum body weight at some 60° latitude and smaller weights both at higher and lower latitudes for New World deer and races of wolves. He found a maximum body size for sheep at some 50° latitude. Ectotherms that stay in the region can 'choose' the lower boundary of the temperature tolerance range such that they switch to the torpor state as soon as the temperature drops to a level at which food becomes scarce. This reduces the growth rate, of course, but not the ultimate body size. Whether the mean body size in a population is affected then depends on harvesting mechanisms.

Population growth rate

Since the (maximum) reproduction rate decreases with a length measure and the juvenile period increases with a length measure, the maximum population growth rate decreases somewhat faster than a length measure, especially for the small species. A crude approximation is the implicit equation obtained from (9.21)

$$\exp(-\dot{r} z a_p^{\rm ref}) = z\dot{r}/\dot{R}_m^{\rm ref} \tag{8.6}$$

For dividing isomorphs, the population growth rate is inversely proportional to the division interval, which corresponds to a juvenile period from an energetics point of view. This gives $\dot{r} = z^{-1}\dot{r}_{\rm ref}$. Fenchel [342] obtained an empirical scaling of weight$^{-1/4}$ for protozoa, and Niklas [837] obtained a value of (gram C)$^{-0.213}$ for cyanobacteria and (gram C)$^{-0.22}$ for unicellular algae. Correction for the contribution of reserves in the size measures gives results very close to the expected scaling relationship.

8.3 Quantitative structure–activity relationships

The concept of one-compartment kinetics has many hidden implications. This section discusses how the parameters covary and determine partition coefficients.

The most obvious property of chemicals for the understanding of toxicokinetics is the n-octanol/water partition coefficient, P_{ow}, which can be estimated from the chemical structure of the compound. Octanol serves as a model for typical lipids of animals, although the model is not always perfect [1059]. It has a density of 827 $\mathrm{g\,dm^{-3}}$, and a molecular weight of 130 dalton, so that 1 $\mathrm{dm^3}$ of octanol contains 6.36 mol. Most comparisons are restricted to the interval $(10^2, 10^6)$ for the P_{ow}. The size of the molecule tends to increase with P_{ow} and, if the P_{ow} is larger than 10^6, the molecules are generally too big to enter cells easily [218].

8.3.1 Kinetics as a function of partition

The molecular details of the transport of a compound between two matrices, such as octanol and water, directly lead to the relationship between the elimination rate and P_{ow}; the P_{ow} has information about the steady state, the elimination rate about the waiting time to reach that steady state. Let us focus on a closed system that evolves to a steady state.

Suppose that N molecules of a compound are distributed over two matrices, and that they can freely travel from one matrix to the other. Both matrices occupy a unit of volume. $N_0(t)$ molecules are present in matrix 0 and $N_1(t) = N - N_0(t)$ molecules in matrix 1 at time t. If first-order kinetics applies, and the total number of molecules $N = N_0(t) + N_1(t)$ is constant, the change of N_0 is given by

$$\frac{d}{dt}\begin{pmatrix} N_0 \\ N_1 \end{pmatrix} = \begin{pmatrix} -\dot{k}_{01} & \dot{k}_{10} \\ \dot{k}_{01} & -\dot{k}_{10} \end{pmatrix} \begin{pmatrix} N_0 \\ N_1 \end{pmatrix} \quad \text{or} \tag{8.7}$$

$$\frac{d}{dt}N_0 = \dot{k}_+(N_0^* - N_0) \tag{8.8}$$

with $\dot{k}_+ = \dot{k}_{01} + \dot{k}_{10}$ and the equilibrium value for N_0 is $N_0^* = N\dot{k}_{10}/\dot{k}_+$. An implicit assumption is that the compound is homogeneously distributed within each matrix.

Suppose now that the exchange rates are proportional to the ratio of the binding forces to the two matrices, i.e. $\dot{k}_{01} = \dot{k}\rho_1/\rho_0$ and $\dot{k}_{10} = \dot{k}\rho_0/\rho_1$, where ρ_i is the binding force of the compound to molecules of matrix i, $i \in \{0, 1\}$, and $\dot{k}$ is a proportionality constant that depends on the properties of the compound, but not on those of the

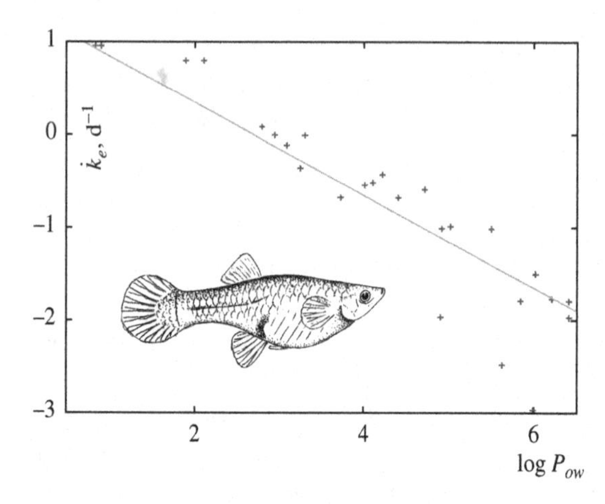

Figure 8.9 The elimination rate in the guppy *Poecilia reticulata* is approximately proportional to $1/\sqrt{P_{ow}}$ for polycyclic hydrocarbons at 23 °C. Data from Voogt *et al.* [1202].

matrix. Although phrased differently, the setting is identical to the concept of fugacity, the escaping tendency of a compound from a phase, that has been successfully used to describe the behaviour of compounds in the environment [730]; it has a simple thermodynamic interpretation [666].

The definition of the partition coefficient is $P_{01} = N_0^*/N_1^*$. Since $N_0^*/N_1^* = N_0^*/(N - N_0^*) = \dot{k}_{10}/\dot{k}_{01} = \rho_0^2/\rho_1^2$, we have that $\rho_0/\rho_1 = \sqrt{P_{01}}$. The result directly follows that $\dot{k}_{01} = \dot{k}\sqrt{P_{10}}$ and $\dot{k}_{10} = \dot{k}\sqrt{P_{01}} = \dot{k}/\sqrt{P_{10}}$, see Figure 8.9.

The bioconcentration coefficient P_{Vd} for fish relates to the octanol/water partition coefficient as $P_{Vd} = 0.048\, P_{ow}$ [729]. Hawker and Connell [475] found the allometric relationships $P_{Vd} = 0.0484\, P_{ow}^{0.898}$ for daphnids and $P_{Vd} = 0.0582\, P_{ow}^{0.844}$ for molluscs in the range $10^2 \leq P_{ow} \leq 10^6$. The scatter in the data is big enough for the relationship $P_{Vd} = 0.02\, P_{ow}$ to apply to both daphnids and molluscs. The proportionality factor directly relates to the fat content. In general we can say that $P_{Vd} = m_o P_{ow}$, where m_o stands for the mass-specific octanol equivalent of the organism, which seems to be taxon-specific. High correlations between P_{Vd} and m_o have been found for fenitrothion in a variety of algae [592], for instance.

Hawker and Connell [474, 475] related the elimination rate $\dot{k}_e$ to P_{ow} and found $\dot{k}_e = 8.851\, P_{ow}^{-0.663}$ d^{-1} for fish, $\dot{k}_e = 113\, P_{ow}^{-0.507}$ d^{-1} for *Daphnia pulex* and $\dot{k}_e = 9.616\, P_{ow}^{-0.540}$ d^{-1} for molluscs. The proportionality factor is inversely proportional to the volumetric length of the animal (see below), which explains the wide range of values. The results for daphnids are most reliable, because they all have the same body size in this case, and confirm the expectation $\dot{k}_e \propto P_{ow}^{-1/2}$, which is based on first-order kinetics. Some workers proposed diffusion layer models where the uptake rate depends hyperbolically on the membrane–water partition coefficient [357], but the derivation neglects the link between diffusion rates and partition coefficients. Others take elimination rates inversely proportional to the animal–water partition coefficient [414, 1149], with the odd implication that the uptake rate is independent of the partition coefficient. This is not consistent with first-order kinetics, where the two media play roles that are exchangeable, which implies a skew-symmetrical relationship between the

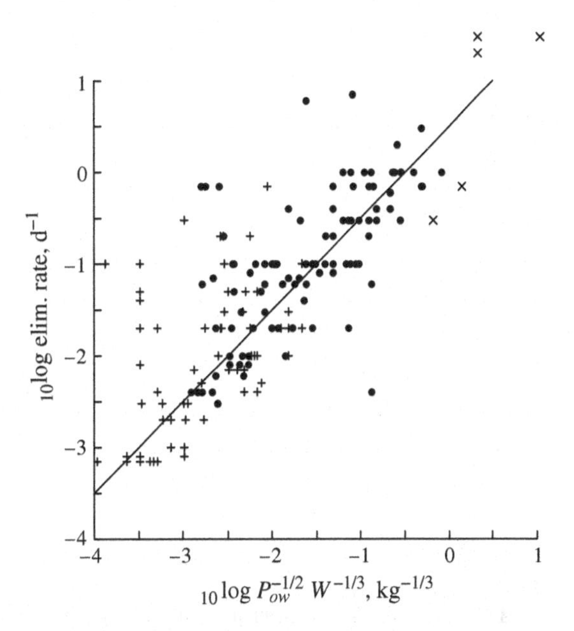

Figure 8.10 The elimination rate depends on the n-octanol/water partition coefficient P_{ow} and the weight W of an organism. It is roughly proportional to $P_{ow}^{-1/2}W^{-1/3}$ with proportionality constant $\sqrt{10}$ d^{-1}kg$^{-1/3}$ for 181 halogenated organic compounds in fish. Data compiled by Hendriks [492]. The marker codes are: $P_{ow} \leq 10^2$ ($\times$), $10^2 \leq P_{ow} < 10^6$ ($\bullet$), $10^6 \leq P_{ow} \leq 10^8$ (+). The range of fish weights is 0.1–900 g. No corrections for differences in temperature have been made, nor for differences in fat content of the fish.

uptake and elimination rates as functions of the partition coefficient; the square root relationship is the only one that satisfies the skew symmetry.

DEB theory predicts that the density of octanol equivalents increases with the body size of the different species of animal because the specific maximum reserve capacity $[E_m]$ increases with a volumetric length, see {300}, and the reserve of some taxa is relatively rich in lipids. These are only general trends and many exceptions occur. Data from Hendriks [492] confirm the general trend; see Figure 8.10.

8.3.2 Film models

The time we have to wait to saturate the tissue of a blank organism to a fraction x of the ultimate level is $t_x = -\dot{k}_e^{-1}\ln(1-x)$. Figure 8.11 illustrates how this time relates to the P_{ow} and is consistent with [1036] for air–water exchange, with [357] for artificial membranes, and with [414] for fish. For low P_{ow} values, the elimination rate hardly depends on the P_{ow} and for large values it is inversely proportional to P_{ow}. As a consequence, the opposite holds for the uptake rate. The result of Thomann [1148] is largely consistent with this relationship, if applied to the proper range of P_{ow} values. Notice that this reasoning does not make use of the considerations for how the exchange rates k_{ij} depend on P_{ij} as presented in the previous section; this is because they do not occur independently in this steady-state flux model, but only in combination as a ratio in the form of P_{ij}.

The steady-flux approximation of two-film model has a one-film model as special case, where e.g. $L_j \to 0$ or $\dot{v}_j \to \infty$. The elimination rate then reduces to $\dot{k}_e = \dot{k}_i(1 - \dot{v}_i/\dot{v}_{ij})^{-1}$, and we must have that $\dot{v}_i < \dot{v}_{ij}$. Since $\dot{v}_{ij} \propto 1/\sqrt{P_{ij}}$, this approximation is only valid in a limited range of P_{ij} values, depending on the value of $\dot{v}_i$.

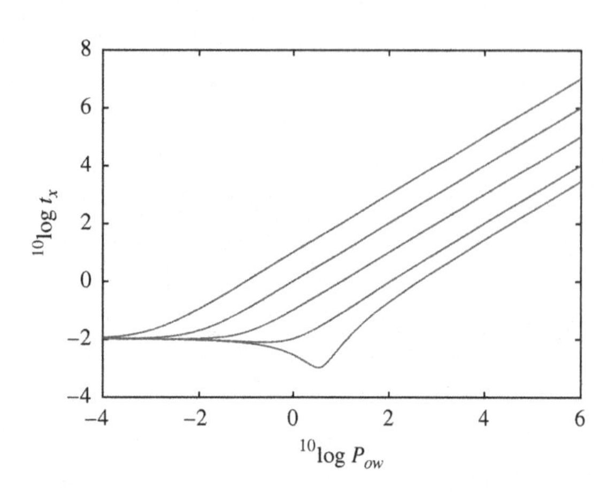

Figure 8.11 A log–log plot of the time to reach an x-level saturation in the tissues of an organism exposed in an environment with a constant concentration of a compound in a two-film model, using the steady-flux approximation. The curves correspond with different values of the velocity $\dot{v}_1$; the upper curve has the lowest velocity. Parameters: $\dot{v}_0 = 1$, $v_1 = .001, .01, .1, 1, 3.6\,\text{mm}\,\text{h}^{-1}$. $\dot{v}_{01} \propto P_{ow}^{-1/2}$, $\dot{v}_{10} \propto P_{ow}^{1/2}$, $\dot{k}_0 = 10\,\text{h}^{-1}$, $x = 0.1$. Notice that the $\dot{v}_1 = 3.6\,\text{mm}\,\text{h}^{-1}$ is close to its maximum value in this parameter combination to ensure positive elimination rates; the steady-flux approximation is probably very poor in this situation.

This illustrates a serious problem with this steady-flux approximation: contrary to the full PDE formulation, we cannot reduce this approximation in a smooth way to the well-mixed special case of a one-compartment model. If we increase the diffusivities $\dot{D}_i$ and/or reduce the thickness of the non-mixed layers L_i, we are forced to increase the exchange rates $\dot{k}_{ij}$ as well to ensure that the transport in the layers is still in pseudo steady state. In other words: the dynamics of the system disappears, and the whole system equilibrates instantaneously. We will see $\dot{k}_{01}$ and $\dot{k}_{10}$ will occur independently in other approximations of the two-film PDE model that do not suffer from this problem. This approximation is popular in situations where the bulk volumes are infinitely large and represent the ocean and the atmosphere, for instance. It then becomes a reasonable assumption to take a constant flux from one medium into the other, without changes in the bulk concentrations. This application is quite different from that in toxicokinetics, where one medium represents a initially blank fish, and the other an aquarium with a compound, and we study the toxicokinetics and effects in transient states.

8.3.3 Bioconcentration coefficient

The bioconcentration coefficient, BC, is an important concept in the kinetics of xenobiotics. It is used among other things as a crude measure to compare xenobiotic compounds and species and to predict effects. For aquatic species and hydrophilic compounds, it is usually defined as the ratio of the concentration in the organism to that in the water, which are both taken to be constant. For terrestrial species and/or lipophilic compounds, it is usually defined as the ratio of the concentration in the organism to that in the food. Applying the BC concept is a bit complicated in the present context, because the concentration in the organism does not become stationary, because of growth and reproduction, even if the concentration in the environment is constant, i.e. in water, food and at constant food density. If the growth rate is low in comparison to the exchange rates, the compound can be in pseudo-equilibrium, but its concentration still depends, generally, on the size of the organism. In addition, reproduction causes a cyclic change in concentration. The oscillations become larger if the organism accumulates its reproductive output over a longer time period. If food density is constant for a long enough period, we have $e = f$ and $\frac{d}{dt}e = 0$. The ultimate concentration on the basis of wet weight then reduces for low growth and reproduction rates to

$$\langle M_Q \rangle_w \to \frac{P_{Vd}c_d + P_{VX} f c_X}{[W_w]} \left(1 + \frac{[M_{Em}]}{[M_V]} P_{EV} f\right) \tag{8.9}$$

This expression can be used to predict how BC depends on body size if species are compared on the basis of the theory presented on {299}. Since P_{VX} is proportional to $\dot{J}_{Xm}$, BC is expected to be linear in the volumetric length. The trend in $[E_m]$ almost cancels out. Figure 8.12 illustrates that the BC for the highly lipophilic compound 2,4,5,2′,4′,5′-hexachlorobiphenyl (PCB153) for aquatic animals is indeed linear in the volumetric length.

This expectation is thus based solely on differences in the uptake of the amount of food. Accumulation in the food chain occurs particularly in terrestrial habitats, and more debatably in aquatic ones. Since top predators tend to have the largest body size, it can be difficult to distinguish food chain effects from body size effects. Food chain effects operate through the partition coefficient for food/water, and body size effects act via the uptake of food.

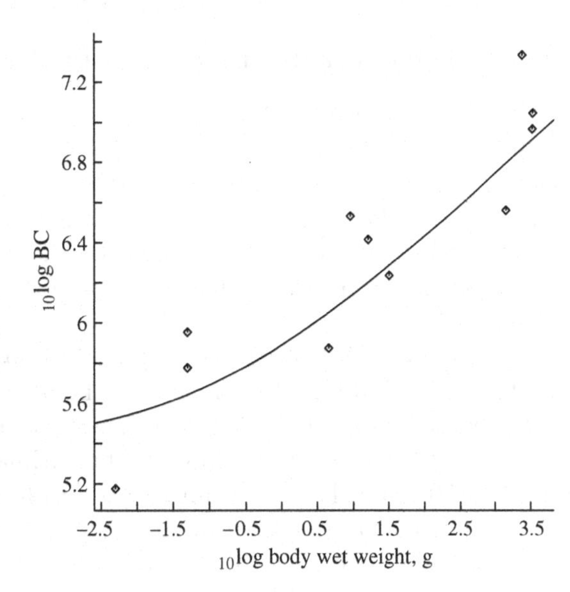

Figure 8.12 Bioconcentration coefficients (BCs) for PCB153 in aquatic organisms in the field, as given in Kooijman and van Haren [656]. Data from Oliver and Niimi [836, 848] and from the Dutch Ministry of Public Works and Transport. The curve represents the least-squares fit of the linear relationship between the BC and the volumetric length. $P_{VX}L_m = 46$ mm.

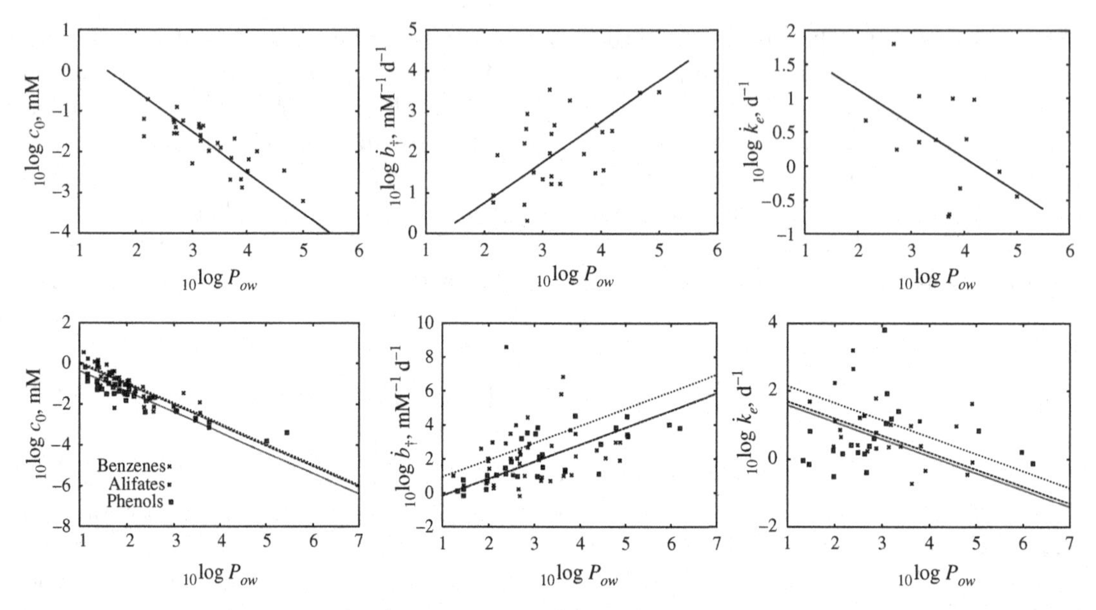

Figure 8.13 The $_{10}\log$ NEC (left), killing rate (middle) and elimination rate (right) of alkyl benzenes (top) and benzenes, aliphatic compounds and phenols (bottom) as a function of the $_{10}\log$ octanol/water partition coefficient. The slopes of the lines, i.e. -1, 1 and -0.5, respectively, follow from simple theoretical considerations. The data in the top panels are from the 4-d bioassays on survival of the fathead minnow *Pimephales promelas*, as presented in [391]. The partition coefficients were obtained from [960] or calculated according to [953]. The data in the bottom panels are from [731] (NECs, killing rates), [474] (elimination rates). The toxicity data originate from [153, 154, 155, 156, 390], as reported in [398]. From Kooijman *et al.* [650].

8.3.4 Effects as a function of partition coefficients

Since the equilibrium tissue concentration is proportional to P_{ow}, we should expect to find that the killing rate $\dot{b}_\dagger \propto P_{ow}$, the tolerance concentration $c_* \propto P_{ow}^{-1}$ and the NEC $c_0 \propto P_{ow}^{-1}$, while the elimination rate $\dot{k}_e \propto P_{ow}^{-0.5}$. Figure 8.13 and the empirical study by de Wolf [1268] support these expectations.

Most of the scatter is due to the fact that octanol is not an ideal chemical model for biomass, and disappears when the log NEC is plotted against the log killing rate, see Figure 8.14.

Könemann [619] observed that the 14 days log LC50 of the guppy *Poecilia reticulata* for 50 'industrial chemicals' is LC50 $= 0.0794\, P_{ow}^{-0.87}$ mol dm^{-3}. To understand this relationship, we have to realise that for a large elimination rate, so a small P_{ow}, the 14 days LC50 is close to the ultimate value, but for a large P_{ow}, the ultimate LC50 is much lower than the LC50.14d. Taking these complexities into account, Figure 8.15 confirms that $\dot{b}_\dagger \propto P_{ow}$ and $\dot{k}_e \propto P_{ow}^{-0.5}$ are indeed consistent with the finding by Könemann. Unfortunately, the data of Figure 8.15 did not allow us to check the relationship for the NEC. Although the NECs had been set to zero,

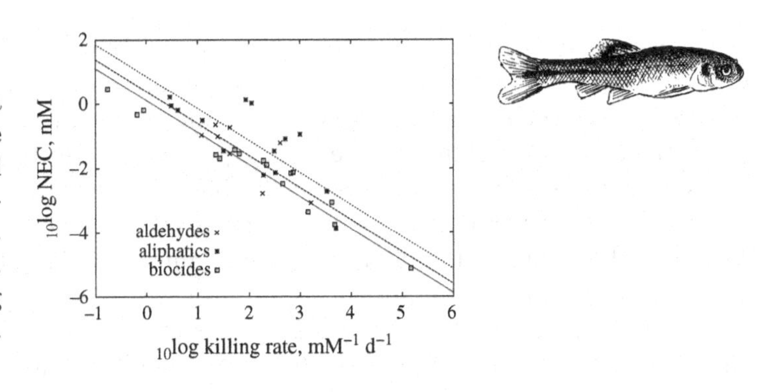

Figure 8.14 The log NEC as a function of the log killing rate for aldehydes, aliphatics and biocides in the fathead minnow *Pimephales promelas*. Data from Gerritsen [398]. The slope is −1, as resulting from theoretical predictions. From Kooijman *et al.* [650].

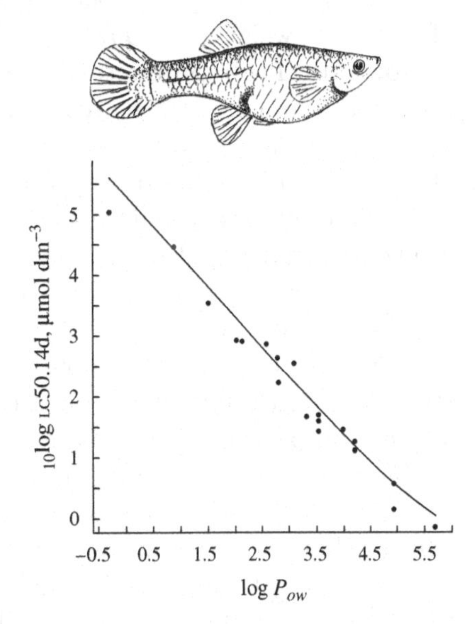

Figure 8.15 The 14 days LC50 values as a function of octanol/water partition coefficient for guppies *Poecilia reticulata* exposed to 21 chlorinated aromatic and other chlorinated hydrocarbons whose P_{ow} ranged from $10^{-0.22}$ (pentachlorobenzene) to $10^{5.21}$ (acetone). Data from Könemann [618]. The calculations are based on the assumptions that the elimination rates equal $50/\sqrt{P_{ow}}\,\mathrm{d}^{-1}$, the killing rates equal $10^{-6.6}P_{ow}\,\mathrm{d}^{-1}\mu\mathrm{mol}^{-1}\mathrm{dm}^3$ and the NECs are zero (see text).

adopting the function NEC $= 10P_{ow}^{-1}$ mmol dm^{-3} hardly changes the result. The conclusion is that the quantitative structure–effects relationships for LC50s follow from first principles.

8.4 Interactions between QSARs and body size scaling relationships

Body size affects chemical kinetics in rather complex ways, so do changes in body size. Since DEB theory is about the dynamics of body size, this directly links to toxicokinetics [649]. It is useful to start with an inventory of the possible uptake and elimination routes of the compounds under consideration, and then consider other chemical and metabolic aspects.

Uptake can be directly from the environment, which is proportional to the surface area of individuals. The implication is that elimination rates are inversely proportional to length. So the time it takes to saturate an organism with a chemical compound is proportional to its (volumetric) length. Uptake can also be via food, and food uptake scales with surface area intra-specifically, but with volume inter-specifically.

Dilution by growth matters, even at low growth rates. The growth rate depends on the size of the individual, relative to the maximum size, so intra- as well as inter-specific scaling relationships contribute.

Elimination can be directly to the environment (involving surface area), and/or to the gut contents (involving the feeding rate), and/or via reproduction or some other species-specific routes. The possible significance of the latter route is obvious from the observation that a female adult daphnid can produce offspring at the rate of 25% of her own weight *per day*. Eggs can represent an important elimination route for compounds. The reproduction rate (in number of offspring per time) is proportional to a weighted sum of surface area and volume intra-specifically, and inversely proportional to a length inter-specifically. Since the mass per offspring is proportional to volume, allocation to reproduction is proportional to surface area inter-specifically. As discussed before, however, the relative size of offspring is a lot more species-specific (so subjected to evolutionary adaptation) than the allocation to reproduction [189, 641, 1185].

The *chemical composition* of biomass also depends on size, since the reserve density (so the ratio of the amounts of reserve and structure) is constant intra-specifically, but proportional to a length inter-specifically. Reserve might be more rich in lipids than structure (depending on the taxa that are studied). This observation obviously matters for the comparison of compounds that differ in P_{ow}.

Chemical transformation in an organism is linked to the metabolic activity of the organism. Lipophilic compounds are frequently transformed into less lipophilic ones, which enhances excretion (elimination). These metabolites are, frequently, more toxic. Moreover, uptake, elimination and transformation frequently involve metabolic activity, which might be linked to the mobilisation rate of reserve.

If effects are receptor mediated, metabolic transformation is even more important [551]; turnover of receptors is possibly linked to metabolic activitiy or to somatic maintenance, in which case the specific turnover rate is independent of body size. The observation that effects are linked to the product of concentration and exposure time motivated many toxicologists to think about the involvement of receptors, although their biochemical identification remained uncertain. This motivation is incorrect, however, if the hazard rate is linear in the (internal) concentration. This is because even without receptors the effect on the survival probability is already via the product of concentration and exposure time. The significance of receptors is in the contribution of the exposure *history* in the effect, rather than of the actual exposure. This requires an in-depth analysis of how effects build up in time and imposes strong constraints on the quality of data. It was only by analysing multiple endpoints simultaneously that we found indications that the effects of organophosphorus esters on fish involve receptors.

These considerations invite for a second thought about effects of chemicals. As long as lipophilic compounds are accumulated in metabolically rather inactive lipids, they are less likely to have metabolic effects. Many animals, and especially mammals, have tissues (the adipose tissue) that are specialised in the storage of such lipids. As soon as these lipids are used, however, effects might show up. This calls for a much more dynamic view on the effects of chemicals, and links up with traditions in pharmacokinetics and medical research on the effects of chemicals.

8.5 Summary

Intra-specific parameter variations have phenotypic and genotypic components; the latter is key to evolutionary change. The Bergmann rule is explained as an adaptation to intra-specific variation patterns in food availability.

The inter-specific covariation of the primary parameters is derived on the basis of the classification of primary parameters into intensive and design parameters, and then that of functions of these parameters, and finally that of quantities that relate more indirectly to primary parameter values and involve interactions between organisms and properties of the environment. They all capture observed scaling relationships very well. The functions of parameters relate to many aspects of life history, and physiological quantities, such as respiration. Many fruitless attempts have been made to explain why respiration scales with body weight to the power 0.75 (approximately). The explanation offered by the DEB theory is that the inter-specific scaling results from an increasing contribution made by the reserves to body weight; the intra-specific scaling results in the decreasing contribution of growth to the energy budget for increasing body weight.

The fugacity argument provides the rules for how kinetic parameters of the 1,1-compartment model covary among compounds with the octanol/water partition coefficient. Film models are variations of the 1,1-compartment model, which allows the evaluation of the covariation of its parameters as well. The covariation of the transport parameters is behind the scaling of effect parameters, so behind the QSARs. Finally the interactions between body size scaling relationships and QSARs are discussed.

9

Living together

The primary purpose of this chapter is to evaluate the consequences of DEB models for individuals at the population and higher levels if extremely simple rules are defined for the interaction between individuals and the energy balance of the whole system. The first section deals with trophic interactions between species, and the constraints on parameter values that ensure a stable coexistence. Then follows a discussion of population dynamics, food chains and (simple) ecosystems.

9.1 Trophic interactions

DEB theory can be used to analyse the dynamics of systems with complex types of mass exchange between the participants in trophic relationships, a rich spectrum ranging from competition to predation. The present aim is to discuss some constraints in these patterns that ensure weak homeostasis of structural masses: the relative abundance of the structural masses of the participating species is independent of the substrate densities in the environment *at steady state*. This matter is taken up again in the discussion on merging {406}.

Trophic relationships are hard to classify; all relationships seem to be unique at close inspection. They are usually based on the judgement of being beneficial for one or both partners, and many different definitions exist for particular inter-species relationships. The oxpecker *Buphagus* feeds on insects that are attracted to wounds of giraffes, antelopes and other bovids; it is not difficult to see why the thin-skinned small antelopes make evident that they do not really appreciate this 'help' from the birds: oxpeckers try to keep wounds attractive for insects. I observed what solution oxpeckers have when wounds are in short supply. I will refrain from a judgement of benefits, and discuss the various relationships purely on the use of substrates. This is not meant to imply, however, that non-trophic relationships are of little importance to population dynamics.

9.1.1 Competition and species diversity

When two species feed on the same substrate in a well-mixed environment, they are said to compete for that substrate. The ratio of the structural masses of two competing V1-morphs is constant, despite variations in the substrate concentration, if the specific population growth rates are identical, so $\dot{r}_1 = \dot{r}_2$, for $\dot{r}_i = \frac{f_i \dot{k}_{Ei} - g_i \dot{k}_{Mi}}{f_i + g_i}$, for $i = 1, 2$, and f_i is the scaled functional response. The growth rates are only equal if $K_1 = K_2$, $[\dot{p}^1_{Am}] = [\dot{p}^2_{Am}]$, $\dot{k}^1_M = \dot{k}^2_M$, $\dot{k}^1_E = \dot{k}^2_E$ and $g_1 = g_2$. In other words, the ratio is only constant if the species are virtually identical in all their energetic properties. The significance of this remark is that syntrophic relationships allow more differences between the species to maintain weak homeostasis. The strict constraints for weak homeostasis explain why pure forms of inter-species competition are rare; most competing species have partially overlapping diets, and differ in preferences. Competition is perhaps most frequent among primary producers, but even they differ in preferences for the various chemical species of nutrients (ammonia versus nitrate, organic nitrogen, etc.).

The literature on population dynamics stresses the competitive exclusion principle: the number of competing species cannot exceed the number of substrates at steady state. The theoretical value of the result is limited, however. Lack of sustainable diversity in community models is only problematic in models with simplistic views on chemical aspects. Mechanisms that maintain diversity (also in community models) are in decreasing order of significance: (1) mutual syntrophy, where the fate of one species is directly linked to that of another; (2) nutritional 'details': the number of substrates is actually large, even if the number of species is small; (3) social interaction, which means that feeding rate is no longer a function of food availability only; (4) spatial structure: extinction is typically local only and followed by immigration from neighbouring patches; (5) temporal structure: complex systems easily have cyclic or chaotic behaviour, even in homogeneous environments, and steady states hardly occur in nature. Changes in feeding conditions come with changes in biomass composition, and so in nutritional value, which feeds back to cause (2). The least important cause, cause (5), has received most attention in the literature.

9.1.2 Syntrophy

Two species have a syntrophic relationship if a recipient species lives off the products of a donor species. The term commensalism is frequently used when the donor does not experience adverse effects. Syntrophy is very common, but the coupling varies from very direct to indirect.

The processing of food requires symbiosis; many animals feed on cellulose-containing phototrophs, but no animal can itself digest cellulose. Most animals have associations with prokaryotes, amoebas and flagellates to digest plant-derived compounds [1076]. These micro-organisms transform cellulose to lipids in the anaerobic intestines of their host animal; the lipids are transported to the aerobic environment of the tissues of the animal for further processing. Attine ants even culture fungi to extract cellulases

[747]. Many symbioses are still poorly understood, such as the Trichomycetes, which live in the guts of a wide variety of arthropods in all habitats [799]; the role of smut fungi (Ustilaginales) in their symbioses with plants also seems more complex than just a parasitic relationship [1183].

When tree leaves fall on the forest floor, fungi release nutrients locked in them by decomposition; the soil fauna accelerates this degradation considerably [1236]. Without this activity by fungi and the soil fauna, trees soon deplete the soil of nutrients, as most leaves last for only one year, even in evergreen species. As mentioned, trees, and plants in general, also need mycorrhizae to release nutrients from their organic matrix.

Faeces, especially those of herbivores, represent nutritious food for other organisms. Organisms specialised on the use of faeces as a resource are known as coprophages. Examples are the bryophyte *Splachnum*, which lives off faeces of herbivores (*S. luteum* actually lives off that of the moose *Alces alces*); the fungus *Coprinus* which lives off mammalian faeces, similar to beetles of the dung beetle family Scarabaeidae and the fly *Sarcophaga*.

The nitrifying bacteria *Nitrosomonas* and *Nitrobacter* oxidise ammonia to nitrite, and nitrite to nitrate, respectively, while other groups (*Pseudomonas*, *Micrococcus*, *Thiobacillus*) convert nitrate to dinitrogen. An even more indirect coupling exists between plants and oceanic diatoms, where plants mobilise silica from rocks [112], which diatoms require to make frustules; terrestrial plants allow diatoms to play a leading role in the plankton of the oceans.

The house dust mite *Glycyphagus domesticus* lives, with help of the fungus *Aspergillus repens*, off human skin flakes (their production is coupled to maintenance); these mites frequently cause allergic reactions in humans, which might stimulate flake production. The moth *Hypochrosis* drinks tears of big mammals, such as Asian elephants and, incidently, humans, but stimulates tear production at the same time. The sucking of mammalian blood by mosquitoes or of plant saps by mistletoes or aphids is only a small step further towards a biotrophic relationship. The honey guider *Indicator* guides mammals (e.g. badgers and humans) to bees' nests, itself feeding on the wax that is left over after the nest has been opened by the guided animal. The birds' activity might increase its average feeding rate as well as that of the guided animal.

A transition to competition is found in sharksuckers *Echeneis*, which feed on fish fragments derived from the shark's meals. Antbirds of the family Formicariidae feed on well-camouflaged locusts that jump to escape from an advancing front of army ants (subfamily Dorylinae); syntrophy here completes the transition to direct food competition.

Methanogens were originally believed to be able to grow on propionate, butyrate and alcohols longer than methanol [1291]. For example *Methanobacillus omelianskii* seemed to oxidise ethanol (C_2H_6O) to acetate ($C_2H_3O_2^-$) and use the electrons to reduce CO_2 to CH_4. This 'species' turned out to consist of two, which use substrates and produce products as follows:

donor	$2\,C_2H_6O + 2\,H_2O + O_2 \rightarrow 2\,C_2H_3O_3^- + 2\,H^+ + 4\,H_2$
recipient	$4\,H_2 + CHO_3^- + H^+ \rightarrow CH_4 + 3\,H_2O$
sum	$2\,C_2H_6O + CHO_3^- + O_2 \rightarrow 2\,C_2H_3O_3^- + H^+ + CH_4 + H_2O$

The donor needs the activities of the recipient to keep the concentration of its product, dihydrogen, down to extremely low levels. This is required to extract energy from the degradation of ethanol. This pair serves as an example of a syntrophic relationship, which will be analysed quantitatively for V1-morphs. What are the constraints on the production of dihydrogen such that the biomass ratio between the species does not change, and the two species behave as a single one, at least in steady state? The interest in the question is to derive evolutionary constraints on the origin of syntrophy.

Direct transfer

The donor obtains its substrate from the environment and the recipient receives product from the donor, which serves as substrate. They grow at specific rates

$$\dot{r}_1 = \frac{f\dot{k}_E^1 - \dot{k}_M^1 g_1}{f + g_1} \quad \text{and} \quad \dot{r}_2 = \frac{\frac{\dot{k}_E^2 j_P}{j_{PAm}^2} \frac{M_V^1}{M_V^2} - \dot{k}_M^2 g_2}{\frac{j_P}{j_{PAm}^2} \frac{M_V^1}{M_V^2} + g_2} \tag{9.1}$$

where f stands for the scaled functional response of the donor, and $j_P = \zeta_{PM}\dot{k}_M^1 g_1 + \zeta_{PA}\dot{k}_E^1 f + \zeta_{PG} g_1 \dot{r}_1$ for the specific flux of product from the donor to the recipient, see (4.86). The flux of product is thus $j_P M_V^1$, while the maximum flux that can be handled by the recipient is $j_{PAm}^2 M_V^2$; the ratio of the two quantifies the scaled reserve density (at steady state). The first assumption is that all product can be handled.

Weak homeostasis is obtained if M_V^1/M_V^2 remains constant, so if $\dot{r}_1 = \dot{r}_2$, independent of the substrate availability of the donor. This leads to $\zeta_{PG} = 0$, $\zeta_{PA} \neq 0$, $\dot{k}_E^1 = \dot{k}_E^2 = \dot{k}_E$ and $\frac{\zeta_{PM}}{\zeta_{PA}} \left(\frac{\dot{k}_M^1}{\dot{k}_E} + \frac{\dot{k}_M^1}{\dot{k}_M^2} \right) = 1 - \frac{\dot{k}_{M1}}{\dot{k}_{M2}}$.

The ratio of the structural masses to the reserves can be expressed as simple functions of parameters at steady state

$$\frac{M_V^1}{M_V^2} = \frac{g_2}{g_1} \frac{1 + \dot{k}_M^2/\dot{k}_E}{1 + \dot{k}_M^1/\dot{k}_E} \frac{j_{PAm}^2}{\zeta_{PA}\dot{k}_E} \quad \text{and} \quad \frac{M_E^1}{M_E^2} = \frac{M_{Em}^1}{M_{Em}^2} \frac{g_1}{g_2} \frac{1 + \dot{k}_M^1/\dot{k}_E}{1 + \dot{k}_M^2/\dot{k}_E} \left(1 + \frac{g_1}{f} \frac{\dot{k}_M^2 - \dot{k}_M^1}{\dot{k}_M^2 + \dot{k}_E} \right)^{-1} \tag{9.2}$$

The ratio of the reserves is only independent of substrate availability for the donor if the maintenance rate coefficients are equal ($\dot{k}_M^2 = \dot{k}_M^1$), that is when $\zeta_{PM} = 0$. The conclusion is that the conditions for weak homeostasis are much less stringent, compared to a competition relationship.

Indirect transfer

Suppose that the donor and the recipient live in a chemostat of throughput rate $\dot{h}$ which is fed with medium containing ethanol in concentration X_{Sr}, and other substrates that might be necessary, except for hydrogen. The donor delivers its product into the well-mixed chemostat. Changes in biomass ratios of donor and recipient are still possible given the constraints of homeostasis if the saturation constant of the recipient for the product is not very small.

The changes in the concentrations of ethanol (substrate S), dihydrogen (product P), and structural biomass of species 1 and 2 are for $f_1 = (1 + K_S/X_S)^{-1}$ and $f_2 = (1 + K_P/X_P)^{-1}$

$$\frac{d}{dt}X_S = (X_{Sr} - X_S)\dot{h} - \zeta_{SA} f_1 \dot{k}_E^1 X_V^1; \quad \frac{d}{dt}e_i = (f_i - e_i)\dot{k}_E^i, \quad i \in \{1, 2\} \tag{9.3}$$

$$\frac{d}{dt}X_P = (\zeta_{PM}\dot{k}_M^1 g_1 + \zeta_{PA}\dot{k}_E^1 f_1 + \zeta_{PG}\dot{r}_1 g_1)X_V^1 - \zeta_{PA} f_2 \dot{k}_E^2 X_V^2 - X_P \dot{h} \tag{9.4}$$

$$\frac{d}{dt}X_V^1 = (\dot{r}_1 - \dot{h})X_V^1 \quad \text{with} \quad \dot{r}_1 = \frac{\dot{k}_E^1 e_1 - \dot{k}_M^1 g_1}{e_1 + g_1} \tag{9.5}$$

$$\frac{d}{dt}X_V^2 = (\dot{r}_2 - \dot{h})X_V^2 \quad \text{with} \quad \dot{r}_2 = \frac{\dot{k}_E^2 e_2 - \dot{k}_M^2 g_2}{e_2 + g_2} \tag{9.6}$$

The expressions for the specific growth rate follow from the DEB theory, as does the production of product (here dihydrogen), which is, generally, a weighted sum of the three basic powers. At steady state we have $\dot{h} = \dot{r}_1 = \dot{r}_2$, and the problem is to find the weight coefficients ζ_{P*} for the production of hydrogen such that X_V^2/X_V^1 does not depend on $\dot{h}$.

The steady-state scaled functional responses are $f_S^1 = g_1 \frac{\dot{k}_M^1 + \dot{h}}{\dot{k}_E^1 - \dot{h}}$ and $f_P^2 = g_2 \frac{\dot{k}_M^2 + \dot{h}}{\dot{k}_E^2 - \dot{h}}$, from which follow the concentrations $X_S = K_S f_S^1/(1 - f_S^1)$ and $X_P = K_P f_P^2/(1 - f_P^2)$. The ratio of the biovolume densities of species 2 and 1 at steady state is

$$\frac{X_V^2}{X_V^1} = \frac{\zeta_{PM}\dot{k}_M^1 g_1 + \zeta_{PA}\dot{k}_E^1 f_S^1 + \zeta_{PG}\dot{h} g_1}{\zeta_{PA}\dot{k}_E^2 f_P^2} + \frac{\zeta_{SA}\dot{k}_E^1 f_S^1}{\zeta_{PA}\dot{k}_E^2 f_P^2} \frac{X_P}{X_{Sr} - X_S} \tag{9.7}$$

The ratio varies within a limited range only, for varying throughput rate $\dot{h}$ and substrate concentration X_{Sr}, if the saturation constant K_P is small, and the constraints apply for weak homeostasis at direct transfer.

9.1.3 Symbiosis

Two species have a symbiotic relationship if the syntrophic one is reciprocal. It is extremely common; think for instance of the micro-flora in digestive tracts of animals, or mycorrhizae in and around plant roots. A discussion of its frequently amazing forms can easily fill a book, and most relationships are probably still unknown.

Products for a favour

The term mutualism is frequently used to indicate a relationship that is reciprocally 'beneficial', without a direct trophic basis that is reciprocal.

The Latin American tree *Cecropia* stores glycogen in specialised plastids in tissue, which is eaten by ants that furiously attack anything that touches the tree. In other ant–plant relationships, plants provide protein granules, and ants take care of plant-eating insects. Another example is the plant–pollinator relationship, where plants provide nectar and pollen, and the pollinator (insects, bats, birds) takes care of directed pollen dispersal. Moreover, most of them also need insects, birds or bats and other animals to be pollinated, e.g. [923, 60], and yet other animals for seed dispersal. Thus, berries, for example of Caprifoliaceae, Solanaceae and Rosaceae, are 'meant' to be eaten [1085]; some seeds have edible appendices (e.g. *Viola*) to promote dispersal, but others have no edible parts in addition to the seed, such as *Adoxa* and *Veronica*, and germinate better after being eaten by snails or birds and ants, respectively. Still other seeds stick to animals (e.g. Boraginaceae, *Arctium*) for dispersal. Fungi, such as the stinkhorn *Phallus* and the truffle *Tuber*, also interact with animals for their dispersal.

These cases represent syntrophic relationships as far as the use of substrates is concerned.

Phototroph–heterotroph associations

Algae frequently go for symbiotic relationships with plants, animals, and other protoctists [951]. Many (tropical) coelenterate species host endosymbiotic dinoflagellates named zooxanthellae [1138], which can still live independently from the host [280]. Scleractinian corals hosting *Symbiodinium* species are the dominant reef-builders. Chloroplasts, including that of *Symbiodinium*, are considered to be endosymbionts themselves. Membrane compositions reveal that chloroplasts of the endosymbiotic dinoflagellate *Amphidinium wigrense* are similar to the cryptomonad endosymbionts of *Gymnodinium acidotum*, which have lost their nucleus [1256]; the chloroplasts of these cryptomonads are possibly derived from a rhodophyte, which encapsulated a prokaryote. Such multiple nestings are frequent [1069]. The view of the eukaryotic cell as an integrated symbiotic community is taking ground [743]. An analysis of trophic interactions in a symbiosis has relevance for cellular biology in general, see {282}. This motivates a more detailed discussion here.

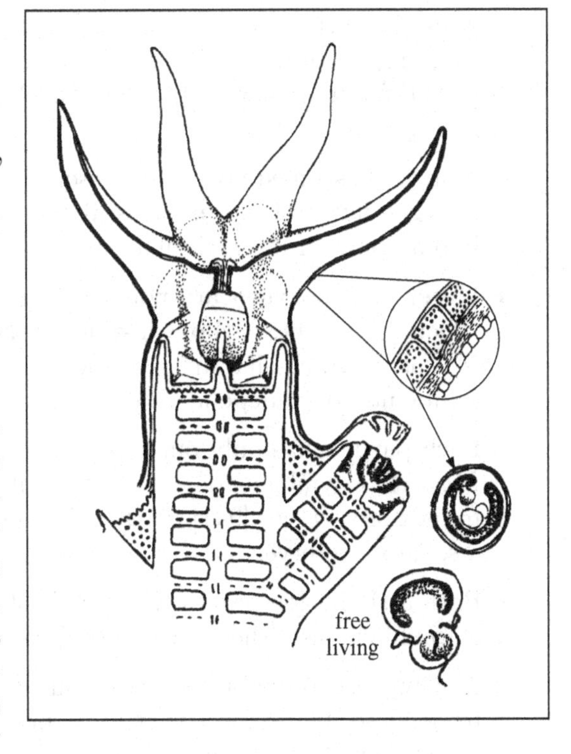

The algal symbionts receive ammonia and carbon dioxide from the host and return lipids and glycerol [64, 1106], which supplement the prey taken by the host. The host can increase the inorganic carbon supply for the symbionts, using the enzyme carbonic anhydrase, which catalyses the interconversion of CO_2 to HCO_3^- [820]. This reduces the carbon dioxide concentration in the host and increases its uptake from the environment. The transfer of bicarbonate to the symbiont is coupled to the calcification process, $Ca^{2+} + 2\,HCO_3^- \rightarrow CaCO_3 + CO_2$, where the carbonate is precipitated by the host and the carbon dioxide is used by the symbiont, see {193}. Symbionts are also found to have heterotrophic abilities [1106] for compounds that are likely to be formed during fermentation processes. In oligotrophic environments, hosts can increase production by one to two orders of magnitude, with the help of symbionts [456].

The stabilising mechanism in the host–symbiont relationship is that the symbiont requires ammonia from the host for growth, and the host requires carbohydrates from the symbiont for *extra* growth that is allowed by food supply. The symbiont cannot grow if it supplies enough carbohydrates to the host to allow the host to use all the ammonia itself; the host generates ammonia, and the symbiont only receives the 'spoils' [456]. This priority in use is reciprocal, and also applies to carbohydrates.

A number of simplifying assumptions are made:

- Bicarbonate is included in carbon dioxide, which in fact stands for inorganic carbon. The complex biochemistry of calcification is simplified to a proportionality with the carbon dioxide that is taken up from the environment. The reported coupling of coral calcification to nitrogen metabolism [233] is taken into account by the full assimilation process, which requires light, inorganic carbon as well as nitrogen.

- Nitrate is assumed not to be present in the environment; it can be included in the nitrogen flux to the symbiosis, if the (variable) nitrate/ammonium ratio is taken into account in the assimilation costs $y_{C_H,ES}$ and $y_{C_H,EH}$.

- Water and dioxygen are non-limiting; the performance of the symbiosis is only affected by light, carbon dioxide, ammonia and food. The composition of food is constant. Self-shading is neglected; light is used proportional to the mass of symbiont, and independent of the surface area of the host.

- Each partner has only one reserve, so the state variables are M_{VH}, M_{EH}, M_{VS}, M_{ES}. The symbiont does not store nitrogen separately; this seems realistic due to lack of vacuoles [941]. It therefore makes little sense to store carbohydrates as a separate reserve.

- Both assimilation processes of the host are parallel; that from carbohydrates and ammonia is fast (these substrate fluxes already have upper boundaries).

- The binding probabilities of all substrates to the Synthesising Units are taken to be close to 1, except that of carbohydrate by the symbiont, which is possibly tunable by the host [286].

- The environment is treated as homogeneous. Flow regimes and diffusive boundary layers usually modify feeding and nutrient uptake [872].

Table 9.1 The chemical compounds of the symbiosis and their transformations and indices. The + signs mean appearance, the − signs disappearance. The signs of the mineral fluxes depend on the chemical indices and parameter values. The labels on rows and columns serve as indices to denote mass fluxes and powers. The table shows the flux matrix $\dot{J}^T$, rather than $\dot{J}$, if the signs are replaced by quantitative expressions presented in Table 9.2.

compounds →			minerals					org. comp.			host		symbiont	
↓ transformations			light	carbon dioxide	water	dioxygen	ammonia	faeces	food	carbohydrate	structure	reserves	structure	reserves
			L	C	H	O	N	P	X	C_H	VH	EH	VS	ES
host	assim. 1	A_1H		+	+	−	+	+	−			+		
host	assim. 2	A_2H		+	+	−	−			−		+		
host	growth	GH		+	+	−	+				+	−		
host	dissip.	DH		+	+	−	+					−		
symb.	assim.	AS	−	−	−	+	−			+				+
symb.	growth	GS		+	+	−	+						+	−
symb.	dissip.	DS		+	+	−	+							−
	carbon	C		1				1	1	1	1	1	1	1
	hydrogen	H			2		3	n_{HP}	n_{HX}	2	n_{HVH}	n_{HEH}	n_{HVS}	n_{HES}
	oxygen	O		2	1	2		n_{OP}	n_{OX}	1	n_{OVH}	n_{OEH}	n_{OVS}	n_{OES}
	nitrogen	N					1	n_{NP}	n_{NX}		n_{NVH}	n_{NEH}	n_{NVS}	n_{NES}

Table 9.2 specifies the fluxes of 11 compounds as indicated in Table 9.1. The fluxes are determined by 19 parameters, and chemical indices, as functions of four environmental variables: light, inorganic carbon, nitrogen and food. The symbiosis can live fully heterotrophically as well as fully autotrophically; as a consequence, it can take up significant amounts of inorganic nitrogen [872]. Figure 9.1 shows that calcification can enhance growth; its measured yearly maximum is 4 $\mathrm{kg\,CaCO_3\,m^{-2}a^{-1}}$, or 3–5 $\mathrm{mm\,a^{-1}}$ [1084]; its daily maximum is three times as high [1083]. The chosen parameter values are just provisional; the figure only serves to illustrate the model structure.

Although the reproduction of the host has been taken into account by the parameter κ_H, the reproduction flux is not listed explicitly. The fluxes associated with assimilation A_2H and AS are given implicitly, and must be obtained numerically. This is hardly a handicap in practice, because a simple Newton–Raphson procedure turns out to be converging rapidly starting from $\dot{J}_{C,A_2H} = \dot{J}_{N,A_2H} = 0$. The change in state is given by $\frac{d}{dt}\mathbf{M} = \dot{\boldsymbol{J}}\mathbf{1}$. The specification allows the following assertions:

Table 9.2 The fluxes in a symbiosis between a heterotrophic isomorphic host H and an autotrophic V1-morphic symbiont S, which experiences the light flux $\dot{J}_{L,F}$, and the densities of inorganic carbon X_C, nitrogen X_N and food X. The reserves enter the fluxes via $m_E^H = M_E^H/M_V^H$ and $m_E^S = M_E^S/M_V^S$. The parameter M_{Vd} just serves as reference for M_V^H to scale $j_{EA_1m}^H$, $j_{NA_2m}^H$, $j_{CA_2m}^H$ and $\dot{k}_E^H$.

$$\dot{J}_{C,A_1H} = -\dot{J}_{X,A_1H} - \dot{J}_{P,A_1H} - \dot{J}_{EH,A_1H}; \quad \dot{J}_{N,A_1H} = -n_{N,X}\dot{J}_{X,A_1H} - n_{N,P} \dot{J}_{P,A_1H} - n_{N,EH}\dot{J}_{EH,A_1H};$$
$$\dot{J}_{P,A_1H} = y_{P,EH}\dot{J}_{EH,A_1H}; \quad \dot{J}_{X,A_1H} = -y_{X,EH}\dot{J}_{EH,A_1H}; \quad \dot{J}_{EH,A_1H} = \frac{j_{EH,A_1Hm}}{1+X_K/X} (M_{VH}/M_{Vd})^{2/3};$$
$$\dot{J}_{N,F} = -\frac{j_{N,A_2Hm}}{1+X_{KN}/X_N}(M_{VH}/M_{V_d})^{2/3}; \quad \dot{J}_{C,F} = -\frac{j_{C,A_2Hm}}{1+X_{KC}/X_C}(M_{VH}/M_{V_d})^{2/3};$$
$$\dot{J}_{C,A_2H} = \dot{J}_{C_H,AS} - \dot{J}_{EH,A_2H}; \quad \dot{J}_{N,A_2H} = -y_{N,EH}\dot{J}_{EH,A_2H}; \quad \dot{J}_{C_H,A_2H} = -y_{C_H,EH}\dot{J}_{EH,A_2H};$$
$$\dot{J}_{EH,A_2H} = \left((y_{EH,N}\dot{J}_{N,A_+H})^{-1} + (y_{EH,C_H}\dot{J}_{C_H,AS})^{-1} - (y_{EH,N}\dot{J}_{N,H} + y_{EH,C_H}\dot{J}_{C_H,AS})^{-1}\right)^{-1};$$
$$\dot{J}_{N,A_+H} = \dot{J}_{N,A_1H} + \dot{J}_{N,GH} + \dot{J}_{N,DH} + (J_{N,+S})_+ - \dot{J}_{N,F}; \quad \dot{J}_{N,+S} = \dot{J}_{N,AS} + \dot{J}_{N,GS} + \dot{J}_{N,DS};$$
$$\dot{J}_{C,GH} = (y_{EH,VH} - 1)\dot{J}_{VH,GH}; \quad \dot{J}_{N,GH} = (n_{N,EH}\, y_{EH,VH} - n_{N,VH})\dot{J}_{VH,GH};$$
$$\dot{J}_{EH,GH} = -y_{EH,VH}\dot{J}_{VH,GH}; \quad \dot{J}_{VH,GH} = M_{VH}\frac{(M_{Vd}/M_{VH})^{1/3}\dot{k}_{EH}m_{EH} - j_{EH,DH}/\kappa_H}{m_{EH} + y_{EH,VH}/\kappa_H};$$
$$\dot{J}_{C,DH} = -\dot{J}_{EH,DH}; \quad \dot{J}_{N,DH} = -n_{N,EH}\dot{J}_{EH,DH}; \quad \dot{J}_{EH,DH} = -j_{EH,DH}M_{VH};$$

$$\dot{J}_{C,AS} = -\dot{J}_{C_H,A_+S}; \quad \dot{J}_{C,A_+S} = \dot{J}_{C,+H} + \dot{J}_{C,GS} + \dot{J}_{C,DS};$$
$$\dot{J}_{C,+H} = -\dot{j}_{C,F} + \dot{J}_{C,A_1H} + \dot{J}_{C,A_2H} + \dot{J}_{C,GH} + \dot{J}_{C,DH};$$
$$\dot{J}_{N,AS} = -n_{N,ES}\dot{J}_{ES,AS}; \quad \dot{J}_{N,A_+S} = (\dot{J}_{N,+H})_+ + \dot{J}_{N,GS} + \dot{J}_{N,DS}$$
$$\dot{J}_{N,+H} = \dot{J}_{N,A_+H} + \dot{J}_{N,A_2H}; \quad \dot{J}_{C_H,AS} = \dot{J}_{C_H,A_+S} - y_{C_H,ES}\dot{J}_{ES,AS};$$
$$\dot{J}_{C_H,A_+S} = \left(\dot{J}_{C,A_+S}^{-1} + (y_{C_H,L}\dot{J}_{L,F})^{-1} - (\dot{J}_{C,A_+S} + y_{C_H,L}\dot{J}_{L,F})^{-1}\right)^{-1};$$
$$\dot{J}_{ES,AS} = \left((y_{ES,N}\dot{J}_{N,A_+S})^{-1} + (y_{ES,C_H}\rho_{C_H}\dot{J}_{C_H,A_+S})^{-1} - (y_{ES,N}\dot{J}_{N,A_+S} + y_{ES,C_H}\rho_{C_H}\dot{J}_{C_H,A_+S})^{-1}\right)^{-1};$$
$$\dot{J}_{C,GS} = (y_{ES,VS} - 1)\dot{J}_{VS,GS}; \quad \dot{J}_{N,GS} = (n_{N,ES}\, y_{ES,VS} - n_{N,VS})\dot{J}_{VS,GS};$$
$$\dot{J}_{ES,GS} = -y_{ES,VS}\dot{J}_{VS,GS}; \quad \dot{J}_{VS,GS} = M_{VS}\frac{\dot{k}_{ES}m_{ES} - j_{ES,DS}}{m_{ES} + y_{ES,VS}};$$
$$\dot{J}_{C,DS} = -\dot{J}_{ES,DS}; \quad \dot{J}_{N,DS} = -n_{N,ES}\dot{J}_{ES,DS}; \quad \dot{J}_{ES,DS} = -j_{ES,DS}M_{VS}$$

- The symbiont does not grow if $\dot{J}_{N,+S} = 0$, in which case $\dot{J}_{N,+H} = 0$ as well; all ammonia released in maintenance is used for assimilation; $\dot{J}_{N,+H} = 0$ if $\dot{J}_{N,A_+H} + \dot{J}_{N,A_2H} = 0$. It also does not grow if $m_{ES} = j_{ES,DS}/\dot{k}_{ES}$; the reserves just cover the maintenance costs. Reserves do not change if $\dot{k}_{ES}M_{ES} = \dot{J}_{ES,AS}$; assimilation equals the catabolic rate.

- The host does not grow if $m_{EH} = \frac{j_{EH,DH}}{\kappa_H \dot{k}_{EH}}\left(\frac{M_{VH}}{M_{Vd}}\right)^{1/3}$; reserves do not change if $\dot{k}_{EH}(M_{Vd}/M_{VH})^{1/3}M_{EH} = \dot{J}_{EH,A_1H} + \dot{J}_{EH,A_2H}$.

- The ratio of the structural masses of symbiont and host does not change if their specific growth rates are equal, so if $j_{VS,GS} = j_{VH,GH}$.

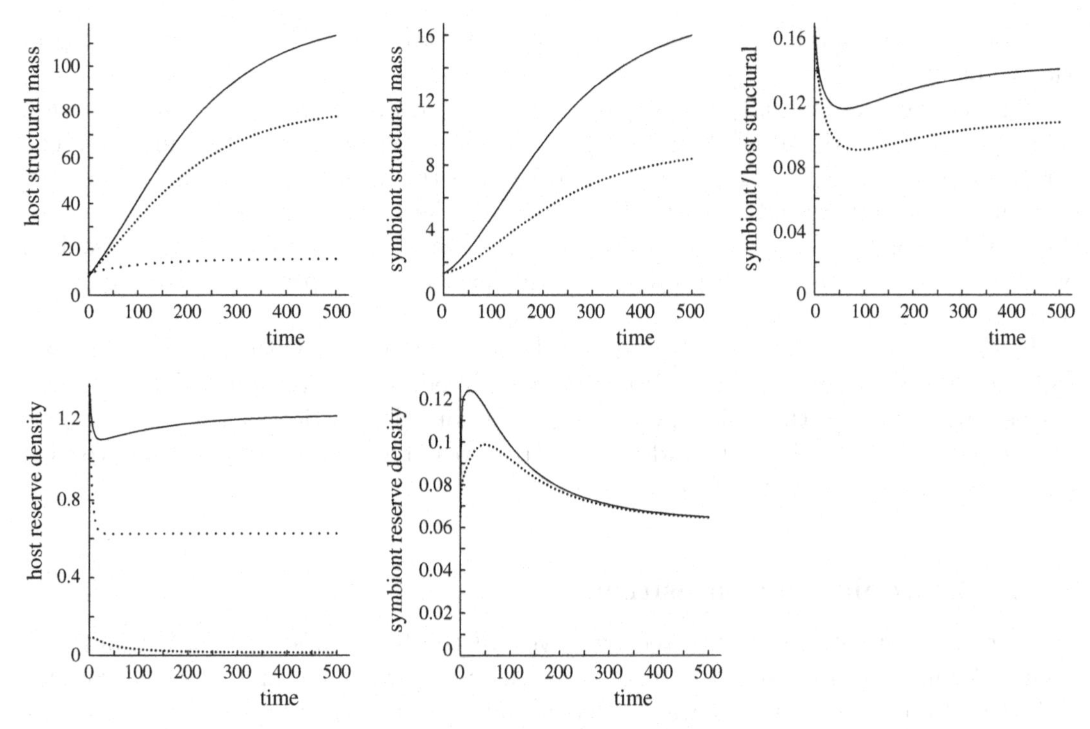

Figure 9.1 The ontogeny of structural masses (in C-moles) and reserve densities of a heterotrophic isomorphic host and an autotrophic V1-morphic symbiont, and their ratio. Light and food are non-limiting and free nitrogen is absent; inorganic carbon is non-limiting (drawn) or absent (finely stippled). Coarsely stippled: host without symbiont.

- The flux ratio $\dot{J}_{EH,A_2H}(\dot{J}_{EH,A_1H}+\dot{J}_{EH,A_2H})^{-1}$ quantifies the photo- versus heterotrophic activity of the host. It is fully phototrophic if $\dot{J}_{X,F} = 0$. The host gains nothing from the symbiont if $\dot{J}_{C_H,AS} = 0$, or if $\dot{J}_{C_H,A_+S} = \dot{J}_{ES,AS}$. Muscatine *et al.* [819] proposed a related measure: the fractional contribution of translocated zooxanthellae C to animal daily respiratory C requirements CZAR $= \dot{J}_{C_H,AS}\dot{J}_{C,+H}^{-1}100\%$.

- The effect of calcification can be evaluated under the various feeding conditions ($\dot{J}_{L,F}$, $\dot{J}_{N,F}$ and $\dot{J}_{X,F}$) with the fraction $\frac{\dot{J}_{EH,A_1H} + \dot{J}_{EH,A_2H} \text{ given } \dot{J}_{C,F} = 0}{\dot{J}_{EH,A_1H} + \dot{J}_{EH,A_2H} \text{ given } \dot{J}_{C,F} = -\infty}$.

- The host is of no use for the symbiont if it captures no prey, and competes with the symbiont for nitrogen.

Figure 9.1 reveals an important implication of the specification of fluxes in Table 9.2: the symbiont/host ratio of structural masses hardly varies. No other regulation seems to be required, other than trophic interactions. The host can tune the population of symbionts via the binding parameter ρ_{C_H} of carbohydrates to the assimilation SU of the symbiont.

Calcifying corals typically consist of a complex carbonate structure covered with a thin layer of living tissue of about constant thickness. Apart from extending the reef, coral colonies in a reef can only grow by increasing the surface area of the supporting structure. The differential growth of biomass and supporting structure controls the shape of the colony. By producing thin axes, the colony can combine a large growth of biomass with little carbonate production, and vice versa by producing solid cerebral structures.

The symbiosis can be simplified smoothly to a mixotroph with a single structure and reserve, by sacrificing the limited degree of freedom in composition. The explicit role of carbohydrates then disappears. A satisfactory description of the diurnal cycle requires separate carbohydrate and nitrogen reserves. The evolutionary perspectives of symbioses are discussed at {406}.

9.1.4 Biotrophy and parasitism

In a biotrophic relationship, the receiver lives off the host's body parts, without necessarily killing the host; it is a transition between a syntrophic and a predatory one. This definition includes most parasites, cows and excludes adult tapeworms (which are competitors). The pearlfish *Carapus* lives inside living sea cucumbers for shelter, where it feeds on the reproductive tissues. Many parasites, such as the avian schistosome *Trichobilharzia* which lives on the reproductive tissues of the pond snail *Lymnaea*, induce their host to increase its investment in growth, by increasing κ, see {42}; this, paradoxically, decreases short-term investment in reproduction, but increases the long-term investment via an increase in body size and so in feeding rate. When the eggs are consumed a little further in their development, and are outside the body of the mother, as in the case of the snake *Dasypeltis* feeding on bird eggs, the relationship is usually called predatory, rather than parasitic. Parasites frequently have intimate metabolic and life history relationships with their host [211, 984, 1247]. DNA sequencing reveals that the feared typhus bacterium *Rickettsia prowazekii* is the closest free-living relative of mitochondria [27], which are completely integrated into almost all eukaryotic cells [743], see {282, 400, 406}.

An intriguing parasitic relationship is the α-proteobacterial symbiont *Wolbachia* of Ecdysozoa (including arthropods and nematodes), which affects sex determination [200, 557, 808, 1237].

9.1.5 Predation and saprotrophy

Although many heterotrophic species eat living prey individuals, most prey–predator models in ecology fail to recognise that the relationship is more complex than just the disappearance of prey individuals and the coupled production of predators. Predators usually have a strong preference for the less healthy prey individuals (possibly in their post-reproductive period where they compete for food with the productive subpopulation), and which are also target for pathogens; once infecting a population, pathogens can attack healthy individuals more easily. Almost all predators are scavengers as well, i.e. they feed on dead biomass, which classifies them as saprotrophs. When a predator dies, a rich supply of substrates and nutrients becomes available in the form of its corpse, which directly or indirectly comes back to the prey. Accidental death or death from ageing by the predator can be considered as maintenance-coupled substrate production processes at the population level. All organisms, therefore, have syntrophic relationships with others at the population level. Predators also provide food for the prey via the release of nutrients (nitrogenous waste, faeces); if the prey happens to be algae or plants, the nutrients can be used directly, or these intermediate organisms will be directly or indirectly food for the prey. These indirect trophic relationships need to be included for a proper understanding of population dynamics.

Most predators also feed on dead prey, other species specialise on sapotrophy, such as burrowing beetles of the family Silphidae. The frequent occurrence of sapotrophy and predation is because the chemical make-up of organisms does not differ that much; because of their great capacity of moving around, animals are often the first to arrive at the feast. Many examples illustrate that it is just a small step from feeding off dead corpses to living off live ones.

Many carnivores have cannibalistic tendencies in periods of low prey abundance. This obviously reduces intra-species competition in the predator population, with more food available to the surviving individuals, while relieving the pressure on the prey. Cannibalism has a strong stabilising effect on population dynamics. DEB models can be used to understand why dwarfs and giants can develop in cannibalistic populations [208].

Prey can develop intricate behavioural and physiological adaptations to avoid predation [1056, 1215], and prey species that share a common predator can develop interrelationships [2]. The predation of insects on plants evolved into intricate relationships [1031].

A low predation pressure on symbiotic partners enhances their stable coexistence [630], whereas coexistence becomes unstable at a high pressure and easily leads to the extinction of both prey and predator. This points to a co-evolution of parameter values quantifying the dynamics in prey–predator systems. The timescale of the effects on fitness is essential; short-term positive effects can go together with long-term negative effects of behavioural traits on fitness. Timescales and indirect side-effects that operate through changes in food availability are important aspects that are usually not included in the literature on evolutionary aspects of life history strategies.

9.2 Population dynamics

The significance of the population level for biological insight at all organisation levels is manifold. It not only sets food availability and predation pressure for each individual, but it also defines the effect of all changes in life history, which is pertinent to evolutionary theory. All other individuals belong to the environment of the particular individual whose fitness is being judged. Fitness, whatever its detailed meaning, relates to the production of offspring, thus it changes the environment of the individual. This is one of the reasons why fitness arguments, which are central to evolutionary theories, should always involve the population level. Feeding on the same resource and being eaten are the major topics in population dynamics, but a real understanding requires analysis of all trophic interactions, as specified in the section on canonical communities, see {377}.

Most models of population dynamics treat individuals as identical objects, so that a population is fully specified by its total number or total biomass. Such populations are called non-structured populations. This obviously leads to attractive simplicity; see e.g. Hastings [471] for an easy introduction. I discuss some doubts about their realism on {349, 358}, doubts that can be removed by turning to structured populations. Structured populations are populations where the individuals differ from each other by one or more characteristics, such as age, which affect feeding, survival and/or reproduction. DEB theory provides an attractive, albeit somewhat complicated, structure. I will show the connection between non-structured and DEB-structured populations step by step.

The differences between individual and population levels fades for V1-morphs, which makes them an ideal paradigm for the connection between non-structured populations and structured ones. The introduction of a structure does not necessarily lead to realistic population models because of the effects of many environmental factors that typically operate at population level: spatial heterogeneity, seasonality, erratic weather, climatic changes, processes of adaptation and selection, subtle species interactions and so on. The occurrence of infectious diseases is perhaps one of the most common causes of decline and extinction of species, which typically operates in a density-dependent way. This means that population dynamics, as discussed in this chapter, still has to be embedded in a wider framework to arrive at realistic descriptions of population dynamics.

Spatial structure can profoundly modify population dynamics, as illustrated here for a compact school of fish, where only the individuals at the front actually feed, while the others starve and frequently interchange position with the individuals in the front. If the school increases its number of individuals, without changing shape, the feeding rate by the school is proportional to the number of individuals to the power 2/3 (based on the same argument as DEB theory uses for individuals), which implies that the feeding rate by the individual decreases for increasing numbers of individuals in the school. This is, however, not a proper population perspective, since this should include rules for the birth and death of schools; the school is here a 'super-individual'.

The interaction between individuals of the same species is here restricted to feeding on the same resource. This point of view might seem a caricature in the eyes of a behavioural ecologist. The general idea, however, is not to produce population models that are as realistic as possible, but to study the consequences of feeding on the same resource. A comparison is then made with non-structured population dynamics and with real-world populations to determine the pay-off between realism and model complexity. If DEB-structured population dynamics predictions are not realistic, while the DEB model is at the individual level, this will give a key to factors that are important in this situation. The basic energetics and trophic interactions must be right before the significance of the more subtle factors can be understood. My fear is that most of the factors shown to be relevant will be specific for a particular species, a particular site and a particular period in time. This casts doubts on the extent to which general theory is applicable and on the feasibility of systems ecology. The main application of population dynamics theory here concerns a mental exercise pertaining to evolutionary theory, with less emphasis on direct testing in real-world populations. The theory should, however, be able to predict population behaviour in simplified environments, such as those found in laboratory set-ups, in bio-reactors and the like, so that it has potential practical applications.

9.2.1 Non-structured populations

The chemostat, a popular device in microbiological research, will be used to make the transition from the intensively studied non-structured populations to DEB-structured populations. In a chemostat, food (substrate) is supplied at a constant rate to a population, which is called a continuous culture. Food density in the inflowing medium is denoted by X_r and the medium is flowing through the chemostat at throughput rate $\dot{h}$ times the volume of the chemostat V_c. Together with the initial conditions (food and biomass density) these controls determine the behaviour of the system, in particular the food (substrate) density X_0 and the biomass density X_1 as functions of time. The index 0 in the notation for food density is added for reasons of symmetry with X_1: the biomass density of predators, i.e. the ratio of the sum of the individual masses and the volume of the chemostat, V_c. So $X_1 = \sum_{i=1}^{N} M_{Vi}/V_c$, if there are N individuals in the population. The reactor is assumed to be spatially homogeneous, and is called a 'continuous flow stirred tank reactor' by engineers.

The chemostat as a model can also be realistic for particular situations outdoors [352, 374]. An important difference between chemostat models and many population dynamic models is that food (substrate) does not propagate in the formulation here, while exponential or logistic growth is the standard assumption in most literature [470, 603, 677, 764]. I do not follow this standard, however, because I want to stick to mass and energy balance equations in a strict way. The growth rate of food should, therefore, depend on its resource levels, which should be modelled as well. In the section on food chains, {372}, higher trophic levels, X_2, $X_3, \ldots$, will be introduced, not lower ones.

Batch cultures, which do not have a supply of food other than that initially present, are a special case of chemostat cultures, where $\dot{h} = 0$. I start with the Lotka–Volterra model, which was and probably still is the standard prey–predator model in ecology. In a sequence of related models, the effect of the stepwise introduction of biological detail that leads to DEB-structured populations will be studied.

Lotka–Volterra model

The Lotka–Volterra model assumes that the predation frequency is proportional to the encounter rate with prey (here substrate), on the basis of what is known as the law of mass action, i.e. the product of the densities of prey and predator. It can be thought of as a linear Taylor approximation of MM-kinetics around food density 0: $f = (1 + K/X_0)^{-1} \simeq X_0/K$ for $X_0 \ll K$. The ingestion rate is taken to be proportional to body volume, as is appropriate for V1-morphs, so that the sum of all ingestion rates by individuals in the population is found by adding the volumes of all individuals and applying the same proportionality constant.

The Lotka–Volterra model for chemostats with throughput rate $\dot{h}$ is

$$\frac{d}{dt}X_0 = \dot{h}X_r - \dot{j}_{XAm}\frac{X_0}{K}X_1 - \dot{h}X_0; \quad \frac{d}{dt}X_1 = Y\dot{j}_{XAm}\frac{X_0}{K}X_1 - \dot{h}X_1 \tag{9.8}$$

where Y stands for the yield factor, i.e. the conversion efficiency from prey to predator biomass; this is taken to be constant here. This model does not account for maintenance or energy reserves, so that in the context of DEB theory we have $Y = \kappa\frac{\mu_{AX}}{\mu_{GV}}$, with $[\dot{p}_M] = 0$ and $[E_m] = 0$. At the individual level, this model implicitly assumes that the feeding rate is proportional to the volume of the individual. This aspect corresponds with V1-morphs. The analysis of the population dynamics can best be done with the dimensionless quantities $\tau \equiv t\dot{h}$, $\jmath_{XAm} \equiv \dot{j}_{XAm}/\dot{h}$, $x_r \equiv X_r/X_K$, $x_0 \equiv X_0/K$, $x_1 \equiv X_1/K$. These substitutions turn (9.8) into

$$\frac{d}{d\tau}x_0 = x_r - \jmath_{XAm}x_0x_1 - x_0; \quad \frac{d}{d\tau}x_1 = Y\jmath_{XAm}x_0x_1 - x_1 \tag{9.9}$$

The equilibrium is found by solving x_0 and x_1 from $\frac{d}{d\tau}x_0 = \frac{d}{d\tau}x_1 = 0$. The positive solutions are $x_0^* = (Y\jmath_{XAm})^{-1}$ and $x_1^* = Yx_r - \jmath_{XAm}^{-1}$. The yield factor in this model has a double interpretation. It stands for the efficiency of converting food into biomass at both the individual and the population levels. To see this, one has to realize that food influx is at rate $\dot{h}Kx_r$ and food output is at rate $\dot{h}Kx_0^* = \frac{\dot{h}K}{Y\jmath_{Xm}}$ at equilibrium. So total food consumption is $\dot{h}K(x_r - \frac{1}{Y\jmath_{Xm}})$. Biomass output is $\dot{h}Kx_1^* = \dot{h}K(Yx_r - \jmath_{XAm}^{-1})$. The conversion efficiency at the population level thus amounts to $\frac{\dot{h}K(Yx_r - \jmath_{XAm}^{-1})}{\dot{h}K(x_r - (Y\jmath_{XAm})^{-1})} = Y$. This is so simple that it seems trivial. That this impression is false soon becomes obvious when we introduce more elements of the DEB machinery; the conversion efficiency at the population level then behaves differently from that at the individual level for non-V1-morphs.

The linear Taylor approximation around the equilibrium of the coupled system (9.9) equals for $\boldsymbol{x}^T \equiv (x_0, x_1)$ and $\boldsymbol{x}^{*T} \equiv (x_0^*, x_1^*)$

$$\frac{d}{d\tau}\boldsymbol{x} \simeq \begin{pmatrix} -\jmath_{Xm}x_1 - 1 & -\jmath_{Xm}x_0 \\ Y\jmath_{Xm}x_1 & Y\jmath_{Xm}x_0 - 1 \end{pmatrix}_{\boldsymbol{x}=\boldsymbol{x}^*} (\boldsymbol{x} - \boldsymbol{x}^*) \tag{9.10}$$

$$\simeq \begin{pmatrix} -Y\jmath_{Xm}x_r & -Y^{-1} \\ Y^2\jmath_{Xm}x_r - Y & 0 \end{pmatrix} \begin{pmatrix} x_0 - \frac{1}{Y\jmath_{Xm}} \\ x_1 - Y(x_r - \frac{1}{Y\jmath_{Xm}}) \end{pmatrix} \tag{9.11}$$

The eigenvalues of the matrix with coefficients, the Jacobian, are -1 and $-Y\jmath_{Xm}x_r + 1$, so that this system does not oscillate. See Edelstein-Keshet [311], and Yodzis [1284] for valuable introductions to this subject, and Hirsch and Smale [511], Ruelle [997] and Arrowsmith and Place [36, 37] for more advanced texts. Mathematical texts on non-linear dynamics systems are now appearing at an overwhelming rate [76, 292, 440, 547, 1151], especially with a focus on 'chaos'. Simple biological problems still seem too complex to analyse analytically, however, and one has to rely on numerical analyses. Figure 9.2 compares the dynamics of the Lotka–Volterra model with other simplifications of the DEB model.

Although this model cannot produce oscillations, with a minor change it can, by feeding the outflowing food (substrate) back into the bio-reactor. This is technically a simple operation. Most microbiologists even neglect the small outflow in open systems in their mass balances. The situation is covered by deleting the third term in (9.9), i.e. $-x_0$. The eigenvalues of the Jacobian then become $-\frac{1}{2}Y\jmath_{XAm}x_r \pm \frac{1}{2}\sqrt{(Y\jmath_{XAm}x_r)^2 - 4Y\jmath_{XAm}x_r}$. For $Y\jmath_{XAm}x_r < 4$, the eigenvalues are complex, thus the system is oscillatory.

Lotka-Volterra, Monod, Marr–Pirt, Droop and DEB models

The Marr–Pirt, Droop and Monod models are special cases of the univariate DEB model for V1-morphs. It reads

$$\frac{d}{d\tau}x_0 = x_r - \jmath_{Xm}fx_1 - x_0; \quad \frac{d}{d\tau}x_1 = Y\jmath_{Xm}fx_1 - x_1 \tag{9.12}$$

with $f = (1 + x_0^{-1})^{-1}$. The yield factor Y is only constant in the Monod model.

The biologically interesting equilibrium values x_0^* and x_1^* can easily be obtained from (9.12), but the result is line filling. The linear Taylor approximation in the equilibrium for the Monod case is:

$$\frac{d}{d\tau}\boldsymbol{x} \simeq \begin{pmatrix} -\frac{x_r + x_0^{*2}}{x_0^* + x_0^{*2}} & -\frac{1}{Y_g} \\ \frac{x_r - x_0^*}{\jmath_{XAm}x_0^{*2}} & 0 \end{pmatrix} (\boldsymbol{x} - \boldsymbol{x}^*) \tag{9.13}$$

The eigenvalues of the Jacobian are -1 and $-\frac{1}{Y_g\jmath_{XAm}}(x_r - \frac{1}{Y_g\jmath_{XAm}-1})(Y_g\jmath_{XAm} - 1)^2$, so the system does not oscillate. The linear Taylor approximation of the functional response is accurate for small equilibrium values of food density, and thus a high value for $Y_g\jmath_{XAm}$, which means that the Monod and the Lotka–Volterra models for the chemostat are very similar. The Monod model has less tendency to oscillate than the Lotka–Volterra model. This becomes visible if the substrate is fed back to the bio-reactor. (Thus we omit the term $-x_0$ in (9.12).) In contrast to the Lotka–Volterra model, the eigenvalues of the Jacobian cannot become complex, so the system cannot oscillate.

Figure 9.2 gives the direction fields of the various simplifications of the DEB model. The functional response in the equilibrium of the Monod model is only 0.4, for the chosen parameter values, which results in a close similarity with the Lotka–Volterra model. The direction fields of the Marr–Pirt and Droop models are rather similar, so the effects of introducing maintenance and reserves are more or less the same. When introduced simultaneously, as in the DEB model, the effect is enhanced. Note that the isocline $\frac{d}{d\tau}x_0 = 0$ hits the axis $x_1 = 0$ at $x_0 = x_r$, which is just outside the frame of the picture for the DEB model, but far outside for the Lotka–Volterra model. For very small initial values for x_0 and x_1, the direction fields show that x_0 will first increase very rapidly to x_r, without a significant increase of x_1, then the $\frac{d}{dt}x_0 = 0$-isocline is crossed and the equilibrium value x_0^*, x_1^* is approached with strongly decreasing speed. This means that x_0 falls back to a very small value for Lotka's model, but much less so for the DEB model. The most obvious difference between the models is in the equilibrium values, where $x_1^* \gg x_0^*$ in Lotka's model, but the (much more realistic) reverse holds in the DEB model. The other models take an intermediate position. The approach of x_0, x_1 to the equilibrium value closely follows the $\frac{d}{dt}x_0 = 0$-isocline if $x_1 > x_1^*$ in all models. The speed in the neighbourhood of the isocline is much less than that further away from the isocline, and the differences in speed are larger for Lotka's model than for the DEB model. These extreme differences in speed mean that the numerical integration of this type of differential equations needs special attention.

Figure 9.3 gives the relative specific growth rates of the Monod, the Marr–Pirt, the Droop and the univariate DEB model for V1-morphs as functions of the scaled substrate density. It further illustrates the effect of maintenance and reserve.

Death

The usefulness of the chemostat in microbiological research lies mainly in the continuous production of cells that are in a particular physiological state. This state depends on the dilution rate. In equilibrium situations, this rate is usually equated to the population growth rate. The implicit assumption being made is that cell death plays a minor role. As long as the dilution rate is high, this assumption is probably realistic, but if the dilution rate is low, its realism is doubtful. Low dilution rates go with low substrate densities and long inter-division intervals. In the section on ageing {214}, the hazard rate for V1-morphs is tied to the respiration rate and so, indirectly, to substrate densities in (6.8). The law of large numbers states that the hazard rate can be interpreted as a mean (deterministic) death rate for large populations. The dynamics for the dead biovolume, $x_\dagger$, reads

$$\frac{d}{dt}x_\dagger = \dot{h}_a x_1 - \dot{h} x_\dagger \tag{9.14}$$

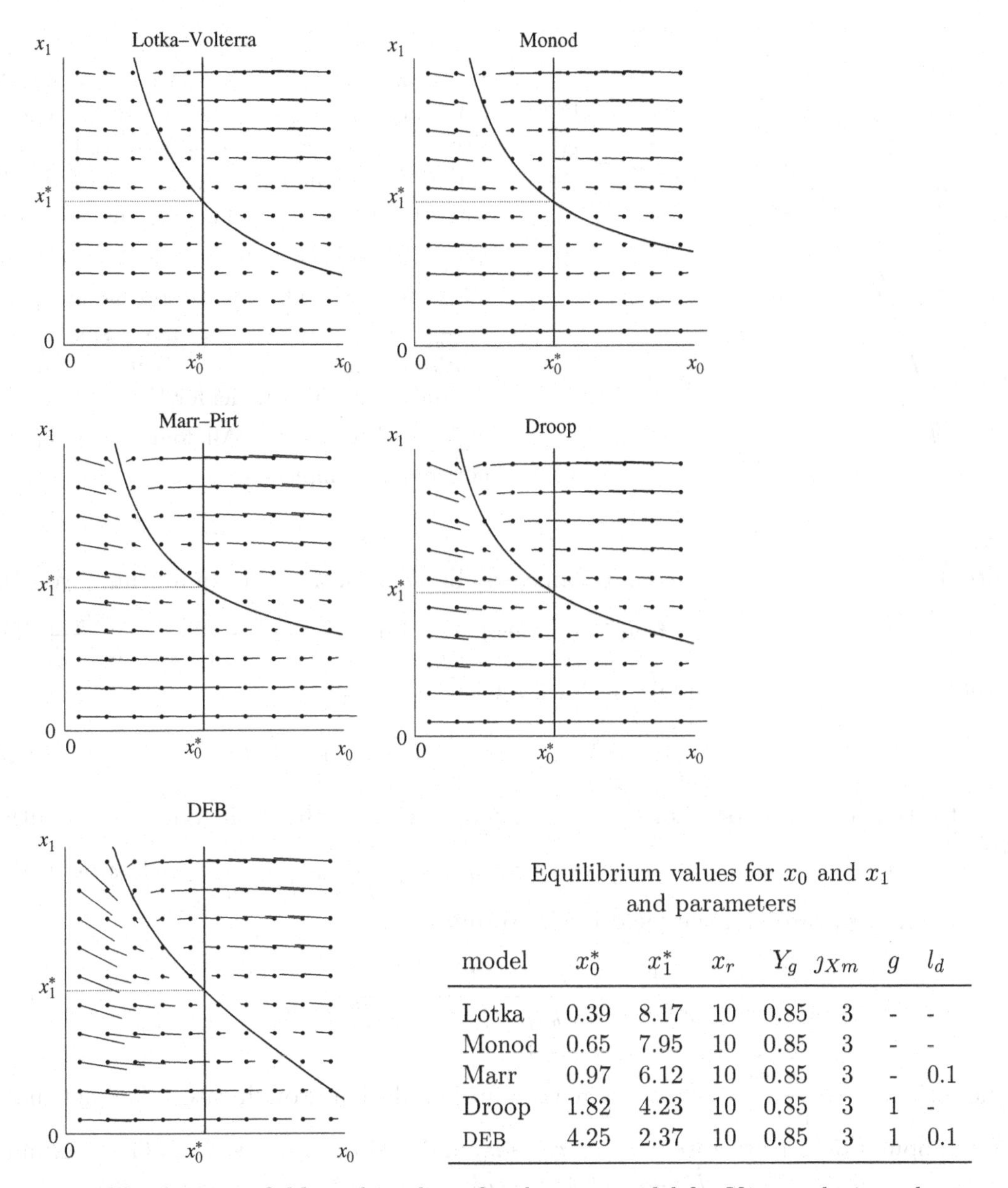

Equilibrium values for x_0 and x_1 and parameters

model	x_0^*	x_1^*	x_r	Y_g	$\jmath_{Xm}$	g	l_d
Lotka	0.39	8.17	10	0.85	3	-	-
Monod	0.65	7.95	10	0.85	3	-	-
Marr	0.97	6.12	10	0.85	3	-	0.1
Droop	1.82	4.23	10	0.85	3	1	-
DEB	4.25	2.37	10	0.85	3	1	0.1

Figure 9.2 The direction fields and isoclines for the DEB model for V1-morphs in a chemostat with reserves at equilibrium, and the various simplifications of this model. The lengths and directions of the line segments indicate the change in scaled food density x_0 and scaled biovolume x_1. The isoclines represent x_0, x_1-values where $\frac{d}{d\tau}x_0 = 0$ or $\frac{d}{d\tau}x_1 = 0$. All parameters and variables are made dimensionless, as indicated in the text.

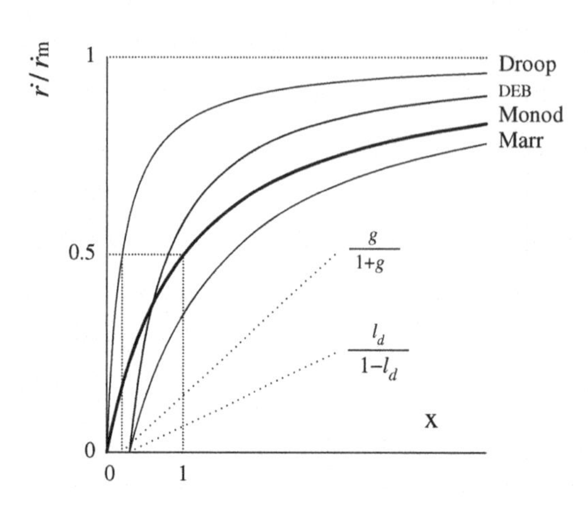

Figure 9.3 In all models, Monod, Marr–Pirt, Droop and DEB, uptake as a fraction of its maximum depends on scaled substrate density x as $f = \frac{x}{1+x}$, as indicated by the thick curve. The Monod model, $\frac{\dot{r}}{\dot{r}_m} = f$, coincides with this curve. The Marr–Pirt model, $\frac{\dot{r}}{\dot{r}_m} = \frac{f/l_d-1}{1/l_d-1}$, includes maintenance and has a translation to the right. The Droop model, $\frac{\dot{r}}{\dot{r}_m} = f\frac{1+g}{f+g}$, includes reserve and has a smaller saturation coefficient (for the $\dot{r}$-curve), whereas the univariate DEB model for V1-morphs $\frac{\dot{r}}{\dot{r}_m} = \frac{1+g}{f+g}\frac{f/l_d-1}{1/l_d-1}$ has both. All four curves have a horizontal asymptote of 1.

with $\dot{h}_a$ denoting the hazard rate. It can easily be seen that, in the equilibrium, we must have that $\dot{h}x_\dagger^* = \dot{h}_a x_1^*$, so the fraction of dead biovolume equals $\frac{x_\dagger^*}{x_1^* + x_\dagger^*} = \frac{\dot{h}_a}{\dot{h} + \dot{h}_a}$. The dynamics of the biomass should account for this loss, thus

$$\frac{d}{dt}x_1 = Y j_{XAm} f x_1 - (\dot{h} + \dot{h}_a)x_1 \tag{9.15}$$

Substitution of the expression for the hazard rate and the yield and the condition $\frac{d}{dt}x_1 = 0$ leads to the equilibrium value for f: $\frac{g(\dot{k}_M + \dot{h})}{\dot{k}_E - \dot{h} - \dot{h}_a(1+g)}$. Back-substitution into the hazard rate and the yield finally results in

$$\frac{x_\dagger^*}{x_1^* + x_\dagger^*} = \frac{\dot{k}_M + \dot{h}}{\dot{k}_M + (\dot{k}_M + \dot{r}_m^\circ)\dot{h}/\dot{h}_a} \tag{9.16}$$

where $\dot{r}_m^\circ = \frac{\dot{k}_E - \dot{k}_M g}{1+g}$ is the gross maximum population growth rate. (The net maximum population growth rate is $\dot{r}_m = \dot{r}_m^\circ - \dot{h}_a$ and $\dot{h} \leq \dot{h}_m \leq \dot{r}_m \leq \dot{r}_m^\circ$.) The maximum throughput rate is $\dot{h}_m = \frac{\dot{k}_E - \dot{h}_a(1+g) - g\dot{k}_M(1+K/X_r)}{1+g(1+K/X_r)}$. Since most microbiological literature does not account for death, and saturation coefficients are usually small, these different maximum rates are usually not distinguished. Figure 9.4 illustrates how the dead fraction depends on the population growth rate.

The significance of the fraction of dead cells is not only of academic interest. Since it is practically impossible to distinguish the living from the dead, it can be used to 'correct' the measured biomass for the dead fraction to obtain the living biomass.

In the section on ageing, {214}, I speculate that prokaryotes might not die instantaneously, but first switch to a physiological state called 'stringent response'. The fraction

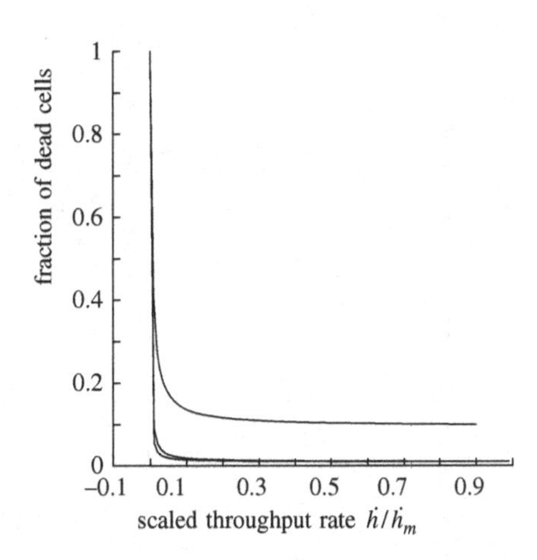

Figure 9.4 The fraction of dead cells depends hyperbolically on the population growth rate, and increases sharply for decreasing population growth rates. The three curves correspond with $\dot{k}_M/\dot{r}_m = 0.05$, $\dot{h}_a/\dot{r}_m = 0.01$ (lower), $\dot{k}_M/\dot{r}_m = 0.1$, $\dot{h}_a/\dot{r}_m = 0.01$ (middle) and $\dot{k}_M/\dot{r}_m = 0.05$, $\dot{h}_a/\dot{r}_m = 0.1$ (upper curve). For high growth rates, the dead fraction is close to $\dot{h}_a/\dot{r}_m$, which will be very small in practice. The curves make it clear that experimental conditions are extremely hard to standardise at low growth rates.

(9.16) can then be interpreted as the fraction of individuals that is in the stringent response. A typical difference between both types of cells is the intracellular concentration of guanosine 4-phosphate (ppGpp), which is usually expressed per gram of total biomass. This quantification implicitly assumes that all cells in the population behave in the same way physiologically, and not that the population can be partitioned into cells that are in the stringent response and those that are not. It remains to be determined which presentation is the more realistic.

Reserves and expo-logistic growth

The univariate DEB model for V1-morphs in chemostats amounts to the following three coupled equations

$$\frac{d}{d\tau}x_0 = x_r - \jmath_{XAm} f x_1 - x_0; \quad \frac{d}{d\tau}e = Y_g \jmath_{XAm} g(f - e)$$
$$\frac{d}{d\tau}x_1 = Y_g \jmath_{XAm} g \frac{e - l_d}{e + g} x_1 - x_1 - h_a \frac{1 + g}{e + g} e x_1 \tag{9.17}$$

A special case of conceptual interest can be solved analytically. This case relates to batch cultures, where no input or output (of substrate or biomass) exists, and the biomass just develops on the substrate that is present at the start of the experiment. If the saturation coefficient, the maintenance costs and ageing rate are small, V1-morphs will grow in a pattern that might be called expo-logistic. Initially they will grow exponentially and after a certain time (which corresponds to the depletion of the substrate) they switch to logistic growth, depleting their reserves. The biomass–time curve is smooth, even at the transition from one mode of growth to the other.

Worked out quantitatively, we get the following results. The functional response f is initially 1, since $K \ll X_0$. If the inoculum is from a culture that has not suffered from substrate depletion, we have $e = 1$ and $X_1(t) = X_1(0) \exp(\dot{r}_m t)$, so the population growth rate is maximal, i.e. $\dot{r}_m = (\dot{k}_E - \dot{k}_M g)(1 + g)^{-1}$. The substrate concentration

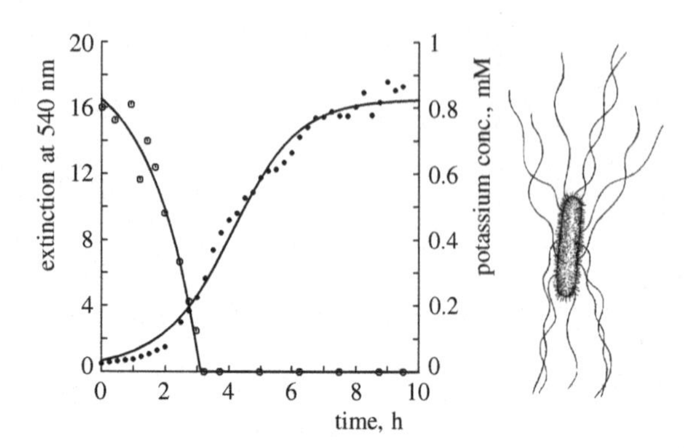

Figure 9.5 The potassium-limited growth of *E. coli* at 30 °C. Data from Mulder [814]. The expo-logistic growth is fully from reserves when potassium is depleted from the environment. Parameters for biomass in dimensionless extinction units: initial potassium concentration $X_0(0) = 0.825$ mM, initial biomass $x_1(0) = 0.657$, maximum specific uptake rate $j_{XAm} = 0.125 \text{ mM h}^{-1}$, investment ratio $g = 0.426$, reserve turnover $\dot{k}_E = 0.925 \text{ h}^{-1}$.

develops as $X_0(t) = X_0(0) - \int_0^t j_{XAm} X_1(t_1)\, dt_1$. It becomes depleted at t_0, say, where $X_0(t_0) = 0$. Substitution gives

$$X_0(t) = X_0(0)(\exp(\dot{r}_m t_0) - \exp(\dot{r}_m t))(\exp(\dot{r}_m t_0) - 1)^{-1} \tag{9.18}$$

where depletion occurs at time $t_0 = \frac{1}{\dot{r}_m} \ln \left\{1 + \frac{X_0(0)}{X_1(0)} \frac{\dot{r}_m}{j_{XAm}}\right\}$. The reserves then decrease exponentially, i.e. $e(t_0 + t) = \exp(-\dot{k}_E t)$. The biovolume thus behaves as $X_1(t_0 + t) = X_1(t_0) \exp\left(\int_0^t \frac{\dot{k}_E e(t_0 + t_1) - \dot{k}_M g}{e(t_0 + t_1) + g}\, dt_1 \right)$. For small maintenance costs, $\dot{k}_M \to 0$, this reduces to $X_1(t_0 + t) = X_1(t_0) \frac{1+g}{\exp(-\dot{k}_E t) + g}$. This is the solution of the well-known logistic growth equation $\frac{d}{dt} X_1 = \dot{k}_E \left(1 - \frac{X_1(t)}{X_1(0)} \frac{g}{1+g}\right) X_1$, see Figure 9.5. The equation originates from Pearl [875] in 1927. If the maintenance costs are not negligibly small, the integral for $X_1(t)$ has to be evaluated numerically. Biomass will first rise to a maximum and then collapse at a rate that depends on the maintenance costs. This behaviour offers the possibility to determine these costs experimentally. The quantitative evaluation can easily be extended to include fed-batch cultures for instance, which have food (substrate) input and no output of food or biomass, but this does involve numerical work.

Similar biovolume–time curves can also arise if the reserve capacity rather than the saturation coefficient is small. If maintenance and ageing are negligible as before, the batch culture can be described by $\frac{d}{dt} X_0 = -j_{XAm} f X_1$ and $\frac{d}{dt} X_1 = Y_g j_{XAm} f X_1$. We must also have $X_1(t) = X_1(0) + Y_g(X_0(0) - X_0(t))$. Substitution and separation of variables gives

$$j_{XAm} Y_g t = \frac{K Y_g}{X_1(\infty)} \ln \frac{X_1(t)(X_1(\infty) - X_1(0))}{X_1(0)(X_1(\infty) - X_1(t))} + \frac{1}{2} \ln \frac{X_1(t)}{X_1(0)} \tag{9.19}$$

Although this expression looks very different from (9.18), the numerical values are practically indistinguishable, as shown in Figure 9.6, where both population growth curves have been fitted to data on *Salmonella*. The only way to distinguish a difference

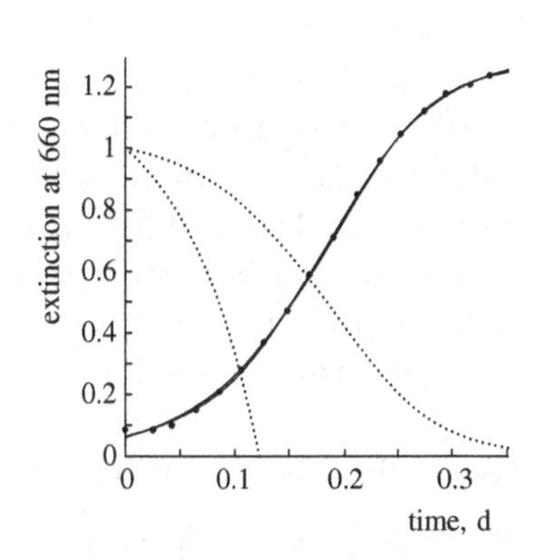

Figure 9.6 A batch culture of *Salmonella typhimurium* strain TA98 at 37 °C in Vogel and Bonner medium with glucose, (excess) histidine and biotin added. Two models have been fitted and plotted: one assumes that the saturation coefficient is negligibly small, but the reserves capacity is substantial, while the other does the opposite. Only the substrate density will tell the difference (stippled curves), but this is not measured. Parameters: $\dot{k}_E = 18.6\,\mathrm{d}^{-1}$, $g = 0.355$, $x_0(0)/j_{XAm} = 0.020\,\mathrm{d}$ or $x_1(\infty) = 1.28$, $YK = 1.31$, $Yj_{XAm} = 23.6\,\mathrm{d}^{-1}$.

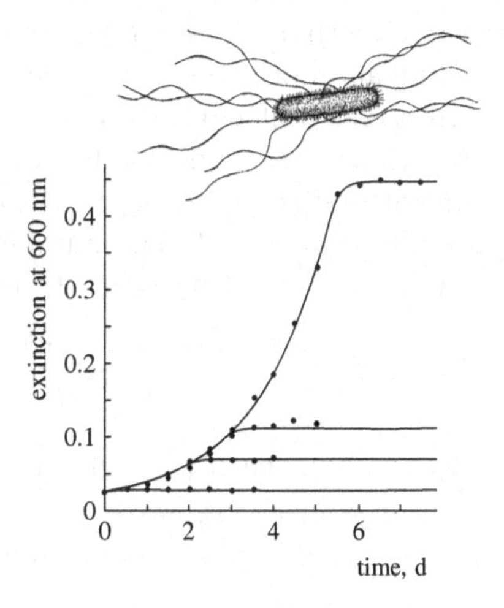

Figure 9.7 Batch cultures of a histidine-deficient strain of *S. typhimurium*, with initially only 0, 0.5, 1 or 5 μg histidine ml^{-1} in the medium, stop growing because of histidine depletion. The fit is based on the assumption of negligible maintenance requirements for histidine, which implies that the extinction plateau is a linear function of the added amount of histidine. The parameters are $j_{XAm} = 8\,\mu\mathrm{g\,His\,ml}^{-1}\,\mathrm{h}^{-1}$, $\dot{k}_E = 5.3\,\mathrm{h}^{-1}$ and $g = 7.958$. One extinction unit corresponds with 7.56×10^8 cells ml^{-1}, so that the yield is $Y_g = \frac{\dot{k}_E}{j_{XAm}g} = 0.0834\,\mathrm{ml}\,\mu\mathrm{g\,His}^{-1}$. This corresponds with $1.1 \times 10^{-10}\mathrm{g\,His\,cell}^{-1} = 3.15 \times 10^5$ molecules His cell^{-1} with a maximum of 4×10^4 molecules histidine in the reserve pool.

is in the simultaneous fit for both biomass and substrate. This illustrates the rather fundamental problem of model identification for populations, even in such a simple case as this with only four free parameters. (To reduce the number of free parameters, maintenance and ageing were taken to be negligible for both special cases.)

If other information is available to allow a choice between various possibilities, such as in the case of very efficient histidine uptake by deficient *Salmonella* strains, see {245}, the growth of batch cultures can be used to estimate the reserve capacity. This has been done in Figure 9.7 to illustrate that under particular circumstances, the DEB model implies mass fluxes, as discussed in more detail on {138}.

9.2.2 Structured populations

It is not my intention to review the rapidly growing literature on structured population dynamics, but, for those who are unfamiliar with the topic, some basic notions are introduced below to help develop intuition. See Heijmans [483], Metz and Diekmann [785], Łomnicki [716], Ebenman and Persson [308], DeAngelis and Gross [254],

Tuljapurkar and Caswell [1173], Gurney and Nisbet [444] and Cushing [237] for reviews of recent developments.

In unstructured models, all individuals are treated as identical, so their state (the i-state) is degenerated, and the population state (the p-state) is simply the number of individuals. This is different in structured models, where a population exists of individuals that differ in their i-state, and the p-state is defined as the frequency distribution of the individuals over the i-state. Individuals with almost identical i-states are thus taken together in a cohort, and counted.

Reproducing neonates

There is no way to prevent neonates from giving rise to new neonates in unstructured populations. This artifact of the formulation can dominate population dynamics at lower growth rates. Comparison with a simple age-structured population, in which individuals reproduce at a constant rate after a certain age a_p, can illustrate this.

In a constant environment, any population grows exponentially given time, structured as well as non-structured. (Real populations will not do so because the environment will soon change because of food depletion.) Let $N(t)$ denote the number of individuals at time t. The numbers follow $N(t) = N(0)\exp(\dot{r}t)$, where the population growth rate $\dot{r}$ is found from the characteristic equation

$$1 = \int_0^\infty \Pr\{\underline{a}_\dagger > a\}\dot{R}(a)\exp(-\dot{r}a)\,da \qquad (9.20)$$

Suppose that death plays a minor role, so $\Pr\{\underline{a}_\dagger > a\} \simeq 1$, and that reproduction is constant after age a_p, so $\dot{R}(a) = (a > a_p)\dot{R}$, where, with some abuse of notation, $\dot{R}$ in the right argument is taken to be a constant. Substitution into (9.20) gives

$$\exp(-\dot{r}a_p) = \dot{r}/\dot{R} \qquad (9.21)$$

This equation ties the population growth rate $\dot{r}$ to the length of the juvenile period and the reproduction rate. It has to be evaluated numerically. For unstructured populations, where $a_p = 0$ must hold, the population growth rate equals the reproduction rate, $\dot{r} = \dot{R}$. For increasing a_p, $\dot{r}$ falls sharply; see Figure 9.8. This means that neonates giving birth to new neonates contribute significantly to unstructured populations.

Discrete individuals

The formulation of the reproduction rate such as $\dot{R}(a) = (a > a_p)\dot{R}$ treats the number of individuals as a continuous variable. Obviously, this is unrealistic, because individuals are discrete units. It would be more appropriate to gradually fill a buffer with energy allocated to reproduction and convert it to a new individual as soon as enough energy has been accumulated. In that case, the reproduction rate becomes $\dot{R}(a) = (a = a_p + i/\dot{R})/da$, for $i = 1, 2, \ldots$. It is zero almost everywhere, but at regular time intervals it switches to ∞ over an infinitesimally small time interval da, such that the mean

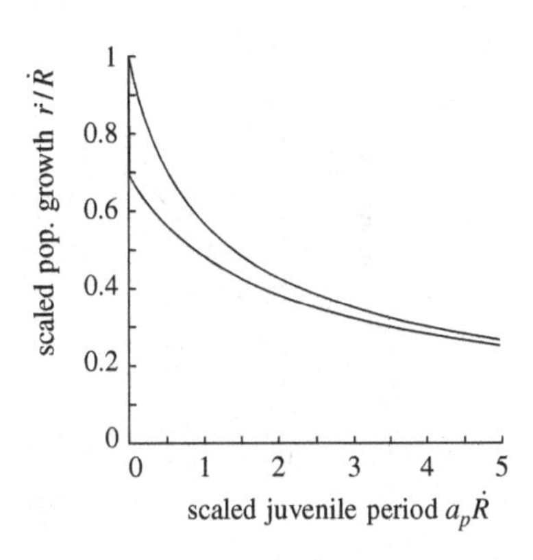

Figure 9.8 For a constant reproduction rate $\dot{R}$ in the adult state, the population growth rate depends sensitively on the length of the juvenile period, as shown in the upper curve. The unit of time is $\dot{R}^{-1}$ and mortality is assumed to be negligible. The lower curve also accounts for the fact that individuals are discrete units of biomass. The required accumulation of reproductive effort to produce such discrete units reduces the population growth rate even further, especially for short juvenile periods. Note that the effect of food availability is not shown in this figure, because it only affects the chosen unit of time.

reproduction rate as an adult over a long period is $\dot{R}$ as before. Giving death a minor role, the characteristic equation becomes

$$\exp(-\dot{r}a_p) = \exp(\dot{r}/\dot{R}) - 1 \tag{9.22}$$

to reveal the effect of individuals being discrete units rather than continuous flows of biomass, cf. (9.21); see Figure 9.8. The effect is most extreme for $a_p = 0$, where $\dot{r} = \dot{R}\ln 2$, which is a fraction of some 0.7 of the continuous biomass case. If young are not produced one by one, but in a litter, which requires longer accumulation times of energy, the discreteness effect is much larger. For a litter size n and a reproduction rate of $\dot{R}(a) = (a = a_p + in/\dot{R})n/da$, the population growth rate is $n^{-1}\ln\{1+n\}$ times the one for continuous biomass with the same mean reproduction rate and negligibly short juvenile period.

The effect of the discrete character of individuals is felt most strongly at low reproduction rates. Since populations tend to grow rapidly in situations where reproduction reduces sharply because of food limitation, this problem is rather fundamental. Reproduction, i.e. the conversion of the energy buffer into offspring, is usually triggered by independent factors (a 2-day moulting cycle in daphnids, seasonal cycles in many other animals). If reproduction is low, details of buffer handling become dominant for population dynamics. Energy that is not sufficient for conversion into the last young dominates population dynamics. Whether it gets lost or remains available for the next litter makes quite a difference and, unfortunately, we know little about what exactly does happen.

Population growth rates and division intervals

The relationship between population growth rate and the division interval can be obtained from a formulation that allows for the production of neonates by letting the mother cell disappear at the moment of division, when two baby cells appear: $\Pr\{\underline{a}_\dagger > a\} = (a \leq a_d)$ and $\dot{R}(a)\,da = 2(a = a_d)$. Substitution into the characteristic equation (9.20) gives $1 = 2\exp(-\dot{r}a_d)$. The division interval a_d is given in (2.23), (4.20) or (4.27) for isomorphs, V1-morphs and rods, respectively.

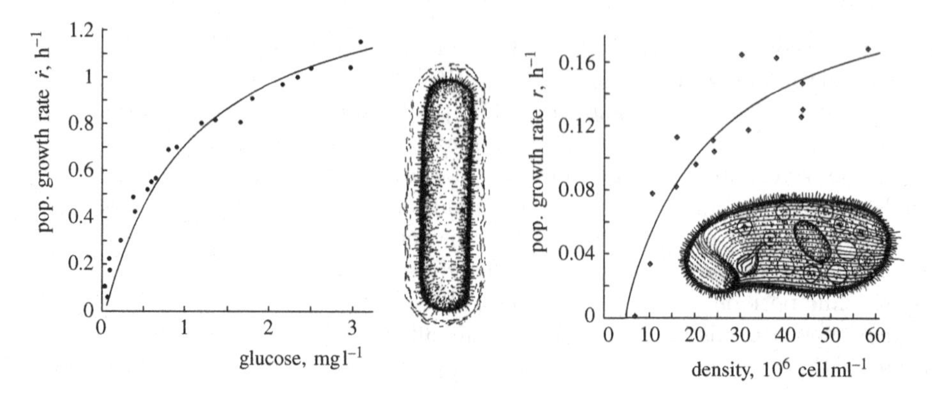

Figure 9.9 The population growth rate as a function of the concentration of substrate or food. The left figure concerns the rod *Klebsiella aerogenes* feeding on glucose at 35 °C. Data from Rutgers *et al.* [1006]. The right figure concerns the isomorphic ciliate *Colpidium campylum* feeding on suspensions of *Enterobacter aerogenes* at 20 °C. Data from Taylor [1144].

The population growth rate is plotted against the substrate concentration for the rods *Escherichia coli* and *Klebsiella aerogenes* in Figures 7.24 and 9.9, and for the isomorph *Colpidium* also in Figure 9.9. The curves closely resemble simple MM-kinetics, which indicates that they contain little information about some of the parameter values of the individual-based DEB model, particularly the energy investment ratio g. Since the goodness of fit is quite acceptable, the modest conclusion can only be that these population responses give little reason to change assumptions about the energy behaviour of individuals. Figure 9.9 also illustrates that the scatter in population responses tends to increase dramatically with body size.

Population structure

For many purposes non-equilibrium situations should be considered, which requires computer simulation studies. Two strategies can be used to follow population dynamics: the family-tree method and the frequency method.

The family-tree method evaluates the changes of the state variables for each individual in the population at each time increment. For this purpose, the individuals are collected in a matrix, where each row represents an individual and each column the value of a state variable. At each time increment rows can be added and/or deleted and at regular time intervals population statistics, such as the total volume of individuals, are evaluated. The amount of required computer time is thus roughly proportional to the number of individuals in the population which must, therefore, be rather limited. This restricts the applicability of this method for analytical purposes, because at low numbers of individuals stochastic phenomena, such as those involved in survival, tend to dominate. The method is very flexible, however, which makes it easy to incorporate differences between individuals with respect to their parameter values. Such differences are realistic and appear to affect population dynamics substantially; see {363}. Kaiser

[570, 571] used the programming environment SIMULA successfully to study the population dynamics of individual dragon-flies, mites and rotifers. Kreft *et al.* [667] simulated the spatial aspects of the individual-based population dynamics of bacteria. Tineke Troost studied the effect of diffusion of parameter values across generations in DEB-structured populations of mixotrophs [1170, 1171, 1172] to study speciation, as done in Adaptive Dynamics, see {365}.

The frequency method is based on bookkeeping in terms of (hyperbolic) partial differential equations, see e.g. (9.24). Several strategies exist to integrate these equations numerically. The method of the escalator boxcar train, perfected by de Roos [988, 989], follows cohorts of individuals through the state space. The border of the state space where individuals appear at birth is partitioned into cells, which are allowed to collect a cohort of neonates for a specified time increment. The reduction of the number of individuals in the cohorts is followed for each time increment, as the cohort moves through the state space. The amount of computer time required is proportional to the number of cohorts, which relates to the volume of the state space as measured by the size of the cells. The number of cells must be chosen by trial and error. The escalator boxcar train is just one method of integrating the partial differential equation, but it appears to be an efficient one compared with methods that use a fixed partitioning of the state space into cells that transfer numbers of individuals among them.

A nasty problem of the (partial) differential equation approach to describe population dynamics is the continuity of the number of neonates if the reproduction rate is very small. This situation occurs in equilibrium situations, if the loss rate is small. The top predators especially are likely to experience very small loss rates. Details of the handling of energy reserves to produce or not produce a single young prove to have a substantial effect on population dynamics.

Synchronisation

Computer simulations of fed-batch cultures of reproducing isomorphs reveal a rather unexpected property of the DEB model. In these simulations the food supply to the population is taken to be constant and the population is harvested by the process of ageing and in a random way. To reduce complicating factors as much as possible, only parthenogenetically reproducing females are considered, using realistic parameter values for *Daphnia magna* feeding on the green alga *Chlorella pyrenoidosa* at 20 °C. Reproduction in daphnids is coupled to moulting, which occurs every 2 to 3 days at 20 °C, irrespective of food availability. Just after moulting, the brood pouch is filled with eggs which hatch just before the next moult. So the intermoult period is beautifully adapted to the incubation time and the buffer for energy allocated to reproduction stays open during the intermoult period. These details are followed in the simulation study because many species produce clutches rather than single eggs.

Figure 9.11 shows a typical result of the population trajectory: the numbers oscillate substantially at low random harvesting rates. Closer inspection reveals that the shape of the number cycles closely follows the survival function of the ageing process. The individuals appear to synchronise their life cycles, i.e. their ages, lengths and energy

Table 9.3 Oscillations can affect crude population statistics. This table compares statistics for computer simulations, assuming that reproduction is by clutches, or by eggs laid one at a time, with statistics that assume the stable age distribution.

statistic	clutch	single egg	stable age
mean scaled functional response, f	0.355	0.340	0.452
mean scaled biomass density, x	1.095	0.99	0.943
mean number of individuals, N	87.0	55.3	18.3
scaled yield coefficient, y	0.214	0.186	0.120

reserve densities, despite the fact that the founder population consists of widely different individuals. This synchronisation is reinforced by the accumulation of reproductive effort in clutches, but it also occurs with single-egg reproduction. The path individuals take in their state space closely follows the no-growth condition. Growth in these populations can only occur via thinning by ageing and the resultant amelioration of the food shortage. After reaching adult volumes, the individuals start to reproduce and mothers are soon outcompeted by their offspring, because they can survive at lower food densities. This has indeed been observed in experimental populations [424, 1146].

Having observed the synchronisation of the individuals, it is not difficult to quantify population dynamics from an individual perspective when we now know that the scaled functional response cycles from $f = l_b$ to l_p. Starting from a maximum $N(0) = \frac{\dot{h}_p X_r g^2 \dot{k}_M^2}{\{\dot{J}_{Xm}\} l_b^3 \dot{v}^2}$ at time 0, the numbers drop according to $N(t) = N(0) \exp(-\int_0^t \dot{h}(t_1)\, dt_1)$, down to $N(t_n) = N(0)(l_b/l_p)^3$. The total biovolume is about constant at $X_1 = \frac{\dot{h}_p X_r V_m^{1/3}}{\kappa\{\dot{J}_{Xm}\}}$; see Table 9.3. At the brief period of takeover by the next generation, the population deviates a little from this regime. It is interesting to note that growth and reproduction are fully determined by the ageing process in this situation. Length-at-age curves do not resemble the saturation curve that is characteristic of the von Bertalanffy growth curve; they are more or less exponential. Biovolume density and the yield are increased by the oscillatory dynamics, compared to expectation on the basis of the stable age distribution.

If the harvesting effort is increased, the population experiences higher food densities and the model details for growth and reproduction become important. The shape of the length-at-age curves switches from 'exponential' to von Bertalanffy, the cycle period shortens, the generations overlap for a longer period because competition between generations becomes less important, and the tendency to synchronise is reduced. All these changes result from the tendency of populations to grow and create situations of food shortage if the harvesting rates drop.

Similar synchronisation phenomena are known for the bakers' yeast *Saccharomyces cerevisiae* [196, 870]. It produces buds as soon the cell exceeds a certain size. This gives a synchronisation mechanism that is closely related to that for *Daphnia*.

Variation between individuals

Although it is not unrealistic to have fluctuating populations at constant food input [1073], the strong tendency of individuals to synchronise their life cycles seems to be unrealistic. Yet the model describes the input–output relationships of individuals rather accurately. A possible explanation is that at the population level some new phenomena play a role, such as slightly different parameter values for different individuals. This gives a stochasticity of a different type than that of the ageing process, which is effectively smoothed out by the law of large numbers. This way of introducing stochasticity seems attractive because the replicability of physiological measurements within one individual generally tends to be better than that between individuals. The exact source of variation in energy parameters, however, is far from obvious. This applies especially to parthenogenetically reproducing daphnids, where recombination is usually assumed not to occur. Hebert [479] however, has reported that (natural) populations of daphnids, which probably originate from a limited number of winter eggs, can have substantial genetic variation. Branta [141] was able to obtain a rapid response to selection in clones of daphnids, which could not be explained by the occurrence of spontaneous mutations. Cytoplasmic factors possibly play a much more important role in gene expression than is recognised at the moment. Koch [611] has discussed individual variability among bacteria.

In principle, it is possible to allow all parameters to scatter independently, but this seems neither feasible nor realistic. High ingestion rates, for instance, usually go with high assimilation rates and storage capacities. The parameter values of the DEB model for different species appear to be linked in a simple way, as discussed in the section on parameter variation {296}. We assume here that the parameters for the different individuals within a species are also linked in this way but vary within a narrow range. The parameters for a particular individual remain constant during its lifetime. In this way, we require only one simple individual-specific multiplier operating on (some of) the original parameters of the DEB model to produce the scatter. The way the scatter appears in the scaled parameters is even simpler [658].

Parameter variation between individuals has interesting effects on population dynamics: a log-normally distributed scatter with even a small coefficient of variation is enough to prevent death by starvation at the takeover of the new generation. Moreover, each generation becomes extinct only halfway through the period of the next generation and the amplitude of the population oscillations is significantly reduced; see Figure 9.11. This may be quantified by its effect on the coefficient of variation for population measures, defined as

$$c_j = \left(\frac{1}{t_n}\int_0^{t_n}\sum_i l_i^j(t)\,dt\right)^{-1}\sqrt{\frac{1}{t_n}\int_0^{t_n}\left(\sum_i l_i^j(t) - \frac{1}{t_n}\int_0^{t_n}\sum_i l_i^j(t_1)\,dt_1\right)^2 dt} \quad (9.23)$$

for $j = 0, 1, 2, 3$. Integration is taken over one typical cycle of length t_n and the summation over all individuals in the population. For values larger than 0.2, the coefficient

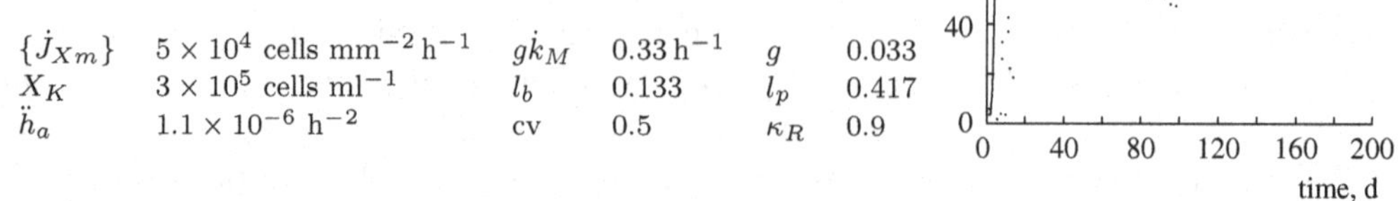

Figure 9.10 Computer simulation of a DEB-structured population of *Daphnia magna*, compared to a real laboratory population at 20 °C with a supply of 5×10^7 cells *Chlorella saccarophila* d^{-1}, starting from five individuals. Data from Fitsch [354]. The parameter values were obtained independently of the observations of individuals. Parameters:

$\{\dot{J}_{Xm}\}$	5×10^4 cells $mm^{-2} h^{-1}$	$g\dot{k}_M$	$0.33\, h^{-1}$	g	0.033
X_K	3×10^5 cells ml^{-1}	l_b	0.133	l_p	0.417
$\ddot{h}_a$	$1.1 \times 10^{-6}\, h^{-2}$	cv	0.5	κ_R	0.9

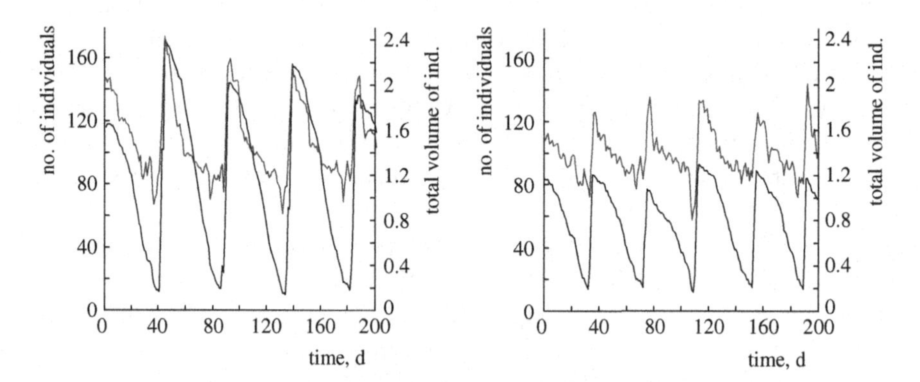

Figure 9.11 The number of individuals (bottom) and the total biovolume (top) in a simulated batch culture of daphnids subjected to ageing as the only method of harvesting. The individuals accumulate reproductive effort during the incubation time in the left figure, while they reproduce egg by egg in the right one. The parameters are $\dot{h}X_r = 7\,\text{units}\,d^{-1}$, $l_b = 0.133$, $l_p = 0.417$, $\dot{J}_{Xm} = 4.99\,\text{units}\,d^{-1}$, $\kappa = 0.3$, $\dot{k}_M = 10\,d^{-1}$, $g = 0.033$, $\ddot{h}_a = 2.5 \times 10^{-5}\,d^{-2}$.

of variation of the scatter parameter barely depresses the variation coefficient of the population measures further; see Figure 9.12.

Figure 9.10 demonstrates that computer simulations of DEB-structured daphnids closely match the dynamics of laboratory populations. The strength of the argument is in the fact that the parameter values for individual performance have been obtained independently.

The oscillations are also likely to be less if one accounts for spatial heterogeneities. This is realistic even for daphnids, because some of the algae adhere to the walls of the vessels and some (but not all) daphnids feed on them [378, 528]. The general features of the dynamics of experimental populations are well captured by the DEB model. Emphasis is given to the competition for food, which Slobodkin [1073] considered to be the only type of interaction operative in his experimental food-limited populations. He suggested that the competition between different age–size categories was responsible

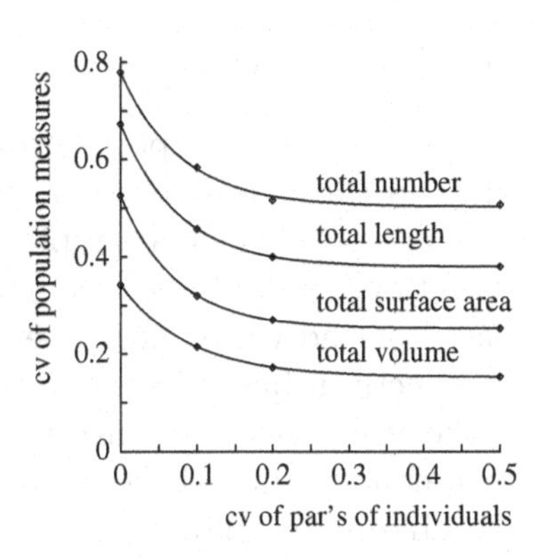

Figure 9.12 The coefficient of variation of the total number, length, surface area and volume of individuals in the population as functions of the coefficient of variation of the scatter parameter that operates on the parameters of individuals. The sharp initial reduction points to the limited realism of strictly deterministic models.

for the observed intrinsic oscillations, which is confirmed by this model analysis. Mrs N. van der Hoeven [517] has concluded, on the basis of a critical survey of the literature on experimental daphnid populations with constant food input, that some fluctuations are caused by external factors. Even populations that tend to stabilise do so, however, by way of a series of damped oscillations, while others seem to fluctuate permanently.

Adaptive dynamics

When parameter values for energy budgets vary among individuals, and rules about how the values carry over to new generations are formulated, selective forces are specified through competition for the same resources. Such selective forces need not be external, the DEB assumptions already imply these forces, namely how differences in feeding translate into differences in reproduction. These rules can obviously be modified as a result of interactions with other populations, such as predators, whose actions directly or indirectly relate to the parameter value of the individuals. Predators can select for particular body sizes, for instance, and body size is determined by DEB parameters. Given a specification of the environment in which the individuals live, the mean parameter values can evolve, and the (multivariate) frequency distribution can become multi-modal, reflecting the process of speciation. Many qualitative properties of this process can be evaluated, even without detailed specification of the models for individuals. This type of problem is called adaptive dynamics [268, 397, 787]. Mrs T. Troost applied it to the affinity of DEB-structured mixotrophs for the autotrophic versus heterotrophic modes, see {404}.

9.2.3 Mass transformation in populations

Mass fluxes in populations are the sum of the individual mass fluxes, but interactions between substrate and biomass densities through the processes of feeding and competition substantially complicate the conversion of substrate to biomass. The results for reproducers and dividers are discussed briefly in the following sections.

Propagation through reproduction

Let us consider a population of parthenogenetically reproducing individuals that develop through embryonic, juvenile and adult stages. Sexually reproducing animals can be included in a simple way, as long as the sex ratio is fixed. The population structure, derived from the collection of individuals that make up that population, is based on individual characteristics. Suppose that there is a maximum for the amount of structural body mass and reserves for individuals, and that we use the scaled length l, the scaled reserve density e and the age a to specify the state of the individual (the techniques to model the dynamics are readily available for an arbitrary number of state variables [452, 453]).

Suppose that a population of individuals lives in a 'black box' and that the individuals only interact through competition for the same food resource. Food is supplied to the black box at a constant rate $\dot{h}_X M_X$, where $\dot{h}_X$ has dimension time^{-1} and M_X is the amount of food (in C-moles per black box volume). Eggs are removed from the black box at a rate $\dot{h}_e$; juveniles and adults are harvested at a rate $\dot{h}$ randomly, i.e. the harvesting process is independent of the state of the individuals (age a, reserves e, size l). Furthermore, the ageing process harvests juveniles and adults at a state-dependent rate $\dot{h}_a$, which is beyond experimental control. Individuals harvested by the ageing process leave the black box instantaneously.

The present purpose is to study how food supplied to the black box converts to body mass and reserves that leave the box in the form of harvested individuals, when the amounts of dioxygen, carbon dioxide, nitrogenous waste and faeces in the black box are kept constant. This implies that these mass fluxes to and from the box equal the use or production by all individuals in the box. Food is not removed, which implies that the amount of food in the box depends on both the food supply and the harvesting rates of individuals.

Before analysing the conversion process in more detail, it is helpful to point out the fundamental difference between the population and the individual level. If no harvesting occurs at all, and food is supplied to the bio-reactor, the population will eventually grow to a size where food input just matches the maintenance needs of the individuals. In this situation no individual is able to grow or reproduce (otherwise we would not have a steady state). The conversion efficiency is then zero. Figure 9.13 illustrates this situation for experimental *Daphnia* populations. By increasing the harvesting rate, the conversion efficiency increases also, at least initially. This illustrates that the conversion process is controlled by the way the population is sandwiched between food input and harvesting. Individual energetics only set the constraints.

In many field situations, the harvesting rate will not be set intentionally. The process of ageing can be considered, for instance, as one of the ways of harvesting through intrinsic causes, but this does not affect the principle. The present aim is to study the behaviour of the yield factor in steady-state situations, so $\dot{r} = 0$, and compare the different life styles: V1-morphs, rods and isomorphs, propagating via division and eggs. For this purpose, let us strip the population of as many details as possible and think of it in terms of inputs and outputs of mass.

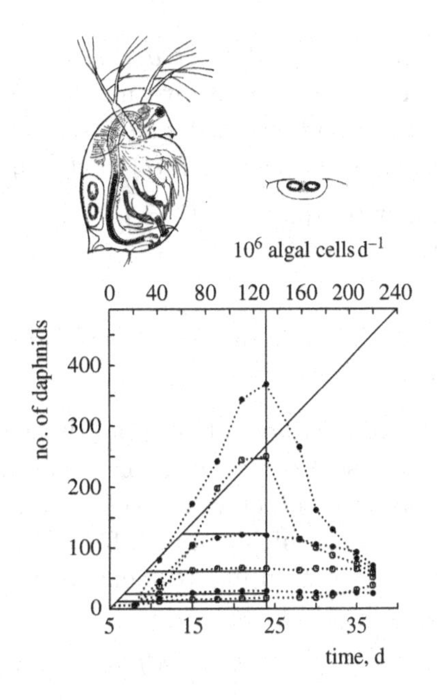

Figure 9.13 Populations of daphnids *Daphnia magna* fed a constant supply of food, the green alga *Chlorella pyrenoidosa*, at 20 °C, grow to a maximum number of individuals that is directly proportional to food input [633]. From this experiment, it can be concluded that each individual requires six algal cells per second just for maintenance. No deaths occurred before day 24. A reduction of food input to 30×10^6 cells day^{-1} after day 24 resulted in almost instant death if the populations were at carrying capacity. The 240×10^6 cells day^{-1} population was still growing when the food supply was suddenly reduced, so the energy reserves were high, and it produced many winter eggs. The daily food supply related to the cumulated number of winter eggs as:

240	120	60	30	12	6	10^6 cells d^{-1}
38	1	3	1	0	0	winter eggs

The conversion process has three control parameters, $\dot{h}_X$, $\dot{h}_e$ and $\dot{h}$, and the aim is now to evaluate all mass fluxes in terms of these three control parameters, given the properties of the individuals. This result is of direct interest to particular biotechnological applications, and to the analysis of ecosystem behaviour, provided that the control parameters are appropriate functions of other populations and the degradation of faeces and dead individuals is specified to recycle the nutrients that are locked in these compounds.

The index + refers to the population, to distinguish fluxes at the population level from those at the individual level. Embryos are treated separately from juveniles and adults, not only because this allows different harvesting rates for both groups, but also because they do not eat, and therefore do not interact with the environment through food.

Given the initial conditions $\phi_e(0, a, e, l)$ and $\phi(0, a, e, l)$, the change in density of embryos and of juveniles plus adults over the state space is given by the McKendrick–von Foerster hyperbolic partial differential equation [359, 1068]:

$$\frac{\partial}{\partial t}\phi_e(t,a,e,l) = -\frac{\partial}{\partial l}\left(\phi_e(t,a,e,l)\frac{d}{dt}l\right) - \frac{\partial}{\partial e}\left(\phi_e(t,a,e,l)\frac{d}{dt}e\right) + \\ - \frac{\partial}{\partial a}\phi_e(t,a,e,l) - \dot{h}_e\phi_e(t,a,e,l) \quad (9.24)$$

$$\frac{\partial}{\partial t}\phi(t,a,e,l) = -\frac{\partial}{\partial l}\left(\phi(t,a,e,l)\frac{d}{dt}l\right) - \frac{\partial}{\partial e}\left(\phi(t,a,e,l)\frac{d}{dt}e\right) + \\ - \frac{\partial}{\partial a}\phi(t,a,e,l) - (\dot{h} + \dot{h}_a(a,e,l))\phi(t,a,e,l) \quad (9.25)$$

where $\int_{a_1}^{a_2} \int_{l_1}^{l_2} \int_{e_1}^{e_2} \phi(t,a,e,l)\, de\, dl\, da$ is the number of individuals (juveniles plus adults) aged somewhere between a_1 and a_2, with a scaled energy density somewhere between e_1 and e_2 and a scaled length somewhere between l_1 and l_2. The total number of juveniles plus adults equals $N(t) = \int_0^\infty \int_{l_b}^1 \int_0^1 \phi(t,a,e,l)\, de\, dl\, da$. The total number of embryos likewise equals $N_e(t) = \int_0^\infty \int_0^{l_b} \int_0^\infty \phi_e(t,a,e,l)\, de\, dl\, da$. The boundary condition at $a = 0$ reads

$$\phi_e(t,0,e_0,l)\frac{d}{dt}a = \delta(l = l_0) \int_0^\infty \int_{l_p}^1 \dot{R}(e,l)\phi(t,a,e,l)\, dl\, da \quad \text{for all } e \tag{9.26}$$

where l_0 denotes the scaled length at $a = 0$, which is taken to be infinitesimally small, l_b the scaled length at birth (i.e. the transition from embryo to juvenile), l_p the scaled length at puberty (i.e. the transition from juvenile to adult), and e_0 the scaled reserve at $a = 0$, which can be a function of e of the mother. The function $\delta(l = l_0)$ is the Dirac delta function in l (dimension: l^{-1}). The boundary condition at $l = l_b$ reads

$$\phi(t,a,e,l_b)\frac{d}{dt}a = \phi_e(t,a,e,l_b)\frac{d}{dt}a \quad \text{for all } a, e \tag{9.27}$$

The individuals can differ at $a = 0$, because e_0 can depend on e, and individuals can make state transitions at different ages and different scaled reserves. The dynamics for food amounts to

$$\frac{d}{dt}M_{X+} = \dot{h}_X M_X + \dot{J}_{X+} \tag{9.28}$$

$$\dot{J}_{X+} \equiv \int_0^\infty \int_{l_b}^1 \int_0^1 \phi(t,a,e,l)\dot{J}_X(e,l)\, de\, dl\, da \tag{9.29}$$

where M_{X+} denotes the food density in C-moles per black box volume, and $\dot{J}_X(e,l)$ the (negative) ingestion rate of an individual of scaled energy reserves e and scaled length l, as discussed in the previous section. The faecal flux $\dot{J}_{P+}$ is simply proportional to the ingestion flux, i.e. $\dot{J}_{P+}/\dot{J}_{X+} = \dot{J}_P/\dot{J}_X$.

The molar fluxes of body mass and reserves ($* = V, E$), are given by

$$\dot{J}_{*+} = \dot{h}_e \int_0^\infty \int_0^{l_b} \int_0^\infty \phi_e(t,a,e,l)M_*(e,l)\, de\, dl\, da + \tag{9.30}$$

$$+ \int_0^\infty \int_{l_b}^1 \int_0^1 (\dot{h} + \dot{h}_a(a))\phi(t,a,e,l)M_*(e,l)\, de\, dl\, da \tag{9.31}$$

The mineral fluxes $\dot{\boldsymbol{J}}_{\mathcal{M}+}$ and the dissipating heat $\dot{p}_{T+}$ follow from (4.35) and (4.82)

$$\dot{\boldsymbol{J}}_{\mathcal{M}+} = -\boldsymbol{n}_{\mathcal{M}}^{-1}\boldsymbol{n}_{\mathcal{O}}\dot{\boldsymbol{J}}_{\mathcal{O}+} \tag{9.32}$$

$$0 = \dot{p}_{T+} + \boldsymbol{\mu}_{\mathcal{M}}^T\dot{\boldsymbol{J}}_{\mathcal{M}+} + \boldsymbol{\mu}_{\mathcal{O}}^T\dot{\boldsymbol{k}}_{\mathcal{O}+}. \tag{9.33}$$

Due to the linear relationships between mass and energy fluxes, the mass fluxes are simple metrics on the densities ϕ_e and ϕ, which are solutions of the partial differential equations (9.24) and (9.25); the determination of the solution generally requires numerical integration.

Steady-state situations for reproduction

At steady state, the easiest approach is to relate the states of the individuals to age, and replace the density $\phi(t, a, e, l)$, by the relative density $\phi^\circ(t, a) = \phi(t, a)/N(t)$. This relative density no longer depends on time at steady state, so we omit the reference to time. $\dot{J}_*(a)$ denotes the flux of compound $*$ with respect to an individual of age a, where a_b is the age at birth and a_p the age at puberty. These ages might be parameters, but the DEB model obtains them from $M_V(a_b) = M_{Vb}$ and $M_V(a_p) = M_{Vp}$.

The characteristic equation applies at steady state

$$M_{E0} = \exp(-\dot{h}_e a_b) \int_{a_p}^{\infty} \exp\left(-\dot{h}a - \int_0^a \dot{h}_a(a_1)\, da_1\right) \dot{J}_{E_R}(a)\, da \tag{9.34}$$

The characteristic equation can be used to solve for the food density M_{X+}, and so the scaled functional response f. Given this food density, the trajectories of the state variables are fixed.

The age distributions of embryos and juveniles plus adults are given by

$$\phi_e^\circ(a) = \frac{\dot{h}_e \exp(-\dot{h}_e a)}{1 - \exp(-\dot{h}_e a_b)} \quad \text{for} \quad a \in [0, a_b] \tag{9.35}$$

$$\phi^\circ(a) = \frac{(\dot{h} + \dot{h}_a(a)) \exp(-\dot{h}a - \int_0^a \dot{h}_a(a_1)\, da_1)}{\int_{a_b}^{\infty} \exp(-\dot{h} - \int_0^a \dot{h}_a(a_1)\, da_1)\, da} \quad \text{for} \quad a \in [a_b, \infty] \tag{9.36}$$

We introduce the expectation operators $\mathcal{E}_e$ and $\mathcal{E}$, i.e. $\mathcal{E}_e Z \equiv \int_0^{a_b} Z(a)\phi_e^\circ(a)\, da$ and $\mathcal{E}Z \equiv \int_{a_b}^{\infty} Z(a)\phi^\circ(a)\, da$, for any function $Z(a)$ of age.

The harvesting rates of organic compounds equal their mass fluxes, i.e.

$$\dot{\boldsymbol{J}}_{\mathcal{O}+} \equiv \begin{pmatrix} \dot{J}_{X+} \\ \dot{J}_{V+} \\ \dot{J}_{E+} \\ \dot{J}_{P+} \end{pmatrix} = \boldsymbol{\eta} N \mathcal{E} \dot{\boldsymbol{p}} = \begin{pmatrix} -\dot{h}_X M_X \\ N_e \mathcal{E}_e \dot{h}_e M_V + N\mathcal{E}(\dot{h} + \dot{h}_a) M_V \\ N_e \mathcal{E}_e \dot{h}_e M_E + N\mathcal{E}(\dot{h} + \dot{h}_a) M_E \\ \dot{h}_X M_X \mu_{AX}/\mu_{AP} \end{pmatrix} \tag{9.37}$$

The numbers of juveniles plus adults in the population, N, and of embryos, N_e, are given by

$$N = \frac{\dot{J}_{X+}}{\mathcal{E}\dot{J}_X} \quad \text{and} \quad N_e = (1 - \exp(-\dot{h}_e a_b)) \frac{N\mathcal{E}\dot{J}_{E_R}}{\dot{h}_e M_{E0}} \tag{9.38}$$

Propagation through division

The ageing rate of dividing organisms is taken to be independent of age and this hazard rate is included in the harvesting rate $\dot{h}$; the state variable age is not used, so the scaled length l and the scaled reserve density e specify the state of the individual. The conversion process of substrate into biomass has two control parameters: $\dot{h}_X$ and $\dot{h}$.

Given the initial condition $\phi(0, e, l)$, the dynamics of density $\phi(t, e, l)$ is then given by

$$\frac{\partial}{\partial t}\phi(t,e,l) = -\frac{\partial}{\partial l}\left(\phi(t,e,l)\frac{d}{dt}l\right) - \frac{\partial}{\partial e}\left(\phi(t,e,l)\frac{d}{dt}e\right) - \dot{h}\phi(t,e,l) \qquad (9.39)$$

with boundary condition

$$\phi(t,e,l_b)\left.\frac{d}{dt}l\right|_{l=l_b} = 2\phi(t,e,l_d)\left.\frac{d}{dt}l\right|_{l=l_d} \quad \text{for all } e \qquad (9.40)$$

where the scaled length at 'birth' relates to the scaled length at division as $l_b = l_d 2^{-1/3}$. This dynamics implies that both daughters are identical.

Suppose now that the dynamics of the scaled reserves is independent of the scaled length, and that the dynamics of the scaled length is proportional to the scaled length. The scaled reserve density then has the property that all individuals will eventually have the same scaled reserve density, which may still vary with time. (The DEB model for V1-morphs is an example of such a model.) For simplicity's sake, we will assume that this also applies at $t = 0$, which removes the need for an individual structure. The consequence is that a population that consists of one giant individual behaves the same as a population of many small ones.

The partial differential equation (9.39) collapses to two ordinary differential equations, one of which is at the population level for the structural body mass

$$\frac{d}{dt}\ln M_{V+} = \frac{d}{dt}\ln l^3 - \dot{h} \qquad (9.41)$$

where the scaled volume kinetics $\frac{d}{dt}l^3 = 3l^2\frac{d}{dt}l$ is given by the model for individuals. The other differential equation is at the individual level for the scaled reserve density kinetics $\frac{d}{dt}e$, which should also be specified by the model for individuals, see e.g. Table 4.1. The scaled reserve density kinetics specifies the (nutritional) state of a random individual.

The expressions for the dissipating heat (9.33) and mineral fluxes (9.32) still apply here, while $\dot{\boldsymbol{J}}_{O+} = \boldsymbol{\eta}\dot{\boldsymbol{p}}_+$, with $\dot{\boldsymbol{p}}_+ = \int_{l_b}^{l_d}\int_0^1 \dot{\boldsymbol{p}}(e,l^3)\phi(t,e,l)\,de\,dl$, and $\dot{\boldsymbol{p}}(e,l^3)$ denotes $\dot{\boldsymbol{p}}$, evaluated at scaled energy reserve e and cubed scaled length l^3. (For V1-morphs it is more convenient to use l^3 as an argument, rather than l.) This result is direct because $\int_{l_b}^{l_d}\int_0^1 l^3\phi(t,e,l)\,de\,dl = M_{V+}/M_{Vm}$, so that

$$\dot{\boldsymbol{p}}_+ = \dot{\boldsymbol{p}}\left(e, \int_{l_b}^{l_d}\int_0^1 l^3\phi(t,e,l)\,de\,dl\right) = \dot{\boldsymbol{p}}(e, M_{V+}/M_{Vm}) = \dot{\boldsymbol{p}}(e,1)M_{V+}/M_{Vm}$$

The latter equality only holds for models such as the DEB model for 1S–V1-morphs, where all powers are proportional to structural body mass. The dynamics for food amounts to

$$\frac{d}{dt}M_{X+} = \dot{h}_X M_X + \dot{J}_{X+} = \dot{h}_X M_X - \frac{\dot{p}_A(e,1)}{\mu_{AX}} \frac{M_{V+}}{M_{Vm}} \tag{9.42}$$

where $\dot{p}_A(e, 1)$ does not depend on the scaled reserves e, in the DEB model.

The environment for the population reduces to the chemostat conditions for the special choice of the harvesting rate $\dot{h}$ relative to the supply rate: $\dot{h}_X M_X = \dot{h}(M_X - M_{X+})$.

Steady-state situations for division

The population growth rate must be zero at steady state. We use this to solve the value of the scaled functional response, i.e. $f = \frac{\dot{k}_M + \dot{h}}{\dot{k}_M/l_d - \dot{h}/g}$ in the case of the DEB model. This model has the nice property that $e = f$ at steady state; it then follows that $M_{X+} = M_K f/(1-f)$, where M_K is the saturation constant of the Holling type II functional response.

The stable age distribution amounts to

$$\phi^\circ(a) = 2\dot{h}\exp(-\dot{h}a) \quad \text{for} \quad a \in [0, \dot{h}^{-1}\ln 2] \tag{9.43}$$

The number of individuals in the population, the total structural body mass and the organic fluxes are given by

$$N = \frac{\dot{J}_{X+}}{\mathcal{E}\dot{J}_X} = \frac{M_{V+}}{l_d^3 M_{Vm}\ln 2} \tag{9.44}$$

$$M_{V+} \equiv N\mathcal{E}M_V = \frac{\dot{h}_X M_X [M_V]}{f[M_X][\dot{J}_{Xm}]} \tag{9.45}$$

$$\dot{\boldsymbol{J}}_{\mathcal{O}+} = \boldsymbol{\eta}\dot{\boldsymbol{p}}_+ = \boldsymbol{\eta}\dot{\boldsymbol{p}}(f,1)M_{V+}/M_{Vm} \tag{9.46}$$

The mean mass per individual is thus $\mathcal{E}M_V = M_{V+}/N$.

The relationship (9.32) for the mineral fluxes still holds. Since the row of $\boldsymbol{n}_{\mathcal{M}}^{-1}$ that corresponds to oxygen, i.e. the third row, can be interpreted as the ratio of the reduction degrees of the elements to that of oxygen if the N substrate is ammonia (see {140}), the third row of $\boldsymbol{n}_{\mathcal{M}}^{-1}\boldsymbol{n}_{\mathcal{O}}$ can be interpreted as the ratio of the reduction degrees of the organic components to that of oxygen. It follows that $-\delta_O \dot{J}_{O+} = \boldsymbol{\delta}_{\mathcal{O}}^T \dot{\boldsymbol{J}}_{\mathcal{O}+}$, when $\boldsymbol{\delta}_{\mathcal{O}}$ denotes the reduction degrees of the organic compounds. If the structural biomass has the same composition as the reserves, and if no products are formed, this further reduces to $-\delta_O \dot{J}_{O+} = \delta_X \dot{J}_{X+} + \delta_W \dot{J}_{W+}$, or $-\delta_O Y_{OW} = \delta_X Y_{XW} + \delta_W$, where index W refers to the total biomass, i.e. the sum of the structural biomass and the reserves, and Y to yield coefficients. This result is well known from the microbiological literature [486, 981] and follows directly from the general assumptions in Table 2.4.

Figure 9.14 compares the measured dioxygen yield with the yield that has to be expected on the basis of this relationship and measured values of biomass yields for a wide variety of organisms and 15 substrates that differ in n_{HX} and n_{OX}, but all have $n_{NX} = 0$. The substantial scatter shows that the error of measurement is large and/or that the biomass composition is not equal for all organisms and is not independent of the growth rate. Generally $n_{*V} \neq n_{*E}$, and n_{*W} depends on the population growth rate $\dot{r}$.

Table 9.4 The possible stochastic events F, S, D_s and D_h, the intensities $\dot{\lambda}_F$, $\dot{\lambda}_S$, $\dot{\lambda}_{D_s}$ and $\dot{\lambda}_{D_h}$ and the steps sizes (dP, dC_s, dC_h), given the state (P, C_s, C_h) of the system at time t. The growth process G is deterministic and continuous. Mass balance restrictions mean that the steps in the three variables are coordinated. The coefficient δ_t varies in time, due to stoichiometric constraints on the growth of the consumers from structure as well as varying reserve of the producers. The system is closed for nutrient, so for producers and consumers as well, while nutrient uptake by the producers is large enough to cause negligibly small concentrations of free nutrient.

event type i intensity $\dot{\lambda}_i$	F feeding $\dot{k}\frac{P}{K}\frac{C_s}{C_\varepsilon}$	S searching $\dot{k}\frac{C_h}{C_\varepsilon}$	D_s dying of C_s $\dot{h}_C\frac{C_s}{C_\varepsilon}$	D_h dying of C_h $\dot{h}_C\frac{C_h}{C_\varepsilon}$	G growing $\dot{r}_P\frac{P}{P_\varepsilon}$
dP	$-P_\varepsilon$	0	0	0	P_ε
dC_s	$-C_\varepsilon$	C_ε	$-C_\varepsilon$	0	0
dC_h	$\delta_t C_\varepsilon$	$-C_\varepsilon$	0	$-C_\varepsilon$	0

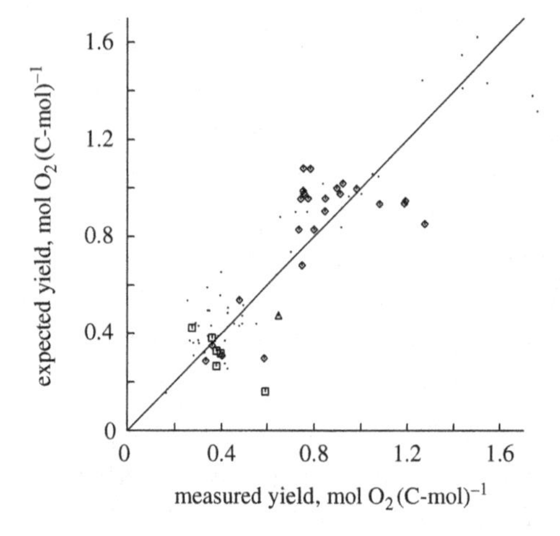

Figure 9.14 The expected molar yield of dioxygen as a function of the measured value based on the assumption of a constant and common biomass composition of $n_{HW_1} = 1.8$, $n_{OW_1} = 0.5$ and $n_{NW_1} = 0.2$ for a wide variety of bacteria (•), yeasts (◇), fungi (□) and the green alga *Chlorella* (△). The expectation is based on measured yields of biomass. Data gathered by Heijnen and Roels [486] from the literature on aerobic growth on a wide variety of substrates without product formation and NH_4^+ as nitrogen substrate.

9.3 Food chains and webs

Many ecosystems have consumers that are linked in a food web; the food chain being its simplest form. Bi- and tri-trophic chains are intensively studied [470, 472, 603, 677, 763, 764]. Most models, however, have growing zero-trophic levels, and are based on implicit assumptions about their food dynamics. This will be avoided here, to allow the application of mass and energy balances. Some popular models are even at odds with conservation principles [623, 625].

A basic problem in the analysis of food web dynamics is the large number of parameters that show up, which reduces the value of the exhausting undertaking of a systematic approach to the analysis of the system's potential behaviour. Several strategies are required to minimise that problem. One of them is to use body size scaling relationships to tie parameter values across species. This reduces the problem of community dynamics in principle to that of particle size distributions in taxon-free communities, as reviewed by Damuth *et al.* [244].

9.3.1 Behaviour of bi-trophic chains

The non-equilibrium dynamics of food chains can be rather complex and sensitively depends on the initial conditions. Figure 9.15 illustrates results for a substrate–bacteria–myxamoebae chain in a chemostat. Bob Kooi has been able to fit the experimental data to the DEB model for V1-morphs with remarkable success. All parameters were estimated on the basis of a weighted least-squares criterion. The fitted system does not account for the digestion of reserves; its incorporation resulted in very similar fits, while the extra parameters were poorly fixed by the data. The main dynamic features are well described by the model. The myxamoebae decrease more rapidly in time than the throughput rate allows by shrinking during starvation. The type of equilibrium of this chain is known as a spiral sink, so that this chain ultimately stabilises, and the period reduces with the amplitude. The numerical integration of the set of differential equations that describe the system was achieved using a fourth-order Runge–Kutta method.

This particular data set was used to illustrate the application of catastrophe theory by Saunders [1014], who concluded that simple generalisations of the Lotka–Volterra model cannot fit this particular data set, because growth is fast when substrate is low. He suggested that the feeding rate for each individual myxamoeba is proportional to the product of the bacteria and the myxamoebae densities. This implies an interaction between the myxamoebae; Bazin and Saunders [71] suggested that the myxamoebae measure their own density via folic acid. Although interactions cannot be excluded, the goodness of fit of the DEB model makes it clear that it is not necessary to include such interactions. The significance of realistic descriptions without interaction is in the extrapolation to other systems; if species-specific interactions do dominate systems behaviour, there can be hardly any hope for the feasibility of community ecology. Reserves cause a time delay in the reaction of the predator to fluctuations in prey and explain why a high growth rate can combine with low substrate densities in these oscillatory systems.

Asymptotic behaviour

Work on the asymptotic behaviour of DEB-structured bi- and tri-trophic chains by Bob Kooi and Martin Boer [119, 120, 121, 458, 624, 628] showed that tri-trophic chains can have quite complex asymptotic behaviour, even in the Marr–Pirt case. It has been extended to include omnivory and symbiotic relationships [122, 621, 629, 630, 676], reserve [626, 647, 675] and stochasticity [655]. Nice overviews of bifurcation analysis for population dynamics can be found in [620, 627]. Chaos in a bivariate prey–predator systems is described in [622].

Figure 9.15 Trajectories of a bi-trophic chain of glucose X_0, graph 1, the bacterium *Escherichia coli* X_1, graph 2, and the cellular slime mould *Dictyostelium discoideum* X_2, graph 3, in a chemostat at 25 °C with throughput rate $\dot{h} = 0.064\,\mathrm{h}^{-1}$ and a glucose concentration of $X_r = 1\,\mathrm{mg\,ml}^{-1}$ in the feed. Graphs 4 and 5 give the mean cell volumes of *Escherichia* and *Dictyostelium*, respectively. Data from Dent *et al.* [264]. The parameter values and equations are

$X_0(0)$	0.433			$\mathrm{mg\,ml}^{-1}$
$X_1(0)$	0.361	$X_2(0)$	0.084	$\mathrm{mm}^3\,\mathrm{ml}^{-1}$
$e_1(0)$	1	$e_2(0)$	1	–
K_1	0.40	K_2	0.18	$\frac{\mu g}{ml}, \frac{mm^3}{ml}$
g_1	0.86	g_2	4.43	–
$\dot{k}_M^1$	0.008	$\dot{k}_M^2$	0.16	h^{-1}
$\dot{k}_E^1$	0.67	$\dot{k}_E^2$	2.05	h^{-1}
$\dot{j}_{XAm}^1$	0.65	$\dot{j}_{XAm}^2$	0.26	$\frac{mg}{mm^3\,h}, \mathrm{h}^{-1}$

$$\frac{d}{dt}e_1 = \dot{k}_E^1 (f_1 - e_1); \quad f_1 = \frac{X_0}{K_1 + X_0}$$

$$\frac{d}{dt}e_2 = \dot{k}_E^2 (f_2 - e_2); \quad f_2 = \frac{X_1}{K_2 + X_1}$$

$$\frac{d}{dt}X_0 = \dot{h}(X_r - X_0) - f_1 \dot{j}_{XAm}^1 X_1$$

$$\frac{d}{dt}X_1 = \left(\frac{\dot{k}_E^1 e_1 - \dot{k}_M^1 g_1}{e_1 + g_1} - \dot{h}\right) X_1 - f_2 \dot{j}_{XAm}^2 X_2$$

$$\frac{d}{dt}X_2 = \left(\frac{\dot{k}_E^2 e_2 - \dot{k}_M^2 g_2}{e_2 + g_2} - \dot{h}\right) X_2$$

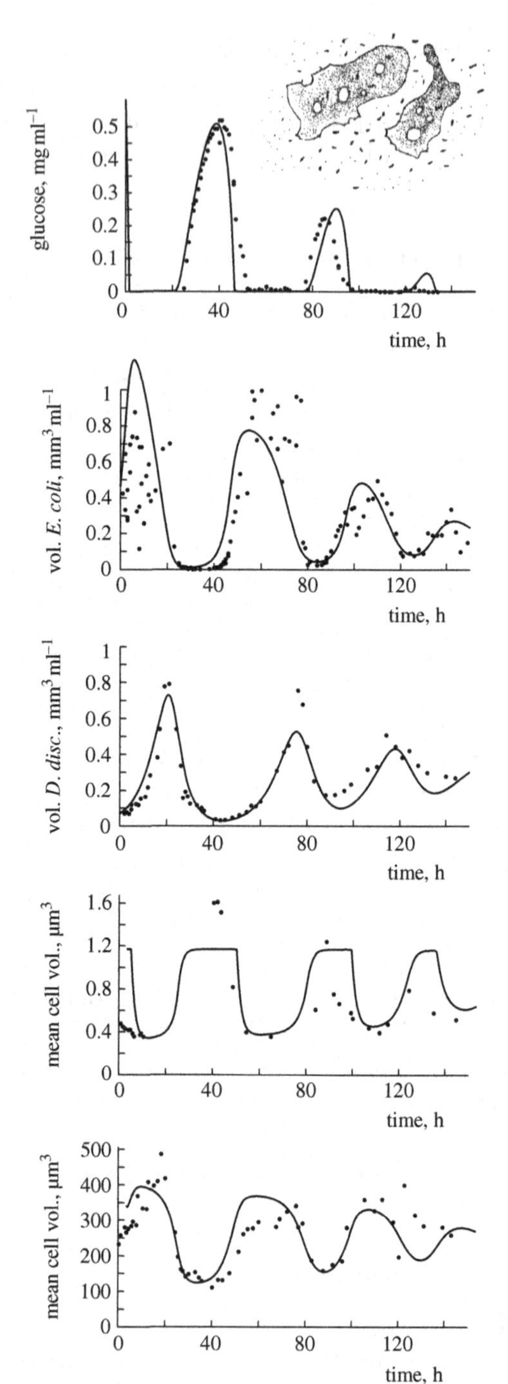

Closed nutrient–producer–consumer system

When one organism eats another one with a chemical composition that can vary, there is a need to deal with conversion efficiencies of prey into predator in a bit more detail than is usual [647]. Suppose that the prey has a single reserve and structure. The assimilation process of the predator then should specify how the two components of its prey, together with nutrients from the environment, transform into predator reserves. Think for instance of daphnids feeding on algae. Algae's main carbon component, cellulose, is of little nutritional value for the daphnid. It is the starch and lipids in the algae's reserves that are daphnids' main energy sources, while they also need ammonia and phosphate, for instance, as building blocks. Daphnids can obtain part of these nutrients from the intracellular reserves of the algae, sometimes they can also obtain them directly from the environment. So the nutritional value of the algae for the daphnids is not a constant, but varies, and depends on environmental conditions.

The implications of a variable nutritional value of the producer (alga) for the consumer (daphnid) can be illustrated with the simple dynamical system

$$m_N = N/P - n_{NC}\, C/P - n_{NP} \tag{9.47}$$

$$\frac{d}{dt}P = \dot{r}_P P - j_{PA} C \quad \text{with} \quad \dot{r}_P = \frac{\dot{k}_N m_N}{y_{NP} + m_N} \quad \text{and} \quad j_{PA} = \frac{j_{PAm} P}{K + P} \tag{9.48}$$

$$\frac{d}{dt}C = (\dot{r}_C - \dot{h})C \quad \text{with} \quad \dot{r}_C = \left(1/\dot{r}_C^P + 1/\dot{r}_C^N - 1/(\dot{r}_C^P + \dot{r}_C^N)\right)^{-1}$$

$$\dot{r}_C^P = y_{CP}\, j_{PA} - \dot{k}_M^P \quad \text{and} \quad \dot{r}_C^N = y_{CN}\, m_N\, j_{PA} - \dot{k}_M^N \tag{9.49}$$

where consumers' reserve density m_N follows from mass conservation, for a total amount of nutrient N; all nutrient that is not in producers' or consumers' structure is in producers' reserve. The chemical indices n_{NP} and n_{NC} stand for producers' and consumers' nutrient content per carbon. The amount of nutrient in the environment is taken to be negligibly small. The consumer has a constant hazard rate $\dot{h}$, and dead producers decompose instantaneously. The producers' reserve turnover rate is $\dot{k}_N$, and producers' maintenance is neglected. Consumers' reserves are not taken into account. These simplifications of the DEB theory amount to Droop's kinetics for the consumer (with a very small half-saturation constant, and a very large specific maximum uptake rate), and Marr–Pirt kinetics for the consumer. The Marr–Pirt kinetics results from the DEB model as a limit for increasing reserve turnover rates. The expression for the growth rate follows from the SU kinetics and the assumption that assimilates from producers' reserve and those from structure are complementary and parallelly processed with a large capacity. There is little need to set a maximum to the capacity here, because that is already set by the maximum specific assimilation rate j_{PAm}. Notice the SU formalism here deals with rates, rather then concentrations, as is basic to its derivation.

Stochastic formulation

The implementation of stochastic events requires the notion of individuals (notably their number), and gives the density of a single producer P_ε and consumer C_ε an explicit and independent role [655].

Table 9.4 gives the possible events F feeding, S searching, D_s dying of C_s and D_h dying of C_h, the intensities $\dot{\lambda}_i$ and the steps sizes at time t. The last process G, the growth of the producers, is supposed to be a deterministic continuous process, not a stochastic point process; the producers continue growing between the Poissonian events, i.e. $\frac{d}{dt}\underline{P} = \dot{r}_P\underline{P}$ where the specific growth rate $\dot{r}_P$ is given in (9.48), producers' reserve density m_N changes as (9.47) and the (variable) yield Y_{CP} is given by $Y_{CP} = \dot{r}_C/j_P$. Between the stochastic jump events m_N, $\dot{r}_P$ and Y_{CP} change smoothly and deterministically, while the consumer densities C_s and C_h remain constant. At a time-incremental basis, $\underline{m}_N$, $\underline{\dot{r}}_P$ and $\underline{Y}_{CP}$ are stochastic, because they are functions of $\underline{P}$ and $\underline{C} = \underline{C}_s + \underline{C}_h$. Together with the initial conditions $P(0)$, $C_s(0)$ and $C_h(0)$, this fully specifies the stochastic dynamics, which we will call the S-model (stochastic model). Again we have the constraint $m_N(0) > 0$ on the initial conditions.

The most important difference with the deterministic formulation is that the Hopf bifurcation point is hardly important for the stochastic model, but the focus point is. Between the focus and the Hopf bifurcation point the deterministic model sports an overshoot behaviour, which lasts longer and has a larger amplitude if closer to the Hopf bifurcation point. Asymptotically, however, the deterministic system settles at the point attractor. The stochastic model, on the contrary, sports irregular semi-oscillatory behaviour in this interval of values for the total amount of nutrient. The oscillations become more regular and the amplitude increases if closer to the Hopf bifurcation points. Around the focus point the model behaviour changes smoothly, but around the Hopf bifurcation point the asymptotic behaviour of the deterministic model changes abruptly, while its transient behaviour and the behaviour of the stochastic model changes smoothly. In summary, the stochastic model responds more smoothly to changes in the total amount of nutrient.

9.3.2 Stability and invasion

Nisbet *et al.* [840] noted that the experimental system appears to be much more stable than is predicted by the bi-trophic Monod model. They concluded that the introduction of maintenance, as proposed by Marr–Pirt, increases the range of operation parameters that give stable chains; however, real-world chains still appear to be more stable. Consistent with the single trophic systems, compared in Figure 9.2, the DEB model for bi- and tri-trophic chains is much more stable than the Monod and the Marr–Pirt model.

A species can invade a trophic system in a chemostat if its per capita growth rate exceeds the throughput rate at an infinitesimally small population size. For most food web models, this occurs when the Lyapunov exponent, which is associated with the dynamics of the invader, is positive at the boundary of the attractor. The transcritical bifurcation point, when the Lyapunov exponent is zero, marks the region where invasion is possible. Using this criterion, numerical studies by Bob Kooi showed that another level-two species can invade in a bi-trophic DEB chain. This means that the level-two species allows escape from the competitive exclusion principle, see [1079], and the two competing species can coexist on a single substrate in the presence of a predator.

Before the 1970s the general insight was that an increase in diversity comes with an increase in stability. May [759] showed that the opposite holds for randomly connected Lotka–Volterra systems. Later, it became evident that the spatial scale is essential, and meta-population theory showed that instability at a small spatial scale can go with stability at a large spatial scale. We are now witnessing a new insight: diversity can go with stability in non-linear systems with more realistic dynamics, even in spatially homogeneous systems.

9.4 Canonical Community

Figure 9.16 illustrates the structure of an idealised, simple, three-species ecosystem. Producers (algae) use light and nutrients to produce organic matter, which is transformed by consumers (grazers), while decomposers (bacteria) release nutrients from the organic matrix [1104]. The system is 'open' to energy flow, but closed to inputs or removal of elemental matter. It might live in a closed bottle, for instance. Exchanges of mass with the rest of the world can be included at a later stage. It is found in many ecosystems, for example, and is very similar to the one used for material turnover in microbial flocs in seawater plankton systems [417, 738].

Microcosms are fairly realistic experimental models for ecosystems [102, 422]. Kawabata and co-workers [383, 584, 825] studied a closed community consisting of the bacterium *Escherichia coli*, the ciliate *Tetrahymena thermophila* and the euglenoid *Euglena gracilis*, for direct and indirect effects of γ-rays. The ciliate grazes on the bacterium, and lives off organic products that are excreted by the euglenoid, which has mixotrophic capabilities. A stable coexistence developed for a period exceeding 130 days. The bacterium did not survive an irradiation of 500 Gy, but the two remaining species continued to exist at lower levels.

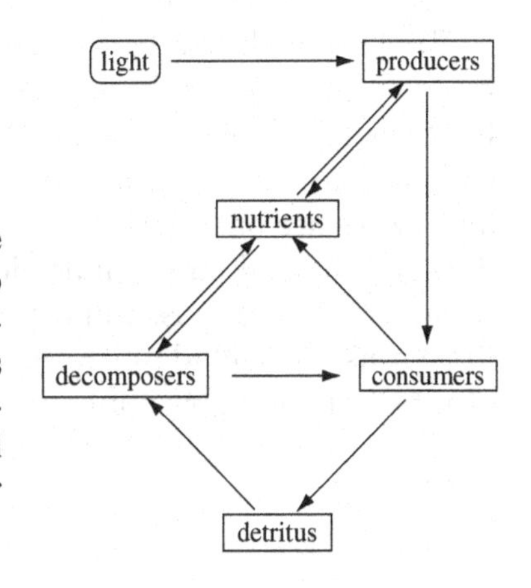

Figure 9.16 The Canonical Community consists of three 'species': producers that gain energy from light and take up nutrients to produce biomass, consumers that feed on producers and decomposers that recycle nutrients from producers and consumers. The community is rather closed for nutrients, but requires a constant supply of energy. Influx and efflux of nutrients largely determine the long-term behaviour of the community.

The results presented here are from [659]. The Canonical Community differs from a prey–predator system by the inclusion of the zero-th trophic level in the dynamics of the system. Prey–predator systems that allow mass balances always require external supply of inert substrate. Many prey–predator systems in the literature, however, assume intrinsic growth of the prey, independent of its food, and, therefore, imply complex dynamics of variables that are excluded from the system. Another difference with a prey–predator system is that all components affect nutrients, which implies more complex trophic interactions between components, as discussed by e.g. Andersen [25].

9.4.1 Mass transformations in communities

The chemical compounds and their transformations in the Canonical Community are presented in Table 9.5. When we replace the signs by model-dependent quantitative expressions, such as in Table 9.6, this turns Table 9.5 into a matrix of fluxes that is known as a scheme matrix [946], which will be indicated by matrix $\dot{\boldsymbol{J}}$; element i, j of matrix $\dot{\boldsymbol{J}}$, called $\dot{J}_{i,j}$, gives the flux of compound i involved in transformation j. We quantify the compounds in terms of moles (for minerals) or C-moles (for organic compounds and biomass), and indicate the vector of moles of all compounds by $\boldsymbol{M}$.

The symbol $\dot{\boldsymbol{J}}_C$ denotes the vector of C-fluxes, while $\dot{J}_{C,GD}$ denotes the C-flux associated with the growth of decomposers. The flux $\dot{J}_C^+$ adds all positive contributions in $\dot{\boldsymbol{J}}_C$, and $\dot{J}_C^-$ all negative ones, so $\dot{J}_C^+ + \dot{J}_C^- = 0$; the quantity $\dot{J}_*^+/M_*$ quantifies the turnover rate of compound $*$ in the system. Index $\mathcal{M}$ collects the four minerals, $\mathcal{O}$ the 11 organic compounds; $\dot{\boldsymbol{J}}_{\mathcal{M},GD}$ denotes the four mineral fluxes that are associated with decomposer growth, $\dot{\boldsymbol{J}}_{\mathcal{O},GB}$ does the same for the 11 organic fluxes; $\boldsymbol{n}_\mathcal{M}$ collects the 4×4 chemical indices for minerals, $\boldsymbol{n}_\mathcal{O}$ is the 4×11 matrix of chemical indices for the organic compounds. Indices C, P and D refer to consumers, producers and decomposers.

When the transformations can be written as functions of the total amount of moles of the various compounds, $\boldsymbol{M}$, the dynamics of $\boldsymbol{M}$ can be written as $\frac{d}{dt}\boldsymbol{M} = \dot{\boldsymbol{J}}\boldsymbol{1}$, which just states that the change in masses equals the sum of the columns of the scheme matrix. The Jacobian $\frac{d}{d\boldsymbol{M}^T}\dot{\boldsymbol{J}}\boldsymbol{1}$ at steady state contains interesting information about the possible behaviour of the system close to the steady state.

Table 9.5 illustrates a case where decomposers and consumers have one type of reserve, and the producers have two, one with and one without nitrogen, to account for their larger metabolic flexibility. Consumers mainly feed on reserves, because they cannot digest cell wall material, which makes up a substantial part of structural mass, and faeces are only derived from structural mass (which implies that their composition does not depend on the nutritional status of the prey). Only ammonia is included, not because it is the most important nutrient, but because organisms excrete it. It makes little sense to include nitrate, for example, without including ammonia; the exclusion of nitrate is just for simplicity's sake.

The system is closed for mass, which means that $\boldsymbol{n}\dot{\boldsymbol{J}} = \boldsymbol{0}$. At steady state, we have $\frac{d}{dt}\boldsymbol{M} = \dot{\boldsymbol{J}}\boldsymbol{1} = \boldsymbol{0}$.

Table 9.5 The chemical compounds of the Canonical Community and their transformations and indices. The + signs mean appearance, the – signs disappearance. The signs of the mineral fluxes depend on the chemical indices and parameter values. The labels on rows and columns serve as indices to denote mass fluxes and powers. The table shows flux matrix $\dot{\boldsymbol{J}}^T$ rather than $\dot{\boldsymbol{J}}$ if the signs are replaced by quantitative expressions presented in Table 9.6.

compounds →			minerals					detritus				consumer		producer			decomp	
← transformations			L light	C carbon dioxide	H water	O dioxygen	N ammonia	PP prod-faeces	PD decomp-faeces	PV dead cons-struc	PE dead cons-res	VC structure	EC reserves	VP structure	E_1P reserves 1	E_2P reserves 2	VD structure	ED reserves
consumer	assim. 1	A_1C		+	+	–	+	+					+	–	–	–		
	assim. 2	A_2C		+	+	–	+		+				+				–	–
	growth	GC		+	+	–	+					+	–					
	dissip.	DC		+	+	–	+						–					
	death	HC								+	+	–	–					
prod.	assim. 1	A_1P	–	–	–	+									+			
	assim. 2	A_2P	–	–	–	+	–									+		
	growth	GP		+	+	–	+							+	–	–		
	dissip. 1	D_1P		+	+	–	+								–			
	dissip. 2	D_2P		+	+	–	+									–		
decomposer	assim. 1	A_1D		+	+	–	+	–										+
	assim. 2	A_2D		+	+	–	+		–									+
	assim. 3	A_3D		+	+	–	+			–								+
	assim. 4	A_4D		+	+	–	+				–							+
	growth	GD		+	+	–	+										+	–
	dissip.	DD		+	+	–	+											–
	carbon	C		1				1	1	1	1	1	1	1	1	1	1	1
	hydrogen	H			2		3	1.6	1.6	1.8	1.8	1.8	1.8	1.6	2	1.6	1.6	1.6
	oxygen	O		2	1	2		0.4	0.4	0.5	0.5	0.5	0.5	0.4	1	0.4	0.4	0.4
	nitrogen	N					1	0.1	0.1	0.2	0.2	0.2	0.2	0.2		0.4	0.2	0.4

Figure 9.17 illustrates that an increase of total nitrogen, starting from a situation where nitrogen is limiting, shifts carbon proportionally from detritus and producers to carbon dioxide, consumers and decomposers, till it ceases to be limiting. A similar increase in carbon also results in a proportional increase in the biomass of all three living components, but ammonia decreases linearly, until it hits a threshold at which the community becomes extinct. An increase of the light level has a more complex effect on biomass. It results in a peak for the consumers and the decomposers, and a dip for

Table 9.6 Fluxes in the Canonical Community of the consumers VC, producers VP and decomposers VD that live in a confined environment, in which all are conceived as V1-morphs. The compounds and transformations are introduced in Table 9.5. Consumers and decomposers have one type of reserves (EC and ED, respectively), the producers have two types (E_1P and E_2P). Detritus includes producer-faeces PP, decomposer-faeces PD (both produced by consumers), and dead consumers (structural mass PV and reserves PE). Carbon dioxide (C) and ammonia (N) are obtained from the balance equation for carbon and nitrogen. The variables x refer to the scaled mass densities: $x_{PP} = M_{PP}/K_{PP}$, $x_{PD} = M_{PD}/K_{PD}$, $x_{PV} = M_{PV}/K_{PV}$, $x_{PE} = M_{PE}/K_{PE}$, $x_P = M_{VP}/K_{VP}$, $x_D = M_{VD}/K_{VD}$, $x_{Li} = \dot{J}_L/\dot{J}_{K,\,Li}$ ($i = 1, 2$), $x_N = M_N/K_{N2}$, $x_{Ci} = M_C/K_{Ci}$ ($i = 1, 2$), where $\dot{J}_L$ is the light flux that is supplied to the system to keep it going.

$$\dot{J}_{VP,\,A_1C} = -M_{VC}\, j_{VP,\,AC,m} \frac{x_P}{1+x_P+x_D}; \quad \dot{J}_{PP,\,A_1C} = -y_{PP,VP}\, \dot{J}_{VP,\,A_1C}$$

$$\dot{J}_{VD,\,A_2C} = -M_{VC}\, j_{VD,\,AC,m} \frac{x_D}{1+x_P+x_C}; \quad \dot{J}_{PD,\,A_2C} = -y_{PD,\,VD} \dot{J}_{VD,\,A_2C}$$

$$\dot{J}_{EC,\,A_iC} = -\sum_* y_{EC,\,*} \dot{J}_{*,\,A_iC} \quad \text{for } (i,*) \in \{(1,VP),(1,E_1P),(1,E_2P),(2,VD),(2,ED)\}$$

$$\dot{J}_{E_iP,\,A_1C} = m_{E_iP} \dot{J}_{VP,\,A_1C} \quad \text{for } i \in \{1,2\}; \quad \dot{J}_{ED,\,A2C} = m_{ED} \dot{J}_{VD,\,A_2C}$$

$$\dot{J}_{VC,\,GC} = M_{VC} \frac{m_{EC}\dot{k}_{EC} - j_{EC,\,MC}}{m_{EC}+y_{EC,\,VC}}; \quad \dot{J}_{EC,\,DC} = -j_{EC,\,MC} M_{VC}; \quad \dot{J}_{PE,\,HC} = m_{EC} \dot{J}_{PV,\,HC}$$

$$\dot{J}_{PV,\,HC} = \dot{h}_a M_{VC} \frac{y_{VC,\,EC}\, m_{EC}}{1+y_{VC,\,EC}\, m_{EC}}; \quad \dot{J}_{VC,\,HC} = -\dot{J}_{PV,\,HC}; \quad \dot{J}_{EC,\,HC} = -\dot{J}_{PE,\,HC}$$

$$\dot{J}_{E_1P,\,A_1P} = M_{VP}\, j_{E_1P,\,AP,m}\, f_{P_1} \quad \text{with} \quad f_{P_1} = \left(1 + \sum_* x_*^{-1} - \left(\sum_* x_*\right)^{-1}\right)^{-1} \quad \text{for } * \in \{L, C\}$$

$$\dot{J}_{E_2P,\,A_2P} = M_{VP}\, j_{E_2P,\,AP,m}\, f_{P_2} \quad \text{for } * \in \{L, N, C\}$$

$$f_{P_2} = \left(1 + \sum_* x_*^{-1} - \left(\sum_{*\notin L} x_*\right)^{-1} - \left(\sum_{*\notin N} x_*\right)^{-1} - \left(\sum_{*\notin C} x_*\right)^{-1} + \left(\sum_* x_*\right)^{-1}\right)^{-1}$$

$$\dot{J}_{VP,\,GP} = \dot{r}_{VP,\,GP} M_{VP} \quad \text{with} \quad \dot{r}_{VP,\,GP} = \left(\sum_i \dot{r}_{E_i}^{-1} - \left(\sum_i \dot{r}_{E_i}\right)^{-1}\right)^{-1} \quad \text{and}$$

$$\dot{r}_{E_i} = \frac{m_{E_iP}(\dot{k}_{E_iP} - \dot{r}_{VP,\,GP}) - j_{E_iP,MP}}{y_{E_iP,\,VP}}; \quad \dot{J}_{E_iP,\,GP} = -y_{E_iP,\,VP} \dot{J}_{VP,\,GP} \quad \text{for } i \in \{1,2\}$$

$$\dot{J}_{E_iP,\,D_iP} = -j_{E_iP,\,MP} M_{VP} - (1-\kappa_{Ei})((\dot{k}_{E_iP} - j_{VP,\,GP}) M_{E_iP} - (j_{E_iP,\,MP} + j_{VP,\,GP}\, y_{E_iP,\,VP}) M_{VP})$$

$$\dot{J}_{*,\,A_iD} = -M_{VD}\, j_{*,\,AD,m} \frac{x_*}{1+x_{PP}+x_{PD}+x_{PV}+x_{PE}}$$

$$\dot{J}_{ED,\,A_iD} = -\dot{J}_{*,\,A_iD}\, y_{ED,\,*} \quad \text{for } (i,*) \in \{(1,PP),(2,PD),(3,PV),(4,PE)\}$$

$$\dot{J}_{VD,\,GD} = M_{VD} \frac{m_{ED}\, \dot{k}_{ED} - j_{ED,\,MD}}{m_{ED}+y_{ED,\,VD}}; \quad \dot{J}_{ED,\,DD} = -j_{ED,\,MD} M_{VD}$$

$$\frac{d}{dt} M_{PP} = \dot{J}_{PP} = \dot{J}_{PP,\,A_1C} + \dot{J}_{PP,\,A_1D}; \quad \frac{d}{dt} M_{PD} = \dot{J}_{PD} = \dot{J}_{PD,\,A_2C} + \dot{J}_{PD,\,A_2D}$$

$$\frac{d}{dt} M_{PV} = \dot{J}_{PV} = \dot{J}_{PV,\,HC} + \dot{J}_{PV,\,A_3D}; \quad \frac{d}{dt} M_{PE} = \dot{J}_{PE} = \dot{J}_{PE,\,HC} + \dot{J}_{PE,\,A_4D}$$

$$\frac{d}{dt} M_{VC} = \dot{J}_{VC} = \dot{J}_{VC,\,GC} + \dot{J}_{VC,\,HC}; \quad \frac{d}{dt} M_{VP} = \dot{J}_{VP} = \dot{J}_{VP,\,GP} + \dot{J}_{VP,\,A_1C}$$

$$\frac{d}{dt} M_{EC} = \dot{J}_{EC} = \dot{J}_{EC,\,GC} + \dot{J}_{EC,\,HC} + \dot{J}_{EC,\,DC} + \dot{J}_{EC,\,A_1C} + \dot{J}_{EC,\,A_2C}$$

$$\frac{d}{dt} M_{VP} = \dot{J}_{VP} = \dot{J}_{VP,\,GP} + \dot{J}_{VP,\,A_1C}; \quad \frac{d}{dt} M_{VD} = \dot{J}_{VD} = \dot{J}_{VD,\,A_2C} + \dot{J}_{VD,\,GD}$$

$$\frac{d}{dt} M_{E_iP} = \dot{J}_{E_iP} = \dot{J}_{E_iP,\,A_1C} + \dot{J}_{E_iP,\,A_iP} + \dot{J}_{E_iP,\,GP} + \dot{J}_{E_iP,\,D_iP} \quad \text{for } i \in \{1,2\}$$

$$\frac{d}{dt} M_{ED} = \dot{J}_{ED} = \dot{J}_{ED,\,A_2D} + \dot{J}_{ED,\,A_1D} + \dot{J}_{ED,\,A_2D} + \dot{J}_{ED,\,A_3D} + \dot{J}_{ED,\,A_4D} + \dot{J}_{ED,\,GD} + J_{ED,\,DD}$$

$$\frac{d}{dt} M_C = \dot{J}_C = -\dot{J}_{PP} - \dot{J}_{PD} - \dot{J}_{PV} - \dot{J}_{PE} - \dot{J}_{VC} - \dot{J}_{EC} - \dot{J}_{VP} - \dot{J}_{E_1P} - \dot{J}_{E_2P} - \dot{J}_{VD} - \dot{J}_{ED}$$

$$\frac{d}{dt} M_N = \dot{J}_N = -0.1\dot{J}_{PP} - 0.1\dot{J}_{PD} - 0.2\dot{J}_{PV} - 0.2\dot{J}_{PE} - 0.2\dot{J}_{VC} - 0.2\dot{J}_{EC} + -0.2\dot{J}_{VP} - 0.4\dot{J}_{E_2P} - 0.2\dot{J}_{VP} - 0.4\dot{J}_{EP}$$

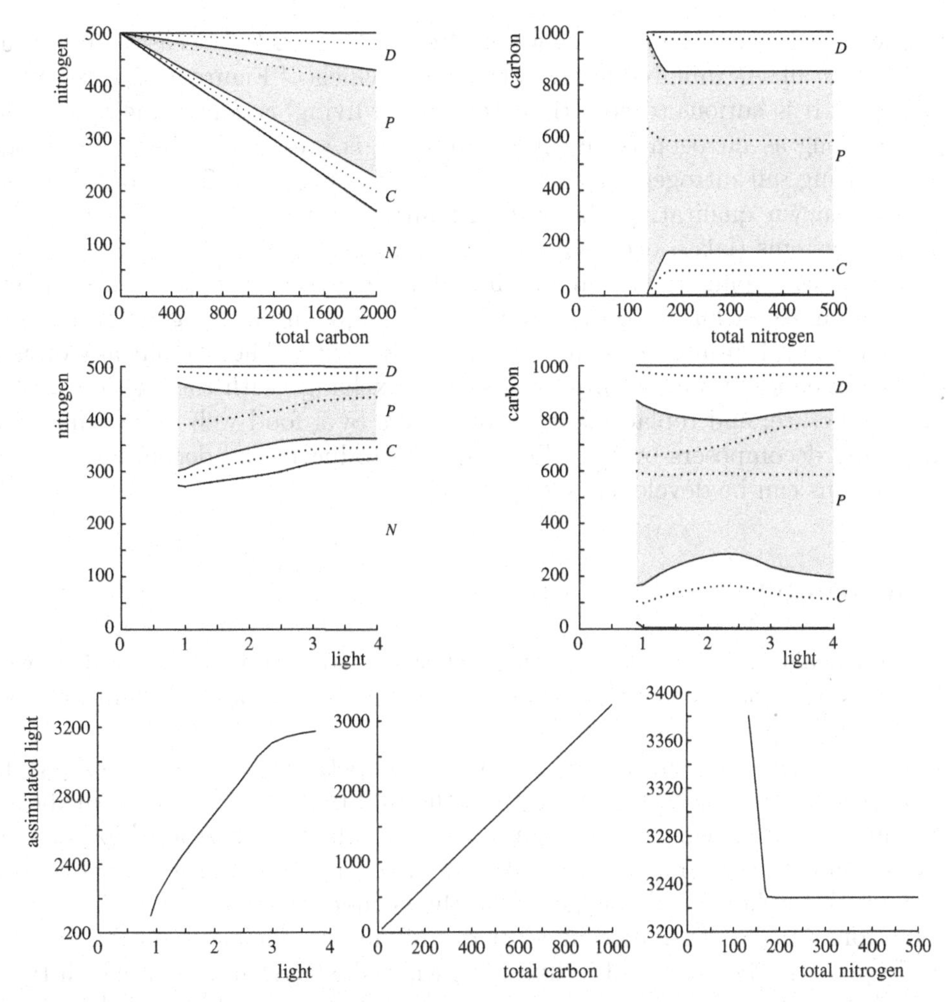

Figure 9.17 The steady-state distribution of carbon and nitrogen in the Canonical Community while increasing the total amount of carbon (upper left), nitrogen (upper right) or light (middle panels), using the DEB model for V1-morphs. The lower panels present the amounts of assimilated light (by the producers), which is proportional to the amount of dissipating heat. The non-changed amounts are 1000 units for carbon, 500 units for nitrogen and 1000 for light. The amounts of carbon and nitrogen are plotted cumulatively, from bottom to top, for the minerals (carbon dioxide, C (very small, not labelled), or ammonia, N), detritus (very small, not labelled), consumers $\mathcal{C}$ (structure and reserve), producers $\mathcal{P}$ (structure, C- and N,C-reserves, grey shaded), decomposers $\mathcal{D}$ (structure and reserve). The producers have three carbon components, and two for nitrogen, because one reserve lacks nitrogen. An increase of light above 4 units has no effect (so all lines are horizontal).

the producers, while an increase beyond the level at which light ceases to be limiting has no effect at all. Assimilated light, in the lower panels of Figure 9.17, quantifies 'the rate of living'. It is curious to note that 'the rate of living' is *de*creasing for *in*creasing nitrogen, as long as nitrogen is limiting. Ammonia is practically absent if nitrogen is strongly limiting, all nitrogen then being fixed into the biota. This corresponds well with widely known qualitative observations: nitrogen minerals are extremely low in oligotrophic systems (lakes, oceans as well as rain forests).

The Canonical Community can be simplified to a two-species, or even a single-species, community of mixotrophs [206, 654]. Since grazing no longer limits life span, ageing has to be taken into account for proper behaviour. The Canonical Community can also be extended in many ways: inclusion of exchange with the outside world and of spatial structure, and replacement of consumers by a food web of consumers, or of producers and decomposers by sets of competing producers and decomposers. Some of these extensions can be developed systematically.

9.5 Summary

This chapter deals with the metabolic interactions between individuals, and shows that individuals depend on each other in many ways; the notion of ecosystem metabolism is developed in steps.

Trophic interactions span a spectrum from competition, via syntrophy, symbiosis and biotrophy, to predation. The strength of the DEB theory is illustrated in the set-up of a full quantitative specification of partners in a symbiotic relationship. The effects of calcification can be evaluated in corals, for instance, and environmental conditions specified where the host does not gain from the symbiont.

Populations can be considered as a set of individuals; their dynamics follows from the ecophysiological behaviour of individuals, when the environment in which they live is specified. Spatial structure is very important, but not considered in this text. The distinction between individuals and populations disappears for V1-morphs in the DEB theory. The univariate DEB model for V1-morphs behaves more stably than that of Lotka–Volterra, Monod, Marr–Pirt and Droop, and has a more realistic prey/predator ratio as equilibrium. Logistic growth in a batch culture can be obtained with the DEB model in two different ways: a small half-saturation coefficient in combination with a large reserve capacity, or vice versa.

The yield of biomass on substrate at population level is controlled by the yield process; the energetics of individuals only sets the upper boundary by a constraint.

Synchronisation of life cycles among individuals can occur spontaneously. Variations in parameter values, in combination with a set of rules that specify how the values carry over to new generations, imply selective forces that lead to speciation.

Population dynamics in food chains experience a delay in response to fluctuations in food availability, and high growth rates can combine (temporarily) with low food levels as as a result. Food chains can show very complex dynamics if the chain length exceeds two. Multiple attractors occur easily, sometimes of the chaotic type. Examples illustrate the application of bifurcation analysis. Contrary to general insight, an increase in diversity can go with an increase in stability in homogeneous environments for more realistic dynamics.

Nutritional 'details', such as the nutritional value of prey's reserve, in prey–predator interactions can affect population dynamics profoundly. In the asymptotic behaviour of stochastic prey–predator systems, the importance of Hopf bifurcation points is taken over by focus points, but the transition is smooth, rather than abrupt.

Canonical Communities serve to illustrate the metabolic interactions between producers, consumers and decomposers as quantified by the DEB theory. If fully closed for mass, the community seems to increase metabolic activity for decreasing nitrogen levels, up to a threshold value, while the activity is proportional to the carbon levels, and satiating in the light levels. The analysis of Canonical Communities unifies the traditionally separated characterisations of ecosystems in terms of structure and function; this separation makes no sense in the context of the DEB theory.

10

Evolution

A proper understanding of metabolic organisation cannot be achieved without exploring its historical roots. The metabolism of individuals has adapted over time to overcome the consequences of changing environmental conditions. Mutation and selection is the well-known evolutionary route since Darwin, but this is a very slow process. It is essential for building up a basic diversity in metabolic performance among the earliest prokaryotes. This explains the slow start of evolution. Much faster is the exchange of plasmids that evolved among prokaryotes [283], which is further accelerated by the process of symbiogenesis, typical for eukaryotes. The latter also duplicate DNA and reshuffle parts of their genome, giving adaptive change even more acceleration. Mutation still continues, of course, but the reshuffling of metabolic modules occurs at rates several orders of magnitude higher. Syntrophy and symbiosis are key to these reshuffling processes and supplement Darwin's notion of survival of the fittest, which is based on competitive exclusion [1009]. The response to changes in the environment is further accelerated by the development of food webs, and therefore of predation, which enhances selection. Owing to their advanced locomotory and sensory systems, animals play an important role in food webs, and so in the acceleration of evolutionary change.

The evolutionary route for individuals as dynamic systems that is discussed below starts from the speculative abiotic origins of life, then deals with the metabolic diversification that evolved in the prokaryotes, and finally leads to the metabolic simplification, coupled to the organisational diversification, of the eukaryotes. I will argue that life became increasingly dependent on itself and that life and climate became increasingly coupled. Then I will discuss the formation of symbioses that are based on mutual syntrophy, evolutionary lines in multicellulars, and finishing with interactions between life and climate. Although some of this material is general textbook knowledge, while other material is rather speculative, my aim is to present it in a way that reveals the role of metabolic organisation on large scales in space and time and the role of DEB theory in this.

10.1 Before the first cells

A possible exergonic process generating energy in the initial stages of life involves the formation of makinawite crusts at the interface of mildly oxidising, iron-rich acidulous ocean water above basaltic floors from which alkaline seepages arose, e.g. [1003]. These crusts consist of FeS layers allowing free electron flow from the reducing environment beneath, generated by the activation of hydrothermal hydrogen. Thus, energy was constantly supplied, which, moreover, could easily be tapped at the steep gradient formed by the crust. FeS can spontaneously form cell-like structures on a solid surface [134, 180, 1001, 1002], and has a high affinity for the ATP ingredients organophosphates and formaldehyde [962], which can form ribulose [79, p 81]. The released energy could stimulate the formation of larger molecules at each inner surface, such as phosphorus or nitrogen compounds. The chemically labile energy-rich inorganic pyrophosphate compounds could have served as energy-transferring molecules [55, 56], whereas the nitrogen-containing molecules on the inner surface of the crust could have developed into nucleic acids or, later, into larger peptides. Of these, the peptides, in turn, could have combined with iron and sulphur complexes in the crust, thus initiating the formation of ferredoxins, or they could have nested themselves within the crust, thus forming the second step in the formation of membranes [1002].

ATP generation via a proton pump across the outer membrane is probably one of the first steps in the evolution of metabolism. The energy for this ATP generation probably came from some extracellular chemoautotrophic process [1001, 1002, 1203].

A possible scenario for the earliest metabolism is presented in Figure 10.1, which may be found in the archaean *Pyrodictium occulatum* [1204]. Few enzymes are required and the substrates are readily available in the deep ocean [289]. Keefe *et al.* [586], however, argue that the oxidation of FeS gives insufficient energy to fix carbon dioxide through the inverse TCA cycle. Yet this fixation may have occurred along other pathways using accumulated ATP. Schoonen *et al.* [1030] demonstrated that the energy of this reaction diminishes sharply at higher temperatures. Contrary to pyrite, greigite ($Fe_5Ni_6S_8$) has structural moieties that are similar to the active centres of certain metallo-enzymes, as well as to electron transfer agents (see, for example, [1002]), and catalyses the transformation

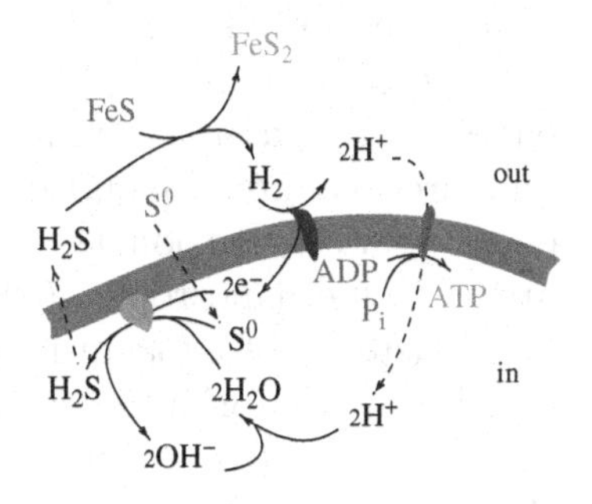

Figure 10.1 A possible early ATP-generating transformation, based on pyrite formation $FeS + S \rightarrow FeS_2$ [1142, 1204], which requires a membrane and only three types of enzyme: proto-hydrogenase, proto-ATP-ase and S^0-reductase. Modified from Madigan *et al.* [732]. Sulphur is imported in exchange for H_2S.

$$2\,CO_2 + CH_3SH + 8\,[H] \rightarrow CH_3COSCH_3 + 3\,H_2O.$$

Irrespective of the biochemical 'details', which are still controversial [855], this transformation rightly places membrane activity central to metabolism, which means that cell size matters. The membranes of membrane-bound vesicles are at the basis of transformations typical for life [1039]. Membranes need membranes (plus genes) for propagation; genes only are not enough [193]. Strong arguments in favour of the hypothesis 'cells before metabolism' include the abiotic abundance of amphiphilic compounds (even on arriving meteorites), the self-organisation of these compounds into membranes and vesicles, and their catalytic properties [251]. This argument only works if amphiphilic compounds tend to accumulate in very specific micro-environments; otherwise they will be too dilute. The modifications of substrates that are taken up from the environment to compounds that function in metabolism were initially probably small, and gradually became substantial. Compartmentalisation is essential for the accumulation of metabolites and for any significant metabolism. Norris and Raine [844] suggest that the RNA world succeeded the lipid world, which is unlikely because the archaean lipids consist of isoprenoid ethers, while eubacterial lipids consist of fatty acids (acyl esters) with completely different enzymes involved in their turnover [573, 580, 1203]. Lipids were probably synthesised first from pyruvate, the end product of the acetyl-CoA pathway and the reverse TCA cycle, before the extensive use of carbohydrates.

Koga *et al.* [614] hypothesised that the eubacterial taxa made the transition from non-cellular ancestors to cellular forms independently from the archaea, see also [751]. This seems unlikely, however, because they are similar in the organisation of their genes (e.g. in operons) and genomes, and in their transcription and translation machinery [192, 849]. Eubacteria do have a unique DNA replicase and replication initiator proteins, however. These properties apply especially to cells, rather than to pre-cellularly existing forms, and are complex enough to make it very unlikely that they evolved twice. Woese [1267] hypothesised that lateral gene transfer could have been intense in proto-cells with a simple organisation; diversification through Darwinian mutation and selection could only occur after a given stage in complexity had been reached, that is when lateral gene transfer could have been much less intense. The eubacteria, archaea and eukaryotes would have crossed this stage independently. Since all eukaryotes once seem to have possessed mitochondria, this origin is unlikely for them. Cavalier-Smith [195] argued that archaea and eukaryotes evolved in parallel from eubacteria since about 850 Ma ago, and that eukaryotes have many properties in common with actinomycetes. However, the differences in, for example, lipid metabolism and many other properties between eubacteria and archaea are difficult to explain in this way. Moreover, carbon isotope differences between carbonates and organic matter of 2.8–2.2 Ga ago are attributed to archaean methanotrophs [608]. Although so far the topic remains speculative, a separate existence of eubacteria and archaea before the initiation of the lipid metabolism and before the origin of eukaryotes through symbiogenesis with mitochondria seems to be the least-problematic sequence explaining metabolic properties among these three taxa.

The ionic strength of cytoplasm of all modern organisms equals that of seawater, which suggests that life arose in the sea.

10.2 Early substrates and taxa

Since genome size might quantify metabolic complexity, it helps to note that some chemoautotrophs have the smallest genome size of all organisms [657]; the togobacterium *Aquifex* is even more interesting since its metabolism might still resemble that of an early cell. Although it is also aerobic, it tolerates only very low dioxygen concentrations, which may have been present when life emerged [22, 522, 579]. Growing optimally at 85 °C in marine thermal vents, it utilises H_2, S^0 or $S_22O_3^-$ as electron donors and O_2 or NO_3^- as electron acceptors. With a genome size of only 1.55 Mbp, its genome amounts to only one-third of that of *E. coli*, which is really small for a non-parasitic prokaryote. One of the smallest known genomes for a non-parasitic bacterium is that of *Nanoarchaeum equitans* with 0.5 Mbp [533], but it lives symbiotically with the H_2-producing and sulphur-reducing archaean *Ignicoccus*, which complicates the comparison. The free-living α-proteobacterium *Pelagibacter ubique*, with a genome size of 1.3 Mbp, is probably phototrophic (using proteorhopsin) and uses organic compounds as carbon and electron source [409, 937]. Its metabolic needs are uncertain, since it is difficult to culture. The phototrophic cyanobacterium *Prochlorococcus* has 1.7 Mbp [380]. These small genome sizes illustrate that autotrophy is metabolically not more complex than heterotrophy.

The early atmosphere was probably rich in carbon dioxide and poor in methane [579], which changed when methanogens started to convert carbon dioxide into methane some 3.7 Ga ago, using dihydrogen as energy substrate. This probably saved the early Earth from becoming deep frozen. Apart from being a product, methane is likely to have been an important substrate (and/or product) during life's origin [477]. Methanogenesis and (anaerobic) methylotrophy are perhaps reversible in some archaea [450]; their metabolic pathways share 16 genes, and are present in some archaeal and eubacterial taxa. The most probable scenario for its evolutionary origin is that it first evolved in the planctobacteria, which transferred it to the proteobacteria and the archaea [205]. This remarkable eubacterial taxon is unique in sporting anaerobic ammonium oxidation (anammox). The anammox clade has ether lipids in their membranes and a proteinaceous cell wall like the archaea [1127]. They have advanced compartmentation and a nuclear membrane like the eukaryotes [379, 709], and are abundant in (living) stromatolites [866]. Fossil stromatolites resemble the living ones closely [270] and date back some 3.5 Ga ago [1214]. Although this points to a key role in early evolution, planctobacteria seem too complex as a contemporary model for an early cell. Moreover, anaerobic methane oxidation (amo) involves sulphate reduction. Isotope data indicate that sulphate reduction originated 3.47 Ga ago [1047]. Sulphate was rare by then [186] and might have been formed photochemically by oxidation of volcanic SO_2 in the upper atmosphere, or phototrophically by green and purple sulphur bacteria (Chlorobiacea, Chromatiacea), see [897].

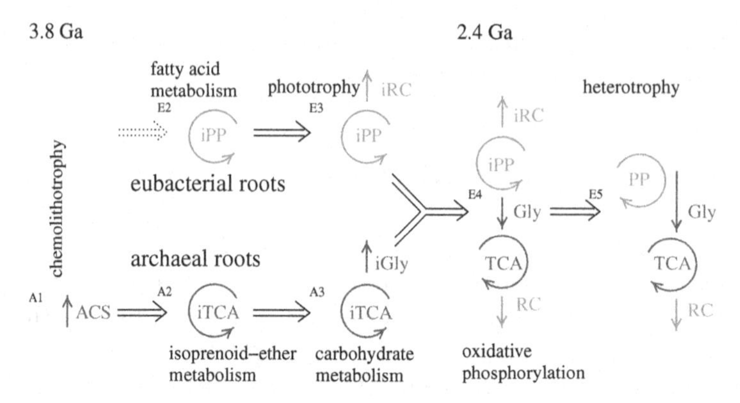

Figure 10.2 Evolution of central metabolism among prokaryotes that formed the basis of eukaryotic organisation of central metabolism. ACS = acetyl-coenzyme A synthase pathway, iPP = inverse pentose phosphate cycle (= Calvin cycle), PP = pentose phosphate cycle, iTCA = inverse tricarboxylic acid cycle, TCA = tricarboxylic acid cycle (= Krebs cycle), iGly = inverse glycolysis, Gly = glycolysis, iRC = inverse respiratory chain, RC = respiratory chain. The arrows indicate the directions of synthesis to show where they reversed; all four main components of eukaryotes' heterotrophic central metabolism originally ran in the reverse direction to store energy and to synthesise metabolites. The approximate timescale is indicated above the scheme (i.e. the origin of life, and that of cyanobacteria and eukaryotes). Contemporary models: A1 *Methanococcus*; A2 *Thermoproteus*; A3 *Sulfolobus*; E2 *Nitrosomonas*; E3 *Chloroflexus*; E4 *Prochlorococcus*; E5 *Escherichia*. Modified from Kooijman and Hengeveld [657].

10.2.1 Evolution of central metabolism

A closer look at the modern central metabolism in an evolutionary perspective might help to get the broad picture, see Figure 10.2. I here suggest that the four main modules of the central metabolism evolved one by one within the prokaryotes, and were recombined, reverted and reapplied. This scenario implies considerable conjugational exchange between the archaea and eubacteria, but given the long history of evolution, such exchanges might have been very rare. The exchange must have been pre-dated by a symbiotic coexistence of archaea and eubacteria to tune their very different metabolic systems. The examples of contemporary models [708, 985] illustrate that the metabolic systems themselves are not hypothetical, but the evolutionary links between these systems obviously are. This is not meant to imply, however, that the taxa also would have these evolutionary links. Some species of *Methanococcus* have most genes of glycolysis; *Thermoproteus* possesses a variant of the reversible Embden–Meyerhof–Parnas and the Entner–Douderoff pathways; *Sulfolobus* has oxidative phosphorylation. These contemporary models are not just evolutionary relicts; the picture is rather complex.

Dioxygen was rare, if not absent, during the time life emerged on Earth which classifies the respiratory chain as an advanced feature. Glucose hardly could have been so central during the remote evolutionary origins of life, since its synthesis and degradation typically involves dioxygen. Like all phototrophic eukaryotes, most chemolithoautotrophic bacteria fix inorganic carbon in the form of carbon dioxide through the Calvin cycle. At present, this cycle is part of the phototrophic machinery, a rather advanced feature in metabolic evolution and not found in any archaea [1027]. It has glucose as its main product, which suggests that the central position of glucose and, therefore, of carbohydrates, evolved only after oxygenic phototrophy evolved. Like the Calvin cycle, eukaryotic and eubacterial glycolysis (the Embden–Meyerhof pathway) is not found in archaea either; hyperthermophilic archaea possess the Embden–Meyerhof pathway in modified form [1027, 1040] and generally do not use the same enzymes [751]. This places the pyruvate processing TCA cycle at the origin of the central metabolism. However, if we leave out the glycolysis as a pyruvate-generating device, what process was generating pyruvate?

Interestingly, the eubacteria *Hydrogenobacter thermophilus* and *Aquifex* use the TCA cycle in reverse, binding and transforming carbon dioxide into building blocks (lipids, see [697]), including pyruvate. Both species are Knallgas bacteria, extracting energy from the oxidation of dihydrogen. The green sulphur bacterium *Chlorobium*, as well as the archaea *Sulfolobus* and *Thermoproteus* [732] also run the TCA cycle in reverse for generating building blocks. Hartman [467], Wächtershäuser [1205] and Morowitz *et al.* [810] hypothesised the reverse TCA cycle to be one of the first biochemical pathways. The interest in hydrogen bacteria relates to the most likely energy source for the first cells on Earth. *Hydrogenobacter* thrives optimally at 70–75 °C in Japanese hot springs. It is an aerobic bacterium, using ammonia and nitrate, but not nitrite, and possesses organelles (mesosomes). Several enzymes of the PP cycle and the glycolytic pathway are present although their activities are low [1096].

The TCA cycle seems to be remarkably efficient, which explains its evolutionary stability. Moreover, it is reversible, which directly relates to its efficiency and the inherent small steps in chemical potential between subsequent metabolites. Yet, with its nine transformations, the TCA cycle is already rather complex, and must have been preceded by simpler CO_2-binding pathways [854], such as the (linear) acetyl-CoA pathway of homoacetogens: $2\,CO_2 + 4\,H_2 + CoASH \rightarrow CH_3COSCoA + 3\,H_2O$ [535, 710]. Apart from H_2, electron donors for acetogenesis include a variety of organic and C_1-compounds. Coenzyme A, which plays an important role in the TCA cycle, is a ribonucleotide and the main substrate for the synthesis of lipids, a relic of the early RNA world [1128]. Several eubacteria and archaea employ the acetyl-CoA pathway; they include autotrophic homoacetogenic and sulphate-reducing bacteria, methanogens, *Closterium*, *Acetobacterium* and others. The RNA world is generally thought to predate the protein/DNA world. RNA originally catalysed all cellular transformations; protein evolved later to support RNA in this role. Many protein enzymes still have RNA-based co-factors (e.g. ribosomes and spliceozomes), while RNA still has catalytic functions. DNA evolved as a chemically more stable archive for RNA, probably in direct connection with the evolution of proteins, and possibly with the intervention of viruses [361, 362]. The step from the RNA to the protein/DNA world came with a need for the regulation of transcription.

The hyperthermophilic methanogens, such as *Methanococcus*, *Methanobacterium* or *Methanopyrus*, have also been proposed as contemporary models for early cells [708]; they have the acetyl-CoA pathway, which they run in both the oxidative and the reductive direction [1066]. Like *Aquifex*, they are thermophilic and taxonomically close to archaea/eubacteria fork (eukaryotes have some properties of both roots), have a small genome (*Methanococcus jannaschii* has 1.66 Mbp, coding for only 1700 genes), and they utilise H_2 as electron donor.

A natural implication of the reversal of the TCA cycle is that the direction of glycolysis was initially reversed as well, and served to synthesise building blocks for e.g. carbohydrates. Comparing the carbohydrate metabolism among various bacterial taxa, Romano and Conway [985] concluded that originally glycolysis must indeed have been reversed. Thus, the reversed glycolytic pathway probably developed as an extension of the reversed TCA cycle, and they both reversed to their present standard direction upon linking to the Calvin cycle, which produces glucose in a phototrophic process. So, what could have been the evolutionary history of phototrophy?

10.2.2 Phototrophy

Phototrophy probably developed more than 3.2 Ga ago [1282]; it seems unlikely that it appeared at the start of evolution, as some authors suggest [113, 114, 191, 468, 1265]. Recent evidence suggests that phototrophy is also possible near hydrothermal vents at the ocean floor [289], where the problem encountered by surface dwellers, namely that of damage by ultraviolet (UV) radiation, is absent. In an anoxic atmosphere, and therefore without ozone, UV damage must have been an important problem for the early phototrophs though, and protection and repair mechanisms against UV damage must have evolved in parallel with phototrophy [271]; possibly phototrophy evolved from UV-protection systems [897].

The green non-sulphur bacterium *Chloroflexus* probably resembles the earliest phototrophs and is unique in lacking the Calvin cycle, as well as the reverse TCA cycle. In the hydroxypropionate pathway, it reduces two CO_2 to glyoxylate, using many enzymes also found in the thermophilic non-phototrophic archaean *Acidianus*. Its photoreaction centre is similar to that of purple bacteria. The reductive dicarboxylic acid cycle of *Chloroflexus* is thought to have evolved into the reductive TCA cycle as found in *Chlorobium*, and further into the reductive pentose phosphate cycle, which is, in fact, the Calvin cycle [468]. Like sulphur and iron-oxidising chemolithotrophs, aerobic nitrifying bacteria use the Calvin cycle for fixing CO_2. The substrate of the first transformation of the monophosphate pathway for oxidising C_1-compounds, such as methane, is very similar to the C_1-acceptor of the Calvin cycle, which suggests a common evolutionary root of these pathways [732]. The first enzyme in the Calvin cycle, Rubisco, is present in most chemolithotrophs and phototrophs and even in some hyperthermophilic archaea. It is the only enzyme of the Calvin cycle of which (some of) the code is found on the genome of chloroplasts. The enzymes that are involved in the Calvin cycle show a substantial diversity among organisms and each has its own rather complex evolutionary history [752]. This complicates the finding of its evolutionary roots; see Figure 10.2.

The thermophilic green sulphur bacterium *Chlorobium tepidum* runs the TCA cycle (and glycolysis) in the opposite direction compared to typical (modern aerobic) organisms [732], indicating an early type of organisation [467, 810, 1205]. In combination with the observations mentioned above, this suggests that the present central glucose-based metabolism evolved when the Calvin cycle became functional in CO_2 binding, and the glycolysis and the TCA cycle reversed to their present standard direction, operating as a glucose and pyruvate processing devices, respectively (see Figure 10.2). Most phototrophs use the Calvin cycle for fixing CO_2 in their cytosol in combination with a pigment system in their membrane for capturing photons. Archaea use a low-efficiency retinal protein and are unable to sustain true autotrophic growth; five of the 11 eubacterial phyla have phototrophy. Bacterio-chlorophyll in green sulphur bacteria is located in chlorosomes, organelles bound by a non-unit membrane, attached to the cytoplasmic membrane. Green non-sulphur and purple bacteria utilise photosystem (PS) II; green sulphur and Gram-positive bacteria utilise PS I, whereas cyanobacteria (including the prochlorophytes) utilise both PS I and II [1297] (which is required for using water as electron donor). In the presence of H_2S as an electron donor, the cyanobacterium *Oscillatoria limnetica* uses only PS I, an ability pointing to the anoxic origin of photosynthesis. Oxygenic photosynthesis is a complex process that requires the coordinated translocation of four electrons. It evolved more than 2.7 Ga ago [109], or perhaps only 2.45–2.32 Ga ago [938]. Based on the observation that bicarbonate serves as an efficient alternative for water as electron donor, Dismukes *et al.* [273] suggested an evolutionary sequence of electron donation: oxalate (by green non-sulphur bacteria), Mn-bicarbonate, bicarbonate and water (by cyanobacteria).

The phototrophic system eventually allowed the evolution of the respiratory chain (the oxidative phosphorylation chain), which uses dioxygen that is formed as a waste product of photosynthesis, as well as the same enzymes in reversed order. If the respiratory chain initially used sulphate, for example, rather than dioxygen as electron acceptor, it could well have evolved simultaneously with the phototrophic system.

The production of dioxygen during phototrophy, which predates the oxidative phosphorylation, changed the Earth, e.g. [273, 681]. It started to accumulate in the atmosphere around 2.3 Ga ago, which shortened the lifetime of atmospheric methane molecules from 10 000 to 10 years with the consequence that the Earth became a 'snowball' [579]. The availability of a large amount of energy and reducing power effectively removed energy limitations; primary production in terrestrial environments is mainly water-limited, that in aquatic environments nutrient-limited. This does not imply, however, that the energetic aspects of metabolism could not be quantified usefully; energy conservation also applies in situations where the energy supply is not rate-limiting. Nutrients may have run short of supplies because of oxidation by dioxygen; this would have slowed down the rate of evolution [22]. First, sulphur precipitated out, followed by iron and toward the end of the Precambrian by phosphate and, since the Cambrian revolution, by calcium as well. Also, under aerobic conditions, nitrogen fixation became difficult, which makes biologically required nitrogen unavailable, despite its continued great abundance as dinitrogen in the environment; see [79, p 41]. Since the Calvin cycle produces fructose 6-phosphate, those autotrophic prokaryotes possessing this cycle are likely to have a glucose-based metabolism. Indeed, the presence of glucose

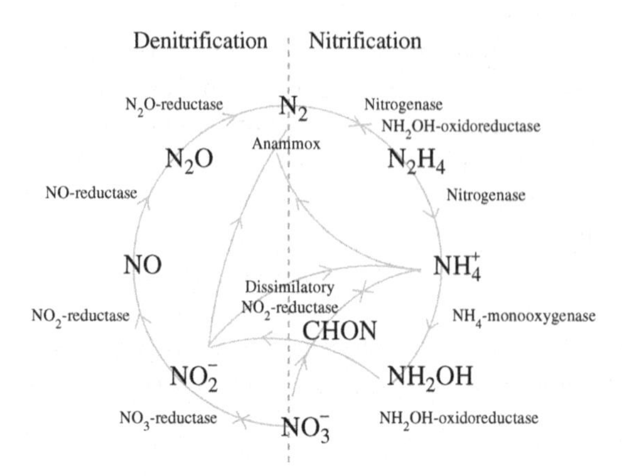

Figure 10.3 Conversions of inorganic nitrogen species by prokaryotes. The compound CHON stands for biomass. Modified from Schalk [1015].

usually suppresses all autotrophic activity. Several obligate chemolithotrophic prokaryotes, such as sulphur-oxidisers, nitrifiers, cyanobacteria and prochlorophytes, contain this cycle in specialised organelles, the carboxysomes, which are tightly packed with Rubisco. Facultative autotrophs, like purple anoxyphototrophs, use the Calvin cycle for fixing CO_2, although they lack the carboxysomes.

10.2.3 Diversification and interactions

The prokaryotes as a group evolved a wide variety of abilities for the processing of substrates, whilst remaining rather specialised as species, e.g. [20]. The nitrogen cycle in Figure 10.3 illustrates this variety, as well as the fact that the products of one group are the substrate of another.

Some of the conversions of inorganic nitrogen species can only be done by a few taxa. The recently discovered anaerobic oxidation of ammonia is only known from the planctobacterium *Brocadia anammoxidans* [1015]. Nonetheless, it might be responsible for the removal of one-half to one-third of the global nitrogen in the deep oceans [242]. The aerobic oxidation of ammonia to nitrite is only known from *Nitrosomonas*, the oxidation of nitrite into nitrate is only known from *Nitrobacter*, and the fixation of dinitrogen can only be done by a few taxa, such as some cyanobacteria, *Azotobacter*, *Azospirillum*, *Azorhizobium*, *Klebsiella*, *Rhizobium* and some others [1094].

Several eukaryotes can respire nitrate non-symbiotically. The ciliate *Loxodes* (Karyorelicta) reduces nitrate to nitrite; the fungi *Fusarium oxysporum* and *Cylindrocarpon tonkinense* reduce nitrate to nitrous oxide; the foraminifera *Globobulimina* and *Nonionella* live in anoxic marine sediments and are able to denitrify nitrate completely to N_2 [975].

The excretion of polysaccharides (carbohydrates) and other organic products by nutrient-limited photosynthesisers (such as cyanobacteria) stimulated heterotrophs to decompose these compounds through the anaerobically operating glycolytic pathway. Thus, other organisms came to use these excreted species-specific compounds as resources, and a huge biodiversity resulted. Apart from the use of each other's products, prokaryotes, such as the proteobacteria *Bdellovibrio* and *Daptobacter*, invented predation on other prokaryotes. When the eukaryotes emerged, many more prokaryote species turned to predation, with transitions to parasitism causing diseases in their eukaryotic hosts. Predators typically have a fully functional metabolism, while parasites use building blocks from the host, reducing their genome with the codes for synthesising these building blocks. The smallest genomes occur in viruses, which probably evolved from their hosts and are not reduced organisms [493, 1129]. Prokaryotic mats on intertidal mudflats and at methane seeps illustrate that the exchange of metabolites between species in a community can be intense [791, 839]. The occurrence of multi-species microbial flocks, such as in sewage treatment plants [137, 139] further illustrates an exchange of metabolites among species. The partners in such syntrophic relationships sometimes live epibiotically, possibly to facilitate exchange. Internalisation further enhances such exchange [647]. The gradual transition of substitutable substrate to become complementary is basic to the formation of obligate syntrophic relationships.

10.3 Evolution of the individual as a dynamic system

The evolution of the organism as a dynamic metabolic system can be described in several steps [662]; some of them are illustrated in Figure 10.4.

10.3.1 Homeostasis induces storage

Variable biomass composition

We start with a living (prokaryotic) cell, surrounded by a membrane. Although it remains hard to define what life is exactly, it represents an activity and, therefore, requires energy. The acquisition of energy and (probably several types of) building blocks to synthesise new structure were separated and the first cells suffered from multiple limitations; they could only flourish if all necessary compounds were present at the same time. Initially there were no reserves and hardly any maintenance costs. A cell's chemical composition varied with the availability of the various substrates. As soon as the membranes were rich in lipids (eubacteria) or isoprenoid ethers (archaea), the accumulation of lipophilic compounds could have been rather passive. The excretion of waste products was not well organised.

Strong homeostasis induces stoichiometric constraints

In a stepwise process, the cells gained control over their chemical composition, which became less dependent on chemical variations in the environment. One mechanism is coupling of the uptake and use of different substrates. How uncoupled uptake of supplementary compounds can gradually change into coupled uptake of complementary compounds is discussed at {406}. With increasing homeostasis, stoichiometric restrictions on growth become more stringent; the cells could only grow if all essential compounds were present at the same time in the direct environment of the cell.

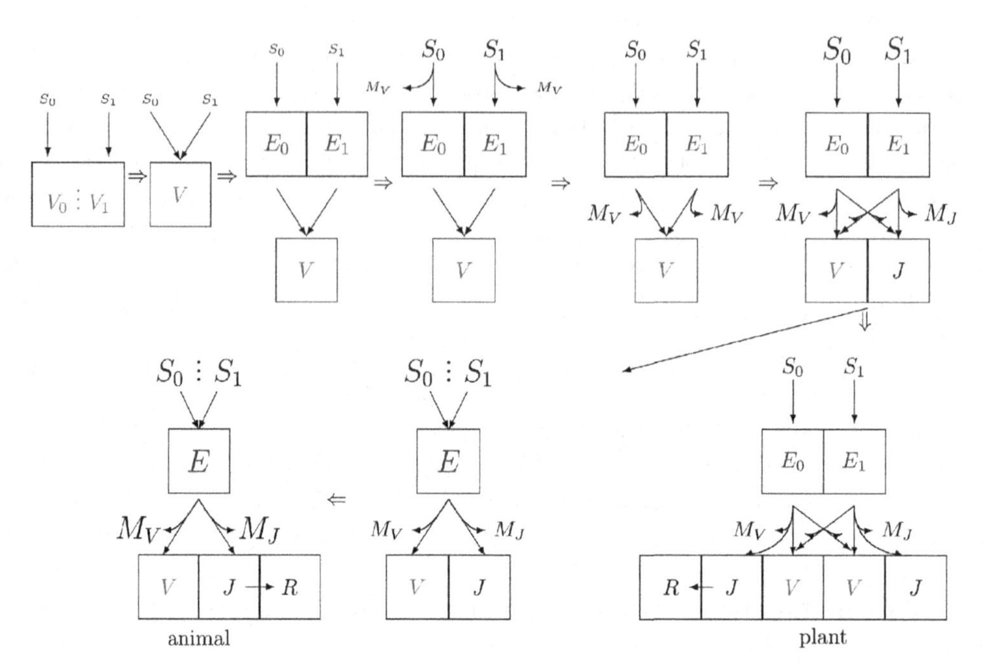

Figure 10.4 Steps in the evolution of the organisation of metabolism of organisms. Symbols: S substrate, E reserve, V structure, J maturity, R reproduction, M_V somatic maintenance, M_J maturity maintenance. Only two of several possible types of E are shown. Font size reflects relative importance. Stacked dots mean loose coupling. The top row shows the development of a prokaryotic system, which bifurcated into a plant, cf. Figure 5.11, and an animal, cf. Figure 2.1, line of development. Modified from Kooijman and Troost [662].

The activity of the cells varied with the environment at a micro-scale, which will typically fluctuate wildly. The reduction in variability of the chemical composition of the cell came with an increased ability to remove waste products, i.e. with a process of production of compounds that are released into the environment.

The mechanism that is described for the reserve dynamics of DEB theory, see {41}, and is based on the control of the number of SUs for growth, might well be an important mechanism for homeostasis in general. Whatever happened, homeostasis gradually became more perfect and biomass can be considered as being composed of a single generalised compound called structure.

Reserves relax stoichiometric constraints

The increased stoichiometric constraints on growth result in a reduction of possible habitats in which the cell can exist. By internalising and storing the essential compounds before use, the cells became less dependent on the requirement for all essential compounds to be present at the same time. In this way, they could smooth out fluctuations in availability at the micro-scale (Figure 10.11). Most substrates are first transported from the environment into the cell across the membrane by carriers before further processing. By reducing the rate of this further processing, storage develops automatically, as will be detailed below. Initially the storage capacity must have been small to avoid osmotic problems, which means that the capacity to process internalised resources is large relative to the capacity to acquire them from the environment. By transforming stored compounds to polymers, these problems could be avoided, and storage capacity could be increased further to smooth out fluctuations more effectively. This can be achieved by increasing the acquisition rate or decreasing the processing rate.

The reason for evolutionary selection toward partitionability might well be in the incremental change in the number of different types of reserves, and thus in the organisational aspects of metabolism. Partitionability and mergeability are mathematically almost the same and to some extend probably reversible in evolution. These changes not only occur within individuals, but also during the internalisation of symbionts.

Excretion: imbalance between availability and requirements

The DEB reserve dynamics implies that the amount of the most limiting reserve covaries with growth, and the amounts of non-limiting reserves can or cannot accumulate under conditions of retarded growth, depending on the excretion of mobilised reserve that is not used; excretion is an essential feature of multiple reserve systems to avoid accumulation without boundary, see Figure 5.4. This is because assimilation does not depend on the amount of reserve, thus also not on the use of reserve; it only depends on the amount of structure and substrate availability. The excretion process can be seen as an enhanced production process of chemical compounds, but its organisation (in terms of the amounts that are excreted under the various conditions) differs from waste production. Waste production is proportional to the source process (assimilation, maintenance, growth). Excretion, on the contrary, reflects an unbalanced availability of resources. The flux is proportional to a fixed fraction of what is rejected by the SU for growth. The theory of SUs quantifies the rejection flux. When the diatom *Pseudonitzschia* becomes silica depleted in an environment that is rich in nitrate, it starts to excrete domoic acids, which drains its nitrogen reserve. Some bacteria produce acetate in environments that are rich in organic compounds, but poor in nutrients, which lowers the pH, which, in turn, has a negative effect on competing species. When only acetates are left, they use these as a substrate.

Empirical evidence has so far [641, Figure 5.5] revealed that the various reserves have the same turnover rate. The reason might be that mobilisation of different reserves involves the same biochemical machinery. This possibly explains why the use of e.g. stored nitrate follows the same dynamics as that of polymers such as carbohydrates and lipids, although the use of nitrate obviously does not involve monomerisation.

Together with waste, excretion products serve an important ecological role as substrate for other organisms [1077]. Most notably polysaccharides that are excreted by phototrophs in response to nutrient limitation provide energy and/or carbon substrate for heterotrophs, so they fuel a production process that is known as the microbial loop. Adaptive dynamics analysis has indicated the importance of syntrophy in evolutionary speciation [277].

Other excretion products are toxic for potential competitors, such as domoic acid produced by the diatom *Pseudonitzschia* in response to nitrogen surplus, which can be highly toxic to a broad spectrum of organisms, including fish. Nitrogen enrichment of the environment by human activity enhances the formation of nitrogen reserves, and so the production of toxicants that contain nitrogen by some algae.

10.3.2 Maintenance enhances storage

Maintenance: increased and internalised

The storing of ions, such as nitrate, creates concentration gradients of compounds across the membrane that have to be maintained. These maintenance costs might originally have been covered by extracellular chemoautotrophic transformations, but this requires the presence of particular compounds (e.g. to deliver energy). Maintenance can only be met in this way if the organism can survive periods without having to meet such costs, i.e. facultative rather than obligatory maintenance. Most maintenance costs are obligatory, however. The next step is to pay the maintenance costs from reserves that are used for energy generation to fuel anabolic work and thus to become less dependent on the local presence of chemoautotrophic substrates. Although extreme starvation, causing exhaustion of reserves, can still affect the ability to meet maintenance costs, such problems will occur much less frequently.

Maintenance requirements were increased further and became less facultative in a number of steps, which we will discuss briefly.

Enhanced uptake and adaptation

Originally carriers (which transport substrate from the environment into the cell across the membrane) were less substrate specific and less efficient, meaning that the cell required relatively high concentrations of substrate. The cell increased the range of habitats in which it could exist by using carriers that are not fully structurally stable, meaning that high-efficiency machinery changes to low efficiency autonomously. The maintenance of high efficiency involves a turnover of carriers.

High-performance carriers are also more substrate specific, which introduces a requirement for regulation of their synthesis and for adaptation to substrate availability in the local environment. The expression of genes coding for the carriers of various substitutable substrates becomes linked to the workload of the carriers, see {291}. The principle that allocation occurs according to relative workload seems to be general and conserved; see the dynamic generalisation of the κ-rule at {204}.

Turnover of structure enhances maintenance

Not only carriers, but many chemical compounds (especially proteins with enzymatic functions) suffer from spontaneous changes that hamper cellular functions. The turnover of these compounds, i.e. breakdown and resynthesis from simple metabolites, restores their functionality [699], but increases maintenance requirements. This mixture of conversion machineries with high and low efficiencies is present in structure and so, due to turnover, to maintenance, it is converted into structure with high-efficiency machinery. The biochemical aspects of the process are reviewed in [605].

These increased requirements made it even more important to use reserves, rather than unpredictable external resources to cover them. When such reserves do not suffice, maintenance costs are met from structure, and cells shrink, see {121}. Paying maintenance from structure is less efficient than from reserve directly, because it involves an extra transformation (namely from reserve to structure). The preference for reserve as the substrate rather than structure would have been weak originally, later becoming stronger. Since the turnover rate of compounds in structure depends on the type of compound (some rates are possibly very low), the metabolites derived from these compounds do not necessarily cover all metabolic needs.

The waste (linked to maintenance and the overhead of growth) and the excreted reserves (linked to stoichiometric restrictions on growth due to homeostasis) serve as substrates for other organisms, so life becomes increasingly dependent on other forms of life even at an early stage. Some of these products were transformed into toxins that suppress competition for nutrients by other species.

Defence systems increases maintenance costs

The invasion of (micro)habitats where toxic compounds are present, and the production of toxic waste and excretion products by other organisms, required the installation of defence systems, which increase maintenance costs, see {50}. Prokaryotes developed a diverse family of defence proteins, called bacteriocins [974]. Phototrophy requires protection against UV radiation. The pathways for anaerobic methane oxidation and methanogenesis possibly evolved from a detoxification system for formaldehyde; this is another illustration of a change in function of a protection system. A general-purpose protection system against toxic compounds consists of proteins that encapsulate toxic molecules. Another general system is to transform lipophilic compounds into more hydrophilic (and so more toxic) ones to enhance excretion, see {233}. The development of a complex double cell membrane in the Didermata (Gram-negative eubacteria) was possibly a response to the excretion of toxic products by other bacteria [443], although the outer membrane is not a typical diffusion barrier [697]. When dioxygen first occurred in the environment as a waste product of oxygenic photosynthesis, it must have been toxic to most organisms [273, 681]; the present core position of carbohydrates in the central metabolism of eukaryotes and its use in energy storage is directly linked to this waste product. The reactive oxygen species (ROS) play an important role in ageing [692], see {214}, and induced the development of defence systems using peroxidase dismutases to fight their effects. While eukaryotes learned not only to protect themselves against dioxygen, but even to make good use of it, they became vulnerable to hydrogen sulphide that is excreted by anaerobically photosynthesising green and purple sulphur bacteria. Biomarkers (isorenieratene) from these bacteria suggest that the great mass extinctions of the Permian, Devonian and Triassic are linked to the toxic effects of hydrogen sulphide and the lack of dioxygen [434, 1217]. Viruses probably arose early in the evolution of life, and necessitated specialised defence systems that dealt with them. These defence systems further increased maintenance costs.

Increase of reserve capacity in response to increase of maintenance

Substrate concentration in the environment is not constant, which poses a problem if there is a continuous need to cover maintenance costs. An increase in maintenance costs therefore requires increased storage capacity in order to avoid situations in which maintenance costs cannot be met. The solution is to further delay the conversion of substrate metabolites to structure, creating a pool of intermediary metabolites. The optimal capacity depends on the variability of substrate availability in the environment and (somatic) maintenance needs. Transformation to polymers (proteins, carbohydrates) and lipids will reduce concentration gradients and osmotic problems, and thus maintenance costs, but involves machinery to perform polymerisation and monomerisation. The development of vacuoles allows spatial separation of ions and cytoplasm to counter osmotic problems. One example is the storage of nitrate in vacuoles of the colourless sulphur bacteria *Thioploca* spp. [567], which use it to oxidise sulphides first to sulphur, for intracellular storage, and then to sulphate for excretion together with ammonium [858]. Organelles like acidocalcisomes also play a role in the storage of cations [276]; vacuoles are discussed at {405}.

A further step to guarantee that obligatory maintenance costs can be met is to catabolise structure. This is inefficient and involves further waste production (so requiring advanced excretion mechanisms), but at least it allows the organism to survive lean periods.

Reserves can contribute considerably to the variability of biomass composition; phytoplankton composition greatly affects the rate at which phytoplankton bind atmospheric carbon dioxide and transport carbon to deep waters [851], known as the biological carbon pump. The activity of the biological carbon pump affects the carbon cycle, and so climate.

10.3.3 Morphological control of metabolism

Morphology influences metabolism for several reasons: assimilation rate is proportional to surface area, maintenance rate to volume and mobilisation rate to the ratio of surface area and volume. This means that surface-area–volume relationships are central to metabolic rates, see {124}. The shape of the growth curve (and so the timing of developmental events) is directly related to the changes in morphology of the cell, see Figure 4.11. Morphology also determines migration capacity, see {122}, which greatly changes the spatial scale that is of metabolic relevance.

κ-rule and emergence of cell cycles

Control of morphology and cell size increased stepwise. Initially the size at division would be highly variable among cells. This variance will be decreased by the installation of a maturation process, where division is initiated as soon as the investment in maturation exceeds a threshold level. This state of maturity creates maturity maintenance costs. Allocation to this maturation programme is a fixed fraction $1-\kappa$ of the mobilised reserve flux, gradually increasing from zero, see {42}. Such an allocation is only simple to achieve if the catabolic flux does not depend on the details of allocation. If the SUs for maturation operate similar to those for somatic maintenance and growth, the fraction κ is constant and depends on the relative abundance and affinity of the maturation SUs.

The metabolic relevance of cell size is in membrane–cytoplasm interactions, see {282}; many catalysing enzymes are only active when bound to membranes [56], and cellular compartmentalisation affects morphology and metabolism. The turnover of reserve decreases with a length measure for an isomorph, see {302}, which comes with the need to reset cell size for unicellulars. Apart from the increase of residence time of compounds in the reserve with a length measure, the cell's surface area to volume ratio decreases with increasing cell size, as does the growth potential. The increase in metabolic performance requires an increase in the amount of DNA and in the time spent on DNA duplication. The trigger for DNA duplication is given when investment into maturation exceeds a given threshold, meaning that a large amount of DNA leads to large cell sizes at division. Prokaryotes partly solved this problem by telescoping generations, i.e. DNA duplication is initiated before the previous duplication cycle is completed, see {279}, and by deleting unused DNA [1120].

10.3.4 Simplification and integration

Reduction of number of reserves

Many eukaryotes started feeding on dead or living biomass with a chemical composition similar to themselves. This covariation in time of all required metabolites for growth removed the necessity to deal with each of those reserves independently. By linking the uptake of various metabolites, the various reserves covary fully in time because their turnover times are equal, as was discussed above. This improved homeostasis, and decreased the need to excrete reserve(s), and allowed further optimisation of enzyme performance; reserve dynamics is mergeable, see {40}. A decrease in the number of reserves also simplifies cells' organisation, which makes it easier to serve as modules in a metabolic structure at a higher level of organisation.

Syntrophy and lateral gene exchange

The evolution of central metabolism as summarised in Figure 10.2 testifies from its syntrophic origins [657, 662]. Some important features are that heterotrophy evolved from phototrophy, which itself evolved from lithotrophy, and that all cycles in the central metabolism of typical modern heterotrophs ran in the opposite direction in the evolutionary past.

This reconstruction suggests that lateral gene exchange between eubacteria and archaea occurred during the evolution of central metabolism. Initially cells could exchange RNA and early strands of DNA relatively easily [1267]; restrictions on exchange became more stringent with increasing metabolic complexity. Many authors suggest that considerable lateral gene exchange occurs in extant prokaryotes [443, 663, 749, 760] by conjugation, plasmids and viruses [1129].

While prokaryotes passed metabolic properties from one taxon to another by lateral exchange of genes, eukaryotes specialised in symbiotic relationships and even internalisation of whole organisms to acquire new metabolic properties. These exchange steps require metabolic preparation, however, both in terms of substrate supply to a new metabolic pathway and of product processing from that pathway for maintenance and/or growth. It seems very likely that these preparation steps include a syntrophic exchange between the donor and the receiver.

First steps in modular recombination: mitochondria

The problem of the origins of mitochondria is not fully resolved. Part of the problem is that mitochondria and hosts exchanged quite a few genes, and the genome of mitochondria reduced considerably, down to 1% of its original bacterial genome [344]. The mitochondrial DNA in kinetoplasts, however, is amplified and can form a network of catenated circular molecules [690]. Mitochondria probably evolved from an α-group purple bacterium [27] in an archaean [52, 748, 751]. However, arguments exist for the existence of mitobionts in the remote past [792], from which mitochondria and prokaryotes developed; the mitobionts differentiated before they associated with various groups of eukaryotes. The amitochondriate pelobiont *Pelomyxa palustris* has intracellular methanogenic bacteria that may have comparable functions. Other members of the α-group of purple bacteria, such as *Agrobacterium* and *Rhizobium* can also live inside cells and spp. sport dinitrogen fixation. *Rickettsia* became parasites, using their host building blocks and reducing their own genome to viral proportions.

It is now widely accepted that all eukaryotes have or once had mitochondria [319, 443, 587, 983, 1062, 1102]. Their internalisation marks the origin of the eukaryotes, which is possibly some 1.5 Ga [608], or 1.7 Ga [938], or 2.0 Ga [944] or 2.7 Ga [151] ago. In fact the eukaryotes may have emerged from the internalisation of a fermenting, facultative anaerobic H_2- and CO_2-producing eubacterium into an autotrophic, obligatory anaerobic H_2- and CO_2-consuming methanogenic archaean [748], the host possibly returning organic metabolites. Once the H_2 production and consumption had been cut out of the metabolism, aerobic environments became available, where the respiratory chain of the symbiont kept the dioxygen concentration in the hosts' cytoplasm at very low levels. The internalisation of (pro)mitochondria might be a response to counter the toxic effects of dioxygen. This hypothesis for the origin of eukaryotes explains why the DNA replication and repair proteins of eukaryotes resemble those of archaea, and not those of eubacteria. Notice that the eukaryotisation, as schematised in Figure 10.5, just represents a recombination and compartmentation of existing modules of the central metabolism (cf. Figure 10.2). Syntrophic associations between methanogens and hydrogenosomes are still abundant; ciliates can have methanogens as endosymbionts and interact in the exchange [345].

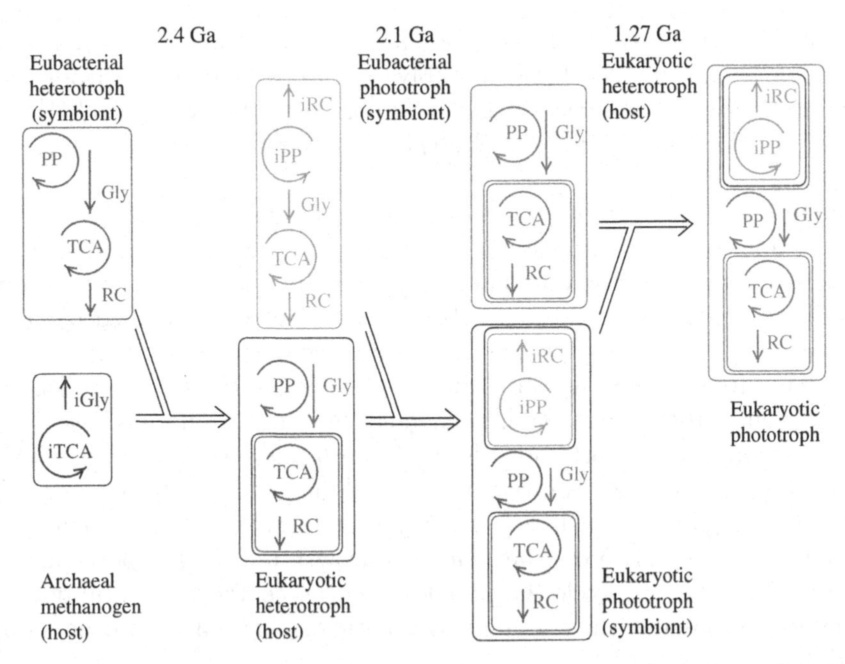

Figure 10.5 Scheme of symbiogenesis events; the first two primary inclusions of prokaryotes (to become mitochondria and chloroplasts respectively) were followed by secondary and tertiary inclusions of eukaryotes. Each inclusion comes with a transfer of metabolic functions to the host. The loss of endosymbionts is not illustrated. See Figure 10.2 for definitions of the modules of central metabolism and for the ancestors of mitochondria and chloroplasts. The outer membrane of the mitochondria is derived from the endosymbiont, and that of the chloroplasts from the host; mitochondria were internalised via membrane rupture, chloroplasts via phagocytosis. Modified from Kooijman and Hengeveld [657]. The scheme explains why all eukaryotes have heterotrophic capabilities.

The penetration of mitochondria into their host required membrane rupture and healing, without causing cell death. The predatory bacterium *Daptobacter* can penetrate the cytoplasm of its bacterial prey *Chromatium*. This may be analogous to early events in the symbiotic acquisition of cell organelles [441]. The outer membrane of the double membrane around mitochondria might be of negibacterial origin [191], which supports the rupture interpretation. At least one example exists of prokaryotic endosymbiosis (β-proteobacteria that harbour γ-proteobacteria [278]) in absence of phagocytosis. More examples exist of penetration through the membrane without killing the victim instantaneously, e.g. [441]. Guerrero's present view, shared by others, is that it happened only once and the logical implication is just in a single individual. If phagocytosis had been well established prior to the entry of a mitochondrion, it is hard to understand why it did not occur more frequently.

This origin of eukaryotes explains their metabolic similarity, compared to the diversity among prokaryotes, and illustrates the narrow borderline between parasitic and symbiotic relationships, see {346}. Since opisthokonts were the first to branch, and animals probably originated in the sea, this internalisation event presumably occurred in a marine environment. The fungi, notably the chytrids, diverged from the animals (unicellular relatives of the choanoflagellates) some 1.6–0.9 Ga ago [1141]. In view of the biology of modern nucleariids and chytrids, this might have occurred in a freshwater environment.

The shape of the cristae of mitochondria is nowadays an important criterion in tracing the evolutionary relationships among protists [871], which points to a slow intracellular evolution. Hydrogenosomes are generally thought to have evolved from mitochondria in anaerobic or micro-aerobic environments. As testified by the presence of mitochondrion-derived genes, some parasitic or commensal groups (Entamoebidae, Microsporidia, diplomonads, Parabasalia, some rumen chytrids, several groups of ciliates) lost their mitochondria [983]. Such genes have not been found (yet) in the amitochondriate oxymonads, retortamonads, *Postgaardi* and *Psalteriomonas*. The absence of mitochondria is perhaps primitive in the pelobionts and the free-living *Trimastrix* [871]. Other organelles, such as peroxysomes and glyoxisomes, probably also have endosymbiotic origins [191, 581]; they follow a growth and fission pattern that is only loosely coupled to the cell cycle. Organelles such as centrioles, undulipodia and the nuclear membrane possibly have an endosymbiotic origin [743], but others oppose this point of view.

A cornerstone of DEB theory is the ability to merge metabolic systems that follow DEB rules such that the merged system again follows DEB rules; the integration of mitochondria into the host system is the first in a long series of similar events; this is further discussed at {406}.

Membrane plasticity

No prokaryote seems to be able to wrap its membranes into vesicles, while membrane transport (including phagocytosis and pinocytosis, vesicle-mediated transport) is basic in eukaryotes [303, 438], and essential for endosymbiotic relationships. Eukaryotes also have ATP-fuelled cytoplasmatic mobility driven by myosin and dynein.

The absence of phagocytosis or pinocytosis in prokaryotes has been used as argument in favour of an independent origin of prokaryotes and eukaryotes [191, 1266]. The protein clathrin plays a key role in membrane invagination, and is not known in prokaryotes. Similarities in the DNA and RNA code and in the whole biochemical and metabolic organisation of prokaryotes and eukaryotes suggest an evolutionary link. Cavalier-Smith [191, 195] argued that eukaryotes descend from some actinobacterium that engulfed a phototrophic posibacterium (an α-proteobacterium) as a mitochondrion, which later lost phototrophy, and used it as a slave to produce ATP. The ability to phagocytose is central to his reasoning. Actomyosin mediates phagocytosis and actinobacteria have proteins somewhat related to myosin, although they do not phagocytose. If he is right that the outer membrane of mitochondria is derived from the original posibacterium, and not from the host, there is little need for the existence of phagocytosis prior to the entry of a posibacterium to become a mitochondrion.

Bell [77, 78] proposed that a lysogenic pox-like DNA virus introduced clatrine-like proteins in an archaean that promoted membrane plasticity. This option also helps to explain the origins of the nuclear membrane, of linear chromosomes with short telometic repeats and of capped mRNA. Viruses might have evolved early [362, 363], but the question remains how viruses got the ability; the fact that membrane-trafficking genes have prokaryotic similarities and many resulted from gene duplications [239] questions a viral origin.

We are beginning to understand the evolutionary roots of cell motility [103], including changes in shape in response to environmental stimuli, and extension of protrusions like lamellipodia and filopodia to allow particles to be enclosed in a phagocytotic cup, which is based on the spatially controlled polymerisation of actin. The eubacterial pathogens *Listeria monocytogenes* and *Shigella flexneri* exhibit actin-based movement in the host cytoplasm [864]. Actin and tubulin have also been isolated from the togobacterium *Thermatoga maritimum* [325]; apart from their role in motility, these proteins also play a key role in the cytoskeleton of eukaryotes, which is used by transporters for the allocation of metabolites to particular destinations. All eukaryotic cytoskeleton elements are presently known from prokaryotes [429].

The evolution of membrane plasticity must have taken place in a time window of some 700 Ma, since biomarker data suggest that the first eukaryotic cells appeared around 2.7 Ga ago [151] (around the time cyanobacteria evolved).

The development of membrane plasticity has been a major evolutionary step, allowing phagocytosis; cells no longer needed to excrete enzymes to split large molecules of substrate into smaller metabolites for uptake with low efficiency, but digestion could be carried out intracellularly, avoiding waste and the necessity for cooperative feeding, see {270}. Fungi possibly never developed this ability and animals evolved from fungi [749] suggesting that the animal lineage developed phagocytosis independently. Recent phylogenetic studies [1107] place the phagocytotic nucleariids at the base of the fungi, however, suggesting that the fungi lost phagocytosis, and that it only developed once. Most animals also excrete enzymes (like their fungal sisters), but since this is in the gut environment, most metabolites arrive at the gut epithelium for uptake. Viridiplantae (glaucophytes, rhodophytes and chlorophytes) gave up phagocytosis, but chromophytes, which received their plastids in the form of rhodophytes, still sport active phagocytosis [24] despite their acquired phototrophic abilities. Phagocytosis allowed the more efficient use of living and dead organisms as a resource. Scavenging, predation and new forms of endosymbioses became widespread.

Membrane plasticity had a huge impact on cellular organisation. The presence of vacuoles increased the capacity to store nutrients [694], and vesicle-mediated intracellular transport reorganised metabolism [303]. By further improving intracellular transport using the endoplasmatic reticulum and further increasing storage capacity, cells could grow bigger and be more motile. Bigger size favours increased metabolic memory, and increased motility allows the organism to search for favourable sites. The eukaryotic endoplasmatic reticulum, build of actin and tubulin networks has a precursor in prokaryotes in the form of MreB proteins [325], so also here, we see gradual improvement.

Plastids

Long after the origin of mitochondria some cyanobacteria evolved into plastids [742, 1139], which made phototrophy available for eukaryotes. Like that of mitochondria, this internalisation event possibly occurred only once in eukaryotic history [195, 259, 260, 768, 980, 1285], see Figure 10.5, but this is controversial [1113].

Sequence data suggest that glaucophytes received the first plastids, and that rhodophytes evolved from them some 2.0 Ga ago [1013, 1135] (or 1.2 Ga according to [608]), while chlorophytes (including plants) diverged from rhodophytes 1.5 Ga ago. The glaucophytes have a poor fossil record, and now consist of a few freshwater species. Where the plastids of glaucophytes retained most of their genome and properties, those of rhodophytes and chlorophytes became progressively reduced by transfer of thousands of genes to the nucleus [750] and by gene loss.

The present occurrence of glaucophytes weakly suggests that the internalisation of a plastid occurred in a freshwater environment. The rhodophytes have their greatest diversity in the sea, and most of their hosts (that possess rhodophyte-derived chloroplasts) are most diverse in the sea, while chlorophytes and their hosts are most diverse in fresh waters. So the habitat in which the internalisation occurred is uncertain.

The secondary endosymbiosis event that seeded the chromophytes was some 1.3 Ga ago [1286] (see Figure 10.5). Rhodophytes became integrated into heterokonts, haptophytes and cryptophytes, while chlorophytes became integrated into euglenophytes and chlorarachniophytes; heterokonts and haptophytes became integrated into dinoflagellates, which themselves (especially *Gymnodinium adriaticum*) engaged into endosymbiotic relationships with animals (corals, other cnidarians and molluscs). Alveolates, to which dinoflagellates, ciliates and sporozoans belong, generally specialised in kleptoplastides (i.e. functional plastids that are acquired by feeding). The presence of plastids in the parasitic kinetoplastids and of cyanobacterial genes in the heterotrophic percolozoans (= Heterolobosea) suggests that secondary endosymbiosis did not take place in the euglenoids, but much earlier in the common ancestor of all excavates, where chloroplasts became lost in the percolozoans [28]. Apart from Dinozoa, cryptophytes (especially *Chrysidiella*) and diatoms engaged in endosymbiotic relationships with radiolarians and foraminiferans, and chlorophytes did so with animals (sponges, *Hydra*, rotifers and platyhelminths). Intracellular chloroplast populations seem to behave more dynamically in kleptomanic and endosymbiotic relationships [1114], compared to fully integrated systems. The coupling of the dynamics of the subsystems can be tight as well as less tight.

Before the arrival of plastids, eukaryotes were heterotrophic. Cyanobacteria are mixotrophic, which makes it likely that their plastids before internalisation were mixotrophs as well. Very few, if any, eukaryotes with plastids became fully specialised on phototrophy, remaining mixotrophic to some extent. Theoretical studies by Tineke Troost on DEB-structured populations of mixotrophs show that the spontaneous evolutionary specialisation into organo- and phototrophs is difficult in spatially homogeneous environments [1172]. In spatially heterogeneous environments, however, such as in the water column where light extinction favours phototrophy at the surface and heterotrophy at the bottom, such specialisation is relatively easy [1171].

Vacuoles and cell structures

Cyanobacteria only develop vacuoles at low pH [1290]; archaea and posibacteria do have gas vacuoles, but their function is totally different from that of eukaryotic vacuoles. Eukaryotes make intensive use of these organelles [694] for storing nutrients in ionic form and carbohydrates; sucrose, a precursor of many other soluble carbohydrates, typically occurs in vacuoles. This organelle probably evolved to solve osmotic problems that came with storing substrates. The storage of water in vacuoles allowed plants to invade the terrestrial environment; almost all other organisms depend on plants in this environment. DEB theory predicts that the storage capacity of energy and building blocks scales with volumetric length to the power of four; since eukaryotic cells are generally larger than prokaryotic ones, storage becomes more important to them. Diatoms typically have extremely large vacuoles, which occupy more than 95% of the cell volume, allowing for a very large surface area (the outer membrane, where the carriers for nutrient uptake are located), relative to their structural mass that requires maintenance. In some species, the large chloroplast wraps around the vacuole like a blanket. Since, according to DEB theory, reserve does not require maintenance, the large ratio of surface area to structural volume explains why diatoms are ecologically so successful, and also why they are the first group of phytoplankton to appear each spring.

The Golgi apparatus, a special set of flat, staked vesicles, called dictyosomes, develops after cell division from the endoplasmic reticulum. They appear and disappear repeatedly in the amitochondriate metamonad *Giardia*. The endoplasmatic reticulum is used by transport proteins to reach their target and deliver their cargo, so it plays a key role in allocation, see {42}, and motivates a flux-based approach to catch transformation, see {100}.

The nuclear envelope can disappear in part of the cell cycle in some eukaryotic taxa and it is also formed by the endoplasmic reticulum. The amitochondriate parabasalid *Trichomonas* does not have a nuclear envelope, while the planctobacterium *Gemmata oscuriglobus* and the poribacteria have one [349]. The possession of a nucleus itself is therefore not a basic requisite distinguishing between prokaryotes and eukaryotes. The situation is quite a bit more complex than molecular biology textbooks suggest; the macronuclei (sometimes more than one) in ciliates are involved in metabolism, for instance, while the micronuclei deal with sexual recombination. Although some prokaryotic cells, such as the planctobacteria, are packed with membranes, eukaryotic cells are generally more compartmentalised, both morphologically and functionally. Compounds can be essential in one compartment, and toxic in another [752]. Eukaryotic cilia differ in structure from the prokaryotic flagella, and are therefore called undulipodia to underline the difference [742]. The microtubular cytoskeleton of eukaryotes is possibly derived from protein constricting the prokaryotic cell membrane during fission, as both use the protein tubulin [325].

Genome reorganisation

Escherichia coli needs 1 hour to duplicates its DNA, while the interdivision interval can be as short as 20 minutes under optimal conditions. It does this using several division forks, see {279}, but this affects it size at division and so its surface-area–volume relationship. Cutting out disused DNA is just one way to reduce the DNA duplication time [1120]. Another possibility is to maintain two chromosomes that are duplicated simultaneously, as in *Rhodobacter sphaeroides* [1132], or, more frequently, to maintain megaplasmids [371, 559, 1118].

The organisation of the eukaryotic genome in chromosomes, with a spindle machinery for genome allocation to daughter cells, enhanced the efficiency of cell propagation by reducing the time needed to duplicate DNA [204], and harnessed plastids, whose duplication is only loosely coupled to the cell cycle in prokaryotes. Since animals such as the ant *Myrmecia croslandi* and the nematode *Parascaris univalens* have only a single chromosome [617], acceleration of DNA duplication is not always vital. It allows more efficient methods of silencing viruses, by changing their genome and incorporating it into that of the host (half of eukaryotic 'junk DNA' consists of these silenced viral genomes). Eukaryotes had to solve the problem of how to couple the duplication cycles of their nuclear genome and that of their mitochondria and chloroplasts. Dynamin-related guanosine triphosphatases (GTPases) seem to play a role in this synchronisation [856]. The nuclear membrane of eukaryotes, poribacteria and planctobacteria possibly allows a better separation of the regulation tasks of gene activity and cellular metabolism by compartmentalisation, which might have been essential to the development of advanced gene regulation mechanisms.

Chromosomes are linked to the evolution of reproduction, which includes cell-to-cell recognition, sexuality and mating systems. Reproduction evolved many times independently, which explains the large diversity. Many eukaryotes have haploid as well as diploid life-stages, and two or more (fungi, rhodophytes) sexes [601]. Although reproduction may seem to have little relevance to metabolism at the level of the individual, metabolic rates at the population level depend on the amount of biomass and, hence, on rates of propagation. Eukaryotes also have a unique DNA topoisomerase I, which is not related to type II topoisomerase of the archaea [364], which questions their origins.

Despite all their properties, the eukaryotic genome size can be small; the genome size of the acidophilic rhodophyte *Cyanidoschyzon* is 8 Mbp, only double the genome size of *E. coli* [204]; the chlorophyte *Ostreococcus tauri* has a genome of only 10 Mbp, and the yeast *Saccharomyces cerivisiae* of 12 Mbp [265]. Typical eukaryotic genome sizes are much larger than that of prokaryotes, however. Apart from silencing of viruses, most of this extension relates to gene regulation functions that are inherent to cell differentiation and the evolution of life stages.

10.4 Merging of individuals in steps

Collaboration in the form of symbioses based on reciprocal syntrophy is basic to biodiversity, and probably to the existence of life [7, 259, 287, 288, 952, 1076]; the frequently complex forms of trophic interactions are discussed at {336 ff}.

The merging of two independent populations of heterotrophs and autotrophs into a single population of mixotrophs occurred frequently in evolutionary history [951, 1147]. This process is known as symbiogenesis [743] and is here discussed following [648].

Endosymbiotic relationships are not always stable on an evolutionary timescale. All of the algal groups have colourless representatives, which implies that they are fully heterotrophic. Half of the species of dinoflagellates, for instance, do not have chloroplasts, probably due to evolutionary loss [192]. The integration is a stepwise process, where plastids' genome size is reduced by gene loss, substitution and transfer to the host's genome, possibly to economise metabolism [191]. Chloroplasts of chlorophytes have a typical genome size of 100 genes, but the genome sizes of chloroplasts of rhodophytes and glaucocystophytes are substantially larger. The chloroplasts of cryptophytes and chlorarachnida still contain a nucleomorph with some chromosomes [736], believed to be derived from their earlier rhodophyte and chlorophyte hosts. The endosymbionts of radiolarians, foraminiferans and animals maintained their full genome. The tightness of the integration is, therefore, reflected at the genome level.

Generally little is known about the population dynamics of intracellular organelles [856]. Mitochondria constitute some 20% of the volume of mammalian cells, but this varies per tissue and the individual's condition. Their number per cell varies between 1000 and 1600 in human liver cells, 500 and 750 in rat myoblasts, some 80 in rabbit peritoneal macrophages, and 1 in mammalian sperm cells [83]. In the case of a single mitochondrion, the growth and division of the mitochondrion must be tightly linked to that of the cell, but generally the dynamics of mitochondria is complex and best described by stochastic models [107]. Mitochondria crawl around in eukaryotic cells [83] and can fuse, resulting in a smaller number of larger mitochondria, and can even form a network, as observed in gametes of the green alga *Chlamydomonas*, for instance. In yeast and many unicellular chlorophytes, a single giant mitochondrion alternates cyclically with numerous small mitochondria. Moreover, the host cell can kill mitochondria and lysosomes can decompose the remains. Likewise, chloroplasts can move through the cell, sometimes in a coordinated way. Chloroplasts can transform reversibly into non-green plastids (proplastids, etioplastids and storage plastids) with other cellular functions.

Apart from changes in numbers, plastids can change in function as well. They can reversibly lose their chlorophyll and fulfil non-photosynthetic tasks, which are permanent in the kinetoplasts (e.g. the endoparasite *Trypanosoma*) and in heterotrophic plants (Petrosaviaceae, Triurdaceae, some Orchidaceae, Burmanniaceae, the prothallium stage of lycopods and ophioglossids, the thalloid liverwort *Cryptothallus mirabilis* [910, p377]), the podocarp *Parasitaxus usta* [334], in parasitic plants (Lennoaceae, Mitrastemonaceae, Cytinaceae, Hydnoraceae, Apodanthaceae, Cynomoriaceae, Orobanchaceae, Rafflesiaceae, Balanophoraceae, some Convolvulaceae), and in predatory plants (some Lentibulariaceae), for instance. This list of exclusively heterotrophic plants suggests that heterotrophy might be more important among plants than is generally recognised. In other taxa, the chloroplasts were lost, such as in the oomycetes like *Phytophthora* [1175], and the would-be phototroph adopted a parasitic life style.

The evolution of organelles strongly suggests an increasingly strong coupling between species that were once more independent. This places eukaryotic cellular physiology

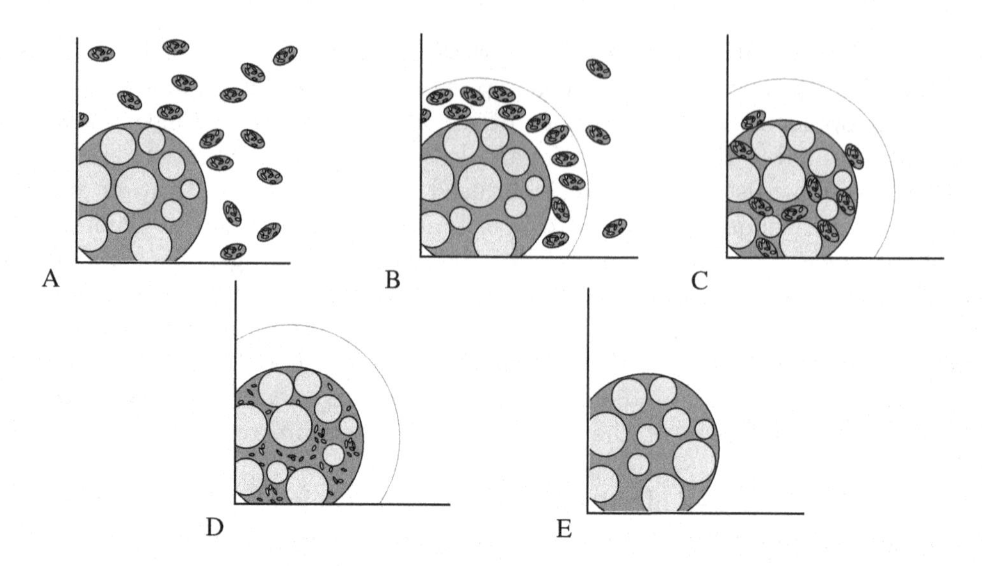

Figure 10.6 Future host (a single individual is indicated) and future symbionts originally live independently (A). When they start exchanging products, an accumulation of symbionts in the hosts' mantle space occurs (B), followed by an internalisation (C). A merging of structures then can occur (D), followed by a merging of reserves (E). A new entity then exists. The light areas in the hosts and symbionts indicate reserves, the dark regions indicate structure; reserves integrate after structures. The mantle space around the hosts' body is indicated where hosts' product accumulates that stimulates symbionts' growth. The text describes these and other steps quantitatively.

firmly in an ecological perspective, and motivates the application of ecological methods to subcellular regulation problems. This mutually dependent dynamics is the focus of the present review, and includes that of intracellular parasites.

The physiological basis of endosymbiosis is probably always reciprocal syntrophy, where each species uses the products of the other species; ammonia–carbohydrate exchange forms the basis of the phototroph–heterotroph interactions, see {341}. Mitochondria receive pyruvate, FAD, GDP, P, and NAD from the cytoplasm, and return FADH, GTP, NADH and intermediary metabolites from the TCA cycle, see {282}. One-way syntrophy, where one species uses the product of the other, but not vice versa, is here treated as a special case of reciprocal syntrophy.

Starting from two free-living populations in the same environment that follow the DEB rules, eight steps of reductions of degrees of freedom can be delineated to arrive at a fully integrated endosymbiotic system that can be treated as a single population for all practical purposes and again follows the DEB rules. All those steps in the asymptotic behaviour of the populations can be made on an incremental basis, i.e. by a continuous and incremental change in (some) parameter values. The general idea is that the parameter values are under evolutionary control. Figure 10.6 illustrates some of the steps.

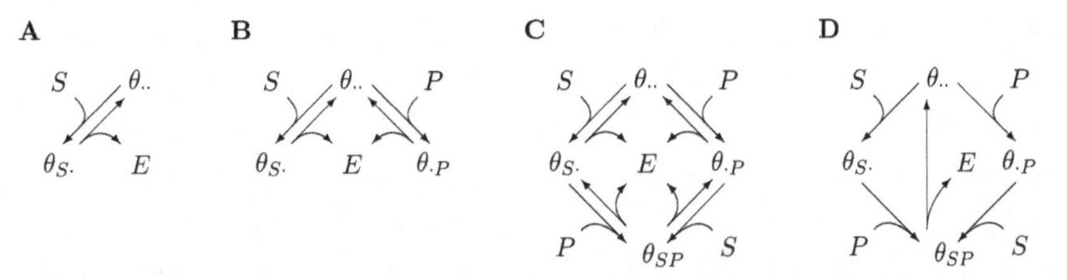

Figure 10.7 The evolution of the transformation from substrate S, and later also from product P, into reserve E. The interaction of substrate and product in the transformation to reserve evolved from sequential-substitutable (B), via parallel-substitutable (C), to parallel-supplementary (D). The symbol θ represents a synthesising unit (SU) that is unbounded $\theta_{..}$, bound to the substrate $\theta_{S.}$, to the product $\theta_{.P}$, or to both θ_{SP}.

10.4.1 Reciprocal syntrophy

Originally two species coexist by living on a different substrate each, so they initially might have little interaction and just happen to live in the same environment. Each species excretes products in a well-mixed environment. A weak form of interaction starts when the products are used by the other species as a substitutable compound for their own substrate, a situation which we can call reciprocal syntrophy. Gradually the nutritional nature of the product changes with respect to the substrate from substitutable to supplementary, and the two species become involved in an obligatory symbiotic relationship; they can no longer live independently of each other; see Figure 10.7. The mechanism can be that the partner's product is a metabolite of an organism's own substrate; eventually the metabolic pathway for that metabolite becomes suppressed and later deleted [1120]. A well-known example is the human inability to synthesise vitamin C, unlike chimpanzees, which is generally interpreted as an adaptation to fruit eating; the genes for coding vitamin C synthesis are still present in the human genome, but they are not expressed. The theory behind the uptake of compounds that make a gradual transition from being substitutable to supplementary is discussed in [137], together with tests against experimental data on co-metabolism.

Figure 10.8 gives the steady-state amounts of structures of hosts and symbionts as functions of the throughput rate of the chemostat for the various steps in symbiogenesis. The throughput rate equals the specific growth rate at steady state. The maximum throughput rate is less than the potential maximum growth rate; equality only holds for infinitely high concentrations of substrate in the inflowing medium. The amounts of structures are zero at a throughput rate of zero because of maintenance. At step 0, where substrates are substitutable to products, the introduction of the partner enhances growth. This is clearly visible in the curve for the host around the maximum throughput rate for the symbiont. Growth stimulation also occurs for the symbiont, of course, but this is less visible in the figure since the host is always present when the symbiont is present with this parameter setting; without the host, the biomass and the maximum throughput rate of the symbiont would be less. The maximum throughput rates for hosts and symbionts can differ as long as substrate and products are substitutable, but

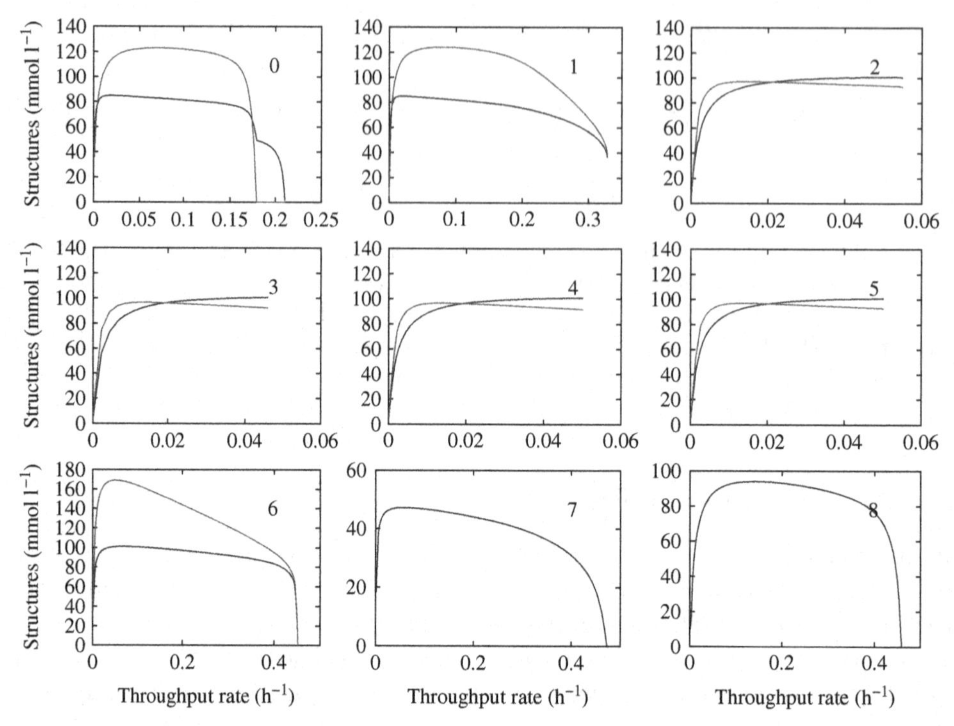

Figure 10.8 Steady-state values of the amounts of structure of hosts (dashed curves) and symbionts (drawn curves) as functions of the throughput rate of the chemostat for the different steps in symbiogenesis. If substrates and products are substitutable (0) symbiosis is facultative and the host can live independently of the symbiont. With this parameter setting the hosts' maximum growth rate is higher. If substrates and products make their transition to become supplementary (1), and especially if they are supplementary (2), symbiosis is obligatory. The environment is homogeneous in steps 0–2 and 5–8, but in step 3 symbionts can live in the free space, in the hosts' mantle space, as well as within the host. The parameter settings are such that the internal population of symbionts outcompetes the mantle and free-living populations; the curve for the symbionts in step 3 corresponds to the internal population; other values for the transport parameters allow the coexistence of all three populations, or any selection from these three. Step 4 has internal symbionts only, but the mantle space can differ from the free space in concentrations of substrates and products. These differences disappear in step 5 (by increasing the transport rates between the spaces). Product transfer is on flux basis in step 6, rather than on concentration basis (steps 0–5); the symbioses can grow much faster and the amounts of structure are again zero at the maximum throughput rate. The transition from step 5 to 6 can be smooth if the host can reduce the leaking of product. The merging of both structures (7) and reserves (8) doesn't have substantial effects at steady state. The single structure in steps 7 and 8 can handle two substrates; each structure in steps 0–6 handles only one.

not if they are supplementary (steps 1 to 6). The amounts of structures then can't decrease gradually to zero for increasingly high growth rates (steps 1 to 5), because product formation, and thus product concentration, will then also decrease gradually to zero, while the equally rapidly growing partner needs a lot of product. Figure 10.8

also shows that the transition from substitutable to supplementary products (from step 1 to 2) comes with a substantial reduction in the maximum growth rate if no other mechanism ensures an easy access to the products; the concentration of products in a well-mixed environment is very low. It is therefore likely that this transition is simultaneous with rather than prior to subsequent steps in symbiogenesis (i.e. spatial clustering, so spatial structure). I will return to this point later.

Examples of product exchange with little spatial clustering can be found among micro-organisms in animal guts, see the discussion on *Methanobacillus* at {338}. *Ruminococcus albus* ferments glucose to acetate, ethanol and dihydrogen, but in the presence of fumarate-fermenting *Vibrio succinogenes*, *R. albus* produces acetate, and not the energetically expensive ethanol; this is only possible when *V. succinogenes* removes dihydrogen [345]. This exchange pattern is typical for methanogen–partner interactions.

The concentration of products in the environment and the biomass ratios of the species can vary substantially in time. In the unlikely case that substrates, products or biota all remain in a given local homogeneous environment, it initially takes an amount of substrate to build up product concentrations, but once these concentrations and the populations settle to a constant value, the environment no longer acts as a sink and the situation is very similar to a direct transfer of product from one species to the other during steady-state. The inefficiency of product transfer in well-mixed environments becomes clear during transient states and if the product decays away (chemically, by physical transport or biologically mediated). A lot of product will not reach the partner, and population levels will be low.

10.4.2 Spatial clustering

Exchange of compounds between the species is enhanced by spatial clustering; most individuals of the small-bodied species live in a narrow mantle around an individual of the large-bodied species. Although the real mantle will not have a sharp boundary with the outer environment, we treat it as a distinct and homogeneous environment that can exchange substrate and products with the outer environment and with the volume inside the host. The individuals of the large-bodied species secrete all products into their mantle; both products and substrates can leave or enter the mantle with certain specific rates according to a generalised diffusion process. We take the mantle's volume (at the population level) just proportional to the structural mass of the large-bodied species (which seems reasonable at the population level). Diffusion and related transport processes mean that the mantle is actually not homogeneous; concentration gradients will build up, see {266}. This level of detail is not required, however, for our present aim.

The emigration rate of the small-bodied species to and from the mantle space will be relatively small. Since the growth rate will be much higher in the mantle, the small-bodied species accumulates in the mantle space of the large-bodied one, depending on the transport rates for substrates and products to the other environment, and on

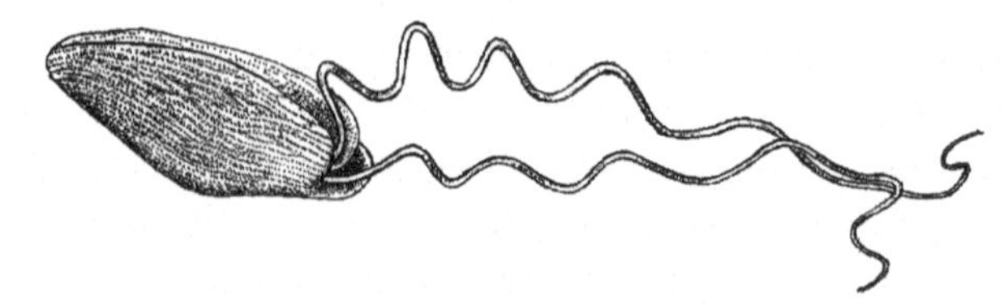

Figure 10.9 The cell surface of the colourless amitochondriate euglenoid *Postgaardi mariagerensis* is fully covered with elongated rod-shaped heterotrophic bacteria. Based on an electron-micrograph in [1063]; the length of the cell is 60 μm.

intra-specific competition. Under rather general conditions for the maximum specific assimilation rate, the specific maintenance and growth costs, and the hazard rate (i.e. the instantaneous death rate), the population inside the mantle space even outcompetes that outside. This spatial clustering does not involve optimisation arguments.

Many species create a special environment to grow their symbionts. Examples are animals' intestines, which harbour the gut flora, or pits in the leaves of the floating fern *Azolla* and *Gunnera*, which harbour cyanobacteria (*Anabaena*) that fix dinitrogen; some flowering plants, such as *Alnus*, Leguminosae, *Hippophaë*, have special structures in the roots for harvesting the bacterium *Rhizobium* for the same purpose. Cockroaches lose their gut flora at moulting, and inoculate from the mother's faecal supplies. Repeated media refreshment for the waterflea *Daphnia magna* reduces daphnids' condition, which is probably caused by a washout of the gut flora; daphnids natural schooling behaviour thus may be related to reinoculation of the gut flora.

10.4.3 Physical contact: epibionts

If partners physically make contact, the exchanging products hardly accumulate in the environment, and compound exchange is directly on the basis of fluxes. Such situations frequently occur; e.g. the surface of the euglenoid *Postgaardi* is completely covered with bacteria (Figure 10.9). Product exchange is probably the reason that many micro-organisms live in flocks, rather than in free suspension [139]. Ascomycetes make physical contact with green algae in lichens. Basidiomycetes do so with vascular plants in ectomycorrhizas. Cyanobacteria, *Prochloron*, live on the outer surface of sea squirts.

If the products decay away, the two species can still improve product exchange by internalisation, where the small-bodied species (endosymbiont) lives inside the large-bodied one (host); this requires phagocytosis. During this internalisation process, the host acquires the product of its symbiont both from the mantle space, as well as from inside its own body. The natural description of the uptake process of product in the mantle space is on the basis of the concentration of product, in combination with a generalised diffusion process that transports the product to the hosts' product-carriers in the outer membrane. While the symbionts' access to the hosts' product is enhanced by internalisation, its access to the substrate can be reduced because that substrate now has to pass through the hosts' outer membrane. From outside the host the endosymbiont is no longer visible and it appears as if the host is now feeding on two substrates, rather than one. The argument becomes subtle if the host transforms the symbionts' substrate before it reaches the symbiont.

The internal population eventually outcompetes the population in the mantle space, and the endosymbiont can lose its capacity to live freely due to adaption to its cytoplasmic environment, which is under the host's homeostatic control. It is curious to note that the chloroplasts of Chromista (which include diatoms, brown algae and many other 'algal' groups) live inside the hosts' endoplasmic reticulum, while they live outside it in other groups [192]. The passage of the extra membrane during the internalisation process is obviously conserved during evolution, which suggests that this process is rare and reveals the evolutionary relationships between these protist taxa.

The numerical studies behind Figure 10.8 did not account for transport mechanisms that enhance the intracellular accumulation of products. A much higher maximum growth rate can be obtained by decreasing the parameters that control the leaking of products and substrates from the cell, for example. The low maximum growth rate shown in steps 2–5 suggests that the control of transport may be a rather essential feature of symbiogenesis. These transport parameters also determine the fate of the three populations of symbionts (free-living, epibiotic and internal). They can all coexist or each of them can outcompete the others, depending on these parameter values. The competitive exclusion principle, which states that the number of species of competitors cannot exceed the number of types of resources at steady state, only holds for homogeneous environments. The values that are used in Figure 10.8 step 3 lead to extinction of the epibiotic and free-living populations. The mantle space still can differ in concentrations of substrates and products in step 4, while in step 5 the transport rates of these compounds from the mantle to the free space and vice versa are so large that the environment is homogeneous again. It is clear that these differences hardly affect the dynamics with this choice of parameter values, the curves in steps 3, 4 and 5 are very similar; the differences clear for smaller transport rates.

The uptake of product evolves from concentration-based to flux-based (see Figure 10.10); this comes with an increase of the maximum throughput rate and a qualitative change in behaviour of the steady-state amounts of structure around this growth rate: the symbiosis becomes independent of the extracellular product concentration. The role of the products partly degrades from an ecological to a physiological one. Figure 10.8 illustrates this in step 6, where the steady-state amount of structure at the maximum throughput rate is (again) zero, while the maximum throughput rate is substantially increased compared with that for step 5.

Many examples are known for endosymbioses. The nitrogen-fixing cyanobacterial symbionts of the diatoms *Rhizosolenia* and *Hemiaulus* live between the cell wall and the cell membrane [1196, 1197]. Pogonophorans (annelids) and *Xyloplax* (echinoderms [50]) do not possess a gut, but harbour chemoautotrophic bacteria inside their tissues. The parabasalian flagellate *Caduceia theobromae* bears two species of ectosymbiotic and one species of endosymbiotic bacteria, which assist wood digestion in termite

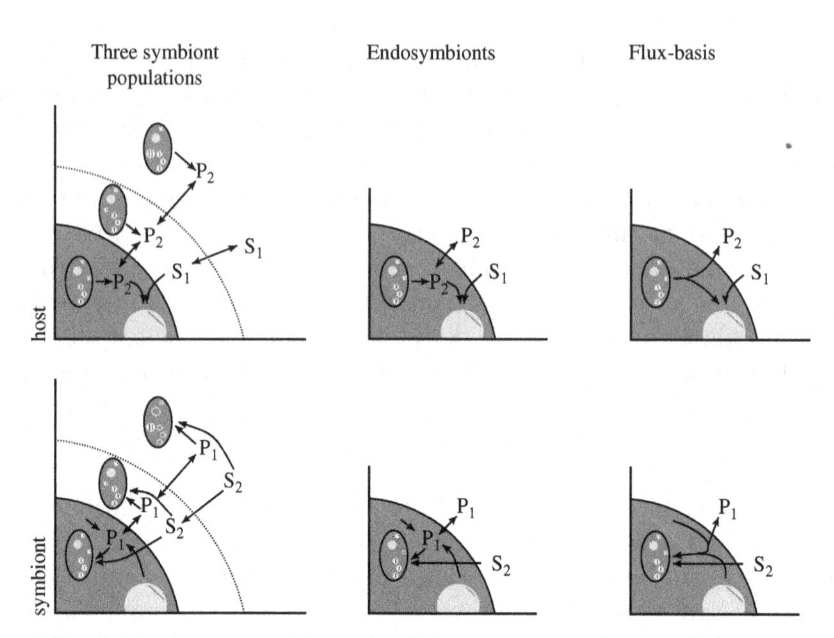

Figure 10.10 The feeding process of host (top) and symbiont (bottom) on substrate (S_1 and S_2) from the environment and product (P_1 and P_2) that is produced by the partner. The first column delineates three populations of symbionts: free, epi- and endosymbionts. The second column only delineates endosymbionts, but still uses intracellular concentrations of products to quantify feeding. With the degradation of intracellular product from an ecological to a physiological variable, feeding is specified in terms of fluxes (third column).

guts. Cyanobacteria live inside the fungus *Geosiphon* and the diatoms *Richelia*, *Hemiaulus* and *Rhopalodia*. Several species of heterotrophic bacteria live endosymbiotically in *Amoeba proteus*.

Associations between the dinitrogen-fixing cyanobacterium *Nostoc* and the fern *Azolla* have been known for some time, but the association with the bryophyte *Pleurozium schreberi* has only recently been discovered [258]; this extremely abundant moss covers most soil in boreal forests and in the taiga. The cyanobacteria are localised in extracellular pockets in these examples, but in some diatoms they live intracellularly.

A close relationship between chlorophytes (or cyanobacteria) and fungi (mainly ascomycetes) evolved relatively recently, i.e. only ca. 450 Ma ago, in the form of lichens and *Geosiphon* [1035]. The fungal partner specialised in decomposing organic matter, which releases nutrients for the algae in exchange for carbohydrates, not unlike the situation in corals. Similarly, mycorrhizae exchange nutrients against carbohydrates with plants, which arose in the same geological period. The endomycorrhizae (presently recognised as a new fungal phylum, the glomeromycetes) evolved right from the beginning of the land plants; the ectomycorrhizae (ascomycetes and basidiomycetes) evolved only during the Cretaceous. These symbioses seemed to have been essential for the invasion of the terrestrial environment [1041].

10.4.4 Weak homeostasis for structure

The ratio of the amounts of structure of the partners varies within a range that becomes increasingly narrow, converging to weak homeostasis, see {204}. The ratio might still depend on the substrate levels at steady state. The importance of products taken up from the environment becomes small; almost all products are exchanged within the body of the host. If the products are fully supplementary to the substrates, some excess product might still leak from host's body, due to stoichiometric restrictions in its use. If part of the product is still substitutable for the substrate, such a leak is unlikely.

Many photosymbionts seem to have a rather constant density in hosts' tissues, although knowledge about digestion of symbionts by hosts as part of a density regulation system is frequently lacking. Product exchange is such a strong regulation mechanism, that other regulation mechanisms are not necessary to explain a relatively constant population density of endosymbionts, but this does not exclude the existence of regulation mechanisms, of course. Regulation mechanisms might affect the coupling of parameters that control product formation.

Heterotrophs not only have syntrophic relationships with photoautotrophs, but also with chemolithotrophs. A nice example concerns the gutless tubificid oligochaete *Olavius algarvensis*, with its sulphate-reducing and sulphide-oxidising endosymbiontic bacteria [298]. These symbionts exchange reduced and oxidised sulphur; the fermentation products of the anaerobic metabolism of the host provide the energy for the sulphate reducers, whereas the organic compounds produced by the sulphide oxidisers fuel the (heterotrophic) metabolism of the host. Taxonomic relationships among hosts can match that among symbionts [289], which suggest considerable co-evolution in syntrophic relationships.

10.4.5 Strong homeostasis for structure

The ratio of the amounts of structure of the partners becomes fixed and independent of the substrate concentrations at steady state; weak homeostasis evolves into strong homeostasis. This might occur by tuning the weight coefficients for how product formation depends on assimilation, maintenance and growth, as discussed quantitatively at {337 ff}. The numerical effect of the merging of the structures is small with our parameter choice (Figure 10.8 from step 6 to step 7). The strong homeostasis condition is usually accompanied by a transfer of (part of) the endosymbionts' DNA to that of host [753]; endosymbionts are now called organelles.

The conditions for strong homeostasis are independent of the details of hosts' assimilation processes. The host might be product as well as substrate limited. The product and the substrate might also be substitutable compounds, such as in the case of algal symbionts of heterotrophs, where the algal carbohydrates serve as an alternative energy source for the host. The significance of these carbohydrates in the hosts' diet might be complex, while the strong homeostasis condition still holds true. If prey is abundant, and the hosts' maximum assimilation capacity is reached, the extra carbohydrates contribute little to the hosts' assimilation. This dynamics is consistent with the rules for sequential processing of substitutable substrates by SUs, {104}, and explains why symbiotic and aposymbiotic hosts grow equally fast at high substrate levels, as has been observed in ciliates and hydras [575, 818]. At low prey abundance (the typical situation in the oligotrophic waters around coral reefs), the extra carbohydrates do contribute to the hosts' diet and propagation.

The fact that the conditions for strong homeostasis are independent of the hosts' assimilation process also implies the independence of details of the endosymbionts' product formation. This simplifies matters considerably, because the excretion of carbohydrates by algal symbionts is not a process covered by fixed associations with assimilation, maintenance and/or growth. It is an active excretion due to stoichiometric constraints of carbohydrates and ammonia (from the host) to form new algal biomass; see next step of integration.

If the host is limited in its growth by endosymbionts' products, similar constraints apply to ensure a constant ratio of structures that is independent of endosymbionts' substrate; the role of host and endosymbionts are just interchanged in this situation. In the case of limitation by substrate of both partners, more stringent constraints apply on energy parameters.

The constraint of small actual assimilation rates might help to explain why symbioses are most frequently found in oligotrophic environments; regulation of relative abundances is more difficult under non-limiting environmental conditions.

Examples of endosymbioses that approach strong homeostasis are mitochondria, hydrogenosomes, chloroplasts and other plastids and peroxisomes. The number of these organelles per host cell depends very much on the species. Diatoms frequently have just a single chloroplast, which implies that the growth and division cycle of the chloroplast must be tightly linked to that of the cell.

The location of mitochondria inside cells further testifies to the optimisation of transport by spatial clustering. Mitochondria cluster close to blood capillaries in mammalian muscle cells and form interdigitating rows with myofibrils to enable peak performance during contraction [83]. The association of mitochondria with the nuclear envelope is thought to relate to the demand of ATP for synthesis in the nucleus, and to the reduction of damage to DNA by reactive oxygen species. The association of mitochondria with the rough endoplasmic reticulum is less well understood but might be related to the movement of mitochondria within cells.

Being able to move independently and over considerable distances, jellyfish, for example, are able to commute between anaerobic conditions at lower water strata for nitrogen intake and higher ones for photosynthesis by their dinozoan endosymbionts supplying them with energy stored in carbohydrates. Dinozoans are engaged in similar relationships with hydropolyps (corals) and molluscs; extensive reefs testify of the evolutionary success of this association.

Some plants can fix dinitrogen with the help of bacteria, encapsulated in specialised tissues. A single receptor seems to be involved in endosymbiotic associations between plants on the one hand and bacteria and fungi on the other [1122], but the recognition process is probably quite complex [868] and not yet fully understood. See Rai *et al.* [934] for a review of symbioses between cyanobacteria and plants.

10.4.6 Coupling of assimilation pathways

The assimilation routes for the organic substrate(s) become coupled, especially in situations where substrate levels covary in time. The reason for the covariation can be purely physical when the substrates originate from a common source (for example another organism with a rather constant chemical composition, or some erosion process of rocks which extracts minerals in fixed ratios). An alternative possibility is that specialisation on a single substrate occurs. Details of product exchange are no longer visible in the dynamics of the integrated system; products now have a strict physiological role, where they still determine the relative importance of the different substructures, and so the substrate uptake capacity. We no longer need to know of their existence to predict how the population responds to environmental factors. If substrates covary in fixed ratios in the environment, the range of the ratio of amounts of reserves becomes increasingly narrow.

In situations where substrate abundances do not covary in time, coupling of the assimilation processes will not occur, and the host will maintain two reserves (see below). Another pattern of development is then likely: part of the unused reserve that is allocated to growth is returned to that reserve, with the result that each reserve can accumulate and reach very high levels when the other reserve limits growth, see {197}.

The functionality of this storing mechanism can be illustrated with algal growth in the sea, where carbohydrate reserves are boosted at the nutrient-poor surface layers where light is plentiful, and the nutrient reserves are filled in the dark bottom layers of the photic zone, which are usually rich in nutrients [654]. Thanks to the uncoupled reserves the alga is able to grow in an environment that would otherwise hardly allow growth. It is the wind, rather than light or nutrients, that is in proximate control of algal growth (Figure 10.11).

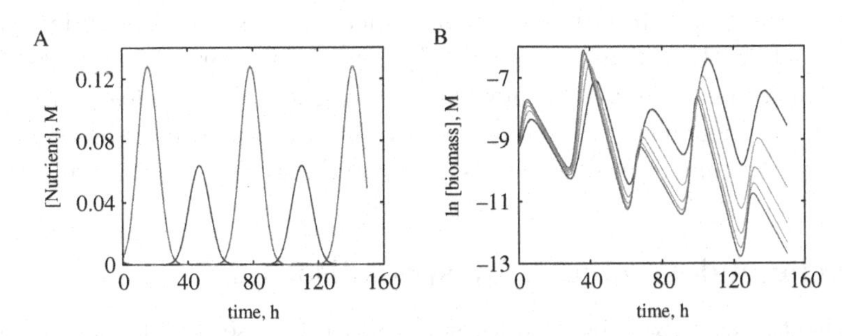

Figure 10.11 (A) A cyclic pattern of two nutrient concentrations, as experienced for example by algae. (B) The expected biomass concentrations in the case of a single structure, and two reserves. The curves correspond to increasing reserve turnover times from top to bottom. The two curves with extreme turnover values are shown with solid lines; the larger this turnover time, the lower the intracellular storage capacity for nutrients. The graphs show that the mean growth rate increases with the storage capacity under these conditions, but that the nutrients must be stored independently.

10.4.7 Weak homeostasis for reserves

Possibly due to their coupling in growth, the turnover rates of reserves seem to be equal, see Figure 5.6, with the consequence that at constant substrate levels, the reserve ratio of the growing populations settles at a constant value and converges to the weak homeostasis condition for reserves. The chemical composition of biomass can still depend on the growth conditions in rather complex ways.

The numerical effect of the merging of the reserves is small using the present parameter values (Figure 10.8 from step 7 to step 8). The merging only requires that the reserve turnover rates are equal, and the ratio of concentrations of substrate remains constant.

Examples of fully coupled single-reserve systems can be found in carnivores, where the rather constant chemical composition of the prey provides the mechanism for the coupling of assimilation of the various nutrients that are required by the carnivore. Parasites also experience a rather constant chemical environment inside their host. Other examples can be found in heterotrophs in eutrophic environments, where a single resource is often limiting, all other resources being available in excess.

10.4.8 Strong homeostasis for reserves

The ratio of intra-host reserves becomes fixed and independent of the substrate levels at steady state by the coupling of the assimilation processes. This causes the ratio of reserve densities to remain fixed during transient states and we arrive at a strong homeostasis for reserves. We can now replace the two reserves by a single combined one. The chemical composition of biomass still depends on growth conditions, but in a less complex way.

The increasingly tight coupling of the dynamics of several types of reserves relates to the situation where maintenance and growth drain reserves in fixed and equal ratios. It is the reason why the metabolic performance of cats can be understood using a single reserve, while that of algae cannot. When a carnivore changes its diet over an evolutionary timescale to become a herbivore, using food with a less constant chemical composition, it frequently continues to be less flexible in its metabolism. This is why the metabolic performance of cows can still be understood using a single reserve.

10.4.9 Cyclic endosymbiosis by specialisation

Symbiogenesis, as described here, allows the host to use substrates, which it could not use without the symbionts. The opposite process is specialisation on a single substrate, where the endosymbiont is no longer functional. This has occurred for instance in aerobic mitochondriate species that invaded anaerobic environments. In some species the mitochondria evolved into hydrogenosomes, but in others the mitochondria were lost. The loss of plastids has occurred in at least some species of all major groups of organisms, e.g. the oomycetes [1175], while in some parasitic groups, such as the kinetoplastids, they assume other functions. Specialisation on metabolic substrates seems to be linked directly to the loss of genes and completes the endosymbiotic cycle.

Some properties of the symbiont might be retained, however, as testified by mitochondrion-derived genes in species that have lost their mitochondria. As mentioned, non-photosynthesising plastids are still functional in plants; such plants can still have arbuscular mycorrhizae, as are found in the orchid *Arachnitis uniflora* [505]. Although the plants cannot transport photosynthetically produced carbohydrate to their fungal partner *Glomus*, it is obviously quite well possible that other metabolites are involved in the exchange. The complex role of plastids shows that the plant is not necessarily parasitising the fungus.

Generally, (endo)symbiosis might be considered to be a process by which metabolic properties are acquired several orders of magnitude faster than by the Darwinian route of mutation and selection [744]. Darwin's mutation/selection route can be particularly cumbersome, because all intermediary stages have to be vital enough to continue the acquisition with incremental steps; almost all metabolic pathways involve several, or even many, enzymes. This provides constraints on the type of properties any particular organism can acquire along the Darwinian route. Such constraints do not apply to acquisition of metabolic traits by endosymbiosis.

10.5 Multicellularity and body size

Although some individual cells can become quite large, with inherent consequences for physiological design and metabolic performance [525, 942], multicellularity can lead to really large body sizes. Multicellularity evolved many times in evolutionary history, even among the prokaryotes. It allows a specialisation of cells to particular functions, and the exchange of products is inherently linked to specialisation. Think, for instance, of filamental chains of cells in cyanobacteria where heterocysts specialise in N_2 fixation. To this end, specialisation requires adaptations for the exclusion of dioxygen and the production of nitrogenase. The existence of dinitrogen fixation in unicellular cyanobacteria shows that all metabolic functions can be combined within a single cell, which is remarkable as its photosynthesis produces dioxygen, inhibiting dinitrogen fixation. A temporal separation of the processes solves the problem, but restricts dinitrogen fixation during darkness; specialisation can be more efficient under certain conditions. The myxobacterium *Chondromyces* and the proteobacteria *Stigmatella* and *Mixococcus* have life cycles that remind us of those of cellular slime moulds, involving a multicellular stage, whereas acetinobacteria such as *Streptomyces* resemble fungal mycelia, e.g. [304]. Pathogens, such as viruses, can kill individual cells without killing the whole organism, which is an important feature of multicellularity, and is basic to the evolution of defence systems.

Cell differentiation is minor in poriferans, reversible in coelenterates and plants, and irreversible in vertebrates. The number of cells of one organism very much depends on the species, and can be up to 10^{17} in whales [976].

Because the maximum reserve density also increases with length, the time to death by starvation will increase which enhances the organism's ability to copy with temporal heterogeneity. At the extreme, the largest whales leave their Antarctic feeding grounds, swim to oligotrophic tropical waters to calve, feed the calf some 600 l of milk per day for several months, and then swim back with their calf to their feeding grounds where they resume feeding. Such factors partly compensate for the disadvantages of a large body size and the associated high minimum food densities.

10.5.1 Differentiation and cellular communication

Multicellularity has many implications. Cells can be organised into tissues and organs, which gives metabolic differentiation once more an extra dimension. The dynamics of organ sizes can be quantified effectively by further partitioning the flux of mobilised reserve (the κ-rule), see {202, 204}. Multicellularity comes with a need for transport processes, see {266}, and regulation of the processes of growth and apoptosis of cells in tissues [995], in which communication between cells plays an important role.

Animals (from cnidarians to chordates) continued the use of (prokaryotic) gap junctions between cells of the same tissue, where a family of proteins called connexins form tissue-specific communication channels. They appear early in embryonic development (in the eight-cell stage in mammals) and are used for nutrient exchange, cell regulation, conduction of electrical impulses, development and differentiation.

Communication by gap junctions allows for limited transport of particular metabolites only, and animals developed both a transport system (blood and lymph) and a (relatively) fast signalling system (the neuronal system). The latter allowed for the development of signal processing from advanced sensors (light, sound, smell, electrical field, pain) in combination with advanced locomotory machinery for food acquisition (mostly other organisms or their products). Advanced methods for food acquisition also came with a requirement for learning and the development of parental care.

Plants use plasmodesmata to interconnect cells, which are tubular extensions of the plasma membrane of 40–50 nm in diameter that traverse the cell wall and interconnect the cytoplasm of adjacent cells into a symplast. Higher fungi form threads of multi-nucleated syncytia, known as mycelia; sometimes septa are present in the hyphae, but they have large pores. Otherwise, the cells of fungi only communicate via the extracellular matrix [804]. Rhodophytes have elaborate pit connections between the cells [275] which have a diameter in the range 0.2–40 μm filled with a plug that projects in the cytoplasm on either side. Their nuclei can move from one cell to the other. Ascomycetes and Basidiomycetes have similar pit connections, but lack the plug structure and the cytoplasm is directly connected unlike the situation in rhodophytes.

Plants differentiated their structure into a root for nutrient uptake linked to water uptake and a shoot for gas exchange, photon acquisition and evaporation of water, see {207}. Roots evolved 410–395 Ma ago, some 50 Ma after the origin of tracheophytes, and probably separately in bryophytes and embryophytes [943]. Leaves typically last 1 year, and fall after recovering (some of) the reserve. This means that plants live syntrophically with the soil biota (especially bacteria and fungi), that feed on this organic rain and release the locked nutrients as waste for renewed uptake by the plants. Moreover, almost all plant species have an endomycorrhiza, i.e. specialised fungi of the phylum Glomeromycetes that are probably involved in drought resistance and nutrient uptake. The Brassicaeae, which are specialists on nutrient-rich soils, do not have endomycorrhizae. Some 30% of plants also have an ectomycorrhiza and many use animals for pollination and dispersal.

10.5.2 Emergence of life-stages: adult and embryo

Multicellularity came with the invention of reproduction by eggs in the form of packages of reserve with a very small amount of structure: the juvenile state thus gave rise to both the adult and the embryo state. Embryos differ from juveniles by not taking up substrates from the environment. That is to say they do not (yet) use the assimilation process for energy and building-block acquisition, although most do take up dioxygen. The spores of endobacteria can be seen as an embryonic stage for prokaryotes. Adults differ from juveniles by allocation to reproduction, rather than further increasing the state of maturity. Unlike dividing juveniles, adults do not reset their state (i.e. the amount of structure, reserve and the state of maturity). Animals, notably vertebrates, and embryophytes, notably the flowering plants, provide the embryo fully with reserve material. Egg size, relative to adult size, has proved highly adaptable in evolutionary history, see {305}.

10.5.3 Further increase in maintenance costs

Multicellular organisation and an active life style, especially in eukaryotes, results in a series of extra maintenance costs. Concentration gradients across the more abundant and dynamic membranes become more important, as well as intracellular transport and movements of the individual. The invasion of the freshwater habitat required a solution to the osmotic condition; many eukaryotes use pulsating vacuoles for this purpose. Invasion of the terrestrial habitat required an answer to the problem of desiccation and of getting rid of nitrogen waste; ammonia is toxic even at low concentrations. Some animals and plants elevate the temperature of parts of their body metabolically to enhance particular physiological functions; birds and mammals have taken this to extremes, see {13}. All these processes increased maintenance requirements further, but also improved the metabolic performance. Such organisms became less dependent on the local chemical and physical conditions.

10.5.4 Ageing and sleeping

When the cyanobacteria eventually enriched the atmosphere with dioxygen, many species adapted to this new situation and energy acquisition from carbohydrates was greatly improved by using dioxygen for oxidation in the respiratory chain. Although means to cope with ROS, or reactive nitrogen species (RNS), were already present, the handling of ROS became important to reduce damage to the metabolic machinery and especially to DNA. This especially holds true for tissues of cells with non-reversible differentiation. Specialised proteins (peroxidase dismutases) were developed and their effectiveness was tuned to compromise between survival of the juvenile period and the use of ROS to generate genetic variability among gametes. The latter is important to allow adaptation to long-term environmental changes that are too large for adaptation within a given genome. Big-bodied species are vulnerable; the length of the juvenile period scales with body length among species {322} whereas the reproductive rate decreases with length {323}. Therefore large-bodied species must have efficient peroxidase dismutases and, therefore, reduce the genetic variability among their gametes, while having few offspring. This makes them vulnerable at the evolutionary timescale.

The neuronal system of animals is sensitive to ROS, and requires sleep for repair [1053, 1054]. Since the required sleeping time tends to be proportional to the specific respiration rate, large-bodied species have more time to search for food. Their speed and the diameter of their home range increases with length, which enhances their ability to cope with spatial heterogeneity.

10.5.5 From supply to demand systems

Plants evolved extreme forms of morphological and biochemical adaptations to the chemical and physical conditions in their direct environment and remained supply systems. Animals, by contrast, especially birds and mammals, excel in behavioural traits designed to meet their metabolic needs. They evolved into demand systems, see {15}. This co-evolved with an increase in the difference between standard and peak metabolic rates, closed circulation systems, advanced forms of endothermy, immune systems and hormonal regulation systems. Since specific food uptake is no longer a function of food density only but also of population density, stable biodiversity is enhanced, see {264}, even in spatially homogeneous and constant environments. Together with the syntrophic basis of coexistence, this can be an important mechanism in the evolution of biodiversity.

10.6 Control over local conditions

By shading and evaporation and littering leaves and branches, trees substantially affect their microclimate and soil properties, and thereby allow other organisms to live there as well. This too can be seen as an aspect of metabolism. *Sphagnum*-dominated peats develop in a particular way, where a series of *Sphagnum* species lower the pH by exchanging ions and hold water, which provides a strong selection force on other species.

Phytoplankters bind nutrients in the photic zone of the oceans, sink below it, die and are degraded by bacteria. Subsequently, a temporary increase in wind speed brings some of the released nutrients back to the photic zone by mixing and enables photosynthesis to continue. The sinking of organic matter is accelerated by grazing zooplankters.

The result of this process is that, over time, phytoplankters build up a nutrient gradient in the water column, that carbon dioxide from the atmosphere becomes buried below the photic zone, and that organic resources are generated for the biota living in the dark waters below this zone and on the ocean floor. Mixing by wind makes phytoplankters commute between the surface, where they can build up and store carbohydrates by photosynthesis, and the bottom of the mixing zone, where they store nutrients. Reserves are essential here for growth, because no single stratum in the water column is favourable for growth; their reserve capacity must be large enough to cover a commuting cycle, which depends on wind speed. Although nutrient availability controls primary production ultimately, wind is doing so proximately. The rain of dead or dying phytoplankters fuels the dark ocean communities, not unlike the rain of plant leaves fuelling soil communities, but then on a vastly larger spatial scale. Little is known about the deep ocean food web; recent studies indicate that cnidarians (jellyfish) form a major component [263].

When part of this organic rain reaches the anoxic ocean floor, the organic matter is decomposed by fermenting bacteria (many species can do this); the produced hydrogen serves as substrate for methanogens (i.e. archaea), which convert carbon dioxide into methane. This methane can accumulate in huge deposits of methane hydrates, which serve as substrate for symbioses between bacteria and a variety of animals, such as the ice worm *Hesiocoeca*, a polychaete. The total amount of carbon in methane hydrates in ocean sediments is more than twice the amount to be found in all known fossil fuels on Earth. If the temperature rises in the deep oceans, the hydrates become unstable and result in a sudden massive methane injection into the atmosphere. This happened e.g. 55 Ma years ago [1288], the Paleocene–Eocene Thermal Maximum (PETM) event, which induced massive extinction. We are just beginning to understand the significance of these communities on ocean floors and deep underground.

The colonisation of the terrestrial environment by plants may in fact have allowed reefs of brachiopods, bryozoans and molluscs (all filter feeders) to flourish in the Silurian and the Devonian (438–360 Ma ago); the reefs in these periods were exceptionally rich [1271]. With the help of their bacterial symbionts, the plants stimulated the conversion from rock to soil, which released nutrients that found their way to the coastal waters, stimulated algal growth, and, hence, the growth of zooplankton, which the reef animals, in turn, filtered out of the water column. Although plant megafossils only appeared in the Silurian, cryptospores, which probably originate from bryophytes, were very abundant in the Ordovician (505–438 Ma ago) [1105]. So the timing of the terrestrial invasion and the reef development supports this link. The reefs degraded gradually during the time Pangaea was formed toward the end of the Permian, which reduced the length of the coastline considerably, and thereby the nutrient flux from the continents to the ocean. Moreover, large continents come with long rivers, and more opportunities for water to evaporate rather than to drain down to the sea; large continents typically have salt deposits. When Pangaea broke up, new coastlines appeared. Moreover, this coincided with a warming of the globe, which brought more rain, more erosion, and high sea levels, which caused large parts of continents to be covered by shallow seas. This combination of factors caused planktonic communities to flourish again in the Cretaceous, and completely new taxa evolved, such as the coccolithophorans and the diatoms. This hypothesis directly links the activities of terrestrial plants to the coastal reef formation through nutrient availability. Although plants reduce erosion on a timescale of thousands of years, they promote erosion on a multi-million years timescale in combination with extreme but very rare physical forces that remove both vegetation and soil. The geological record of the Walvis Ridge suggests that the mechanism of physical–chemical forces that remove the vegetation, followed by erosion and nutrient enrichment of coastal waters in association with recolonisation of the rocky environment by plants might also have been operative in e.g. the 0.1 Ma recovery period following the PETM event (D. Kroon, personal communication).

10.7 Control over global conditions

Climate modelling mainly deals with energy (temperature) and water balances. Heat and water transport and redistribution, including radiation and convection in atmospheres and oceans, depend on many chemical aspects which means that climate modelling cannot be uncoupled from modelling biogeochemical cycling. I here focus on radiation, as affected via albedo and absorption by greenhouse gases [643].

10.7.1 Water

Because of its abundance, water is by far the most important greenhouse gas. Most climate models keep the mean global relative humidity constant at 50%, e.g. [234], but this assumption can be questioned. The origin of water is still unclear; some think it originates from degassing of the hot young planet [666], others think from meteoric contributions in the form of carbonaceous chondrites [91], which possibly continues today.

Plants modify water transport in several ways. Although plants can extract foggy water from the atmosphere particularly in arid environments (by condensation at their surface as well as via the emission of condensation kernels), they generally pump water from the soil into the atmosphere, and increase the water capacity of terrestrial environments by promoting soil formation in bare environments (chemically, with help of bacteria [96]) thereby reducing water runoff to the oceans. This became painfully clear during the flooding disasters in Bangladesh that followed the removal of Himalayan forests in India. On a short timescale, plants greatly reduce erosion; their roots prevent or reduce soil transport by common mild physical forces. In combination with rare strong and usually very temporary physical forces that remove vegetation (fires in combination with hurricanes or floods, for instance), however, plants increase erosion on a longer timescale, because plants enhance soil formation in rocky environments. Because such 'catastrophes' are rare, they have little impact on short timescales. The effects of plants on climate and geochemistry were perhaps most dramatic during their conquest of dry environments in the middle Devonian. It came with a massive discharge of nutrients and organic matter into the seas, which led to anoxia and massive extinctions in the oceans [16].

Plants, therefore, affect the nutrient (nitrate, phosphate, silica, carbonates) supply to the oceans in complex ways, and thus the role of life in the oceans in the carbon cycle. Plants pump water from the soil into the atmosphere much faster in the tropics than in the temperate regions because of temperature (high temperature comes with large evaporation), seasonal torpor (seasons become more pronounced toward the poles, so plants are active during a shorter period in the year toward the poles) and nutrients in the soil (plants pump to get nutrients, which are rare in tropical soils). Plant-produced cellulose accumulates in soils and increases its water retention capacity profoundly.

10.7.2 Carbon dioxide

Carbon dioxide is the second most important greenhouse gas. Its dynamics involves the global carbon cycle, which is still poorly known quantitatively. This is partly due to the coupling with other cycles.

Carbon dioxide is removed from the atmosphere by chemical weathering of silicate rocks, which couples the carbon and silica cycles. This weathering occurs via wet deposition, and gives a coupling between the carbon and the water cycles. When ocean downwashed calcium carbonate and silica oxide precipitate and become deeply buried by continental drift in Earth's mantle, segregation occurs into calcium silicate and carbon dioxide; volcanic activity puts carbon dioxide back into the atmosphere. Geochemists generally hold this rock cycle to be the main long-term control of the climate system.

Westbroek [1240] argued that the role of life in the precipitation processes of carbonates and silica oxide became gradually more important during evolution. Mucus formers (by preventing spontaneous precipitation of supersaturated carbonates) and calcifiers have controlled carbonates since the Cretaceous. Diatoms (and radiolarians) have controlled silicates since the Jurassic [671]. Corals and calcifying plankton (coccolithophores and foraminiferans) have an almost equal share in calcification. In fresh water, charophytes are in this guild. For every pair of bicarbonate ions, one is transformed into carbon dioxide for metabolism, and one into carbonate. Planktonic-derived carbonate partly dissolves, and contributes to the build-up of a concentration gradient of inorganic carbon in the ocean. This promotes the absorption of carbon dioxide from the atmosphere by seawater.

The dry deposition of carbon dioxide in the ocean is further enhanced by the organic carbon pump, where inorganic carbon is fixed into organic carbon, which travels down to deep layers by gravity. This process is accelerated by predation where unicellular algae are compacted into faecal pellets, and partial microbial decomposition recycles nutrients to the euphotic zone, boosting primary production. The secondary production also finds its way to the deep layers. Chitin plays an important role in the organic carbon pump. This is because it is difficult to degrade and occurs in relatively large (and heavy) particles and is produced by zooplankton in the upper ocean layer (zooplankton). When dying (or moulting), chitin sinks which causes a drain of (organic) carbon and nitrogen to the deep in a C : N ratio of 8:1. This ratio controls the ability of phytoplankton to fix atmospheric carbon dioxide [850]. This process is of importance on a timescale in the order of millennia (the cycle time for ocean's deep water), and so is relevant for assessing effects of an increase of atmospheric carbon by humans. It is less important on much longer timescales.

In nutrient-rich shallow water, organic matter can accumulate fast enough to form anaerobic sediments, where decomposition is slow and incomplete and fossilisation into mineral oil occurs. Although textbooks on marine biogeochemistry do not always fully recognise the role of plants in the global carbon cycle, see [702, p 139], coal deposits in freshwater marshes are substantial enough to affect global climate. Oil formed by plankton and coal by plants mainly occurs on continental edges, and affects climate on the multi-million years timescale.

10.7.3 Methane

Methane is the third most important greenhouse gas; 85% of all emitted methane is (presently) produced by methanogens (in syntrophic relationships with other organisms, sometimes endosymbiotic) in anaerobic environments (sediments, guts) [348, 734]. The flux is presently enhanced by large-scale deforestation by humans via termites. Apart from accumulation in the atmosphere, and in fossilised gas, big pools ($2\,10^3$–$5\,10^6$ Pg) of methane hydrates rest on nearshore ocean sediments. Since methane can capture infrared radiation 25 times better than carbon dioxide, on a molar basis, a release of the methane hydrates can potentially destabilise the climate system [684]. Oxidation of methane is a chief source of water in the stratosphere [176], where it interferes with radiation.

Like carbon dioxide, the methane balance is part of the global carbon cycle. Since most of life's activity is limited by nutrients, the carbon cycle cannot be studied without involving other cycles. Nitrogen (nitrate, ammonia) is the primary limiting nutrient, but iron might be limiting as well in parts of the oceans [42, 197]. After assuming that dinitrogen-fixing cyanobacteria could eventually relieve nitrogen limitation, Tyrell [1177] came to the conclusion that nitrogen was proximately limiting primary production, and phosphate was ultimately doing so. The question remains, however, are cyanobacteria active enough? Many important questions about the nitrogen cycle are still open, even whether oceans represent a sink or a source of ammonia, nitrates and nitrous oxide [549]. The latter is, after methane, the next most important greenhouse gas, which can intercept infrared radiation 200 times better than carbon dioxide.

Most nutrients enter the oceans via rivers from terrestrial habitats, which couples both systems and makes coastal zones very productive. The surface area of this habitat has obviously been under control by continental drift and seawater level changes, and therefore with ice formation and temperature. These remarks serve to show the link between climate and biochemical cycles.

10.7.4 Dioxygen

Complex relationships exist between the carbon and oxygen cycles. Dioxygen results from photosynthesis, so there is a direct relationship between dioxygen in the atmosphere and buried fossil carbon. The latter probably exceeds dioxygen on a molar basis, because of e.g. the oxidation of iron and other reduced pools in the early history of the Earth. Photorespiration links dioxygen to carbon dioxide levels {191}. This effect of dioxygen is possibly an evolutionary accident that resulted from the anoxic origins of Rubisco. Spontaneous fires require at least 75% of present-day dioxygen levels, and dioxygen probably now sets an upper boundary to the accumulation of organic matter in terrestrial environments, and so partly controls the burial of fossilised carbon [1274]. The extensive coal fires in China at 1 km depth, that have occurred since human memory, illustrate the importance of this process. Model calculations by Berner [90] suggest, however, that dioxygen was twice the present value during the Carboniferous. If true, this points to the control of fossil carbon accumulation by dioxygen being weak. The big question is, of course, to what extent humans are perturbing the climate system by enhancing the burning of biomass and fossil carbon. The massive burning of the world's rain forests after the latest El Niño event makes it clear that their rate of disappearance is accelerating, despite worldwide concern.

10.7.5 Albedo

Apart from greenhouse gases, the radiation balance is affected by albedo. Ice and clouds are the main controlling components. Cloud formation is induced by micro-aerosols, which result from combustion processes, volcanoes and ocean-spray-derived salt particles. Phytoplankton (diatoms, coccolithophorans) affects albedo via the production of dimethyl sulfide (DMS), which becomes transformed to sulphuric acid in the atmosphere, acting as condensation nuclei. The production is associated with cell death, because the precursor of DMS is mainly used in cell's osmoregulation. Plants, and especially conifers, which dominate in taiga and on mountain slopes, produce isoprenes and terpenes [142], which, after some oxidation transformations, also result in condensation nuclei. Since plants cover a main part of the continents, plants change the colour, and so the albedo of the Earth, in a direct way. Condensation nuclei derived from human-mediated sulphate emissions now seem to dominate natural sources, and possibly counterbalance the enhanced carbon dioxide emissions [201].

Ice affects the climate system via the albedo and ocean level. If temperature drops, ice grows and increases the albedo, which makes it even colder. It also lowers the ocean level, however, which enhances weathering of fossil carbon and increases atmospheric carbon dioxide. This affects temperature in the opposite direction, and illustrates a coupling between albedo, and the carbon and water cycles.

10.8 Effects of climate on life

Climate affects life mainly through temperature, and in terrestrial environments, by precipitation and humidity. Nutrient supply and drain is usually directly coupled to water transport. The transport of organisms themselves in water and in air can also be coupled to climate. The effects are in determining both geographical distribution patterns, and abundance and activity rates.

For effects of body temperature on metabolic rates, see {17}. Many species of organism that do not switch to the torpor state escape bad seasons by migration, see {122}, some of them travelling on a global scale. Endotherms (birds and mammals) are well known examples of spectacular migrations; their energy budgets are tightly linked to the water balance {153}. The capacity to survive periods of starvation has close links with body size {323}.

Plant production increases in an approximately linear way with annual precipitation, which illustrates the importance of water availability in terrestrial environments. Plants use water for several purposes, one of them being the transport of nutrients from the soil to their roots. This is why the ratio of the surface areas of shoots and roots enters in the saturation constants for nutrient uptake by plants {154}. Precipitation also affects nutrient availability via leakage.

Because multiple reserve systems have to deal with excretion {195}, assimilation is much more loosely coupled to maintenance and growth compared to single reserve systems. The way temperature affects photosynthesis differs from how it affects growth (synthesis of structure) {22}, with the consequence that the excretion of carbohydrates (mobilised from its reserve, but rejected by the SUs for growth) depends on temperature. This means that the importance of the microbial loop is temperature dependent. Single-reserve systems, by contrast, do not excrete in this way and so do not have this degree of freedom {18}.

Extensive pampas and savannah ecosystems, as well as the recently formed fynbos vegetation in southern Africa, require regular fires of a particular intensity for existence. Many plant species require fire to trigger germination.

Local differences between seasons in temperate and polar areas are large with respect to global climate changes during the evolution of the Earth, which complicates the construction of simple models that aim to be realistic.

10.9 Summary

Evolution accelerated considerably, starting with a long period where mutation and selection created an initial diversity, followed by a period during which recombination of metabolic modules enhanced diversity. The evolution of the central metabolism testifies for the significance of the exchange of metabolic modules, that of symbiogenesis even more. DEB theory is unique in its ability to capture why and how systems that follow DEB rules can merge such that the resulting merged structure again follows DEB rules.

Life started with the increase of the ability to maintain homeostasis, which induced stoichiometric constraints on production, and necessitated the use of reserves. The animal evolutionary line reduced the number of reserves, while increasing their acquisition ability and enhancing mobility. The plant evolutionary line increased metabolic versatility and morphological flexibility.

I summarise observations that lead to the conclusion that life did become increasingly dependent on life, which is in harmony with the stepwise improvement of various forms of homeostasis, which is key to DEB theory {8}. The increased ability to control the internal physical–chemical conditions is matched by the ability to do that externally.

The interactions between life and climate are discussed, which are of substantial quantitative importance. DEB theory can be used to quantify these interactions.

11

Evaluation

The aim of this short chapter is to place the DEB model in the context of research in eco-energetics. I first summarise the collection of empirical models that turn out to be special cases of DEB theory. Then I discuss some problems that are inherent to biochemical models for metabolism of individuals. Finally I discuss static energy budgets and their time-dependent relatives, the net production models, in comparison with the DEB approach.

11.1 Empirical models that are special cases of DEB theory

The emphasis of DEB theory is on mechanisms. This implies a radical rejection of the standard application of allometric functions, which I consider to be a blind alley that hampers understanding. Although it has never been my objective to glue existing ideas and models together into one consistent framework, many aspects and special cases of the DEB theory turned out to be identical or very similar to classic models, see Table 11.1. DEB theory not only shows how and why these models are related, it also specifies the conditions under which these models might be realistic, and it extends the scope from the thermodynamics of subcellular processes to population dynamics.

11.2 A weird world at small scales

Many attempts exist to build models for cellular metabolism following the fate of some important compounds. Apart from the limited usefulness of models with many parameters and variables, these models typically suffer from lack of spatial structure, and the transformations are typically based on classic chemical kinetics. The following considerations question the validity of this approach.

Basic to chemical kinetics is the law of mass action: transformation rates are proportional to meeting frequencies, which are taken proportional to the product of

Table 11.1 Empirical models that turn out to be special cases of DEB models, or very good numerical approximations to them.

author	year	page	model
Lavoisier	1780	{162}	multiple regression of heat against mineral fluxes
Gompertz	1825	{219}	survival probability for ageing
Arrhenius	1889	{17}	temperature dependence of physiological rates
Huxley	1891	{202}	allometric growth of body parts
Henri	1902	{103}	Michaelis–Menten kinetics
Blackman	1905	{267}	bi-linear functional response
Hill	1910	{206}	Hill's functional response
Thornton	1917	{160}	heat dissipation
Pütter	1920	{52}	von Bertalanffy growth of individuals
Pearl	1927	{355}	logistic population growth
Fisher and Tippitt	1928	{218}	Weibull ageing
Kleiber	1932	{303}	respiration scales with body weight$^{3/4}$
Mayneord	1932	{136}	cube root growth of tumours
Monod	1942	{131}	growth of bacterial populations
Emerson	1950	{136}	cube root growth of bacterial colonies
Huggett and Widdas	1951	{63}	foetal growth
Weibull	1951	{218}	survival probability for ageing
Best	1955	{266}	diffusion limitation of uptake
Smith	1957	{149}	embryonic respiration
Leudeking and Piret	1959	{165}	microbial product formation
Holling	1959	{33}	hyperbolic functional response
Marr and Pirt	1962	{131}	maintenance in yields of biomass
Droop	1973	{131}	reserve (cell quota) dynamics
Rahn and Ar	1974	{306}	water loss in bird eggs
Hungate	1975	{278}	digestion
Beer and Anderson	1997	{59}	development of salmonid embryos

concentrations of substrate. This rests on transport by diffusion or convection. A few observations might help to reveal that the application of classic chemical kinetics in cellular metabolism is problematic.

This even holds for the concept 'concentration' of a compound inside cells. The concept 'concentration' is rather problematic in spatially highly structured environments, such as in growing cells, where many transformations are mediated by membrane-bound enzymes. Use of concentrations should be restricted to well-mixed local environments, such as the idealised environment outside organisms. Ratios of amounts, called densities, can play a role in transformations. Densities resemble concentrations, but the compounds are not necessarily well mixed at a molecular level.

Consider a typical bacterial cell of volume $0.25\,\mu m^3$ and an internal pH of 7 at 25 °C [642]; the intracellular compartments of eukaryotic cells are about the same size. If the cell consisted of pure water, it would have $N = 8\,10^9$ water molecules. For a dissociation rate of water $\dot{k}_1 = 2.4\,10^{-5}\,s^{-1}$ and an association rate of the ions $\dot{k}_2 = 10^3\,ion^{-1}s^{-1}$

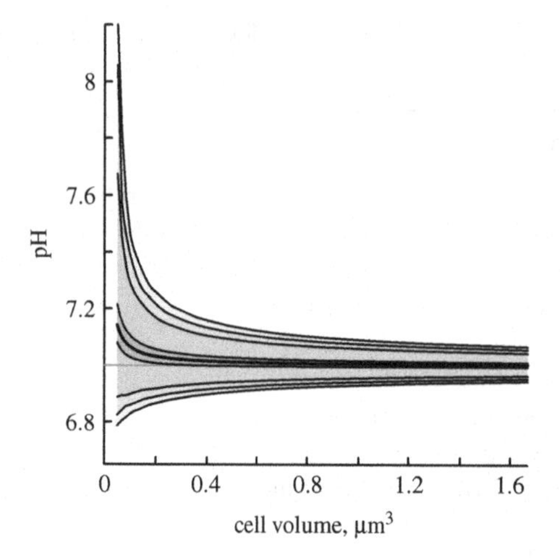

Figure 11.1 If the process $H_2O \rightleftharpoons H^+ + OH^-$ occurs randomly and the mean number of free protons is m, the probability of n free protons equals $P_n = (m^n/n!)^2/I_0(2m)$, where $I_0(x) = \sum_{i=0}^{\infty}(x/2)^{2i}(i!)^{-2}$ is the modified Bessel function. The 95%, 90%, 80% and 20% confidence intervals of pH in cells of pure water with pH 7 are plotted as a function of the cell size. They increase dramatically for decreasing cell sizes for cells (or cell compartments) less than $0.5\,\mu m^3$. The thick curve represents the mean pH, which goes up sharply for very small cell sizes.

(in ice, $\dot{k}_2$ is faster), the equilibrium number of 'free' protons is $m = \sqrt{N\dot{k}_1/\dot{k}_2} = 13.9$. Random dissociation of water, and association of protons and hydroxyl ions make this number fluctuate wildly [775]. Figure 11.1 shows that the (asymptotic) frequency distribution of the number of protons, and so of pH, dramatically increases in variance for decreasing cell sizes for volumes smaller that $0.5\,\mu m^3$. We have to think in terms of pH *distributions* rather than pH *values*. Many chemical properties of compounds depend on the pH, which makes matters really complex.

The relaxation time is given by $\tau = 1/\sqrt{4N\dot{k}_1\dot{k}_2}$ and amounts to $36\,\mu s$. A water molecule is created, by association of a proton and a hydroxyl ion, and is annihilated by dissociation about twice a day at 25 °C. Brownian motion transports a water molecule about 3 cm between creation and annihilation, while protons and hydroxyl ions are transported some 3 μm, on average. However, these distances do not fit into a cell (or cell compartment), which must lead to the conclusion that undisturbed diffusion does not occur in cells. These expectations are based on pure water, but a more realistic cytoplasm composition does not eliminate the problem.

Water in very small volumes behaves as a liquid crystal [163, 54], rather than as a liquid, which has substantial consequences for kinetics. Electrical potentials reveal the crystalline properties. They decay exponentially as a function of distance L, so they are proportional to $\exp(-L/L_D)$. The parameter L_D, called the Debye distance, is about 0.1 μm for water at 25 °C [1233], which means the electrical potential of a proton would be felt through most of the cell, even if it did not move.

Macromolecular crowding can affect transformation rates substantially [10]. Organelles crawl around in cells, divide, merge and may be destroyed. Substrates are delivered to enzymes by transporter proteins, that follow paths along the endoplasmic reticulum; products are removed from their site of origin in a similar way. Active allocation of substrates to particular transformations is common. The small number of molecules of any type requires stochastic rather than deterministic specification of their dynamics. Many enzymes are only active if bound to membranes. Many substrates

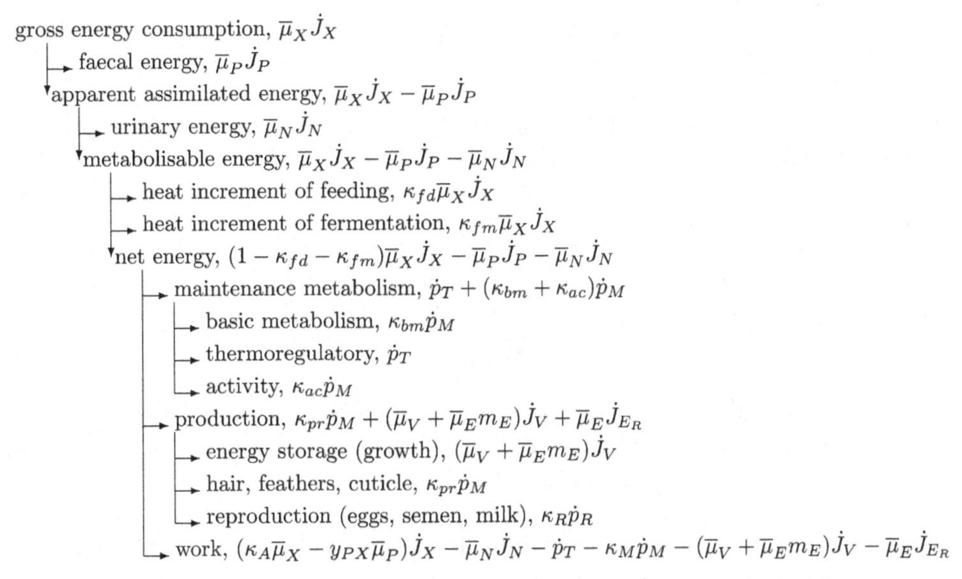

Figure 11.2 A typical Static Energy Budget [1262, p 87]. The symbols refer to the powers in the Dynamic Energy Budget, and reveal the links; see text for explanation.

play a role in a metabolic network, which gives complex connections. The cell follows a cell cycle, and many subcellular structures and activities have complex links to this cycle. All these aspects mean that transformations that are based on the law of mass action are problematic for living cells.

11.3 Static Energy Budgets

Most of the literature on animal energetics concerns Static Energy Budgets (SEBs). The term 'budget' refers to the conservation of energy, i.e. the various allocated powers add to the power input. SEBs can only be compared to DEBs at steady state, by averaging over a sufficient number of meals, but not so many that size changes. Figure 11.2 gives the relationship between both approaches. The following differences exist:

- SEB deals with energies that are fixed in the different products, while DEB deals with energies allocated to assimilation, maintenance and growth; the difference is in the overhead costs. The reconstruction assumes that the SEB balance is complete, so no products are formed coupled to growth, and also that we are dealing with an adult, and are using a combustion frame of reference (the energy content of oxygen, carbon dioxide, water and ammonia is set to zero). Some of the mapping depends on details of how the quantities are actually measured. The total balance sheet amounts to $\overline{\mu}_X \dot{J}_X = \overline{\mu}_P \dot{J}_P + (\overline{\mu}_V + \overline{\mu}_E m_E) \dot{J}_V + \kappa_R \dot{p}_R + \overline{\mu}_N \dot{J}_N + \dot{p}_{T+}$, or $\dot{p}_A = \dot{p}_M + \dot{p}_J + \dot{p}_T + (1 + m_E y_{VE}) \dot{p}_G + \dot{p}_R$.

- 'faecal energy' represents a fixed fraction of 'gross energy consumption' in DEB, so $\overline{\mu}_P \dot{J}_P = \overline{\mu}_P y_{PX} \dot{J}_X$.

- 'urinary energy' is decomposed in DEB into contributions from assimilation, maintenance and growth: $\dot{J}_N = \dot{J}_{NA} + \dot{J}_{NM} + \dot{J}_{NG}$. Subtraction from the 'apparent assimilation energy' complicates the mapping of the remaining energy to maintenance and (re)production.

- 'heat increment of feeding' and of 'fermentation' are included in the overhead costs of assimilation and, therefore, fractions of the 'gross energy consumption'. The fractions κ_{fd} and κ_{fm} are constants.

- 'net energy' equals $\dot{p}_A - \overline{\mu}_N(\dot{J}_{NM} + \dot{J}_{NG})$, assuming that both heat increments cover all assimilation overheads, except for the assimilation part of urinary energy. It follows that $\kappa_{fd} + \kappa_{fm} = \frac{1 - y_{PX}\overline{\mu}_P/\overline{\mu}_X}{1 + \overline{\mu}_N \eta_{NA}}$.

- 'basic metabolism', 'activity', and 'hair, feather, cuticle' are all fractions of somatic maintenance costs in DEB, so $\kappa_{bm} + \kappa_{ac} + \overline{\mu}_N \dot{J}_{NM}/\dot{p}_M = 1$. This mapping includes the overhead costs of maintenance in 'basic metabolism', the correctness depending on the way it is measured.

- 'energy storage' includes the energy fixed in new reserves and new structural mass. Notice that $\overline{\mu}_V \dot{J}_V < \dot{p}_G$; the overhead costs of growth in DEB go into 'work' in SEB.

- 'milk production' (of female mammals) comes with a temporal change in the parameters $\{\dot{J}_{Xm}\}$ and $\{\dot{p}_{Am}\}$ in DEB.

- 'work' includes part of the overhead costs of growth and maintenance in SEB's balance sheet. The abbreviations $\kappa_A = 1 - \kappa_{fd} - \kappa_{fm}$ and $\kappa_M = \kappa_{bm} + \kappa_{ac} + \kappa_{pr}$ have been made. Alternative expressions are: $\dot{p}_A - \dot{p}_M - \dot{p}_T - \kappa_R \dot{p}_R - \overline{\mu}_N \dot{J}_{NG} - (\overline{\mu}_V + \overline{\mu}_E m_E)\dot{J}_V = \dot{p}_G(1 + m_E y_{VE}) + \dot{p}_R(1 - \kappa_R) + \dot{p}_J - \overline{\mu}_X \dot{J}_X + \overline{\mu}_p \dot{J}_P + \kappa_R \dot{p}_R + \dot{p}_{T+} = \dot{p}_{T+} + \dot{p}_J + \dot{p}_R + \dot{p}_G(1 + y_{VE} m_E) + \overline{\mu}_P \dot{J}_P - \overline{\mu}_X \dot{J}_X$.

The significance of including certain energy allocations in the overhead of others is in the comparison of energy budgets, both between different organisms and with respect to changes in time. Such inclusions greatly simplify the structure of energy budgets, and reduce the flexibility, i.e. reveal patterns of covariation of allocations. It boils down to the sharp distinction that DEBs make between the power allocated to, for example, growth, and the power that is actually fixed in new biomass; SEBs can only handle the energy that is fixed in new biomass, because the energy allocated to growth can only be assessed indirectly via changes of the budget in time. The reconstruction beautifully shows that 'work' has many contributions in SEB, and cannot be interpreted easily. The term is misleading, by suggesting that the individual can spend it freely.

Practical applications of SEBs to individuals of different sizes and different environmental conditions reveal that the resulting coefficients vary in ways that are difficult to understand; the scaling exponent of allometric functions to account for size differences turn out to depend on size [1184].

Von Bertalanffy [95] related the respiration rate to the rate of anabolism. I cannot follow this reasoning. At first sight, synthesis processes are reducing by nature, which makes catabolism a better candidate for seeking a relationship with respiration. In the standard SEB studies, respiration rates are usually identified with maintenance metabolism. These routine metabolic costs are a lump sum, including the maintenance of concentration gradients across membranes, protein turnover, regulation, transport (blood circulation, muscle tonus), and an average level of movement. The Scope For Growth (SFG) concept rests on this identification. The idea behind this concept is that energy contained in faeces and the energy equivalent of respiration are subtracted from energy derived from food, the remainder being available for growth [70]. The SFG concept is built on SEBs, {433}. In the DEB model, where energy derived from food is added to the reserves, the most natural candidate for a relationship with respiration is the rate at which the reserves are used. This is underpinned on {147}.

Although respiration rates are measured over short periods (typically a couple of minutes) and the actual growth of the body is absolutely negligible, the energy investment in growth can still be substantial. Parry [869] estimates the cost of growth between 17% and 29% of the metabolism of an 'average' ectotherm population. The respiration rate includes routine metabolic costs as well as costs of growth [998]. This interpretation is, therefore, incompatible with the SFG concept. Since the DEB model does not use respiration rates as a primary variable, the interpretation problems concerning respiration rates only play a role in testing the model.

11.4 Net production models

The DEB model assumes that assimilates are added to reserves, and reserves are used to fuel other metabolic processes (maintenance, growth, maturation, reproduction). Roger Nisbet [839] proposed the term assimilation models for models based on this assumption, to distinguish them from net production models. The latter models first subtract maintenance costs from assimilates, before allocation to other metabolic processes occurs; they represent a kind of time-dependent SEB models. Several net production models have been worked out [25, 451, 706, 841, 992]. It will not always be easy to use experimental data to choose between assimilation and net production models.

The choice of an assimilation structure rather than a net production structure is primarily motivated by simplicity in several respects, including mechanistic arguments with respect to metabolic control. The first argument is that embryos do not feed, but nevertheless have to pay maintenance costs. Net production models then suffer from the choice of letting embryos differ from juveniles by allowing embryos to pay maintenance from reserves, or treating yolk as a new state variable that is typical for embryos [706]. The second argument is that if feeding is not sufficient to pay maintenance costs, they have to be paid from reserves; it does not seem realistic to assume that an animal dies from starvation while it has lots of reserves. Most animals feed on meals anyway, while the storage in the gut cannot explain survival between the meals. Net production models must, therefore, contain elements of assimilation models, and switches have to be installed to pay maintenance from assimilates and/or from reserves. The analysis of the mathematical properties of models with switches rapidly becomes more problematic with the number of switches. The feeding and reproduction switch seems to be unavoidable for all energetic models. This set of arguments relates to the organisation of metabolism, which is much more independent of the environment in assimilation models than in net production models. This allows a simpler regulation system for metabolism, which is mainly driven by signals from the nutritional state of the organism itself, rather than from signals directly taken from the environment.

The net production models that are presently available have many more parameters than the assimilation model in this book; the simplest and best comparable production model is that formulated by Dina Lika and Roger Nisbet [706]. Besides structural mass, its state variables include yolk (in the embryonic stage), reserves and the maximum experienced reserve density in the juvenile and adult stages. Although it is possible to simplify net production models and reduce the number of parameters, I am convinced that they need more parameters and state variables than assimilation models with a comparable amount of detail. This is because they have to handle switches, and specify growth investment in a more complex way. The parameter κ of the assimilation model in this book specifies the investment in growth (plus somatic maintenance) versus reproduction (plus maturity maintenance); growth ceases automatically when the energy allocated to growth plus somatic maintenance is required for somatic maintenance. Reproduction can continue, while growth ceases. Net production models need at least one extra parameter to obtain this type of behaviour. Maintenance in net production models is paid for by food if possible, but from reserves if necessary, which requires an extra parameter for the maintenance costs. Maintenance is always paid from reserves in assimilation models (except in extreme starvation during shrinking). The number of parameters and state variables is a measure of the complexity of a model.

The mechanism for reserve dynamics and weak homeostasis that is proposed here, structural homeostasis {10}, does not apply to net production models. I expect that it is difficult to implement weak homeostasis mechanistically in net production models. If true, this means that biomass composition is changing, even at steady state, and structure always has to be disentangled from reserves in tests against experimental data. I also expect that it is difficult to derive realistic body size scaling relationships and to explain the method of indirect calorimetry on the basis of net production models.

Since reserves are wired prior to allocation to reproduction in production models, and not used for growth, they are hard to apply to dividing organisms, such as micro-organisms. The growth of plant biomass from tubers, and growth during starvation (see Figure 4.3), for instance, are also hard to implement; they need an extra state variable similar to yolk in the embryo.

Toxic compounds, parasites or the light regime can change the value of κ, which has complex consequences (growth is reduced and development lasts for a shorter time and/or reproduction is greater because of the higher investment, which is partially cancelled by the reduction of assimilation due to reduced growth). These complex changes can be described realistically with effects on a single parameter in the present assimilation model, while net production models need a more complex description that involves effects on more parameters. This type of perturbation of metabolism is perhaps the strongest argument in favour of assimilation models.

11.5 Summary

Many empirical models turn out to be special cases of DEB theory. We now know the mechanisms behind these models, and can better understand under what conditions they can be applied. DEB theory also reveals the coherence behind these models, which seem to be completely unrelated at first sight. Using the empirical support of these empirical models, which accumulated over the ages, DEB theory is presently the best tested theory in biology.

I briefly discuss alternative approaches to the subject of energetics, starting with that by molecular biology. The detailed molecular dynamics in cells is complex because of their spatial structure and the role of membranes; the small size of cell compartments gives them properties of liquid crystals and classic chemical dynamics hardly applies. Static Energy Budget models are (still) most popular in ecophysiology, and time-dependent Static Energy Budget models, net production models, are typically used to capture the energetics of individuals during their life cycle. This approach has led, in the past, to a particular interpretation of respiration rates that makes it difficult to understand why respiration approximately scales with weight to the power 3/4. All scaling relationships are difficult to understand with this method. I discuss some consistency issues that need further clarification in the context of Static Energy Budgets.

References

[1] K. Aagaard. Profundal chironimid populations during a fertilization experiment in Langvatn, Norway. *Holarct. Ecol.*, **5**:325–331, 1982.

[2] P. Abrams. Indirect interactions between species that share a predator: varieties of indirect effects. In W. C. Kerfoot and A. Sih, editors, *Predation: Direct and Indirect Impacts on Aquatic Communities*, pages 38–54. Univ. of New England Press, Hanover, 1987.

[3] R. A. Ackerman. Growth and gas exchange of embryonic sea turtles (*Chelonia, Caretta*). *Copeia*, **1981**:757–765, 1981.

[4] R. A. Ackerman. Oxygen consumption by sea turtle (*Chelonia, Caretta*) eggs during development. *Physiol. Zool.*, **54**:316–324, 1981.

[5] R. A. Ackerman, G. C. Whittow, C. V. Paganelli, and T. N. Pettit. Oxygen consumption, gas exchange, and growth of embryonic wedge-tailed shearwaters (*Puffinus pacificus chlororhynchus*). *Physiol. Zool.*, **53**:201–221, 1980.

[6] W. Agosta. *Bombardier Beetles and Fever Trees*. Addison Wesley, Reading, 1995.

[7] V. Ahmadjian and S. Paracer. *Symbiosis: An Introduction to Biological Associations*. Univ. of New England Press, Hanover, 1986.

[8] D. L. Aksnes and J. K. Egge. A theoretical model for nutrient uptake in phytoplankton. *Mar. Ecol. Prog. Ser.*, **70**:65–72, 1991.

[9] A. H. Y. Al-Adhub and A. B. Bowers. Growth and breeding of *Dichelopandalus bonnieri* in Isle of Man Waters. *J. Mar. Biol. Assoc. U.K.*, **57**:229–238, 1977.

[10] M. Al-Habori. Macromolecular crowding and its role as intracellular signalling of cell volume regulation. *Int. J. Biochem. Cell Biol.*, **33**:844–864, 2001.

[11] A. W. H. Al-Hakim, M. I. A. Al-Mehdi, and A. H. J. Al-Salman. Determination of age, growth and sexual maturity of *Barbus grypus* in the Dukan reservoir of Iraq. *J. Fish Biol.*, **18**:299–308, 1981.

[12] O. Alda Alvarez, T. Jager, E. Marco Redondo, and J. E. Kammenga. Assessing physiological modes of action of toxic stressors with the nematode *Acrobeloides nanus*. *Environ. Toxicol. Chem.*, **25**:3230–3237, 2006.

[13] M. Aleksiuk and I. M. Cowan. The winter metabolic depression in Arctic beavers (*Castor canadensis* Kuhl) with comparisons to California beavers. *Can. J. Zool.*, **47**:965–979, 1969.

[14] R. McN. Alexander. *Exploring Biomechanics: Animals in Motion*. Scientific American Library, New York, 1992.

[15] V. Y. Alexandrov. *Cells, Molecules and Temperature*. Springer-Verlag, Berlin, 1977.

[16] T. J. Algeo, S. E. Scheckler, and J. B. Maynard. Effects of the middle to late devonian spread of vascular land plants on weathering regimes, marine biotas, and global climate. In P. C. Gensel and D. Edwards, editors, *Plants Invade the Land: Evolutionary and Environmental Perspectives*. Columbia University Press, New York, 2001.

[17] A. L. Alldredge. Abandoned larvacean houses: a unique food source in the pelagic environment. *Science*, **177**:885–887, 1972.

[18] A. L. Alldredge. Discarded appendicularian houses as sources of food, surface habitats and particulate organic matter in planktonic environments. *Limnol. Oceanogr.*, **21**:14–23, 1976.

[19] A. L. Alldredge. House morphology and mechanisms of feeding in *Oikopleuridae* (*Tunicata, Appendicularia*). *J. Zool. Lond.*, **181**:175–188, 1977.

[20] J. P. Amend and E. L. Shock. Energetics of overall metabolic reactions of thermophilic and hyperthermophilic archaea and bacteria. *FEMS Microbiol. Rev.*, **25**:175–243, 2001.

[21] B. N. Ames, J. McCann, and E. Yamasaki. Methods for detecting carcinogens and mutagens with *Salmonella*/mammalian microsome mutagenicity test. *Mutat. Res.*, **31**:347–364, 1975.

[22] A. D. Anbar and A. H. Knoll. Proterozoic ocean chemistry and evolution: a bioinorganic bridge? *Science*, **1137**:1137–1142, 2002.

[23] J. S. Andersen, J. J. M. Bedaux, S. A. L. M. Kooijman, and H. Holst. The influence of design parameters on statistical inference in non-linear estimation; a simulation study based on survival data and hazard modelling. *J. Agric. Biol. Environ. Stat.*, **5**:323–341, 2000.

[24] R. A. Andersen. Biology and systematics of heterokont and haptophyte algae. *Am. J. Bot.*, **91**:1508–1522, 2004.

[25] T. R. Andersen. *Pelagic Nutrient Cycles: Herbivores as Sources and Sinks.* Springer-Verlag, Berlin, 1997.

[26] A. J. Anderson and E. A. Dawes. Occurence, metabolism, metabolic control, and industrial uses of bacterial polyhydroxylkanoates. *Microbiol. Rev.*, **54**:450–472, 1990.

[27] S. G. E. Anderson, A. Zomorodipour, J. O. Andersson, T. Sicheritz-Pontén, U. C. M. Alsmark, R. M. Podowski, A. K. Näslund, A.-S. Eriksson, H. H. Winkler, and C. G. Kurland. The genome sequence of *Rickettsia prowazekii* and the origin of mitochondria. *Nature*, **396**:133–140, 1998.

[28] J. O. Andersson and A. J. Roger. A cyanobacterial gene in nonphotosybthetic protests an early chloroplast acquisition in eukaryotes? *Current Biol.*, **12**:115–119, 2002.

[29] A. D. Ansell. Reproduction, growth and mortality of *Venus striatula* (Da Costa) in Kames Bay, Millport. *J. Mar. Biol. Assoc. U.K.*, **41**:191–215, 1961.

[30] R. S. Appeldoorn. Variation in the growth rate of *Mya arenaria* and its relationship to the environment as analyzed through principal components analysis and the ω parameter of the von Bertalanffy equation. *J. Fish Biol.*, **81**:75–84, 1983.

[31] A. Ar and H. Rahn. Interdependence of gas conductance, incubation length, and weight of the avian egg. In J. Piiper, editor, *Respiration Function in Birds.* Springer-Verlag, Berlin, 1978.

[32] K. Arbaĉiauskas and Z. Gasiūnaitė. Growth and fecundity of *Daphnia* after diapause and their impact on the development of a population. *Hydrobiologia*, **320**:209–222, 1996.

[33] R. Arking. *Biology of Aging.* Sinauer, Sunderland, 1998.

[34] E. A. Armstrong. *The Wren.* The New Naturalist. Collins, London, 1955.

[35] N. Armstrong, R. E. Chaplin, D. I. Chapman, and B. Smith. Observations on the reproduction of female wild and park Fallow deer (*Dama dama*) in southern England. *J. Zool. Lond.*, **158**:27–37, 1969.

[36] D. K. Arrowsmith and C. M. Place. *An Introduction to Dynamical Systems.* Cambridge University Press, 1990.

[37] D. K. Arrowsmith and C. M. Place. *Dynamical Systems: Differential Equations, Maps and Chaotic Behaviour.* Chapman & Hall, London, 1992.

[38] P. W. Atkins. *Physical Chemistry.* Oxford University Press, 1989.

[39] P. W. Atkins. *Physical Chemistry* 2nd edn. Oxford University Press, Oxford, 1990.

[40] M. Atlas. The rate of oxygen consumption of frogs during embryonic development and growth. *Physiol. Zool.*, **11**:278–291, 1938.

[41] H. J. W. de Baar. Von Liebig's law of the minimum and plankton ecology. *Prog. Oceanogr.*, **33**:347–386, 1994.

[42] H. J. W. de Baar and P. W. Boyd. The role of iron in plankton ecology and carbon dioxide transfer of the global oceans. In R. B. Hanson, H. W. Ducklow, and J. G. Field, editors, *The Changing Ocean Carbon Cycle*, pages 61–140. Cambridge University Press, 2000.

[43] J. Baas, B. P. P. van Houte, C. A. M. van Gestel, and S. A. L. M. Kooijman. Modelling the effects of binary mixtures on survival in time. *Environ. Toxicol. Chem.*, **26**:1320–1327, 2007.

[44] J. Baas, T. Jager, and S. A. L. M. Kooijman. Estimation of no effect concentrations from exposure experiments when values scatter among individuals. *Ecotoxicol. Modell.*, **220**:411–418, 2009.

[45] J. Baas, T. Jager, and S. A. L. M. Kooijman. A model to analyze effects of complex mixtures on survival. *Ecotoxicol. Environ. Saf.*, **72**:669–676, 2009.

[46] T. Baekken. Growth patterns and food habits of *Baetis rhodani*, *Capnia pygmaea* and *Diura nanseni* in a west Norwegian river. *Holarct. Ecol.*, **4**:139–144, 1981.

[47] T. B. Bagenal. The growth rate of the hake, *Merluccius merluccius* (L.), in the Clyde and other Scottish sea areas. *J. Mar. Biol. Assoc. U.K.*, **33**:69–95, 1954.

[48] T. B. Bagenal. Notes on the biology of the schelly *Coregonus lavaretus* (L.) in Haweswater and Ullswater. *J. Fish Biol.*, **2**:137–154, 1970.

[49] J. E. Bailey and D. F. Ollis. *Biochemical Engineering Fundamentals.* McGraw-Hill, London, 1986.

[50] A. N. Baker, F. W. E. Rowe, and H. E. S. Clark. A new class of Echinodermata from New Zealand. *Nature*, **321**:862–864, 1986.

[51] B. Bakker. Control and regulation of glycolysis in *Trypanosoma brucei.* PhD thesis, Vrije Universiteit, Amsterdam, 1998.

[52] S. L. Baldauf, D. Bhattachary, J. Cockrill, P. Hugenholtz, A. G. B. Pawlowski, an J. Simpson. The tree of life. In J. Cracraft and M. J. Donoghue, editors, *Assembling the Tree of Life*, pages 43–75. Oxford University Press, 2004.

[53] J. H. van Balen. A comparative study of the breeding ecology of the great tit *Parus major* in different habitats. *Ardea*, **61**:1–93, 1973.

[54] P. Ball. *A Biography of Water.* Weidenfeld & Nicolson, London, 1999.

[55] H. Baltscheffsky. Energy conversion leading to the origin and early evolution of life: did inorganic pyrophosphate precede adenosine triphosphate? In H. Baltscheffsky, editor, *Origin and Evolution of Biological Energy Conversion*, pages 1–9. VCH Publishers, Cambridge, 1996.

[56] M. Baltscheffsky, A. Schultz, and H. Baltscheffsky. H^+-PPases: a tightly membrane-bound family. *FEBS Letters*, **457**:527–533, 1999.

[57] C. A. M. Baltus. De effekten van voedselregime en lood op de eerste reproduktie van *Daphnia magna.* Technical report, Dienst DBW/RIZA, 1990.

[58] M. C. Barber, L. A. Suárez, and R. R. Lassiter. Modeling bioconcentration of nonpolar organic pollutants by fish. *Environ. Toxicol. Chem.*, **7**:545–558, 1988.

[59] J. Barcroft, R. H. E. Elliott, L. B. Flexner, F. G. Hall, W. Herkel, E. F. McCarthy, T. McClurkin, and M. Talaat. Conditions of foetal respiration in the goat. *J. Physiol. (Lond.)*, **83**:192–214, 1935.

[60] F. G. Barth. *Insects and Flowers.* Princeton University Press, 1991.

[61] G. A. Bartholomew and D. L. Goldstein. The energetics of development in a very large altricial bird, the brown pelican. In R. S. Seymour, editor, *Respiration and Metabolism of Embryonic Vertebrates*, pages 347–357. Dr. W. Junk Publishers, Dordrecht, 1984.

[62] I. V. Basawa and B. L. S. P. Rao. *Statistical Inference for Stochastic Processes.* Academic Press, London, 1980.

[63] G. C. Bateman and R. P. Balda. Growth, development, and food habits of young Pinon jays. *Auk*, **90**:39–61, 1973.

[64] J. F. Battey. Carbon metabolism in zooxanthellae-coelenterate symbioses. In W. Reisser, editor, *Algae and Symbioses*, pages 153–187. Biopress, Bristol, 1992.

[65] E. H. Battley. Calculation of entropy change accompanying growth of *Escherichia coli* K-12 on succinic acid. *Biotechnol. Bioeng.*, **41**:422–428, 1993.

[66] B. Baule. Zu Mitscherlichs Gesetz der physiologischen Beziehungen. *Landw. Jahrb.*, **51**:363–385, 1917.

[67] J. L. Baxter. A study of the yellowtail *Seriola dorsalis* (Gill). *Fish. Bull. (Dublin)*, **110**:1–96, 1960.

[68] B. L. Bayne, A. J. S. Hawkins, and E. Navarro. Feeding and digestion by the mussel *Mytilus edulis* L. (*Bivalvia, Mollusca*) in mixtures of silt and algal cells at low concentrations. *J. Exp. Mar. Biol. Ecol.*, **111**:1–22, 1987.

[69] B. L. Bayne, A. J. S. Hawkins, E. Navarro, and I. P. Iglesias. Effects of seston concentration on feeding, digestion and growth in the mussel *Mytilus edulis. Mar. Ecol. Prog. Ser.*, **55**:47–54, 1989.

[70] B. L. Bayne and R. C. Newell. Physiological energetics of marine molluscs. In

A. S. M. Saleuddin and K. M. Wilbur, editors, *Physiology*, vol. 4 of *The Mollusca*, ch. 1. Academic Press, London, 1983.

[71] M. J. Bazin and P. T. Saunders. Determination of critical variables in a microbial predator–prey system by catastrophe theory. *Nature*, **275**:52–54, 1978.

[72] R. C. Beason and E. C. Franks. Development of young horned larks. *Auk*, **90**:359–363, 1983.

[73] J. R. Beddington. Mutual interference between parasites or predators and its effect on searching efficiency. *J. Anim. Ecol.*, **44**:331–340, 1975.

[74] W. N. Beer and J. J. Anderson. Modelling the growth of salmonid embryos. *J. Theor. Biol.*, **189**:297–306, 1997.

[75] D. Beerling. *The Emerald Planet*. Oxford University Press, 2007.

[76] D. J. Bell. *Mathematics of Linear and Nonlinear Systems*. Clarendon Press, Oxford, 1990.

[77] P. J. L. Bell. Viral eukaryogenesis: was the ancestor of the nucleus a complex DNA virus? *J. Mol. Evol.*, **53**:251–256, 2001.

[78] P. J. L. Bell. Sex and the eukaryotic cell cycle is consistent with a viral ancestry for the eukaryotic nucleus. *J. Theor. Biol.*, **243**:54–63, 2006.

[79] S. Bengtson. *Early Life on Earth*. Columbia University Press, New York, 1994.

[80] D. B. Bennett. Growth of the edible crab (*Cancer pagurus* L.) off south-west England. *J. Mar. Biol. Assoc. U.K.*, **54**:803–823, 1974.

[81] A. A. Benson and R. F. Lee. The role of wax in oceanic food chains. *Sci. Am.*, **3**:77–86, 1975.

[82] T. G. Benton, S. J. Plaistow, A. P. Beckerman, C. T. Lapsley, and S. Littlejohns. Changes in maternal investment in eggs can affect population dynamics. *Proc. R. Soc. Lond. B Biol. Sci.*, **272**:1351–1356, 2005.

[83] J. Bereiter-Hahn. Behaviour of mitochondria in the living cell. *Int. Rev. Cytol.*, **122**:1–63, 1990.

[84] T. G. O. Berg and B. Ljunggren. The rate of growth of a single yeast cell. *Biotechnol. Bioeng.*, **24**:2739–2741, 1982.

[85] C. Bergmann. Über die Verhältnisse der Wärmeökonomie der Tiere zu ihrer Grösse. *Goett. Stud.*, **1**:595–708, 1847.

[86] H. Bergmann and H. Helb. *Stimmen der Vögel Europas*. BLV Verlagsgesellschaft, München, 1982.

[87] H. H. Bergmann, S. Klaus, F. Muller, and J. Wiesner. *Das Haselhuhn*. A. Ziemsen-Verlag, Wittenberg Lutherstadt, 1978.

[88] C. Bernard, A. G. B. Simpson, and D. J. Patterson. Some free-living flagellates (Protista) from anoxic habitatats. *Ophelia*, **52**:113–142, 2000.

[89] J. Bernardo. Determinants of maturation in animals. *Trends Ecol. Evol.*, **8**:166–173, 1993.

[90] R. A. Berner. Atmospheric oxygen, tectonics and life. In S. H. Schneider and P. J. Boston, editors, *Scientists on Gaia*, pages 161–173. MIT Press, Cambridge, 1991.

[91] R. A. Berner, and E. K. Berner. *Global Environment: Water, Air, and Geochemical Cycles*. Prentice Hall, Englewood Cliffs, 1996.

[92] D. R. Berry, editor. *Physiology of Industrial Fungi*. Blackwell Sci. Publ., Oxford, 1988.

[93] H. H. Berry. Hand-rearing lesser flamingos. In J. Kear and N. Duplaix-Hall, editors, *Flamingos*. T. & A. D. Poyser, Berkhamsted, 1975.

[94] L. von Bertalanffy. A quantitative theory of organic growth. *Hum. Biol.*, **10**:181–213, 1938.

[95] L. von Bertalanffy. Quantitative laws in metabolism and growth. *Q. Rev. Biol.*, **32**:217–231, 1957.

[96] J. Berthelin. Microbial weathering. In W. E. Krumbein, editor, *Microbial Geochemistry*, pages 233–262. Blackwell Sci. Publ., Oxford, 1983.

[97] P. Berthold. Über den Einfluss der Nestlingsnahrung auf die Jugendentwickling, insbesondere auf das Flugelwachstum, bei der Mönchsgrasmücke (*Sylvia atricapilla*). *Vogelwarte*, **28**:257–263, 1976.

[98] D. F. Bertram and R. R. Strathmann. Effects of maternal and larval nutrition on growth and form of planktotrophic larvae. *Ecology*, **79**:315–327, 1998.

[99] J. Best. The influence on intracellular enzymatic properties for kinetics data obtained from living cells. *J. Cell. Comp. Physiol.*, **46**:1–27, 1955.

[100] J. J. Beun. PHB metabolism and N-removal in sequencing batch granular sludge reactors. PhD thesis, University of Delft, 2001.

[101] R. J. H. Beverton and S. J. Holt. On the dynamics of exploited fish populations. *Fish. Invest. Ser. I*, **2**, 1957.

[102] R. J. Beyers. The metabolism of twelve aquatic laboratory microecosystems. *Ecol. Monogr.*, **33**:281–306, 1963.

[103] J. Bigay, P. Guonon, S. Robineau, and B. Antonny. Lipid packing sensed by arfgap1 couples copi coat disassembly to membrane bilayer curvature. *Nature*, **426**:563–566, 2003.

[104] R. Bijlsma. *De Boomvalk.* Kosmos vogelmonografieën. Kosmos, Amsterdam, 1980.

[105] R. J. Bijlsma. Modelling whole-plant metabolism of carbon and nitrogen: a basis for comparative plant ecology and morphology. PhD thesis, Vrije Universiteit, Amsterdam, 1999.

[106] T. Birkhead. *Great Auk Islands: A Field Biologist in the Arctic.* T. & A. D. Poyser, London, 1993.

[107] C. W. Birky. The inheritance of genes in mitochondria and chloroplasts: laws, mechanisms, and models. *Annu. Rev. Gen.*, **35**:125–148, 2001.

[108] E. C. Birney and D. D. Baird. Why do some mammals polyovulate to produce a litter of two? *Am. Nat.*, **126**:136–140, 1985.

[109] C. J. Bjerrum and D. E. Canfield. Ocean productivity before about 1.9 gyr ago limited by phosphorus adsorption onto iron oxides. *Nature*, **417**:159–162, 2002.

[110] D. G. Blackburn. Convergent evolution of viviparity, matrotrophy and specializations for fetal nutrition in reptiles and other vertebrates. *Am. Zool.*, **32**:313–321, 1992.

[111] F. F. Blackman. Optima and limiting factors. *Ann. Bot.*, **19**:281–295, 1905.

[112] W. Bland and D. Rolls. *Weathering: An Introduction to the Scientific Principles.* Arnold, London, 1998.

[113] R. E. Blankenship and H. Hartman. Origin and early evolution of photosynthesis. *Photosynth. Res.*, **33**:91–111, 1992.

[114] R. E. Blankenship and H. Hartman. The origin and evolution of oxygenic photosynthesis. *Trends Biochem. Sci.*, **23**:94–97, 1998.

[115] J. H. S. Blaxter and J. R. Hunter. The biology of the clupoid fishes. In J. H. S. Blaxter, F. S. Russell, and M. Yonge, editors, *Advances in Marine Biology*, vol. 20, pages 1–223. Academic Press, London, 1982.

[116] K. Blaxter. *Energy Metabolism in Animals and Man.* Cambridge University Press, 1989.

[117] C. I. Bliss. The method of probits. *Science*, **79**:38–39, 1934.

[118] H. Blumel. *Die Rohrammer.* A. Ziemsen-Verlag, Wittenberg Lutherstadt, 1982.

[119] M. P. Boer. The dynamics of tri-trophic food chains. PhD thesis, Vrije Universiteit, Amsterdam, 2000.

[120] M. P. Boer, B. W. Kooi, and S. A. L. M. Kooijman. Food chain dynamics in the chemostat. *Math. Biosci.*, **150**:43–62, 1998.

[121] M. P. Boer, B. W. Kooi, and S. A. L. M. Kooijman. Homoclinic and heteroclinic orbits in a tri-trophic food chain. *J. Math. Biol.*, **39**:19–38, 1999.

[122] M. P. Boer, B. W. Kooi, and S. A. L. M. Kooijman. Multiple attractors and boundary crises in a tri-trophic food chain. *Math. Biosci.*, **169**:109–128, 2001.

[123] P. D. Boersma, N. T. Wheelwright, M. K. Nerini, and E. S. Wheelwright. The breeding biology of the fork-tailed storm-petrel (*Oceanodroma furcata*). *Auk*, **97**:268–282, 1980.

[124] S. Bohlken and J. Joosse. The effect of photoperiod on female reproductive activity and growth of the fresh water pulmonate snail *Lymnaea stagnalis* kept under laboratory conditions. *Int. J. Invertebr. Reprod.*, **4**:213–222, 1982.

[125] C. Bohr and K. A. Hasselbalch. Über die Kohlensäureproduction des Hühnerembryos. *Skand. Arch. Physiol.*, **10**:149–173, 1900.

[126] J. T. Bonner. *Size and Cycle: An Essay on the Structure of Biology.* Princeton University Press, 1965.

[127] T. A. Bookhout. Prenatal development of snowshoe hares. *J. Wildl. Manage.*, **28**:338–345, 1964.

[128] T. Borchardt. Influence of food quantity on the kinetics of cadmium uptake and loss via food and seawater in *Mytilus edulis. Mar. Biol.*, **76**:67–76, 1983.

[129] T. Borchardt. Relationships between carbon and cadmium uptake in *Mytilus edulis. Mar. Biol.*, **85**:233–244, 1985.

[130] F. van den Bosch. The velocity of spatial population expansion. PhD thesis, Rijksuniversiteit Leiden, 1990.

[131] F. van den Bosch, J. A. J. Metz, and O. Diekmann. The velocity of spatial population expansion. *J. Math. Biol.*, **28**:529–565, 1990.

[132] Y. Bourles, M. Alunno-Bruscia, S. Pouvreau, G. Tollu, D. Leguay, C. Arnaud, P. Goulletquer, and S. A. L. M. Kooijman. Modelling growth and reproduction of the pacific oyster *Crassostrea gigas*: application of the oyster-DEB model in a coastal pond. *J. Sea Res.*, **62**:2009.

[133] A. B. Bowers. Breeding and growth of whiting (*Gadus merlangus* L.) in Isle of Man waters. *J. Mar. Biol. Assoc. U.K.*, **33**:97–122, 1954.

[134] A. J. Boyce, M. L. Coleman, and M. J. Russell. Formation of fossil hydrothermal chimneys and mounds from silvermines, Ireland. *Nature*, **306**:545–550, 1983.

[135] C. R. Boyden. Trace element content and body size in molluscs. *Nature*, **251**:311–314, 1974.

[136] A. E. Brafield and M. J. Llewellyn. *Animal Energetics.* Blackie, Glasgow, 1982.

[137] B. W. Brandt. Realistic characterizations of biodegradation. PhD thesis, Vrije Universiteit, Amsterdam, 2002.

[138] B. W. Brandt, F. D. L. Kelpin, I. M. M. van Leeuwen, and S. A. L. M. Kooijman. Modelling microbial adaptation to changing availability of substrates. *Water Res.*, **38**:1003–1013, 2004.

[139] B. W. Brandt and S. A. L. M. Kooijman. Two parameters account for the flocculated growth of microbes in biodegradation assays. *Biotechnol. Bioeng.*, **70**:677–684, 2000.

[140] B. W. Brandt, I. M. M. van Leeuwen, and S. A. L. M. Kooijman. A general model for multiple substrate biodegradation. Application to co-metabolism of non structurally analogous compounds. *Water Res.*, **37**:4843–4854, 2003.

[141] A. M. Branta. *Studies on the Physiology, Genetics and Evolution of some* Cladocera. Technical report, Carnegie Institute, Washington, 1939.

[142] G. B. Brasseur, J. J. Orlando, and G. S. Tyndall. *Atmospheric Chemistry and Global Change.* Oxford University Press, 1999.

[143] D. Brawand, W. Wahli, and H. Kaessmann. Loss of egg yolk genes in mammals and the origin of lactation and placentation. *PLoS Biol.*, **6**:0507–0517, 2008.

[144] P. R. Bregazzi and C. R. Kennedy. The biology of pike, *Esox lucius* L., in southern eutrophic lakes. *J. Fish Biol.*, **17**:91–112, 1980.

[145] W. W. Brehm. Breeding the green-billed toucan at the Walsrode Bird park. *Int. Zoo Yearb.*, **9**:134–135, 1969.

[146] H. Bremer and P. P. Dennis. Modulation of chemical composition and other parameters of the cell by growth rate. In F. C. Neidhardt, editor, Escherichia coli *and* Salmonella typhimurium, pages 1527–1542. Am. Soc. Microbiol., Washington, 1987.

[147] H. Brendelberger and W. Geller. Variability of filter structures in eight *Daphnia* species: mesh sizes and filtering areas. *J. Plankton Res.*, **7**:473–486, 1985.

[148] J. R. Brett. The relation of size to rate of oxygen consumption and sustained swimming speed of sockeye salmon (*Oncorhynchus nerka*). *J. Fish. Res. Board Can.*, **22**:1491–1497, 1965.

[149] F. H. van den Brink. *Zoogdierengids.* Elsevier, Amsterdam, 1955.

[150] I. L. Brisbin Jr and L. J. Tally. Age-specific changes in the major body components and caloric value of growing Japanese quail. *Auk*, **90**:624–635, 1973.

[151] J. J. Brocks, G. A. Logan, R. Buick, and R. E. Summons. Archean molecular fossils and the early rise of eukaryotes. *Science*, **285**:1033–1036, 1999.

[152] S. Broekhuizen. Hazen in Nederland. PhD thesis, Landbouw Hogeschool, Wageningen, 1982.

[153] L. T. Brooke, D. J. Call, and L. T. Brooke. *Acute Toxicities of Organic Chemicals to Fathead Minnow (*Pimephlus promelas*)*, vol. 4 of *Center for Lake Superior Environmental Studies.* University of Wisconsin, Madison, 1988.

[154] L. T. Brooke, D. J. Call, and L. T. Brooke. *Acute Toxicities of Organic Chemicals to Fathead Minnow (*Pimephlus promelas*)*, vol. 5 of *Center for Lake Superior Environmental Studies.* University of Wisconsin, Madison, 1990.

[155] L. T. Brooke, D. L. Geiger, D. J. Call, and C. E. Northcott. *Acute Toxicities of Organic Chemicals to Fathead Minnow (*Pimephlus promelas*)*, vol. 1 of *Center for Lake Superior Environmental Studies.* University of Wisconsin, Madison, 1984.

[156] L. T. Brooke, S. H. Poirrier, and D. J. Call. *Acute Toxicities of Organic Chemicals to Fathead Minnow (*Pimephlus promelas*)*, vol. 3 of *Center for Lake Superior Environmental Studies.* University of Wisconsin, Madison, 1986.

[157] M. Brooke. *The Manx Shearwater.* T. & A. D. Poyser, London, 1990.

[158] J. L. Brooks and S. I. Dodson. Predation, body size, and composition of plankton. *Science*, **150**:28–35, 1965.

[159] S. P. J. Brooks, A. de Zwaan, G. van den Thillart, O. Cattani, and A. Corti. Differential survival of *Venus gallina* and *Scapharca inaequivalvis* during anoxic stress: covalent modification of phosphofructokinase and glycogen phosphorylase during anoxia. *J. Comp. Physiol.*, **161**:207–212, 1991.

[160] D. J. Brousseau. Analysis of growth rate in *Mya arenaria* using the von Bertalanffy equation. *Mar. Biol.*, **51**:221–227, 1979.

[161] E. Brouwer. On simple formulae for calculating the heat expenditure and the quantities of carbohydrate and fat metabolized in ruminants from data on gaseous exchange and urine N. In *1st Symposium on Energy Metabolism*, pages 182–194. European Association for Animal Production, Rome, 1958.

[162] R. Brouwer and C. T. de Wit. A simulation model of plant growth with special attention to root growth and its consequences. In W. J. Whittington, editor, *Root Growth*, pages 224–279. Butterworth, London, 1969.

[163] G. H. Brown and J. J. Wolken. *Liquid Crystals and Biological Structures.* Academic Press, New York, 1979.

[164] W. Y. Brown. Growth and fledging age of sooty tern chicks. *Auk*, **93**:179–183, 1976.

[165] W. A. Bruggeman, L. B. J. M. Martron, D. Kooiman, and O. Hutzinger. Accumulation and elimination kinetics of di-, tri- and tetra chlorobiphenyls by goldfish after dietary and aqueous exposure. *Chemosphere*, **10**:811–832, 1981.

[166] L. W. Bruinzeel and T. Piersma. Cost reduction in the cold: heat generated by terrestrial locomotion partly substitutes for thermoregulation costs in knot *Calidris canutus. Ibis*, **140**:323–328, 1998.

[167] B. Bruun. *Gids voor de vogels van Europa.* Elsevier, Amsterdam, 1970.

[168] C. Bryant. *Metazoan Life without Oxygen.* Chapman & Hall, London, 1991.

[169] M. M. Bryden. Growth of the south elephant seal *Mirounga leonina* (Linn.). *Growth*, **33**:69–82, 1969.

[170] T. L. Bucher. Parrot eggs, embryos, and nestlings: patterns and energetics of growth and development. *Physiol. Zool.*, **56**:465–483, 1983.

[171] L. J. Buckley. RNA–DNA ratio: an index of larval fish growth in the sea. *Mar. Biol.*, **80**:291–298, 1984.

[172] J. J. Bull. Sex determination mechanisms: an evolutionary perspective. In S. C. Stearns, editor, *The Evolution of Sex and its Consequences*, pages 93–114. Birkhäuser, Basel, 1987.

[173] P. C. Bull and A. H. Whitaker. The amphibians, reptiles, birds and mammals. In G. Kuschel, editor, *Biogeography and Ecology in New Zealand*, pages 231–276. Dr. W. Junk, Dordrecht, 1975.

[174] B. A. Bulthuis. Stoichiometry of growth and product-formation by *Bacillus licheniformis.* PhD thesis, Vrije Universiteit, Amsterdam, 1990.

[175] C. W. Burns. The relationship between body size of filterfeeding cladocera and the maximum size of particles ingested. *Limnol. Oceanogr.*, **13**:675–678, 1968.

[176] S. S. Butcher, R. J. Charlson, G. H. Orians, and G. V. Wolfe. *Global Biochemical Cycles.* Academic Press, London, 1992.

[177] D. K. Button. Biochemical basis for whole-cell uptake kinetics: specific affinity, oligotrophic capacity, and the meaning of Michaelis constant. *Appl. Environ. Microbiol.*, **57**:2033–2038, 1991.

[178] M. Byrne. Viviparity and intragonadal cannibalism in the diminutive asterinid sea stars *Patiriella vivipara* and *P. parvivipara. Mar. Biol.*, **125**:551–567, 1996.

[179] P. L. Cadwallader and A. K. Eden. Food and growth of hatchery-produced chinook-salmon, *Oncorhynchus tschawytscha* (Walbaum), in landlocked lake Purrumbete, Victoria, Australia. *J. Fish Biol.*, **18**:321–330, 1981.

[180] A. G. Cairns-Smith, A. J. Hall, and M. J. Russell. Mineral theories of the origin of life and an iron sulphide example. *Origins Life Evol. Biosphere*, **22**:161–180, 1992.

[181] W. A. Calder III. The kiwi and egg design: evolution as a package deal. *BioScience*, **29**:461–467, 1979.

[182] W. A. Calder III. *Size, Function and Life History.* Harvard University Press, Cambridge, 1984.

[183] P. Calow and C. R. Townsend. Resource utilization in growth. In C. R. Townsend and P. Calow, editors, *Physiological Ecology*, pages 220–244. Blackwell Sci. Publ., Oxford, 1981.

[184] N. Cameron. *The Measurement of Human Growth*. Croom Helm, London, 1984.

[185] R. N. Campbell. The growth of brown trout, *Salmo trutta* L. in northern Scottish lochs with special reference to the improvement of fisheries. *J. Fish Biol.*, **3**:1–28, 1971.

[186] D. E. Canfield, K. S. Habicht, and B. Thamdrup. The archean sulfur cycle and the early histroy of atmospheric oxygen. *Science*, **288**:658–661, 2000.

[187] H. Canter-Lund and J. W. G. Lund. *Freshwater Algae: Their Microscopic World Explored*. Biopress, Bristol, 1995.

[188] J. F. M. F. Cardoso. Growth and reproduction in bivalves. PhD thesis, Groningen University, 2007.

[189] J. F. M. F. Cardoso, H. W. van der Veer, and S. A. L. M. Kooijman. Body size scaling relationships in bivalves: a comparison of field data with predictions by dynamic energy budgets (DEB theory). *J. Sea Res.*, **56**:125–139, 2006.

[190] M. Cashel and K. E. Rudd. The stringent response. In F. C. Neidhardt, editor, Escherichia coli *and* Salmonella typhimurium, vol. 2, pages 1410–1438. Am. Soc. Microbiol., Washington, 1987.

[191] T. Cavalier-Smith. The origin of cells, a symbiosis between genes, catalysts and membranes. *Cold Spring Harbor Symp. Quant. Biol.*, **52**:805–824, 1987.

[192] T. Cavalier-Smith. A revised six-kingdom system of life. *Biol. Rev. Camb. Philos. Soc.*, **73**:203–266, 1998.

[193] T. Cavalier-Smith. Membrane heredity and early chloroplast evolution. *Trends Plant Sci.*, **5**:174–182, 2000.

[194] T. A. Cavalier-Smith. Chloroplast evolution: secondary symbiogenesis and multiple losses. *Current Biol.*, **12**:R62–R64, 2002.

[195] T. A. Cavalier-Smith. The phagotrophic origin of eukaryotes and phylogenetic classification of protozoa. *Int. J. Syst. Evol. Microbiol.*, **52**:297–354, 2002.

[196] L. Cazzador. Analysis of oscillations in yeast continuous cultures by a new simplified model. *Bull. Math. Biol.*, **53**:685–700, 1991.

[197] F. Chai, S. T. Lindley, J. R. Toggweiler, and R. R. Barber. Testing the importance of iron and grazing in the maintenance of the high nitrate condition in the equatorial pacific ocean: a physical-biological model study. In R. B. Hanson, H. W. Ducklow, and J. G. Field, editors, *The Changing Ocean Carbon Cycle*, pages 155–186. Cambridge University Press, 2000.

[198] G. Chapelle and L. Peck. Polar gigantism dictated by oxygen availability. *Nature*, **399**:114–115, 1999.

[199] D. J. Chapman and J. W. Schopf. Biological and chemical effects of the development of an aerobic environment. In J. W. Schopf, editor, *Earth's Earliest Biosphere: Its Origin and Evolution*, pages 302–320. Princeton University Press, 1983.

[200] S. Charlat and H. Mercot. *Wolbachia*, mitochondria and sterility. *Trends Ecol. Evol.*, **16**:431–432, 2001.

[201] R. J. Charlson. The vanishing climatic role of dimethyl sulfide. In G. M. Woodwell and F. T. Mackenzie, editors, *Biotic Feedbacks in the Global Climatic System*, pages 251–277. Oxford University Press, 1995.

[202] E. L. Charnov and D. Berrigan. Dimensionless numbers and the assembly rules for life histories. *Phil. Trans. R. Soc. Lond. B Biol. Sci.*, **332**:41–48, 1991.

[203] F. F. Chehab, K. Mounzih, R. Lu, and M. E. Lim. Early onset of reproductive function in normal female mice treated with leptin. *Science*, **275**:88–90, 1997.

[204] J. Chela-Flores. First step in eukaryogenesis: physical phenomena in the origin and evolution of chromosome structure. *Origins Life Evol. Biosphere*, **28**:215–225, 1998.

[205] L. Chistoserdova, C. Jenkins, M. G. Kalyuzhnaya, C. J. Marx, A. Lapidus, J. A. Vorholt, J. T. Stanley, and M. E. Lidstrom. The enigmatic Planctomycetes may hold a key to the origins of methanogenesis and methylotrophy. *Mol. Biol. Evol.*, **21**:1234–1241, 2004.

[206] D. Chivian, E. L. Brodie, *et al.* Environmental genomics reveals a single-species ecosystem deep within earth. *Science*, **322**:275–278, 2008.

[207] K. Christoffersen and A. Jespersen. Gut evacuation rates and ingestion rates of *Eudiaptomus*

graciloides measured by means of gut fluorescence method. *J. Plankton Res.*, **8**:973–983, 1986.

[208] D. Claessen, A. M. de Roos, and L. Persson. Dwarfs and giants: cannibalism and competition in size-structured populations. *Am. Nat.*, **155**:219–237, 2000.

[209] B. M. Clark, N. F. Mangelson, L. L. St. Clair, L. B. Rees, G. S. Bench, and J. R. Southon. Measurement of age and growth rate in the crustose saxicolous lichen *Caloplaca trachyphylla* using ^{14}C accelerator mass spectrometry. *Lichenologist*, **32**:399–403, 2000.

[210] A. Clarke. Temperature and embryonic development in polar marine invertebrates. *Int. J. Invertebr. Reprod.*, **5**:71–82, 1982.

[211] D. H. Clayton and J. Moore. *Host–Parasite Evolution: General Principles and Avian Models.* Oxford University Press, 1997.

[212] G. A. Clayton, C. Nixey, and G. Monaghan. Meat yield in turkeys. *Br. Poult. Sci.*, **19**:755–763, 1978.

[213] J. H. L. Cloethe. Prenatal growth in the merino sheep. *Onderstepoort J. Vet. Sci. Anim. Ind.*, **13**:417–558, 1939.

[214] J. F. Collins and M. H. Richmond. Rate of growth of *Bacillus cereus* between divisions. *J. Gen. Microbiol.*, **28**:15–33, 1962.

[215] A. Comfort. *The Biology of Senescence.* Churchill Livingstone, Edinburgh, 1979.

[216] A. Comfort. Effect of delayed and resumed growth on the longevity of a fish (*Lebistes reticulatus* Peters) in captivity. *Gerontologia (Basel)*, **8**:150–155, 1983.

[217] J. D. Congdon, D. W. Tinkle, and P. C. Rosen. Egg components and utilization during development in aquatic turtles. *Copeia*, **1983**:264–268, 1983.

[218] D. W. Connell and D. W. Hawker. Use of polynomial expressions to describe the bioconcentration of hydrophobic chemicals by fish. *Ecotoxicol. Environ. Saf.*, **16**:242–257, 1988.

[219] R. J. Conover. Transformation of organic matter. In O. Kinne, editor, *Marine Ecology*, vol. 4, pages 221–499. J. Wiley, Chichester, 1978.

[220] G. D. Constanz. Growth of nestling rufous hummingbirds. *Auk*, **97**:622–624, 1980.

[221] C. L. Cooney, D. I. C. Wang, and R. I. Mateles. Measurement of heat evolution and correlation with oxygen consumption during microbial growth. *Biotechnol. Bioeng.*, **11**:269–281, 1968.

[222] S. Cooper. The constrained hoop: an explanation of the overscoot in cell length during a shift-up of *Escherichia coli*. *J. Bacteriol.*, **171**:5239–5243, 1989.

[223] J. D. Coronios. Development of behaviour in fetal cat. *Genet. Psychol. Monogr.*, **14**:283–330, 1933.

[224] R. R. Coutts and I. W. Rowlands. The reproductive cycle of the Skomer vole (*Clethrionomys glareolus skomerensis*). *J. Zool. Lond.*, **158**:1–25, 1969.

[225] D. R. Cox and D. Oakes. *Analysis of Survival Data.* Chapman & Hall, London, 1984.

[226] R. C. Crabtree and F. A. Bazzaz. Seedling response of four birch species to simulated nitrogen deposition – ammonium versus nitrate. *Ecol. Appl.*, **3**:315–321, 1993.

[227] D. Cragg-Hine and J. W. Jones. The growth of dace *Leuciscus leuciscus* (L.), roach *Rutilus rutilus* (L.) and chub *Sqalius cephalus* (L.) in Willow Brook, Northamptonshire. *J. Fish Biol.*, **1**:59–82, 1969.

[228] J. F. Craig. A note on growth and mortality of trout, *Salmo trutta* L., in afferent streams of Windermere. *J. Fish Biol.*, **20**:423–429, 1982.

[229] G. Creutz. *Der Weiss-storch.* A. Ziemsen-Verlag, Wittenberg Lutherstadt, 1985.

[230] E. G. Crichton. Aspects of reproduction in the genus *Notomys* (*Muridae*). *Austral. J. Zool.*, **22**:439–447, 1974.

[231] R. E. Criss. *Principles of Stable Isotope Distribution.* Oxford Univerity Press, 1999.

[232] J. R. Cronmiller and C. F. Thompson. Experimental manipulation of brood size in red-winged blackbirds. *Auk*, **97**:559–565, 1980.

[233] C. J. Crossland and D. J. Barnes. The role of metabolic nitrogen in coral calcification. *Mar. Biol.*, **28**:325–332, 1974.

[234] T. J. Crowley and G. R. North. *Palaeoclimatology.* Oxford University Press, 1991.

[235] H. Crum. *A Focus on Peatlands and Peat Mosses.* University of Michigan Press, Ann Abor, 1992.

[236] E. H. Curtis, J. J. Beauchamp, and B. G. Blaylock. Application of various mathematical models to data from the uptake of methyl mercury in bluegill sunfish (*Lepomis macrochirus*). *Ecol. Modell.*, **3**:273–284, 1977.

[237] J. M. Cushing. *An Introduction to Structured Population Dynamics.* SIAM, Philadelphia, 1998.

[238] H. Cyr and M. L. Pace. Allometric theory: extrapolations from individuals to communities. *Ecology*, **74**:1234–1245, 1993.

[239] J. B. Dacks and M. C. Field. Evolution of the eukaryotic mambrane-trafficking system: origin, tempo and mode. *J. Cell Sci.*, **120**:2977–2985, 2007.

[240] W. B. Dade, P. A. Jumars, and D. L. Penry. Supply-side optimization: maximizing absorptive rates. In R. N. Hughes, editor, *Behavioural Mechanisms of Food Selection*, pages 531–555. Springer-Verlag, Berlin, 1990.

[241] M. J. Dagg. Complete carbon and nitrogen budgets for the carnivorous amphipod, *Calliopius laeviusculus* (Kroyer). *Int. Rev. Gesamten Hydrobiol.*, **61**:297–357, 1976.

[242] T. Dalsgaard, D. E. Canfield, J. Petersen, B. Thamdrup, and J. Acuna-Gonzalez. N-2 production by the anammox reaction in the anoxic water column of Golfo Dulce, Costa Rica. *Nature*, **422**:606–608, 2003.

[243] J. D. Damuth. Of size and abundance. *Nature*, **351**:268–269, 1991.

[244] J. D. Damuth. Taxon-free characterization of animal communities. In A. K. Behrensmeyer, J. D. Damuth, W. A. DiMichele, R. Potts, H.-D. Sues, and S. L. Wing, editors, *Terrestrial Ecosystems through Time: Evolutionary Paleoecology of Terrestrial Plants and Animals*, pages 183–203. University of Chicago Press, 1992.

[245] N. Dan and S. E. R. Bailey. Growth, mortality, and feeding rates of the snail *Helix aspersa* at different population densities in the laboratory, and the depression of activity of helicid snails by other individuals, or their mucus. *J. Molluscan Stud.*, **48**:257–265, 1982.

[246] M. C. Dash and A. K. Hota. Density effects on the survival, growth rate, and metamorphosis of *Rana tigrina* tadpoles. *Ecology*, **61**:1025–1028, 1980.

[247] H. A. David and M. L. Moeschberger. *The Theory of Competing Risks*, vol. 39 of *Griffin's Statistical Monographs and Courses.* C. Griffin, London, 1978.

[248] P. J. Davoll and M. W. Silver. Marine snow aggregates: life history sequence and microbial community of abandoned larvacean houses from Monterey Bay, California. *Mar. Ecol. Prog. Ser.*, **33**:111–120, 1986.

[249] E. A. Dawes. The constrained hoop: an explanation of the overschoot in cell length during a shift-up of *Escherichia coli.* In T. R. Gray and J. R. Postgate, editors, *The Survival of Vegetative Microbes*, pages 19–53. Cambridge University Press, 1976.

[250] W. R. Dawson. Evaporative losses of water by birds. *Biochem. Physiol.*, **71**:495–509, 1982.

[251] D. W. Deamer and R. M. Pashley. Amphiphilic components of the murchison carbonaceous chondrite: surface properties and membrane formation. *Orig. Life Evol. Biosphere*, **19**:21–38, 1989.

[252] J. Dean. *Lange's Handbook of Chemistry*, 12th edn. McGraw-Hill, New York, 1979.

[253] D. L. DeAngelis, R. A. Goldstein, and R. V. O'Neill. A model for trophic interaction. *Ecology*, **56**:881–892, 1975.

[254] D. L. DeAngelis and L. J. Gross, editors. *Individual-Based Models and Approaches in Ecology.* Chapman & Hall, London, 1992.

[255] D. C. Deeming and M. W. J. Ferguson. Environmental regulation of sex determination in reptiles. *Phil. Trans. R. Soc. Lond. B Biol. Sci.*, **322**:19–39, 1988.

[256] D. C. Deeming and M. W. J. Ferguson. Effects of incubation temperature on growth and development of embryos of *Alligator mississippiensis. J. Comp. Physiol. B*, **159**:183–193, 1989.

[257] T. E. DeLaca, D. M. Karl, and J. H. Lipps. Direct use of dissolved organic carbon by agglutinated benthic foraminifera. *Nature*, **289**:287–289, 1981.

[258] T. H. DeLuca, O. Zackrisson, M.-C. Nilsson, and A. Sellstedt. Quantifying nitrogen-fixation in feather moss carpets of boreal forests. *Nature*, **429**:917–920, 2002.

[259] C. F. Delwiche. Tracing the thread of plastid diversity through the tapestry of life. *Am. Nat.*, **154**:S164–S177, 1999.

[260] C.F. Delwiche, R.A. Andersen, D. Bhattacharya, B.D. Mishler, and R.M. McCourt. Algal evolution and the early radiation of green plants. In J. Cracraft and M.J. Donoghue, editors, *Assembling the Tree of Life*, pages 121–167. Oxford University Press, 2004.

[261] M.W. Demment and P.J. van Soest. A nutritional explanation for body size patterns of ruminant and nonruminant herbivores. *Am. Nat.*, **125**:641–672, 1985.

[262] W.R. Demott. Feeding selectivities and relative ingestion rates of *Daphnia* and *Bosmina*. *Limnol. Oceanogr.*, **27**:518–527, 1982.

[263] C. Dennis. Close encounters of the jelly kind. *Nature*, **426**:12–14, 2003.

[264] V.E. Dent, M.J. Bazin, and P.T. Saunders. Behaviour of *Dictyostelium discoideum* amoebae and *Escherichia coli* grown together in chemostat culture. *Arch. Microbiol.*, **109**:187–194, 1976.

[265] E. Derelle, C. Ferraz, P. Lagoda, S. Eycheni, *et al.* DNA libraries for sequencing the genome of *Ostreococcus tauri* (*Chlorophyta, Prasinophyceae*): the smallest free-living eukaryotic cell. *J. Phycol.*, **38**:1150–1156, 2002.

[266] A.W. Diamond. The red-footed booby on Aldabra atoll, Indian Ocean. *Ardea*, **62**:196–218, 1974.

[267] J.M. Diamond and R.K. Buddington. Intestinal nutrient absorption in herbivores and carnivores. In P. Dejours, L. Bolis, C.R. Taylor, and E.R. Weibel, editors, *Comparative Physiology: Life in Water and on Land*, pages 193–203. Liviana Press, Padua, 1987.

[268] U. Dieckmann and R. Law. The mathematical theory of coevolution: a derivation from stochastic processes. *J. Math. Biol.*, **34**:579–612, 1996.

[269] C. Dijkstra. Reproductive tactics in the kestrel *Falco tinnunculus*: a study in evolutionary biology. PhD thesis, University of Groningen, 1988.

[270] R.F. Dill, E.A. Shinn, A.T. Jones, K. Kelly, and R.P. Steinen. Giant subtidal stromatolites forming in normal salinity waters. *Nature*, **324**:55–58, 1986.

[271] J.G. Dillon and R.W. Castenholz. Scytonemin, a cyanobacterial sheath pigment, protects against UVC radiation: implications for early photosynthetic life. *J. Phycol.*, **35**:673–681, 1999.

[272] F.A. Dipper, C.R. Bridges, and A. Menz. Age, growth and feeding in the ballan wrasse *Labrus bergylta* Ascanius 1767. *J. Fish Biol.*, **11**:105–120, 1977.

[273] G.C. Dismukes, V.V. Klimov, S.V. Baranov, Dasguptam J. Kozlov, Yu. N., and A. Tyryshkin. The origin of atmospheric oxygen on earth: the innovation of oxygenic photosynthesis. *Proc. Natl. Acad. Sci. U.S.A.*, **98**:2170–2175, 2001.

[274] H. Dittberner and W. Dittberner. *Die Schafstelze*. A. Ziemsen-Verlag, Wittenberg Lutherstadt, 1984.

[275] P.S. Dixon. *Biology of the Rhodophyta*. Oliver & Boyd, Edinburgh, 1973.

[276] R. Docampo, W. de Souza, K. Miranda, P. Rohloff, and S.N.J. Moreno. Acidocalcisomes: conserved from bacteria to man. *Nature Rev. Microbiol.*, **3**:251–261, 2005.

[277] M. Doebeli. A model for the evolutionary dynamics of cross-feeding polymorphisms in microorganisms. *Pop. Ecol.*, **44**:59–70, 2002.

[278] C.D. von Dohlen, S. Kohler, S.T. Alsop, and W.R. McManus. Mealybug β-proteobacterial endosymbionts contain γ-proteobacterial symbionts. *Nature*, **412**:433–436, 2001.

[279] D. Dolmen. Growth and size of *Triturus vulgaris* and *T. cristatus* (*Amphibia*) in different parts of Norway. *Holarct. Ecol.*, **6**:356–371, 1983.

[280] S.L. Domotor and C.F. D'Elia. Nutrient uptake kinetics and growth of zooxanthellae. *Mar. Biol.*, **80**:93–101, 1984.

[281] W.D. Donachie. Relationship between cell size and time of initiation of DNA replication. *Nature*, **219**:1077–1079, 1968.

[282] W.D. Donachie, K.G. Begg, and M. Vicente. Cell length, cell growth and cell division. *Nature*, **264**:328–333, 1976.

[283] W.F. Doolittle. Phylogenetic classification and the universal tree. *Science*, **284**:2124–2128, 1999.

[284] Q. Dortch. The interaction between ammonium and nitrate uptake in phytoplankton. *Mar. Ecol. Prog. Ser.*, **61**:183–201, 1990.

[285] P.G. Doucet and N.M. van Straalen. Analysis of hunger from feeding rate observations. *Anim. Behav.*, **28**:913–921, 1980.

[286] A. Douglas, P. J. McAuley, and P. S. Davies. Algal symbiosis in cnidaria. *J. Zool. Lond.*, **231**:175–178, 1993.

[287] A. E. Douglas. *Symbiotic Interactions.* Oxford University Press, 1994.

[288] S. E. Douglas. Plastid evolution: origin, diversity, trends. *Curr. Top. Gen. Dev.*, **8**:655–661, 1998.

[289] C. L. van Dover. *The Ecology of Deep-Sea Hydrothermal Vents.* Princeton University Press, 2000.

[290] W. L. Downing and J. A. Downing. Molluscan shell growth and loss. *Nature*, **362**:506, 1993.

[291] R. L. Draper. The prenatal growth of the guinea-pig. *Anat. Rec.*, **18**:369–392, 1920.

[292] P. G. Drazin. *Nonlinear Systems.* Cambridge University Press, 1992.

[293] R. H. Drent. Functional aspects of incubation in the herring gull. *Behaviour, Suppl.*, **17**:1–132, 1970.

[294] M. R. Droop. Some thoughts on nutrient limitation in algae. *J. Phycol.*, **9**:264–272, 1973.

[295] M. R. Droop. The nutrient status of algal cells in continuous culture. *J. Mar. Biol. Assoc. U.K.*, **54**:825–855, 1974.

[296] M. R. Droop. 25 years of algal growth kinetics. *Bot. Mar.*, **26**:99–112, 1983.

[297] G. L. Dryden. Growth and development of *Suncus murinus* in captivity on Guam. *J. Mammal.*, **49**:51–62, 1968.

[298] N. Dubilier, C. Mulders, T. Felderman, D. de Beer, A. Pernthaler, M. Klein, M. Wagner, C. Erséus, F. Thiermann, J. Krieger, O. Giere, and R. Rudolf Amann. Endosymbiotic sulphate-reducing and sulphide-oxidizing bacteria in an oligochaete worm. *Nature*, **411**:298–302, 2001.

[299] W. E. Duellman. Reproductive strategies of frogs. *Sci. Am.*, **7**:58–65, 1992.

[300] W. E. Duellman and L. Trueb. *Biology of Amphibians.* McGraw-Hill, New York, 1986.

[301] G. C. A. Duineveld and M. I. Jenness. Difference in growth rates of the sea urchin *Echiocardium cordatum* as estimated by the parameter ω of the von Bertalanffy equation to skeletal rings. *Mar. Ecol.*, **19**:65–72, 1984.

[302] E. H. Dunn. Growth, body components and energy content of nestling double-crested cormorants. *Condor*, **77**:431–438, 1975.

[303] C. de Duve. *A Guided Tour of the Living Cell.* Scientific American Library, New York, 1984.

[304] M. Dworkin. *Developmental Biology of the Bacteria.* Benjamin-Cummings, Menlo Park, 1985.

[305] E. M. Dzialowski and P. R. Sotherland. Maternal effects of egg size on emu *Dromaius novaehollandiae* egg composition and hatchling phenotype. *J. Exp. Biol.*, **207**:597–606, 2004.

[306] O. N. Eaton. Weight and length measurements of fetuses of karakul sheep and of goats. *Growth*, **16**:175–187, 1952.

[307] Muller E. B. and Nisbet R. M. Survival and production in variable resource environments. *Bull. Math. Biol.*, **62**:1163–1189, 2000.

[308] B. Ebenman and L. Persson. *Size-Structured Populations: Ecology and Evolution.* Springer-Verlag, Berlin, 1988.

[309] K. J. Eckelbarger, P. A. Linley, and J. P. Grassle. Role of ovarian follicle cells in vitellogenesis and oocyte resorption in *Capitella* sp. I (Polychaeta). *Mar. Biol.*, **79**:133–144, 1984.

[310] R. Eckert, D. Randall, and G. Augustine. *Animal Physiology.* W. H. Freeman, New York, 1988.

[311] L. Edelstein-Keshet. *Mathematical Models in Biology.* Random House, New York, 1988.

[312] D. A. Egloff and D. S. Palmer. Size relations of the filtering area of two *Daphnia* species. *Limnol. Oceanogr.*, **16**:900–905, 1971.

[313] E. Eichinger. Bacterial degradation of dissolved organic carbon in the water column: an experimental and modelling approach. PhD thesis, Université de Mediterranée & Vrije Universiteit, Marseille and Amsterdam, 2008.

[314] S. Einum and I. A. Fleming. Maternal effects of egg size in brown trout (*Salmo trutta*): norms of reaction to environmental quality. *Proc. R. Soc. Lond. B Biol. Sci.*, **266**:2095–2097, 1999.

[315] J. F. Eisenberg. *The Mammalian Radiations: An Analysis of Trends in Evolution, Adaptation, and Behaviour.* University of Chicago Press, 1981.

[316] R. C. Elandt-Johnson and N. L. Johnson. *Survival Models and Data Analysis.* J. Wiley, Chichester, 1980.

[317] R. H. E. Elliott, F. G. Hall, and A. St. G. Huggett. The blood volume and oxygen capacity of the foetal blood in the goat. *J. Physiol. (Lond.)*, **82**:160–171, 1934.

[318] G. W. Elmes, J. A. Thomas, and J. C. Wardlaw. Larvae of *Maculinea rebeli*, a large-blue butterfly, and their *Myrmica* host ants: wild adoption and behaviour in ant-nests. *J. Zool. Lond.*, **223**:447–460, 1991.

[319] T. M. Embley and R. P. Hirt. Early branching eukaryotes? *Curr. Opin. Gen. Dev.*, **8**:624–629, 1998.

[320] S. Emerson. The growth phase in *Neurospora* corresponding to the logarithmic phase in unicellular organisms. *J. Bacteriol.*, **60**:221–223, 1950.

[321] L. H. Emmons and F. Feer. *Neotropical Rainforest Mammals.* University of Chicago Press, 1990.

[322] J. Emsley. Potent painkiller from poisonous frog. *New Sci.*, **30**(May):14, 1992.

[323] R. H. Emson and P. V. Mladenov. Studies of the fissiparous holothurian *Holothuria parvula* (Selenka) (Echinodermata: Holothuroidea). *J. Exp. Mar. Biol. Ecol.*, **111**:195–211, 1987.

[324] H. Engler. *Die Teichralle.* A. Ziemsen-Verlag, Wittenberg Lutherstadt, 1983.

[325] F. van den Ent, L. A. Amos, and J. Löwe. Prokaryotic origin of the actin cytoskeleton. *Nature*, **413**:39–44, 2001.

[326] G. Ernsting and J. A. Isaaks. Accelerated ageing: a cost of reproduction in the carabid beetle *Notiophilus biguttatus* F. *Funct. Ecol.*, **5**:229–303, 1991.

[327] G. Ernsting and J. A. Isaaks. Effects of temperature and season on egg size, hatchling size and adult size in *Notiophilus biguttatus. Ecol. Entomol.*, **22**:32–40, 1997.

[328] A. A. Esener, J. A. Roels, and N. W. F. Kossen. Dependence of the elemental composition of *K. pneumoniae* on the steady-state specific growth rate. *Biotechnol. Bioeng.*, **24**:1445–1449, 1982.

[329] A. A. Esener, J. A. Roels, and N. W. F. Kossen. Theory and applications of unstructured growth models: kinetic and energetic aspects. *Biotechnol. Bioeng.*, **25**:2803–2841, 1983.

[330] E. Evers and S. A. L. M. Kooijman. Feeding and oxygen consumption in *Daphnia magna*: a study in energy budgets. *Neth. J. Zool.*, **39**:56–78, 1989.

[331] N. Fairall. Prenatal development of the impala *Aepyceros melampus. Koedoe*, **12**:97–103, 1969.

[332] U. Falkengren-Grerup and H. Lakkenborg-Kristensen. Importance of ammonium and nitrate to the performance of herb-layer species from deciduous forests in southern Sweden. *Environ. Exp. Bot.*, **34**:31–38, 1994.

[333] P. G. Falkowski and J. A. Raven. *Aquatic Photosynthesis.* Blackwell Sci. Publ., Oxford, 1997.

[334] A. Farjon. *A Natural History of Conifers.* Timber Press, Portland, 2008.

[335] J. O. Farlow. A consideration of the trophic dynamics of a late Cretaceous large-dinosaur community (Oldman Formation). *Ecology*, **57**:841–857, 1976.

[336] G. D. Farquhar. Models describing the kinetics of ribulose riphosphate carboxylase-oxygenase. *Arch. Biochem. Biophys.*, **193**:456–468, 1979.

[337] G. D. Farquhar, S. von Caemmerer, and J. A. Berry. A biochemical model of photosynthetic CO_2 assimilation in leaves of C_3 species. *Planta*, **149**:78–90, 1980.

[338] H. S. Fawcett. The temperature relations of growth in certain parasitic fungi. *Univ. Calif. Berkeley Publ. Agr. Sci.*, **4**:183–232, 1921.

[339] M. A. Fedak and H. J. Seeherman. Reappraisal of energetics of locomotion shows identical cost in bipeds and quadrupeds including ostrich and horse. *Nature*, **282**:713–716, 1979.

[340] R. Fenaux. Rhythm of secretion of oikopleurid's houses. *Bull. Mar. Sci.*, **37**:498–503, 1985.

[341] R. Fenaux and G. Gorsky. Cycle vital et croissance de l'appendiculaire *Oikopleura longicauda. Ann. Inst. Oceanogr.*, **59**:107–116, 1983.

[342] T. Fenchel. Intrinsic rate of natural increase: the relationship with body size. *Oecologia (Berlin)*, **14**:317–326, 1974.

[343] T. Fenchel. Eukaryotic life: anaerobic physiology. In D. McL. Roberts, P. Sharp, G. Alderson, and M. Collins, editors, *Evolution of Microbial Life*, pages 185–203. Cambridge University Press, 1996.

[344] T. Fenchel. *Origin and Early Evolution of Life.* Oxford University Press, 2002.

[345] T. Fenchel and B. L. Finlays. *Ecology and Evolution in Anoxic Worlds.* Oxford University Press, 1995.

[346] M. W. J. Ferguson. Palatal shelf elevation in the Wistar rat fetus. *J. Anat.*, **125**:555–577, 1978.

[347] S. Fernandez-Baca, W. Hansel, and C. Novoa. Embryonic mortality in the alpaca. *Biol. Reprod.*, **3**:243–251, 1970.

[348] J. G. Ferry. *Methanogenesis: Ecology, Physiology, Biochemistry and Genetics.* Chapman & Hall, New York, 1993.

[349] L. Fieseler, M. Horn, M. Wagner, and U. Hentschel. Discovery of the novel candidate phylum "*Poribacteria*" in marine sponges. *Appl. Environ. Microbiol.*, **70**:3724–3732, 2004.

[350] C. E. Finch. *Longevity, Senescence, and the Genome.* University of Chicago Press, 1990.

[351] M. S. Finkler, J. B. van Orman, and P. R. Sotherland. Experimental manipulation of egg quality in chickens: influence of albumen and yolk on the size and body composition of near-term embryos in a precocial bird. *J. Comp. Physiol. B*, **168**:17–24, 1998.

[352] B. J. Finlay and T. Fenchel. An anaerobic ciliate as a natural chemostat for the growth of endosymbiotic methanogens. *Eur. J. Protistol.*, **28**:127–137, 1992.

[353] R. A. Fisher and L. H. C. Tippitt. Limiting forms of the frequency distribution of the largest or the smallest memeber of a sample. *Proc. Camb. Phil. Soc.*, **24**:180–190, 1928.

[354] V. Fitsch. Laborversuche und Simulationen zur kausalen Analyse der Populationsdynamik von *Daphnia magna*. PhD thesis, Rheinisch–Westfälischen Technischen Hochschule Aachen, 1990.

[355] L. B. Flexner and H. A. Pohl. The transfer of radioactive sodium across the placenta of the white rat. *J. Cell. Comp. Physiol.*, **18**:49–59, 1941.

[356] P. R. Flood. Architecture of, and water circulation and flow rate in, the house of the planktonic tunicate *Oikopleura labradoriensis*. *Mar. Biol.*, **111**:95–111, 1991.

[357] G. L. Flynn and S. H. Yalkowsky. Correlation and prediction of mass transport across membranes. 1. Influence of alkyl chain length on flux-determining properties of barrier and diffusant. *J. Pharm. Sci.*, **61**:838–852, 1972.

[358] K. J. Flynn, M. J. R. Fasham, and C. R. Hipkin. Modelling the interactions between ammonium and nitrate in marine phytoplankton. *Phil. Trans. R. Soc. Lond. B Biol. Sci.*, **352**:1625–1645, 1997.

[359] H. von Foerster. Some remarks on changing populations. In F. Stohlman, editor, *The Kinetics of Cellular Proliferation*, pages 382–407. Grune & Stratton, New York, 1959.

[360] J. Forsberg and C. Wiklund. Protandry in the green-veined white butterfly, *Pieris napi* L. (*Lepidoptera; Pieridae*). *Funct. Ecol.*, **2**:81–88, 1988.

[361] P. Forterre. Genomics and the early cellular evolution. the origin of the DNA world. *C. R. Acad. Sci.*, **324**:1067–1076, 2001.

[362] P. Forterre. The origin of DNA genomes and DNA replication proteins. *Curr. Opin. Microbiol.*, **5**:525–532, 2002.

[363] P. Forterre. The origin of viruses and their possible roles in major evolutionary transitions. *Virus Res.*, **117**:5–16, 2006.

[364] P. Forterre, A. Bergerat, and P. Lopez-Garvia. The unique DNA topology and DNA topoisomerase of hyperthermophilic archaea. *FEMS Microbiol. Rev.*, **18**:237–248, 1996.

[365] B. L. Foster Smith. The effect of concentration of suspension and inert material on the assimilation of algae by three bivalves. *J. Mar. Biol. Assoc. U.K.*, **55**:411–418, 1975.

[366] B. L. Foster Smith. The effect of concentration of suspension on the filtration rates and pseudofaecal production for *Mytilus edulis* L., *Cerastoderma edule* L. and *Venerupsis pullastra* (Montagu). *J. Exp. Mar. Biol. Ecol.*, **17**:1–22, 1975.

[367] P. da Franca. Determinacao da idade em *Gambusia holbrookii* (Girard). *Arq. Mus. Bocage*, **24**:87–93, 1953.

[368] D. R. Franz. Population age structure, growth and longevity of the marine gastropod *Uropalpinx cinerea* Say. *Biol. Bull. (Woods Hole)*, **140**:63–72, 1971.

[369] J. J. R. Fraústo da Silva and R. J. P. Williams. *The Biological Chemistry of the Elements: The Inorganic Chemistry of Life.* Clarendon Press, Oxford, 1993.

[370] A. G. Fredrickson, R. D. Megee III, and H. M. Tsuchiy. Mathematical models for fermentation processes. *Adv. Appl. Microbiol.*, **13**:419–465, 1970.

[371] B. Friedich. Genetics of energy converting systems in aerobic chemolithotrophs. In H. G. Schlegel and E. Bowien, editors,

Autotrophic Bacteria, pages 415–436. Springer-Verlag, Berlin, 1989.

[372] E. Frommhold. *Die Kreuzotter.* A. Ziemsen-Verlag, Wittenberg Lutherstadt, 1969.

[373] B. W. Frost. Effects of size and concentration of food particles on the feeding behaviour of the marine copepod *Calanus pacificus*. *Limnol. Oceanogr.*, **17**:805–815, 1972.

[374] B. W. Frost and N. C. Franzen. Grazing and iron limitation in the control of phytoplankton stock and nutrient concentration: a chemostat analogue of the pacific equatorial upwelling zone. *Mar. Ecol. Prog. Ser.*, **83**:291–303, 1992.

[375] W. E. Frost and C. Kipling. The growth of charr, *Salvinus willughbii* Gunther, in Windermere. *J. Fish Biol.*, **16**:279–289, 1980.

[376] B. Fry. *Stable Isotope Ecology*. Springer-Verlag, New York, 2006.

[377] F. E. J. Fry. The effect of environmental factors on the physiology of fish. In W. S. Hoar and D. J. Randall, editors, *Fish Physiology*, vol. 6, pages 1–87. Academic Press, London, 1971.

[378] G. Fryer. Evolution and adaptive radiation in *Macrothricidae* (*Crustacea*: *Cladocera*): a study in comparative functional morphology and ecology. *Proc. R. Soc. Lond. B Biol. Sci.*, **269**:142–385, 1974.

[379] J. A. Fuerst. Intracellular compartmentation in planctomycetes. *Rev. Microbiol.*, **59**:299–328, 2005.

[380] J. Fuhrman. Genome sequences from the sea. *Nature*, **424**:1001–1002, 2003.

[381] E. Fuller, editor. *Kiwis*. Swan Hill Press, Shrewsbury, 1991.

[382] T. W. Fulton. *The Sovereignty of the Seas*. Blackwood, Edinburgh, 1911.

[383] S. Fuma, H. Takeda, K. Miyamoto, K. Yanagisawa, Y. Inoue, N. Sato, M. Hirano, and Z. Kawabata. Effects of γ-rays on the populations of the steady state ecological microcosm. *Int. J. Radiat. Biol.*, **74**:145–150, 1998.

[384] R. W. Furness. *The Skuas*. T. & A. D. Poyser, Calton, 1987.

[385] V. F. Gallucci and T. J. Quinn. Reparameterizing, fitting, and testing a simple growth model. *Trans. Am. Fish. Soc.*, **108**:14–25, 1979.

[386] L. Garby and P. Larsen. *Bioenergetics: Its Thermodynamic Foundations*. Cambridge University Press, 1995.

[387] A. J. Gaston. *The Ancient Murrelet*. T. & A. D. Poyser, London, 1992.

[388] R. E. Gatten, K. Miller, and R. J. Full. Energetics at rest and during locomotion. In M. E. Feder and W. W. Burggren, editors, *Environmental Physiology of the Amphibians*, pages 314–377. University of Chicago Press, 1992.

[389] M. Gatto, C. Ricci, and M. Loga. Assessing the response of demographic parameters to density in a rotifer population. *Ecol. Modell.*, **62**:209–232, 1992.

[390] D. L. Geiger, C. E. Northcott, D. J. Call, and L. T. Brooke. *Acute Toxicities of Organic Chemicals to Fathead Minnow (*Pimephales promelas*)*, vol. 1 of *Center for Lake Superior Environmental Studies*. University of Wisconsin Press, 1985.

[391] D. L. Geiger, L. T. Brooke, and D. J. Call. *Acute Toxicities of Organic Chemicals to Fathead Minnow (*Pimephales promelas*)*, vols. 2–5 of *Center for Lake Superior Environmental Studies*. Univerity of Wisconsin Press, 1985–1990.

[392] V. Geist. Bergmann's rule is invalid. *Can. J. Zool.*, **65**:1035–1038, 1987.

[393] W. Geller. Die Nahrungsaufname von *Daphnia pulex* in Aghängigkeit von der Futterkonzentration, der Temperatur, der Köpergrösse und dem Hungerzustand der Tiere. *Arch. Hydrobiol./ Suppl.*, **48**:47–107, 1975.

[394] W. Geller and H. Müller. Seasonal variability in the relationship between body length and individual dry weight as related to food abundance and clutch size in two coexisting *Daphnia*. *J. Plankton Res.*, **7**:1–18, 1985.

[395] R. B. Gennis. *Biomembranes: Molecular Structure and Function*. Springer-Verlag, Berlin, 1989.

[396] M. Genoud. Energetic strategies of shrews: ecological constraints and evolutionary implications. *Mammal Rev.*, **4**:173–193, 1988.

[397] S. A. H. Geritz, É. Kisdi, G. Meszna, and J. A. J. Metz. Evolutionarily singular strategies and the adaptive growth and branching of the evolutionary tree. *Evol. Ecol.*, **12**:35–57, 1998.

[398] A. A. M. Gerritsen. The influence of body size, life stage, and sex on the toxicity of alkylphenols to *Daphnia magna*. PhD thesis, University of Utrecht, 1997.

[399] R. D. Gettinger, G. L. Paukstis, and W. H. N. Gutzke. Influence of hydric environment on oxygen consumption by embryonic turtles *Chelydra serpentina* and *Trionyx spiniferus*. *Physiol. Zool.*, **57**:468–473, 1984.

[400] W. Gewalt. *Der Weisswal (*Delphinapterus leucas*)*. A. Ziemsen-Verlag, Wittenberg Lutherstadt, 1976.

[401] A. G. Gibbs. Biochemistry at depth. In D. J. Randall and A. P. Farrell, editors, *Deep-Sea Fishes.*, pages 239–278. Academic Press, San Diego, 1997.

[402] R. N. Gibson. Observations on the biology of the giant goby *Gobius cobitis* Pallas. *J. Fish Biol.*, **2**:281–288, 1970.

[403] R. N. Gibson and I. A. Ezzi. The biology of a Scottish population of Fries' goby, *Lesueurigobius friesii*. *J. Fish Biol.*, **12**:371–389, 1978.

[404] R. N. Gibson and I. A. Ezzi. The biology of the scaldfish, *Arnoglossus laterna* (Walbaum) on the west coast of Scotland. *J. Fish Biol.*, **17**:565–575, 1980.

[405] R. N. Gibson and I. A. Ezzi. The biology of the Norway goby, *Pomatoschistus norvegicus* (Collett), on the west coast of Scotland. *J. Fish Biol.*, **19**:697–714, 1981.

[406] A. Gierer. *Hydra* as a model for the development of biological form. *Sci. Am.*, **12**:44–54, 1974.

[407] M. A. Gilbert. Growth rate, longevity and maximum size of *Macoma baltica* (L.). *Biol. Bull. (Woods Hole)*, **145**:119–126, 1973.

[408] U. Gille and F. V. Salomon. Heart and body growth in ducks. *Growth Dev. Aging*, **58**:75–81, 1994.

[409] S. J. Giovannoni, H. J. Tripp, S. Givan, M. Podar, K. L. Vergin, D. Baptista, L. Bibbs, J. Eads, T. H. Richardson, M. S. Noordewier, M. Rappé, J. M. Short, J. C. Carrington, and E. J. Mathur. Genome streamlining in a cosmopolitan oceanic bacterium. *Science*, **309**:1242–1245, 2005.

[410] S. Glasstone, K. J. Laidler, and H. Eyring. *The Theory of Rate Processes*. McGraw-Hill, London, 1941.

[411] D. S. Glazier. Effects of food, genotype, and maternal size and age on offspring investment in *Daphnia Magna*. *Ecology*, **73**:910–926, 1992.

[412] Z. M. Gliwicz. Food thresholds and body size in cladocerans. *Nature*, **343**:638–640, 1990.

[413] Z. M. Gliwicz and C. Guisande. Family planning in *Daphnia*: resistance to starvation in offspring born to mothers grown at different food levels. *Oecologia (Berlin)*, **91**:463–467, 1992.

[414] F. A. P. C. Gobas and A. Opperhuizen. Bioconcentration of hydrophobic chemicals in fish: relationship with membrane permeation. *Environ. Toxicol. Chem.*, **5**:637–646, 1986.

[415] K. Godfrey. *Compartment Models and their Application*. Academic Press, New York, 1983.

[416] L. Godfrey, M. Sutherland, D. Boy, and N. Gomberg. Scaling of limb joint surface areas in anthropoid primates and other mammals. *J. Zool. Lond.*, **223**:603–625, 1991.

[417] J. C. Goldman. Conceptual role for microaggregates in pelagic waters. *Bull. Mar. Sci.*, **35**:462–476, 1984.

[418] J. C. Goldman and E. J. Cerpenter. A kinetic approach of temperature on algal growth. *Limnol. Oceanogr.*, **19**:756–766, 1974.

[419] C. R. Goldspink. A note on the growth-rate and year-class strength of bream, *Abramis brama* (L.), in three eutrophic lakes, England. *J. Fish Biol.*, **19**:665–673, 1981.

[420] B. Gompertz. On the nature of the function expressive of the law of mortality, and on a new method of determining the value of life contingencies. *Phil. Trans. R. Soc. Lond. B Biol. Sci.*, **27**:513–585, 1825.

[421] M. Gophen and W. Geller. Filter mesh size and food particle uptake by *Daphnia*. *Oecologia (Berlin)*, **64**:408–412, 1984.

[422] R. W. Gordon, R. J. Beyers, E. P. Odum, and R. G. Eagon. Studies of a simple laboratory microecosystem: bacterial activities in a heterotrophic succession. *Ecology*, **50**:86–100, 1969.

[423] S. J. Gould. *Ontogeny and Phylogeny*. Belknap Press, Cambridge, 1977.

[424] C. E. Goulden, L. L. Henry, and A. J. Tessier. Body size, energy reserves, and competitive ability in three species of cladocera. *Ecology*, **63**:1780–1789, 1982.

[425] C. E. Goulden and L. L. Hornig. Population oscillations and energy reserves in planktonic cladocera and their consequences to competition. *Proc. Natl. Acad. Sci. U.S.A.*, **77**:1716–1720, 1980.

[426] B. W. Grant and W. P. Porter. Modelling global macroclimatic constraints on ectotherm energy budgets. *Am. Zool.*, **32**:154–178, 1992.

[427] C. J. Grant and A. V. Spain. Reproduction, growth and size allometry of *Liza vaigiensis* (Quoy & Gaimard) (*Pisces: Mugilidae*) from North Queensland inshore waters. *Austral. J. Zool.*, **23**:475–485, 1975.

[428] C. J. Grant and A. V. Spain. Reproduction, growth and size allometry of *Valamugil seheli* (Forskal) (*Pisces: Mugilidae*) from North Queensland inshore waters. *Austral. J. Zool.*, **23**:463–474, 1975.

[429] P. J. Graumann. Cytoskeletal elements in bacteria. *Rev. Microbiol.*, **61**:589–618, 2007.

[430] J. Gray. The growth of fish. 1. The relationship between embryo and yolk in *Salmo fario*. *J. Exp. Biol.*, **4**:215–225, 1926.

[431] R. H. Green. *Birds: The Fauna of Tasmania.* Potoroo Publishing, Launceston, 1995.

[432] W. Greve. Ökologische Untersuchungen an *Pleurobrachia pileus*. 1. Freilanduntersuchungen. *Helgol. Wiss. Meeresunters.*, **22**:303–325, 1971.

[433] W. Greve. Ökologische Untersuchungen an *Pleurobrachia pileus*. 2. Laboratoriumuntersuchungen. *Helgol. Wiss. Meeresunters.*, **23**:141–164, 1972.

[434] K. Grice, C. Cao, G. D Love, M. E. Böttcher, R. J. Twitchett, E. Grosjean, R. E. Summons, S. C. Turgeon, W. Dunning, and Y. Jin. Photic zone euxinia during the Permian–Triassic superanoxic event. *Science (New York)*, **307**:706–709, 2005.

[435] P. Groeneveld. Control of specific growth rate and physiology of the yeast *Kluyveromyces marxianus*: a biothermokinetic approach. PhD thesis, Vrije Universiteit, Amsterdam, 1999.

[436] W. T. de Groot. Modelling the multiple nutrient limitation of algal growth. *Ecol. Modell.*, **18**:99–119, 1983.

[437] R. K. Grosberg. Life-history variation within a population of the colonial ascidian *Botryllus schlosseri*. 1: The genetic and environmental control of seasonal variation. *Evolution*, **42**:900–920, 1988.

[438] J. Gruenberg. The endocytic pathway: a mosaic of domains. *Nature Rev.*, **2**:721–730, 2001.

[439] R. Grundel. Determinants of nestling feeding rates and parental investment in the mountain chickadee. *Condor*, **89**:319–328, 1987.

[440] J. Guckenheimer and P. Holmes. *Nonlinear Oscillations, Dynamical Systems, and Bifurcations of Vector Fields.* Springer-Verlag, Berlin, 1983.

[441] R. Guerrero. Predation as prerequisite to organelle origin: *Daptobacter* as example. In L. Margulis and R. Fester, editors, *Symbiosis as a Source of Evolutionary Innovation*, pages 106–117. MIT Press, Cambridge, 1991.

[442] R. Günther. *Die Wasserfrösche Europas.* A. Ziemsen-Verlag, Wittenberg Lutherstadt, 1990.

[443] R. S. Gupta. What are archaebacteria: life's third domain or monoderm prokaryotes related to gram-positive bacteria? A new proposal for the classification of prokaryotic organisms. *Mol. Microbiol.*, **29**:695–707, 1998.

[444] W. S. C Gurney and R. M. Nisbet. *Ecological Dynamics.* Oxford University Press, 1998.

[445] W. C. Gurney and R. M. Nisbet. Resource allocation, hyperphagia and compensatory growth. *Bull. Math. Biol.*, **66**:1731–1753, 2004.

[446] R. Haase. *Thermodynamics of Irreversible Processes.* Dover Publications, New York, 1990.

[447] P. Haccou and E. Meelis. *Statistical Analysis of Behavioural Data: An Approach Based on Time-Structured Models.* Oxford University Press, 1992.

[448] U. Halbach. Einfluss der Temperatur auf die Populationsdynamik der planktischer Radertieres *Brachionus calyciflorus* Pallas. *Oecologia (Berlin)*, **4**:176–207, 1970.

[449] D. O. Hall and K. K. Rao. *Photosynthesis.* Cambridge University Press, 1999.

[450] S. J. Hallam, N. Putnam, C. M. Preston, J. C. Detter, D. Rokhsar, P. M. Richardson, and E. F. DeLong. Reverse methanogenesis: testing the hypothesis with environmental genomics. *Science (New York)*, **305**:1457–1462, 2004.

[451] T. G. Hallam, R. R. Lassiter, and S. A. L. M. Kooijman. Effects of toxicants on aquatic populations. In S. A. Levin, T. G. Hallam, and L. F. Gross, editors, *Mathematical Ecology*, pages 352–382. Springer-Verlag, London, 1989.

[452] T. G. Hallam, R. R. Lassiter, J. Li, and W. McKinney. An approach for modelling populations with continuous structured models. In D. L. DeAngelis and L. J. Gross, editors, *Individual-Based Approaches and Models in Ecology*, pages 312–337. Springer-Verlag, Berlin, 1992.

[453] T. G. Hallam, R. R. Lassiter, J. Li, and L. A. Suarez. Modelling individuals employing an integrated energy response: application to *Daphnia*. *Ecology*, **71**:938–954, 1990.

[454] T. G. Hallam and J. L. de Luna. Effects of toxicants on population: a qualitative approach. 3. Environmental and food chain pathways. *J. Theor. Biol.*, **109**:411–429, 1984.

[455] B. Halliwell and J. M. C. Gutteridge. *Free Radicals in Biology and Medicine*. Oxford University Press, 1999.

[456] P. Hallock. Algal symbiosis: a mathematical analysis. *Mar. Biol.*, **62**:249–255, 1981.

[457] P. P. F. Hanegraaf. Mass and energy fluxes in microorganisms according to the Dynamic Energy Budget theory for filaments. PhD thesis, Vrije Universiteit, Amsterdam, 1997.

[458] P. P. F. Hanegraaf, B. W. Kooi, and S. A. L. M. Kooijman. The role of intracellular components in food chain dynamics. *C. R. Acad. Sci.*, **323**:99–111, 2000.

[459] R. J. F. van Haren and S. A. L. M. Kooijman. Application of the dynamic energy budget model to *Mytilus edulis* (L). *Neth. J. Sea Res.*, **31**:119–133, 1993.

[460] D. Harman. Role of free radicals in mutation, cancer, aging and maintenance of life. *Radiat. Res.*, **16**:752–763, 1962.

[461] D. Harman. The aging process. *Proc. Natl. Acad. Sci. U.S.A.*, **78**:7124–7128, 1981.

[462] J. L. Harper. *Population Biology of Plants*. Academic Press, London, 1977.

[463] M. P. Harris. The biology of an endangered species, the dark-rumped petrel (*Pterodroma phaeopygia*), in the Galapagos Islands. *Condor*, **72**:76–84, 1970.

[464] C. Harrison. *A Field Guide to the Nests, Eggs and Nestlings of European Birds*. Collins, London, 1975.

[465] R. Harrison and G. G. Lunt. *Biological Membranes: Their Structure and Function*. Blackie, Glasgow, 1980.

[466] M. W. Hart. What are the costs of small egg size for a marine invertebrate with feeding planktonic larvae? *Am. Nat.*, **146**:415–426, 1995.

[467] H. Hartman. Speculations on the origin and evolution of metabolism. *J. Mol. Evol.*, **4**:359–370, 1975.

[468] H. Hartman. Photosynthesis and the origin of life. *Orig. Life Evol. Biosphere*, **28**:515–521, 1998.

[469] K. A. Hasselbalch. Über den respiratorischen Stoffechsel des Hühnerembryos. *Skand. Arch. Physiol.*, **10**:353–402, 1900.

[470] A. Hastings. What equilibrium behaviour of Lotka–Volterra models does not tell us about food webs. In G. A. Polis and K. O. Winemiller, editors, *Food Webs*, pages 211–217. Chapman & Hall, London, 1996.

[471] A. Hastings. *Population Biology: Concepts and Models*. Springer-Verlag, New York, 1997.

[472] A. Hastings and T. Powell. Chaos in a three-species food chain. *Ecology*, **72**:896–903, 1991.

[473] F. Haverschmidt. Notes on the life history of *Amazilia fimbriata* in Surinam. *Wilson Bull.*, **61**:69–79, 1952.

[474] D. W. Hawker and D. W. Connell. Relationships between partition coefficient, uptake rate constant, clearance rate constant and time to equilibrium for bioaccumulation. *Chemosphere*, **14**:1205–1219, 1985.

[475] D. W. Hawker and D. W. Connell. Bioconcentration of lipophilic compounds by some aquatic organisms. *Ecotoxicol. Environ. Saf.*, **11**:184–197, 1986.

[476] A. J. S. Hawkins and B. L. Bayne. Seasonal variations in the balance between physiological mechanisms of feeding and digestion in *Mytilus edulis* (Bivalvia, Mollusca). *Mar. Biol.*, **82**:233–240, 1984.

[477] J. M. Hayes. Global methanotrophy at the archeon-proterozoic transition. In S. Bengtson, editor, *Early Life on Earth.*, pages 220–236. Columbia University Press, New York, 1994.

[478] D. D. Heath, C. W. Fox, and J. W. Heath. Maternal effects on offspring size: variation through early development of chinook salmon. *Evolution*, **53**:1605–1611, 1999.

[479] P. D. N. Hebert. The genetics of *Cladocera*. In W. C. Kerfoot, editor, *Evolution and Ecology of Zooplankton Communities*, pages 329–336. Univ. of New England Press, Hanover, 1980.

[480] P. D. N. Hebert. Genotypic characteristic of cyclic parthenogens and their obligately asexual derivates. In S. C. Stearns, editor, *The Evolution of Sex and its Consequences*, pages 175–195. Birkhäuser, Basel, 1987.

[481] H. Heesterbeek. R_0. PhD thesis, Rijksuniversiteit Leiden, 1992.

[482] H. J. A. M. Heijmans. Holling's 'hungry mantid' model for the invertebrate functional response considered as a Markov process. 3. Stable satiation distribution. *J. Math. Biol.*, **21**:115–1431, 1984.

[483] H. J. A. M. Heijmans. Dynamics of structured populations. PhD thesis, University of Amsterdam, 1985.

[484] J. J. Heijnen. A new thermodynamically based correlation of chemotrophic biomass yields. *Antonie van Leeuwenhoek*, **60**:235–256, 1991.

[485] J. J. Heijnen and J. P. van Dijken. In search of a thermodynamic description of biomass yields for the chemotrophic growth of micro organisms. *Biotechnol. Bioeng.*, **39**:833–858, 1992.

[486] J. J. Heijnen and J. A. Roels. A macroscopic model describing yield and maintenance relationships in aerobic fermentation. *Biotechnol. Bioeng.*, **23**:739–761, 1981.

[487] B. Heinrich. *The Hot-Blooded Insects: Strategies and Mechanisms of Thermoregulation.* Harvard Universitiy Press, Cambridge, 1993.

[488] R. Heinrich and S. Schuster. *The Regulation of Cellular Systems.* Chapman & Hall, New York, 1996.

[489] J. M. Hellawell. Age determination and growth of the grayling *Thymallus thymallus* (L.) of the River Lugg, Herefordshire. *J. Fish Biol.*, **1**:373–382, 1969.

[490] M. A. Hemminga. Regulation of glycogen metabolism in the freshwater snail *Lymnaea stagnalis*. PhD thesis, Vrije Universiteit, Amsterdam, 1984.

[491] A. M. Hemmingsen. Energy metabolism as related to body size and respiratory surfaces, and its evolution. *Rep. Steno. Mem. Hosp. Nordisk Insulinlaboratorium*, **9**:1–110, 1969.

[492] A. J. Hendriks. Modelling non-equilibrium concentrations of microcontaminants in organisms: comparative kinetics as a function of species size and octanol–water partitioning. *Chemosphere*, **30**:265–292, 1995.

[493] R. W. Hendrix, M. C. Smith, R. N. Burns, M. E. Ford, and G. F. Hatfull. Evolutionary relationships among diverse bacteriophages: all the world's a phage. *Proc. Natl. Acad. Sci. U.S.A.*, **96**:2192–2197, 1999.

[494] V. Henri. Théorie générale de l'action de quelques diastases. *C. R. Acad. Sci.*, **135**:916–919, 1902.

[495] W. G. Heptner and A. A. Nasimowitsch. *Der Elch (*Alces alces*).* A. Ziemsen-Verlag, Wittenberg Lutherstadt, 1974.

[496] S. L. Herendeen, R. A. VanBogelen, and F. C. Neidhardt. Levels of major proteins of *Escherichia coli* during growth at different temperatures. *J. Bacteriol.*, **138**:185, 1979.

[497] C. F. Herreid II and B. Kessel. Thermal conductance in birds and mammals. *Comp. Biochem. Physiol.*, **21**:405–414, 1967.

[498] H. C. Hess. The evolution of parental care in brooding spirorbid polychaetes: the effect of scaling constraints. *Am. Nat.*, **141**:577–596, 1993.

[499] D. J. S. Hetzel. The growth and carcass characteristics of crosses between alabio and tegal ducks and muscovy and pekin drakes. *Br. Poult. Sci.*, **24**:555–563, 1983.

[500] E. H. W. Heugens, T. Jager, R. Creyghton, M. H. S. Kraak, A. J. Hendriks, N. M. van Straalen, and W. Admiraal. Temperature dependent effects of cadmium on *Daphnia magna*: accumulation versus sensitivity. *Environ. Sci. Technol.*, **37**:2145–2151, 2003.

[501] H. R. Hewer and K. M. Backhouse. Embryology and foetal growth of the grey seal, *Halichoerus grypus*. *J. Zool. Lond.*, **155**:507–533, 1968.

[502] C. J. Hewett. Growth and moulting in the common lobster (*Homarus vulgaris* Milne-Edwards). *J. Mar. Biol. Assoc. U.K.*, **54**:379–391, 1974.

[503] P. S. Hewlett and R. L. Plackett. *The Interpretation of Quantal Responses in Biology.* Edward Arnold, London, 1979.

[504] L. C. Hewson. Age, maturity, spawning and food of burbot, *Lota lota*, in Lake Winnipeg. *J. Fish. Res. Board Can.*, **12**:930–940, 1955.

[505] D. S. Hibbet. When good relationships go bad. *Nature*, **419**:345–346, 2002.

[506] R. W. Hickman. Allometry and growth of the green-lipped mussel *Perna canaliculus* in New Zealand. *Mar. Biol.*, **51**:311–327, 1979.

[507] R. Hile. Age and growth of the cisco *Leuchichthys artedi* (Le Sueur) in lake of the north-eastern High Land Eisconsian. *Bull. Bur. Fish., Wash.*, **48**:211–317, 1936.

[508] H. V. Hill. The possible effects of aggregation on the molecules of haemoglobin on its dissociation curves. *J. Physiol. (Lond.)*, **40**:IV–VII, 1910.

[509] R. Hill and C. P. Whittingham. *Photosynthesis.* Methuen, London, 1955.

[510] T. L. Hill. *An Introduction to Statistical Thermodynamics.* Dover Publications, New York, 1986.

[511] M. W. Hirsch and S. Smale. *Differential Equations, Dynamical Systems, and Linear Algebra.* Academic Press, London, 1974.

[512] P. W. Hochachka. *Living without Oxygen.* Harvard University Press, Cambridge, 1980.

[513] P. W. Hochachka and M. Guppy. *Metabolic Arrest and the Control of Biological Time.* Harvard University Press, Cambridge, 1987.

[514] P. A. R. Hockey. Growth and energetics of the African black oystercatcher *Haematopus moquini*. *Ardea*, **72**:111–117, 1984.

[515] J. K. M. Hodasi. The effects of different light regimes on the behaviour biology of *Achatina (Achatina) achatina* (Linne). *J. Molluscan Stud.*, **48**:283–293, 1982.

[516] C. van den Hoek, D. G. Mann, and H. M. Jahn. *Algae: An Introduction to Phycology.* Cambridge University Press, 1995.

[517] N. van der Hoeven. The population dynamics of daphnia at constant food supply: a review, re-evaluation and analysis of experimental series from the literature. *Neth. J. Zool.*, **39**:126–155, 1989.

[518] N. van der Hoeven, S. A. L. M. Kooijman, and W. K. de Raat. Salmonella test: relation between mutagenicity and number of revertant colonies. *Mutat. Res.*, **234**:289–302, 1990.

[519] M. Hoffman. Yeast biology enters a surprising new phase. *Science (New York)*, **255**:1510–1511, 1992.

[520] M. A. Hofman. Energy metabolism, brain size and longevity in mammals. *Q. Rev. Biol.*, **58**:495–512, 1983.

[521] M. J. Holden. The growth rates of *Raja brachyura*, *R. clavata* and *R. montaqui* as determined from tagging data. *J. Physiol. (Lond.)*, **34**:161–168, 1972.

[522] H. D. Holland. Early proterozoic atmospheric change. In *Early Life on Earth*, S. Bengtson, editor, pages 237–244. Columbia University Press, New York, 1994.

[523] C. S. Holling. Some characteristics of simple types of predation and parasitism. *Can. Entomol.*, **91**:385–398, 1959.

[524] B. Holmes. The perils of planting pesticides. *New Sci.*, **28**:34–37, 1993.

[525] A. B. Hope and N. A. Walker. *The Physiology of Giant Algal Cells.* Cambridge University Press, 1975.

[526] N. J. Horan. *Biological Wastewater Treatment Systems.* J. Wiley, Chichester, 1990.

[527] H. J. Horstmann. Sauerstoffverbrauch und Trockengewicht der Embryonen von *Lymnaea stagnalis* L. *Z. Vgl. Physiol.*, **41**:390–404, 1958.

[528] R. A. Horton, M. Rowan, K. E. Webster, and R. H. Peters. Browsing and grazing by cladoceran filter feeders. *Can. J. Zool.*, **57**:206–212, 1979.

[529] M. A. Houck, J. A. Gauthier, and R. E. Strauss. Allometric scaling in the earliest fossil bird, *Archaeopteryx lithographica*. *Science*, **247**:195–198, 1990.

[530] S. E. L. Houde and M. R. Roman. Effects of food quality on the functional ingestion response of the copepod *Acartia tonsa*. *Mar. Ecol. Prog. Ser.*, **48**:69–77, 1987.

[531] J. Hovekamp. On the growth of larval plaice in the North Sea. PhD thesis, University of Groningen, 1991.

[532] R. A. How. Reproduction, growth and survival of young in the mountain possum, *Trichosurus caninus* (*Marsupialia*). *Austral. J. Zool.*, **24**:189–199, 1976.

[533] H. Huber, M. J. Hohn, R. Rachel, T. Fuchs, V. C. Wimmer, and K. O. Stetter. A new phylum of Archaea represented by nanosized hyperthermophilic symbiont. *Nature*, **417**:63–67, 2002.

[534] M. F. Hubert, P. Laroque, J. P. Gillet, and K. P. Keenan. The effects of diet, ad libitum feeding, and moderate and severe dietary restriction on body weight, survival, clinical pathology parameters, and cause of death in control Sprague–Dawley rats. *Toxicol. Sci.*, **58**:195–207, 2000.

[535] J. Hugenholtz and L. G. Ljungdahl. Metabolism and energy generation in homoacetogenic clostridia. *FEMS Microbiol. Rev.*, **87**:383–389, 1990.

[536] A. St. G. Huggett and W. F. Widdas. The relationship between mammalian foetal weight and conception age. *J. Physiol. (Lond.)*, **114**:306–317, 1951.

[537] R. E. Hungate. The rumen microbial ecosystem. *Annu. Rev. Ecol. Syst.*, **6**:39–66, 1975.

[538] P. C. Hunt and J. W. Jones. Trout in Llyn Alaw, Anglesey, North Wales. 2. Growth. *J. Fish Biol.*, **4**:409–424, 1972.

[539] R. von Hutterer. Beobachtungen zur Geburt und Jugendentwicklung der Zwergspitzmaus, *Sorex minutus* L. (*Soricidae – Insectivora*). *Z. Sauegetierkd.*, **41**:1–22, 1976.

[540] J. S. Huxley. *Problems of Relative Growth*. Methuen, London, 1932.

[541] H. L. Ibsen. Prenatal growth in guinea-pigs with special reference to environmental factors affecting weight at birth. *J. Exp. Zool.*, **51**:51–91, 1928.

[542] T. D. Iles and P. O. Johnson. The correlation table analysis of a sprat (*Clupea sprattus* L.) year-class to separate two groups differing in growth characteristics. *J. Cons. Int. Explor. Mer.*, **27**:287–303, 1962.

[543] H. Ineichen, U. Riesen-Willi, and J. Fisher. Experimental contributions to the ecology of *Chironomus* (*Diptera*). 2. Influence of the photoperiod on the development of *Chironomus plumosus* in the 4th larval instar. *Oecologia (Berlin)*, **39**:161–183, 1979.

[544] L. Ingle, T. R. Wood, and A. M. Banta. A study of longevity, growth, reproduction and heart rate in *Daphnia longispina* as influenced by limitations in quantity of food. *J. Exp. Zool.*, **76**:325–352, 1937.

[545] J. L. Ingraham. Growth of psychrophilic bacteria. *J. Bacteriol.*, **76**:75–80, 1958.

[546] R. Ivell. The biology and ecology of a brackish lagoon bivalve, *Cerastoderma glaucum* Bruguiere, in Lago Lungo, Italy. *J. Molluscan Stud.*, **45**:364–382, 1979.

[547] E. A. Jackson. *Perspectives of Nonlinear Dynamics*, vols. I and II. Cambridge University Press, 1991.

[548] J. A. Jacquez. *Compartment Analysis in Biology and Medicine*. Elsevier, Amsterdam, 1972.

[549] D. A. Jaffe. The nitrogen cycle. In S. S. Butcher, R. J. Charlson, G. H. Orians, and G. V. Wolfe, editors, *Global Biogeochemical Cycles*, pages 263–284. Academic Press, London, 1992.

[550] T. Jager, T. Crommentuijn, C. A. M. van Gestel, and S. A. L. M. Kooijman. Simultaneous modelling of multiple endpoints in life-cycle toxicity tests. *Environ. Sci. Technol.*, **38**:2894–2900, 2004.

[551] T. Jager and S. A. L. M. Kooijman. Modeling receptor kinetics in the analysis of survival data for organophosphorus pesticides. *Environ. Sci. Technol.*, **39**:8307–8314, 2005.

[552] M. R. James. Distribution, biomass and production of the freshwater mussel, *Hyridella menziesi* (Gray), in Lake Taupo, New Zealand. *Freshwater Biol.*, **15**:307–314, 1985.

[553] M. J. W. Jansen and J. A. J. Metz. How many victims will a pitfall make? *Acta Biotheor.*, **28**:98–122, 1979.

[554] G. Janssen. On the genetical ecology of springtails. PhD thesis, Vrije Universiteit, Amsterdam, 1985.

[555] D. Jardine and M. K. Litvak. Direct yolk sac volume manipulation of zebrafish embryos and the relationship between offspring size and yolk sac volume. *J. Fish Biol.*, **63**:388–397, 2003.

[556] X. Ji, W.-G. Du, and W.-Q. Xu. Experimental manipulation of eggs size and hatchling size in the cobra, *Naja naja atra* (Elapidae). *Neth. J. Zool.*, **49**:167–175, 1999.

[557] F. M. Jiggins, J. K. Bentley, M. E. N. Majerus, and G. D. D. Hurst. How many species are infected with *Wolbachia*? Cryptic sex ratio distorters revealed to be common by intensive sampling. *Proc. R. Soc. Lond. Biol. Sci.*, **268**:1123–1126, 2001.

[558] M. Jobling. Mathematical models of gastric emptying and the estimation of daily rates of food consumption for fish. *J. Fish Biol.*, **19**:245–257, 1981.

[559] A. W. B. Johnston, J. L. Firmin, and L. Rosen. On the analysis of symbiotic genes in *Rhizobium*. *Symp. Soc. Gen. Microbiol.*, **42**:439–455, 1988.

[560] A. Jones. Studies on egg development and larval rearing of turbot, *Scophthalmus maximus* L. and brill, *Scophthalmus rhombus* L., in the laboratory. *J. Mar. Biol. Assoc. U.K.*, **52**:965–986, 1972.

[561] G. P. Jones. Contribution to the biology of the redbanded perch, *Ellerkeldia huntii* (Hector), with a discussion on hermaphroditism. *J. Fish Biol.*, **17**:197–207, 1980.

[562] H. G. Jones. *Plant and Microclimate: A Quantitative Approach to Environmental Plant Physiology.* Cambridge University Press, 1992.

[563] J. W. Jones and H. B. N. Hynes. The age and growth of *Gasterosteus aculeatus*, *Pygosteus pungitius* and *Spinachia vulgaris*, as shown by their otoliths. *J. Anim. Ecol.*, **19**:59–73, 1950.

[564] M. de Jong-Brink. Parasites make use of multiple strategies to change energy budgetting in their host. In H. J. Th. Goos, R. K. Rastogi, H. Vandry, and R. Pierantoni, editors, *Perspective in Comparative Endocrinology: Unity and Diversity*, pages 341–348. Monduzzi Editore, 2001.

[565] E. N. G. Joosse and J. B. Buker. Uptake and excretion of lead by litter-dwelling Collembola. *Environ. Pollut.*, **18**:235–240, 1979.

[566] E. N. G. Joosse and E. Veltkamp. Some aspects of growth, moulting and reproduction in five species of surface dwelling Collembola. *Neth. J. Zool.*, **20**:315–328, 1970.

[567] B. B. Jørgensen and V. A. Gallardo. *Thioploca* spp.: filamentous sulfur bacteria with nitrate vacuoles. *FEMS Microbiol. Rev.*, **28**:301–313, 1999.

[568] D. M. Joubert. A study of pre-natal growth and development in the sheep. *J. Agric. Sci.*, **47**:382–428, 1956.

[569] J. Juget, V. Goubier, and D. Barthélémy. Intrisic and extrinsic variables controlling the productivity of asexual populations of *Nais* spp. (Naididae, Oligochaeta). *Hydrobiologia*, **180**:177–184, 1989.

[570] H. Kaiser. Populationsdynamik und Eigenschaften einzelner Individuen. *Verh. Ges. Ökol. Erlangen*, **1974**:25–38, 1974.

[571] H. Kaiser. The dynamics of populations as result of the properties of individual animals. *Fortschr. Zool.*, **25**:109–136, 1979.

[572] J. D. Kalbfleisch and R. L. Prentice. *The Statistical Analysis of Failure Time Data.* J. Wiley, Chichester, 1980.

[573] O. Kandler. The early diversification of life and the origin of the three domains: a proposal. In J. Wiegel and M. W. W. Adams, editors, *Thermophiles: The Keys to Molecular Evolution and the Origin of Life.*, pages 19–31. Taylor & Francis, Washington, 1998.

[574] G. Kapocsy. *Weissbart- und Weissflugelseeschwalbe.* A. Ziemsen-Verlag, Wittenberg Lutherstadt, 1979.

[575] S. J. Karakashian. Growth of *Paramecium bursaria* as influenced by the presence of algal symbionts. *Physiol. Zool.*, **36**:52–68, 1963.

[576] W. H. Karasov. Daily energy expenditure and the cost of activity in mammals. *Am. Zool.*, **32**:238–248, 1992.

[577] K. B. Karsten, L. N. Andriamandimbiarisoa, S. F. Fox, and C. J. Raxworthy. A unique life history among tetrapods: an annual chameleon living mostly as an egg. *Proc. Natl. Acad. Sci. U.S.A.*, **105**:8980–8984, 2008.

[578] V. V. Kashkin. Heat exchange of bird eggs during incubation. *Biophysics*, **6**:57–63, 1961.

[579] J. F. Kasting. Earth's early atmosphere. *Science*, **259**:920–925, 2001.

[580] M. Kates. The phytanyl ether-linked polar lipids and isopreniod neutral lipids of extremely halophilic bacteria. *Lipids*, **15**:301–342, 1979.

[581] L. A. Katz. The tangled web: gene genealogies and the origin of eukaryotes. *Am. Nat.*, **154**:S137–S145, 1999.

[582] L. Kaufman. Innere und Äussere Wachstumsfactoren: Untersuchungen an Hühnern und Tauben. *Wilhelm Roux Arch. Entwicklungsmech. Org.*, **6**:395–431, 1930.

[583] N. Kautsky. Quantitative studies on gonad cycle, fecundity, reproductive output and recruitment in a baltic *Mytilus edulis* population. *Mar. Biol.*, **67**:143–160, 1982.

[584] Z. Kawabata, K. Matsui, K. Okazaki, M. Nasu, N. Nakano, and T. Sugai. Synthesis of a species-defined microcosm with protozoa. *J. Protozool. Res.*, **5**:23–26, 1995.

[585] C. Kayser and A. Heusner. Étude comparative du métabolisme énergétiques dans la série animale. *J. Physiol. (Paris)*, **56**:489–524, 1964.

[586] A. D. Keefe, S. L. Miller, and G. Bada. Investigation of the prebiotic synthesis of amino acids and RNA bases from CO_2 using FES/H_2S as a reducing agent. *Proc. Natl. Acad. Sci. U.S.A.*, **92**:11904–11906, 1995.

[587] P. J. Keeling. A kingdom's progress: Archaezoa and the origin of eukaryotes. *BioEssays*, **20**:87–95, 1998.

[588] B. D. Keller. Coexistence of sea urchins in seagrass meadows: an experimental analysis of competition and predation. *Ecology*, **64**:1581–1598, 1983.

[589] C. M. Kemper. Growth and development of the Australian musid *Pseudomys novaehollandiae*. *Austral. J. Zool.*, **24**:27–37, 1976.

[590] S. C. Kendeigh. Factors affecting the length of incubation. *Auk*, **57**:499–513, 1940.

[591] M. Kennedy and P. Fitzmaurice. Some aspects of the biology of gudgeon *Gobio gobio* (L.) in Irish waters. *J. Fish Biol.*, **4**:425–440, 1972.

[592] R. A. Kent and D. Currie. Predicting algal sensitivity to a pesticide stress. *Environ. Toxicol. Chem.*, **14**:983–991, 1995.

[593] K. Kersting and W. Holterman. The feeding behaviour of *Daphnia magna*, studied with the coulter counter. *Mitt. Int. Ver. Theor. Angew. Limnol.*, **18**:1434–1440, 1973.

[594] M. Kessel and Y. Cohen. Ultrastructure of square bacteria from a brine pool in southern Sinai. *J. Bacteriol.*, **150**:851–860, 1982.

[595] A. Keve. *Der Eichelhaher*. A. Ziemsen-Verlag, Wittenberg Lutherstadt, 1985.

[596] P. Kindlmann and A. F. G. Dixon. Developmental constraints in the evolution of reproductive strategies: telescoping of generations in parthenogenetic aphids. *Funct. Ecol.*, **3**:531–537, 1989.

[597] C. M. King. *The Natural History of Weasels and Stoats*. Croom Helm, London, 1989.

[598] P. Kingston. Some observations on the effects of temperature and salinity upon the growth of *Cardium edule* and *Cardium glaucum* larvae in the laboratory. *J. Mar. Biol. Assoc. U.K.*, **54**:309–317, 1974.

[599] F. C. Kinsky. The yearly cycle of the northern blue penguin (*Eudyptula minor novaehollandiae*) in the Wellington Harbour area. *Rec. Dom. Mus. (Wellington)*, **3**:145–215, 1960.

[600] T. Kiørboe, F. Møhlenberg, and O. Nøhr. Effect of suspended bottom material on growth and energetics in *Mytilus edulis*. *Mar. Biol.*, **61**:283–288, 1981.

[601] M. Kirkpatrick, editor. *The Evolution of Haploid Diploid Life Cycles*. American Mathematical Society, Providence, 1993.

[602] M. Klaassen, G. Slagsvold, and C. Beek. Metabolic rate and thermostability in relation to availability of yolk in hatchlings of black-legged kittiwake and domestic chicken. *Auk*, **104**:787–789, 1987.

[603] A. Klebanoff and A. Hastings. Chaos in a three-species food chain. *J. Math. Biol.*, **32**:427–451, 1994.

[604] O. T. Kleiven, P. Larsson, and A. Hobæk. Sexual reproduction in *Daphnia magna* requires three stimuli. *Oikos*, **64**:197–206, 1992.

[605] D. J. Klionsky. Regulated self-cannibalism. *Nature*, **431**:31–32, 2004.

[606] C. Klok and A. M. de Roos. Population level consequences of toxicological influences on individual growth and reproduction in *Lumbricus rubellus* (Lumbricidae, Oligochaeta). *Ecotoxicol. Environ. Saf.*, **33**:118–127, 1996.

[607] H. N. Kluyver. Food consumption in relation to habitat in breeding chickadees. *Auk*, **78**:532–550, 1961.

[608] A. H. Knoll. *Life on a Young Planet: The First Three Billion Years of Evolution on Earth*. Princeton University Press, 2003.

[609] A. L. Koch. Overall controls on the biosynthesis of ribosomes in growing bacteria. *J. Theor. Biol.*, **28**:203–231, 1970.

[610] A. L. Koch. The macroeconomics of bacterial growth. In M. Fletcher and G. D. Floodgate, editors, *Bacteria in their Natural Environment*, pages 1–42. Academic Press, London, 1985.

[611] A. L. Koch. The variability and individuality of the bacterium. In F. C. Neidhardt, editor, Escherichia coli *and* Salmonella typhimurium*: Cellular and Molecular Biology*, pages 1606–1614. American Society for Microbiology, Washington, 1987.

[612] A. L. Koch. Diffusion. the crucial process in many aspects of the biology of bacteria. *Adv. Microb. Ecol.*, **11**:37–70, 1990.

[613] A. L. Koch. Quantitative aspects of cellular turnover. *Antonie van Leeuwenhoek*, **60**:175–191, 1991.

[614] Y. Koga, T. Kyuragi, M. Nishhihara, and N. Sone. Did archaeal and bacterial cells arise independently from noncellular precursors? A hypothesis stating that the advent of membrane phospholipids with enantiomeric glycerophosphate backbones caused the separation of the two lines of descent. *J. Mol. Evol.*, **46**:54–63, 1998.

[615] A. C. Kohler. Variations in the growth of Atlantic cod (*Gadus morhua* L.). *J. Fish. Res. Board Can.*, **21**:57–100, 1964.

[616] M. Konarzewski. A model of growth in altricial birds based on changes in water content of the tissues. *Ornis Scand.*, **19**:290–296, 1988.

[617] A. S. Kondrashov. Evolutionary genetics of life cycles. *Annu. Rev. Ecol. Syst.*, **28**:391–435, 1997.

[618] W. H. Könemann. Quantitative structure–activity relationships for kinetics and toxicity of aquatic pollutants and their mixtures in fish. PhD thesis, Utrecht University, 1980.

[619] W. H. Könemann. Quantitative structure–activity relationships in fish toxicity studies. 1. Relationship for 50 industrial pollutants. *Toxicology*, **19**:209–221, 1981.

[620] B. W. Kooi. Numerical bifurcation analysis of ecosystems in a spatially homogeneous environment. *Acta Biotheor.*, **51**:189–222, 2003.

[621] B. W. Kooi and M. P. Boer. Bifurcations in ecosystem models and their biological interpretation. *Appl. Anal.*, **77**:29–59, 2001.

[622] B. W. Kooi and M. P. Boer. Chaotic behaviour of a predator–prey system. *Dyn. Cont. Discr. Impul. Syst., Ser. B, Appl. Algor.*, **10**:259–272, 2003.

[623] B. W. Kooi, M. P. Boer, and S. A. L. M. Kooijman. Mass balance equation versus logistic equation in food chains. *J. Biol. Syst.*, **5**:77–85, 1997.

[624] B. W. Kooi, M. P. Boer, and S. A. L. M. Kooijman. Consequences of population models on the dynamics of food chains. *Math. Biosci.*, **153**:99–124, 1998.

[625] B. W. Kooi, M. P. Boer, and S. A. L. M. Kooijman. On the use of the logistic equation in food chains. *Bull. Math. Biol.*, **60**:231–246, 1998.

[626] B. W. Kooi and P. P. F. Hanegraaf. Bi-trophic food chain dynamics with multiple component populations. *Bull. Math. Biol.*, **63**:271–299, 2001.

[627] B. W. Kooi and F. D. L. Kelpin. Structured population dynamics: a modeling perspective. *Comm. Theor. Biol.*, **8**:125–168, 2003.

[628] B. W. Kooi and S. A. L. M. Kooijman. Invading species can stabilize simple trophic systems. *Ecol. Modell.*, **133**:57–72, 2000.

[629] B. W. Kooi, L. D. J. Kuijper, M. P. Boer, and S. A. L. M. Kooijman. Numerical bifurcation analysis of a tri-trophic food web with omnivory. *Math. Biosci.*, **177**:201–228, 2002.

[630] B. W. Kooi, L. D. J. Kuijper, and S. A. L. M. Kooijman. Consequences of symbiosis for food web dynamics. *J. Math. Biol.*, **3**:227–271, 2004.

[631] B. W. Kooi and T. A. Troost. Advantages of storage in a fluctuating environment. *Theor. Pop. Biol.*, **70**:527–541, 2006.

[632] S. A. L. M. Kooijman. Statistical aspects of the determination of mortality rates in bioassays. *Water Res.*, **17**:749–759, 1983.

[633] S. A. L. M. Kooijman. Toxicity at population level. In J. Cairns, editor, *Multispecies Toxicity Testing*, pages 143–164. Pergamon Press, New York, 1985.

[634] S. A. L. M. Kooijman. Population dynamics on the basis of budgets. In J. A. J. Metz and O. Diekmann, editors, *The Dynamics of Physiologically Structured Populations*, pages 266–297. Springer-Verlag, Berlin, 1986.

[635] S. A. L. M. Kooijman. What the hen can tell about her egg: egg development on the basis of budgets. *J. Math. Biol.*, **23**:163–185, 1986.

[636] S. A. L. M. Kooijman. The von Bertalanffy growth rate as a function of physiological parameters: a comparative analysis. In

T. G. Hallam, L. J. Gross, and S. A. Levin, editors, *Mathematical Ecology*, pages 3–45. World Scientific Press, Singapore, 1988.

[637] S. A. L. M. Kooijman. *Dynamic Energy Budgets in Biological Systems: Theory and Applications in Ecotoxicology*. Cambridge University Press, 1993.

[638] S. A. L. M. Kooijman. The stoichiometry of animal energetics. *J. theor. Biol.*, **177**:139–149, 1995.

[639] S. A. L. M. Kooijman. An alternative for NOEC exists, but the standard model has to be replaced first. *Oikos*, **75**:310–316, 1996.

[640] S. A. L. M. Kooijman. The synthesizing unit as model for the stoichiometric fusion and branching of metabolic fluxes. *Biophys. Chem.*, **73**:179–188, 1998.

[641] S. A. L. M. Kooijman. *Dynamic Energy and Mass Budgets in Biological Systems*, 2nd edn. Cambridge University Press, 2000.

[642] S. A. L. M. Kooijman. Quantitative aspects of metabolic organization; a discussion of concepts. *Phil. Trans. R. Soc. Lond. B Biol. Sci.*, **356**:331–349, 2001.

[643] S. A. L. M. Kooijman. On the coevolution of life and its environment. In J. Miller, P. J. Boston, S. H. Schneider, and E. Crist, editors, *Scientists Debate Gaia: The Next Century*, pages 343–351. MIT Press, Cambridge, 2004.

[644] S. A. L. M. Kooijman. Pseudo-faeces production in bivalves. *J. Sea Research*, **56**:103–106, 2006.

[645] S. A. L. M. Kooijman. What the egg can tell about its hen: embryo development on the basis of dynamic energy budgets. *J. Math. Biol.*, **58**:377–394. 2009.

[646] S. A. L. M. Kooijman. Social interactions can affect feeding behaviour of fish in tanks. *J. Sea Res.*, **62**:(in press), 2009.

[647] S. A. L. M. Kooijman, T. R. Andersen, and B. W. Kooi. Dynamic energy budget representations of stoichiometric constraints to population models. *Ecology*, **85**:1230–1243, 2004.

[648] S. A. L. M. Kooijman, P. Auger, J. C. Poggiale, and B. W. Kooi. Quantitative steps in symbiogenesis and the evolution of homeostasis. *Biol. Rev.*, **78**:435–463, 2003.

[649] S. A. L. M. Kooijman, J. Baas, D. Bontje, M. Broerse, C. A. M. van Gestel, and T. Jager. Ecotoxicological applications of dynamic energy budget theory. In J. Devillers, editor, *Ecotoxicology Modeling*. Springer-Verlag, Berlin, 2009.

[650] S. A. L. M. Kooijman, J. Baas, D. Bontje, M. Broerse, T. Jager, C. van Gestel, and B. van Hattum. Scaling relationships based on partition coefficients and body sizes have similarities and interactions. *SAR and QSAR in Environ. Res.*, **18**:315–330, 2007.

[651] S. A. L. M. Kooijman and J. J. M. Bedaux. *The Analysis of Aquatic Toxicity Data*. Unje Universiteit Press, Amsterdam, 1996.

[652] S. A. L. M. Kooijman and J. J. M. Bedaux. Analysis of toxicity tests on *Daphnia* survival and reproduction. *Water Res.*, **30**:1711–1723, 1996.

[653] S. A. L. M. Kooijman and J. J. M. Bedaux. Dynamic effects of compounds on animal energetics and their population consequences. In J. E. Kammenga and R. Laskowski, editors, *Demography in Ecotoxicology*, pages 27–41. J. Wiley, New York, 2000.

[654] S. A. L. M. Kooijman, H. A. Dijkstra, and B. W. Kooi. Light-induced mass turnover in a mono-species community of mixotrophs. *J. theor. Biol.*, **214**:233–254, 2002.

[655] S. A. L. M. Kooijman, J. Grasman, and B. W. Kooi. A new class of non-linear stochastic population models with mass conservation. *Math. Biosci.*, **210**:378–394, 2007.

[656] S. A. L. M. Kooijman and R. J. F. van Haren. Animal energy budgets affect the kinetics of xenobiotics. *Chemosphere*, **21**:681–693, 1990.

[657] S. A. L. M. Kooijman and R. Hengeveld. The symbiotic nature of metabolic evolution. In T. A. C. Reydon and L. Hemerik, editors, *Current Themes in Theoretical Biology: A Dutch Perspective.*, pages 159–202. Springer-Verlag, Dordrecht, 2005.

[658] S. A. L. M. Kooijman, N. van der Hoeven, and D. C. van der Werf. Population consequences of a physiological model for individuals. *Funct. Ecol.*, **3**:325–336, 1989.

[659] S. A. L. M. Kooijman and R. M. Nisbet. How light and nutrients affect life in a closed bottle. In S. E. Jørgensen, editor, *Thermodynamics and Ecological Modeling*, pages 19–60. CRC Publ., Boca Raton, 2000.

[660] S. A. L. M. Kooijman and L. A. Segel. How growth affects the fate of cellular substrates. *Bull. Math. Biol.*, **67**:57–77, 2005.

[661] S. A. L. M. Kooijman, T. Sousa, L. Pecquerie, J. van der Meer, and T. Jager. From food-dependent statistics to metabolic parameters, a practical guide to the use of dynamic energy budget theory. *Biol. Rev.*, **83**:533–552, 2008.

[662] S. A. L. M. Kooijman and T. A. Troost. Quantitative steps in the evolution of metabolic organisation as specified by the dynamic energy budget theory. *Biol. Rev.*, **82**:1–30, 2007.

[663] E. V. Koonin, K. S. Makarova, and L. Aravind. Horizontal gene transfer in prokaryotes: quantification and classification. *Annu. Rev. Microbiol.*, **55**:709–742, 2001.

[664] A. Kowald and T. B. L. Kirkwood. Towards a network theory of ageing: a model combining the free radical theory and the protein error theory. *J. theor. Biol.*, **168**:75–94, 1994.

[665] A. Kowald and T. B. L. Kirkwood. A network theory of ageing: the interactions of defective mitochondria, aberrant proteins, free radicals and scavengers in the ageing process. *Mutat. Res.*, **316**:209–235, 1996.

[666] K. B. Krauskopf and D. K. Bird. *Introduction to Geochemistry.* McGraw-Hill, New York, 1995.

[667] J.-U. Kreft, G. Booth, and W. T. Wimpenny. BacSim, a simulator for individual-based modelling of bacterial colony growth. *Microbiology*, **144**:3275–3287, 1998.

[668] F. Kreit and W. Z. Black. *Basic Heat Transfer.* Harper & Row, 1980.

[669] H. J. Kreuzer. *Nonequilibrium Thermodynamics and its Statistical Foundations.* Clarendon Press, Oxford, 1981.

[670] H. J. Kronzucker, M. Y. Siddiqi, and A. D. M. Glass. Conifer root discrimination against soil nitrate and the ecology of forest succession. *Nature*, **385**:59–61, 1997.

[671] W. E. Krumbein and D. Werner. The microbial silica cycle. In W. E. Krumbein, editor, *Microbial Geochemistry*, pages 125–157. Blackwell Sci. Publ., Oxford, 1983.

[672] M. Kshatriya and R. W. Blake. Theoretical model of migration energetics in the blue whale, *Balaenoptera musculus*. *J. Theor. Biol.*, **133**:479–498, 1988.

[673] H. E. Kubitschek. Cell growth and abrupt doubling of membrane proteins in *Escherichia coli* during the division cycle. *J. Gen. Microbiol.*, **136**:599–606, 1990.

[674] S. W. Kuffler, J. G. Nicholls, and A. R. Martin. *From Neuron to Brain.* Sinauer, Sunderland, 1984.

[675] L. D. J. Kuijper, T. R. Anderson, and S. A. L. M. Kooijman. C and N gross efficiencies of copepod egg production studies using a dynamic energy budget model. *J. Plankton Res.*, **26**:1–15, 2003.

[676] L. D. J. Kuijper, B. W. Kooi, C. Zonneveld, and S. A. L. M. Kooijman. Omnivory and food web dynamics. *Ecol. Modell.*, **163**:19–32, 2003.

[677] Yu. A. Kuznetsov and S. Rinaldi. Remarks on food chain dynamics. *Math. Biosci.*, **134**:1–33, 1996.

[678] D. Lack. *Ecological Adaptations for Breeding in Birds.* Methuen, London, 1968.

[679] A. K. Laird. Dynamics of tumor growth. *Br. J. Cancer*, **18**:490–502, 1964.

[680] H. Lambers, F. S. Chapin III, and T. L. Pons. *Plant Physiological Ecology.* Springer-Verlag, New York, 1998.

[681] N. Lane. *Oxygen: The Molecule that Made the World.* Oxford University Press, 2002.

[682] A. R. G. Lang and J. E. Begg. Movements of *Helianthus annuus* leaves and heads. *J. Appl. Ecol.*, **16**:299–306, 1979.

[683] E. Lanting. *Madagascar: A World Out of Time.* Aperture, New York, 1990.

[684] D. A. Lashof. Gaia on the brink: biogeochemical feedback processes in global warming. In S. H. Schneider and P. J. Boston, editors, *Scientists on Gaia*, pages 393–404. MIT Press, Cambridge, 1991.

[685] R. R. Lassiter and T. G. Hallam. Survival of the fattest: a theory for assessing acute effects of hydrophobic, reversibly acting chemicals on populations. *Ecology*, **109**:411–429, 1988.

[686] D. M. Lavigne, W. Barchard, S. Innes, and N. A. Oritsland. *Pinniped Bioenergetics*, vol. IV of *FAO Fisheries Series No. 5*. FAO, Rome, 1981.

[687] R. M. Laws. The elephant seal (*Mirounga leonina* Linn). 1. Growth and age. *RIDS Sci. Rep.*, **8**:1–37, 1953.

[688] R. M. Laws. Age criteria for the African elephant, *Loxodonta a. africana*. *E. Afr. Wildl.*, **4**:1–37, 1966.

[689] M. LeCroy and C. T. Collins. Growth and survival of roseate and common tern chicks. *Auk*, **89**:595–611, 1972.

[690] J. J. Lee, G. F. Leedale, and P. Bradbury. *An Illustrated Guide to the Protozoa*. Society of Protozoologists, Lawrence, KS, 2000.

[691] I. M. M. van Leeuwen. Mathematical models in cancer risk assessment. PhD thesis, Vrije Universiteit, Amsterdam, 2003.

[692] I. M. M. van Leeuwen, F. D. L. Kelpin, and S. A. L. M. Kooijman. A mathematical model that accounts for the effects of caloric restriction on body weight and longevity. *Biogerontology*, **3**:373–381, 2002.

[693] I. M. M. van Leeuwen, C. Zonneveld, and S. A. L. M. Kooijman. The embedded tumor: host physiology is important for the interpretation of tumor growth. *Br. J. Cancer*, **89**:2254–2263, 2003.

[694] R. A. Leigh and D. Sanders. *The Plant Vacuole*. Academic Press, San Diego, 1997.

[695] W. A. Lell. The relation of the volume of the amniotic fluid to the weight of the fetus at different stages of pregnancy in the rabbit. *Anat. Rec.*, **51**:119–123, 1931.

[696] V. Lemesle and L. Mailleret. Mechanistic investigation of the algae growth "Droop" model. *Acta Biotheor.*, **56**:87–102, 2008.

[697] J. W. Lengeler, G. Drews, and H. G. Schlegel, editors. *Biology of the Prokaryotes*. Thieme, Stuttgart, 1999.

[698] R. Leudeking and E. L. Piret. A kinetic study of the lactic acid fermentation. *J. Biochem. Microbiol. Technol. Eng.*, **1**:393, 1959.

[699] B. Levine and D. J. Klionsky. Development by self-digestion: molecular mechanisms and biological functions of autophagy. *Devel. Cell*, **6**:463–477, 2004.

[700] G. Levy. Demonstration of Michaelis–Menten kinetics in man. *J. Pharm. Sci.*, **54**:496, 1965.

[701] G. Levy, A. W. Vogel, and L. P. Amsel. Capacity-limited salicylurate formation during prolonged administration of aspirin to healthy human subjects. *J. Pharm. Sci.*, **58**:503–504, 1969.

[702] S. M. Libes. *An Introduction to Marine Biogeochemistry*. J. Wiley, New York, 1992.

[703] J. von Liebig. *Chemistry in its Application to Agriculture and Physiology*. Taylor & Walton, London, 1840.

[704] N. Lifson and R. McClintock. Theory of use of the turnover rates of body water for measuring energy and material balance. *J. Theor. Biol.*, **12**:46–74, 1966.

[705] J. D. Ligon. Still more responses of the poor-will to low temperatures. *Condor*, **72**:496–498, 1970.

[706] K. Lika and R. M. Nisbet. A dynamic energy budget model based on partitioning of net production. *J. Math. Biol.*, **41**:361–386, 2000.

[707] K. Lika and N. Papandroulakis. Modeling feeding processes: a test of a new model for sea bream (*Sparus aurata*) larvae. *Can. J. Fish. Aquat. Sci.*, **99**:1–11, 2004.

[708] P. A. Lindahl and B. Chang. The evolution of acetyl-CoA synthase. *Orig. Life Evol. Biosphere*, **31**:403–434, 2001.

[709] M. R. Lindsay, R. J. Webb, M. Strouss, M. S. M. Jetten, M. K. Butler, R. J. Forde, and J. A. Fuerst. Cell compartmentalisation in planctomycetes: novel types of structural organisation for the bacterial cell. *Arch. Microbiol.*, **175**:413–429, 2001.

[710] L. G. Ljungdahl. The acetyl-CoA pathway and the chemiosmotic generation of ATP during acetogenesis. In H. L. Drake, editor, *Acetogenesis*, pages 63–87. Chapman & Hall, New York, 1994.

[711] C. Lockyer. Growth and energy budgets of large baleen whales from the southern hemisphere. In *Mammals in the Seas*, vol. III of *FAO Fisheries series No. 5*. FAO, Rome, 1981.

[712] B. E. Logan and J. R. Hunt. Bioflocculation as a microbial response to substrate limitations. *Biotechnol. Bioeng.*, **31**:91–101, 1988.

[713] B. E. Logan and D. L. Kirchman. Uptake of dissolved organics by marine bacteria as a function of fluid motion. *Mar. Biol.*, **111**:175–181, 1991.

[714] H. Lohrl. *Die Tannemeise*. A. Ziemsen-Verlag, Wittenberg Lutherstadt, 1977.

[715] J. Loman. Microevolution and maternal effects on tadpole *Rana temporaria* growth and development rate. *J. Zool. (Lond.)*, **257**:93–99, 2002.

[716] A. Łomnicki. *Population Ecology of Individuals*. Princeton University Press, 1988.

[717] J. A. Long, K. Trinajstic, G. C. Young, and T. Senden. Live birth in the Devonian period. *Nature*, **453**:650–651, 2008.

[718] B. G. Loughton and S. S. Tobe. Blood volume in the African migratory locust. *Can. J. Zool.*, **47**:1333–1336, 1969.

[719] G. Louw. *Physiological Animal Ecology.* Longman, Harlow, 1993.

[720] B. Lovegrove. *The Living Deserts of Southern Africa.* Fernwood Press, Vlaeberg, 1993.

[721] B. G. Lovegrove and C. Wissel. Sociality in mole-rats: metabolic scaling and the role of risk sensitivity. *Oecologia (Berlin)*, **74**:600–606, 1988.

[722] Y. Loya and L. Fishelson. Ecology of fish breeding in brackish water ponds near the Dead Sea (Israel). *J. Fish Biol.*, **1**:261–278, 1969.

[723] A. G. Lyne. Observations on the breeding and growth of the marsupial *Pelametes nasuta* geoffroy, with notes on other bandicoots. *Austral J. Zool.*, **12**:322–339, 1964.

[724] J. W. MacArthur and W. H. T. Baillie. Metabolic activity and duration of life. 1. Influence of temperature on longevity in *Daphnia magna. J. Exp. Zool.*, **53**:221–242, 1929.

[725] R. H. MacArthur and E. O. Wilson. *The Theory of Island Biogeography.* Princeton University Press, 1967.

[726] D. MacDold. *European Mammals: Evolution and Behaviour.* HarperCollins, London, 1995.

[727] B. A. MacDonald and R. J. Thompson. Influence of temperature and food availability on ecological energetics of the giant scallop *Placopecten magellanicus.* 1. Growth rates of shell and somatic tissue. *Mar. Ecol. Prog. Ser.*, **25**:279–294, 1985.

[728] E. C. MacDowell, E. Allen, and C. G. Macdowell. The prenatal growth of the mouse. *J. Gen. Physiol.*, **11**:57–70, 1927.

[729] D. MacKay. Correlation of bioconcentration factors. *Environ. Sci. Technol.*, **16**:274–278, 1982.

[730] D. MacKay and S. Paterson. Calculating fugacity. *Environ. Sci. Technol.*, **15**:1006–1014, 1981.

[731] D. Mackay, W. Y. Shiu, and K. C. Ma. *Illustrated Handbook of Physical–Chemical Properties and Environmental Fate for Organic Chemicals.* Lewis Publishers, Chelsea, 1992.

[732] M. T. Madigan, J. M. Martinko, and J. Parker. *Brock Biology of Microorganisms.* Prentice-Hall, Englewood Cliffs, 2000.

[733] T. H. Mague, E. Friberg, D. J. Hughes, and I. Morris. Extracellular release of carbon by marine phytoplankton; a physiological approach. *Limnol. Oceanogr.*, **25**:262–279, 1980.

[734] R. A. Mah, D. M. Ward, L. Baresi, and T. L. Glass. Biogenesis of methane. *Annu. Rev. Microbiol.*, **31**:309–341, 1977.

[735] S. P. Mahoney and W. Threlfall. Notes on the eggs, embryos and chick growth of the common guillemots *Uria aalge* in Newfoundland. *Ibis*, **123**:211–218, 1981.

[736] U.-G. Maier, S. E. Douglas, and T. Cavalier-Smith. The nucleomorph genomes of cryptophytes and chlorarachniophytes. *Protist*, **151**:103–109, 2000.

[737] S. E. Manahan. *Environmental Chemistry.* Lewis Publishers, 1994.

[738] K. H. Mann and J. R. N. Lazier. *Dynamics of Marine Ecosystems.* Blackwell Sci. Publ., Oxford, 1996.

[739] R. H. K. Mann. The growth and reproductive stategy of the gudgeon, *Gobio gobio* (L.), in two hard-water rivers in southern England. *J. Fish Biol.*, **17**:163–176, 1980.

[740] S. C. Manolis, G. J. W. Webb, and K. E. Dempsey. Crocodile egg chemistry. In G. J. W. Webb, S. C. Manolis, and P. J. Whitehead, editors, *Wildlife Management: Crocodiles and Alligators*, pages 445–472. Beatty, Sydney, 1987.

[741] D. A. Manuwal. The natural history of Cassin's auklet (*Ptychoramphus aleuticus*). *Condor*, **76**:421–431, 1974.

[742] L. Margulis. *Origins of Eukaryotic Cells.* W. H. Freeman, San Fransisco, 1970.

[743] L. Margulis. *Symbiosis in Cell Evolution.* W. H. Freeman, New York, 1993.

[744] L. Margulis and R. Fester, editors. *Symbiosis as a Source of Evolutionary Innovation.* MIT Press, Cambridge, 1991.

[745] D. M. Maron and B. N. Ames. Revised methods for the *Salmonella* mutagenicity test. *Mutat. Res.*, **113**:173–215, 1983.

[746] A. G. Marr, E. H. Nilson, and D. J. Clark. The maintenance requirement of *Escherichia coli. Ann. N.Y. Acad. Sci.*, **102**:536–548, 1963.

[747] M. M. Martin. *Invertebrate–Microbial Interactions: Ingested Fungal Enzymes in Arthropod Biology.* Comstock, Ithaca, 1987.

[748] W. Martin and M. Muller. The hydrogen hypothesis for the first eukaryote. *Nature*, **392**:37–41, 1998.

[749] W. Martin, C. Rotte, M. Hoffmeister, U. Theissen, G. Gelius-Dietrich, A. Ahr, and K. Henze. Early cell evolution, eukaryotes, anoxia, sulfide, oxygen, fungi first (?), and a three of genomes revisited. *Life*, **55**:193–204, 2003.

[750] W. Martin, T. Rujan, E. Richly, A. Hansen, Cornelsen, S., T. Lins, D. Leister, B. Stoebe, M. Hasegawa, and D. Penny. Evolutionary analysis of *Arabidopsis*, cyanobacterial, and chloroplast genomes reveals plastid phylogeny and thousands of cyanobacterial genes in the nucleus. *Proc. Natl. Acad. Sci. U.S.A.*, **99**:12246–12251, 2002.

[751] W. Martin and M. Russel. On the origin of cells: a hypothesis for the evolutionary transitions from abiotic geochemistry to chemoautotrophic prokaryotes and from prokaryotes to nucleated cell. *Phil. Trans. R. Soc. Lond. B Biol. Sci.*, **358**:59–85, 2003.

[752] W. Martin and C. Schnarrenberger. The evolution of the Calvin cycle from prokaryotic to eukaryotic chromosomes: a case study of functional redundancy in ancient pathways through endosymbiosis. *Curr. Genet.*, **32**:1–18, 1997.

[753] W. Martin, B. Stoebe, V. Goremykin, S. Hansmann, M. Hasegawa, and K. V. Kowallik. Gene transfer to the nucleus and the evolution of chloroplasts. *Nature*, **393**:162–165, 1998.

[754] D. Masman. The annual cycle of the kestrel *Falco tinnunculus*: a study in behavioural energetics. PhD thesis, University of Groningen, 1986.

[755] G. L. C. Matthaei. On the effect of temperature on carbon-dioxide assimilation. *Phil. Trans. R. Soc. Lond. B Biol. Sci.*, **197**:47–105, 1905.

[756] A. G. Matthysse, K. Deschet, M. Williams, M. Marry, A. R. White, and W. C. Smith. A functional cellulose synthase from ascidian epidermis. *Proc. Natl. Acad. Sci. U.S.A.*, **101**:986–991, 2004.

[757] J. Mauchline and L. R. Fisher. *The Biology of Euphausids*. Academic Press, London, 1969.

[758] J. E. Maunder and W. Threlfall. The breeding biology of the black-legged kittiwake in Newfoundland. *Auk*, **89**:789–816, 1972.

[759] R. M. May. *Stability and Complexity in Model Ecosystems*. Princeton University Press, 1973.

[760] J. Maynard Smith, N. H. Smith, M. O'Rourke, and B. G. Spratt. How clonal are bacteria? *Proc. Natl. Acad. Sci. U.S.A.*, **90**:4384–4388, 1993.

[761] W. V. Mayneord. On a law of growth of Jensen's rat sarcoma. *Am. J. Cancer*, **16**:841–846, 1932.

[762] G. M. Maynes. Growth of the Parma wallaby, *Macropus parma* Waterhouse. *Austral J. Zool.*, **24**:217–236, 1976.

[763] K. McCann and P. Yodzis. Nonlinear dynamics and population disappearances. *Am. Nat.*, **144**:873–879, 1994.

[764] K. McCann and P. Yodzis. Bifurcation structure of a tree-species food chain model. *Theor. Pop. Biol.*, **48**:93–125, 1995.

[765] E. McCauley, W. W. Murdoch, and R. M. Nisbet. Growth, reproduction, and mortality of *Daphnia pulex* Leydig: life at low food. *Funct. Ecol.*, **4**:505–514, 1990.

[766] P. McCullagh and J. A. Nelder. *Generalized Linear Models*. Chapman & Hall, London, 1983.

[767] E. H. McEwan. Growth and development of the barren-ground caribou. 2. Postnatal growth rates. *Can. J. Zool.*, **46**:1023–1029, 1968.

[768] G. I. McFadden. Primary and secondary endosymbiosis and the origin of plastids. *J. Phycol.*, **37**:951–959, 2001.

[769] M. D. McGurk. Effects of delayed feeding and temperature on the age of irreversible starvation and on the rates of growth and mortality of pacific herring larvae. *Mar. Biol.*, **84**:13–26, 1984.

[770] G. S. McIntyre and R. H. Gooding. Egg size, contents, and quality: maternal-age and-size effects on house fly eggs. *Can. J. Zool.*, **78**:1544–1551, 2000.

[771] J. W. McMahon. Some physical factors influencing the feeding behavior of *Daphnia magna* Straus. *Can. J. Zool.*, **43**:603–611, 1965.

[772] T. A. McMahon. *Size and Shape in Biology*. Freeman, New York, 1973.

[773] T. A. McMahon and J. T. Bonner. *On Size and Life*. Scientific American Library, New York, 1983.

[774] B. K. McNab. On the ecological significance of Bergmann's rule. *Ecology*, **52**:845–854, 1971.

[775] D. A. McQuarrie. Stochastic approach to chemical kinetics. *J. Appl. Prob.*, **4**:413–478, 1967.

[776] J. van der Meer. Metabolic theories in biology. *Trends Ecol. Evol.*, **21**:136–140, 2006.

[777] E. M. Meijer, H. W. van Verseveld, E. G. van der Beek, and A. H. Stouthamer. Energy conservation during aerobic growth in *Paracoccus denitrificans*. *Arch. Microbiol.*, **112**:25–34, 1977.

[778] T. H. M. Meijer. Reproductive decisions in the kestrel *Falco tinnunculus*: a study in physiological ecology. PhD thesis, University of Groningen, 1988.

[779] T. H. M. Meijer, S. Daan, and M. Hall. Family planning in the kestrel (*Falco tinnuculus*): the proximate control of covariation of laying data and clutch size. *Behaviour*, **114**:117–136, 1990.

[780] E. Meléndez-Hevia. The game of the pentose phosphate cycle: a mathematical approach to study the optimization in design of metabolic pathways during evolution. *Biomed. Biochem. Acta*, **49**:903–916, 1990.

[781] E. Meléndez-Hevia and A. Isidoro. The game of the pentose phosphate cycle. *J. Theor. Biol.*, **117**:251–263, 1985.

[782] M. L. Mendelsohn. Cell proliferation and tumor growth. In L. F. Lamerton and R. J. M. Fry, editors, *Cell Proliferation*, pages 190–210. Blackwell Sci. Publ., Oxford, 1963.

[783] J. A. J. Metz and F. H. D. van Batenburg. Holling's 'hungry mantid' model for the invertebrate functional response considered as a Markov process. Part I: The full model and some of its limits. *J. Math. Biol.*, **22**:209–238, 1985.

[784] J. A. J. Metz and F. H. D. van Batenburg. Holling's 'hungry mantid' model for the invertebrate functional response considered as a Markov process. 2. Negligible handling time. *J. Math. Biol.*, **22**:239–257, 1985.

[785] J. A. J. Metz and O. Diekmann. *The Dynamics of Physiologically Structured Populations*. Springer-Verlag, Berlin, 1986.

[786] J. A. J. Metz and O. Diekmann. Exact finite dimensional representations of models for physiologically structured populations. 1. The abstract foundations of linear chain trickery. In J. A. Goldstein, F. Kappel, and W. Schappacher, editors, *Differential Equations with Applications in Biology, Physics, and Engineering*, pages 269–289. Marcel Dekker, New York, 1991.

[787] J. A. J. Metz, S. A. H. Geritz, G. Meszna, F. J. A. Jacobs, and J. S. van Heerwaarden. Adaptive dynamics, a geometrical study of the consequences of nearly faithful reproduction. In S. J. van Strien and S. M. Verduyn Lunel, editors, *Stochastic and Spatial Structures of Dynamical Systems*, pages 183–231. North-Holland, Amsterdam, 1996.

[788] E. A. Meuleman. Host–parasite interrelations between the freshwater pulmonate *Biomphalaria pfeifferi* and the trematode *Schistosoma mansoni*. *Neth. J. Zool.*, **22**:355–427, 1972.

[789] H. Meyer and L. Ahlswede. Über das intrauterine Wachstum und die Körperzusammensetzung von Fohlen sowie den Nährstoffbedarf tragender Stuten. *Übersicht. Tierernähr.*, **4**:263–292, 1976.

[790] D. G. Meyers. Egg development of a chydorid cladoceran, *Chydorus sphaericus*, exposed to constant and alternating temperatures: significance to secondary productivity in fresh water. *Ecology*, **61**:309–320, 1984.

[791] W. Michaelis, R. Seifert, K. Nauhaus, *et al.* Microbial reefs in the black sea fuelled by anaerobic oxidation of methane. *Science (New York)*, **297**:1013–836, 2002.

[792] R. Mikelsaar. A view of early cellular evolution. *J. Mol. Evol.*, **25**:168–183, 1987.

[793] H. Mikkola. *Der Bartkauz*. A. Ziemsen-Verlag, Wittenberg Lutherstadt, 1981.

[794] T. T. Milby and E. W. Henderson. The comparative growth rates of turkeys, ducks, geese and pheasants. *Poult. Sci.*, **16**:155–165, 1937.

[795] J. S. Millar. Post partum reproductive characteristics of eutherian mammals. *Evolution*, **35**:1149–1163, 1981.

[796] P. J. Miller. Age, growth, and reproduction of the rock goby, *Gobius paganellus* L., in the Isle of Man. *J. Mar. Biol. Assoc. U.K.*, **41**:737–769, 1961.

[797] P. Milton. Biology of littoral blenniid fishes on the coast of south-west England. *J. Mar. Biol. Assoc. U.K.*, **63**:223–237, 1983.

[798] A. Mira. Why is meiosis arrested? *J. Theor. Biol.*, **194**:275–287, 1998.

[799] J. K. Misra and R. W. Lichtwardt. *Illustrated Genera of Trichomycetes: Fungal Symbionts of Insects and Other Arthropods*. Science Publishers, Enfield, NH, 2000.

[800] J. M. Mitchison. The growth of single cells. 3. *Streptococcus faecalis*. *Exp. Cell Res.*, **22**:208–225, 1961.

[801] G. G. Mittelbach. Predation and resource partitioning in two sunfishes (Centrarchidae). *Ecology*, **65**:499–513, 1984.

[802] J. Monod. *Recherches sur la croissance des cultures bactériennes*, 2nd edn. Hermann, Paris, 1942.

[803] J. L. Monteith and M. H. Unsworth. *Principles of Environmental Physics*. E. Arnold, London, 1990.

[804] D. Moore. *Fungal Morphogenesis*. Cambridge University Press, 1998.

[805] N. Moran and P. Baumann. Phylogenetics of cytoplasmically inherited microorganisms of arthropods. *Trends Ecol. Evol.*, **9**:15–20, 1994.

[806] F. M. M. Morel. Kinetics of nutrient uptake and growth in phytoplankton. *J. Phycol.*, **23**:137–150, 1987.

[807] S. Moreno, P. Nurse, and P. Russell. Regulation of mitosis by cyclic accumulation of $p80^{cdc25}$ mitotic inducer in fission yeast. *Nature*, **344**:549–552, 1990.

[808] Y. Moret, P. Juchault, and T. Rigaud. *Wolbachia* endosymbiont responsible for cytoplasmic incompatibility in a terrestrial crustacean: effects in natural and foreign hosts. *Heredity*, **86**:325–332, 2001.

[809] S. A. Morley, R. S. Batty, P. Geffen, and A. J. Tytler. Egg size manipulation: a technique for investigating maternal effects on the hatching characteristics of herring. *J. Fish Biol.*, **55 (suppl. A)**:233–238, 1999.

[810] H. J. Morowitz, J. D. Kosttelrik, J. Yang, and G. D. Cody. The origin of intermediary metabolism. *Proc. Natl. Acad. Sci. U.S.A.*, **97**:7704–7708, 2000.

[811] J. A. Morrison, C. E. Trainer, and P. L. Wright. Breeding season in elk as determined from known-age embryos. *J. Wildl. Manage.*, **23**:27–34, 1959.

[812] G. J. Morton, D. E. Cummings, D. G. Baskin, G. S. Barsh, and M. W. Schwartz. Central nervous system control of food intake and body weight. *Nature*, **443**:289–294, 2006.

[813] B. S. Muir. Comparison of growth rates for native and hatchery-stocked populations of *Esox masquinongy* in Nogies Creek, Ontario. *J. Fish. Res. Board Can.*, **17**:919–927, 1960.

[814] M. M. Mulder. Energetic aspects of bacterial growth: a mosaic non-equilibrium thermodynamic approach. PhD thesis, University of Amsterdam, 1988.

[815] E. B. Muller. Bacterial energetics in aerobic wastewater treatment. PhD thesis, Vrije Universiteit, Amsterdam, 1994.

[816] M. T. Murphy. Ecological aspects of the reproductive biology of eastern kingbirds: geographic comparisons. *Ecology*, **64**:914–928, 1983.

[817] A. W. Murray and M. W. Kirschner. What controls the cell cycle? *Sci. Am.*, **3**:34–41, 1991.

[818] L. Muscatine and H. M. Lenhoff. Symbiosis of hydra and algae. 2. Effects of limited food and starvation on growth of symbiotic and aposymbiotic hydra. *Biol. Bull. (Woods Hole)*, **129**:316–328, 1965.

[819] L. Muscatine, L. R. McCloskey, and R. E. Marian. Estimating the daily contribution of carbon from zooxanthellae to coral animal respiration. *Limnol. Oceanogr.*, **26**:601–611, 1981.

[820] L. Muscatine and V. Weis. Productivity of zooxanthellae and biogeochemical cycles. In P. G. Falkowski and A. D. Woodhed, editors, *Primary Productivity and Biochemical Cycles in the Sea*, pages 257–271. Plenum Press, New York, 1992.

[821] S. Nagasawa. Parasitism and diseases in chaetognaths. In Q. Bone, H. Kapp, and A. C. Pierrot-Bults, editors, *The Biology of Chaetognaths*, pages 76–85. Oxford University Press, 1991.

[822] S. Nagasawa and T. Nemoto. Presence of bacteria in guts of marine crustaceans on their fecal pellets. *J. Plankton Res.*, **8**:505–517, 1988.

[823] R. G. Nager, P. Monaghan, and D. C. Houston. Within-clutch trade-offs between the number and quality of eggs: experimental manipulations in gulls. *Ecology*, **81**:1339–1350, 2000.

[824] R. G. Nager, P. Monaghan, D. C. Houston, K. E. Arnold, J. D. Blount, and N. Berboven. Maternal effects through the avian egg. *Acta Zool. Sinica*, **52 (suppl.)**:658–661, 2006.

[825] H. Nakajima and Z. Kawabata. Sensitivity analysis in microbial communities. In R. R. Colwell, U. Simidiu, and K. Ohwada, editors, *Microbial Diversity in Time and Space*, pages 85–91. Plenum Press, New York, 1996.

[826] A. Narang, A. Konopka, and D. Ramkrishna. New patterns of mixed substrate growth in batch cultures of *Escherichia coli* K12. *Biotechnol. Bioeng.*, **55**:747–757, 1997.

[827] R. D. M. Nash. The biology of Fries' goby, *Lesueurigobius friesii* (Malm) in the Firth of Clyde, Scotland, and a comparison with other stocks. *J. Fish Biol.*, **21**:69–85, 1982.

[828] S. Nee, A. F. Read, J. D. Greenwood, and P. H. Harvey. The relationship between abundance and body size in British birds. *Nature*, **351**:312–313, 1991.

[829] F. C. Neidhardt, J. L. Ingraham, and M. Schaechter. *Physiology of the Bacterial Cell: A Molecular Approach.* Sinauer, Sunderland, 1990.

[830] J. A. Nelder. The fitting of a generalization of the logistic curve. *Biometrics*, **17**:89–110, 1961.

[831] B. Nelson. *The Gannet.* T. & A. D. Poyser, Berkhamsted, 1978.

[832] E. Nestaas and D. I. C. Wang. Computer control of the penicillum fermentation using the filtration probe in conjunction with a structured model. *Biotechnol. Bioeng.*, **25**:781–796, 1983.

[833] M. C. Newman and D. K. Doubet. Size dependence of mercury (II) accumulation in the mosquitofish *Gambusia affinis* (Baird and Girard). *Arch. Environ. Contam. Toxicol.*, **18**:819–825, 1989.

[834] M. C. Newman and S. V. Mitz. Size dependence of zinc elimination and uptake from water by mosquitofish *Gambusia affinis* (baird and girard). *Aquat. Toxicol.*, **12**:17–32, 1988.

[835] I. Newton. *The Migration Ecology of Birds.* Elsevier, Amsterdam, 2008.

[836] A. J. Niimi and B. G. Oliver. Distribution of polychlorinated biphenyl congeners and other hydrocarbons in whole fish and muscle among Ontario salmonids. *Environ. Sci. Technol.*, **23**:83–88, 1989.

[837] K. J. Niklas. *Plant Allometry: The Scaling of Form and Process.* Chicago University Press, 1994.

[838] B. Nisbet. *Nutrition and Feeding Strategies in Protozoa.* Croom Helm, London, 1984.

[839] E. G. Nisbet and C. M. R. Fowler. Archaean metabolic evolution of microbial mate. *Proc. R. Soc. Lond. B Biol. Sci.*, **266**:2375–2382, 1999.

[840] R. M. Nisbet, A. Cunningham, and W. S. C. Gurney. Endogenous metabolism and the stability of microbial prey–predator systems. *Biotechnol. Bioeng.*, **25**:301–306, 1983.

[841] R. M. Nisbet, A. H. Ross, and A. J. Brooks. Empirically-based dynamic energy budget models: theory and application to ecotoxicology. *Nonlinear World*, **3**:85–106, 1996.

[842] P. S. Nobel. *Physicochemical and Environmental Plant Physiology.* Academic Press, San Diego, 1999.

[843] D. R. Nobles and R. M. Brown. The pivotal role of cyanobacteria in the evolution of cellulose synthase and cellulose synthase-like proteins. *Cellulose*, **11**:437–448, 2004.

[844] V. Norris and D. J. Raine. A fission–fusion origin for life. *Orig. Life Evol. Biosphere*, **29**:523–537, 1998.

[845] T. H. Y. Nose and Y. Hiyama. Age determination and growth of yellowfin tuna, *Thunnus albacares* Bonnaterre by vertebrae. *Bull. Jpn. Soc. Sci. Fish.*, **31**:414–422, 1965.

[846] W. J. O'Brian. The dynamics of nutrient limitation of phytoplankton algae: a model reconsidered. *Ecology*, **55**:135–141, 1974.

[847] A. Okubo. *Diffusion and Ecological Problems: Mathematical Models.* Springer-Verlag, Berlin, 1980.

[848] B. G. Oliver and A. J. Niimi. Trophodynamic analysis of polychlorinated biphenyl congeners and other chlorinated hydrocarbons in Lake Ontario ecosystem. *Environ. Sci. Technol.*, **22**:388–397, 1988.

[849] G. J. Olsen and C. R. Woese. Lessons from an archaeal genome: what are we learning from *Methanococcus jannaschii*? *Trends Genet.*, **12**:377–379, 1996.

[850] A. W. Omta. Eddies and algal stoichiometry: physical and biological influences on the organic carbon pump. PhD thesis, Vrije Universiteit, Amsterdam, 2009.

[851] A. W. Omta, J. Bruggeman, S. A. L. M. Kooijman, and H. A. Dijkstra. The biological carbon pump revisited: feedback mechanisms between climate and the redfield ratio. *Geophys. Res. Lett.*, 2006.

[852] R. V. O'Neill, D. L. DeAngelis, J. J. Pastor, B. J. Jackson, and W. M. Post. Multiple nutrient limitations in ecological models. *Ecol. Modell.*, **46**:147–163, 1989.

[853] A. Opperhuizen. Bioconcentration in fish and other distribution processes of hydrophobic chemicals in aquatic environments. PhD thesis, University of Amsterdam, 1986.

[854] L. E. Orgel. Self-organizing biochemical cycles. *Proc. Natl. Acad. Sci. U.S.A.*, **97**:12503–12507, 2000.

[855] R. Österberg. On the prebiotic role of iron and sulfur. *Orig. Life Evol. Biosphere*, **27**:481–484, 1997.

[856] K. W. Osteryoung and J. Nunnari. The division of endosymbiotic organelles. *Science*, **302**:1698–1704, 2003.

[857] A. Y. Ota and M. R. Landry. Nucleic acids as growth rate indicators for early developmental stages of *Calanus pacificus* Brodsky. *J. Exp. Mar. Biol. Ecol.*, **80**:147–160, 1984.

[858] S. Otte, J. G. Kuenen, L. P. Nielsen, H. W. Pearl, J. Zopfi, H. N. Schulz, A. Teska, B. Strotmann, V. A. Gallardo, and B. B. Jørgenssen. Nitrogen, carbon, and sulfur metabolism in natural *Tioploca* samples. *Appl. Environ. Microbiol.*, **65**:3148–3157, 1999.

[859] M. J. Packard, G. C. Packard, and W. H. N. Gutzke. Calcium metabolism in embryos of the oviparous snake *Coluber constrictor*. *J. Exp. Biol.*, **110**:99–112, 1984.

[860] M. J. Packard, G. C. Packard, J. D. Miller, M. E. Jones, and W. H. N. Gutzke. Calcium mobilization, water balance, and growth in embryos of the agamid lizard *Amphibolurus barbatus*. *J. Exp. Zool.*, **235**:349–357, 1985.

[861] M. J. Packard, T. M. Short, G. C. Packard, and T. A. Gorell. Sources of calcium for embryonic development in eggs of the snapping turtle *Chelydra serpentina*. *J. Exp. Zool.*, **230**:81–87, 1984.

[862] R. T. Paine. Growth and size distribution of the brachiopod *Terebratalia transversa* Sowerby. *Pac. Sci.*, **23**:337–343, 1969.

[863] J. E. Paloheimo, S. J. Crabtree, and W. D. Taylor. Growth model for *Daphnia*. *Can. J. Fish. Aquat. Sci.*, **39**:598–606, 1982.

[864] D. Pantaloni, C. Le Clainche, and M.-F. Carlier. Mechanism of actin-based motility. *Science*, **292**:1502–1506, 2001.

[865] S. Papa and V. P. Skulacheve. Reactive oxygen species, mitochondria, apoptosis and aging. *Mol. Cell Biochem.*, **174**:305–319, 1997.

[866] D. Papineau, J. J. Walker, S. J. Moizsis, and N. R. Pace. Composition and structure of microbial communities from stromatolites of Hamelin Pool in Shark Bay, Western Australia. *Appl. Environ. Microbiol.*, **71**:4822–4832, 2005.

[867] J. R. Parks. *A Theory of Feeding and Growth of Animals*. Springer-Verlag, Berlin, 1982.

[868] M. Parniske and J. A. Downie. Lock, keys and symbioses. *Nature*, **425**:569–570, 2003.

[869] G. D. Parry. The influence of the cost of growth on ectotherm metabolism. *J. Theor. Biol.*, **101**:453–477, 1983.

[870] S. J. Parulekar, G. B. Semones, M. J. Rolf, J. C. Lievense, and H. C. Lim. Induction and elimination of oscillations in continuous cultures of *Saccharomyces cerevisiae*. *Biotechnol. Bioeng.*, **28**:700–710, 1986.

[871] D. J. Patterson. The diversity of eukaryotes. *Am. Nat.*, **154**:S96–S124, 1999.

[872] M. R. Patterson. A chemical engineering view of cnidarian symbioses. *Am. Zool.*, **32**:566–582, 1992.

[873] D. Pauly and P. Martosubroto. The population dynamics of *Nemipterus marginatus* (Cuvier & Val.) off Western Kalimantan, South China Sea. *J. Fish Biol.*, **17**:263–273, 1980.

[874] P. R. Payne and E. F. Wheeler. Growth of the foetus. *Nature*, **215**:849–850, 1967.

[875] R. Pearl. The growth of populations. *Q. Rev. Biol.*, **2**:532–548, 1927.

[876] L. Pecquerie. Bioenergetic modelling of the growth, development and reproduction of a small pelagic fish: the Bay of Biscay anchovy. PhD thesis, Agrocampus Rennes and Vrije Universiteit, Amsterdam, 2008.

[877] L. Pecquerie, P. Petitgas, and S. A. L. M. Kooijman. Environmental impact on anchovy spawning duration in the context of the dynamic energy budget theory. *J. Sea Res.*, **62**:(in press), 2009.

[878] T. J. Pedley. *Scale Effects in Animal Locomotion*. Academic Press, London, 1977.

[879] D. L. Penry and P. A. Jumars. Chemical reactor analysis and optimal digestion. *BioScience*, **36**:310–315, 1986.

[880] D. L. Penry and P. A. Jumars. Modeling animal guts as chemical reactors. *Am. Nat.*, **129**:69–96, 1987.

[881] B. Peretz and L. Adkins. An index of age when birthdate is unknown in *Aplysia californica*: shell size and growth in long-term maricultured animals. *Biol. Bull. (Woods Hole)*, **162**:333–344, 1982.

[882] A. R. R. Péry, J. J. M. Bedaux, C. Zonneveld, and S. A. L. M. Kooijman. Analysis of bioassays with time-varying concentrations. *Water Res.*, **35**:3825–3832, 2001.

[883] A. R. R. Péry, P. Flammarion, B. Vollat, J. J. M. Bedaux, S. A. L. M. Kooijman, and J. Garric. Using a biology-based model (DEBtox) to analyse bioassays in ecotoxicology: opportunities and recommendations. *Environ. Toxicol. Chem.*, **21**:459–465, 2002.

[884] R. H. Peters. *The Ecological Implications of Body Size.* Cambridge University Press, 1983.

[885] T. N. Pettit, G. S. Grant, G. C. Whittow, H. Rahn, and C. V. Paganelli. Respiratory gas exchange and growth of white tern embryos. *Condor*, **83**:355–361, 1981.

[886] T. N. Pettit, G. S. Grant, G. C. Whittow, H. Rahn, and C. V. Paganelli. Embryonic oxygen consumption and growth of laysan and black-footed albatross. *Am. J. Physiol.*, **242**:121–128, 1982.

[887] T. N. Pettit and G. C. Whittow. Embryonic respiration and growth in two species of noddy terns. *Physiol. Zool.*, **56**:455–464, 1983.

[888] T. N. Pettit, G. C. Whittow, and G. S. Grant. Caloric content and energetic budget of tropical seabird eggs. In G. C. Whittow and H. Rahn, editors, *Seabird Energetics*, pages 113–138. Plenum Press, New York, 1984.

[889] H. G. Petzold. *Die Anakondas.* A. Ziemsen-Verlag, Wittenberg Lutherstadt, 1984.

[890] J. Phillipson. Bioenergetic options and phylogeny. In C. R. Townsend and P. Calow, editors, *Physiological Ecology*, pages 20–45. Blackwell Sci. Publ., Oxford, 1981.

[891] E. R. Pianka. On r and K selection. *Am. Nat.*, **104**:592–597, 1970.

[892] E. R. Pianka. *Evolutionary Ecology.* Harper & Row, London, 1978.

[893] E. R. Pianka. *Ecology and Natural History of Desert Lizards: Analyses of the Ecological Niche of Community Structure.* Princeton University Press, 1986.

[894] T. Piersma. Estimating energy reserves of great crested grebes *Podiceps cristatus* on the basis of body dimensions. *Ardea*, **72**:119–126, 1984.

[895] T. Piersma, N. Cadée, and S. Daan. Seasonality in basal metabolic rate and thermal conductance in a long-distance migrant shorebird, the knot (*Calidris canutus*). *J. Comp. Physiol. B*, **165**:37–45, 1995.

[896] T. Piersma and R. I. G. Morrison. Energy expenditure and water turnover of incubating ruddy turnstones: high costs under high arctic climatic conditions. *Auk*, **111**:366–376, 1994.

[897] B. K. Pierson, H. K. Mitchell, and A. L. Ruff-Roberts. *Chloroflexus aurantiacus* and ultraviolet radiation: implications for Archean shallow-water stromatolites. *Orig. Life Evol. Biosphere*, **23**:243–260, 1993.

[898] J. Pilarska. Eco-physiological studies on *Brachionus rubens* Ehrbg (*Rotatoria*). 1. Food selectivity and feeding rate. *Pol. Arch. Hydrobiol.*, **24**:319–328, 1977.

[899] E. M. del Pino. Marsupial frogs. *Sci. Am.*, **5**:76–84, 1989.

[900] S. J. Pirt. The maintenance energy of bacteria in growing cultures. *Proc. R. Soc. Lond. B Biol. Sci.*, **163**:224–231, 1965.

[901] S. J. Pirt. *Principles of Microbe and Cell Cultivation.* Blackwell Sci. Publ., Oxford, 1975.

[902] S. J. Pirt and D. S. Callow. Studies of the growth of *Penicillium chrysogenum* in continuous flow culture with reference to penicillin production. *J. Appl. Bacteriol.*, **23**:87–98, 1960.

[903] G. Pohle and M. Telford. Post-larval growth of *Dissodactylus primitivus* Bouvier, 1917 (*Brachyura: Pinnotheridae*) under laboratory conditions. *Biol. Bull. (Woods Hole)*, **163**:211–224, 1982.

[904] J. S. Poindexter. Oligotrophy: fast and famine existence. *Adv. Microb. Ecol.*, **5**:63–89, 1981.

[905] A. Polle. Mehler reaction: friend or foe in photosynthesis? *Bot. Acta*, **109**:84–89, 1996.

[906] D. E. Pomeroy. Some aspects of the ecology of the land snail, *Helicella virgata*, in South Australia. *Austral. J. Zool.*, **17**:495–514, 1969.

[907] R. W. Pomeroy. Infertility and neonatal mortality in the sow. 3. Neonatal mortality and foetal development. *J. Agric. Sci.*, **54**:31–56, 1960.

[908] W. E. Poole. Breeding biology and current status of the grey kangaroo, *Macropus fuliginosus fuliginosus*, of Kangaroo Island, South Australia. *Austral. J. Zool.*, **24**:169–187, 1976.

[909] D. Poppe and B. Vos. *De Buizerd.* K. Kosmos, Amsterdam, 1982.

[910] R. Porley and N. Hodgetts. *Mosses and Liverworts.* Collins, London, 2005.

[911] K. G. Porter, J. Gerritsen, and J. D. Orcutt. The effect of food concentration on swimming patterns, feeding behavior, ingestion, assimilation and respiration by *Daphnia. Limnol. Oceanogr.*, **27**:935–949, 1982.

[912] K. G. Porter, M. L. Pace, and J. F. Battey. Ciliate protozoans as links in freshwater planktonic food chains. *Nature*, **277**:563–564, 1979.

[913] R. K. Porter and M. D. Brand. Body mass dependence of H^+ leak in mitochondria and its relevance to metabolic rate. *Nature*, **362**:628–630, 1993.

[914] L. Posthuma, R. F. Hogervorst, E. N. G. Joosse, and N. M. van Straalen. Genetic variation and covariation for characteristics associated with cadmium tolerance in natural populations of the springtail *Orchesella cincta* (L.). *Evolution*, **47**:619–631, 1993.

[915] L. Posthuma, R. F. Hogervorst, and N. M. van Straalen. Adaptation to soil pollution by cadmium excretion in natural populations of *Orchesella cincta* (L.) (*Collembola*). *Arch. Environ. Contam. Toxicol.*, **22**:146–156, 1992.

[916] E. Postma, W. A. Scheffers, and J. P. van Dijken. Kinetics of growth and glucose transport in glucose-limited chemostat cultures of *Saccharomyces cerevisiae* cbs 8086. *Yeast*, **5**:159–165, 1989.

[917] E. Postma, C. Verduyn, W. A. Scheffers, and J. P. van Dijken. Enzyme analysis of the crabtree effect in glucose-limited chemostat cultures of *Saccharomyces cerevisiae. Appl. Environ. Microbiol.*, **55**:468–477, 1989.

[918] S. Pouvreau, Y. Bourles, S. Lefebvre, A. Gangnery, and M. Alunno-Bruscia. Application of a dynamic energy budget model to the pacific oyster *Crassostrea gigas*, reared under various environmental conditions. *J. Sea Res.*, **56**:156–167, 2006.

[919] D. M. Prescott. Relations between cell growth and cell division. In D. Rudnick, editor, *Rythmic and Synthetic Processes in Growth*, pages 59–74. Princeton University Press, 1957.

[920] G. D. Prestwich. The chemical defences of termites. *Sci. Am.*, **249**:68–75, 1983.

[921] H. H. Prince, P. B. Siegel, and G. W. Cornwell. Embryonic growth of mallard and pekin ducks. *Growth*, **32**:225–233, 1968.

[922] W. G. Pritchard. Scaling in the animal kingdom. *Bull. Math. Biol.*, **55**:111–129, 1993.

[923] M. Proctor and P. Yeo. *The Pollination of Flowers.* Collins, London, 1973.

[924] D. R. Prothero and W. A. Berggren, editors. *Eocene–Oligocene Climatic and Biotic Evolution.* Princeton University Press, 1992.

[925] D. M. Purdy and H. H. Hillemann. Prenatal growth in the golden hamster (*Cricetus auratus*). *Anat. Rec.*, **106**:591–597, 1950.

[926] A. Pütter. Studien über physiologische Ähnlichkeit. 6. Wachstumsähnlichkeiten. *Arch. Gesamte Physiol. Mench. Tiere*, **180**:298–340, 1920.

[927] P. A. Racey and S. M. Swift. Variations in gestation length in a colony of pipistrelle bats (*Pipistrellus pipistrellus*) from year to year. *J. Reprod. Fertil.*, **61**:123–129, 1981.

[928] E. C. Raff, E. M. Popodi, B. J. Sly, F. R. Turner, J. T. Villinski, and R. A. Raff. A novel ontogenetic pathway in hybrid embryos between species with different modes of development. *Development*, **126**:1937–1945, 1999.

[929] U. Rahm. *Die Afrikanische Wurzelratte.* A. Ziemsen-Verlag, Wittenberg Lutherstadt, 1980.

[930] H. Rahn and A. Ar. The avian egg: incubation time and water loss. *Condor*, **76**:147–152, 1974.

[931] H. Rahn, A. Ar, and C. V. Paganelli. How bird eggs breathe. *Sci. Am.*, **2**:38–47, 1979.

[932] H. Rahn and C. V. Paganelli. Gas fluxes in avian eggs: driving forces and the pathway for exchange. *Comp. Biochem. Physiol. A*, **95**:1–15, 1990.

[933] H. Rahn, C. V. Paganelli, and A. Ar. Relation of avian egg weight to body weight. *Auk*, **92**:750–765, 1975.

[934] A. N. Rai, E. Soderback, and B. Bergman. Cyanobacterium–plant symbioses. *New Phytol.*, **147**:449–481, 2000.

[935] K. Raman. *Transport Phenomena in Plants.* Narosa, London, 1997.

[936] J. E. Randall, G. R. Allen, and R. C. Steen. *Fishes of the Great Barrier Reef and Coral Sea.* University of Hawaii Press, Honolulu, 1990.

[937] M. S. Rappé, S. A. Connon, K. L. Vergin, and S. J. Giovannoni. Cultivation of the ubiquitous SAR11 marine bacterioplankton clade. *Nature*, **438**:630–633, 2002.

[938] B. Rasmussen, I. R. Fletcher, J. J. Brocks, and M. R. Kilburn. Reassessing the first appearance of eukaryotes and cyanobacteria. *Nature*, **455**:1101–1104, 2008.

[939] C. Ratledge. Biotechnology as applied to the oils and fats industry. *Fette Seifen Anstrichm.*, **86**:379–389, 1984.

[940] J. A. Raven. *Energetics and Transport in Aquatic Plants.* Alan R. Liss, New York, 1984.

[941] J. A. Raven. The vacuole: a cost–benefit analysis. In R. A. Leigh and D. Sanders, editors, *The Plant Vacuole*, pages 59–86. Academic Press, San Diego, 1997.

[942] J. A. Raven and C. Brownlee. Understanding membrane function. *J. Phycol.*, 2001.

[943] J. A. Raven and D. Edwards. Roots: evolutionary origins and biogeochemical significance. *J. Exp. Bot.*, **52**:381–401, 2001.

[944] J. A. Raven and Z. H. Yin. The past, present and future of nitrogenous compounds in the atmosphere, and their interactions with plants. *New Phytol.*, **139**:205–219, 1998.

[945] P. J. Reay. Some aspects of the biology of the sandeel, *Ammodytes tobianus* L., in Langstone Harbour, Hampshire. *J. Mar. Biol. Assoc. U.K.*, **53**:325–346, 1973.

[946] C. Reder. Metabolic control theory: a structural approach. *J. Theor. Biol.*, **135**:175–201, 1988.

[947] A. C. Redfield. The biological control of chemical factors in the environment. *Am. Sci.*, **46**:205–221, 1958.

[948] M. R. Reeve. The biology of *Chaetognatha*. 1. Quantitative aspects of growth and egg production in *Sagitta hispida*. In J. H. Steele, editor, *Marine Food Chains*, pages 168–189. University of California Press, Berkeley, 1970.

[949] M. R. Reeve and L. D. Baker. Production of two planktonic carnivores (Chaetognath and Ctenophore) in south Florida inshore waters. *Fish. Bull. (Dublin)*, **73**:238–248, 1975.

[950] M. J. Reiss. *The Allometry of Growth and Reproduction.* Cambridge University Press, 1989.

[951] W. Reisser. *Algae and Symbioses.* Biopress, Bristol, 1992.

[952] W. Reisser, R. Meier, and B. Kurmeier. The regulation of the endosymbiotic algal population size in ciliate–algae associations: an ecological model. In H. E. A. Schenk and W. Schwemmler, editors, *Endocytobiology*, vol. 2, pages 533–543. W. de Gruyter, Berlin, 1983.

[953] R. F. Rekker. The hydrophobic fragmental constant, its derivation and application: a means of characterizing membrane systems. In W. Th. Nauta and R. F. Rekker, editors, *Pharmacochemistry Library.* Elsevier, Amsterdam, 1977.

[954] H. Remmert. *Arctic Animal Ecology.* Springer-Verlag, Berlin, 1980.

[955] D. Reznick, H. Callahan, and R. Llauredo. Maternal effects on offspring quality in poeciliid fishes. *Am. Zool.*, **36**:147–156, 1996.

[956] D. N. Reznick, M. Mateos, and M. S. Springer. Independent origins and rapid evolution of the placenta in the fish genus *Poeciliopsis*. *Science*, **298**:1018–1020, 2002.

[957] F. J. Richards. A flexible growth function for empirical use. *J. Exp. Bot.*, **10**:290–300, 1959.

[958] S. W. Richards, D. Merriman, and L. H. Calhoun. Studies on the marine resources of southern New England. 9. The biology of the little skate, *Raja erinacea* Mitchill. *Bull. Bingham Oceanogr. Collect. Yale Univ.*, **18**:5–67, 1963.

[959] D. Richardson. *The Vanishing Lichens: Their History, Biology and Importance.* David and Charles, Newton Abbot, 1975.

[960] M. L. Richardson and S. Gangolli. *The Dictionary of Substances and their Effects.* Royal Society of Chemistry, Cambridge, 1995.

[961] S. Richman. The transformation of energy by *Daphnia pulex*. *Ecol. Monogr.*, **28**:273–291, 1958.

[962] D. Rickard, I. B. Butler, and A. Olroyd. A novel iron sulphide switch and its implications for earth and planetary science. *Earth Planet. Sci. Let.*, **189**:85–91, 2001.

[963] R. E. Ricklefs. Patterns of growth in birds. *Ibis*, **110**:419–451, 1968.

[964] R. E. Ricklefs. Patterns of growth in birds. 3. Growth and development of the cactus wren. *Condor*, **77**:34–45, 1975.

[965] R. E. Ricklefs. Adaptation, constraint, and compromise in avian postnatal development. *Biol. Rev. Camb. Phil. Soc.*, **54**:269–290, 1979.

[966] R. E. Ricklefs and A. Scheuerlein. Biological implications of the Weibull and Gompertz models of aging. *J. Gerontol. A*, **57**:B69–B76, 2002.

[967] R. E. Ricklefs and T. Webb. Water content, thermoregulation, and the growth rate of skeletal muscles in the European starling. *Auk*, **102**:369–377, 1985.

[968] R. E. Ricklefs, S. White, and J. Cullen. Postnatal development of Leach's storm-petrel. *Auk*, **97**:768–781, 1980.

[969] U. Riebesell, D. A. Wolf-Gladrow, and V. Smetacek. Carbon dioxide limitation of marine phytoplankton growth rates. *Nature*, **361**:249–251, 1993.

[970] F. H. Rigler. Zooplankton. In W. T. Edmondson, editor, *A Manual on Methods for the Assessment of Secondary Productivity in Fresh Waters*, pages 228–255. Bartholomew Press, Dorking, 1971.

[971] H. U. Riisgård, E. Bjornestad, and F. Møhlenberg. Accumulation of cadmium in the mussel *Mytilus edulis*: kinetics and importance of uptake via food and sea water. *Mar. Biol.*, **96**:349–353, 1987.

[972] H. U. Riisgård and F. Møhlenberg. An improved automatic recording apparatus for determining the filtration rate of *Mytilus edulis* as a function of size and algal concentration. *Mar. Biol.*, **66**:259–265, 1979.

[973] H. U. Riisgård and A. Randløv. Energy budgets, growth and filtration rates in *Mytilus edulis* at different algal concentrations. *Mar. Biol.*, **61**:227–234, 1981.

[974] M. A. Riley and J. E. Wertz. Bacteriochins: evolution, ecology and application. *Annu. Rev. Microbiol.*, **56**:117–137, 2002.

[975] N. Risgaard-Petersen, A. M. Langezaal, S. Ingvardsen, M. C. Schmid, M. S. M. Jetten, H. J. M. op den Camp, J. W. M. Derksen, E. Piña Ochoa, S. P. Eriksson, L. P. Nielsen, N. P. Revsbech, T. Cedhagen, and G. J. van der Zwaan. Evidence for complete denitrification in a benthic foraminifer. *Nature*, **443**:93–96, 2006.

[976] M. Rizzotti. *Early Evolution.* Birkhäuser, Basel, 2000.

[977] C. T. Robbins and A. N. Moen. Uterine composition and growth in pregnant white-tailed deer. *J. Wildl. Manage.*, **39**:684–691, 1975.

[978] J. R. Robertson and G. W. Salt. Responses in growth, mortality, and reproduction to variable food levels by the rotifer *Asplanchna girodi*. *Ecology*, **62**:1585–1596, 1981.

[979] P. G. Rodhouse, C. M. Roden, G. M. Burnell, M. P. Hensey, T. McMahon, B. Ottway, and T. H. Ryan. Food resource, gametogenesis and growth of *Mytilus edulis* on the shore and in suspended culture: Killary Harbour, Ireland. *Mar. Biol.*, **64**:513–529, 1984.

[980] N. Rodriguez-Ezpeleta, H. Brinkmann, S. C. Burey, G. Roure, B. Burger, W. Loffelhardt, H. J. Bohnert, H. Philippe, and B. F. Lang. Monophyly of primary photosynthetic eukaryotes: green plants, red algae, and glaucophytes. *Curr. Biol.*, **15**:1325–1330, 2005.

[981] J. A. Roels. *Energetics and Kinetics in Biotechnology.* Elsevier, Amsterdam, 1983.

[982] D. A. Roff. *The Evolution of Life Histories.* Chapman & Hall, New York, 1992.

[983] A. J. Roger. Reconstructing early events in eukaryotic evolution. *Am. Nat.*, **154**:S146–S163, 1999.

[984] K. Rohde. *Ecology of Marine Parasites.* CAB International, Wallingford, 1993.

[985] A. H. Romano and T. Conway. Evolution of carbohydrate metabolic pathways. *Res. Microbiol.*, **147**:448–455, 1996.

[986] A. L. Romanov. *The Avian Embryo.* Macmillan, New York, 1960.

[987] C. Romijn and W. Lokhorst. Foetal respiration in the hen. *Physiol. Zool.*, **2**:187–197, 1951.

[988] A. de Roos. Numerical methods for structured population models: the escalator boxcar train. *Num. Meth. Part. Diff. Eq.*, **4**:173–195, 1988.

[989] A. M. de Roos. Escalator boxcar train package. http://staff.science.uva.nl/ aroos/Ebt.htm, 2007.

[990] M. R. Rose. Laboratory evolution of postponed senescence in *Drosophila melanogaster*. *Evolution*, **38**:1004–1010, 1984.

[991] J. L. Roseberry and W. D. Klimstra. Annual weight cycles in male and female bobwhite quail. *Auk*, **88**:116–123, 1971.

[992] A. H. Ross and R. M. Nisbet. Dynamic models of growth and reproduction of the mussel *Mytilus edulis* L. *Funct. Ecol.*, **4**:777–787, 1990.

[993] M. C. Rossiter. Environmentally based maternal effects: a hidden force in insect population dynamics? *Oecologia*, **87**:288–294, 1991.

[994] M. C. Rossiter. Maternal effects generate variation in life history: consequences of egg weight plasticity in the gypsy moth. *Funct. Ecol.*, **5**:386–393, 1991.

[995] M. E. Rothenberg and Y.-N. Jan. The hyppo hypothesis. *Nature*, **425**:469–470, 2003.

[996] W. Rudolph. *Die Hausenten*. A. Ziemsen-Verlag, Wittenberg Lutherstadt, 1978.

[997] D. Ruelle. *Elements of Differentiable Dynamics and Bifurcation Theory*. Academic Press, San Diego, 1989.

[998] P. C. de Ruiter and G. Ernsting. Effects of ration on energy allocation in a carabid beetle. *Funct. Ecol.*, **1**:109–116, 1987.

[999] J. B. Russel and G. M. Cook. Energetics of bacterial growth: balance of anabolic and catabolic reactions. *Microbiol. Rev.*, **59**:48–62, 1995.

[1000] R. W. Russel and F. A. P. C. Gobas. Calibration of the freshwater mussel *Elliptio complanata*, for quantitative monitoring of hexachlorobenzene and octachlorostyrene in aquatic systems. *Bull. Environ. Contam. Toxicol.*, **43**:576–582, 1989.

[1001] M. J. Russell and A. J. Hall. The emergence of life from iron monosulphide bubbles at a submarine hydrothermal redox and pH front. *J. Geol. Soc. Lond.*, **154**:377–402, 1997.

[1002] M. J. Russell and A. J. Hall. From geochemistry to biochemistry: chemiosmotic coupling and transition element clusters in the onset of life and photosynthesis. *Geochem. News*, **133**(Oct):6–12, 2002.

[1003] M. J. Russell, R. M. Daniel, A. J. Hall, and J. A. Sherringham. A hydrothermally precipitated catalytic iron sulfide membrane as a first step toward life. *J. Mol. Evol.*, **39**:231–243, 1994.

[1004] R. L. Rusting. Why do we age? *Sci. Am.*, **12**:86–95, 1992.

[1005] M. Rutgers. Control and thermodynamics of microbial growth. PhD thesis, University of Amsterdam, 1990.

[1006] M. Rutgers, M. J. Teixeira de Mattos, P. W. Postma, and K. van Dam. Establishment of the steady state in glucose-limited chemostat cultures of *Kleibsiella pneumoniae*. *J. Gen. Microbiol.*, **133**:445–453, 1987.

[1007] A. J. Rutter and F. H. Whitehead. *The Water Relations of Plants*. Blackwell, London, 1963.

[1008] I. Ružić. Two-compartment model of radionuclide accumulation into marine organisms. 1. Accumulation from a medium of constant activity. *Mar. Biol.*, **15**:105–112, 1972.

[1009] F. Ryan. *Darwin's Blind Spot*. Texere, New York, 2003.

[1010] S. I. Sandler and H. Orbey. On the thermodynamics of microbial growth processes. *Biotechnol. Bioeng.*, **38**:697–718, 1991.

[1011] J. R. Sargent. Marine wax esters. *Sci. Prog.*, **65**:437–458, 1978.

[1012] J. Sarvala. Effect of temperature on the duration of egg, nauplius and copepodite development of some freshwater benthic copepoda. *Freshwater Biol.*, **9**:515–534, 1979.

[1013] G. W. Saunders and M. H. Hommersand. Assessing red algal supraordinal diversity and taxonomy in the context of contemporary systematic data. *Am. J. Bot.*, **91**:1494–1507, 2004.

[1014] P. T. Saunders. *An Introduction to Catastrophe Theory*. Cambridge University Press, 1980.

[1015] J. Schalk. A study of the metabolic pathway of anaerobic ammonium oxidation. PhD thesis, University of Delft, 2000.

[1016] H. Schatzmann. Anaerobes Wachstum von *Saccharomyces cerevisiae*: Regulatorische Aspekte des glycolytischen und respirativen Stoffwechsels. PhD thesis, Techn. Hochschule, Zurich, 1975.

[1017] H. G. Schegel. *Algemeine Mikrobiologie*. Thieme, Stuttgart, 1981.

[1018] H. Scheufler and A. Stiefel. *Der Kampflaufer*. A. Ziemsen-Verlag, Wittenberg Lutherstadt, 1985.

[1019] D. W. Schindler. Feeding, assimilation and respiration rates of *Daphnia magna* under various environmental conditions and their relation to production estimates. *J. Anim. Ecol.*, **37**:369–385, 1968.

[1020] I. I. Schmalhausen and E. Syngajewskaja. Studien über Wachstum und Differenzierung. 1. Die individuelle Wachtumskurve von *Paramaecium caudatum*. *Roux's Arch. Dev. Biol.*, **105**:711–717, 1925.

[1021] U. Schmidt. *Vampirfledermäuse*. A. Ziemsen-Verlag, Wittenberg, 1978.

[1022] K. Schmidt-Nielsen. *Scaling: Why Is Animal Size so Important?* Cambridge University Press, 1984.

[1023] R. Schneider. Polychlorinated biphenyls (PCBs) in cod tissues from the Western Baltic: significance of equilibrium partitioning and lipid composition in the bioaccumulation of lipophilic pollutants in gill-breathing animals. *Meeresforschung*, **29**:69–79, 1982.

[1024] S. A. Schoenberg, A. E. Maccubbin, and R. E. Hodson. Cellulose digestion by freshwater microcrustacea. *Limnol. Oceanogr.*, **29**:1132–1136, 1984.

[1025] P. F. Scholander. Evolution of climatic adaptation in homeotherms. *Evolution*, **9**:15–26, 1955.

[1026] M. Schonfeld. *Der Fitislaubsanger*. A. Ziemsen-Verlag, Wittenberg Lutherstadt, 1982.

[1027] P. Schönheit and T. Schafer. Metabolism of hyperthermophiles. *World J. Microbiol. Biotechnol.*, **11**:26–57, 1995.

[1028] S. Schonn. *Der Sperlingskauz*. A. Ziemsen-Verlag, Wittenberg Lutherstadt, 1980.

[1029] R. M. Schoolfield, P. J. H. Sharpe, and C. E. Magnuson. Non-linear regression of biological temperature-dependent rate models based on absolute reaction-rate theory. *J. Theor. Biol.*, **88**:719–731, 1981.

[1030] M. A. A. Schoonen, Y. Xu, and J. Bebie. Energetics and kinetics of the prebiotic synthesis of simple organic and amino acids with the FeS-H_2S/FeS_2 redox couple as a reductant. *Orig. Life Evol. Biosphere*, **29**:5–32, 1999.

[1031] L. M. Schoonhoven, T. Jermy, and J. J. A. van Loon. *Insect-Plant Biology: From Physiology to Ecology*. Chapman & Hall, London, 1998.

[1032] H. N. Schulz, T. Brinkhoff, T. G. Ferdelman, M. Hernández Marineé, A. Teske, and B. B. Jørgensen. Dense populations of a giant sulfur bacterium in Namibian shelf sediments. *Science*, **284**:493–495, 1999.

[1033] H. N. Schulz and B. B. Jørgensen. Big bacteria. *Annu. Rev. Microbiol.*, **55**:105–137, 2001.

[1034] K. L. Schulze and R. S. Lipe. Relationship between substrate concentration, growth rate, and respiration rate of *Escherichia coli* in continuous culture. *Arch. Mikrobiol.*, **48**:1–20, 1964.

[1035] A. Schüssler. Molecular phylogeny, taxonomy, and evolution of *Geosiphon pyriformis* and arbuscular mycorrhizal fungi. *Plant and Soil*, **244**:75–83, 2002.

[1036] R. P. Schwarzenbach, P. M. Gschwend, and D. M. Imboden. *Environmental Organic Chemistry*. J. Wiley, New York, 1993.

[1037] W. A. Searcy. Optimum body sizes at different ambient temperatures: an energetics explanation of Bergmann's rule. *J. Theor. Biol.*, **83**:579–593, 1980.

[1038] P. Sébert. Pressure effects on shallow-water fishes. In D. J. Randall and A. P. Farrell, editors, *Deep-Sea Fishes*, pages 279–324. Academic Press, San Diego, 1997.

[1039] D. Segré, D. Ben-Eli, D. W. Deamer, and D. Lancet. The lipid world. *Orig. Life Evol. Biosphere*, **31**:119–145, 2001.

[1040] M. Selig, K. B. Xavier, H. Santos, and P. Schnheit. Comparative analysis of Embden–Meyerhof and Entner–Doudoroff glycolytic pathways in hyperthermophilic archaea and the bacterium *Thermotoga*. *Arch. Microbiol.*, **167**:217–232, 1997.

[1041] M.-A. Selosse and F. Le Tacon. The land flora: a phototroph–fungus partnership? *Trends Ecol. Evol.*, **13**:15–20, 1998.

[1042] H. P. Senn. Kinetik und Regulation des Zuckerabbaus von *Escherichia coli* ML 30 bei tiefen Zucker Konzentrationen. PhD thesis, Techn. Hochschule, Zurich, 1989.

[1043] M. Shafi and P. S. Maitland. The age and growth of perch (*Perca fluviatilis* L.) in two Scottish lochs. *J. Fish Biol.*, **3**:39–57, 1971.

[1044] J. A. Shapiro. Bacteria as multicellular organisms. *Sci. Am.*, **6**:62–69, 1988.

[1045] P. J. H. Sharpe and D. W. DeMichele. Reaction kinetics of poikilotherm development. *J. Theor. Biol.*, **64**:649–670, 1977.

[1046] J. E. Shelbourne. A predator–prey relationship for plaice larvae feeding on *Oikopleura*. *J. Mar. Biol. Assoc. U.K.*, **42**:243–252, 1962.

[1047] Y. Shen and R. Buick. The antiquity of microbial sulfate reduction. *Earth Sci. Rev.*, **64**:243–272, 2004.

[1048] J. M. Shick. *A Functional Biology of Sea Anemones.* Chapman & Hall, London, 1991.

[1049] R. Shine. Why is sex determined by nest temperature in many reptiles? *Trends Ecol. Evol.*, **14**:186–189, 1999.

[1050] R. Shine and E. L. Charnov. Patterns of survival, growth, and maturation in snakes and lizards. *Am. Nat.*, **139**:1257–1269, 1992.

[1051] Y. Shlesinger and Y. Loya. Coral community reproductive pattern: Red Sea versus the Great Barrier Reef. *Science*, **228**:1333–1335, 1985.

[1052] R. V. Short. Species differences in reproductive mechanisms. In C. R. Austin and R. V. Short, editors, *Reproductive Fitness*, vol. 4 of *Reproduction in Mammals*, pages 24–61. Cambridge University Press, 1984.

[1053] J. M. Siegel. The REM sleep–memory consolidation hypothesis. *Science*, **294**:1058–1063, 2001.

[1054] J. M. Siegel. Why we sleep. *Sci. Am.*, **289**:72–77, 2003.

[1055] L. Sigmund. Die postembryonale Entwicklung der Wasserralle. *Sylvia*, **15**:85–118, 1958.

[1056] A. Sih. Predators and prey lifestyles: an evolutionary and ecological overview. In W. C. Kerfoot and A. Sih, editors, *Predation: Direct and Indirect Impacts on Aquatic Communities*, pages 203–224. Univ. of New England Press, Hanover, 1987.

[1057] D. T. H. M. Sijm, K. W. Broersen, D. F. de Roode, and P. Mayer. Bioconcentration kinetics of hydrophobic chemicals in different densities of *Chlorella pyrenoidosa*. *Environ. Toxicol. Chem.*, **17**:1695–1704, 1998.

[1058] D. T. H. M. Sijm and A. van der Linde. Size-dependent bioconcentration kinetics of hydrophobic organic chemicals in fish based on diffusive mass transfer and allometric relationships. *Environ. Sci. Technol.*, **29**:2769–2777, 1995.

[1059] D. T. H. M. Sijm, G. Schüürmann, P. J. Vries, and A. Opperhuizen. Aqueous solubility, octanol solubility, and octanol/water partition coefficient of nine hydrophobic dyes. *Environ. Toxicol. Chem.*, **18**:1109–1117, 1999.

[1060] D. T. H. M. Sijm, W. Seinen, and A. Opperhuizen. Life-cycle biomagnification study in fish. *Environ. Sci. Technol.*, **26**:2162–2174, 1992.

[1061] D. T. H. M. Sijm, M. E. Verberne, W. J. de Jonge, P. Pärt, and A. Opperhuizen. Allometry in the uptake of hydrophobic chemicals determined *in vivo* and in isolated perfused gills. *Toxicol. Appl. Pharmacol.*, **131**:130–135, 1995.

[1062] A. G. Simpson and A. J. Roger. Eukaryotic evolution: getting to the root of the problem. *Curr. Biol.*, **12**:R691–R693, 2002.

[1063] A. G. B. Simpson, J. van den Hoff, C. Bernard, H. R. Burton, and D. J. Patterson. The ultrastructure and systematic position of the euglenozoon *Postgaardi mariagerensis*, Fenchel et al. *Arch. Protistenkd.*, **147**:213–225, 1996.

[1064] G. G. Simpson. *Penguins: Past and Present, Here and There.* Yale University Press, New Haven, 1976.

[1065] M. R. Simpson and S. Boutin. Muskrat life history: a comparison of a northern and southern population. *Ecography*, **16**:5–10, 1993.

[1066] P. G. Simpson and W. B. Whitman. Anabolic pathways in methanogens. In J. G. Ferry, editor, *Methanogenesis*, pages 445–472. Chapman & Hall, New York, 1993.

[1067] B. Sinervo. The evolution of maternal investment in lizard: an experimental and comparative analysis of egg size and its effect on offspring performance. *Evolution*, **44**:279–294, 1990.

[1068] J. W. Sinko and W. Streifer. A model for populations reproducing by fission. *Ecology*, **52**:330–335, 1967.

[1069] P. Sitte, S. Eschbach, and M. Maerz. The role of symbiosis in algal evolution. In W. Reisser, editor, *Algae and Symbioses*, pages 711–733. Biopress, Bristol, 1992.

[1070] J. G. Skellam. The formulation and interpretation of mathematical models of diffusionary processes in population biology. In M. S. Bartlett and R. W. Hiorns, editors, *The Mathematical Theory of the Dynamics of Biological Populations*, pages 63–85. Academic Press, San Diego, 1973.

[1071] R. O. Slatyer. *Plant–Water Relationships.* Academic Press, London, 1967.

[1072] W. Slob and C. Janse. A quantitative method to evaluate the quality of interrupted animal cultures in aging studies. *Mech. Ageing Dev.*, **42**:275–290, 1988.

[1073] L. B. Slobodkin. Population dynamics in *Daphnia obtusa* Kurz. *Ecol. Monogr.*, **24**:69–88, 1954.

[1074] J. F. Sluiters. Parasite–host relationship of the avian schistosome *Trichobilharzia ocellata* and the hermaphrodite gastropod *Lymnea stagnalis*. PhD thesis, Vrije Universiteit, Amsterdam, 1983.

[1075] G. L. Small. *The Blue Whale*. Columbia Univesity Press, New York, 1971.

[1076] D. C. Smith and A. E. Douglas. *The Biology of Symbiosis*. Arnold, London, 1987.

[1077] D. J. Smith and G. J. C. Underwood. The production of extracellular carbohydrates by estuarine benthic diatoms: the effects of growth phase and light and dark treatment. *J. Phycol.*, **36**:321–333, 2000.

[1078] F. A. Smith and N. A. Walker. Photosynthesis by aquatic plants: effects of unstirred layers in relation to assimilation of CO_2 and HCO_3^- and to carbon isotopic discrimination. *New Phytol.*, **86**:245–259, 1980.

[1079] H. L. Smith and P. Waltman. *The Theory of the Chemostat*. Cambridge University Press, 1994.

[1080] M. L. Smith, J. N. Bruhn, and J. B. Anderson. The fungus *Armillaria bulbosa* is among the largest and oldest living organisms. *Nature*, **356**:428–431, 1992.

[1081] S. Smith. Early development and hatching. In M. E. Brown, editor, *The Physiology of Fishes*, vol. 1, pages 323–359. Academic Press, San Diego, 1957.

[1082] S. M. Smith. *The Black-Capped Chickadee: Behavioral Ecology and Natural History*. Comstock, Ithaca, 1991.

[1083] S. V. Smith. The Houtman Abrolhos Islands: carbon metabolism of coral reefs at high latitude. *Limnol. Oceanogr.*, **26**:612–621, 1981.

[1084] S. V. Smith and D. W. Kinsey. Calcium carbonate production, coral reef growth, and sea level change. *Science*, **194**:937–939, 1976.

[1085] B. Snow and D. Snow. *Birds and Berries*. T. & A. D. Poyser, Calton, 1988.

[1086] D. W. Snow. The natural history of the oilbird, *Steatornis caripensis*, in Trinidad, W.I. 1. General behaviour and breeding habits. *Zoologica (New York)*, **46**:27–48, 1961.

[1087] T. Sousa, T. Domingos, and S. A. L. M. Kooijman. From empirical patterns to theory: a formal metabolic theory of life. *Philo. Trans. R. Soc. Lond. B Biol. Sci.*, **363**:2453–2464, 2008.

[1088] T. Sousa, R. Mota, T. Domingos, and S. A. L. M. Kooijman. The thermodynamics of organisms in the context of DEB theory. *Physi. Rev. E*, **74**:1–15, 2006.

[1089] G. M. Southward. A method of calculating body lengths from otolith measurements for pacific halibut and its application to Portlock-Albatross Grounds data between 1935 and 1957. *J. Fish. Res. Board Can.*, **19**:339–362, 1962.

[1090] A. L. Spaans. On the feeding ecology of the herring gull *Larus argentatus* Pont. in the northern part of the Netherlands. *Ardea*, **59**:75–188, 1971.

[1091] G. Spitzer. Jahreszeitliche Aspekte der Biologie der Bartmeise (*Panurus biarmicus*). *J. Ornithol.*, **11**:241–275, 1972.

[1092] J. R. Spotila and E. A. Standora. Energy budgets of ectothermic vertebrates. *Am. Zool.*, **25**:973–986, 1985.

[1093] J. B. Sprague. Measurement of pollutant toxicity to fish. 1. Bioassay methods for acute toxicity. *Water Res.*, **3**:793–821, 1969.

[1094] J. I. Sprent. *The Ecology of the Nitrogen Cycle*. Cambridge University Press, 1987.

[1095] M. Sprung. Physiological energetics of mussel larvae (*Mytilus edulis*). 1. Shell growth and biomass. *Mar. Ecol. Prog. Ser.*, **17**:283–293, 1984.

[1096] J. T. Staley, M. P. Bryant, N. Pfennig, and J. G. Holt. *Bergey's Manual of Systematic Bacteriology*. Williams & Wilkins, Baltimore, 1989.

[1097] M. W. Stanier, L. E. Mount, and J. Bligh. *Energy Balance and Temperature Regulation*. Cambridge University Press, 1984.

[1098] R. Y. Stanier, E. A. Adelberg, and J. L. Ingraham. *General Microbiology*. Macmillan, London, 1976.

[1099] R. Y. Stanier, J. L. Ingraham, M. L. Wheelis, and P. R. Painter. *The Microbial World*. Prentice-Hall, Englewood Cliffs, 1986.

[1100] P. Starke-Reed. Oxygen radicals and aging. In C. E. Thomas and B. Kalyanaman, editors, *Oxygen Radicals and the Disease Process*, pages 65–83. Harwood, Amsterdam, 1997.

[1101] S. C. Stearns. *The Evolution of Life Histories*. Oxford University Press, 1992.

[1102] A. Stechmann and T. Cavalier-Smith. Rooting the eukaryote tree by using a derived gene fusion. *Science*, **297**:89–91, 2002.

[1103] G. G. Steel. *Growth Kinetics of Tumours.* Clarendon Press, Oxford, 1977.

[1104] J. H. Steele. *The Structure of Marine Ecosystems.* Oxford University Press, 1993.

[1105] P. Steemans and C. H. Wellman. Miospores and the emergence of land plants. In B. Webby, F. Paris, M. L. Droser, and I. C. Percival, editors, *The Great Ordovician Biodiversification Event*, pages 361–366. Columbia University Press, New York, 2004.

[1106] R. G. Steen. The bioenergetics of symbiotic sea anemones (*Anthozoa: Actinaria*). *Symbiosis*, **5**:103–142, 1988.

[1107] E. T. Steenkamp, J. Wright, and S. L. Baldauf. The protistan origins of animals and fungi. *Mol. Biol. Evol.*, **23**:93–106, 2006.

[1108] R. S. Stemberger and J. J. Gilbert. Body size, food concentration, and population growth in planktonic rotifers. *Ecology*, **66**:1151–1159, 1985.

[1109] M. N. Stephens. *The Otter Report.* Universal Federation of Animal Welfare, Hertfordshire, 1957.

[1110] R. W. Sterner and J. J. Elser. *Ecological Stoichiometry.* Princeton University Press, 2002.

[1111] J. D. Stevens. Vertebral rings as a means of age determination in the blue shark (*Prionace glauca* L.). *J. Mar. Biol. Assoc. U.K.*, **55**:657–665, 1975.

[1112] J. R. Steward and M. B. Thompson. Evolution of placentation amoung squamate reptiles: recent research and future directions. *Comp. Biochem. Physiol.*, **127**:411–431, 2000.

[1113] J. W. Stiller, D. C. Reel, and J. C. Johnson. A single origin of plastids revisited: convergent evolution in organellar genome content. *J. Phycol.*, **39**:95–105, 2003.

[1114] D. K. Stoecker and M. W. Silver. Replacement and aging of chloroplasts in *Strombidium capitatum* (*Ciliophora: Oligaotrichida*). *Mar. Biol.*, **107**:491–502, 1990.

[1115] B. Stonehouse. The emperor penguin. 1. Breeding behaviour and development. *F.I.D.S. Scientific Reports* 6, 1953.

[1116] B. Stonehouse. The brown skua of South Georgia. *F.I.D.S. Scientific Reports* 14, 1956.

[1117] B. Stonehouse. The king penguin of South Georgia. 1. Breeding behaviour and development. *F.I.D.S. Scientific Reports* 23, 1960.

[1118] A. H. Stouthamer. Metabolic pathways in *Paracoccus denitrificans* and closely related bacteria in relation to the phylogeny of prokaryotes. *Antonie van Leeuwenhoek*, **61**:1–33, 1992.

[1119] A. H. Stouthamer, B. A. Bulthuis, and H. W. van Verseveld. Energetics of growth at low growth rates and its relevance for the maintenance concept. In R. K. Poole, M. J. Bazin, and C. W. Keevil, editors, *Microbial Growth Dynamics*, pages 85–102. IRL Press, Oxford, 1990.

[1120] A. H. Stouthamer and S. A. L. M. Kooijman. Why it pays for bacteria to delete disused DNA and to maintain megaplasmids. *Antonie van Leeuwenhoek*, **63**:39–43, 1993.

[1121] N. M. van Straalen, T. B. A. Burghouts, M. J. Doornhof, G. M. Groot, M. P. M. Janssen, E. N. G. Joosse, J. H. van Meerendonk, J. P. J. J. Theeuwen, H. A. Verhoef, and H. R. Zoomer. Efficiency of lead and cadmium excretion in populations of *Orchesalla cincta* (Collembola) from various contaminated forest soils. *J. Appl. Ecol.*, **24**:953–968, 1987.

[1122] S. Stracke, C. Kistner, S. Yoshida, *et al.* A plant receptor-like kinase required for both bacterial and fungal symbiosis. *Nature*, **417**:959–962, 2002.

[1123] R. Strahan, editor. *The Complete Book of Australian Mammals.* Angus & Robertson, London, 1983.

[1124] R. R. Strathmann and M. F. Strathmann. The relationship between adult size and brooding in marine invertebrates. *Am. Nat.*, **119**:91–1011, 1982.

[1125] J. R. Strickler. Calanoid copepods, feeding currents and the role of gravity. *Science*, **218**:158–160, 1982.

[1126] T. Strömgren and C. Cary. Growth in length of *Mytilus edulis* L. fed on different algal diets. *J. Mar. Biol. Assoc. U.K.*, **76**:23–34, 1984.

[1127] M. Strous and M. S. M. Jetten. Anaerobic oxidation of mathane and ammonium. *Annu. Revi. Microbiol.*, **58**:99–117, 2004.

[1128] L. Stryer. *Biochemistry.* W. H. Freeman, New York, 1988.

[1129] M. R. Sullivan, J. B. Waterbury, and S. W. Chisholm. Cyanophages infecting the oceanic cyanobacterium *Prochlorococcus*. *Nature*, **424**:1047–1050, 2003.

[1130] W. C. Summers. Age and growth of *Loligo pealei*, a population study of the common Atlantic coast squid. *Biol. Bull. (Woods Hole)*, **141**: 189–201, 1971.

[1131] D. W. Sutcliffe, T. R. Carrick, and L. G. Willoughby. Effects of diet, body size, age and temperature on growth rates in the amphipod *Gammarus pulex*. *Freshwater Biol.*, **11**:183–214, 1981.

[1132] A. Suwanto and S. Kaplan. Physical and genetic mapping of the *Rhodobacter sphaeroides* 2.4.1. genome: presence of two unique circular chromosomes. *J. Bacteriol.*, **171**:5850–5859, 1989.

[1133] C. Swennen, M. F. Leopold, and M. Stock. Notes on growth and behaviour of the American razor clam *Ensis directus* in the Wadden Sea and the predation on it by birds. *Helgol. Wiss. Meeresunters*, **39**:255–261, 1985.

[1134] E. Syngajewskaja. The individual growth of protozoa: *Blepharisma lateritia* and *Actinophrys sp. Trav. Inst. Zool. Biol. Acad. Sci. Ukr.*, **8**:151–157, 1935.

[1135] H. Tappan. Possible eucaryotic algae (*Bangiophycidae*) among early Proterozoic microfossils. *Geol. Soc. Am. Bull.*, **87**:633–639, 1976.

[1136] W. M. Tattersall and E. M. Sheppard. Observations on the asteriod genus *Luidia*. In *James Johnstone Memorial Volume*, pages 35–61. Liverpool University Press, 1934.

[1137] C. R. Taylor, K. Schmidt-Nielsen, and J. L. Raab. Scaling of energetic costs of running to body size in mammals. *Am. J. Physiol.*, **219**:1104–1107, 1970.

[1138] D. L. Taylor. Symbiotic marine algae: taxonomy and biological fitness. In W. B. Vernberg, editor, *Symbiosis in the Sea*, pages 245–262. University of South Carolina Press, Columbia, 1974.

[1139] F. J. R. Taylor. Implications and extensions of the serial endosymbiosis theory of the origin of eukaryotes. *Taxon*, **23**:229–258, 1974.

[1140] J. M. Taylor and B. E. Horner. Sexual maturation in the Australian rodent *Rattus fuscipes assimilis*. *Austral. J. Zool.*, **19**:1–17, 1971.

[1141] J. W. Taylor, J. Spatafora, K. O'Donnell, F. Lutzoni, T. James, D. S. Hibbert, D. Geiser, T. D. Bruns, and M. Blackwell. The fungi. In J. Cracraft and M. J. Donoghue, editors, *Assembling the Tree of Life*, pages 171–194. Oxford University Press, 2004.

[1142] P. Taylor, T. E. Rummery, and D. G. Owen. Reactions of iron monosulfide solids with aqueous hydrogen sulfide up to 160 °C. *J. Inorgani. Nucl. Chem.*, **41**:1683–1687, 1979.

[1143] R. H. Taylor. Growth of adelie penguin (*Pygoscelis adeliae* Hombron and Jacquinot) chicks. *New Zealand J. Sci.*, **5**:191–197, 1962.

[1144] W. D. Taylor. Growth responses of ciliate protozoa to the abundance of their bacterial prey. *Microb. Ecol.*, **4**:207–214, 1978.

[1145] H. Tennekes. *De Wetten van de Vliegkunst: Over Stijgen, Dalen, Vliegen en Zweven.* Aramith Uitgevers, Bloemendaal, 1992.

[1146] A. J. Tessier and C. E. Goulden. Cladoceran juvenile growth. *Limnol. Oceanogr.*, **32**:680–685, 1987.

[1147] T. F. Thingstad, H. Havskum, K. Garde, and B. Riemann. On the strategy of "eating your competitor": a mathematical analysis of algal mixotrophy. *Ecology*, **77**:2108–2118, 1996.

[1148] R. V. Thomann. Bioaccumulation model of organic chemical distribution in aquatic food chains. *Environ. Sci. Technol.*, **23**:699–707, 1989.

[1149] T. W. Thomann and J. A. Mueller. *Principles of Surface Water Quality Modeling and Control.* Harper & Row, New York, 1987.

[1150] P. G. Thomas and T. Ikeda. Sexual regression, shrinkage, re-maturation and growth of spent female *Euphausia superba* in the laboratory. *Mar. Biol.*, **95**:357–363, 1987.

[1151] J. M. T. Thompson and H. B. Stewart. *Nonlinear Dynamics and Chaos.* J. Wiley, Chichester, 1986.

[1152] K. S. Thompson. *Living Fossil: The Story of the Coelacanth.* Hutchinson, London, 1991.

[1153] M. B. Thompson. Patterns of metabolism in embryonic reptiles. *Respir. Physiol.*, **76**:243–256, 1989.

[1154] W. M. Thornton. The relation of oxygen to the heat of combustion of organic compounds. *Phil. Mag.*, **33**(Ser 6):196–203, 1917.

[1155] S. T. Threlkeld. Starvation and the size structure of zooplankton communities. *Freshwater Biol.*, **6**:489–496, 1976.

[1156] E. V. Thuesen. The tetrodotoxin venom of chaetognaths. In Q. Bone, H. Kapp, and A. C. Pierrot-Bults, editors, *The Biology of*

Chaetognaths, pages 55–60. Oxford University Press, 1991.

[1157] R. R. Tice and R. B. Setlow. DNA repair and replication in aging organisms and cells. In C. E. Finch and E. L. Schneider, editors, *Handbook of the Biology of Aging*, pages 173–224. Van Nostrand, New York, 1985.

[1158] W. L. N. Tickell. *The Biology of the Great Albatrosses,* Diomedea exulans *and* Diomedea epomophora. American Geophysical Union, Washington, 1968.

[1159] A. G. M. Tielens. Energy generation in parasitic helminths. *Parasitol. Today*, **10**:346–352, 1994.

[1160] A. G. M. Tielens and J. J. van Hellemond. Differences in energy metabolism between *Trypanosomatidae*. *Parasitol. Today*, **14**:265–271, 1998.

[1161] A. G. M. Tielens and J. J. van Hellemond. The electron transport chain in anaerobically functioning eukaryotes. *Biochim. Biophys. Acta*, **1365**:71–78, 1998.

[1162] A. G. M. Tielens, C. Rotte, J. J. van Hellemond, and W. Martin. Mitochondria as we don't know them. *Trends Biochem. Sci.*, **27**:564–572, 2002.

[1163] K. Tiews. Biologische Untersuchungen am Roten Thun (*Thunnus thynnus* [Linnaeus]) in der Nordsee. *Ber. Dtsch. Wiss. Komm. Meeresforsch.*, **14**:192–220, 1957.

[1164] C. Tolla. Bacterial populations dynamics modelling applied to the study of bioturbation effects on the nitrogen cycle in marine sediments. PhD thesis, University of Marseille and Vrije Universiteit, Amsterdam, 2006.

[1165] C. Tolla, S. A. L. M. Kooijman, and J. C. Poggiale. A kinetic inhibition mechanism for the maintenance process. *J. Theor. Biol.*, **244**:576–587, 2007.

[1166] H. Topiwala and C. G. Sinclair. Temperature relationship in continuous culture. *Biotechnol. Bioeng.*, **13**:795–813, 1971.

[1167] N. R. Towers, J. K. Raison, G. M. Kellerman, and A. W. Linnane. Effects of temperature-induced phase changes in membranes on protein synthesis by bound ribosomes. *Biochim. Biophys. Acta*, **287**:301–311, 1972.

[1168] A. P. J. Trinci. A kinetic study of the growth of *Aspergillus nidulans* and other fungi. *J. Gen. Microbiol.*, **57**:11–24, 1969.

[1169] A. P. J. Trinci, G. D. Robson, M. G. Wiebe, B. Cunliffe, and T. W. Naylor. Growth and morphology of *Fusarium graminearum* and other fungi in batch and continuous culture. In R. K. Poole, M. J. Bazin, and C. W. Keevil, editors, *Microbial Growth Dynamics*, pages 17–38. IRL Press, Oxford, 1990.

[1170] T. Troost. Evolution of community metabolism. PhD thesis, Vrije Universiteit, Amsterdam, 2006.

[1171] T. A. Troost, B. W. Kooi, and S. A. L. M. Kooijman. Ecological specialization of mixotrophic plankton in a mixed water column. *Am. Nat.*, **166**:E45–E61, 2005.

[1172] T. A. Troost, B. W. Kooi, and S. A. L. M. Kooijman. When do mixotrophs specialize? Adaptive dynamics theory applied to a dynamic energy budget model. *Math. Biosci.*, **193**:159–182, 2005.

[1173] S. Tuljapurkar and H. Caswell. *Structured-Population Models in Marine, Terrestrial, and Freshwater Systems.* Chapman & Hall, New York, 1996.

[1174] A. W. H. Turnpenny, R. N. Bamber, and P. A. Henderson. Biology of the sand-smelt (*Atherina presbyter* Valenciennes) around Fawlay power station. *J. Fish Biol.*, **18**:417–427, 1981.

[1175] B. M. Tyler, S. Tripathy, *et al. Phytophthora* genome sequences uncover evolutionary origins and mechanisms of pathogenesis. *Science*, **313**:1261–1266, 2006.

[1176] H. Tyndale-Biscoe. *Life of Marsupials.* E. Arnold, London, 1973.

[1177] T. Tyrell. The relative influences of nitrogen and phosphorous on oceanic primary production. *Nature*, **400**:525–531, 1999.

[1178] D. E. Ullrey, J. I. Sprague, D. E. Becker, and E. R. Miller. Growth of the swine fetus. *J. Anim. Sci.*, **24**:711–717, 1965.

[1179] F. A. Urquhart. *The Monarch Butterfly: International Traveler.* Nelson Hall, Chicago, 1987.

[1180] E. Ursin. A mathematical model of some aspects of fish growth, respiration, and mortality. *J. Fish. Res. Board Can.*, **24**:2355–2453, 1967.

[1181] K. Uvnäs-Moberg. The gastrointestinal tract in growth and reproduction. *Sci. Am.*, **7**:60–65, 1989.

[1182] H. A. Vanderploeg, G.-A. Paffenhöfer, and J. R. Liebig. Concentration–variable interactions between calanoid copepods and particles of different food quality: observations and hypothesis. In R. N. Hughes, editor, *Behavioural Mechanisms of Food Selection*, pages 595–613. Springer-Verlag, Berlin, 1990.

[1183] K. Vánky. *Illustrated Genera of Smut Fungi.*, vol. 1 of *Cryptogamic Studies.* Gustav Fischer, Stuttgart, 1987.

[1184] H. van der Veer, J. F. M. F. Cardoso, M. A. Peck, and S. A. L. M. Kooijman. Physiological performance of plaice *Pleuronectes platessa* (L.): from Static to Dynamic Energy Budgets. *J. Sea Res.*, **62**:(in press), 2009.

[1185] H. W. van der Veer, S. A. L. M. Kooijman, W. C. Leggett, and J. van der Meer. Body size scaling relationships in flatfish as predicted by dynamic energy bugets (DEB theory): implications for recruitment. *J. Sea Res.*, **50**:255–270, 2003.

[1186] H. W. van der Veer, S. A. L. M. Kooijman, and J. van der Meer. Intra- and interspecific comparison of energy flow in North Atlantic flatfish species by means of dynamic energy budgets. *J. Sea Res.*, **45**:303–320, 2001.

[1187] N. Verboven and T. Piersma. Is the evaporative water loss of knot *Calidris canutus* higher in tropical than in temperate climates? *Ibis*, **137**:308–316, 1995.

[1188] C Verduyn. Energetic aspects of metabolic fluxes in yeasts. PhD thesis, Technical Univertsity, Delft, 1992.

[1189] K. Vermeer. The importance of plankton to Cassin's auklets during breeding. *J. Plankton Res.*, **3**:315–329, 1981.

[1190] H. W. van Verseveld, M. Braster, F. C. Boogerd, B. Chance, and A. H. Stouthamer. Energetic aspects of growth of *Paracoccus denitrificans*: oxygen-limitation and shift from anaerobic nitrate-limitation to aerobic succinate-limitation. *Arch. Microbiol.*, **135**:229–236, 1983.

[1191] H. W. van Verseveld and A. H. Stouthamer. Oxidative phosphorylation in *Micrococcus denitrificans*: calculation of the P/O ration in growing cells. *Arch. Microbiol.*, **107**:241–247, 1976.

[1192] H. W. van Verseveld and A. H. Stouthamer. Growth yields and the efficiency of oxidative phosphorylation during autotrophic growth of *Paracoccus denitrificans* on methanol and formate. *Arch. Microbiol.*, **118**:21–26, 1978.

[1193] H. W. van Verseveld and A. H. Stouthamer. Two-(carbon) substrate-limited growth of *Paracoccus denitrificans* on mannitol and formate. *FEMS Microbiol. Lett.*, **7**:207–211, 1980.

[1194] J. Vidal and T. E. Whitledge. Rates of metabolism of planktonic crustaceans as related to body weight and temperature of habitat. *J. Plankton Res.*, **4**:77–84, 1982.

[1195] J. Vijverberg. Effect of temperature in laboratory studies on development and growth of *Cladocera* and *Copepoda* from Tjeukemeer, the Netherlands. *Freshwater Biol.*, **10**:317–340, 1980.

[1196] T. A. Villareal. Evaluation of nitrogen fixation in the diatom genus *Rhizosolenia* Ehr. in the absence of its cyanobacterial symbiont *Richelia intracellularis* Schmidt. *J. Plankton Res.*, **9**:965–971, 1987.

[1197] T. A. Villareal. Nitrogen fixation of the cyanobacterial symbiont of the diatom genus *Hemiaulus*. *Mar. Ecol. Prog. Ser.*, **76**:201–204, 1991.

[1198] C. M. Vleck, D. F. Hoyt, and D. Vleck. Metabolism of embryonic embryos: patterns in altricial and precocial birds. *Physiol. Zool.*, **52**:363–377, 1979.

[1199] C. M. Vleck, D. Vleck, and D. F. Hoyt. Patterns of metabolism and growth in avian embryos. *Am. Zool.*, **20**:405–416, 1980.

[1200] D. Vleck, C. M. Vleck, and R. S. Seymour. Energetics of embryonic development in the megapode birds, mallee fowl *Leipoa ocellata* and brush turkey *Alectura lathami*. *Physiol. (Zool.)*, **57**:444–456, 1984.

[1201] N. J. Volkman and W. Trivelpiece. Growth in pygoscelid penguin chicks. *J. Zool. Lond.*, **191**:521–530, 1980.

[1202] P. de Voogt, B. van Hattum, P. Leonards, J. C. Klamer, and H. Govers. Bioconcentration of polycyclic hydrocarbons in the guppy (*Poecilia reticulata*). *Aquat. Toxicol.*, **20**:169–194, 1991.

[1203] G. Wächtershäuser. Before enzymes and templates: theory of surface metabolism. *Microbiol. Rev.*, **52**:452–484, 1988.

[1204] G Wächtershäuser. Pyrite formation, the first energy source for life: A hypothesis. *System. Appl. Microbiol.*, **10**:207–210, 1988.

[1205] G Wächtershäuser. Evolution of the 1st metabolic cycles. *Proc. Natl. Acad. Sci. U.S.A.*, **87**:200–204, 1990.

[1206] T. G. Waddell, P. Repovic, E. Meléndez-Hevia, R. Heinrich, and F. Montero. Optimization of glycolysis: a new look at the efficiency of energy coupling. *Biochem. Educ.*, **25**:204–205, 1997.

[1207] J. G. Wagner. The kinetics of alcohol elimination in man. *Acta Pharmacol. Toxicol.*, **14**:265–289, 1958.

[1208] J. G. Wagner. A modern view in pharmacokinetics. *J. Pharmacokinet. Biopharm.*, **1**:363–401, 1973.

[1209] J. G. Wagner. Do you need a pharmacokinetic model, and, if so, which one? *J. Pharmacokinet. Biopharm.*, **3**:457–478, 1975.

[1210] J. G. Wagner. Time to reach steady state and prediction of steady-state concentrations for drugs obeying Michaelis–Menten elimination kinetics. *J. Pharmacokinet. Biopharm.*, **6**:209–225, 1978.

[1211] C. H. Walker. Kinetic model to predict bioaccumulation of pollutant. *Funct. Ecol.*, **4**:295–301, 1990.

[1212] E. P. Walker. *Mammals of the World.* Johns Hopkins University Press, Baltimore, 1975.

[1213] G.M. Walker. *Yeast: Physiology and Biotechnology.* J. Wiley, Chichester, 1998.

[1214] M. R. Walker. Stromatolites: the main geological source of information on the evolution of the early benthos. In S. Bengtson, editor, *Early Life on Earth*, pages 270–286. Columbia University Press, New York, 1994.

[1215] M. Walls, H. Caswell, and M. Ketola. Demographic costs of Chaoborus-induced defences in *Daphnia pulex. Oecologia*, **87**:43–50, 1991.

[1216] A.E. Walsby. A square bacterium. *Nature*, **283**:69–73, 1980.

[1217] P. Ward. Precambrian strikes back. *New Sci.*, (9 Feb):40–43, 2008.

[1218] J. Warham. The crested penguins. In B. Stonehouse, editor, *The Biology of Penguins*, pages 189–269. Macmillan, London, 1975.

[1219] B.L. Warwick. Prenatal growth of swine. *J. Morphol.*, **46**:59–84, 1928.

[1220] I. Watanabe and S. Okada. Effects of temperature on growth rate of cultured mammalian cells (l5178y). *J. Cell Biol.*, **32**:309–323, 1967.

[1221] L. Watson. *Whales of the World.* Hutchinson, London, 1981.

[1222] E. Watts and S. Young. Components of *Daphnia* feeding behaviour. *J. Plankton Res.*, **2**:203–212, 1980.

[1223] J. van Waversveld, A. D. F. Addink, and G. van den Thillart. The anaerobic energy metabolism of goldfish determined by simultaneous direct and indirect calorimetry during anoxia and hypoxia. *J. Comp. Physiol. B*, **159**:263–268, 1989.

[1224] J. van Waversveld, A. D. F. Addink, G. van den Thillart, and H. Smit. Heat production of fish: a literature review. *Comp. Biochem. Physiol. A*, **92**:159–162, 1989.

[1225] H. Wawrzyniak and G. Sohns. *Die Bartmeise.* A. Ziemsen-Verlag, Wittenberg Lutherstadt, 1986.

[1226] G. J. W. Webb, D. Choqeunot, and P. J. Whitehead. Nests, eggs, and embryonic development of *Carettochelys insculpta* (*Chelonia*: *Carettochelidae*) from Northern Australia. *J. Zool. (Lond.) B*, **1**:521–550, 1986.

[1227] G. J. W. Webb, S. C. Manolis, K. E. Dempsey, and P. J. Whitehead. Crocodilian eggs: a functional overview. In G. J. W. Webb, S. C. Manolis, and P. J. Whitehead, editors, *Wildlife Management: Crocodiles and Alligators*, pages 417–422. Beatty, Sydney, 1987.

[1228] T. P. Weber and T. Piersma. Basal metabolic rate and the mass of tissues differing in metabolic scope: migration-related covariation between individual knots *Calidris canutus. J. Avian Biol.*, **27**:215–224, 1996.

[1229] W. Weibull. A statistical distribution of wide applicability. *J. Appl. Mech.*, **18**:293–297, 1951.

[1230] A. P. Weinbach. The human growth curve. 1. Prenatal. *Growth*, **5**:217–233, 1941.

[1231] R. Weindruch, R. L. Walford, S. Fligiel, and D. Guthrie. The retardation of aging in mice by dietary restriction: longevity, cancer, immunity and lifetime energy intake. *J. Nutr.*, **116**:641–654, 1986.

[1232] J. Weiner. Physiological limits to sustainable energy budgets in birds and mammals: ecological implications. *Trends Ecol. Evol.*, **7**:384–388, 1992.

[1233] T. F. Weis. *Cellular Biophysics.*, vol. 1, *Transport.* MIT Press, Cambridge, 1996.

[1234] E. Weitnauer-Rudin. *Mein Vogel aus dem Leben der Mauerseglers* Apus apus. Basellandschaftlicher Natur- und Vogelschutzverband, Liestal, 1983.

[1235] M. J. Wells. Cephalopods do it differently. *New Sci.*, **3**:333–337, 1983.

[1236] J. van Wensum. Isopods and pollutants in decomposing leaf litter. PhD thesis, Vrije Universeit, Amsterdam, 1992.

[1237] J. H. Werren. Evolution and consequences of *Wolbachia* symbioses in invertebrates. *Am. Zool.*, **40**:1255–1255, 2000.

[1238] D. F. Werschkul. Nestling mortality and the adaptive significance of early locomotion in the little blue heron. *Auk*, **96**:116–130, 1979.

[1239] I. C. West. *The Biochemistry of Membrane Transport.* Chapman & Hall, London, 1983.

[1240] P. Westbroek. *Life as a Geological Force: Dynamics of the Earth.* W. W. Norton, New York, 1991.

[1241] K. Westerterp. The energy budget of the nestling starling *Sturnus vulgaris*, a field study. *Ardea*, **61**:137–158, 1973.

[1242] D. F. Westlake. Some effects of low-velocity currents on the metabolism of aquatic macrophytes. *J. Exp. Bot.*, **18**:187–205, 1967.

[1243] R. Wette, I. N. Katz, and E. Y. Rodin. Stochastic processes for solid tumor kinetics: surface-regulated growth. *Math. Biosci.*, **19**:231–225, 1974.

[1244] T. C. R. White. *The Inadequate Environment: Nitrogen and the Abundance of Animals.* Springer-Verlag, Berlin, 1993.

[1245] P. J. Whitehead. Respiration of *Crocodylus johnstoni* embryos. In G. J. W. Webb, S. C. Manolis, and P. J. Whitehead, editors, *Wildlife Management: Crocodiles and Alligators*, pages 473–497. Beatty, Sydney, 1987.

[1246] P. J. Whitehead, G. J. W. Webb, and R. S. Seymour. Effect of incubation temperature on development of *Crocodylus johnstoni* embryos. *Physiol. Zool.*, **63**:949–964, 1990.

[1247] P. J. Whitfield. *The Biology of Parasitism: An Introduction to the Study of Associating Organisms.* E. Arnold, London, 1979.

[1248] J. N. C. Whyte, J. R. Englar, and B. L. Carswell. Biochemical composition and energy reserves in *Crassostrea gigas* exposed to different levels of nutrition. *Aquaculture*, **90**:157–172, 1990.

[1249] A. P. Wickens. *The Causes of Aging.* Harwood, Amsterdam, 1998.

[1250] J. Widdows, P. Fieth, and C. M. Worral. Relationship between seston, available food and feeding activity in the common mussel *Mytilus edulis. Mar. Biol.*, **50**:195–207, 1979.

[1251] E. Widmark and J. Tandberg. Über die Bedingungen für die Akkumulation indifferenter narkoliken theoretische Bereckerunger. *Biochem. Z.*, **147**:358–369, 1924.

[1252] P. Wiersma and T. Piersma. Effects of microhabitat, flocking, climate and migratory goal on energy expenditure in the annual cycle of red knots. *Condor*, **96**:257–279, 1994.

[1253] H. Wijnandts. Ecological energetics of the long-eared owl (*Asio otus*). *Ardea*, **72**:1–92, 1984.

[1254] M. Wilbrink, M. Treskes, T. A. de Vlieger, and N. P. E. Vermeulen. Comparative toxicokinetics of 2,2′- and 4,4′-dichlorobiphenyls in the pond snail *Lymnaea stagnalis* (L.). *Arch. Environ. Contam. Toxicol.*, **19**:69–79, 1989.

[1255] H. M. Wilbur. Interactions of food level and population density in *Rana sylvatica. Ecology*, **58**:206–209, 1977.

[1256] L. W. Wilcox and G. J. Wedemayer. Dinoflagellate with blue-green chloroplasts derived from an endosymbiotic eukaryote. *Science*, **227**:192–194, 1985.

[1257] D. I. Williamson. *Larvae and Evolution.* Chapman & Hall, London, 1992.

[1258] P. Williamson and M. A. Kendall. Population age structure and growth of the trochid *Monodonta lineata* determined from shell rings. *J. Mar. Biol. Assoc. U.K.*, **61**:1011–1026, 1981.

[1259] D. B. Wingate. First successful hand-rearing of an abondoned Bermuda petrel chick. *Ibis*, **114**:97–101, 1972.

[1260] J. E. Winter. The filtration rate of *Mytilus edulis* and its dependence on algal concentrations, measured by a continuous automatic recording apparatus. *Mar. Biol.*, **22**:317–328, 1973.

[1261] L. M. Winters, W. W. Green, and R. E. Comstock. Prenatal development of the bovine. *Techn. Bull. Minn. Agric. Exp. Station*, **151**:1–50, 1942.

[1262] P. C. Withers. *Comparative Animal Physiology.* Saunders College Publishing, Fort Worth, 1992.

[1263] P. C. Withers and J. U. M. Jarvis. The effect of huddling on thermoregulation and oxygen consumption for the naked mole-rat. *Comp. Biochem. Physiol.*, **66**:215–219, 1980.

[1264] M. Witten. A return to time, cells, systems, and aging. 3. Gompertzian models of biological aging and some possible roles for critical elements. *Mech. Ageing Dev.*, **32**:141–177, 1985.

[1265] C. R. Woese. A proposal concerning the origin of life on the planet earth. *J. Mol. Evol.*, **12**:95–100, 1979.

[1266] C. R. Woese. Archaebacteria. *Sci. Am.*, **6**:98–122, 1981.

[1267] C. R. Woese. On the evolution of cells. *Proc. Natl. Acad. Sci. U.S.A.*, **99**:8742–8747, 2002.

[1268] W. de Wolf, J. H. Canton, J. W. Deneer, R. C. C. Wegman, and J. L. M. Hermens. Quantitative structure–activity relationships and mixture–toxicity studies of alcohols and chlorohydrocarbons: reproducibility of effects on growth and reproduction of *Daphnia magna*. *Aquat. Toxicol.*, **12**:39–49, 1988.

[1269] D. Wolf-Gladrow and U. Riebesell. Diffusion and reactions in the vicinity of plankton: a refined model for inorganic carbon transport. *Mar. Chem.*, **59**:17–34, 1997.

[1270] D. A. Wolf-Gladrow, J. Bijma, and R. E. Zeebe. Model simulation of the carbonate chemistry in the micro-environment of symbiont bearing foraminifera. *Mar. Chem.*, **94**:181–198, 1999.

[1271] R. Wood. *Reef Evolution*. Oxford University Press, 1999.

[1272] F. I. Woodward. *Climate and Plant Distribution*. Cambridge University Press, 1987.

[1273] R. J. Wootton. *Ecology of Teleost Fishes*. Chapman & Hall, London, 1990.

[1274] T. R. Worsley, R. D. Nance, and J. B. Moody. Tectonics, carbon, life, and climate for the last three billion years: a unified system? In S. H. Schneider and P. J. Boston, editors, *Scientists on Gaia*, pages 200–210. MIT Press, Cambridge, 1991.

[1275] J. P. Wourms. Viviparity: the maternal–fetal relationship in fishes. *Am. Zool.*, **21**:473–515, 1981.

[1276] J. P. Wourms and J. Lombardi. Reflections on the evolution of piscine viviparity. *Am. Zool.*, **32**:276–293, 1992.

[1277] G. A. Wray. Punctuated evolution of embryos. *Science (New York)*, **267**:1115–1116, 1995.

[1278] G. A. Wray and A. E. Bely. The evolution of echinoderm development is driven by several distinct factors. *Development* (**suppl.**):97–106, 1994.

[1279] G. A. Wray and R. A. Raff. The evolution of developmental strategy in marine invertebrates. *Trends Ecol. Evol.*, **6**:45–50, 1991.

[1280] J. R. Wright and R. G. Hartnoll. An energy budget for a population of the limpet *Patella vulgata*. *J. Mar. Biol. Assoc. U.K.*, **61**:627–646, 1981.

[1281] I. Wyllie. *The Cuckoo*. Batsford, London, 1981.

[1282] J. Xiong, W. M. Fisher, K. Inoue, *et al.* Molecular evidence for the early evolution of photosynthesis. *Science*, **289**:1724–1730, 2000.

[1283] A. Ykema. Lipid production in the oleaginous yeast *Apiotrichum curvantum*. PhD thesis, Vrije Universiteit, Amsterdam, 1989.

[1284] P. Yodzis. *Introduction to Theoretical Ecology*. Harper & Row, New York, 1989.

[1285] S. H. Yoon, J. D. Hacket, C. Ciniglia, G. Pinto, and D. Bhattacharya. The single, ancient origin of chromist plastids. *Proc. Natl. Acad. Sci. U.S.A.*, **99**:15507–15512, 2002.

[1286] S. H. Yoon, J. D. Hacket, C. Ciniglia, G. Pinto, and D. Bhattacharya. A molecular timeline for the origin of photosynthetic eukaryotes. *Mol. Biol. Evol.*, **21**:809–818, 2004.

[1287] T. Yoshinaga, A. Hagiwara, and K. Tsukamoto. Effect of periodical starvation on the survival of offspring in the rotifer *Brachionus plicatilis*. *Fish. Sci.*, **67**:373–374, 2001.

[1288] J. C. Zachos, M. W. Wara, S. Bohaty, *et al.* A transient rise in tropical sea surface temperature during the Paleocene–Eocene thermal maximum. *Science (New York)*, **302**:1551–1554, 2003.

[1289] C. Zhang and A. M. Cuervo. Restoration of chaperone-mediated autophagy in aging liver improves cellular maintenance and hepatic function. *Nature Med.*, **14**:959–965, 2008.

[1290] Y. J. Zhao, H. Y. Wu, H. L. Guo, M. Xu, K. Cheng, and H. Y. Zhu. Vacuolation induced by unfavorable pH in cyanobacteria. *Prog. Nat. Sci.*, **11**:931–936, 2001.

[1291] S. H. Zinder. Physiological ecology of methanogens. In J. G. Ferry, editor, *Methanogenesis: Ecology, Physiology, Biochemistry and Genetics*, pages 128–206. Chapman & Hall, London, 1993.

[1292] C. Zonneveld. A cell-based model for the chlorophyll *a* to carbon ratio in phytoplankton. *Ecol. Modell.*, **113**:55–70, 1998.

[1293] C. Zonneveld. Photoinhibition as affected by photoacclimation in phytoplankton: a model approach. *J. Theor. Biol.*, **193**:115–123, 1998.

[1294] C. Zonneveld and S. A. L. M. Kooijman. Application of a general energy budget model to *Lymnaea stagnalis*. *Funct. Ecol.*, **3**:269–278, 1989.

[1295] C. Zonneveld and S. A. L. M. Kooijman. Body temperature affects the shape of avian growth curves. *J. Biol. Syst.*, **1**:363–374, 1993.

[1296] C. Zonneveld and S. A. L. M. Kooijman. Comparative kinetics of embryo development. *Bull. Math. Biol.*, **3**:609–635, 1993.

[1297] G. Zubay. *Origins of Life on Earth and in the Cosmos.* Academic Press, San Diego, 2000.

Glossary

acidity The negative logarithm, with base 10, of the proton concentration expressed in mole dm^{-3}. It is known as the pH.

alga An autotrophic (or mixotrophic) protoctist.

allometry The group of analyses based on a linear relationship between the logarithm of some physiological or ecological variable and the logarithm of the body weight of individuals.

allometric function A function of the type $y(x) = ax^b$, where a and b are parameters and $x > 0$.

altricial A mode of development where the neonate is still in an early stage of development and requires attention from the parents. Typical altricial birds and mammals are naked and blind at birth. The opposite of altricial is precocial.

anabolism The collection of biochemical processes involved in the synthesis of structural body mass.

animal Metazoan, ranging from sponges to chordates, which feeds on organisms or their products.

Arrhenius temperature The value of the slope of the linear graph one gets if the logarithm of a physiological rate is plotted against the inverse absolute temperature. It has dimension temperature, but it does not relate to a temperature that exists at a site.

aspect ratio The dimensionless ratio between the length and the diameter of an object with the shape of a cylinder (filaments, rods). The length of rods includes both hemispheres.

assimilation Generation of reserves from substrates (food); see also intake and uptake.

ATP Adenosine triphosphate is a chemical compound that is used by all cells to store or retrieve energy via hydrolysis of one or two phosphate bonds.

Avogadro constant The number of C-atoms in 12 g of ^{12}C, which is $6.022\,05 \times 10^{23}$ mol^{-1}.

C-mole Ratio of the number of carbon atoms of a compound to the Avogadro's constant, where the frequencies of non-carbon elements are expressed relative to carbon.

canonical Relating to the simplest form to which various equations and schemata can be reduced without loss of generality.

catabolism The collection of biochemical processes involved in the decomposition of compounds for the generation of energy and/or source material for anabolic processes; here used for the use of reserves for metabolism (maintenance and growth).

chemical potential The change in the total free energy of a mixture of compounds per mole of substance when an infinitesimal amount of a substance is added, while temperature, pressure and all other compounds are constant.

coefficient of variation The dimensionless ratio of the (sample) standard deviation and the mean. It is a useful measure for the scatter of realisations of a random variable that has a natural origin. The measure is useless for temperatures measured in degrees Celsius, for example.

combustion reference In this frame of reference, the chemical potentials of H_2O, HCO_3^-, NH_4^+, H^+ and O_2 are taken to be 0. The chemical potentials of organic compounds in the standard thermodynamic frame of reference (pH=7, 298 K, unit molarity) are corrected for this setting by equating the dissipation free energy in both frames of reference, when the compound is fully oxidised. The chemical potential of compound $CH_xO_yN_z$ in the combustion frame of reference is expressed in the standard frame of reference as $\overline{\mu}_{CH_xO_yN_z} = \overline{\mu}^\circ_{CH_xO_yN_z} + \frac{1}{2}(2 - x + 3z)\overline{\mu}^\circ_{H_2O} - \overline{\mu}^\circ_{HCO_3^-} - (1 - z)\overline{\mu}^\circ_{H^+} - z\overline{\mu}^\circ_{NH_4^+} + \frac{1}{4}(4 + x - 2y - 3z)\overline{\mu}^\circ_{O_2}$.

compound parameter A function of original parameters. It is usually a simple product and/or ratio.

cubic spline function A function consisting of a number of third-degree polynomials glued together in a smooth way for adjacent intervals of the argument. This is done by requiring that polynomials which meet at a particular argument value x_i have the same value y_i, and the same first two derivatives at that point. The points x_i, y_i, for $i = 1, 2, \ldots, n$ with $n \geq 4$ are considered as the parameters of the cubic spline. For descriptive purposes, splines have the advantage over higher-order polynomials because their global behaviour is much less influenced by local behaviour.

DEB Dynamic Energy Budget considers the energy budget of an individual during its life cycle, and includes changes in properties, in contrast to the Static Energy Budget. DEB theory, which is discussed in this book, is the particular theory on the organisational structure of the DEB.

density The ratio of two masses; but these masses are not necessarily homogeneously mixed, contrary to the concept 'concentration'.

dissipation The collection of transformations that convert reserve into (mineral) products in ways not linked to growth.

dissociation constant The negative logarithm, with base 10, of the ratio of the product of the proton and the ion concentration, to the molecule concentration. It is known as the pK.

DNA Deoxyribonucleic acid, the carrier of genetic information in all living cells.

Ecdysozoa A rather recently delineated taxon of moulting evertebrates to which arthropods and nematodes belong.

eclosion Hatching of imago from pupa (of a holometabolic insect).

ectotherm An organism that is not an endotherm.

eigenvalue If a special vector, an eigenvector, is multiplied by a square matrix, the result is the same as multiplying that vector by a scalar value, known as the eigenvalue. Each square matrix has a number of different independent eigenvectors. This number is less than or equal to the number of rows (or columns). Each eigenvector has its own eigenvalue, but some of the eigenvalues may be equal.

endotherm An animal that usually keeps its body temperature within a narrow range by producing heat. Birds and mammals do this for most of time that they are active. Some other species (insects, tuna fish) have endothermic tendencies.

enthalpy Heat content with dimension energy mole^{-1}. The enthalpy of a system increases by an amount equal to the energy supplied as heat if the temperature and pressure do not change.

entropy The cumulative ratio of heat capacity to temperature of a body when its temperature is gradually increased from zero (absolute) temperature to the temperature of observation. Its dimension is energy$\times$(temperature mole)$^{-1}$. The equivalent definition of the ratio of enthalpy minus free energy to temperature is more useful in biological applications.

estimation The use of measurements to assign values to one or more parameters of a model. This is usually done in some formalised manner that allows evaluation of the uncertainty of the result.

eukaryote An organism that has a nucleus; it contrasts with prokaryote, and includes protoctists, plants and animals.

expectation The theoretical mean of a function of a random variable. For a function g of a random variable $\underline{x}$ with probability density $\phi_{\underline{x}}$, its formal definition is $\mathcal{E}g(\underline{x}) \equiv \int_x g(x)\phi_{\underline{x}}(x)\,dx$. For $g(\underline{x}) = \underline{x}$, the expectation of $\underline{x}$ is the theoretical mean.

exponential distribution The random variable $\underline{t}$ is exponentially distributed with parameter $\dot{r}$ if the probability density is $\phi_{\underline{t}}(t) = \dot{r}\exp\{-\dot{r}t\}$. The mean of $\underline{t}$ equals $\dot{r}^{-1}$. An exponential function has the form $y(x) = y(o)\exp\{\dot{r}t\}$; during exponential growth, body or population size follows an exponential function.

first-order process A process that can be described by a differential equation where the change of a quantity is linear in the quantity itself.

flux An amount of mass or energy per unit of time. An energy flux is physically known as a power.

free energy The maximum amount of energy of a system that is potentially available for 'work'. In biological systems, this 'work' usually consists of driving chemical reactions against the direction of their thermodynamic decay.

functional response The ingestion rate of an organism as a function of food density.

generalised compound Mixture of chemical compounds that does not change in composition: fixed stoichiometries for synthesis (organic substrate, reserves and structural mass are generalised compounds).

growth Increase in structural body mass, measured as an increase in volume in most organisms. I do not include anabolic processes that are part of maintenance.

hazard rate The probability per time increment that death strikes at a certain age, given survival up to that age.

heat capacity The mole-specific amount of heat absorbed by a substance to increase one Kelvin in temperature. Heat capacity typically depends on temperature and has dimension energy mole^{-1}.

heterotroph An organism that uses organic compounds as a source of energy.

homeostasis The ability to run metabolism independent of the (fluctuating) environment. Most organisms can keep the chemical composition of their body (rather) constant, despite changes in the chemical composition of the environment, they remain isomorphic during the life cycle and some maintain a constant body temperature or a constant feeding rate rather independent of food availability.

intake Food intake is the process of removal of food from the environment; food typically enters the digestive system in animals, where it is transformed by enzymes assisted by gut flora. This process is an important at the ecosystem level, where competition controls abundance. See also uptake.

isomorph An organism that does not change its shape during growth.

iteroparous Able to reproduce several times, rather than just once.

large number law The strong law of large numbers states that the difference between the mean of a set of random variables and its theoretical mean is small, with an overwhelming probability, given that the set is large enough.

maintenance A rather vague term denoting the collection of energy-demanding processes that life seems to require to keep going, excluding all production processes.

mass action law The law that states that the meeting frequency of two types of particles is proportional to the product of their densities, i.e. number of particles per unit of volume.

mitochondrion The organelle of cells of most eukaryotes that houses the tricarboxylic acid (TCA) cycle and the respiratory chain and has a double membrane and some DNA. The population dynamics of mitochondria in cells can be complex; they can divide and fuse and crawl through the cells.

morph Organism in which surface area that is involved in uptake grows proportional to volume^0 (V0-morph), to volume$^{2/3}$ (isomorph) or to volume^1 (V1-morph).

NADPH Nicotinamide adenine dinucleotide phosphate is a chemical compound that is used by all cells to accept pairs of electrons.

nutrients Inorganic substrates used for the synthesis of reserves; carbon dioxide and ammonia are examples, and light is also included for convenience.

ODE Ordinary differential equation, which is an equation of the type $\frac{d}{dt}y = f(t, y)$, for some function f of t and y.

opisthokont A rather recently delineated taxon to which animals and fungi belong.

ovoviviparous Having embryos that develop energetically independent from, but inside, the mother.

parameter A quantity in a model that describes the behaviour of state variables. It is usually assumed to be a constant and its value is typically estimated from data using explicit criteria.

parthenogenesis The mode of reproduction where females produce eggs that hatch into new females without the interference of males.

partition coefficient The ratio of the equilibrium concentrations of a compound dissolved in two immiscible solvents, which is taken to be independent of the actual concentrations. The concentrations are here expressed per unit of weight of solvent (not per unit of volume or per mole of solvent).

PDE Partial differential equation, which is an equation of the type $\frac{d}{ds}y + \frac{d}{dt}y = f(s, t, y)$, for some function f of s, t and y.

phylum A taxon that collects organisms with the same body plan.

plant Embryophyte, which includes mosses, ferns and relatives, gymnosperms and flowering plants.

Poisson distribution A random integer-valued variable $\underline{X}$ is Poisson distributed with parameter (mean) λ if $\Pr\{\underline{X} = x\} = \frac{\lambda^x}{x!} \exp\{-\lambda\}$. If intervals between independent events are exponentially distributed, the number of events in a fixed time period will be Poisson distributed.

polynomial A polynomial of degree n of argument x is a function of the type $\sum_{i=0}^{n} c_i x^i$, where $c_0, c_0, \ldots, c_n$ and $c_n \neq 0$ are fixed coefficients.

POM Particulate organic matter.

precocial A mode of development where the neonate is in an advanced state of development and usually does not require attention from the parents. Typical precocial birds and mammals have feathers or hair and gather food by themselves. The opposite of precocial is altricial.

probability density function A non-negative function, here called ϕ, belonging to a continuous random variable, $\underline{x}$ for instance, with the property that $\int_{x_1}^{x_2} \phi_{\underline{x}}(x)\, dx = \Pr\{x_1 < \underline{x} < x_2\}$.

prokaryote An organism that does not have a nucleus, i.e. a eubacterium or archaebacterium; it contrasts with a eukaryote.

protoctist A eukaryote that is not a plant or animal.

QSAR Quantitative structure–activity relationship, see {327}.

reduction degree A property of a molecule. Its value equals the sum of the valences of the atoms minus the electrical charge.

relaxation time A characteristic time that indicates how long a dynamic system requires to return to its equilibrium after perturbation. It is a compound parameter with the dimension time, standing for the first term of the Taylor expansion of the differential equation that describes the dynamics of the system, evaluated in its equilibrium.

respiration quotient The ratio between carbon dioxide production and oxygen consumption, expressed on a molar basis.

RNS Reactive nitrogen species.

rod A bacterium with the shape of a croquette or sausage, that grows in length only. It is here idealised by a cylinder with hemispheres at both ends.

ROS Reactive oxygen species, which include free radicals.

state variable A variable which determines, together with other state variables, the behaviour of a system. The crux of the concept is that the collection of state variables, together with the input, determines the behaviour of the system completely.

SEB Static Energy Budget considers the individual during a short period only, in which it does not change in any property. Respiration is typically interpreted as loss that is subtracted from assimilation to arrive at a scope for growth. This contrasts with respiration in DEB theory, where it includes overhead costs of growth and reproduction.

survivor function A rather misleading term standing for the probability that a given random variable exceeds a specified value. All random variables have a survivor function, even those without any connection to life span. It equals one minus the distribution function. The term is sometimes synonymous with upper tail probability.

taxon A systematic unit, which is used in the classification of organisms. It can be species, genus, family, order, class, phylum, kingdom.

Taylor expansion The approximation of a function by a polynomial of a certain degree that is thought to be accurate for argument values around a specified value. The coefficients of the polynomial are obtained by equating the function value and its first n derivatives at the specified value to that of the n degree polynomial.

TCA cycle Tricarboxylic acid cycle, which is part of the central metabolism.

uptake Food (or substrate) uptake is the passing of the gut wall (or cell membrane) of food-derived metabolites. See also intake and assimilation.

volumetric length The cubic root of the volume of an object. It has dimension length.

weighted sum The sum of terms that are multiplied with weight coefficients before addition. If the terms do not have the same dimension, the dimensions of the different weight coefficients convert the dimensions of weighted terms to the same dimension.

zero-th order process A process that can be described by a differential equation where the change of a quantity is constant.

zooplankter An individual belonging to the zooplankton, i.e. a group of usually small aquatic animals that live in free suspension and do not actively move far in the horizontal direction.

Notation and symbols

Notation differences between the second and third editions

The heating length is now called L_T rather than L_h to make a better link to $\dot{p}_T$. The (half) saturation coefficient is now called K, rather than X_K to simplify the notation.

L now means volumetric structural length, and L_w some physical length, in analogy with V_w.

Since the theory has been substantially extended, and new variables needed to be considered, quite a few new symbols appeared.

Notation rules

Some readers will be annoyed by the notation, which sometimes differs from the one usual in a particular specialisation. One problem is that conventions in e.g. microbiology differ from those in ecology, so not all conventions can be observed at the same time. The symbol D, for example, is used by microbiologists for the dilution rate in chemostats, but by chemists for diffusivity. A voluminous literature on population dynamics exists, where it is standard to use the symbol l for survival probability. This works well as long as one does not want to use lengths in the same text! Another problem is that most literature does not distinguish structural biomass from energy reserves, which both contribute to e.g. dry weight. So the conventional symbols actually differ in meaning from the ones used here.

Few texts deal with such a broad spectrum of phenomena as this book. A consequence is that any symbol table is soon exhausted if one carelessly assigns new symbols to all kinds of variables that show up.

The following conventions are used to reduce this problem and to aid memory.

Symbols

- Variables denoted by symbols that differ only in indices, have the same dimensions. For example M_E and M_V are both moles.
- The interpretation of the leading character does not relate to that of the index character. For example, the M in M_E stands for mass in moles, but in $\dot{k}_M$ it stands for maintenance.
- Some lowercase symbols relate to uppercase ones via scaling: $\{e, E\}$, $\{m, M\}$, $\{j, J\}$, $\{l, L\}$, $\{u, U\}$, $\{w, W\}$ and $\{x, X\}$.

- Structure V has a special role in DEB notation. The structural volume V_V is abbreviated as V. Many quantities are expressed per structural mass, volume or surface area. Likewise the energy of reserve E_E is abbreviated as E.

- Analogous to the tradition in chemistry, quantities that are expressed per unit of structural volume have square brackets, [], so $[M_*] = M_*/V = M_* L^{-3}$. Quantities per unit of structural surface area have braces, {}, so $\dot{p}_* = \{\dot{p}_*\}L^2 = [\dot{p}_*]L^3 = [\dot{p}_*]V$. Quantities per unit of weight have angles, $\langle\rangle$, (with indices w and d for wet and dry weight). Likewise $m_* = M_*/M_V$ is used for the amount of compound $*$ relative to the amount of structure, all expressed in C-moles. This notation is chosen to stress that these symbols refer to relative quantities, rather than absolute ones. They do not indicate concentrations in the chemical sense, because well-mixedness at the molecular level is not assumed. Likewise $j_{*_1 *_2} = \dot{J}_{*_1 *_2}/M_V$, where $*_1$ refers to the type of compound and $*_2$ to the process.

- Parentheses, square brackets and braces around numbers refer to equations, references and pages respectively.

- Rates have dots, which merely indicate the dimension 'per time'. Dots (and primes) do *not* stand for the derivative as in some mathematical and physical texts (see the subsection 'Expressions'). Dots, brackets and braces allow an easy test for some dimensions, and reduce the number of different symbols for related variables. If time has been scaled, i.e. the time unit is some particular value making scaled time dimensionless, the dot has been removed from the rate that is expressed in scaled time.

- Molar values have an overbar.

- Random variables are underscored. The notation $\underline{x}|\underline{x} > x$ means: the random variable $\underline{x}$ given that it is larger than the value x. It can occur in expressions for the probability, Pr{}, or for the probability density function, $\phi()$, of distribution function, $\Phi()$.

- Vectors and matrices are printed in bold face. A bold number represents a vector or matrix of elements with that value; so $\dot{\boldsymbol{J}}\boldsymbol{1}$ is the summation of matrix $\dot{\boldsymbol{J}}$ across columns and $\boldsymbol{1}^T\dot{\boldsymbol{J}}$ across rows; $\boldsymbol{x} = \boldsymbol{0}$ means that all elements of $\boldsymbol{x}$ are 0.

- Organic compounds are quantified in C-mol, which stand for the number of C-atoms as a multiple of Avogadro's number. So 6 C-mol of glucose equals 1 mol of glucose. Notice that for simple compounds, such as glucose we have both the option to express it in mole or C-mole, but for generalised compounds we can only express them in C-mole. So we always use C-mole.

- Mass–mass couplers y, also called yield coefficients, are constant , but yield coefficients Y can vary in time. E.g. Y_{WX} stands for the C-moles of biomass W that is formed per consumed C-mole of substrate X; it is not constant and depends on the specific growth rate. Moreover, y is taken to be non-negative, while Y can be negative, if one compound is appearing, and the other disappearing. They represent ratios of molar fluxes, so $Y_{VE} = \dot{J}_{EG}/\dot{J}_{VG}$ is the ratio of the flux of reserve E (here meant to be a type) that is allocated to growth G (here meant to be a process) and the flux of structure V that is synthesised in the growth process. As a consequence we have $y_{EV} = y_{VE}^{-1}$.

- Energy–mass couplers $\mu_{*_1*_2} = \dot{p}_{*_1}/\dot{J}_{*_2*_1}$ for process $*_1$ and mass of type $*_2$ are inverse to the mass–energy couplers $\eta_{*_2*_1} = \mu_{*_1*_2}^{-1}$. Notice that the sequence of indices changed. The mass–mass couplers $\zeta_{*_1*_2} = \frac{\overline{\mu}_E m_{Em}}{\mu_{*_2*_1}}$ are scaled energy–mass couplers, but now relative to the maximum reserve energy density $\overline{\mu}_E m_{Em}$.

Indices

Indices are catenated, the first subscript frequently specifying the variable to which the symbol relates. For example M_V stands for a mole of structural biomass, where V is structural biovolume. Some indices have a specific meaning:

$*$	indicates that several other symbols can be substituted. It is known as 'wildcard' in computer science. As superscript it denotes the equilibrium value of the variable.
$'$	indicates a scaling as superscript.
i, j	are counters that refer to types or species; they can take the values $1, 2, \ldots$
m	stands for 'maximum'. For example $\dot{p}_{Am}$ is the maximum value that $\dot{p}_A$ can attain.
$+$	can refer to the sum of elements, such as $V_+ = \sum_i V_i$, or to addition, such as X_{i+1}.

Indices for compounds refer to

C	carbon dioxide	$C-$	bicarbonate	E	reserve	E_R	reprod. reserve
H	water, maturity	$\mathcal{M}$	minerals	N_H	ammonia	N_O	nitrate
O	dioxygen	$\mathcal{O}$	org. compounds	P	product (faeces)	Q	toxic compound
V	structural mass	X	food				

Indices for processes refer to

F	feeding	A	assimilation	C	mobilisation	D	dissipation
H	maturation	G	growth	J	mat. maint.	M	vol. linked som. maint.
R	reproduction	S	som. maint.	$T+$	dissipating heat	T	surf. linked som. maint.

Expressions

- An expression between parentheses with an index '+' means: take the maximum of 0 and that expression, so $(x-y)_+ \equiv \max\{0, x-y\}$. The symbol '$\equiv$' means 'is per definition'. It is just another way of writing, you are not supposed to understand that the equality is true.

- Although the mathematical standard for notation should generally be preferred over that of any computer language, I make one exception: the logic boolean, e.g. $(x < x_s)$. It always comes with parentheses and has value 1 if true or value 0 if false. It appears as part of an expression. Simple rules apply, such as:

$$(x \leq x_s)(x \geq x_s) = (x = x_s)$$
$$(x \leq x_s) = (x = x_s) + (x < x_s) = 1 - (x > x_s)$$
$$\int_{x_1=-\infty}^{x} (x_1 = x_s)\, dx_1/dx = (x \geq x_s)$$
$$\int_{x_1=-\infty}^{x} (x_1 \geq x_s)\, dx_1 = (x - x_s)_+$$

- The following operators occur:

$\frac{d}{dt}X\vert_{t_1}$	derivative of X with respect to t evaluated at $t = t_1$
$\frac{\partial}{\partial t}X\vert_{t_1}$	partial derivative of X with respect to t evaluated at $t = t_1$
$\mathcal{E}g(\underline{x})$	expectation of a function g of the random variable $\underline{x}$
var $\underline{x}$	variance of the random variable $\underline{x}$: $\mathcal{E}(\underline{x} - \mathcal{E}\underline{x})^2$
cv $\underline{x}$	coefficient of variation of the random variable $\underline{x}$: $\sqrt{\text{var } \underline{x}}/\mathcal{E}\underline{x}$
$\boldsymbol{x}^T$	transpose of vector or matrix $\boldsymbol{x}$ (interchange rows and columns)
$\vdots$ or ,	catenation of matrices across columns: $\boldsymbol{n} = (\boldsymbol{n}_\mathcal{M} \vdots \boldsymbol{n}_\mathcal{O})$
;	catenation across rows: $(\dot{\boldsymbol{J}}_1^T, \dot{\boldsymbol{J}}_2^T)^T = (\dot{\boldsymbol{J}}_1; \dot{\boldsymbol{J}}_2)$

Signs

Fluxes of appearing compounds at the level of the individual plus its environment are typically taken to be positive, and of disappearing compounds negative. Such fluxes are indicated with a single index for the compound. If the process is also indicated, so two indices are used, such as the mobilisation flux $\dot{J}_{EC}$ the flux typically taken positive. The sign-problem is complex, however, and depends on the level of observation and the choice of state variables (i.e. pools). Where the sign is not obvious, I mention it explicitly. Parameters are always positive, and yield coefficients written with a lower case y are taken as parameters, but yield coefficients written with an upper case Y are ratios of fluxes (so they are variables, which might vary in time) and can be negative. The yield of structure on reserve in the growth process is $Y_{VE}^G = -y_{VE}$, with primary parameter $y_{VE} > 0$.

The mass-specific fluxes, $\dot{j}$, hazard rates, $\dot{h}$, and energy fluxes, $\dot{p}$, are always taken to be positive.

Units, dimensions and types

The SI system is used to present units of measurements; the symbol 'a' stands for year.

In the description of the dimensions in the list of symbols, the following symbols are used:

–	no dimension	L	length (of individual)	e	energy ($\equiv ml^2t^{-2}$)
t	time	l	length (of environment)	T	temperature
#	number (mole)	m	mass (weight)		

These dimension symbols just stand for an abbreviation of the dimension, and differ in meaning from symbols in the symbol column. A difference between the dimensions l and L is that the latter involves an arbitrary choice of the length to be measured (e.g. including or excluding a tail). The morph interferes with the choice. The dimensions differ because the sum of lengths of objects for which l and L apply does not have any useful meaning. The list below does not include symbols that are used in a brief description only. The page number refers to the page where the symbol is introduced.

The choice of symbols relates to dimensions, and not to types. Three types are specified in the description in the list: constant, c, variable, v, and function, f. This classification cannot be rigorous, however. The temperature T, for example, is indicated to be a constant, but it can also be considered as a function of time, in which case all rate constants are functions of time as well. On the other hand, variables such as food density X can be held constant in particular situations. Variables such as structural biovolume V are constant during a short period, such as is relevant for the study of the process of digestion, but not during a longer period, such as is relevant for the study of life cycles. The choice of type can be considered as a default, deviations being mentioned in the text.

List of frequently used symbols

symbol	dim	type	page	interpretation
a	t	v	{54}	age, i.e. time since gametogenesis of fertilisation
a_b	t	v	{54}	age at birth (hatching), i.e. end of embryonic stage
a_p	t	v	{67}	age at puberty, i.e. end of juvenile stage
$a_\dagger$	t	v	{215}	age at death (life span)
$\dot{b}_\dagger$	$l^3\#^{-1}t^{-1}$	c	{238}	killing rate by toxicant
$\dot{B}_\dagger^{*_1*_2}$	$l^6\#^{-2}t^{-1}$	c	{248}	interaction parameter for compounds $*_1$ and $*_2$ in the hazard rate
$B_*^{*_1*_2}$	$l^6\#^{-2}$	v	{248}	interaction parameter for compounds $*_1$ and $*_2$ on target parameter $*$
$B_x(a,b)$	–	f	{61}	incomplete beta function
c_0	$\#l^{-3}$	c	{238}	no-effect concentration (NEC) of toxicant in the environment
c_d	$\#l^{-3}$	v	{223}	concentration of toxicant in the water (dissolved)
c_e	$\#l^{-3}$	v	{238}	scaled internal conc. of toxicant above the NEC: $(c_V - c_0)_+$
c_X	$\#l^{-3}$	v	{230}	concentration of toxicant in food

symbol	dim	type	page	interpretation
c_V	$\# l^{-3}$	v	{223}	scaled internal concentration of toxicant: $[M_Q]P_{dV}$
c_*	$\# l^{-3}$	v	{242}	scaled tolerance conc. of toxicant for target parameter $*$
d_*	$m\,L^{-3}$	c	{11}	density of compound $*$
$\dot{D}$	$l^2 t^{-1}$	c	{267}	diffusivity
e	–	v	{39}	scaled energy density: $[E]/[E_m] = m_E/m_{Em}$
e_b	–	v	{54}	scaled energy density at birth
e_H	–	v	{55}	scaled maturity density: gu_H/l^3
e_H^b	–	v	{54}	scaled maturity density at birth: gu_H^b/l_b^3
e_R	–	v	{72}	scaled energy allocated to reproduction: $E_R E_m^{-1}$
E	e	v	{37}	non-allocated energy in reserve
E_0	e	v	{54}	energy costs of one egg/foetus
E_H	e	v	{49}	accumulated energy investment into maturation
E_H^b	e	c	{54}	maturation threshold for feeding (birth)
E_H^p	e	c	{49}	maturation threshold for reproduction (puberty)
E_R	e	v	{81}	energy in reserve that is allocated to reproduction
$[E]$	$e\,L^{-3}$	v	{37}	energy density: E/V
$[E_b]$	$e\,L^{-3}$	v	{54}	energy density at birth
$[E_G]$	$e\,L^{-3}$	c	{37}	volume-specific costs of structure
$[E_m]$	$e\,L^{-3}$	c	{39}	maximum energy density: $\{\dot{p}_{Am}\}/\dot{v}$
f	–	v	{33}	scaled functional response: $f = \frac{X}{K+X} = \frac{x}{1+x}$
$\dot{F}$	$l^3 t^{-1}$	v	{34}	filtering rate
$\dot{F}_m$	$l^3 t^{-1}$	c	{34}	maximum filtering rate
$\{\dot{F}_m\}$	$l^3 L^{-2} t^{-1}$	c	{34}	specific searching rate
g	–	c	{51}	energy investment ratio: $\frac{[E_G]}{\kappa[E_m]}$
$\dot{h}$	t^{-1}	v	{215}	number-specific predation probability rate (hazard rate)
$\dot{h}_a$	t^{-1}	c	{220}	ageing rate for unicellulars: $\frac{[E_G]}{\kappa\mu_{QC}} \frac{\dot{k}_E+\dot{k}_M}{g+1}$
$\dot{h}_G$	t^{-1}	c	{218}	Gompertz ageing rate
$\dot{h}_W$	t^{-1}	c	{218}	Weibull ageing rate
$\ddot{h}_a$	t^{-2}	c	{216}	Weibull ageing acceleration
$\dot{h}_m$	t^{-1}	c	{354}	max. throughput rate in a chemostat without complete washout
$\overline{h}_*$	$e\,\#^{-1}$	v	{160}	molar enthalpy of compound $*$
i_Q	$l^3 L^{-3} t^{-1}$	c	{223}	uptake rate of toxicant
j_*	$\#\,\#^{-1} t^{-1}$	v	{92}	structure-specific flux of compound $*$: $\dot{J}_*/M_V$
$\dot{J}_*$	$\# t^{-1}$	v	{139}	flux of compound $*$
$\dot{J}_{*_1,*_2}$	$\# t^{-1}$	v	{92}	flux of compound $*_1$ associated with process $*_2$
$\dot{\boldsymbol{J}}$	$\# t^{-1}$	v	{139}	matrix of fluxes of compounds $\dot{J}_{*_1,*_2}$
$\{\dot{J}_{XAm}\}$	$\# L^{-2} t^{-1}$	c	{34}	surface-area-specific max ingestion rate
$[\dot{J}_{Xm}]$	$\# L^{-3} t^{-1}$	c	{127}	volume-specific maximum ingestion rate: $\{\dot{J}_{Xm}\}V_d^{-1/3}$
k	–	c	{50}	maintenance ratio: $\dot{k}_J/\dot{k}_M$
$\dot{k}_e$	t^{-1}	c	{223}	elimination rate of toxicant

symbol	dim	type	page	interpretation
$\dot{k}_E$	t^{-1}	c	{128}	specific-energy conductance: $\dot{v}V_d^{-1/3}$
$\dot{k}_J$	t^{-1}	c	{50}	maturity maintenance rate coefficient
$\dot{k}_M$	t^{-1}	c	{45}	somatic maintenance rate coefficient: $[\dot{p}_M]/[E_G]$
K_*	$\# l^{-3 \text{ or } -2}$	c	{26}	(half) saturation coefficient of compound $*$; default: food
l	–	v	{51}	scaled body length: $(V/V_m)^{1/3} = L/L_m$
l_b	–	c	{54}	scaled body length at birth: $(V_b/V_m)^{1/3} = L_b/L_m$
l_d	–	c	{280}	scaled cell length at division: $(V_d/V_m)^{1/3} = \dot{k}_M g/\dot{k}_E$
l_T	–	c	{46}	scaled heating length: L_T/L_m
l_p	–	c	{72}	scaled body length at puberty: $(V_p/V_m)^{1/3} = L_p/L_m$
L	L	v	{11}	volumetric structural length: $V^{1/3}$
L_b	L	v	{52}	volumetric length at birth: $V_b^{1/3}$
L_d	L	v	{132}	volumetric length at cell division
L_m	L	c	{51}	maximum volumetric length: $V_m^{1/3} = \dot{v}/g\dot{k}_M$
L_p	L	v	{66}	volumetric length at puberty: $V_p^{1/3}$
L_T	L	c	{46}	volumetric heating length: $V_T^{1/3} = \{\dot{p}_T\}/[\dot{p}_M]$
L_w	L	v	{11}	physical length: $L/\delta_{\mathcal{M}}$
m_*	$\#\#^{-1}$	v	{83}	mass of compound $*$ in moles relative to M_V: M_*/M_V
m_{Em}	$\#\#^{-1}$	v	{94}	max molar reserve density: $M_{Em}/M_V = [M_{Em}]/[M_V]$
M_*	$\#$	v	{83}	mass of compound $*$ in moles
$\mathcal{M}(V)$	–	f	{125}	shape (morph) correction function: $\frac{\text{real surface area}}{\text{isomorphic surface area}}$
$[M_{Em}]$	$\# L^{-3}$	c	{83}	maximum reserve density in non-embryos in C-moles $[E_m]/\overline{\mu}_E$
$[M_{sm}]$	$\# L^{-3}$	c	{274}	maximum volume-specific capacity of the stomach for food
$[M_V]$	$\# L^{-3}$	c	{84}	number of C-atoms per unit of structural body volume V
$[M_Q^0]$	$\# L^{-3}$	c	{238}	(internal) no-effect concentration of compound Q; cf. c_0
$n_{*_1 *_2}$	$\#\#^{-1}$	c	{138}	number of atoms of element $*_1$ present in compound $*_2$
$n^0_{*_1 *_2}$	$\#\#^{-1}$	c	{158}	number of isotopes 0 of element $*_1$ present in a pool of comp. $*_2$
$n^{0k}_{*_1 *_2}$	$\#\#^{-1}$	c	{96}	number of isotopes 0 of element $*_1$ present in comp. $*_2$ in process k
$\boldsymbol{n}$	$\#\#^{-1}$	c	{138}	matrix of chemical indices $n_{*_1 *_2}$
N	$\#$	v	{369}	(total) number of individuals: $\int_a \phi_N(a)\, da$
$\dot{p}_*$	et^{-1}	v	{26}	energy flux (power) of process $*$
$\dot{p}_{T+}$	et^{-1}	v	{160}	total dissipating heat
$\dot{p}_{TT}$	et^{-1}	v	{161}	radiation and convection heat

symbol	dim	type	page	interpretation
$\dot{\mathbf{p}}$	$e\,t^{-1}$	v	{139}	vector of basic powers: $(\dot{p}_A\,\dot{p}_D\,\dot{p}_G)$
$\{\dot{p}_{Am}\}$	$e\,L^{-2}t^{-1}$	c	{36}	surface-area-specific maximum assimilation rate
$[\dot{p}_{Am}]$	$e\,L^{-3}t^{-1}$	c	{128}	volume-specific maximum assimilation rate: $\{\dot{p}_{Am}\}V_d^{-1/3}$
$[\dot{p}_M]$	$e\,L^{-3}t^{-1}$	c	{44}	specific volume-linked somatic maintenance rate: $\dot{p}_M/V$
$[\dot{p}_S]$	$e\,L^{-3}t^{-1}$	v	{37}	volume-specific somatic maintenance rate: $\dot{p}_S/V = [\dot{p}_M] + \{\dot{p}_T\}/L$
$\{\dot{p}_T\}$	$e\,L^{-2}t^{-1}$	c	{44}	specific surface area-linked somatic maintenance rate: $\dot{p}_T\,V^{-2/3}$
$P_{*_1*_2}$	–	c	{230}	partition coeff. of a compound in matrix $*_1$ and $*_2$ (moles per volume)
P_{ow}	–	c	{327}	octanol/water partition coefficient of a compound
P_{PX}	–	c	{230}	faeces/food partition coefficient of a compound
P_{Vd}	l^3L^{-3}	c	{223}	biomass/water (dissolved fraction) partition coefficient of a compound
P_{VW}	–	c	{231}	structural/total body mass partition coefficient of a compound
$q(c,t)$	–	v	{238}	survival probability to a toxic compound
$\dot{r}$	t^{-1}	c	{37}	number-specific population growth rate
$\dot{r}_B$	t^{-1}	c	{52}	von Bertalanffy growth rate: $(3/\dot{k}_M + 3fV_m^{1/3}/\dot{v})^{-1} = \dot{k}_M g/3(f+g)$
$\dot{r}_m$	t^{-1}	c	{354}	(net) maximum number-specific population growth rate
$\dot{r}_m^{\circ}$	t^{-1}	c	{354}	gross maximum number-specific population growth rate
$\dot{R}$	$\#t^{-1}$	v	{69}	reproduction rate, i.e. number of eggs or young per time
$\dot{R}_m$	$\#t^{-1}$	c	{71}	max reproduction rate
s	–	v	{242}	stress value
s_0	–	c	{243}	stress value without effect
s_G	–	c	{216}	Gompertz stress coefficient
$\overline{s}_*$	$e\,T^{-1}\#^{-1}$	v	{160}	molar entropy of compound $*$
t	t	v	{37}	time
t_d	t	v	{281}	inter division period
t_D	t	c	{279}	DNA duplication time
t_g	t	v	{275}	gut residence time
t_R	t	v	{73}	time at spawning
t_s	t	v	{274}	mean stomach residence time
T	T	c	{17}	temperature
T_A	T	c	{17}	Arrhenius temperature
T_b	T	c	{14}	body temperature
T_e	T	c	{14}	environmental temperature
u_E	–	v	{54}	scaled reserve: $U_E\frac{g^2\dot{k}_M^3}{\dot{v}^2} = \frac{E}{g[E_m]V_m} = \frac{el^3}{g}$

symbol	dim	type	page	interpretation
u_E^0	–	v	{54}	initial scaled reserve: $U_E^0 \frac{g^2 \dot{k}_M^3}{\dot{v}^2} = \frac{E_0}{g[E_m]V_m}$
u_H	–	v	{54}	scaled maturity: $U_H g^2 \frac{\dot{k}_M^3}{\dot{v}^2} = \frac{E_H}{g[E_m]V_m}$
u_H^b	–	v	{54}	scaled maturity at birth: $U_H^b g^2 \frac{\dot{k}_M^3}{\dot{v}^2} = \frac{E_H^b}{g[E_m]V_m}$
u_H^p	–	v	{66}	scaled maturity at puberty: $U_H^p g^2 \frac{\dot{k}_M^3}{\dot{v}^2} = \frac{E_H^p}{g[E_m]V_m}$
U_E	tL^2	v	{37}	scaled reserve: $M_E/\{\dot{J}_{EAm}\}$
U_H	tL^2	v	{75}	scaled maturity: $M_H/\{\dot{J}_{EAm}\}$
U_H^b	tL^2	v	{75}	scaled maturity at birth: $M_H^b/\{\dot{J}_{EAm}\}$
U_H^p	tL^2	v	{66}	scaled maturity at puberty: $M_H^p/\{\dot{J}_{EAm}\}$
$\dot{v}$	Lt^{-1}	c	{39}	energy conductance (velocity): $\{\dot{p}_{Am}\}/[E_m]$
v_H^b	–	c	{54}	scaled maturity volume at birth: $\frac{M_H^b g^2 \dot{k}_M^3}{\dot{v}^2\{\dot{J}_{EAm}\}(1-\kappa)} = \frac{u_H^b}{1-\kappa}$
V	L^3	v	{81}	structural body volume
V_b	L^3	c	{55}	structural body volume at birth (transition embryo/juvenile)
V_d	L^3	c	{132}	structural cell volume at division
V_T	L^3	c	{46}	structural volume reduction due to heating: $\{\dot{p}_T\}^3[\dot{p}_M]^{-3} = L_T^3$
V_m	L^3	c	{51}	maximum structural body volume: $(\kappa\{\dot{p}_{Am}\})^3[\dot{p}_M]^{-3} = (\dot{v}/\dot{k}_M g)^3$
V_p	L^3	c	{66}	structural body volume at puberty (transition juvenile/adult)
V_w	L^3	c	{81}	physical volume
V_∞	L^3	c	{133}	ultimate structural body volume
w_*	$m\,\#^{-1}$	c	{84}	molar weight of compound $*$
W_d	m	v	{81}	dry weight of (total) biomass
W_w	m	v	{11}	wet weight of (total) biomass
x	–	v	{350}	scaled biomass density in environment: X/K
X_*	$\# l^{-3 \text{ or } -2}$	v	{26}	biomass density of compound $*$ in environment; default: food
X_r	$\# l^{-3}$	c	{349}	substrate density in feed of chemostat
$y_{*_1 *_2}$	$\#\#^{-1}$	c	{41}	coefficient that couples mass flux $*_1$ to mass flux $*_2$
$Y_{*_1 *_2}^k$	$\#\#^{-1}$	v	{94}	yield that couples flux $*_1$ to flux $*_2$ in process k: $\dot{J}_{*_1 k}/\dot{J}_{*_2 k}$
z	–	v	{300}	zoom factor to compare body sizes
$\alpha_{*_1 *_2}^{*_3}$	–	c	{96}	reshuffle coefficient for element $*_1$ of compound $*_2$ in process $*_3$
$\beta_{*_1 *_2}^{0 *_3}$	–	c	{98}	odds ratio of isotope 0 of element $*_1$ of compound $*_2$ in process $*_3$
$\gamma_{*_1 *_2}^0$	–	c	{95}	$\frac{\text{number of isotopes 0 of element } *_1 \text{ in comp. } *_2}{\text{number of atoms of element } *_1}$
$\Gamma(x)$	–	f	{219}	gamma function
δ	–	c	{132}	aspect ratio
δ_l	–	c	{175}	shape parameter of generalised logistic growth
$\delta_{\mathcal{M}}$	–	c	{11}	shape (morph) coefficient: $V^{1/3}/L$

symbol	dim	type	page	interpretation
$\eta_{*_1 *_2}$	$\# e^{-1}$	c	{139}	coefficient that couples mass flux $*_1$ to energy flux $*_2$: $\mu^{-1}_{*_2 *_1}$
$\boldsymbol{\eta}$	$\# e^{-1}$	c	{139}	matrix of coefficients that couple mass to energy fluxes
θ	–	v	{102}	fraction of a number of items: $0 \leq \theta \leq 1$
κ	–	c	{25}	fraction of mobilised reserve allocated to soma
κ_A	–	c	{187}	fraction of assimilation that originates from well-fed-prey reserves
κ_E	–	c	{196}	fraction of rejected flux of reserves that returns to reserves
κ_R	–	c	{51}	fraction of reproduction energy fixed in eggs
κ_X	–	c	{36}	fraction of food energy fixed in reserve
$\overline{\mu}_*$	$e \#^{-1}$	c	{168}	specific chemical potential of compound $*$
$\mu_{*_1 *_2}$	$e \#^{-1}$	c	{139}	coefficient that couples energy flux $*_1$ to mass flux $*_2$: $\eta^{-1}_{*_2 *_1}$
$\overline{\boldsymbol{\mu}}_{\mathcal{M}}$	$e \#^{-1}$	c	{161}	vector of specific chemical potentials of 'minerals'
$\overline{\boldsymbol{\mu}}_{\mathcal{O}}$	$e \#^{-1}$	c	{161}	vector of specific chemical potentials of organic compounds
$\overline{\mu}_*$	$e \#^{-1}$	c	{170}	molar Gibbs energy of compound $*$
ρ	–	c	{105}	binding probability of substrate
$\dot{\sigma}$	$e T^{-1} t^{-1}$	v	{162}	rate of entropy production
τ	–	v	{115}	scaled time or age: typically $t\dot{k}_M$ or $a\dot{k}_M$
τ_b	–	v	{54}	scaled age at birth: $a_b \dot{k}_M$
ω_*	–	c	{81}	contribution of reserve to body weight or physical volume
$\zeta_{*_1 *_2}$	$\# \#^{-1}$	c	{166}	coefficient that couples mass flux $*_1$ to energy flux $*_2$: $\overline{\mu}_E m_{Em} \mu^{-1}_{*_2 *_1}$

Taxonomic index

Index

Printed in the United States
By Bookmasters